A Research Companion

to

Principles and Standards for School Mathematics

A Research Companion to *Principles and Standards for School Mathematics*

Edited by
Jeremy Kilpatrick
W. Gary Martin
Deborah Schifter

The National Council of Teachers of Mathematics, Inc.
1906 Association Drive, Reston, VA 20191-1502
(703) 620-9840; (800) 235-7566; www.nctm.org

ISBN 0-87353-537-5

Second printing 2003

Printed in the United States of America

Table of Contents

Preface

This book originated in late 1996 when members of the Commission on the Future of the Standards of the National Council of Teachers of Mathematics (NCTM) recognized that the document they were tentatively calling *Standards 2000* would need to give more attention to the scholarly literature than previous NCTM Standards documents had done. In two lively, well-attended sessions at the research presession of the annual meeting of the NCTM in Minneapolis in April 1997, Mary Lindquist (chair of the Commission on the Future of the Standards), Joan Ferrini-Mundy (chair of the Standards 2000 Writing Group), and Jeremy Kilpatrick (a member of the Commission) led a discussion in which teachers and researchers shared their thoughts about perspectives from theory, research, and practice that ought to guide *Standards 2000.*

Although the presession participants were virtually unanimous in asking for some articulation of those perspectives, most seemed to consider explicit statements of theory or reviews of research as content that should not appear in the document but that could be referenced there. In response, Jeremy Kilpatrick proposed to the National Science Foundation (NSF) a project entitled "Foundations for School Mathematics" to convene a small working conference of mathematics educators, mathematicians, teachers, and educational theorists and researchers, together with the Standards 2000 Writing Group leaders, to lay out the various perspectives and consider how they ought to be portrayed. The papers produced for the conference and subsequently revised in light of discussions there would both guide the writers of *Standards 2000* and yield a document that would serve as a companion to it. Funding was received from NSF, and the conference was held in March 1998 in Atlanta.

During the first half of the three-day conference, the participants, who had received draft papers in advance, heard a brief presentation from each paper's author and then discussed in plenary their reactions to each paper. The papers were presented by Anna Sfard, Jim Greeno, Ellice Forman, and Bob Siegler, each of whom laid out theoretical foundations for the standards effort. The participants also wrote individual reflections on that theme. During the second half of the conference, the participants broke into groups to discuss changes needed in school mathematics and how those changes might be addressed in *Standards 2000.* They discussed the proposed companion document on research, identifying underdeveloped themes and relevant literature. The writers of *Standards 2000* had already met in summer 1997 to delineate the project and discuss the issues involved, and they were grateful to have draft papers from the conference to use when they met in summer 1998 to produce a discussion draft.

During the months in which the conference was being planned and held, the NCTM Research Advisory Committee (RAC), chaired by Paul Cobb, was commissioning some so-called white papers on specific research topics that would also guide and assist the writers of *Standards 2000.* When the writers met in summer 1998 to draft the updated document, they had drafts of white papers and the papers from the Atlanta conference.

In 1999, with encouragement from NSF to use available project funds to expand the proposed research companion document, Jeremy Kilpatrick began working with Gary Martin, the Standards 2000 project director, and Deborah Schifter, the new RAC chair, to obtain reviews and revisions of the existing papers as well as to commission and get reviews and revisions of additional papers to fill in some gaps in the collection. Reviewers were drawn primarily but not exclusively from the RAC membership as well as from the membership of NCTM's Instructional Issues Advisory Committee and its Professional Development and Status Advisory Committee. This book contains the complete set of revised papers.

The editors are grateful to the many people who worked to bring this project to fruition: the authors, some of whom had to wait patiently while other chapters were being written, reviewed, and revised; the participants at the Atlanta conference (listed below), whose comments were instrumental in shaping the final versions of the conference papers; the reviewers, who offered numerous useful suggestions for improving the papers; the RAC, for encouraging NCTM to publish this volume; and the editorial staff at NCTM headquarters, particularly Harry Tunis, Jean Carpenter, Debra Kushner, Ann Butterfield, Glenn Fink, and Rebecca Totten, who worked diligently to make the book both readable and attractive.

Jeremy Kilpatrick
W. Gary Martin
Deborah Schifter
Editors

Participants in the Conference on Foundations for School Mathematics

March 1998, Atlanta

Sadie Bragg
Gail Burrill
Alverna Champion
Jere Confrey
Joan Ferrini-Mundy
Ellice Forman
Enrique Galindo
James Greeno
Gila Hanna
Patricio Herbst
James Hiebert
Jeane Joyner
Jeremy Kilpatrick
Glenda Lappan
Mary Lindquist
Gary Martin
Douglas McLeod
David Pimm
Henry Pollak
Barbara Reys
Alan Schoenfeld
Anna Sfard
Robert Siegler
Edward Silver
Judy Sowder
George Stanic
James Stigler

Chapter 1

Introduction

Jeremy Kilpatrick, University of Georgia

When, during the 1980s, the nation was deemed "at risk" because its schoolchildren seemed ill prepared to meet the challenges of the modern age, the term *standards* entered the parlance of U.S. education. The National Commission on Excellence in Education (1983), deploring what it saw as a drift toward minimum standards in education, recommended that schools and colleges set higher, more rigorous standards for academic achievement. The National Council of Teachers of Mathematics (NCTM) was the first group to formulate sets of standards—understood as statements of values—issuing standards for curriculum and evaluation in 1989, for teaching in 1991, and for assessment in 1995. The NCTM *Standards* documents, particularly the first, were met with acclaim in the United States and Canada, and they influenced subsequent efforts in other school subjects (for a detailed account of NCTM's efforts at setting standards, see McLeod, Stake, Schappelle, Mellissinos, & Gierl, 1996).

As seems inevitable when statements are made about reforming school mathematics, even when drafts have been widely circulated and discussed, the first NCTM *Standards* documents eventually attracted considerable criticism. Some mathematicians objected to what they considered an overemphasis on modeling and applications and an underemphasis on proof. Some psychologists and educators took issue with a view of teaching and learning that encouraged students to work in small groups and to explore mathematical ideas using computers, calculators, and physical objects. Calls went out for schools not to make the changes advocated by the *Standards* documents until those changes had been "validated" by research, and politicians began to pass legislation that called for instructional materials, programs, and assessments to be based on research (Kilpatrick, 2001).

Even as the standards documents were being produced, NCTM leaders recognized that their message would need to be recast for the twenty-first century. In 1995, a Commission on the Future of the Standards was appointed to oversee the creation, publication, and distribution of an updated document through a project called Standards 2000. A document was drafted in summer 1998, reviewed extensively during the subsequent academic year, and revised in summer 1999. The result, *Principles and Standards for School Mathematics*, was published in 2000.

The Commission on the Future of the Standards had not met many times before the members realized that whatever form the update would take, it would need to do a better job of acknowledging research and other scholarship in mathematics education. In the first three *Standards* documents, underlying assumptions about the nature of mathematics, student learning, instructional practice, assessment practice, and teacher development had not always been explicit and had rarely been linked to the scholarly literature. The documents were not well anchored in either research or theory. The purpose of this book is to provide that anchoring for *Principles and Standards* (NCTM, 2000).

In chapter 2, James Hiebert provides a comprehensive analysis of what research should be expected to do in setting standards for school mathematics, of what is known in general from the relevant research, and of why teachers have not acted on what is known from that research. The subsequent three sections of the book explore the research literature in detail, provide a variety of theoretical perspectives on school mathematics, and examine the process of setting standards from a policy perspective.

The first section, constituting the bulk of the book, contains chapters on research relevant to topics in the NCTM (2000) *Principles and Standards* document. The section opens with three chapters that address teaching and assessment in light of the new principles and standards. Chapter 3, by Deborah Loewenberg Ball and Hyman Bass, is on mathematical reasoning and its development in school classrooms; they offer a conception of teaching based on commitment to the subject matter, the learner, and the classroom community. Arguing that mathematical reasoning cannot be separated from knowing mathematics with understanding, they identify steps that teachers might take to support the development of students' reasoning. Research on teaching, teachers' knowledge, and the professional development of teachers are treated in chapter 4 by Denise Mewborn, who looks at the literature supporting NCTM's vision of mathematics teaching and how teachers might learn to teach according to that vision. She concludes that although research provides no single portrait of "good" teaching, it does present general guidelines for teachers of mathematics. In chapter 5, Linda Wilson and Patricia Kenney address the topic of mathematics assessment, considering both small-scale assessments used by teachers to guide their instruction and large-scale assessments given to large numbers of students. Relevant research is examined in terms of the standards appearing in *Assessment Standards for School Mathematics* (NCTM, 1995), with the research appearing to support both the standards themselves and the capacity of teachers and researchers to find ways to meet those standards.

The next nine chapters in the section survey the literature on curriculum topics associated with the content standards. In chapter 6, Karen Fuson reviews recent research on the development of fluency in whole-number computation, arguing that a more sophisticated view of the process is both justified by the literature and demanded by current goals for school mathematics. Chapter 7 is by Patrick Thompson and Luis Saldanha, who examine how fractions might be treated in the curriculum, some of the difficulties fractions present for teaching and learning, and what it means to understand and reason with fractions. The teaching of the so-called facts of arithmetic and of the algorithms for performing arithmetic operations is treated by Koeno Gravemeijer and Frans van Galen in chapter 8 with an approach they term *algorithmetizing*, an alternative to teaching facts and algorithms by using physical materials. They argue for placing facts and algorithms at the center of school mathematics, making them grounds for students to develop useful and meaningful knowledge on their own.

In chapter 9, Daniel Chazan and Michal Yerushalmy review cognitive research on the learning of algebra, pointing out that *algebra* is a term used with many meanings in research, in the curriculum, and in discussions of reforming school mathematics. They contend that algebra instruction will be successful only if it acknowledges the cognitive complexity of what is being taught. To illustrate some of that complexity, Erick Smith offers another perspective on algebra in chapter 10, positing the theme of "stasis and change" as a means of integrating the curriculum strands of patterns, functions, and algebra. Smith then reviews research literature supporting that theme, illustrating its implications for teaching and calling for further research, consideration, and exploration.

The curriculum strand of geometry is addressed by Douglas Clements in chapter 11. He summarizes the status of geometry—the "forgotten strand"—in school mathematics and some theories of geometric thinking, learning, and teaching. Reviewing research on a variety of issues, he concludes that the research is available to help rescue geometry curriculum and teaching practices from their current "abominable" state. In chapter 12, Richard Lehrer tackles the measurement strand, concentrating, as the research literature does, on children's conceptions of spatial measure and using developmental frameworks to summarize children's evolving understanding. He finds no prescriptions for instruction other than the need for children to go beyond procedural competence to reason about and understand measurement.

The topic of statistics in school mathematics is dealt with in chapter 13 by Clifford Konold and Traci Higgins, who review the literature on reasoning about data, concentrating on the learning of descriptive statistics in the early elementary school years and concluding that students need multiple experiences with real data if they are to understand data representation and analysis. In chapter 14, Michael Shaughnessy discusses the mushrooming literature on students' conceptions of probability. He also discusses some trends in students' performance on assessments and offers some suggestions for teaching, arguing that both research and practice should address the interaction of probability and statistics.

The remaining five chapters in the section deal with topics that cut across the curriculum. In chapter 15, Erna Yackel and Gila Hanna review the research literature on reasoning and proof, pointing at some limitations of that literature, stressing the variation that students show in sophistication of reasoning, and concluding that much effort is required to create a classroom environment in which mathematics is understood as reasoning. Magdalene Lampert and Paul Cobb, in chapter 16 on communication and the use of language, use the metaphor of participation in learning, in contrast with acquisition of learning, to frame their review. They illustrate with several so-called design experiments how mathematical communication might be taught to all students. The final three chapters in the section deal with different facets of the new Representation Standard for school mathematics. In chapter 17, Stephen Monk considers what it means to use graphs as tools for understanding, concentrating on graphs that portray either change over time or data for analysis and observing that graphs can not only communicate meaning but also generate it. In chapter 18, Stephen Smith looks at how children represent mathematical problems, examining how children's views of mathematics are reflected in the representations they produce; he uses the idiosyncratic representations made by four children to argue that, without instruction, children do not view representations as part of a system, using them instead to solve particular problems and not as tools for general use. Gerald Goldin then offers in chapter 19 a unifying perspective on the topic of representations that reconciles some disparate ideas from the literature. He proposes a theory of representational systems that accommodates both external and internal representations, thereby providing an inclusive approach to research.

The second section of the book contains four chapters, each of which approaches the question of principles and standards for school mathematics from a different perspective on teaching and learning. In chapter 20, Robert Siegler offers a cognitive science approach, taking a set of topics relevant to school mathematics and summarizing the empirical findings and theoretical conclusions related to each of them. He points out that although cognitive science has much to say about sources of learning difficulty in mathematics and about potentially effective instructional procedures, translation into specific classroom contexts requires much additional effort from teachers and schools. In chapter 21, James Greeno expounds a situative perspective that uses research and theory on individual cognition and on social communication to examine the systems in which teachers participate as they interact with other people in the classroom and elsewhere in their workplace. He views learning and doing mathematics as involving sustained participation in mathematical practices and calls for research that will explore varieties of teaching practices rather than seek to identify uniform sets of conditions and practices. Sociocultural theory is examined in chapter 22 by Ellice Forman, who demonstrates its relevance for reform in school mathematics by showing how it links instructional practices to learning outcomes and how it models mathematics learning as occurring through communication in complex, goal-directed activity systems. She argues that social and cultural activity should supplant impersonal rationality as the basis for studying reform in school mathematics if that reform is to reach all students. Anna Sfard takes an eclectic approach in chapter 23, arguing for a plurality of outlooks and addressing the tension between the needs of the child and the needs of mathematics by using both "acquisitionist" (learning as compiling ideas into cognitive structures) theories and "participationist" (learning as reorganizing activity in a community of practice) theories to support her case. She uses ten needs of learners to frame her argument, concluding that although no final resolution has been attained to the dilemmas in school mathematics she has identified, near-term, approximate solutions can be found.

Chapter 24 constitutes the final section of the book. Joan Ferrini-Mundy and Gary Martin discuss the role and value of educational research in establishing policy, using NCTM's efforts to update its standards for school mathematics as an illustrative case. Adopting Silver's (1990) formulation of how research—including theoretical perspectives, research methods, and research findings—might influence practice, they examine how it influenced the development and content of the updated document.

Taken together, the chapters in this volume should serve as a useful companion to the NCTM (2000) *Principles and Standards for School Mathematics* for anyone seeking to explore its underpinnings in the scholarly literature. Not every claim made in the *Principles and Standards* document is supported by the research cited herein, nor does every assertion in these chapters have an echo in the document. The authors of *Principles and Standards* were setting forth a vision for what school mathematics ought to become; consequently, they could not always produce evidence that what they envisioned would work out in practice. If teachers were forced to wait until research supported each proposed change in school mathematics, it would never occur. Although tens of thousands of studies in mathematics education have been generated over the years, the scholarly literature is silent on some issues and cannot say much about others. Moreover, despite our attempt to be reasonably comprehensive, this book does not cover the entire terrain of school mathematics. Whole domains are missing, and few chapter authors were able to undertake comprehensive reviews. Nonetheless, the book synthesizes a sizable portion of the literature to provide valuable insight into current thinking about school mathematics.

REFERENCES

Kilpatrick, J. (2001). Where's the evidence? *Journal for Research in Mathematics Education, 32*, 421–427.

McLeod, D. B., Stake, R. E., Schappelle, B. P., Mellissinos, M., & Gierl, M. J. (1996). Setting the standards: NCTM's role in the reform of mathematics education. In S. A. Raizen & E. D. Britton (Eds.), *Bold ventures: Vol. 3. Case studies of U.S. innovations in mathematics education* (pp. 13–132). Dordrecht, The Netherlands: Kluwer.

National Commission on Excellence in Education. (1983). *A nation at risk: The imperative for educational reform.* Washington, DC: U.S. Government Printing Office.

National Council of Teachers of Mathematics. (1989). *Curriculum and evaluation standards for school mathematics.* Reston, VA: Author.

National Council of Teachers of Mathematics. (1991). *Professional standards for teaching mathematics.* Reston, VA: Author.

National Council of Teachers of Mathematics. (1995). *Assessment standards for school mathematics.* Reston, VA: Author.

National Council of Teachers of Mathematics. (2000). *Principles and standards for school mathematics.* Reston, VA: Author.

Silver, E. A. (1990). Contributions of research to practice: Applying findings, methods, and perspectives. In T. Cooney (Ed.), *Teaching and learning mathematics in the 1990s* (pp. 1–11). Reston, VA: National Council of Teachers of Mathematics.

CHAPTER 2

What Research Says About the NCTM Standards

James Hiebert
University of Delaware

What does research say about the Standards contained in the documents published by the National Council of Teachers of Mathematics ([NCTM], 1989, 1991, 1995, 2000)?[1] Most people have just one concern when they ask that question: Does research support the Standards? That is a fair question, and it has two answers. One is simple but incomplete; the other is longer but more informative. The simple answer is yes. The Standards are consistent with the best and most recent evidence on teaching and learning mathematics. Said another way, the Standards do not violate the major findings from research and, where relevant results are available, the Standards are consistent with those results. Of course, in matters as complex as connecting educational standards with research, simple answers cannot tell the whole story. In fact, it would be silly to stop here and presume that we now know whether research supports the Standards.

The full answer is more complicated than yes for two reasons: (1) standards in any field are rarely based solely on research, so the connection between research and standards is never straightforward; and (2) research in mathematics education does not shine equally brightly on all aspects of the NCTM Standards, so one cannot make a blanket statement that applies to all the connections. The more complicated story takes a little longer to tell but is worth recounting. By tracing the connections between research and the Standards, one becomes more informed about what to expect from research, how to make sense of the public debates surrounding the Standards, and how to raise the level of these debates. By understanding the research base, one also gets a better sense of what is possible for students to learn and what teachers and researchers might do to help students realize those possibilities.

The story begins by setting appropriate expectations for research. Understanding the current contributions of research requires understanding what research can and cannot do. Once appropriate expectations are in place, the story continues by examining the effects of the current program of school mathematics in the United States. Research can say a great deal about how the current program is working. The story then moves to the promise of alternatives. How much research support is available for teachers to change what they are doing? And if change is indicated, in what direction should they move? Even when teachers wish to change and know where they want to go, change does not happen automatically. Research also says a good deal about the process of change. The story concludes by examining this process of changing classroom practice.

What Should We Expect From Research?

How nice it would be if we could look at the research evidence and decide whether the Standards are right or wrong. That ability would make decisions simple and bring an end to the debates about the direction of mathematics education in the United States. Is that outcome altogether impossible? After all, can't people do the same in other professions? Actually, they can't. Standards and research rarely have a clear relationship. The health professions, for example, propose standards for living a healthy life—diet, exercise, and so on—but medical research does not prove that those standards are the best

[1]The phrase "NCTM Standards," or just "Standards," is used to capture the recommendations for K–12 curriculum, teaching, and assessment contained in the initial three-volume set (NCTM, 1989, 1991, 1995) and in the revised volume *Principles and Standards for School Mathematics* (NCTM, 2000).

ones. Is meat good for you or not? Is it better to use butter or margarine? Should you have exactly seven servings of fruits and vegetables every day, or would six be enough? These simple-sounding questions do not have simple answers. Indeed, health standards themselves have changed over time, driven by changes in culture and values as well as by research findings. In education—a field at least as complex as physical health—we should not be surprised that research has not answered all our questions. But research can still be useful. To understand its rightful contribution, we must get a sense of what it cannot do as well as what it can do.

Some Things Research Cannot Do

Research cannot select standards. Standards in mathematics education represent the goals we set for our students. They are value judgments about what we would like our students to know and be able to do. They are chosen through a complex process that is fed by societal expectations, past practice, research information, and visions of the professionals in the field. The process is similar to that which operates in selecting standards in other professional fields, such as law or agriculture. The role that research plays in selecting standards is difficult to pin down, but in the end, research does not select standards. Standards are, ultimately, statements about what we value most. They represent our priorities.[2]

That standards are value laden does not mean that research has nothing to say about them. To sharpen the claim that research does not select standards, a distinction must be made between influence and selection. Mathematics education is filled with examples of ways that research can *influence* the nature of standards and the way they are defined. For example, in the early 1900s, mathematics was viewed as a valuable subject because learning mathematics was believed to exercise the mind, and the mind, like a muscle, needed exercise to become strong. The research of E. L. Thorndike (1922; Thorndike & Woodworth, 1901), however, suggested that the idea of mind as muscle was a poor analogy. Students' minds did not appear to become stronger from studying mathematics; that is, they did not become smarter in other areas but simply learned mathematics. Standards today, therefore, rarely prescribe mathematical activity to exercise the mind. Thorndike's research encouraged a move away from those kinds of standards.

Research on learning can also have the opposite effect: It can show what is possible and draw attention toward certain standards. Research demonstrating young children's ability to solve simple arithmetic story problems *before* instruction provides one example (Carpenter, Moser, & Romberg, 1982). Standards increasingly emphasize young students' inventions of arithmetic procedures because, in part, we know that young students are capable of such inventions.

Research in the subject itself also can shape the kinds of standards that are selected. For example, research and development in mathematics have opened up vast new areas of study in non-Euclidean geometries, such as transformation and projective geometry. These topics have found their way into the elementary and secondary curricula and are identified in the NCTM Standards.

Consequently, research in a variety of fields related to mathematics education can influence the standards that are selected and the way in which they are expressed. But that influence occurs only when the implications of the research are valued. Mathematical inventions by students are not included in a set of standards simply because students are capable of inventing; they are included because an additional value judgment has been made: that invention is an important mathematical process. Non-Euclidean geometries are included not just because they exist but also because a judgment has been made about their importance in the field of mathematics. Research by itself does not select standards. But it can influence the process by questioning assumptions, uncovering deficiencies, revealing possibilities, and creating new needs.

Research cannot prove what is best. There is increasing pressure to prove scientifically the best course to take in making curricular and pedagogical decisions in mathematics. Should we teach in this way or that way? Should we use this textbook series or that textbook series? The stakes are rising because mathematics is increasingly seen as important for all students. The decisions we make have consequences for everyone, but the decisions are becoming more difficult because more and more curricular and pedagogical options are available. Scientific research is looked to as the solution because, after all, science has taken us to Mars with the Pathfinder and has healed painful backs with incredibly precise arthroscopic surgeries.

Looking to scientific research is a good thing; the more reliable information we have, the better our decisions will be. But in every field, science has its limits. To make rational decisions, one must have some sense of those limits as well as of the potential contributions.

---

[2]See NCTM's "Statement of Beliefs" (posted on their website, www.nctm.org) for a description of basic values and priorities that underlie the NCTM Standards.

Consider again the requirements for healthy living. How many servings of fruits and vegetables should we have each day? We probably do not expect science to prove the optimal amounts for everyone, because there are just too many factors that influence the answer. How much exercise do we get? How much do we weigh? What are our genetic make-ups and our past histories? What are our metabolism rates? And exactly which fruits and vegetables do we have in mind? It would be impossible to control all of these factors to *prove* that a certain amount of fruits and vegetables is best.

The situation in education is similar. Most outcomes are influenced by more factors than anyone can identify, let alone control. Does this realization mean that research is a waste of time? Not at all. Just because research cannot prove whether a particular decision is the best one does not mean that research is irrelevant. In complex environments, such as our bodies and school classrooms, a special relationship exists between research and decision making. Decisions are probability estimates, and research data help in estimating the likelihood of success. The clearer the results, the more confident we are that we are making good decisions. But we usually make decisions with levels of confidence and not with certainty.

Here is an example of a conjecture that might be true or false: *It is better to use calculators than not to use calculators in elementary school.* That conjecture is simple enough and is receiving a lot of heated debate. Is it possible to prove or disprove the conjecture? Suppose we try to prove it by comparing classrooms in which calculators are and are not used. First, we must decide what *better* means and how to measure it. Does *better* mean that students, at the end, understand mathematics more deeply, solve challenging problems more effectively, execute written computation procedures more quickly, or like mathematics more? Deciding what *better* means is not a trivial task. That decision also requires being clear about values and priorities. Suppose, for the sake of argument, that we take "It is better to use calculators than not to use calculators" to mean "Students can execute written computation procedures more accurately and quickly when they use calculators than when they do not." Many people would guess that, if that interpretation expressed the valued outcome, the no-calculator classroom would be better.

How could we test that hypothesis? How would we set up a fair comparison between the calculator and the no-calculator treatments? A reasonable approach would be to develop, with our desired learning goal in mind, the best imaginable instructional program with the calculator and the best imaginable instructional program without the calculator. That approach would mean that students in the two programs would probably be completing different tasks and engaging in different activities because what is possible with the calculator is very different from what is possible without it. But now we have a problem because we will not know what caused the differences in students' learning. Was it the calculator, the other differences between the instructional programs, or both? Maybe we could solve this problem by keeping the instructional programs identical and just plopping the calculators into one set of classrooms and not the others. But into which instructional program should the calculators be plopped: the one that maximizes the benefits of the calculator or the one that assumes they are not available? Neither choice is good, because both would lead to difficult problems of interpretation. Maybe we should split the differences between the instructional programs. But then we would have a program that no one would intentionally design.

Do these problems mean that all the studies on using calculators—and there have been many—are uninterpretable? No. But these problems do mean that no single study will prove once and for all whether students should use calculators. The best way to draw conclusions regarding issues like this is to review the many studies done under a variety of conditions and look for patterns in the results. Perhaps studies in the early grades show one kind of pattern and studies in the later grades show another pattern. Or perhaps studies using calculators in one way show one pattern of results and studies using calculators in another way show another pattern. As it happens, this kind of review of calculator use has been done, and so we can provide a partial and tentative answer to the question of calculator use (Hembree & Dessart, 1986). The results suggest that using calculators along with common pencil-and-paper activities does not harm students' skill development and supports increased problem-solving skills and better attitudes toward mathematics. That conclusion does not mean, by the way, that such results will be found in every classroom, but it does suggest two things: (1) that teachers can decide to use calculators wisely during mathematics instruction with some confidence, and (2) that when calculators are blamed for damaging students' mathematical competence, it would be useful to check the full instructional program. The problem is likely to be due to a poor use of calculators or a feature of instruction unrelated to calculators and not to the calculators themselves.

If research cannot prove that one course of action is *the* best one, it follows that research cannot prescribe a curriculum and a pedagogical approach for all students and for all time. Decisions about curriculum and pedagogy are always tentative and made with some level of confidence that changes over time with new information and changing conditions. Research can and should play a critical role in making informed decisions and setting the levels of confidence, but we cannot look to research for clear prescriptions.

Things Research Can Do

Research is not just filled with limitations; it also holds enormous potential. To set appropriate expectations for research, understanding its potential benefits is as important as understanding its limitations.

Prevent us from making unwitting mistakes. Research can serve as a brake for the educational juggernaut. It can slow down the tendency to endorse ideas based on speculations and on the enthusiastic opinions of charismatic leaders by showing that some goals, although highly valued by many, are unattainable. For example, it would be nice if studying mathematics actually did exercise the mind and make people smarter in some general way. As noted earlier, however, research suggests that this goal is unrealistic. Until there is evidence to the contrary, it would be unrealistic to shape standards with this goal in mind.

Research can also probe beneath the surface to show why some goals are difficult to achieve and why particular programs do not work as well as expected. This probing allows us to develop a much clearer understanding of the factors we must consider in making decisions. An example was recently reported by investigators of the QUASAR project, a large-scale effort to improve the mathematics education program of inner-city middle schools. In some QUASAR schools, students' achievement was not rising as expected. It would have been easy to conclude that the program simply was not effective for some students. But the investigators took a second look, comparing schools in which students' achievement was increasing with schools in which it was not (Parke & Smith, 1998). They found major differences in the staffing situations in the two kinds of schools. In the less successful schools, the rate of teacher and principal turnover was very high. This turnover resulted in a relatively weak implementation plan and fewer, more superficial changes in classroom instruction. It would be a mistake, therefore, to conclude that the program itself is ineffective; instead, one can conclude only that weak implementations are ineffective and that such implementations can occur when staff members do not have the time to learn new practices.

The QUASAR example reminds us that useful research data on curriculum and instruction programs include more than student achievement scores. Information is also needed on what is actually happening in classrooms. Educators and policy makers in the United States can and do make many mistakes because they do not have sufficient information of this kind. An example from California is illustrative (Stigler, 1998). In 1995, faced with falling mathematics achievement scores, the state superintendent of public instruction appointed a task force to study the situation and propose solutions. Why, if California's recently accepted standards had received so much acclaim, were students' achievement scores so low? Discussion at the task force meetings soon turned to the standards. Were they to blame? Some members thought so; some members disagreed and defended the standards. The debate that ensued signaled the kinds of debates that continue today.

Lost in those early debates in California was that fact that no information was available on the extent to which the standards were influencing mathematics instruction in the state's classrooms. Without knowing what was happening in classrooms, how could the effectiveness of the standards be assessed? How could wise decisions be made? As it turned out, fewer than one third of the elementary school teachers reported having abandoned rather traditional methods of teaching mathematics (Cohen & Hill, 1998). And teachers tend to overreport the reform-minded changes they make (Stigler & Hiebert, 1997). The presumption that the new standards were causing the decline appears to have been mistaken.

That story is not meant to single out California; few, if any, states regularly collect information on what is happening inside their classrooms. This situation is unfortunate because it blocks one way in which research can prevent mistakes. Informed policy decisions often require knowing how earlier recommendations—including the issuance of standards—are influencing classroom practice. Such information can help to prevent the pendulum swings that are often evident in successive educational recommendations.

Show what is possible and what looks promising. In addition to applying the brakes, research can also step on the accelerator. It can demonstrate what is possible for students—what they can learn under specific kinds of conditions. Research can show that students can reach certain goals and that some kinds of instruction are especially effective in helping them get there. For example, a research study might show that given appropriate instruction, students at particular ages can learn more about probability, or engage in more deductive reasoning, than they are now. Research of this kind can help verify that improvements in particular areas of school mathematics are feasible, that specific visions of the professionals in the field are reasonable.

Because several subsequent sections in this chapter are devoted to reviewing this kind of information, only a brief caveat is added here. In any professional field, research into what might be done is necessarily incomplete. Such research is generated as new questions are asked, new possibilities envisioned. And in a growing field, new questions are asked faster than research can address them, especially during times of changing cultures, values, and priorities—times that we are witnessing now in mathematics and science education. The consequence is that research can never answer all the questions about which directions are most promising. Where such research has been conducted, however, the information can be extremely useful.

What Do We Know?

Armed with some guidelines for what to expect from research, we are in a better position to interpret the research findings that are relevant for the NCTM Standards. Turning to the literature, however, we are faced with an immediate dilemma: Which studies should we include? There are hundreds of them. Some seem to be more relevant than others; some seem to be of higher quality than others; some present results that conflict with others. How do we make sense of the vast array of reports? There is no simple answer. We could say that we will include only reports of traditional scientific experiments,[3] studies in which one treatment is compared with another and researchers try to ensure that all features not being compared are similar in both treatments. That criterion sounds reasonable. The problem is that these studies are useful for answering only some kinds of questions. Other questions do not lend themselves to studies requiring scientific manipulation in the traditional sense. Indeed, holding constant the non-manipulated features is often impossible in classroom research. Recall the conjecture about calculators discussed in the previous section. Not only can a restriction to experimental studies lead to a misinterpretation of their results, it also indicates that other kinds of research methods must be considered.

Three criteria are helpful for judging the usefulness of research in mathematics education. First, the research must be of high quality. That means it must be educationally significant and must have scientific merit. Educationally significant studies are those that ask important questions; they address significant problems in teaching and learning mathematics. Studies with scientific merit are studies that present trustworthy results. Trustworthiness can be demonstrated by researchers in many ways, including by giving full descriptions of the methods so that readers can draw their own conclusions from the results.[4]

The second criterion is that, across the set of studies addressing a particular topic, a variety of perspectives and methodologies should be represented. Because there is rarely a single, critical study that answers an educational question, we must look for convergence from a number of studies. Our confidence rises if we find convergence from multiple directions and from studies that have used different research methods.

A third criterion is that research should be directed toward understanding a phenomenon. Why do some things work better than others? Under what conditions can we expect certain approaches to work? Understanding why things work provides the information we need to anticipate the ways in which they might need to be modified to work in other contexts.

In addition to using these three criteria in selecting research, the following discussion focuses on research directed toward (1) issues of teaching and learning, especially the empirical connection between teaching methods or curriculum materials and students' learning, and (2) assessing the degree to which particular kinds of instructional practices are prevalent in classrooms. Both kinds of research are relevant for the NCTM Standards, either for the nature of the Standards themselves or for policy decisions that require information on the extent to which the Standards are influencing classroom practice. Of special interest are research domains in which enough work has been done for results to be accumulating, forming patterns, and suggesting conclusions that can be drawn with some confidence. Finally, most of the studies cited were conducted in the United States. Although many studies that fit the previous criteria have been conducted in other countries, the question always arises of whether something that works well in one culture can be

[3] A team of researchers made this decision in their report to the California State Board of Education (Dixon, Carnine, Lee, Wallin, & Chard, 1998).

[4] The issue of what constitutes high-quality research is, itself, a complicated question. See, for example, the presentations in Part V ("Evaluation of Research in Mathematics Education") in Sierpinska and Kilpatrick (1998), including chapters by Lester and Lambdin and by Hanna; see also Kilpatrick (1993).

imported into another culture. To avoid that concern, this review focuses on research from U.S. classrooms.

Baseline Conclusion: Students Learn What They Have an Opportunity to Learn

One of the most reliable findings from research on teaching and learning is that students learn what they are given opportunities to learn. "Opportunity to learn" is a significant phrase. It means more than just receiving information. Providing an opportunity to learn means setting up the conditions for learning that take into account students' entry knowledge, the nature and purpose of the tasks and activities, the kind of engagement required, and so on. Sometimes, one hears an alternative version of that finding: Students learn what they are taught. The reality, however, is not quite that simple. Imagine, for example, first graders' being taught about calculus by listening to a traditional lecture. They will probably not learn much about calculus, given their relative lack of preparation, the inaccessibility of the ideas, and so on. But they might learn something, like how to sit quietly and listen to discourse they do not understand. Why will they learn that behavior and not calculus? Because it is what they have an opportunity to learn. It may not have been what was intended, but it was what the conditions afforded. Providing an opportunity to learn what is intended means providing the conditions in which students are likely to *engage* in tasks that involve the relevant content. Such engagement might include listening, talking, writing, reasoning, and a variety of other intellectual processes.

The consequence of this fact is obvious, and it can be stated in two ways. The first is descriptive: When one finds that students have acquired particular kinds of knowledge and skills, one can presume that they have had opportunities to learn these kinds of knowledge and skills. In other words, the kinds of knowledge and skills students acquire say something about the opportunities they have had. The second way of stating the consequence is prescriptive: If one wants students to acquire particular kinds of knowledge and skill, then one must provide them opportunities to do so. If one wants them to *know* certain things, such as the fact that common fractions represent exact measures and decimal fractions represent approximate measures, then one must engage them in tasks that address this issue. If one wants them to do certain things, such as reason deductively, then one must engage them in doing those things.

We can use this rather simple, baseline conclusion to interpret much of the research in mathematics teaching and learning. The review begins with what we know about the outcomes of traditional classroom practice and then moves to the results from alternative practices. An effort is made to explore the reasons for the research findings.

Student Learning With Traditional Curricula and Pedagogy

What is traditional pedagogy? Some people might be surprised to learn that a quite consistent, predictable way of teaching mathematics prevails in the United States. Here is an often-cited account from a researcher's observations of mathematics lessons:

> First, answers were given for the previous day's assignment. A brief explanation, sometimes none at all, was given of the new material, and problems were assigned for the next day. The remainder of the class was devoted to students working independently on the homework while the teacher moved about the room answering questions. The most noticeable thing about math classes was the repetition of this routine. (Welch, 1978, p. 6)

Readers might, as many do, recognize their own school mathematics experience in this description.

A good deal of data, collected over many years, describes classroom mathematics teaching in very similar ways. In 1975, the National Advisory Committee on Mathematical Education (NACOME) commissioned a study of elementary school mathematics instruction. They concluded the following:

> The [median] mathematics period is about 43 minutes long, and about half of this time is written work. A single text is used in whole-class instruction. The text is followed fairly closely. . . . Teachers are essentially teaching the same way they were taught in school. (p. 77)

One year later, the National Science Foundation funded a series of studies on classroom practice, including a national survey of teaching practices (Weiss, 1978) and a series of case studies (Stake & Easley, 1978). The quote above from Welch is taken from one of the case studies. The full set of findings is not easy to summarize, but James Fey (1979), who was asked to synthesize the results, noted that the most common pattern of classroom practice was "extensive teacher-directed explanation and questioning followed by student seatwork on paper-and-pencil assignments" (p. 494).[5] This kind of instruction, often called recitation, has been common for a long time, certainly since the turn of the twentieth century (Hoetker & Ahlbrand, 1969; Tharp & Gallimore, 1988).

Today, teachers continue to teach much like their forebears did. The best information on current practice comes from the Third International Mathematics and Science Study (TIMSS), a large-scale international study that included a "video study" of classroom teaching and an analysis of the curricula commonly used in the United States. The video study yielded a familiar sounding description of teaching:

> The typical eighth-grade mathematics lesson in the U.S. is organized around two phases: an acquisition phase and an application phase. In the acquisition phase, the teacher demonstrates or leads a discussion on how to solve a sample problem. The aim is to clarify the steps in the procedure so that students will be able to execute the same procedure on their own. In the application phase, students practice using the procedure by solving problems similar to the sample problem. (Stigler & Hiebert, 1997, p. 18).[6]

The emphasis on teaching procedures, especially computation procedures, is evident in the TIMSS video study. For example, teachers give little attention during the lesson to helping students develop conceptual ideas or even connect the procedures they are learning with the concepts that show why those procedures work. In fact, in 78% of the topics covered during the U.S. lessons, teachers stated procedures and ideas but did not develop them. The teacher, for example, might simply state that the area of a triangle is found by multiplying one half times the base times the height rather than discuss why this formula might work or show how it could be developed from finding the areas of parallelograms. Given the demonstration-practice routine of most lessons, we might not be surprised to learn that 96% of the time students were doing seatwork, they were practicing procedures. Less than 1% of the time were they doing creative work, such as inventing new procedures or analyzing new problems (Stigler & Hiebert, 1999).

Coupled with this information on teaching practices, the TIMSS curriculum analysis data also show that the traditional U.S. curriculum is relatively repetitive, unfocused, and undemanding (Schmidt, McKnight, & Raizen, 1996; Silver, 1998). Compared with curricula in other countries, the U.S. curriculum provides few opportunities for students to solve challenging problems and to engage in mathematical reasoning, communicating, conjecturing, justifying, and proving. Much of the curriculum deals with calculating and defining, and much of that activity is carried out in a rather simplistic way. Such a curriculum fits well with the traditional method of teaching.

What do students learn from these kinds of curricula and pedagogy? Using the baseline conclusion presented in the preceding section, we are tempted to guess the kinds of learning that result from these opportunities. But we can do better than guess. Extensive data sets are available that describe the mathematics achievement of students in the United States.

What do students learn in traditional programs? What do students learn from the mathematics curriculum and the instructional methods that have characterized most U.S. classrooms over the past years? The best information comes from the National Assessment of Educational Progress (NAEP), the TIMSS, and various statewide assessments. NAEP is especially useful because the items are matched to the U.S. curriculum and because the sampling design ensures a representative sample of U.S. students. More focused research projects can help illuminate particular subject areas or grade levels.

What do we know? From the NAEP administered in 1996, we know that almost all students learn to add, subtract, multiply, and divide whole numbers, most learn to do very simple arithmetic with fractions, decimals, and percents, and many are reasonably successful in solving one-step problems that involve these skills. For example, in eighth grade, 91% of the students surveyed added three-digit numbers with regrouping, 80% completed a long division problem, 83% rounded a decimal number to the nearest whole number, 58% found the percentage of a number, and 51% solved a problem like the following:

---

[5] These findings are especially interesting because they were collected shortly after an intensive effort to change mathematics education in the United States, an effort commonly called the "new math." In the 1960s and early 1970s, a number of projects produced curriculum materials and textbooks that significantly changed the nature of what was to be taught in Grades 1 to 12. But classroom practice seemed to be largely unaffected by these curriculum efforts. As the 1975 NACOME report goes on to say, "Almost none of the concepts, methods, or big ideas of modern mathematics programs have appeared in the median classroom" (p. 77).

[6] This description matches well what others have reported from a variety of observations, surveys, and literature reviews. For example, Dixon et al. (1998) summarize the conventional method of mathematics instruction, often used as control treatments in experimental mathematics education studies, in much the same way.

"Josie has 2/3 of a cup of cocoa. She is making chocolate cookies. Each batch of cookies uses 1/6 of a cup of cocoa. How many batches can she make?" (Kouba & Wearne, 2000; Wearne & Kouba, 2000).

Students' performance on most simple computation items has risen slightly or remained the same over the past decade or so (Wearne & Kouba, 2000). In fact, that statement seems to characterize the story one could tell looking back over the past 50 to 60 years. In 1967, during a time when some were concerned that the "new math" had eroded students' computational skills, Johnson and Rising noted that students' computation performance had never been as high as expected. For example, they cite a large study in 1932 that found that only 20% of 12th graders correctly computed 2.1 percent of 60 (p. 22).

In other areas, such as measurement and geometry, most students today also learn the simplest knowledge and skills, such as reading measuring instruments, calculating perimeters and areas of rectangles, and labeling simple polygons. For example, in eighth grade, 79% of the students surveyed correctly read a weighing scale even though not all increments were labeled, 66% drew a rectangle of 12 square units on square grid paper, and 78% used manipulatives to form a specified polygon (Kenney & Kouba, 1997; Strutchens & Blume, 1997).

We also know, however, that students' knowledge and skills are very fragile and apparently were learned without much depth or conceptual understanding. This fragility and shallowness become evident by studying performance on related items that require students to extend these skills, reason about them, or explain why they work. For example, in eighth grade, only 35% of the students correctly completed the following problems (Kouba & Wearne, 2000; Wearne & Kouba, 2000): Identify how many pieces are left if 65 pieces of candy are divided equally among 15 bags with each bag having as many as possible. Order fractions, like $\frac{1}{2}$, $\frac{2}{2}$, and $\frac{6}{7}$, by size (with calculators available). Find the number that is 7% more than 50. Multistep problems pose an even greater challenge. For example, 8% of the eighth graders solved a multistep problem on planning a trip that required adding miles, finding distance from miles per gallon, and calculating a fractional part of the trip (Wearne & Kouba, 2000).

Two conclusions can be drawn from these results. First, students' knowledge of the foundational topics of elementary mathematics is limited. It is evident only in the simplest contexts and is neither robust nor rich enough to extend to more complex or novel situations. A considerable research base has been building over the past 20 years to show that one reason for this limitation is the separation between procedural and conceptual knowledge (Hiebert, 1986, Hiebert & Carpenter, 1992). Given traditional instructional approaches, many students acquire skills for manipulating symbols and develop some understandings of the concepts, but these skills and understandings are unconnected. For example, Hiebert and Wearne (1986) report that only 33% of the seventh graders they interviewed could explain why they "lined up the decimal points" when they added and subtracted decimals. Many students follow rules and execute procedures they do not understand, making it impossible for them to modify or extend their skills to fit new situations or to monitor their performance and catch errors when they occur. After several years of executing procedures they do not understand, students' behavior is so rule-governed and so little affected by conceptual understanding that one can model their behavior and predict the errors they will make by looking only at the symbol manipulation rules they have been taught and pretending that they are following these rules like robots with poor memories (Brown & VanLehn, 1982; Hiebert & Wearne, 1985; Matz, 1980; VanLehn, 1983).

A second conclusion from the evidence on students' current mathematical performance is that students are more proficient with the processes of calculating, labeling, and defining than with other mathematical processes, such as reasoning, communicating, conjecturing, and justifying. Although students clearly are limited in using the former processes, they show considerably more expertise with them than with the latter. Students' performance on items that make demands on the latter processes is uniformly low.

Reasons for the outcomes. Although drawing direct connections between classroom practice and students' learning is always complicated and tricky, the massive amounts of converging data suggest that the baseline conclusion is, once again, confirmed. Students in the United States are learning best the kinds of mathematics that they are having the most opportunities to learn. Research on classroom practice indicates that students have more opportunities to learn simple calculation procedures, terms, and definitions than either to learn more complex procedures and why they work or to engage in mathematical processes other than calculation and memorization. Achievement data indicate that students are indeed learning simple calculation procedures, terms, and definitions. They are not learning what they have few opportunities to learn: how to adjust procedures to solve new problems or how to engage in other mathematical processes.

Before reviewing what we know about the effects of alternative instructional programs, we should note that the current debate about the future of mathematics education in the United States often casts the issue as a comparison between the traditional "proven" approaches and the new "experimental" approaches. Arguments against change sometimes claim that implementing unproven new programs is poor policy, and even unethical. Several years ago, Lee Hochberg, a reporter for Oregon Public Broadcasting, had this to say during a story on reform-minded mathematics teaching for the PBS NewsHour with Jim Lehrer: "Although there never was any scientific research conducted on the effectiveness of this style of teaching, the NCTM hoped that it would better prepare American students for the modern adult workplace" (11 May 1998).

The commendable part of these arguments is that they claim to promote research-based decision making. That effort is certainly appropriate and is, in fact, the reason for this chapter. But presuming that traditional approaches have proven to be successful is ignoring the largest database we have. The evidence suggests that the traditional curriculum and instructional methods in the United States are not serving students well. The evidence suggests that the long-running experiment we have been conducting with traditional methods has serious deficiencies and that we should attend carefully to the research findings that are accumulating regarding alternative programs.

Student Learning in Alternative Programs

Decisions about which instructional programs schools should adopt in place of traditional programs would be easy if we could compare them directly. But such comparison is often difficult. Why? For one thing, the different programs often have different goals and are designed to facilitate different kinds of learning. That point is crucial for interpreting research findings, so it needs to be elaborated.

Changing goals of alternative programs. In a general sense, the NCTM Standards call for a broadening of the goals of mathematics education. The Standards documents argue that the traditional focus on facts and skills should be expanded to include conceptual understanding and engagement in a range of mathematical processes. Mathematical expertise, they say, is found not only in students' factual knowledge but also in the strategic decisions they make while solving problems and in their ability to communicate these ideas with others.

Standards often are interpreted as statements about "what students should know and be able to do." What does that mean? An obvious, and appropriate, interpretation is that competence in mathematics comes from both knowing *and* doing. Traditional programs usually focus on knowing, and traditional assessments focus on knowledge that students have acquired. But what about doing? What does doing mathematics mean? Most mathematicians would say that it means engaging in the intellectual processes that are essential to mathematics: solving problems through reasoning, conjecturing, inferring, deducing, justifying, and so on. These processes are valued not only because they produce a certain kind of knowledge but also because an ability to engage in them is important in itself. Being able to do mathematics is an important part of mathematical competence.[7]

Research related to changing goals. Although decisions about goals are based on values and beliefs, several sources of information address the context and wisdom of these decisions. The information can be sorted into two statements: One sets the choice of goals in an international context, and the other pertains to whether the goals are too ambitious.

1. The "basics," as we have long defined them in the United States, are relatively limited and shallow.

As noted, the TIMSS analyses of curriculum and classroom instruction suggest that current U.S. instructional programs cover many topics but focus on a relatively narrow band of skills within these topics (Schmidt et al., 1996; Stigler & Hiebert, 1999). According to the curriculum documents and surveys of instructional programs, the basic goals in the United States apparently consist of learning terms, definitions, and computational processes. In eighth grade, for example, the basics are arithmetic for whole numbers, fractions, decimals, and percents, together with introductory algebra. In contrast, many other countries identify a richer and deeper set of goals for students, including engagement in mathematical processes, such as inductive and deductive reasoning, and conceptual understanding of the topics that students study.

If students are to become good mathematical problem solvers, they need more than definitions and computational skills. Good problem solvers analyze the meaning of problems, decide what approaches are most promising, select from the procedures they already know or adjust them for the situation, execute the procedures and monitor how well they are working, and check the

[7] See the chapters in this volume by Forman, by Greeno, and by Sfard for a further development of this argument.

reasonableness of answers (Charles & Silver, 1988; Schoenfeld, 1985). These skills do not develop automatically; students must have opportunities to develop and practice them. Such opportunities are offered more frequently in many other countries than in the United States.

2. Almost all children can learn more mathematics than they are learning now. In other words, it is reasonable to set more ambitious learning goals for students.

Several sources of information indicate that the goals set by the NCTM Standards are not overly ambitious. In fact, the traditional emphasis on skill learning appears to be overly cautious; it seems to underestimate the capabilities of U.S. students. Students can learn more than just skills, and focusing only on skills places unnecessary limits on students' growth.

The evidence for this claim converges from a number of research areas. One set of results shows that most young children when they enter school can already solve simple addition, subtraction, multiplication, and division story problems (Carpenter, Ansell, Franke, Fennema, & Weisbeck, 1993; Carpenter et al., 1982; Fuson, 1988; Gelman & Gallistel, 1978; Siegler & Robinson, 1981). Especially striking is the finding that children's invented problem-solving methods are based on their analyses of the meaning of the problems and do not show the common nonsensical kinds of errors made frequently by older children. For children who enter school without the counting skills and number sense of their peers and are unable to solve these problems, instructional programs can be designed to enrich their knowledge to a level equal to that of their peers (Griffin, Case, & Siegler, 1994). These results mean, among other things, that early instructional programs can be recalibrated to take into account children's entry competencies.

A second set of results shows that some people learn mathematics outside of school—on the street or in places of employment—and that they can learn with high levels of skill, low error rates, and sensible problem-solving methods. Much of this work has been done in other cultures and has documented the kinds of mathematics used by those engaged in commercial activities (Ginsburg, 1978; Nunes, Schliemann, & Carraher, 1993; Saxe, 1991; Scribner, 1984). Sometimes the mathematics learned in these settings demands a higher level of proficiency than the mathematics learned by the same people in school (Nunes et al., 1993). Although this work does not address directly how one might teach mathematics in school, it does suggest that, given the need and the appropriate context, people *can* learn and apply mathematics sensibly and proficiently.

A third kind of evidence supporting the claim for more ambitious expectations comes from international comparisons. Many such comparisons show that students in a number of other countries are learning more mathematics, and learning mathematics more deeply, than students in the United States (National Center for Education Statistics, 1996, 1997, 1998; Stevenson & Stigler, 1992 ; Travers & McKnight, 1985). There are, of course, many reasons for between-country differences in student performance. Differences in cultures, in home lives and home-school relationships, and in opportunities to learn both in and out of school all may influence how much mathematics students learn. But two points can be drawn from the cross-national data: (1) Given the level of learning evident in other countries, no cognitive or developmental constraints would appear to be preventing U.S. students from learning more than they do now; and (2) at least some of the national differences in students' performance might be due to such school and classroom factors as the nature of the curricula in use and the common instructional methods.

The final source of evidence for believing that U.S. students can learn much more than they are learning now comes from a wide range of studies showing that when extra attention is paid to designing classroom instruction with specific learning goals in mind, students usually improve their achievement of these goals (Carpenter, Fennema, Peterson, Chiang, & Loef, 1989; Cobb et al., 1991; Cognition & Technology Group at Vanderbilt [CTGV], 1997; Fawcett, 1938; Good, Grouws, & Ebmeier, 1983; Griffin et al., 1994; Heid, 1988; Hiebert & Wearne, 1993; Markovits & Sowder, 1994; Stein & Lane, 1996; Wearne & Hiebert, 1989; Wood & Sellers, 1996). Many of the studies cited here focused on learning goals consistent with those of the NCTM Standards, but studies from earlier eras that focused on somewhat different goals could be cited to make the same point. Additional effort in designing curriculum and instruction with learning goals in mind often pays off with improved student learning of the kind intended. This argument, of course, is an instance of the baseline conclusion proposed at the outset: Students' learning is related to opportunity to learn. But the specific point here is that students are not the problem; they can learn more if we can enrich their opportunities to do so.

But what are the specific effects of alternative instructional programs inspired by the NCTM Standards that share some of these more ambitious goals? What do students learn from such programs?

What, exactly, is the pedagogy of alternative programs? The media are filled with stories of the successes and failures of reform-minded instructional programs in mathematics. Indeed, the wars being fought over the future direction of mathematics education are filled with claims and counterclaims about whether new programs work. Here is an excerpt from the 26 April 26 1998 edition of the *Riverside Press-Enterprise* newspaper:

> High failure rates and concerns that students are not learning the math skills they need has prompted a third of Inland area high schools trying a new college-prep program to drop it. Riverside's Poly High School discontinued College Preparatory Mathematics in June after only 27 percent of the algebra I students earned a C or better. One semester after scrapping the program, the passing rate went up to 42 percent. (Sharma, 1998)

As the story continues, we see that there is no consensus about whether the new program is a failure or about why it is having the reported effect. Many opinions are expressed, such as that NCTM-inspired programs like this one are doomed to fail, but no clear conclusions are drawn. Of course, there can be no clear conclusions, because no information was collected systematically about what was going on in classrooms. We do not know how the program was being implemented, so we have no way to evaluate its effectiveness. Debates may revolve around the intent of curriculum, but they are debates about values. If we want to know about the effects of the program, we need to know how the program was being practiced, in classrooms.[8] Unfortunately, many of the claims and counterclaims about the effects of alternative programs are based on these kinds of stories, without the benefit of real information.

To answer questions about the effects of alternative programs, we first need to describe what the programs look like in practice. We need to consult research studies that have described, in detail, the nature of the classroom practice being evaluated. What kinds of learning opportunities are provided in these programs? That question is almost impossible to answer in a brief review because so many different alternative programs have been developed. Even if we stick with programs that are trying to translate the NCTM Standards into practice, too many versions exist to lump them into one description. To make matters more complicated, in many research domains (particular grade levels and mathematical topics), individual studies have reported intriguing results, but the domains do not yet show a convergence of findings from multiple research programs. Although results in these areas may be suggestive, we cannot draw conclusions with high levels of confidence.

One area in which considerable work has been done in designing and testing alternative instructional programs is the arithmetic of the primary grades (Carpenter et al., 1989; Cobb et al., 1991; Fennema et al., 1996; Fuson & Briars, 1990; Hiebert & Wearne, 1992, 1993, 1996; Kamii, 1985; Kamii & Joseph, 1989; Villasenor & Kepner, 1993; Wood & Sellers, 1996). Because many of the investigators have been engaged in independent research programs, differences are evident in the alternative instructional programs that have been implemented in classrooms. But some significant similarities are also apparent, and these similarities are of particular interest.

The following features characterize many of the alternative programs in primary-grade arithmetic:

- Building directly on students' entry knowledge and skills. Many students enter school being able to count and solve simple arithmetic problems. Alternative programs take advantage of this proficiency by gradually increasing the range of problem types and the size of the numbers.
- Providing opportunities for both invention and practice. Classroom activity often revolves around solving problems that require some creative work by students and some practice of already learned skills. For example, second graders may have been doing subtraction problems like 345 – 127 = __ but then are asked to work out a method for 403 – 265 = __, a problem with a 0 in the minuend.
- Focusing on the analysis of (multiple) methods. Classroom discussion usually centers on the methods for solving problems, methods that have been presented by the students or the teacher. Methods are compared for similarities and differences, advantages and disadvantages.
- Asking students to provide explanations. Students are expected to present solutions to problems, to describe the methods they use, and to explain why the methods work.

Although these features are distilled from the arithmetic programs for young children, they are not limited

---

[8]Beyond the absence of information about classroom practice, there are other missing elements in this story, elements that are needed to interpret the "facts." For example, what does it mean for the passing grades a teacher assigns to move from 27% to 42%? Are students learning more? Maybe they are, but maybe they are being tested on easier material.

to such programs. Research reports of alternative instructional programs in other areas share these features. These programs include the comprehensive problem-solving program for middle school students commonly referred to as the Jasper Project (CTGV, 1997), as well as smaller-scale research programs on students' learning of common fractions (Behr, Wachsmuth, Post, & Lesh, 1984; Mack, 1990), decimal fractions (Wearne & Hiebert, 1988, 1989), and calculus (Heid, 1988; Palmiter, 1991). In these studies, both the general features of the instructional programs and the student outcomes are consistent with the findings of the studies on arithmetic in the early grades. That consistency is apparent even though some features of the programs, such as the large-scale problems students solve in the Jasper Project or the technology-enhanced nature of the calculus instruction, make them look quite different on the surface.

What do students learn in alternative programs? The conclusions that we can draw about students' learning in these instructional programs are shaped continuously by new research findings. Much, but not all, of the research comes from the programs focusing on elementary school students' learning of arithmetic, where the evidence is converging rapidly. A current set of conclusions we can draw about students' learning in these instructional programs would include the following:

- Instructional programs that emphasize conceptual development, with the goal of understanding, can facilitate significant mathematics learning without sacrificing skill proficiency.

In light of the previous discussion, it should come as no surprise that instruction can be designed to promote deeper conceptual understanding. If students have more opportunity to construct mathematical understandings, they will construct them more often and more deeply. The question is, At what cost? Will they fail to master other knowledge or skills that we value? That question is essential given the public debates about the effects of reform-minded instruction. The claim above says that students can learn both, that well-designed and implemented instructional programs can facilitate both conceptual understanding and procedural skill. What is the evidence for that claim? Once again, it comes from several sets of converging information.

The most direct source is the collection of studies that have compared primary-grade students' performance in arithmetic after some students participated in a year or more of instruction designed to promote conceptual understanding (Carpenter et al., 1989; Cobb et al., 1991; Kamii, 1985; Kamii & Joseph, 1989; Hiebert & Wearne, 1993, 1996; Wood & Sellers, 1996). Overall, the results showed that students in the alternative programs learned the standard arithmetic skills at least as well as their traditionally taught peers and, more than that, constructed a deeper understanding of the concepts, such as place value, that underlie the procedures. This understanding showed itself in a variety of ways, including students' ability to invent new procedures or modify old ones to solve new problems. When individual students were followed over time, it was clear that more students in the alternative program connected concepts, like place value, with procedures, like regrouping to subtract, and that such understanding *enhanced* skill development instead of detracting from it (Hiebert & Wearne, 1996).

Although we have no reason to think that findings in the upper grades would show significantly different results, less data are available that address the question directly. Many of the studies with older students have implemented alternative instruction of this kind for shorter periods of time—weeks rather than years (Mack, 1990; Wearne & Hiebert, 1988). Although the results show trends that are consistent with the results for younger students, longer-term studies must be conducted to confirm these results. When such studies have been done, the results fit the pattern. After one year of participation in the middle school Jasper Project, students showed the same level of skill proficiency as their traditionally taught peers but higher levels of problem-solving ability and better attitudes toward mathematics (CTGV, 1997). As another example, Heid (1988) compared two semester-long calculus classes. One class emphasized the conceptual ideas of the course, using many of the instructional features listed above (adjusted for calculus students). They practiced computational procedures only during the last few weeks, after the students and the instructor had developed them. The other class practiced the procedures in a traditional way for the full 15 weeks. The results were much like those of primary-grade programs: The two classes showed similar levels of skill proficiency, but the conceptual development class showed a deeper understanding of the ideas.

An important point to emphasize is that the alternative programs reviewed here did not choose conceptual development to the exclusion of practice. They did not concentrate only on conceptual understanding and ignore skills. Because the roles of conceptual development and practice have been the flash point for a good deal of misunderstanding, some clarification is worthwhile. Practice clearly is essential for acquiring cognitive skills of almost any kind (Anderson, Reder, & Simon, 1996). What distinguishes the alternative programs is that other mathematical processes, such as invention and justification, are equally visible. Because these processes often are absent from traditional programs, they have received the attention in the public debates. But practice and the development of skills are nonetheless a part of the alternative programs reviewed here.

The finding that conceptual understanding does not detract from skill proficiency and may even enhance the recall and use of skills is not surprising to psychologists who study learning (e.g., Brownell, 1935; Bruner, 1973; Hilgard, 1957; Siegler, this volume). That relationship has been known for years and makes good sense. If students understand how and why a procedure works, they will probably remember it better and might be able to adjust it to solve a new problem; if students have memorized a procedure and have no clue how it works, they have little chance of using it flexibly. That phenomenon is exactly what is reported when the progress of individual students is traced over several years as they acquire the concepts and skills of arithmetic (Carpenter, Franke, Jacobs, Fennema, & Empson, 1998; Hiebert & Wearne, 1996).

The significance of the relationship between conceptual understanding and skill proficiency has not been lost on mathematics educators who have demonstrated over the years that more effective instructional programs pay more attention to conceptual development. These results have been found even when the goals of mathematics learning were restricted to the standard achievement measures. In one of the largest and most systematic research programs on mathematics teaching and learning, Good et al. (1983) demonstrated that when mathematics teachers devoted a substantive portion of each lesson to developing the topic conceptually, students' performance increased on standard achievement measures, most of which were skill oriented.

The difference in the more recent alternative programs lies in instructional features that aim to develop both concepts and skills and in research studies reporting their effects that give detailed descriptions of how those features interact. The following conclusions capture two important findings.

1. Problems to be solved can be used effectively as a context for students to learn new concepts *and* skills, not just as applications of previously learned skills.

The traditional approach to solving problems in U.S. classrooms is to teach a procedure and then assign students problems on which to practice the procedure. Problems are viewed as applications of already learned procedures. This view fits the acquisition-application pattern of U.S. lessons. It is so ingrained in our thinking that teachers begin to believe it is unfair to present students with a problem for which they have not already learned a solution method.

The alternative instructional programs cited earlier stand these assumptions on their head. The programs were designed from different theories of how students best learn to use skills effectively. Acquiring skills does not need to be separated from using them. Students can acquire skills *as* they develop them to solve problems. In fact, the development of the skill can, itself, be treated as a problem for students to solve. For example, suppose that students knew the meaning of $\frac{2}{3}$ and $\frac{3}{5}$. They might be given the following problem: Find the best method for adding the two numbers. The students could reason through this problem, develop several different methods, and then analyze the advantages and disadvantages of each. Practice of the new methods then would occur through solving problems that extended these methods to even more challenging situations. Treating the development of procedures as problems to be solved is a characteristic of many of the alternative programs cited above.

The results of these programs indicate that this approach can help students become proficient skill users—equally proficient as, and sometimes more proficient than, their traditionally taught peers. And these results are found with less time devoted to practicing. This phenomenon is not surprising. If one understands how and why a method works, one does not need to practice it as much to use it correctly and remember it. The data indicate that is what happens when students develop procedures in the context of solving problems.

2. If students overpractice procedures before they understand them, it is more difficult to make sense of them later.

A long-running debate has been whether students should practice procedures first and then try to understand them or understand the procedures first before practicing them. Carefully developed theoretical arguments have been launched on both sides of the issue: practice before understanding (Nolan, 1973) and understanding before practice (Brownell, 1954; Brueckner, 1939; Kaput, 1987). Many of the alternative instructional programs cited above finesse the issue by mixing the two so that both are occurring simultaneously. But evidence from other studies suggests that the traditional approach, which often encourages students to practice rules and procedures they do not understand, might make it difficult for them to go back later and construct an understanding (Brownell & Chazal, 1935; Mack, 1990; Resnick

& Omanson, 1987; Wearne & Hiebert, 1988). For example, Wearne and Hiebert (1988) found that fifth and sixth graders who had already practiced rules for adding and subtracting decimal fractions were less inclined than fourth graders, with no such experience, to construct an understanding of related decimal concepts.

Cognitive science provides some explanations for why it is difficult to make sense of an overpracticed procedure. As humans are introduced to new information, such as a rule for adding decimal fractions, the information is stored as separate statements, perhaps corresponding to the steps in the procedure. As the procedure is practiced, the individual statements or steps become increasingly connected and eventually lose their identity, becoming stored as a single procedure (Anderson, 1983). This condensation is useful for fast execution but not for reflecting on the meaning of each step and why the steps fit together the way they do. Hatano (1988) observed, "This process of acceleration of calculation speed results in a sacrifice of understanding and of the construction of conceptual knowledge. It is hard to unpack a merged specific rule to find the meaning of any given step" (p. 64). The moral of the story is that, if students are to develop both proficiency and understanding of skills, the most efficient instructional approach is to build understanding into students' experience from the beginning.

Summary

The United States has been engaged in a long running experiment with a particular instructional method and curriculum. Instruction has focused on teachers' demonstrating procedures and providing time for students to practice them. The curriculum emphasizes procedures and definitions. With the changing, more ambitious learning goals for students and the increasing belief that all students should become mathematically competent, the deficiencies of the traditional approach are becoming more apparent. Students still are learning best what they have an opportunity to learn, but the opportunities are unnecessarily limited.

Alternative instructional programs designed with more ambitious learning goals in mind have demonstrated that students can acquire both concepts and skills at higher levels than in comparative traditional programs. This conclusion is significant, and it sounds like a conclusion that could end this chapter. But now we come to a curious fact, a fact that can be phrased as a question: If these research findings are trustworthy, why have we not seen more such alternative programs around the country? Why have we not acted on what we know?

Why Have We Not Acted on What We Know?

If it is true that instructional programs can be designed to facilitate more ambitious learning goals for students, why do we not see them more often?[9] Why do we read stories of failed programs, like the story carried in the *Riverside Press-Enterprise* (Sharma, 1998)? Perhaps the research data are flawed. Maybe the alternative programs are not very effective. But the topics reviewed above were chosen to minimize this possibility—they represent areas that have been heavily researched by a variety of independent programs. The convergence of findings suggests that flawed data is an unlikely explanation.

Another possibility, one that is receiving considerable empirical support, is that the alternative programs, which show great promise in research settings, might not be implemented effectively when adopted by schools and districts. The reason for this ineffectiveness is simple but underappreciated: It is difficult to change the way we teach. The new, more ambitious instructional programs require teachers to make substantial changes. They are asked to adopt broader learning goals, to think differently about mathematics and how students learn it, and to change their instructional methods. Such changes do not happen automatically; they require learning. And learning for teachers, just as for students, requires an opportunity to learn. Unless such opportunities are provided, teachers are asked to do the impossible—teach in new ways without having had a chance to learn them. It would be foolish to frame an entire analysis of high-quality mathematics instruction around the opportunities for students to learn, as I have done here, and then, when wondering about the problems with effective implementation, overlook the opportunities for teachers to learn.

Recent research has not overlooked the link between teachers' learning and effective implementation of new programs, and it has uncovered three significant findings.

1. Most teachers have relatively few opportunities to learn new methods of teaching.

This finding has been reported mostly for elementary school teachers (Cohen & Hill, 1998; Lord, 1994; O'Day & Smith, 1993; Weiss, 1994), but we have no reason to believe it is not equally true for secondary teachers. For example, in California, the site of a great deal of reform

[9] Since this chapter was written, many such curriculum and instructional programs have been initiated, and the results of these efforts deserve careful monitoring. The conclusions identified in this section, however, are still relevant.

activity, fewer than 5% of elementary school teachers participated in more than two weeks of special in-service education during the year of the survey (Cohen & Hill, 1998). Most teachers had one day or less of learning opportunities.

The absence of learning opportunities for teachers is especially striking when compared with other occupations concerned about improvement. General Motors, for example, provides 92 hours per year for continuing education for every Saturn employee to improve his or her design and production skills. In the Cohen & Hill (1998) survey, only about 4% of California elementary school teachers had this much time for improving their skills. Albert Shanker (1993), former president of the American Federation of Teachers, said this about the Saturn program:

> It is ironic that a bunch of people whose business is building cars understand so well the importance of educating their employees, whereas people in education seem to assume that teachers and other school staff will be able to step right into a new way of doing things with little or no help. If it takes ... 92 hours a year per employee to make a better automobile, it will take that and more to make better schools. (p. 34)

2. The amount and kinds of opportunities to learn make a difference in teachers' classroom practice and in students' learning.

Do learning opportunities for teachers really make a big difference? A rapidly building research base says, unequivocally, yes. Opportunities for teachers to learn yields big dividends in improved classroom practices, and, in turn, in increased student learning (Brown, Smith, & Stein, 1996; CTVG, 1997;Cohen & Hill, 1998; Fennema et al., 1996; Parke & Smith, 1998; Saxe, 1996; Swafford, Jones, & Thornton, 1997). In fact, some reports show that investing in teacher learning is the most effective way to improve student learning (Greenwald, Hedges, & Laine, 1996).

The relationship between opportunity to learn and changes in classroom practice works in two ways. One is that rich opportunities to learn result in classroom practice that is consistent with the intent of the alternative programs that teachers are trying to implement. This change, in turn, yields the kinds of learning gains cited earlier. The other way is that if opportunities are weak or irrelevant to the new program, teachers' practice changes only superficially and sometimes in unintended ways. For example, teachers might ask students to work in small groups but retain the same goals and mathematical tasks. That approach leaves students' opportunities to learn essentially unchanged. Hiebert and Stigler (2000) report that many eighth-grade mathematics teachers who said that they were implementing the NCTM recommendations showed only superficial similarities with the recommendations. That claim is what one would expect if these teachers have had few opportunities to acquire new teaching skills.

3. Rich opportunities to learn new teaching practices share several core features.

The third finding parallels what we know about effective learning opportunities for students—effective teacher in-service programs share a number of core features (CTGV, 1997; Cohen & Hill, 1998; Elmore, Peterson, & McCarthey, 1996; Fennema et al., 1996; Little, 1982, 1993; Schifter & Fosnot, 1993; Stein, Silver, & Smith, 1998; Stigler & Hiebert, 1997; Swafford et al., 1997). These features are (1) ongoing collaboration—measured in years—of teachers for purposes of planning, with (2) the explicit goal of improving students' achievement of clear learning goals, (3) anchored by attention to students' thinking, the curriculum, and pedagogy, with (4) access to alternative ideas and methods, and opportunities to observe these in action.

If most teachers have one day or less of learning opportunities during the year, they will obviously experience none of these features. And opportunities that do not include these features cannot be expected to help teachers improve their practice significantly. No wonder, then, that the alternative programs are not spreading wildly throughout the country. Research shows that instructional programs aligned with the NCTM Standards are effective, but it also shows that lots of hard work will be needed to scale up from research sites to all classrooms.

Conclusions

We can now more easily see why the short answer given to the opening question of this chapter is incomplete. Answering yes to whether the NCTM Standards are supported by research vastly oversimplifies the issues. Oversimplifications can confuse public discussion and damage efforts to achieve the goal that most people share: improved mathematics learning in our schools.

The longer, more complete answer that has been spun out in the preceding pages can be summarized as follows: The Standards proposed by NCTM are, in many ways, more ambitious than those of traditional programs. Based on beliefs about what students should know and be able to do, the Standards address conceptual understanding and the use of key mathematical processes as well as skill proficiency. The best evidence we have indicates that most traditional programs do not provide students with many opportunities to achieve these additional goals and, not surprisingly, that most students do not achieve them. Alternative programs can be designed to provide these opportunities and, where these programs have been implemented with fidelity for reasonable lengths of time, students have learned more and learned more deeply than in traditional programs. Although the primary evidence comes from elementary school, especially the primary grades, none of the evidence is inconsistent. That is, no programs at any level that share the core instructional features and have been implemented as intended for reasonable lengths of time, show that students perform more poorly than their traditionally taught peers.

But that is not the end of the story. Alternative programs consistent with the NCTM Standards often require considerable learning by the teacher. Without new opportunities to learn, teachers must either stick with their traditional approaches or add on a feature or two of the new programs (e.g., small-group activity) while retaining the same goals and lesson designs. From the available evidence, one can reasonably presume that it is these practices that are often being critiqued as not producing higher achievement.

What we have learned from research now brings us back to an issue of values. We now know that we *can* design curriculum and pedagogy to help students meet the ambitious learning goals outlined by the NCTM Standards. The question is whether we value these goals enough to invest in opportunities for teachers to learn to teach in the ways the goals require. That is a policy decision, but it is a decision that can be guided by research.

ACKNOWLEDGMENTS

A brief version of this chapter appeared as "Relationships Between Research and the NCTM Standards" in the January 1999 issue of the *Journal for Research in Mathematics Education*. I thank Bent Schmidt-Nielsen, and Brent McClain for their helpful comments on an earlier draft.

REFERENCES

Anderson, J. R. (1983). *The architecture of cognition*. Cambridge, MA: Harvard University Press.

Anderson, J. R., Reder, L. M., & Simon, H. A. (1996). Situated learning and education. *Educational Researcher, 25*(4), 5–11.

Behr, M. J., Wachsmuth, I., Post, T. R., & Lesh, R. (1984). Order and equivalence of rational numbers: A clinical teaching experiment. *Journal for Research in Mathematics Education, 15*, 323–341.

Brown, C. A., Smith, M. S., & Stein, M. K. (1996, April). *Linking teacher support to enhanced classroom instruction*. Paper presented at the meeting of the American Educational Research Association, New York.

Brown, J. S., & VanLehn, K. (1982). Towards a generative theory of "bugs." In T. P. Carpenter, J. M. Moser, & T. A. Romberg (Eds.), *Addition and subtraction: A cognitive perspective* (pp. 117–135). Hillsdale, NJ: Erlbaum.

Brownell, W. A. (1935). Psychological considerations in the learning and teaching of arithmetic. In W. E. Reeve (Ed.), *The teaching of arithmetic* (Tenth yearbook of the National Council of Teachers of Mathematics, pp. 1–31). New York: Bureau of Publications, Teachers College, Columbia University.

Brownell, W. A. (1954). The revolution in arithmetic. *Arithmetic Teacher, 1*, 1–5.

Brownell, W. A., & Chazal, C. B. (1935). The effects of premature drill in third-grade arithmetic. *Journal of Educational Research, 29*, 17–28.

Brueckner, L. J. (1939). The development of ability in arithmetic. In G. M. Whipple (Ed.), *Child development and the curriculum* (Thirty-eighth yearbook of the National Society for the Study of Education, Part 1, pp. 275–298). Bloomington, IL: Public School Publishing.

Bruner, J. S. (1973). *Beyond the information given*. New York: Norton.

Carpenter, T. P., Ansell, E., Franke, M. L., Fennema, E., & Weisbeck, L. (1993). Models of problem solving: A study of kindergarten children's problem-solving processes. *Journal for Research in Mathematics Education, 24*, 428–441.

Carpenter, T. P., Fennema, E., Peterson, P. L., Chiang, C. P., & Loef, M. (1989). Using knowledge of children's mathematical thinking in classroom teaching: An experimental study. *American Educational Research Journal, 26*, 499–531.

Carpenter, T. P., Franke, M. L., Jacobs, V. R., Fennema, E., & Empson, S. B. (1998). A longitudinal study of invention and understanding in children's multidigit addition and subtraction. *Journal for Research in Mathematics Education, 29*, 3–20.

Carpenter, T. P., Moser, J. M., & Romberg, T. A. (Eds.). (1982). *Addition and subtraction: A cognitive perspective*. Hillsdale, NJ: Erlbaum.

Charles, R. I., & Silver, E. A. (Eds.). (1988). *The teaching and assessing of mathematical problem solving*. Reston, VA: National Council of Teachers of Mathematics.

Cobb, P., Wood, T., Yackel, E., Nicholls, J., Wheatley, G., Trigatti, B., & Perlwitz, M. (1991). Assessment of a problem-centered second-grade mathematics project. *Journal for Research in Mathematics Education, 22*, 3–29.

Cognition and Technology Group at Vanderbilt (1997). *The Jasper project: Lessons in curriculum, instruction, assessment, and professional development.* Mahwah, NJ: Erlbaum.

Cohen, D. K., & Hill, H. C. (1998). *State policy and classroom performance: Mathematics reform in California* (CPRE Policy Briefs, RB-23). Philadelphia: University of Pennsylvania, Consortium for Policy Research in Education.

Dixon, R. C., Carnine, D. W., Lee, D.-S., Wallin, J., & Chard, D. (1998). *Review of high quality experimental mathematics research* (Report to the California State Board of Education). Eugene: University of Oregon, National Center to Improve the Tools of Educators.

Elmore, R. F., Peterson, P. L., & McCarthey, S. J. (1996). *Restructuring in the classroom: Teaching, learning, and school organization.* San Francisco: Jossey-Bass

Fawcett, H. P. (1938). *The nature of proof* (Thirteenth yearbook of the National Council of Teachers of Mathematics). New York: Bureau of Publications, Teachers College, Columbia University.

Fennema, E., Carpenter, T. P., Franke, M. L., Levi, L., Jacobs, V. R., & Empson, S. B. (1996). A longitudinal study of learning to use children's thinking in mathematics instruction. *Journal for Research in Mathematics Education, 27*, 403–434.

Fey, J. (1979). Mathematics teaching today: Perspectives from three national surveys. *Mathematics Teacher, 72*, 490–504.

Fuson, K. C. (1988). *Children's counting and concepts of number.* New York: Springer-Verlag.

Fuson, K. C., & Briars, D. J. (1990). Using a base-ten blocks learning/teaching approach for first- and second-grade place-value and multidigit addition and subtraction. *Journal for Research in Mathematics Education, 21*, 180–206.

Gelman, R., & Gallistel, C. R. (1978). *The child's understanding of number.* Cambridge, MA: Harvard University Press.

Ginsburg, H. (1978). Poor children, African mathematics, and the problem of schooling. *Educational Research Quarterly, 2*, 26–43.

Good, T. L., Grouws, D. A., & Ebmeier, H. (1983). *Active mathematics teaching.* New York: Longman.

Greenwald, R., Hedges, L. V., & Laine, R. D. (1996). The effect of school resources on student achievement. *Review of Educational Research, 66*, 361–396.

Griffin, S. A., Case, R., & Siegler, R. S. (1994). Rightstart: Providing the central conceptual prerequisites for first formal learning of arithmetic to students at risk for school failure. In K. McGilly (Ed.), *Classroom lessons: Integrating cognitive-theory and classroom practice* (pp. 25–49). Cambridge, MA: MIT Press.

Hanna, G. (1998). Evaluating research papers in mathematics education. In A. Sierpinska & J. Kilpatrick (Eds.), *Mathematics education as a research domain: A search for identity* (New ICMI Studies Series, Vol. 4, pp. 399–407). Dordrecht, The Netherlands: Kluwer.

Hatano, G. (1988). Social and motivational bases for mathematical understanding. In G. B. Saxe & M. Gearhart (Eds.), *Children's mathematics* (pp. 55–70). San Francisco: Jossey-Bass.

Heid, M. K. (1988). Resequencing skills and concepts in applied calculus using the computer as a tool. *Journal for Research in Mathematics Education, 19*, 3–25.

Hembree, R., & Dessart, D. J. (1986). Effects of hand-held calculators in precollege mathematics education: A meta-analysis. *Journal for Research in Mathematics Education, 17*, 83–99.

Hiebert, J. (Ed.). (1986). *Conceptual and procedural knowledge: The case of mathematics.* Hillsdale, NJ: Erlbaum.

Hiebert, J., & Carpenter, T. P. (1992). Learning and teaching with understanding. In D. A. Grouws (Ed.), *Handbook of research on mathematics teaching and learning* (pp. 65–97). New York: Macmillan.

Hiebert, J., & Stigler, J. W. (2000). A proposal for improving classroom teaching: Lessons from the TIMSS Video Study. *Elementary School Journal, 101*, 3–20.

Hiebert, J., & Wearne, D. (1985). A model of students' decimal computation procedures. *Cognition and Instruction, 2*, 175–205.

Hiebert, J., & Wearne, D. (1986). Procedures over concepts: The acquisition of decimal number knowledge. In J. Hiebert (Ed.), *Conceptual and procedural knowledge: The case of mathematics* (pp. 199–223). Hillsdale, NJ: Erlbaum.

Hiebert, J., & Wearne, D. (1992). Links between teaching and learning place value with understanding in first grade. *Journal for Research in Mathematics Education, 23*, 98–122.

Hiebert, J., & Wearne, D. (1993). Instructional tasks, classroom discourse, and students' learning in second-grade arithmetic. *American Educational Research Journal, 30*, 393–425.

Hiebert, J., & Wearne, D. (1996). Instruction, understanding and skill in multidigit addition and subtraction. *Cognition and Instruction, 14*, 251–283.

Hilgard, E. R. (1957). *Introduction to psychology* (2nd ed.). New York: Harcourt Brace.

Hoetker, J., & Ahlbrand, W. (1969). The persistence of the recitation. *American Educational Research Journal, 6*, 145–167.

Johnson, D. A., & Rising, G. R. (1967). *Guidelines for teaching mathematics.* Belmont, CA: Wordsworth.

Kamii, C. K. (1985). *Young children reinvent arithmetic.* New York: Teachers College Press.

Kamii, C., & Joseph. L. L. (1989). *Young children continue to reinvent arithmetic.* New York: Teachers College Press.

Kaput, J. J. (1987). Toward a theory of symbol use in mathematics. In C. Janvier (Ed.), *Problems of representation in the teaching and learning of mathematics* (pp. 159–195). Hillsdale, NJ: Erlbaum.

Kenney, P. A., & Kouba, V. L. (1997). What do students know about measurement? In P. A. Kenney & E. A. Silver (Eds.),

Results from the sixth mathematics assessment of the National Assessment of Educational Progress (pp. 141–163). Reston, VA: National Council of Teachers of Mathematics.

Kilpatrick, J. (1993). Beyond face value: Assessing research in mathematics education. In G. Nissen & M. Blomhøj (Eds.), *Criteria for scientific quality and relevance in the didactics of mathematics* (pp. 15–34). Roskilde, Denmark: Danish Research Council for the Humanities.

Kouba, V. L., & Wearne, D. (2000). Whole number properties and operations. In E. A. Silver & P. A. Kenney (Eds.), *Results from the seventh mathematics assessment of the National Assessment of Educational Progress* (pp. 141–161). Reston, VA: National Council of Teachers of Mathematics.

Lester, F. K., Jr., & Lambdin, D. V. (1998). The ship of Theseus and other metaphors for thinking about what we value in mathematics education research. In A. Sierpinska & J. Kilpatrick (Eds.), *Mathematics education as a research domain: A search for identity* (New ICMI Studies Series, Vol. 4, pp. 415–425). Dordrecht, The Netherlands: Kluwer.

Little, J. (1982). Norms of collegiality and experimentation: Workplace conditions of school success. *American Educational Research Journal, 19*, 325–340.

Little, J. W. (1993). Teachers' professional development in a climate of educational reform. *Educational Evaluation and Policy Analysis, 15*, 129–151.

Lord, B. (1994). Teachers' professional development: Critical colleagueship and the role of professional communities. In N. Cobb (Ed.), *The future of education: Perspectives on national standards in education* (pp. 175–204). New York: College Entrance Examination Board.

Mack, N. K. (1990). Learning fractions with understanding: Building on informal knowledge. *Journal for Research in Mathematics Education, 21*, 16–32.

Markovits, Z., & Sowder, J. (1994). Developing number sense: An intervention study in grade 7. *Journal for Research in Mathematics Education, 25*, 4–29.

Matz, M. (1980). Towards a computational theory of algebraic competence. *Journal of Mathematical Behavior, 3*, 93–166.

National Advisory Committee on Mathematics Education. (1975). *Overview and analysis of school mathematics, grades K–12.* Washington, DC: Conference Board of the Mathematical Sciences.

National Center for Education Statistics. (1996). *Pursuing excellence: A study of U.S. eighth-grade mathematics and science teaching, learning, curriculum, and achievement in international context.* Washington, DC: U.S. Department of Education.

National Center for Education Statistics. (1997). *Pursuing excellence: A study of U.S. fourth-grade mathematics and science achievement in international context.* Washington, DC: U.S. Department of Education.

National Center for Education Statistics. (1998). *Pursuing excellence: A study of U.S. twelfth-grade mathematics and science achievement in international context.* Washington, DC: U.S. Department of Education.

National Council of Teachers of Mathematics. (1989). *Curriculum and evaluation standards for school mathematics.* Reston, VA: Author.

National Council of Teachers of Mathematics. (1991). *Professional standards for teaching mathematics.* Reston, VA: Author.

National Council of Teachers of Mathematics. (1995). *Assessment standards for school mathematics.* Reston, VA: Author.

National Council of Teachers of Mathematics. (2000). *Principles and standards for school mathematics.* Reston, VA: Author.

Nolan, J. D. (1973). Conceptual and rote learning in children. *Teachers College Record, 75*, 251–258.

Nunes, T., Schliemann, A. D., & Carraher, D. W. (1993). *Street mathematics and school mathematics.* New York: Cambridge University Press.

O'Day, J., & Smith, M. (1993). Systemic reform and educational opportunity. In S. Fuhrman (Ed.), *Designing coherent policy* (pp. 250-312). San Francisco: Jossey-Bass.

Palmiter, J. R. (1991). Effects of computer algebra systems on concept and skill acquisition in calculus. *Journal for Research in Mathematics Education, 22*, 151–156.

Parke, C. S., & Smith, M. (1998, April). *Examining student learning outcomes in the QUASAR project and comparing features at two sites that may account for their differential outcomes.* Paper presented at the meeting of the American Educational Research Association, San Diego.

Resnick, L. B., & Omanson, S. F. (1987). Learning to understand arithmetic. In R. Glaser (Ed.), *Advances in instructional psychology* (Vol. 3, pp. 41–95). Hillsdale, NJ: Erlbaum.

Saxe, G. B. (1991). *Cultural and cognitive development: Studies in mathematical understanding.* Hillsdale, NJ: Erlbaum.

Saxe, G. (1996, April). *Integrating findings across datasets: The classroom contexts of changes in children's understanding of fractions.* Paper presented at the meeting of the American Educational Research Association, New York.

Schifter, D, & Fosnot, C. T. (1993). *Reconstructing mathematics education: Stories of teachers meeting the challenge of reform.* New York: Teachers College Press.

Schmidt, W. H., McKnight, C. C., & Raizen, S. A. (1996). *A splintered vision: An investigation of U.S. science and mathematics education.* Boston: Kluwer.

Schoenfeld, A. H. (1985). *Mathematical problem solving.* Orlando, FL: Academic Press.

Scribner, S. (1984). Studying working intelligence. In B. Rogoff & J. Lave (Eds.), *Everyday cognition: Its development in social context* (pp. 9–40). Cambridge, MA: Harvard University Press.

Shanker, A. (1993, January 24). Where we stand: Ninety-two hours. *The New York Times*, pp. 33–34.

Sharma, A. (1998, April 26). Math not working for some. *Riverside Press-Enterprise.*

Siegler, R. S., & Robinson, M. (1981). The development of numerical understandings. In H. W. Reese & L. P. Lipsitt (Eds.), *Advances in child development and behavior* (Vol. 16, pp. 241–311). New York: Academic Press.

Sierpinska, A., & Kilpatrick, J. (Eds.). (1998). *Mathematics education as a research domain: A search for identity* (New ICMI Studies Series, Vol. 4). Dordrecht, The Netherlands: Kluwer.

Silver, E. A. (1998). *Improving mathematics in middle school: Lessons from TIMSS and related research.* Washington, DC: U.S. Department of Education.

Stake, R., & Easley, J. (Eds.). (1978). *Case studies in science education.* Urbana: University of Illinois.

Stein, M. K., & Lane, S. (1996). Instructional tasks and the development of student capacity to think and reason: An analysis of the relationship between teaching and learning in a reform mathematics project. *Educational Research and Evaluation, 2*(1), 50–80.

Stein, M. K., Silver, E. A., & Smith, M. S. (1998). Mathematics reform and teacher development: A community of practice perspective. In J. Greeno & S. Goldman (Eds.), *Thinking practices in mathematics and science learning* (pp. 17–52). Mahwah, NJ: Erlbaum.

Stevenson, H. W., & Stigler, J. W. (1992). *The learning gap.* New York: Simon & Schuster.

Stigler, J. W. (1998). Video surveys: New data for the improvement of classroom instruction. In S. G. Paris & H. M. Wellman (Eds.), *Global prospects for education: Development, culture, and schooling* (pp. 129–168). Washington, DC: American Psychological Association.

Stigler, J. W., & Hiebert, J. (1997). Understanding and improving classroom mathematics instruction: An overview of the TIMSS video study. *Phi Delta Kappan, 79*(1), 14–21.

Stigler, J. W., & Hiebert, J. (1999). *The teaching gap: Best ideas from the world's teachers for improving education in the classroom.* New York: Free Press.

Strutchens, M. E., & Blume, G. W. (1997). What do students know about geometry? In P. A. Kenney & E. A. Silver (Eds.), *Results from the sixth mathematics assessment of the National Assessment of Educational Progress* (pp. 165–193). Reston, VA: National Council of Teachers of Mathematics.

Swafford, J. O., Jones, G. A., & Thornton, C. A. (1997). Increased knowledge in geometry and instructional practice. *Journal for Research in Mathematics Education, 28*, 467–483.

Tharp. R. G., & Gallimore, R. (1988). *Rousing minds to life: Teaching, learning, and schooling in social context.* New York: Cambridge University Press.

Thorndike, E. L. (1922). *The psychology of arithmetic.* New York: Macmillan.

Thorndike, E. L., & Woodworth, R. S. (1901). The influence of improvement in one mental function upon the efficiency of other functions. *Psychological Review, 8*, 247–261, 384–395, 553–564.

Travers, K. J., & McKnight, C. C. (1985). Mathematics achievement in U.S. schools: Preliminary findings from the Second IEA Mathematics Study. *Phi Delta Kappan, 66*, 407–413.

VanLehn, K. (1983). On the representation of procedures in repair theory. In H. P. Ginsburg (Ed.), *The development of mathematical thinking* (pp. 201–252). New York: Academic Press.

Villasenor, A., Jr., & Kepner, H. S., Jr. (1993). Arithmetic from a problem-solving perspective: An urban implementation. *Journal for Research in Mathematics Education, 24*, 62–69.

Wearne, D., & Hiebert, J. (1988). A cognitive approach to meaningful mathematics instruction: Testing a local theory using decimal numbers. *Journal for Research in Mathematics Education, 19*, 371–384.

Wearne, D., & Hiebert, J. (1989). Cognitive changes during conceptually based instruction on decimal fractions. *Journal of Educational Psychology, 81*, 507–513.

Wearne, D., & Kouba, V. L. (2000). Rational numbers. In E. A. Silver & P. A. Kenney (Eds.), *Results from the seventh mathematics assessment of the National Assessment of Educational Progress* (pp. 163–191). Reston, VA: National Council of Teachers of Mathematics.

Weiss, I. (1978). *Report of the 1977 national survey of science, mathematics, and social studies education.* Research Triangle Park, NC: Research Triangle Institute.

Weiss, I. (1994). *A profile of science and mathematics education in the United States:* 1993. Chapel Hill, NC: Horizon Research.

Welch, W. (1978). Science education in Urbanville: A case study. In R. Stake & J. Easley (Eds.), *Case studies in science education* (pp. 5-1–5-33). Urbana: University of Illinois.

Wood, T., & Sellers, P. (1996). Assessment of a problem-centered mathematics program: Third grade. *Journal for Research in Mathematics Education, 27*, 337–353.

Section 1

Research on Principles and Standards

CHAPTER 3

Making Mathematics Reasonable in School

Deborah Loewenberg Ball and Hyman Bass
University of Michigan

Benny, a sixth grader whose teacher regards him as one of her best pupils, is interviewed by a classroom visitor (Erlwanger, 1973). The interviewer probes his knowledge of decimals:

Interviewer: What would you get if you add .3 + .4?

Benny: That would be... oh, seven [07]... point oh seven [.07].

Interviewer: How do you decide where to put the point?

Benny: Because there's two points: at the front of the 4 and the front of the 3. So you have to have two numbers after the decimal, because... you know... two decimals. Now like if I had .44, .44 (i.e., .44 + .44), I have to have four numbers after the decimal (i.e., .0088). (p. 4)

He does the same thing when he multiplies decimals.

Interviewer: What about .7 × .5?

Benny: That would be .35.

Interviewer: And how do you decide on the point?

Benny: Because there's two points, one in front of each number; so you have to add both of the numbers left... 1 and 1 is 2; so there has to be two numbers left for the decimal. (p. 5)

Using these methods, Benny produces such answers as 4 + 1.6 = 2.0 and 7.48 – 7 = 7.41 and yet is unaware that his answers are wrong. Because Benny knows that numbers can be represented in different ways, if his answer is different from the teacher's answer or from the answer in the answer key, he simply assumes that his answer is just another way to write the correct one. "It's like a wild goose chase," he explains emotionally . Mathematics is anything but reasonable for Benny. His experience and the conclusions he draws about mathematics are also not uncommon. Contrast his experience in school with that of the children in the following excerpt from Deborah Ball's third-grade class (transcript of class, 19 January 1990). In this example, all names are pseudonyms, standardized across published analyses of these data and selected to be culturally similar to the children's real names. Near the end of a class, the children are concluding a discussion with their teacher:

Riba: (to the class) So Sean is saying that some even numbers, in a pattern, can be even and odd, and some can't. Four can't, because it's two groups. Six can. Eight can't. Ten can. *(Pointing at the number line above the chalkboard, she uses a pointer to mark off consecutive even numbers.)* Can't. Can. Can't. Can....

Ofala: Well, I just think that just because twenty-two is eleven groups, that doesn't mean it's an odd number. My conjecture, I think it's always true, is that if all twos are circled in a number, then it's an even number.

Sean: What conjecture?

Ball: Ofala, tell him what you're talking about when you talk about your conjecture. He's not sure what you're referring to.

Ofala: That conjecture I already...

Sean: That's not a conjecture. That's a *definition.*

These children and their classmates are struggling with concepts of evenness and oddness as a consequence of one child's claim that six could be even, and it could also be odd because you have an odd number of groups of two. Not unlike Benny, they are confronting puzzling mathematics. They had thought that they understood even and odd numbers. How could the number six be both even and odd? But rather than simply accept the student's notion, appeal to the teacher, or dismiss what might be seen as nonsense, the student's classmates are actively reasoning about elaborations of the claim and about the student's reasons for it. They have developed resources for inspecting and judging mathematical claims and arguments and for revising and developing mathematical ideas. In the process, they are solidifying their understanding of the definitions of even and odd. For them, unlike Benny, mathematics is reasonable, that is, something about which one can reason.

Mathematical Reasoning and Proof: Essential as Both End and Means

NCTM's *Principles and Standards for School Mathematics* (2000) makes a strong statement about the centrality of mathematical reasoning by including a major standard on reasoning and proof for all grades (p. 56):

> **Reasoning and Proof**
>
> Instructional programs pre-kindergarten through grade 12 should enable all students to—
>
> - recognize reasoning and proof as fundamental aspects of mathematics;
> - make and investigate mathematical conjectures;
> - develop and evaluate mathematical arguments and proofs;
> - select and use various types of reasoning and methods of proof.

Some may regard a standard on reasoning and proof as a nice, but esoteric, embellishment to the main curricular goals in mathematics. Some might even consider the standard expendable. Quite the contrary. Mathematical reasoning is no less than a basic skill. Why do we make this claim?

First, the notion of *mathematical understanding* is meaningless without a serious emphasis on reasoning. What, after all, would mathematical "understanding" mean if it were not founded on mathematical reasoning? Take, for example, understanding multiplication of decimals. Benny, like many adults, counted decimal places to determine the number of places in an answer, but he had no idea what it meant. Not understanding the reasons underlying the procedure meant that he made senseless mistakes. Why does multiplying $.7 \times .5$ produce an answer with two decimal places—.35—whereas adding the same numbers, $.7 + .5$, yields an answer with just one decimal place—1.2? Unjustified knowledge is unreasoned and, hence, easily becomes unreasonable.

A second reason for claiming that mathematical reasoning is a basic skill is that such reasoning is fundamental to using mathematics. Knowing particular mathematical ideas and procedures as mere fact or routine is insufficient for using those ideas flexibly in diverse cases. First graders who have learned to use the equals sign to signal the result of an operation on two numbers are baffled when presented with $8 = \_\_ + 5$. They think that the 8 does not "tell" them to "do" anything, and so they often say there is no number to write in the blank. Their thinking results from using the equals sign in number sentences without reasoning about the concept of equality (Carpenter & Franke, 2001; Falkner, Levi, & Carpenter, 1999). Or consider people who know that the probability of two independent events can be calculated by multiplying the probability of the first event by the probability of the second. For flips of two fair coins, they thus may correctly calculate the probability of two heads, or of two tails, as $1/4$ ($= 1/2 \times 1/2$) and yet say that the probability of one head and one tail is $1/3$, seeing this case as one of three possibilities (two heads, two tails, or mixed), failing to take account of the fact that the mixed case has two ways of occurring.

Third, mathematical reasoning is fundamental to reconstructing faded knowledge when a demand for it arises. A person who once knew how to divide fractions but has forgotten the algorithm can rebuild a reasonable procedure if he can use the meaning of division and of fractions to reason about dividing one fraction by another. Analyzing the basic meaning of division allows him to see dividing fractions as not essentially different from dividing any whole number by another: Dividing $4/5$ by $2/3$ is conceptually like dividing 6 by 3, even though the numbers involved in the former example have more complex descriptions. Or, consider a person who learned—but has since forgotten—the formulas to calculate probabilities of independent and of mutually exclusive events. When asked what the probability is of tossing two coins and getting two the same (i.e., two heads or two tails), she may be unsure what operations to perform. Should she multiply the probabilities of each outcome? Add them? What are the outcomes? If she can reason, however, about the logic of a probabilistic situation to analyze whether the outcomes are independent or mutually exclusive, she will likely be able to figure out that she must first multiply the probability of flipping one head times the probability of flipping a second and must do the same for two tails; the successive tosses are independent. Each of these probabilities is $1/2 \times$

1/2, or 1/4. But to calculate the probability of getting two coins the same, she will realize that she then must add the chance of tossing two heads to that of tossing two tails; the two-heads and two-tails outcomes are mutually exclusive. Being able to reason mathematically allows her to recapture a way to work successfully with the problem.

Our point is that mathematical reasoning is as fundamental to knowing and using mathematics as comprehension of text is to reading. Readers who can only decode words can hardly be said to know how to read. Reading competently depends on being able to understand the structures of texts and nuances of language; to interpret authors' ideas; and to visualize, evaluate, and infer meanings. Likewise, merely being able to operate mathematically does not assure being able to do and use mathematics in useful ways. Procedural operations are fundamental to reasonable mathematical activity but are by themselves little more than the analog of reciting text based on the phonetic and structural analysis of words. Making mathematics reasonable means making it reasoned and, therefore, known in useful and usable ways.

This chapter examines what is entailed by mathematical reasoning, and what this looks like as it develops in students in classrooms. Drawing from work on elementary school teaching and learning, and on the practices of mathematics, we discuss a framework for what we call the reasoning of justification—or proof. Next, we turn to what it might take to make mathematics reasonable—that is, what can support the reasoning about mathematics in school.

What Do We Mean by Mathematical Reasoning?

Making mathematics reasonable is more than individual sense making. Making sense refers to making mathematical ideas sensible, or perceptible, and allows for understanding based only on personal conviction. Reasoning, as we use it, comprises a set of practices and norms that are collective, not merely individual or idiosyncratic, and rooted in the discipline. Making mathematics reasonable entails making it subject to, and the result of, such reasoning. That an idea makes sense to me is not the same as reasoning toward understandings that are shared by others with whom I discuss and critically examine that idea toward a shared conviction.

The desire to know and to understand has led people to develop disciplined means of reasoning, of exploring and verifying, of hypothesizing and justifying, in many arenas of human activity. Historians reason about evidence from the past, physicians reason about patients' symptoms, chefs reason about composing ingredients under particular conditions, and pilots reason about instrument readings. In none of these examples do individuals make sense in whatever ways they choose. Instead, in each of these arenas, people have developed methods of reliable thinking that afford inspection, analysis, judgment, and conclusions. These methods of reasoning are the particular means of constructing and evaluating knowledge in a domain.

Much has been written in recent years about constructivist theories of learning and their implications for instruction. Indeed, *constructivism* has arguably been one of the most influential—and most multiply interpreted—ideas in mathematics education. Our research analyzes classroom mathematics learning and teaching in light of ideas about constructing knowledge that are rooted in mathematics as a discipline. Lampert (1990, 1992) has been exploring similar resonances between the practices of knowing mathematics in school and those of knowing mathematics in the discipline. When students are at work in a mathematics class, for example, we see them as constructing mathematical knowledge. Looking at the development of students' knowledge in this way highlights the fundamentally mathematical nature of their—and hence, their teachers'—work. The ways in which students seek to justify claims, convince their classmates and teacher, and participate in the collective development of publicly accepted mathematical knowledge have powerful resonances with mathematicians' work. As students explore problems, make and inspect claims, and seek to prove their validity, even as young children, they engage in substantial forms of mathematical reasoning and make use of mathematical resources. Smith (1999) provides a vivid portrait of this through his close analyses of four nine-year-olds' individual mathematical reasoning. His account focuses on their use of language and representations as they draw together and use mathematical resources to solve problems. This mathematical perspective makes visible some fundamental aspects of mathematics teaching and learning that are hidden when instruction is viewed from a purely cognitive or sociocultural perspective. In particular, this analysis allows for and explores a subject-specific view of learning.

This work finds company in recent advances in other fields (e.g., Wilson, 2001; Wineburg, 1996). Shari Levine Rose (1999), in a study of fourth graders' learning of history in her own classroom, distinguishes between what she was able to see in her students' work when she viewed it from the perspective of generic theories of learning and when she later began to view their work using a lens of historical reasoning. Initially, she explains, she was "influenced by constructivist theories of learning,... [believing] that children drew upon knowledge, values, and beliefs in actively *making sense* of new information." She argues, however, that the generic perspective did not help her see how the children were constructing meaning of historical events. But with the historical lens, she was struck by the "historical nature" of children's sense making. They repeatedly sought understanding through constructing stories, much as historians fashion narratives, embedding meaning and interpretation in context. Rose writes about how the historical perspectives that she brought to bear in hearing and interpreting her students made visible how the children came to know the past and constructed meaning of historical events in ways that were much more rooted in the nature of historical reasoning.

Viewed from the perspective of the practicing mathematician, reasoning is one of the principal instruments for developing mathematical understanding and for constructing new mathematical knowledge. Mathematical reasoning can serve as an instrument of inquiry in discovering and exploring new ideas, a process that we call the *reasoning of inquiry*. Mathematical reasoning also functions centrally in justifying or proving mathematical claims, a process that we call the *reasoning of justification*, the focus of this chapter.

Historically, this sort of mathematical reasoning has primarily been found in the high school geometry curriculum in the context of constructing two-column proofs, sometimes treated more as ritual than as an instrument of sense making. What might be entailed by a broader conception and practice of mathematical reasoning in school, as called for by the NCTM's standard on reasoning and proof? In this chapter, we offer and illustrate a conceptual framework for learning and teaching mathematical reasoning.

Teaching Commitments to Mathematics, Students, and Community

Our study of mathematical reasoning is framed by a conception of teaching founded on three specific commitments—to the integrity of the discipline, to taking individual students' thinking seriously, and to the collective as an intellectual community. These commitments orient, but do not determine, practice. First is a commitment to draw from mathematics as a discipline in intellectually sound and honest ways (Ball, 1993; Ball & Bass, 2000a, 2000b; Bruner 1960; Lampert, 1990, 1992, 2001). So in the case of this analysis, we ask, What mathematical habits and dispositions are crucial to doing and learning mathematics? What is the basis of mathematical reasoning? How is mathematical knowledge constructed in the discipline? How do mathematicians interact as a community over knowledge claims? These and other questions have guided our study of making mathematics reasonable, by and for children, in school.

In counterpoint, we assume that teaching demands a sensitivity and responsiveness to students' ideas, interests, lives, and trajectories. Teachers strive to hear their students, to work with them as they investigate and interpret their worlds. Respecting students means attending to who they are and what they bring as well as helping them grow beyond where they are now or where they think they can go. But attending to individual students' interests and proclivities is not enough. In school, teachers must be concerned with "covering" the mandated curriculum so that each student is prepared for the next grade and for the standardized tests used to chart his or her progress. And while many seek to redefine what "*covering the curriculum*" might mean (Lampert 1992, 2001), caring for students means also being responsible to current definitions of progress and learning (Delpit, 1985). Understanding teaching as centrally guided by students' ideas and thinking has led our work on mathematical reasoning to close examination of how students come to hold and believe in mathematical knowledge (Ball & Bass, 2000b).

Finally, the teaching in which we are interested aims to create a classroom community in which differences are valued; in which students learn to care about and respect one another; and in which commitments to a just, democratic, and rational society are embodied and learned (Dewey, 1916; Schwab, 1976). Care and respect for others includes listening to, hearing, and being able to represent others' ideas, even those with which one disagrees. Respect also means taking others' ideas seriously, appraising them critically, and evaluating their validity. In this work, we consider mathematical reasoning as producing more than individual conviction: as generating public knowledge that is usable by the collective.

A Framework for Mathematical Reasoning

The reasoning of justification in mathematics rests on two foundations. One foundation is a body of public knowledge on which to stand as a point of departure and that defines the granularity of acceptable mathematical reasoning within a given context or community. The second foundation of mathematical reasoning is language—symbols, terms, and other representations and their definitions—and rules of logic and syntax for their meaningful use in formulating claims and the networks of relationships used to justify them.

We first discuss the *base of public knowledge*, the term we use to refer to the knowledge on which claims and arguments are based within some context. Yackel and Cobb (1996) use the label *taken-as-shared* to refer to the meanings, norms, and ideas that are negotiated and used as common within a classroom. Edwards and Mercer (1987; 1989) also write about the development of common knowledge in teaching and learning and focus particularly on the discourse patterns whereby teachers establish such common knowledge. With Yackel and Cobb, we are interested in normative aspects of mathematics discussions specific to students' mathematical activity, such as agreements about what counts as mathematically different solutions or what counts as an acceptable mathematical explanation. And like Edwards and Mercer, we are interested in the development of common knowledge. In our framework, we focus further on the specific mathematical knowledge that is available for public use by a particular community in constructing mathematical claims and in seeking to justify those claims to others. This knowledge is of particular ideas, accepted procedures, defined terms, and methods of mathematical investigation and verification. This knowledge is already assumed or developed—part of the record of the children's prior experience or the class's past work. By identifying it as public, we seek to avoid implying that each member of the community knows it individually in the same way; ascertaining how mathematical knowledge is known individually is an empirical question beyond the scope of our analysis. We mean, rather, to call attention to the knowledge that can comfortably be assumed and used publicly without additional explanation. We contrast such knowledge with ideas or procedures that are not shared and must therefore be established before they can be used to justify claims in the collective discourse of a community.

This base of public knowledge is defined relative to a particular community of reasoners. For professional mathematicians, the base of public knowledge might consist of an axiom system for some mathematical structure (e.g., Euclidean geometry or group theory), simply admitted as given, plus a body of previously developed and publicly accepted mathematical knowledge derived from those axioms. We argue that this idea of a base of public knowledge is useful in understanding the work of a class of elementary school students as well, where this base of public knowledge comprises the expanding set of publicly established ideas and shared knowledge that can be used by the class in explanation or justification.

Take a rudimentary example. Early in September one day, in Ball's third-grade class, the children were working on this problem: "Write number sentences for 10." In the early stages of their work, most students were writing simple equations: $4 + 6 = 10$, $3 + 7 = 10$, and $8 + 2 = 10$. Pressing them gently into more complex solutions, Ball gave the children the following challenge (transcript of class, 18 September 1989):

> I wonder if someone can think of a number sentence that uses more than two numbers here. Just so we have a bunch of ideas of how we could do this. Who can make a number sentence that equals 10 but has more than two numbers adding up to 10?

Quickly, Tembe began, "One plus one plus one plus one—..."

Ball wrote on the board as Tembe rattled off numbers: "One plus one plus one plus one plus one plus one plus one. Plus three."

"Why does that equal ten?" asked Ball, who then called on Harooun to justify Tembe's claim. Harooun, repeating, said, "That's just one plus one plus one plus one...." Ball asked the students, "How do we know that that equals ten?" Riba, eager to respond, explained, "Because one plus one plus one plus one plus one plus one plus one and plus three equals ten." Ball, still not satisfied, said, "You're just sort of reading it. How could you prove it to somebody who wasn't sure?"

Riba: Because I counted it.

Ball: What did you count? What did you find out?

Riba: There's one and the next one is two and the next one is three, next one is four, next one is five, next one is six, next one is seven, next one is—seven, and then three more, eight, nine, ten."

In this simple segment, Tembe's offered solution, the first that included more than two terms, was not something automatically presumed to be within the common knowledge of the class, and Ball asked for justification. If this exchange had taken place in a fifth-grade class, one would likely presume differently. A young child's initial sense of addition comes from counting, which is adding one at a time. Adding many terms at once or adding two numbers larger than one in a single step are higher-order operations, not only for young children but also mathematically. When Riba is called on to explain why the string of terms equals 10, she first just recites the equation. When Ball presses her, "But how could you prove it to somebody who wasn't sure?" Riba replies that she "counted," and Ball then encourages her further to make this counting public. In response, Riba expands her explanation:

> There's one and the next one is two and the next one is three, next one is four, next one is five, next one is six, next one is seven, next one is—seven, and then three more, eight, nine, ten.

Perhaps it is through this counting that Riba first proves to herself that Tembe's formulation was valid. Her teacher is requiring her to make this reasoning public to persuade the class as well. Her teacher then publicly validates Riba's work, underlining early in the year a standard for explanation and justification that is more than simple restating of the assertion:

> Do you see the difference in Riba's second explanation? Did you see how she really showed us how it equals ten? The first time you just read it. And the second time you explained it. That was really nice.

Here the teacher does more than praise Riba. She points explicitly to Riba's work—the mathematical explanation she has constructed—and comments on the difference between repeating a statement and explaining it.

A process of reasoning typically consists of a sequence of steps, each of which has the form of justifying one claim by invocation of another, to which the first claim is logically reduced. This process, which merely transforms one claim into another, is not a vicious circle, because the reduced claim is typically of a more elementary or accessible nature and, in a finite number of such steps, one arrives at a claim that requires no further warrant because of being part of the base of publicly shared knowledge and therefore universally persuasive within a particular community of reasoners. Thus, the base of public knowledge both constrains and enables the stepping-stones of an argument. In that sense, publicly shared knowledge defines the granularity of acceptable mathematical reasoning within a given context. In the example above, the addition represented in the equation $1 + 1 + 1 + 1 + 1 + 1 + 1 + 3 = 10$ was not, at that moment, presumed to be part of the base of established knowledge of that class, and so Riba was pressed to reduce the claim to an iterated counting, keeping track of the total as she counted. At that point, Tembe's assertion was sufficiently reduced to a level that relied on knowledge common to the class—counting by ones—and required no further justification.

The crucial issue is how to justify a mathematical claim. One way to justify a claim is to state the claim and to undergird its truth by the sheer force of authority. Students often receive mathematical knowledge in school that is justified by little else than the textbook's or the teacher's assertion. By default, the book has epistemic authority: Teachers explain assignments to pupils by saying, "This is what they want you to do here," and the right answers are found in the answer key. According to Davis (1967), learning to "play by the rules" often involves a "suspension of sense-making" in school mathematics. But that route is the antithesis of warranting claims through a process of mathematical reasoning.

The base of public knowledge consists of knowledge of certain facts and concepts; of the meanings of mathematical terms and expressions; and of procedures and resources for calculation, for problem solving, and even for reasoning. The base of public knowledge is always present, in both latent and active forms, although it may be tacit and only implicit in the discourse of the community, whether mathematicians or third graders.

Whether a particular piece of knowledge is in fact commonly shared is an empirical question, one that a teacher must often assess. Did everyone in the class understand and agree that Riba's elaborated explanation satisfactorily proved Tembe's claim? And, further, how many children were not already adequately convinced by his initial statement? These questions are difficult to resolve fully; they are also not completely within the teacher's or the students' view as a class discussion proceeds. Still, reasoning within a community depends on the presumption of common knowledge and shared established methods. This presumption is most often represented by the use of knowledge already used and established publicly. Arguments that do not build on publicly shared knowledge are unlikely to produce grounded conviction in others. At the same time, the process of reasoning can in fact help build and extend a group's common knowledge. As claims are proved, and ideas developed, the claims may become part of the legacy of public knowledge on which subsequent claims may depend and build.

In our analysis, mathematical language is the foundation of mathematical reasoning that is complementary to the base of publicly shared knowledge. Language is used here expansively, comprising the entire linguistic infrastructure that supports mathematical communication with its requirements for precision, clarity, and economy of expression. Language is essential for mathematical reasoning and for communicating about mathematical ideas, claims, explanations, and proofs. Language is a medium in which mathematics is enacted, used, and created.

In our framework, language includes the nature and role of definitions in mathematics; the nature of, and rules for, manipulating symbolic notation; and the conceptual compression afforded by timely use of such notation. Definitions and terms play a crucial role: Not simply delivered names to be memorized, definitions and terms originate in, and emerge from, new ideas and concepts and develop through active investigation and reflection. Definitions and terms facilitate reasoning about those new ideas by naming and specification. Decisions about what to name, when to name it, and how to specify that which is being named are important components of mathematical sensibility and discrimination central to the construction of mathematical knowledge. Using symbolic and other representations to encode ideas, as well as decoding ideas represented in symbolic or other forms, are essential communicative tools for the construction of mathematical knowledge. Precise language is also needed to articulate the correspondences between equivalent representations of the same mathematical entity or concept. Notation can be used to compress ideas into forms that, when done skillfully, can reduce computation and manipulation to manageable proportions; how and when to do this is an important skill of mathematical representation useful in reasoning.

Mathematical language is central to constructing mathematical knowledge; it provides resources with which claims are developed, made, and justified. Lampert writes,

> Mathematical discourse is about figuring out what is true, once the members of the discourse community agree on their definitions and assumptions. These definitions and assumptions are not given, but are negotiated in the process of figuring out what is true. (1990, p. 42)

Some disagreements stem from divergent or unreconciled uses of terminology, whereas others are rooted in substantive and conflicting mathematical claims (Crumbaugh, 1998; Lampert, 1998). The ability to distinguish between issues of terminology and issues of mathematical claims requires sensitivity to the nature and role of language in mathematics. We return to this aspect of language in the examples analyzed in the next section of this chapter.

Developing Mathematical Reasoning in a Third-Grade Class

In this section, we take a close look at two classroom episodes in which elementary school students are learning to engage in mathematical reasoning. These classroom episodes are based on Ball's third-grade class records for the school year from 1989 to 1990 (see Lampert & Ball, 1998). We have chosen two excerpts of instruction from the same classroom, one less than 5 months after the first. In fact, the first episode is from the first day of mathematics class in September, and the second, from a day late in January. Our purpose in choosing these two segments is to examine the evolution of the reasoning of justification. What were the students and teacher doing as they sought to reason about mathematics in September, and in what ways were those approaches the same or different in January, less than 5 months later? What do the students seem to be learning? Comparing these two points in time helps establish that mathematical reasoning is something that students can learn to do and, hence, that teachers can teach. Our analysis leads to the final section of the chapter, in which we turn to considering what approaches teachers might adopt to help make mathematics reasonable in school.

In the first episode, early in September, the students are working on an arithmetic problem that has multiple, but finitely many—six—solutions. The task is to find all the solutions and then to show that all solutions have been found. In the later episode, in January, students are working on a conjecture about addition of odd numbers: that an odd number plus an odd number equals an even number. The task is to determine whether this conjecture is true. In both instances, the mathematical work for students calls for justification: How do you know that you have found all the answers? Can you prove that this statement is always true? What the students do in January entails more complex reasoning in that it concerns a claim about all the infinitely many pairs of odd numbers; this work shows substantial development of reasoning skills and sensibility since September.

September: How Do You Know That You Have Them All?

In the first regular mathematics period of the school year, the students are working on the following problem:

> I have pennies, nickels, and dimes in my pocket. Suppose I pull out two coins. How much money might I have?

In setting up the task, Ball reads the problem with the students and they try an example together. She asks the class, "I'm going to pull out two coins. How much money could I pull out? Like this, like I'm not even looking and I'm going to reach in and pull out one coin and reach in and pull out another. How much money might I have?" "Ten cents," proposes Lucy.

Ball asks for more explanation: "How could I pull out 10 cents?"

Lucy: Two 5s.

Ball: What do other people think about that? If I pulled out two nickels, would I have 10 cents?

Students: Yeah.

Ball: How do you know that? How do you know that that would be 10 cents? Ofala?

Ofala: Because five plus five is ten.

The students work on the task for about 20 minutes while Ball circulates, noting what different students are doing and asking and answering an occasional question. A few times during class, Ball calls the group together to share bits of work or to discuss how the work is proceeding. We zoom in on the class discussion of solutions to the problem because it affords a close look at the group's early efforts to reason mathematically, as individuals and as a group.

About 20 minutes before the end of class, Ball brings them back together. She elicits solutions to the problem from different students. The class discusses and verifies each proposed solution. Ball records the answers on the board in a list:

15¢
20¢
6¢
11¢
2¢
10¢

Ball then reads the solutions off the board:

Ball: We have 15 cents, 20 cents, 6 cents, 11 cents, 2 cents, and 10 cents. Any more? Look at your lists, and see if you have anything that we didn't put on the board. (pause) No, Ofala? You don't have anything else in your notebook? (pause) Jeannie, do you have anything else in your notebook? Riba, do you? Does anybody? How many different possible answers did we find for this problem? How many answers did we come up with here?

Latifa: Six.

Latifa is right. They have come up with six solutions for the problem. However, Ball does not affirm Latifa's answer. Instead, she presses a bit, pushing for justification.

Ball: Six answers. How do we know that we have them all, though? How do we know there isn't a seventh one? Or an eighth one that we didn't find yet?

Ofala, noting a constraint in the given statement of the problem, says she thinks that six is all they can make because they cannot use quarters. Mei comments, "I think we have them all 'cause we had lots of them already and all the people had six."

Lisa speaks next, but her voice is almost inaudible. Ball interrupts to tell her to "talk so that other people can hear you," and directs the rest of the class to listen to what Lisa is saying. Lisa declares, more audibly, "We can only pick up two coins, and if we pick up seven, then we would be picking up three or four." Ball, seeking to understand Lisa's argument, asks for clarification, "So to get a seventh answer, we'd have to pick up three coins?" Lisa assents, and Ball asks her how she knows. Lisa says something about a nickel and a penny, "and if we add another penny, it will be seven and three coins." She seems to talking about a solution for seven cents, but, because it requires three coins, she may be trying to show that the production of a seventh solution—Ball's original question—would require more than two coins.

At this juncture, Sheena raises her hand. She says that she has been working on the question, and she keeps trying to find more solutions but keeps "getting the same answers." Ball repeats her idea to the class: "If you keep picking them up, you'll get the same answers? How many people solved this problem by picking up coins until they got the same ones again?... How many people reached into the box or onto their pile and kept picking them up until they started to repeat?" Many students raise their hands. "What's another way to do the problem?" she asks, looking around.

Mei offers another method: "You first think of what you can make from, you can make out of nickels, dimes, and pennies; and then you take, and then you write them down and you think about it some more until you're—(pause) then you'll get them all."

The children seem reasonably satisfied that they have found all the solutions. They believe they have found them all because they cannot find any more. This argument is empirical or inductive, not deductive. It is the kind of empirical reasoning that can increase confidence in a scientific hypothesis, but it is not a mathematical proof. Still, the children find it persuasive, and they do not seem to conceive any method by which one could confirm more definitively that they have all solutions. Ball ends class by urging them to "think very hard before math tomorrow" to see whether they can find any more answers to this problem. By not affirming the completeness of the solution—with six answers—she seeks to maintain some need to show more firmly that they found all the answers, something more than trying many solutions or being personally convinced.

At this early stage, the children have not witnessed or been formally introduced to the notion of mathematical proof. But we see here that the imperative to mathematically justify is being seeded in their work, before they even know what a proof is or looks like. The challenges to justify their own conjectures serve to motivate them to construct, through reflection and analysis, some of the intellectual architecture of mathematical reasoning.

January: Can We Prove That Betsy's Conjecture Is Always True?

We revisit the class almost five months later. The students have been working on problems that involve patterns with sums. For example, they have worked on such problems as the following:

> Erasers cost 2¢, and pencils cost 7¢. How many different combinations of erasers and pencils can you buy if you want to spend exactly 30¢?

The numbers in this problem have been chosen deliberately so that the children might notice that an even number of pencils must be bought if the total has to be 30—an even number—because 7 is an odd number. Indeed, solving such problems has generated a series of conjectures about even and odd numbers:

Even + even = even
Even + odd = odd
Odd + odd = even

Ball has helped the children formulate these conjectures and understand them and has challenged the children to see whether they could prove that these statements are always true. The students have generated long lists of examples in their notebooks, seeking to confirm the conjectures or to find examples that do not work and therefore would show that a conjecture is not (always) true. They are working in small groups on a particular conjecture.

In the course of this work, something new happens. For the first time, the thought occurs to some of the children that to prove such a conjecture is something they have not done before and that doing so presents challenges they have not previously appreciated. Jeannie and Sheena, who are working on the conjecture that an odd number plus an odd number equals an even number, report on their work:

> *Jeannie:* Me and Sheena were working together, but we didn't find one that didn't work. We were trying to prove that... you can't prove that Betsy's conjecture (*odd + odd = even*) always works. Because, um, there's, um, like numbers go on and on forever, and that means odd numbers and even numbers go on forever, so you couldn't prove that *all* of them work.

Ofala protests. She declares, looking closely into her notebook, that she has tried "almost eighteen" of them and even some special cases, and they have all "worked," so she thinks that "it can always work."

Mei then offers a very different kind of objection. "I think it could always work because with those conjectures [motioning to several previously discussed and widely agreed-on conjectures posted above the chalkboard], we haven't even tried them with all the numbers there is, so why do you say that those work? We haven't tried those with all the numbers that there ever could be."

Ball asks whether this statement means that she is disagreeing with Jeannie or agreeing with her. "I disagree," replies Mei, emphatically. Ball asks her to clarify what she is saying, and she repeats and amplifies her point that the class has already agreed to accept other conjectures even though they were not able to check them with every number. But Jeannie and Sheena are not to be deterred. Jeannie says that she never said that those other conjectures were true all the time. "Then why didn't you disagree when everyone agreed with those conjectures?" presses Mei. Sheena explains that they had never thought about any of this before, and now they see that a problem exists because numbers go on forever. Class ends with this issue unresolved, and Ball tells the students to think some more about what Jeannie and Sheena are claiming: "What do other people think? Do you think we can't prove that it's always true, or do you think that we can prove that it's always true?"

A few days later, the class takes a big step. Betsy, together with a few classmates, presents a proof of the conjecture, which they first illustrate with 7 + 7:

> What we figured out how it's always true is that we would have seven dots, or lines, plus seven lines (draws fourteen hash marks on the board, seven at a time)

> . . . and then (counts the lines), and then we said that we had to circle them by twos (she starts circling twos from right to left)...
>
> 
>
> and also we said that... an even number is... just a second (finishes circling groups of two lines) that if you added another even one to an odd number, or another 1 to an odd number, then it would equal an even number 'cause all odd numbers if you circle them, what we found out, all odd numbers if you circle them by twos, there's 1 left over; so if you... plus 1, um, or *if you plus another odd number, then the two 1's left over will group together, and it will make an even number* (italics added).

And then after a couple of questions by other children, Betsy elaborates:

> If an odd number plus an odd number, if you add another number with an odd number, then it equals an even number because an even number plus 1 equals an odd number, so *if you added two odd numbers together, you can add the 1's left over and it would always equal an even number* (italics added).

Ball turns to the class:

> Does anybody want to comment either on the example or on Betsy's proof? Do people think that does prove that an odd plus an odd would always be even?

Mei, seeming to focus mainly on the illustrated example, 7 + 7, questions whether this proves that the conjecture would always be true. She wonders about checking all numbers. She says, "I don't think so, because you don't know about, like, in the thousands, and you don't know the numbers, like, you don't even know how you pronounce it, or how you say it." Jeannie disagrees, arguing that Betsy is showing that this would work for all numbers: "It could be any two odd numbers because, um, there's always one left."

Mei still does not accept this idea: "I think what I am trying to say... I'm trying to say that that's [the diagram that Betsy has drawn on the board] only one example. You can't really say that it will work for every odd number, even... I don't think it would work for numbers that we can't say or figure out what they are."

Riba wonders, too: "How does Betsy know that it will always work? She never tried all the numbers." Betsy herself concedes the last point:

> *Betsy:* Mathematicians can't even do that. You would die before you counted every number.
>
> *Riba:* I know, that's why I mean, that's what I mean... you don't know if it always works.
>
> *Ball:* Anyone else have a comment? Sheena?
>
> *Sheena:* I agree sort of with Riba, but you don't have to try all the numbers, because you would die before you tried all the numbers, because there are some numbers you can't even pronounce and some numbers you don't even know that they're there. But still, I still think it's true because, um, we've tried like enough of, of examples already. It proves it's true.
>
> *Riba:* I don't think so. Like if... even though we don't know how to pronounce it, we could write it down and say, like, "this number."

The discussion continues for a few more minutes. The children wrestle with whether Betsy's argument shows that any two odd numbers will always add to an even number. Unlike the discussion in September, here the lack of agreement on a proof prevents consensus in the group that the conjecture is true. Struggling with whether they can prove the conjecture, they are still not willing to believe something is true without being able to prove it. Some children claim that they have tried enough examples to be convinced; others remind them that "there are numbers we have not checked" that might show the conjecture to be false. But the idea of a general proof is beginning to capture many of the students. And Riba, struggling to articulate some way of referring to the idea of "some number that we don't know what it is," unwittingly highlights the need for new mathematical language.

Ball sends the children back into their groups again, seeking to prove or disprove the conjectures. A bit later in that same class, she asks groups to report on their work and progress.

Mark, who has been quiet throughout the preceding discussions, reports on his work with Nathan: "Well, first, me and Nathan, we were just getting answers and we weren't thinking about proof. And we were still getting answers and [then] we were trying to prove, and Betsy came and she had proved it, and then we all agreed that it would work." Mark and Nathan's work has been transformed by what has happened. No longer content to produce lists of examples, they glimpse what it might mean to prove a statement, and they see the possibility of doing so. Moreover, they are convinced by Betsy's argument that her conjecture is always true.

The work of this class is a long way from what they were doing just less than 5 months earlier. We turn in the next section to consider the nature of the children's mathematical reasoning and explore the teaching and learning of reasoning in class.

September: What Is It Going to Take to "Do Mathematics" in Third Grade?

Our first episode takes place at the beginning of the year. Many students do not know one another, and the teacher is unacquainted with what they know. Norms for mathematical work have yet to be developed, and the language and ideas on which the class's work can be based are just emerging. If students are to engage collectively in substantial mathematical work, they need to learn to listen attentively, and critically, to the views of others. They need to learn to offer, justify, and critically evaluate mathematical claims, both their own and those of others. These skills are likely not ones that they already possess. Neither are these skills ones they are likely to develop simply by being expected to behave or think in such ways. How will they develop these practices of mathematical reasoning and acquire the knowledge and skills that support them?

One site for such learning is situated in the nature of the mathematical tasks on which students are asked to work. Consider the problem on this first day of class:

> I have pennies, nickels, and dimes in my pocket. Suppose I pull out two coins. How much money might I have?

As stated, this problem could be solved by providing a single answer—for example, 6¢. In this class, however, the enacted task involves several levels of mathematical work beyond this outcome. First, students are expected to justify their answers. How is six cents an answer to this problem? Why is it an answer to this problem? Second, students are directed to find multiple answers to the problem and, together, to find all the possible solutions. Finally, in the whole-group discussion, the teacher asks the class to prove that they have found all such solutions. These layers of additional mathematical work, part of the task in its enacted version, create contexts for mathematical reasoning implied but not automatically entailed by the original problem statement.

Although this problem's opportunities for mathematical reasoning are not explicit in its basic formulation, the problem has a substantial curriculum potential. *Curriculum potential*, as Ben-Peretz (1990) develops it, is a characteristic of curriculum that describes the opportunities for learning and teaching offered by the structure and scope of the tasks. The problem's possibilities emerge and are developed as the students enact the problem in class. The opportunities that arise for learning to reason mathematically depend on how the problem is used. Enactment can either extend or degrade the mathematics learning potential of a given task (Stein, Grover, & Henningsen, 1996).

A second site for the development of mathematical reasoning is embedded in how the task is used in class and can be seen in the class discussion of solutions to the problem. Whether and how these solutions are discussed can hide or make visible the underlying reasoning. For example, in this discussion, we see the teacher ask, "How much money might I have?" When a child offers 10 cents as a solution, she does not simply affirm this answer or ask others whether they agree. Instead, she asks for justification: How could ten cents be made under the given conditions—two coins, using pennies, nickels, and dimes? Translating nickels to their numeric value, the child responds, "Two fives." Ball then asks the other children to verify this answer by asking a question: Do two nickels form ten cents? Several say yes, and she presses for an explanation: "How do you *know* that?" The unpacking of what underlies the solution "ten cents" becomes part of the work and opens up opportunities to develop practices of mathematical reasoning.

Later in the lesson, we see the teacher engage the class in another aspect of mathematical reasoning: whether the solutions identified and listed on the board cover all possible solutions. The process of recording on the board the various student solutions, which might otherwise remain unnoticed by some of the students, makes individuals' work available for public discussion and helps set up the resources for their collective work. First Ball asks them to look at their own work and see whether they have other solutions not listed on the board. That they can easily view all the solutions makes work on this question more accessible. When no one can produce other solutions, Ball asks whether they know that they have found them all. The children's responses reveal that they are relying exclusively on empirical information to determine the solution's completeness: They have tried over and over again but keep getting the same answers. For this problem, despite the teacher's repeated questions—Are there other answers? How do we know that there is not a seventh solution?—the children are not yet able to prove that exactly six solutions are possible. Merely questioning students is not sufficient to help them learn this particular aspect of mathematical reasoning. Still, although these students seem satisfied that they have found all possible solutions, their teacher does not permit the problem to be finally complete. She directs the students to think more about this question, leaving some doubt about the current solution.

Consider this: This problem does have exactly six solutions, and well before the end of class the children have successfully identified and justified them. This work has engaged the students in reasoning as a substantial component of the task. Some might say that the class has finished the work. But the mathematical reasoning possible—and necessary—is not complete. Obtaining six correct answers is not equivalent to showing that six is the total number of answers to this problem. The teacher is the one who moves this remaining mathematical issue into the public workspace, who presses it, and who maintains it. And more work—more opportunities to engage in this sort of reasoning, more teaching, and more learning—will be necessary to develop the students' ability to prove the completeness of a finite solution set. Students can, for example, be encouraged to imagine drawing first one coin and then a second, each of which can be a penny (*p*), nickel (*n*), or dime (*d*), leading to exactly nine (3 × 3) possibilities: *pp*, *pn*, *pd*, *np*, *nn*, *nd*, *dp*, *dn*, and *dd*. Of course, some of them, for example, *pn* and *np*, yield the same amount of money, 6 cents, so altogether only six different amounts of money are represented. Alternatively, students could be encouraged to use a more direct route that does not pay attention to the order in which the coins are drawn. Thus one might first consider those cases that draw at least one penny: *pp*, *pn*, and *pd*. With no more pennies, if at least one nickel is drawn, the possibilities are *nn* and *nd*. Finally, the only case involving neither pennies nor nickels is *dd*. Thus we have six possibilities, and each one gives a different amount of money. Even without this extra work, the teacher may have helped seed the children's awareness of an important mathematical question: Do I have all the solutions, and how do I know?"

That this class is observed from the very beginning of the school year is important. How does a foundation begin to be laid for the development of mathematical reasoning? First, the problem with its layers of possible mathematical engagement can likely be made accessible to every student. The arithmetic called for, that of adding combinations of 1, 5, and 10, is within reach of these third graders. And although some students are unfamiliar with the values of U.S. coins, the teacher talks about their worth and makes actual coins available for the students. This accessibility of the task helps create a common space for collective mathematical reasoning.

Second, for a base of knowledge to be used in class, an approach that is acceptably usable by this group of children must be made visible and shared. To that end, the work that students do on this problem, individually and in group discussion, reveals what students know and can do, as well as aspects of their mathematical habits and dispositions. Finally, practices of mathematical work are begun during this class. Children are called on to communicate their solutions, not only to the teacher but also to the class. When Lisa offers a comment in class, Ball asks her to speak so that others can hear and directs the others to listen. Children are also asked to evaluate the productions and claims of the other students. The work of the class is recorded both in student notebooks for future reference and on the chalkboard for public discussion.

Late January: Examining the Foundations for Building Mathematical Knowledge in Third Grade

In the previous class example, we saw the foundations beginning to be laid for mathematical reasoning. We saw the initial stages of work that could serve to identify and make public the mathematical knowledge usable for making and justifying claims. And we saw practices of reasoning being first developed.

Four and a half months later, we glimpse the more established foundation for their work and the evolution of these practices. For example, in September the children were not used to providing justification, whereas in January they assumed this need. In the fall, we saw them unconcerned with one another's ideas, but at midyear we see them focused on, and attentive to, the collective work.

This episode provides a glimpse of the children using tools of mathematical reasoning to grapple with a problem. First, the task has a sophisticated quality: to prove (or disprove) "Betsy's conjecture," that "odd + odd = even." In other words, Betsy is suggesting that the sum of two odd numbers is even. Whereas the two-coin problem called for a finite number of solutions with the task of exhibiting them and showing that all solutions had been found, Betsy's conjecture, in contrast, asserts a property of any pair of odd numbers, and there are infinitely many odd numbers.

Here the task is not to seek or rule out some undiscovered example, and exhibiting all the examples is impossible because the claim is universally quantified over an infinite set. The challenge here, rather, is to confirm a property for all the examples. Because this confirmation requires reasoning about examples that cannot all be tried individually, two of the students, Jeanie and

Sheena, make the revolutionary proposal that Betsy's conjecture cannot be proved. They have taken the class to a new precipice that challenges the very foundations of their work together: If you cannot try all the numbers, or even know what all of them are, then how can you ever say that something about all of them is really true? This sudden realization of the implications of the infinitude of numbers takes Jeannie and Sheena aback. They have encountered the force of making a claim about a property of an infinite set and the corresponding burden of trying to prove it.

The mathematical task—seeking to prove whether a conjecture is true—suggests that the students' base of mathematical knowledge and skills has developed since September. The students are familiar with and are able to use definitions of even and odd numbers. Their testing out of various pairs of odd numbers, including large cases, shows that they are now more able to add whole numbers quickly and accurately. They know to arrange and keep track of their empirical work as they experiment with these combinations, and they have acquired the skills to do so. They also are developing a language—*conjecture*, *definition*, and *proof*—about mathematical knowledge. Although their meanings for these terms are as yet informal, this vocabulary has emerged in the context of doing mathematical work and provides the students with terms of reference for knowing and doing mathematics. The students' work on the conjecture shows, too, that their mathematical reasoning skills have also developed. At the outset, the children verify many numerical cases of Betsy's conjecture. They seem to conclude that the conjecture must be true because they have verified a large number of diverse cases and not found a counterexample. Listing many examples suffices to convince many of the students that the conjecture is true. This stage of the work is, like the earlier work on the two-coin problem, empirical or inductive but not mathematically conclusive. Still, the children are more inclined to question and seek to prove the truth of a mathematical claim.

But for two children, Jeannie and Sheena, a new insight emerges. In working on this already challenging task, they begin to realize a new problem: They are trying to establish the truth of a claim for all numbers, but they cannot check all numbers. They observe, "You can't prove that Betsy's conjecture (*odd* + *odd* = *even*) always works. Because... , like, numbers go on and on forever and that means odd numbers and even numbers go on forever, so you couldn't prove that all of them work."

Jeannie and Sheena are making the dramatic claim that Betsy's conjecture cannot be proved, because, essentially, it represents an infinite number of claims—one for each pair of odd numbers—and verifying infinitely many cases is beyond human capacity. This insight into the nature of claims involving universal quantification—all—over an infinite set and the seeming paradox of having the capacity to prove such claims with a procedure of finite extent is profound.

These two students not only reach an important mathematical insight but are confident enough to share it with the class. Both were rather quiet and shy in September. Both were also then convinced that checking examples was an adequate means of justification.

That they reach this insight is not the only aspect to notice here, however. As the teacher turns their idea to the class, the other students take up on it immediately. They spontaneously consider the girls' claim and are both able and willing to respond. Lucy agrees, but Ofala objects, citing the long list of examples she has recorded in her notebook. And Mei, invoking a principle of logical consistency, points out all the other conjectures that have already been accepted as true without verifying every possible number.

The students have acquired habits of mathematical reasoning: They assume that a classmate's posited claim is theirs to respond to, and they draw on past work and knowledge as they respond. Even students who do not talk in the whole-group discussion on this day, such as Mark and Nathan, use this discussion to continue and amend their own work. Mark remarks later, "Well, first, me and Nathan, we were just getting answers and we weren't thinking about proof." He describes how they went back to work after this discussion and began trying to prove the conjecture, not just list more examples.

Over the next couple of classes the situation advances dramatically. The students manage to construct a convincing mathematical proof of Betsy's conjecture: "All odd numbers if you circle them by twos, there's one left over, so if you... plus one, um, or if you plus another odd number, then the two left over will group together, and it will make an even number."

This argument is a general one, about any two odd numbers, not a particular numerical example. It invokes a general property that derives from one of the definitions of odd numbers, namely, that a number is odd if, when it is grouped into twos, there is one left over. It also uses a definition of even numbers, namely, a number is even if nothing is left over when it is grouped into twos. Betsy's argument is that when combining two odd numbers, one combines the two ones left over to form a pair, and so the combination of the two odd numbers can be grouped into pairs with nothing left over and hence is even. Consequently, the proof is able to transcend the objection raised by Jeannie and Sheena—that if each pair of odd numbers cannot be inspected directly, determining the truth of the claim is impossible.

The proof relies critically on the definition of odd numbers, a definition that itself contains an assertion quantified over the infinite set of all odd numbers and so it has the capacity to support logical conclusions of similarly infinite purview.

We can distill here the compelling logic of this argument, which Betsy haltingly struggled to articulate but of which not all the students were immediately convinced. Sheena spoke for more than herself when she said, "I agree sort of with Riba, but you don't have to try all the numbers, because you would die before you tried all the numbers, because there are some numbers you can't even pronounce and some numbers you don't even know that they're there. But, still, I still think it's true because, um, we've tried like enough of, of examples already. It proves it's true." Riba responds, "I don't think so. Like if...even though we don't know how to pronounce it, we could write it down and say, like, 'this number.'" We can see in this proposal from Riba the conceptual germ of algebraic notation for a variable as representing any number in some class. Part of Betsy's and Riba's struggle is the lack of a developed mathematical language with which to designate an indeterminate quantity ranging over some infinite set of numbers—for example, the idea of "any odd number," which can be represented using algebraic notation in the form $2x + 1$, where x is any whole number. Riba's effort to name "this number" can be heard as a plea for this linguistic tool, which the work on this problem of proof naturally invites.

In this episode, we see among the children the first emergence of some fundamental mathematical constructs: the need and challenge, from Jeannie and Sheena; the possibility, from Betsy, of structural mathematical proof grounded in precise definitions; norms for logical consistency, offered by Mei; and quantified language for discussing properties of all elements of an infinite collection, stated by Betsy and Riba. And we see evidence of assimilation of these ideas by other students, here, Mark and Nathan. This episode provides a glimpse of how far the students have come in developing a base of public knowledge that they can use to reason about mathematical claims as well as a host of skills and habits, language, and practices for engaging in reasoning, individually and collectively. We see the migration of newly constructed student ideas and productions through the class as they become rejected or else assimilated and consensual, thereby becoming eventually a part of the base of public knowledge. That the sum of two odd numbers will always be an even number was to become one such item.

Teaching Reasonable Mathematics in School

One common reaction to such accounts as the ones in this chapter is "My students could never do this" or "These students must be gifted." We disagree. Students can reason in this way, but only when they have learned to do so. It would be peculiar to argue that students could never learn division or fractions or concepts of probability, but neither is it usually questioned that these things need to be taught and learned if they are to be known. Our exploration of mathematical reasoning as it began to develop in this third-grade class makes plain that such reasoning—just as anything else we teach in mathematics—is something that can be and is learned. The base of public knowledge that can be used to establish new ideas and the mathematical language for their expression and use lay the foundation for practices of constructing mathematics. We saw both the base and the language develop from September to January. Whereas in September, the students replied to questions when asked and focused on the teacher, by January, they engaged mathematical claims themselves and responded to and used one another's statements. We saw signs of the children's learning to ask such questions as "How can I prove that this solution is valid?" and "Do I have all the solutions to this problem?" We glimpsed them learning to expect explicitness from a definition and working with an agreed-on definition to prove a claim.

We argue in this chapter that mathematical reasoning is not auxiliary to basic goals of mathematics education but rather is fundamental to knowing and being proficient with mathematics—that mathematical reasoning is itself basic. If that argument is true, then the work of teaching includes helping students learn to engage in mathematical reasoning. What does this work entail for teachers? What are some of the things that teachers might do to enable the development of mathematical reasoning? We offer a preliminary set of ideas about practices of teaching mathematics in ways that support the learning of mathematics in ways that are reasonable, that is, subject to reasoning by students and their teachers.

Designing and Using Mathematical Tasks

A first arena is to select and adapt tasks in order to create the need and opportunity for substantial mathematical reasoning. Many problems exist that, by design, create considerable space for mathematical reasoning and are worth seeking out and learning to use well. But good opportunities can also be created by enhancing ordinary mathematical work. The two-coin problem discussed above provides one example. Taken as originally worded and then enacted by letting students find one or more possible answers would have afforded much less space for exploring the problem's mathematical territory. Students might have been able to record 20¢ or 11¢ on their papers and handed them in. The task would have remained one that supplied application and some basic addition practice and little else. So one thing teachers can do is make justification a part of the tasks they use, adding such questions as "How do you know that you have all the solutions?" to problems, or "Suppose someone challenged your solution. How could you prove to them that your answer is right?" Teachers can ask students to evaluate solutions by giving them solved problems and asking them whether the solutions are valid. Problems can be adapted or designed to precipitate student conjectures. Carpenter and Franke (2001) are engaged in designing such sequences of tasks with teachers and are finding that even young students of elementary age can begin to notice patterns or generate conjectures from such work. For example, teachers might ask students first to calculate

$$\begin{array}{r} 68 \\ +57 \\ \hline \end{array}$$

and then the following:

$$\begin{array}{r} 67 \\ +58 \\ \hline \end{array} \qquad \begin{array}{r} 69 \\ +56 \\ \hline \end{array} \qquad \begin{array}{r} 70 \\ +55 \\ \hline \end{array}$$

Teachers might then ask students what should be added to 65 to yield the same sum as the original problem. Asking whether students notice a pattern might precipitate some useful general ideas about arithmetic expressions, such as that changing the numbers you are adding so that one goes up the same amount that the other goes down results in the same sum. These claims could also be examined and proved in their own right. This example illustrates how mathematical reasoning can be naturally integrated into basic work on addition computation, a site for important student practice. Teachers can also seek opportunities to augment such ordinary mathematical tasks. For example, the work above could be extended: What if the sum (125) is multiplied by 2? How might you change the terms you are adding to make the equation true? What if you multiplied 125 by 5?

$$\begin{array}{r} + \quad ? \\ ? \\ \hline 2 \times 125 \end{array} \qquad \begin{array}{r} 68 \\ +\ 57 \\ \hline 125 \end{array} \qquad \begin{array}{r} + \quad ? \\ ? \\ \hline 5 \times 125 \end{array}$$

Such explorations of operations could be further extended to subtraction, multiplication, and division.

Making Records of Mathematical Knowledge

As we saw in this chapter, mathematical tasks themselves do not comprise the work that teachers and students do. How they are used is equally important. This section on making records to support teaching and learning draws on, and has benefited especially from, our research with Ben-Peretz and Cohen. If mathematical reasoning is to be central to the work that teachers and students do, then teachers play an important role in making mathematical knowledge and language public and in scaffolding their use. One element concerns making records of the mathematical work of the class: how the chalkboard is used; when the overhead projector, with its transitory record, is not used; what is recorded on pieces of paper and preserved on classroom walls; and what students are helped to record in notebooks. Doing this recording makes mathematical work public and available for collective development, scrutiny, and subsequent use. Definitions that have been worked out can be posted in the room. Students' conjectures may be provisionally recorded while they are being tested; established conclusions may later replace such conjectures. For example, once the students completed their work on the conjectures about adding odd and even numbers, the agreed-on conjectures were printed on colored paper and taped to the wall:

Odd + odd = even

Even + even = even

Odd + even = odd

Even during a class period, having available different solutions, diagrams, and ideas that teacher and students can use to reason about the problems at hand supports the work. Without such records, the "texts" of class work are remote from individual and collective attention and the history of an idea's or a problem's development is obscured. Being able to look at three distinct methods for solving a given problem permits such questions as "Are these all correct?" "Are these really different or the 'same'?" and "In what way are these the same?" Seeing arrayed the six solutions to the two-coin problem made it easier to consider whether the class had found all possible solutions. And having previous conjectures posted above the board gave Mei the grounds to remind the class that they had already agreed on these earlier claims. Obviously, what teachers decide to display on the board or in a permanent spot, who places it there, and how such public records are used are all matters worthy of careful attention.

Naming

Making records of mathematical work includes considering how to name or refer to ideas, methods, problems, solutions, and so on. *Betsy's conjecture* referred to the claim, suggested by Betsy, that an odd number plus an odd number equals an even number. Naming it as such provided a compact means of referring to this provisional idea. Naming it after Betsy reflected a common practice in mathematics of identifying mathematical ideas by their authors. As teachers choose to name ideas for students, sensitivity to whose ideas are named and to the possibility that several students may have come up with the same idea matters. Mathematical ideas, methods, and solutions can also be given other sorts of names. *Our working definitions of even and odd numbers* might be the label for a provisional articulation of these definitions. Worth considering, however, are ways to make knowledge easily retrievable and its referents collectively recognized.

Making Ideas Public

Making mathematical knowledge and language public also requires deftly moving individuals' ideas into the collective space. When individual students offer ideas or solutions, these often become bilateral exchanges between them and their teachers. For students to take note of and use one another's ideas requires deliberate work. First, students' contributions must be comprehensible to their peers. This step requires that students speak loudly enough to be heard and learn to listen closely to others' talk. When Lisa offered an idea in the September class, Ball asked her to speak up and told others to consider what she was saying. Teachers' moves may be as simple as that. But teachers may also have to help students articulate their ideas in ways that are both audible and understandable. They may have to ask individual students to repeat what they have said or ask them questions about what they are saying. These tactics help make the ideas that individual students are conveying more explicit, so their contributions do not remain private, vague, half-developed, and weakly articulated statements to which others cannot usefully respond. Making ideas public involves helping make them accessible for others' consideration. Once ideas are more clearly expressed, teachers may ask students to respond directly to another student's point, may ask students to explain what a classmate has said, or may ask students whether they can articulate how a classmate reached a conclusion. For example, in September, Ball asked Haroun to justify a claim that Tembe made. In January, Ball asked,

> What do other people think about this argument that Jillian and Lin are having? Lin's saying that she is still not convinced that this could apply for all numbers—big, big numbers, and Jillian is saying that it would always be true because there'd always be one left over from each odd number. Rania, what do you think about that?

She might also have asked Betsy's classmates whether they could see how she was using the definition of odd numbers to make her proof. As students' ideas become regular sources of the class's work, students will both speak more clearly and ask one another to speak more audibly.

Teachers also play an important role in modeling the use of others' ideas, using public mathematical knowledge, and using language carefully. Teachers can make references to "Lucy's method" or remark on uses of established ideas: "How is this idea related to Tembe's conjecture?" They can expect connections with public knowledge by asking such questions as "Are you using the definition that we agreed on for even numbers?" or "How does what we figured out about multiplying by 10 or 100 or 1000 help with this problem?"

Conclusion

We have argued that mathematical reasoning is inseparable from knowing mathematics with understanding, and that mathematical reasoning is itself a basic mathematical skill. Making mathematics reasonable in school is not a frill. Doing so, however, requires more than slogans and exhortation. Simply posing open-ended mathematical problems that require mathematical reasoning is not sufficient to help students learn to reason mathematically. Neither is merely asking students to explain their thinking.

In this chapter, we characterized two foundations of the mathematical reasoning of justification: a base of public knowledge and mathematical language. The reasoning of justification requires pedagogy and design, just as the teaching of any other aspect of mathematics. Students must learn to use publicly established ideas, methods, and language to make, inspect, validate, improve, and extend the mathematical knowledge of the class. And teachers must provide the resources for, and create an environment that encourages and makes possible, complex student work.

In the last section of the chapter, we identified some initiatives and moves that teachers might make to support the development of students' mathematical reasoning. One is the design or enhancement of mathematical tasks to create a demand for mathematical reasoning. Another is to create lasting records of established as well as currently negotiated mathematical knowledge. Such records afford both reference to prior work and a shared sense of the class's base of public knowledge. This record making, and the classroom discourse that it enables, includes the development of mathematical language—such as working definitions of basic concepts, and names for important ideas. Putting students' mathematical productions into public space and helping make those productions comprehensible so that other students can critically examine them entail norms of communication, of both careful and audible articulation of mathematical ideas, and of attentive listening.

All these steps require learning. In the episodes analyzed above, we have tried not only to illustrate what substantial mathematical reasoning by third graders can look like but also to show that these skills and practices were learned across time, from the first day of class in September. We also have made explicit some of the pedagogical moves and designs that contributed to that learning, teacher practices that play an important role in making mathematics reasonable for students. But these practices are not ones that teachers can do simply because it is suggested that they do so. Just as students need to learn to reason mathematically, so, too, must teachers develop and learn practices to support such learning. If students are to acquire the basic skills to build mathematical knowledge by reasoning, then teachers must have opportunities to develop the knowledge and practice necessary to make mathematics reasonable in school.

ACKNOWLEDGMENTS

The preparation of this paper was supported, in part, by the Spencer Foundation for the project Crossing Boundaries: Probing the Interplay of Mathematics and Pedagogy in Elementary Teaching, Major Grant no. 199800202. The paper draws, in part, on Ball and Bass (2000b). Our work draws on data collected during the 1989–1990 school year under a National Science Foundation grant to Deborah Loewenberg Ball and Magdalene Lampert for a project in which we set out to investigate the potential of using new technologies together with extensive records of practice to design new approaches to the pedagogy and curriculum of teacher education. Daily records were made of Ball's third-grade and Lampert's fifth-grade mathematics classes, including videotapes and audiotapes of lessons, photocopies of students' work, teacher's journal and plans, tests, quizzes, homework, and the mathematics problems and tasks on which the students worked. For a discussion of this project and its results, see Lampert and Ball (1998).

The ideas in this chapter were vetted with members of the Mathematics Teaching and Learning to Teach Project at the University of Michigan, and we acknowledge gratefully their many contributions: Merrie Blunk, Mark Hoover, Deidre LeFevre, Jennifer Lewis, Geoffrey Phelps, Ed Wall, and Raven Wallace. We also acknowledge the careful reading by, and comments of, Deborah Schifter, Denise Mewborn, and Patrick Callahan. Our discussion of the work of teaching to support mathematical reasoning has been helpfully guided by our work with others on the Mathematics Teaching and Learning to Teach Project. In particular, our consideration of the nature and uses of mathematical tasks, the significance of naming ideas, and the teacher's role in moving individuals' ideas into the public space have benefited from our collective work. We acknowledge and thank especially Mark Hoover, Jennifer Lewis, and Ed Wall for their contributions to these particular ideas.

REFERENCES

Ball, D. L. (1993). With an eye on the mathematical horizon: Dilemmas of teaching elementary school mathematics. *Elementary School Journal, 93*, 373–397.

Ball, D. L., & Bass, H. (2000a). Interweaving content and pedagogy in teaching and learning to teach: Knowing and using mathematics. In J. Boaler (Ed.), *Multiple perspectives on the teaching and learning of mathematics* (pp. 83–104). Westport, CT: Ablex.

Ball, D. L., & Bass, H. (2000b). Making believe: The collective construction of public mathematical knowledge in the elementary classroom. In D. Phillips (Ed.), *Constructivism in education* (99th yearbook of the National Society for the Study of Education, Part 1, pp. 193–224). Chicago: University of Chicago Press.

Ben-Peretz, M. (1990). *The teacher-curriculum encounter: Freeing teachers from the tyranny of texts.* Albany, NY: SUNY Press.

Bruner, J. (1960). *The process of education.* Cambridge: Harvard University Press.

Carpenter, T. P., & Franke, M. L. (2001). Developing algebraic reasoning in the elementary school: Generalization and proof. In H. Chick, K. Stacey, J. Vincent, & J. Vincent (Eds.), *The future of the teaching and learning of algebra* (Proceedings of the 12th ICMI Study Conference). Melbourne, Australia: University of Melbourne.

Crumbaugh, C. (1998). *"Yeah but I thought it would still make a square": A study of fourth-graders' disagreement during whole-group mathematics discussion.* Unpublished doctoral dissertation, Michigan State University, East Lansing.

Davis, R. (1967). Mathematics teaching—With special reference to epistemological problems. *Journal of Research and Development in Education* (Monograph No. 1).

Delpit, L. (1985). The silenced dialogue: Power and pedagogy in educating other people's children. *Harvard Educational Review, 58,* 280–298.

Dewey, J. (1916). *Democracy and education.* Chicago: University of Chicago Press.

Edwards, D., & Mercer, N. (1987). Reconstructing context: The conventionalization of classroom knowledge. *Discourse Processes, 12,* 91–104.

Edwards, D., & Mercer, N. (1989). *Common knowledge.* London: Routledge.

Erlwanger, S. (1973). Benny's conceptions of rules and answers in IPI mathematics. *Journal of Children's Mathematical Behavior, 1*(3), 157–283.

Falkner, K., Levi, L., & Carpenter, T. P. (1999). Early childhood corner: Children's understanding of equality: A foundation *for algebra. Teaching Children Mathematics, 6,* 232–236.

Lampert, M. (1990). When the problem is not the question and the answer is not the solution. *American Educational Research Journal, 27,* 29–63.

Lampert, M. (1992). Practices and problems in teaching authentic mathematics. In F. Oser, D. Andreas, J. Patry (Eds.), *Effective and responsible teaching: The new synthesis* (pp. 295–314). San Francisco: Jossey-Bass.

Lampert, M. (1998). *Talking mathematics in school.* Cambridge, MA: Cambridge University Press.

Lampert, M. (2001). *Teaching problems and the problems of teaching.* New Haven, CT: Yale University Press.

Lampert, M., & Ball, D. L. (1998). *Mathematics, teaching, and multimedia: Investigations of real practice.* New York: Teachers College Press.

National Council of Teachers of Mathematics. (2000). *Principles and standards for school mathematics.* Reston, VA: Author.

Rose, S. L (1999). *Understanding children's historical sense-making.* Unpublished doctoral dissertation, Michigan State University, East Lansing.

Schwab, J. J. (1976). Education and the state: Learning community. In *Great ideas today* (pp. 234–271). Chicago: Encyclopedia Britannica.

Smith, S. (1999). *Children, learning theory, and mathematics: An analysis of the language and representations in children's mathematical reasoning.* Unpublished doctoral dissertation, Michigan State University, East Lansing.

Stein, M. K., Grover, B., & Henningsen, M. (1996). Building student capacity for mathematical thinking and reasoning: An analysis of mathematical tasks used in reform classrooms. *American Educational Research Journal, 33,* 455–488.

Wilson, S. (2001). Research on history teaching. In V. Richardson (Ed.). *Handbook for research on teaching* (4th ed., pp. 527–565). Washington, DC: American Educational Research Association.

Wineburg, S. (1996). The psychology of teaching and learning history. In D. Berliner & R. Calfee (Eds.), *Handbook of educational psychology* (pp. 423–437). New York: Simon & Schuster.

Yackel, E., & Cobb, P. (1996). Sociomathematical norms, argumentation, and intellectual mathematics. *Journal for Research in Mathematics Education, 27,* 458–477.

Chapter 4

Teaching, Teachers' Knowledge, and Their Professional Development

Denise S. Mewborn, University of Georgia

One year after the release of the draft of *Curriculum and Evaluation Standards for School Mathematics* (National Council of Teachers of Mathematics [NCTM], 1989), Cooney (1988), drawing on work by Freudenthal, asserted that "reform is not a matter of paper but a matter of people" (p. 355). The people to whom Cooney referred are teachers and those who support teachers in their efforts to implement instruction that is consistent with the vision of *Principles and Standards for School Mathematics* (NCTM, 2000). The crucial role of teachers is highlighted in some of the vignettes in *Principles and Standards* (see, e.g., pp. 120–121). The purpose of this chapter is to present some of the research findings that support NCTM's vision of mathematics teaching and to illuminate what is known about what it takes to teach in the manner described by *Principles and Standards.*

Impact of Teachers on Students' Learning

Principles and Standards asserts, "Students' understanding of mathematics, their ability to use it to solve problems, and their confidence in, and disposition toward, mathematics are all shaped by the teaching they encounter in school" (NCTM, 2000, pp. 16–17). NCTM is not alone in its claim that teachers are essential to students' learning and achievement. For example, the American Council on Education (ACE) asserted, "The success of the student depends most of all on the quality of the teacher" (1999, p. 5). A series of studies by Sanders and his colleagues using the Tennessee Value-Added Assessment System (e.g., Sanders & Horn, 1998; Sanders & Rivers, 1996) supports the statements by NCTM, ACE, and others. These studies showed that teachers are the main determinant of student academic achievement as measured by standardized test scores. Teachers were deemed to be "effective" based on their students' gain scores on a standardized test. Students assigned to effective teachers had significantly higher gains from the previous year than did students assigned to ineffective teachers. While these studies controlled for such variables as students' socioeconomic status, they did not identify the specific classroom practices that characterize "effectiveness" in teaching.

A multitude of studies have been undertaken in the past 40 years in an attempt to determine which teacher characteristics affect student achievement, and the results largely have been inconclusive. In 1999, the Center for the Study of Teaching and Policy released a 50-state report (Darling-Hammond, 1999) that summarized the research on the impact on student achievement of teachers' general academic ability and intelligence, subject matter knowledge, knowledge of teaching and learning, teaching experience, certification status, and teaching behaviors and practices. Darling-Hammond attributed the surprisingly contradictory and inconclusive results, in part, to the methodological complexity of measuring such variables.

Perhaps the most widely cited example of researchers' attempts to link teacher characteristics and student learning in mathematics was a study conducted as part of the National Longitudinal Study of Mathematical Abilities [NLSMA] and summarized by Begle in his 1979 book *Critical Variables in Mathematics Education: Findings from a Survey of the Empirical Literature*. Begle found no evidence to suggest a significant positive relationship between student achievement in mathematics and the variables of teachers' number of years of teaching, highest academic degree, academic credits beyond a BA, mathematics credits beginning with calculus, credits in mathematics methods, in-service or extension courses, other preparation in the past 5 years, and majoring or minoring in mathematics. Begle (1979) concluded,

> There is no doubt that teachers play an important role in the learning of mathematics by their students. However, the specific ways in which teachers' understanding, attitudes, and characteristics affect their students are not widely understood. (p. 27)

Begle's statement is still true today. We lack a detailed understanding of which teacher characteristics affect student learning and how they affect student learning.

In contrast to Begle's work, some recent studies have documented modest connections between specific teacher characteristics and student achievement. Monk (1994) found that the number of mathematics education courses a teacher had taken was significantly related to gains in student achievement. In particular, he found that teachers' coursework in mathematics education had a greater impact on student achievement than did teachers' coursework in mathematics. Interestingly, Begle (1979) found a similar result, but that result is rarely cited. Not surprisingly, in cases in which there does seem to be a connection between teacher characteristics and student achievement, the relationship is nonlinear, and threshold effects seem to occur (Darling-Hammond, 1999; Monk, 1994).

Fennema et al. (1996) reported on a longitudinal study of primary-grade teachers using the Cognitively Guided Instruction (CGI) program. CGI is a teacher development project in which teachers learn about research-based models of students' mathematical thinking. The teachers then investigate the implications of these models for their classroom practices. Fennema et al. found that student achievement in the areas of concepts and problem solving improved in the classrooms of CGI teachers. In many cases, student achievement increased by a standard deviation. In addition to examining student achievement scores, Fennema and her colleagues summarized teachers' classroom practices that affected student achievement. Instructional practices that seemed to be associated with higher student achievement included (a) providing time for students to work intensively with mathematical ideas in a problem-solving context, (b) providing opportunities for students to engage in conversations with one another about their mathematical ideas, and (c) adapting instruction to the problem-solving level of the students. Similarly, a study of mathematics teaching in California after the adoption of the *California Mathematics Framework* showed that children learned more mathematics in classrooms where instruction (a) was based on students' ways of thinking, (b) engaged students in problem solving with rich problems, and (c) assisted students in seeing the underlying links among various mathematical concepts and symbols (Gearhart et al., 1999).

Studies such as those cited above provide convincing evidence that students' learning of mathematics content is influenced by what teachers do in the classroom. Other studies provide evidence that teachers' practices also influence what students have the opportunity to learn about mathematics as a discipline. Ball and her colleagues have studied her mathematics teaching in a third-grade classroom over the course of a school year. One particular example from her work illustrates how a teacher can affect students' opportunity to learn mathematics. In a lesson involving even and odd numbers, one of Ball's students, Shea, said that 6 was both an even and an odd number. He explained that 6 was even because 6 objects could be "split in half without having to use halves" (Ball, 1992, p. 14); he was using the definition of evenness that the students had constructed. He explained that 6 was also odd because when 6 objects were placed in piles of 2, there were 3 piles and 3 is an odd number. This definition of odd numbers was not the one that had been agreed upon by the classroom community. However, rather than tell Shea that he was not being consistent in his application of definitions, Ball asked the other students in the class to comment on Shea's claim and engaged them in determining whether other numbers had the property that Shea noticed. Ball was able to use her own knowledge of mathematics content to determine that Shea's observation about the number 6 was not a random occurrence but rather that other numbers fit Shea's description and that those numbers had a pattern. Thus, she decided that engaging the students in the investigation, although not the lesson she had planned for the day, would be a worthwhile task. The students eventually developed a general rule for such numbers, and, in keeping with how things are often done in the larger mathematics community, they named these numbers *Shea numbers* in honor of the mathematician who first advanced the idea.

In this lesson, Ball used her knowledge of mathematics as a discipline to guide student learning. As evidenced by classroom discussions and a subsequent written task completed by the children individually, the children learned the mathematics of how to determine whether a number is even or odd and whether a number is a Shea number. Further, through this and other lessons, the children learned something about what it means to participate in a mathematical community. They learned how a conjecture is scrutinized by peers, how it is stated formally, and how it becomes an accepted part of the culture of a mathematical community.

Ball (1991) studied three experienced teachers as they taught long multiplication to fourth graders and analyzed the ways in which their knowledge about mathematics as a discipline influenced their classroom practice and, hence, students' opportunities to learn mathematics in the ways advocated by NCTM. Two of the teachers focused their instruction on helping children get correct answers through the proper use of procedures. They used mnemonic devices to help students remember the steps of the procedure and "placeholders" to minimize students' errors. Students were given lots of opportunities to practice procedures and get feedback from the teacher. Although one teacher did show evidence that she understood the mathematics behind the placement of partial products and the role of place value, she chose not to explain that mathematics to her students because she did not think it was important to their success in obtaining correct answers. Thus, the students in these classrooms did not have the opportunity to grapple with the meaning of multiplication and how it might be represented with larger numbers. They did not have the opportunity to make sense of the procedure they were being taught by linking it to what they already knew.

In contrast, the third teacher in the study, mathematics educator Magdalene Lampert, had a goal of developing students' mathematical and reasoning skills to enable them to validate their own and others' thinking. Thus, she engaged her students in the intellectual activities of inventing procedures, justifying the validity of their procedures, and explaining their procedures to peers. Like the students of the other two teachers, Lampert's students learned how to get correct answers to multiplication problems, but her students also had the opportunity to learn something about how mathematics is created, canonized, and communicated to others.

The question of how teaching affects student learning is still largely unanswered by research. Sanders and Horn (1998) provided evidence that teachers do figure prominently in children's achievement and that so-called good teaching can generally meet the challenges of economics, administration, and curriculum. However, a complicated relationship exists among teachers' knowledge, their teaching practices, and student learning that research has yet to untangle.

Teachers' Knowledge

Principles and Standards asserts that "teachers must know and understand deeply the mathematics they are teaching and be able to draw on that knowledge with flexibility in their teaching tasks" (NCTM, 2000, p. 17). An abundance of research tells us that many teachers do not possess this deep and rich knowledge of mathematics. Much of this research has been done with preservice teachers (e.g., Ball, 1988), but some examples exist in the literature on in-service teachers as well (e.g., Putnam, Heaton, Prawat, & Remillard, 1992). Further, most of this research has been done with elementary school teachers (e.g., Leinhardt & Smith, 1985), although some has involved middle school teachers (e.g., Post, Harel, Behr, & Lesh, 1991) and secondary school teachers (e.g., Even & Tirosh, 1995). This body of research overwhelmingly paints a dismal picture of teachers' conceptual knowledge of the mathematics they are expected to teach. By and large, teachers have a strong command of the procedural knowledge of mathematics, but they lack a conceptual understanding of the ideas that underpin the procedures.

Putnam et al. (1992) studied four fifth-grade teachers during mathematics instruction and identified aspects of the teachers' mathematical knowledge that affected their instruction. They noted that "the limits of [the teachers'] knowledge of mathematics became apparent and their efforts fell short of providing students with powerful mathematical experiences" (p. 221). For example, one teacher's lack of familiarity with area and perimeter led her to encourage her students to multiply the dimensions of a park to determine how much fencing would be required to enclose it; the situation called for a measure of perimeter, but the teacher directed her students to compute a measure of area. In another lesson, the same teacher failed to recognize the complexities her students would face in attempting to determine how many cubic feet of sand would be required to fill a sandbox. The students measured two dimensions of the sandbox in yards and one dimension in feet. The students were then allowed to erroneously multiply measurements given in feet by measurements given in yards, resulting in an incorrect and unreasonable answer (Heaton, 1992).

Perhaps, however, it is more instructive to examine the research that shows what teachers who have a strong command of mathematics content can do in the classroom. Fernández (1997) observed lessons taught by nine secondary mathematics teachers who held bachelor's degrees in mathematics or mathematics-related fields and Master of Arts in Teaching degrees in secondary mathematics. She focused on the teachers' responses to unexpected student answers to see how the teachers' knowledge affected their teaching practice. She noted situations in which the teachers' strong content knowledge enabled them to do any of a variety of things: provide a counterexample to uncover an error in students' thinking, follow through on a student's comment to lead to a contradiction or a viable solution, apply a student's method to a simpler or related problem, understand a student's alternative method, and incorporate a student's alternative method into instruction.

Sowder, Philipp, Armstrong, and Schappelle (1998) reported on a 2-year teacher enhancement project with five middle school teachers teaching rational numbers, quantity, and proportional reasoning. The researchers found that teachers' practices changed as their content knowledge increased and deepened. For example, the teachers were more willing to try new mathematics with their students and became less dependent on prescribed curricula. They saw their students as more capable mathematically and thus encouraged and expected more conceptual explanations of material. Sowder et al. also found that students' knowledge of fractions increased as their teachers learned more mathematics. Further, the students showed evidence of deeper mathematical understanding than is typical of other students in their grade level.

In another study of the effects of professional development focused on specific mathematical content, Swafford, Jones, and Thornton (1997) studied the classroom practices of eight middle school teachers who completed a summer course in geometry. As a result of the course, the teachers made significant gains in their geometry content knowledge, which affected their classroom practices. Similar to the results found by Sowder et al., the teachers in Swafford et al.'s study spent more time on geometry instruction and incorporated new instructional tasks (such as tessellations) and methods (such as encouraging student discourse) into their practices. The teachers reported that they were more confident in their abilities to elicit and respond to higher levels of geometric thinking and more likely to take risks to enhance student learning as a result of their experience in the geometry course.

One study, in particular, shows that teachers need more than strong mathematical knowledge in order to be effective teachers. Thompson and Thompson (1994, 1996) studied a middle school teacher as he worked one-on-one with a middle school student on the concept of rate. They found that although the teacher had a reasonably robust conception of rate, he was unable to articulate that conception in a way that helped the student learn about rate conceptually. In fact, the teacher's inclination to describe rate in terms of whole numbers and whole-number operations actually reinforced the student's incorrect additive (rather than multiplicative) way of thinking. Thompson and Thompson noted that because the teacher had a strong conception of rate, he tended to automatically lay his own understanding on top of the child's explanations rather than listen carefully to her reasoning. This tendency to interpret the child's explanations in light of his own mathematical understanding led him to misjudge what the student was thinking. On some occasions the teacher and student seemed to be having a conversation but were actually talking past each other because the teacher was unaware that the child did not possess an understanding of rate that was identical to his. The teacher, although having a strong conception of rate, was not able to "step outside" his own thinking to really listen to the child.

This study shows that teachers need more than strong mathematics content knowledge to be able to enact the vision of teaching set forth in *Principles and Standards*. They also need knowledge of students, curriculum, learning theory, and pedagogy. *Principles and Standards* acknowledges the importance of all these types of knowledge.

Professional Development and the Conditions That Promote Change

Teaching, as envisioned by the writers of *Principles and Standards*, is a continual journey. There is no one right way to teach, and one does not "master" teaching. There is always room to grow and change. Heaton (1994, 2000), an experienced teacher who engaged in a year-long self-study as she tried to change her teaching practice to be consistent with current calls for reform, characterized teaching as "inherently under construction" (p. 341) and "continuous invention" (p. 341). She noted that teaching is very situation-specific and that therefore teachers are constantly tinkering with what they are doing. Further, as teachers seek to make changes in practice, they must continue teaching at the same time. Thus, a teacher's classroom is a "learning laboratory" (Cobb, Wood, & Yackel, 1990, p. 131). To understand and embrace the reforms envisioned by NCTM, however, teachers need professional development and support. *Principles and Standards* notes, "The work and time of teachers must be structured to allow and support professional development that will benefit them and their students" (NCTM, 2000, p. 19). In recent years, a number of well-researched professional development projects have provided some useful insights into the conditions that promote change.

The design of effective professional development opportunities should be grounded in sound theories about learning in general, and adult learning in particular. Lappan and Briars (1995) characterized the view of student learning supported by NCTM:

- Learning is contextual: What students learn is fundamentally connected with how they learn it.
- Learning occurs best through dialogue, discussion, and interaction.
- Learners must be actively involved in the process.
- A variety of models must be used to meet the [needs] of all learners (e.g., working individually, in pairs, and in cooperative groups).
- Learners benefit from reviewing, critiquing, and revising one another's work. (p. 133)

Just as one cannot expect students to learn something simply by being told that it is so, one cannot expect teachers to change their teaching practice simply because they have been told to do so. Cohen (1990) presented the case of a teacher, Mrs. Oublier, who was taught about contemporary reform issues in a manner that was antithetical to the reform itself. "She was told to do something, like students in many traditional math classrooms. She was told that it was important. Brief explanations were offered, and a synopsis of what she was to learn was provided in a text" (p. 263). In contrast, Schifter (1998) provided examples of teachers who participated in professional development sessions in which they were actively engaged in learning new ideas. These teachers were able to translate what they learned into classroom practice. Schifter's research suggests that teachers need to have opportunities to learn mathematics in the ways in which they are expected to teach it to students. They need to struggle with important mathematical ideas, justify their thinking to peers, investigate alternative solutions proposed by others, and reconsider their conceptions of what it means to do mathematics. In short, teachers' thinking needs to be at the center of professional development sessions just as children's thinking needs to be at the center of mathematics instruction.

A second important aspect of professional development, which is alluded to above, is that teachers need to revisit the mathematics they are teaching to gain insights into the conceptual underpinnings of topics and the interconnections among topics. Enhancing teachers' mathematical knowledge has been a component of a number of professional development projects, and the evidence overwhelmingly suggests that this is a crucial part of learning to teach differently (Brown, Smith, & Stein, 1996; Campbell & White, 1997; Heaton, 1994, 2000; Kilpatrick, Hancock, Mewborn, & Stallings, 1996; Swafford, Jones, & Thornton, 1997).

As the study by Thompson and Thompson (1994, 1996) demonstrates, it is not sufficient for teachers to have a robust knowledge of mathematics. They must also have a propensity to listen to students' reasoning and the knowledge base to make sense of students' mathematical thinking. Thus, teachers need professional development that affords them the opportunity to cultivate their listening skills and their ability to analyze children's ideas. Several projects have incorporated a focus on children's mathematical thinking—Cognitively Guided Instruction (Fennema et al., 1996), Integrated Mathematics Assessment Program (Gearhart et al., 1999), Mathematics Case Methods Project (Barnett, Goldenstein, & Jackson, 1994), and Teaching to the Big Ideas (Schifter, 1998). Many of these projects use artifacts of practice (e.g., videocases, written cases, and written student work) as a source of professional development curricula.

Professional development opportunities for teachers need to occur in a context in which teachers can try what they have learned in their classrooms. Fennema et al. (1996) found that teachers who participated in the CGI program used their classrooms as testing grounds to evaluate the information they were receiving in workshops. As they were able to validate this information in their classrooms, the teachers began to take ownership of the models that were presented and to interpret them in light of their classroom experience. Schifter (1998) reported that although the teachers in her study gained a great deal from reading transcripts and watching videotapes of children's mathematical thinking, they needed to hear *their own* students' mathematical thinking. Thus, it was crucial that the professional development project spanned a school year so that teachers could try out what they were learning in their own classroom.

Similarly, Kazemi and Franke (2000) studied teachers in an urban elementary school as they worked together to reflect on and change their classroom practice by sharing student work with one another. Those teachers who were more engaged in the project consistently applied what they learned from their peers to their classroom practice. These teachers saw their classrooms as natural extensions of the professional development work. In contrast, those teachers who were less engaged in the project viewed the professional development activity as separate from their classroom practice. They were less inclined to adapt mathematical problems used in the working groups for use in their classrooms, and they were less likely to bring examples of student work to the working group meetings.

Teachers need collegial support as they engage in changing their practices. This support includes being provided the time to meet with colleagues to discuss the implementation of new ideas and to reflect individually on their teaching. The QUASAR (Quantitative Understanding Amplifying Student Achievement and Reasoning) Project (Silver, Smith, & Nelson, 1995; Stein, Silver, & Smith, 1998) provides an example of a sustained professional development project that focused on school-based change in urban middle schools. Researchers in this project found that teachers needed safe, supportive environments in which to discuss issues of content and pedagogy with peers. Having colleagues with whom to discuss successful and exciting experiences as well as difficult and frustrating experiences was essential to teachers' development. However, Gearhart et al. (1999) found that professional development that was primarily aimed at providing collegial support without a concomitant focus on mathematics, children's mathematical thinking, or curriculum was not as effective in helping teachers change their practices as professional development with an explicit focus on such topics.

Teachers at the North Carolina School of Science and Mathematics worked together to revamp their precalculus curriculum and eventually wrote a textbook (Barrett et al., 1992). Although writing a textbook is not a typical form of professional development, it afforded these teachers opportunities to grapple with unfamiliar mathematics and to talk openly about pedagogical issues. The teachers reported that they likely would not have undertaken, or been successful with, the changes they made if each individual teacher had acted alone. Instead, they formed a collaborative learning community, changed their individual teaching practices, and influenced a number of other teachers through their book and workshops (Kilpatrick et al., 1996).

Even when individual teachers seek to change their practice on their own, a supportive local community is essential to their success. Romagnano (1994) and Heaton (1994, 2000) reported the centrality of particular colleagues in their personal quests to teach mathematics differently. Their colleagues helped them grapple with the uncertainties they faced, assisted them in learning new mathematical ideas, and provided a mirror to help them look reflectively at their practices.

A number of scholars have reported on models for continuous improvement of teaching used in Japan and China (Ma, 1999; Stigler & Hiebert, 1999). These models seem to incorporate all the components of professional development identified above: long-term, school-based reform in a community of learners with opportunities to grapple with significant mathematical ideas and to consider how students engage with these mathematical ideas. In these countries, teachers spend a significant portion of their school day in collaboration with one another. Chinese teachers spend substantial amounts of time studying the government's framework, the textbooks, and the teacher's manual to understand how topics are sequenced, why particular examples are used, and how best to make use of the materials to accomplish the stated objectives. Ma noted that this preparation time "occupies a significant status in Chinese teachers' work" (p. 135) and that much of this studying is done with colleagues. Teachers meet weekly in teaching research groups for formal discussion of, and reflection on, their teaching. During both individual and group study of teaching materials, Chinese teachers likely encounter both the opportunity and the necessity to deepen their knowledge of the mathematics content they are teaching. To provide time for this intensive study with colleagues, Chinese teachers teach only three or four 45-minute class periods each day. Further, 80% of the Chinese elementary school teachers in Ma's study taught only mathematics. U.S. elementary school teachers, by contrast, are expected to understand and teach science, social studies, reading, language arts, health, and a host of other subjects.

The Asian model of continuous improvement highlights the importance of viewing professional development as part of the daily work of teachers. A study of the mathematics reform movement in California conducted by Cohen and Hill (1998) shows that professional development also needs to be aligned with all other aspects of teachers' work. Cohen and Hill found that professional development was most successful in changing teachers' practices and in enhancing student achievement when the professional development was designed to help teachers understand student curriculum materials that reflected the intended reform. Further, they demonstrated a link between teachers' professional development and student achievement when the curriculum, teachers' professional development work, and student assessment measures were aligned with the prevailing policies of the reform.

The Limits of Research

Although much research has been conducted on teachers and teaching, some questions remain that research cannot answer definitively. One such question is "What constitutes 'good' teaching?" A number of studies cited in this chapter paint portraits of "good" teachers. However, it is unclear whether there is such as thing as "good" teaching, what it looks like, and how one becomes a "good" teacher. The research does show clearly that there is no single answer to what counts as good teaching or how one becomes a good teacher. Teaching is an enormously complex activity that cannot be reduced to a list of do's and don'ts. No particular instructional approach is necessarily ruled out, and no particular approach is necessarily always good. Context plays a crucial role in instructional decision-making, and each situation is unique. Thus, *Principles and Standards* does not attempt to provide prescriptions whereby teachers can follow a formula that will lead to effective mathematics teaching. Rather, it provides general guidelines, based on sound research, to guide and shape the work of mathematics teachers.

ACKNOWLEDGMENT

The ideas presented in this chapter were influenced substantially by my work with Deborah Loewenberg Ball.

REFERENCES

American Council on Education. (1999). *To touch the future: Transforming the way teachers are taught. An action agenda for college and university presidents.* Washington, DC: ACE.

Ball, D. L. (1992). Magical hopes: Manipulatives and the reform of math education. *American Educator, 16*(2), 14–18, 46–47.

Ball, D. L. (1991). Research on teaching mathematics: Making subject matter knowledge part of the equation. In J. Brophy (Ed.), *Advances in research on teaching, Vol. 2: Teachers' knowledge of subject matter as it relates to their teaching practice* (pp. 1–48). Greenwich, CT: JAI Press.

Ball, D. L. (1988). *The subject matter preparation of prospective mathematics teachers: Challenging the myths* (Research Report 88-3). East Lansing, MI: Michigan State University, National Center for Research on Teacher Learning.

Barnett, C., Goldenstein, D., Jackson, B. (1994). Fractions, decimals, ratios & percents: *Hard to teach and hard to learn?* Portsmouth, NH: Heinemann.

Barrett, G. B., Bartkovich, K. G., Compton, H. L., Davis, S., Doyle, D., Goebel, J. A., et al. (1992). *Contemporary precalculus through applications*: Functions, data analysis, and matrices. Dedham, MA: Janson.

Begle, E. G. (1979). *Critical variables in mathematics education: Findings from a survey of the empirical literature.* Washington, DC: Mathematical Association of America and National Council of Teachers of Mathematics.

Brown, C. A., Smith, M. S., & Stein, M. K. (1996, April). *Linking teacher professional development to enhanced classroom instruction.* Paper presented at the meeting of the American Educational Research Association, New York.

Campbell, P. F., & White, D. Y. (1997). Project IMPACT: Influencing and supporting teacher change in predominantly minority schools. In E. Fennema & B. S. Nelson (Eds.), *Mathematics teachers in transition* (pp. 309–355). Mahwah, NJ: Erlbaum.

Cobb, P., Wood, T., & Yackel, E. (1990). Classrooms as learning environments for teachers and researchers. In R. Davis, C. Maher, & N. Noddings (Eds.), *Constructivist views on the teaching and learning of mathematics* (JRME Monograph 4, pp. 125–146). Reston, VA: National Council of Teachers of Mathematics.

Cohen, D. K. (1990). A revolution in one classroom: The case of Mrs. Oublier. *Educational Evaluation and Policy Analysis, 12*, 327–345.

Cohen, D. D., & Hill, H. C. (1998). *State policy and classroom performance: Mathematics reform in California* (CPRE Policy Brief). Philadelphia: Consortium for Policy Research in Education.

Cooney, T. J. (1988). The issue of reform: What have we learned from yesteryear? *Mathematics Teacher, 81*, 352–363.

Darling-Hammond, L. (1999). *Teacher quality and student achievement: A review of state policy evidence.* Seattle: University of Washington, Center for Teaching and Policy.

Even, R., & Tirosh, D. (1995). Subject-matter knowledge and knowledge about students as sources of teacher presentations of the subject matter. *Educational Studies in Mathematics, 29*, 1–20.

Fennema, E., Carpenter, T. P., Franke, M. L., Levi, L., Jacobs, V. R., & Empson, S. B. (1996). A longitudinal study of learning to use children's thinking in mathematics instruction. *Journal for Research in Mathematics Education, 27*, 403–434.

Fernández, E. (1997). *The "'Standards'-like" role of teachers' mathematical knowledge in responding to unanticipated student observations.* Paper presented at the meeting of the American Educational Research Association, Chicago.

Gearhart, M., Saxe, G. B., Seltzer, M., Schlackman, J., Ching, C. C., Nasir, N., et al. (1999). Opportunities to learn fractions in elementary mathematics classrooms. *Journal for Research in Mathematics Education 30*, 286–315.

Heaton, R. M. (2000). *Teaching mathematics to the new standards: Relearning the dance.* New York: Teachers College Press.

Heaton, R. M. (1994). Creating and studying a practice of teaching elementary mathematics for understanding (Doctoral dissertation, Michigan State University, 1994). *Dissertation Abstracts International, 55-07A*, 1860.

Heaton, R. M. (1992). Who is minding the mathematics content? A case study of a fifth-grade teacher. *Elementary School Journal, 93*, 153–162.

Kazemi, E., & Frank, M. L. (2000). Understanding teacher learning as changing participation in communities of practice. In M. L. Fernández (Ed.), *Proceedings of the Twenty-Second Annual Meeting of the North American Chapter of the International Groups for the Psychology of Mathematics Education* (pp. 561–566). Columbus, OH: ERIC Clearinghouse for Science, Mathematics, and Environmental Education.

Kilpatrick, J., Hancock, L., Mewborn, D. S., & Stallings, L. (1996). Teaching and learning cross-country mathematics: A story of innovation in precalculus. In S. A. Raizen & E. D. Britton (Eds.), *Bold ventures: Vol. 3. Case studies of U.S. innovations in mathematics education* (pp. 133–243). Dordrecht, The Netherlands: Kluwer.

Lappan, G., & Briars, D. (1995). How should mathematics be taught? In I. M. Carl (Ed.), *Prospects for school mathematics* (pp. 131–156). Reston, VA: National Council of Teachers of Mathematics.

Leinhardt, G., & Smith, D. A. (1985). Expertise in mathematics instruction: Subject matter knowledge. *Journal of Educational Psychology*, 77, 247–271.

Ma, L. (1999). *Knowing and teaching elementary mathematics: Teachers' understanding of fundamental mathematics in China and the United States.* Mahwah, NJ: Erlbaum.

Monk, D. H. (1994). Subject area preparation of secondary mathematics and science teachers and student achievement. *Economics of Education Review, 13*, 125–145.

National Council of Teachers of Mathematics (NCTM). (1989). *Curriculum and Evaluation Standards for School Mathematics.* Reston, VA: Author.

National Council of Teachers of Mathematics. (2000). *Principles and standards for school mathematics.* Reston, VA: Author.

Post, T. R., Harel, G., Behr, M. J., & Lesh, R. (1991). Intermediate teachers' knowledge of rational number concepts. In E. Fennema, T. P. Carpenter, & S. J. Lamon (Eds.), *Integrating research on teaching and learning mathematics* (pp. 177–198). Albany: State University of New York Press.

Putnam, R. T., Heaton, R. M., Prawat, R. S., & Remillard, J. (1992). Teaching mathematics for understanding: Discussing case studies of four fifth-grade teachers. *Elementary School Journal, 93*, 213–228.

Romagnano, L. (1994). *Wrestling with change: The dilemmas of teaching real mathematics.* Portsmouth, NH: Heinemann.

Sanders, W. L., & Horn, S. P. (1998). Research findings from the Tennessee Value-Added Assessment System (TVAAS) Database: Implications for educational evaluation and research. *Journal of Personnel Evaluation in Education, 12*, 247–256.

Sanders, W. L., & Rivers, J. C. (1996). *Cumulative and residual effects of teachers on future student academic achievement.* Knoxville: University of Tennessee Value-Added Research and Assessment Center.

Schifter, D. (1998). Learning mathematics for teaching: From a teachers' seminar to the classroom. *Journal of Mathematics Teacher Education, 1*, 55–87.

Silver, E. A., Smith, M. S., & Nelson, B. S. (1995). The QUASAR Project: Equity concerns meet mathematics education reform in the middle school. In W. G. Secada, E. Fennema, & L. B. Adajian (Eds.), *New directions for equity in mathematics education* (pp. 9–56). New York: Cambridge University Press.

Sowder, J. T., Philipp, R. A., Armstrong, B. E., & Schappelle, B. P. (1998). *Middle-grade teachers' mathematical knowledge and its relationship to instruction: A research monograph.* Albany: State University of New York Press.

Stein, M. K., Silver, E. A., & Smith, M. S. (1998). Mathematics reform and teacher development from the community of practice perspective. In J. Greeno & S. Goldman (Eds.), *Thinking practices: A symposium on mathematics and science learning* (pp. 17–52). Hillsdale, NJ: Erlbaum.

Stigler, J., & Hiebert, J. (1999). *The teaching gap: Best ideas from the world's teachers for improving education in the classroom.* New York: Free Press.

Swafford, J. O., Jones, G. A., & Thornton, C. A. (1997). Increased knowledge in geometry and instructional practice. *Journal for Research in Mathematics Education, 28*, 467–483.

Thompson, A. G., & Thompson, P. W. (1996). Talking about rates conceptually, Part 2: Mathematical knowledge for teaching. *Journal for Research in Mathematics Education, 27*, 2–24.

Thompson, P. W., & Thompson, A. G. (1994). Talking about rates conceptually, Part 1: A teacher's struggle. *Journal for Research in Mathematics Education, 25*, 279–303.

CHAPTER 5

Classroom and Large-Scale Assessment

Linda Dager Wilson, University of Maryland
Patricia Ann Kenney, University of Michigan

The activities of assessment are central to the acts of teaching and learning mathematics. The central questions of assessment in mathematics are, What do our students know about mathematics? and What are they able to do? The answers to these questions are vital to each constituency involved in educating students in mathematics. Teachers need the information to plan for instruction; students, to monitor their own learning; parents, to provide necessary supports for learning; and administrators and policymakers, to make informed decisions. Yet, as central as assessment is to teaching and learning, the education research community is only just beginning to compile evidence about assessment activities in mathematics.

What are the most effective methods of assessment in the mathematics classroom? How can teachers ensure that the information they gather is valid and reliable? How can assessment tasks and instruments be designed so as to measure important mathematics, support learning, and be equitable? What forms of assessment are appropriate for different content areas of mathematics or for different developmental levels of students? What should be the teacher's role in large-scale assessment? What should be the student's role in classroom assessment?

These are but a few of the questions that could be asked about assessment in mathematics. In this chapter, we attempt to shed some light on some of these questions, insofar as research evidence is available to support various claims. We also raise issues about classroom and large-scale assessment that must be addressed, even if the research base is not yet well developed for some of them.

Overview

We have divided this chapter into two major forms of assessment—classroom assessment and large-scale assessment—given that some of the issues and questions about these two kinds of assessment are necessarily different. We deal first with classroom assessment, using the six assessment standards from *Assessment Standards for School Mathematics* (National Council of Teachers of Mathematics, 1995) as our organizer. We look at recent research that sheds light on classroom assessment in relation to each of the six standards. Next, we examine issues in large-scale assessment.

Definition of Terms

To clarify what we mean by the terms we use, we will adopt as our definition of *assessment* the one used in the assessment standards document (NCTM, 1995). That is, assessment is the process of gathering evidence about a student's knowledge of, ability to use, and disposition toward, mathematics—and of making inferences from that evidence for a variety of purposes (p. 3). This broad definition includes all the activities that educators (or students themselves) use to learn what students know and can do in mathematics. From classroom questioning and discourse to standardized achievement tests, assessment is seen as a process that is all about data gathering and analysis. One cannot know what is inside another's mind, so educators and others must gather evidence of that knowledge and then infer what the student knows.

Classroom assessment is designed or used by classroom teachers for making instructional decisions, monitoring students' progress, or evaluating students' achievement. In this chapter, we emphasize assessment that incorporates actions from the teacher in response to the evidence gathered from learners, in other words, formative assessment as opposed to summative.

Large-scale assessment refers to tests that are summative in nature and administered to large numbers of students. These tests are external to the course of regular classroom activities and are usually designed without much input from teachers. They include norm-referenced standardized tests (e.g., TerraNova [formerly the California Achievement Tests], Iowa Tests of Basic Skills [ITBS], Stanford Achievement Tests); local and state tests developed for accountability purposes (e.g., Maryland State Performance Assessment Program, North Carolina End-of-Grade tests, Texas Assessment of Academic Skills [TAAS]); and large-scale surveys given at the national or international level (e.g., National Assessment of Educational Progress [NAEP], Third International Mathematics and Science Study [TIMSS]).

Classroom teachers often do not have direct influence on large-scale assessment programs, but such programs often have a profound influence on what happens in classrooms. The latter influence is particularly strong for tests used for accountability purposes and that have consequences for students, teachers, schools, and districts. Although this chapter primarily focuses on research studies in classroom assessment, it also is important to mention issues concerning the current state of large-scale assessment programs as well as studies documenting evidence of their influence on mathematics teaching and learning.

Issues in Classroom Assessment

Mathematics: Assessment Should Reflect the Mathematics That All Students Need to Know and Be Able to Do

The nature and use of teacher tasks. What kinds of tasks should teachers use with students in order to enhance mathematics learning and assessment? The most important feature of good tasks is that they embody mathematics that is important for students to know and be able to do. Whether the tasks chosen are to be used for instructional purposes or for assessment, the primary consideration must be the mathematics in the task. As the Mathematical Sciences Education Board (MSEB, 1993a, p.6) expressed it, "The mathematics in an assessment must never be distorted or trivialized for the convenience of assessment." In two publications, MSEB developed a conceptual framework for designing, developing, implementing, and interpreting assessment tasks in mathematics, and it developed prototypes of such tasks for use by classroom teachers (MSEB, 1993a; 1993b).

The nature of the tasks that teachers use is an important feature of a successful classroom, defined as one in which students are taught mathematics for understanding. According to Ames (1992), tasks should be novel and varied in interest, offer reasonable challenge, help students develop short-term goals, focus on meaningful aspects of learning, and support the development and use of effective learning strategies. As Blumenfeld (1992) points out, a task that is too challenging can lead to students' avoidance of the risks involved. For students who are far behind, it is difficult to encourage their efforts without making them aware of how far behind they are. Yet, tasks can be meaningful for various reasons, and teachers should emphasize those meanings that might be productive for learning.

Do teacher tasks assess important mathematics—and when teachers use tasks at a high cognitive level, are they able to sustain that level during instruction? Stein, Grover, and Henningsen (1996) investigated a sample of 144 mathematics tasks used during reform-oriented instruction, analyzing their features (e.g., number of solution strategies, communication requirements) and cognitive demands (e.g., memorization, "doing of mathematics"). The findings suggest that teachers were selecting the kinds of tasks that should have led students to engage in higher-order thinking, but the cognitive demands of some high-level tasks tended to decline to the procedural level during classroom implementation. Some factors contributing to the decline were students' pressing the teacher to specify explicit procedures; teachers' performing the more challenging aspects of tasks; and students' failing to engage in the tasks because of lack of interest, motivation, or prior knowledge.

One way to categorize the difficulty level of assessment tasks is to consider the extent to which a task is like others that the students have experienced. In other words, a task could be nearly identical to one that students have done before and require merely the exact replication of a previously learned procedure. Or it might be a problem that is typical of, but not identical to, the one studied and require that students identify and use the appropriate algorithm. Or the task might be quite different, calling for new reasoning and construction of a new approach. Dumas-Carre and Larcher (1987) found that students need special and explicit training for tackling tasks of the last type. All three types of tasks are needed, and teachers could use a scheme like this to organize the tasks they use and to plan how to react to students' responses to them.

Teacher knowledge. How do teachers interpret the assessment of understanding? Cooney, Badger, and Wilson (1993) report on a study of 201 middle and high school mathematics teachers who were asked to write or draw first a typical problem for students testing minimal understanding and then a problem testing deep and thorough understanding of mathematics. The responses were categorized into four levels. The majority of teachers created items at lower levels of understanding, even though they were ostensibly testing for deep and thorough understanding. The problems they constructed typically required a simple computation or were one-step word problems. The conclusion was that teachers did not have a good understanding of how to assess more complex mathematics.

Does teachers' content knowledge affect their ability to interpret student responses? Wilson and Kenney (1997) gave an open-ended fourth-grade task to undergraduate education majors and then asked them to interpret a set of student responses to the task. About one fourth of the participants could not adequately respond to the initial task; the lack of a conceptual understanding of the task appeared to hinder participants' ability to give valid justifications for the scores they gave to student work.

Learning: Assessment Should Enhance Mathematics Learning

Evidence that formative assessment affects learning. Formative assessment, or assessment that focuses on teachers' responses to the student learning data they encounter on a daily basis, can profoundly affect learning. Black and Wiliam (1998) conclude from an examination of 250 research studies on classroom assessment that "formative assessment does improve learning"—and that the achievement gains are "among the largest ever reported for educational interventions." The effect size of 0.7, on average, illustrates just how large these gains are. If this effect size could be achieved on a nationwide scale, the outcome would be equivalent to raising the mathematics attainment score of an 'average' country like the United States into the "top five" after the Pacific rim countries of Singapore, Korea, Japan, and Hong Kong (Beaton, Mullis, Martin, Gonzalez, Kelly, & Smith, 1996). In other words, if mathematics teachers were to focus their efforts on classroom assessment that is primarily formative in nature, students' learning gains would be impressive. These efforts would include gathering data through classroom questioning and discourse, using a variety of assessment tasks, and attending primarily to what students know and understand.

Discourse and questions. One of the important ways that teachers assess students' learning is through ongoing, everyday classroom use of questioning. As teachers and students engage in classroom discourse, teachers are obtaining important assessment information about where students are in their understanding. They can use this information to make decisions about the lesson, about the assignments, or about longer-range planning. Strong research evidence suggests that the nature of the discourse that occurs in a mathematics classroom can have a significant impact on students' learning. Collaborative discourse is particularly vital to students' learning. In other words, significant student learning is likely to occur when students are encouraged and trained to engage the teacher in conversation about mathematics, especially at the level of understanding a concept (e.g., Clarke, 1988; Johnson & Johnson, 1990). When students were trained to generate higher-order questions with peers, they outperformed students who lacked such training (King, 1990, 1992, 1994; Foos, Mora, & Tkacz, 1994; King & Rosenshine, 1993, 1996).

Unfortunately, as Stiggins and Griswold (1989) found, teachers' questioning is dominated by recall questions. Further, although teachers who were trained to teach higher-order thinking skills asked more relevant questions, their use of higher-order questions was still infrequent. Thus, teachers need more encouragement to engage their students in discourse at a higher level of thinking. The question of how the discourse operates in the classroom also remains to be examined. Bromme and Steinberg (1994) studied novice mathematics teachers' classroom strategies and found that these teachers tended to treat students' questions as if they were individual concerns, whereas the responses of expert teachers tended to be directed more to a "collective" student.

When students respond to teachers' questions, they do so within the social context of the classroom. Each classroom embodies a set of expectations that influence the way students participate in discourse. For example, if a teacher's questions are primarily of a "lower order" (e.g., questions asking for recall of a correct procedure), students come to expect those kinds of questions. If the teacher then were to ask questions demanding understanding or the application of a concept, students could view those questions as unfair, illegitimate, or even meaningless (Schoenfeld, 1985).

When a teacher questions a student, the teacher's beliefs, knowledge, and experience influence both the questions asked and the way that answers are interpreted. The extent to which a teacher's questions may lead to informative assessment data depends on the richness of the teacher's questions. For example, if the teacher limits questions to a narrow band of procedural questions, the answers given may not be sufficient for the teacher to make informed inferences about the breadth or depth of students' understanding. That is, the teacher may take a series of correct answers by a student as evidence of understanding, when in fact it is very limited evidence merely of the student's ability to give the correct answers (for a broader discussion of this issue, see von Glaserfeld, 1987, p. 13).

To enhance students' learning, classroom assessment must include adequate dialogue between teachers and students. When that dialogue is maintained at a high level of thinking, when students operate under the expectation that mathematical understanding is the primary goal, and when teachers are attuned to the kinds of questions they ask and the responses they give so as to make adjustments to instruction, then formative assessment is working well.

Equity: Assessment Should Promote Equity

Formative assessment and expectations. Formative assessment, when done effectively, relies on inductive thinking, that is, the teacher gathers data about students' learning and uses those data to draw conclusions and make decisions. Often, children with special needs are given prescriptions for learning in the form of individual learning programs. These prescriptions and their application tend to be based on a more deductive approach. That is, given that the student has "this" profile, we would expect "this" kind of learning to be effective. When teachers rely solely on such information rather than on formative data, their deductive conclusions may limit students' achievement. A meta-analysis of 21 studies of this phenomenon found that when teachers worked to set rules about how to review formative data and what actions might follow, the mean effect size on student achievement was 0.92, compared with 0.42 for students of teachers who worked from a priori prescriptions for individualized learning programs (Fuchs & Fuchs, 1986). This disparity gives strong evidence of the striking effects of formative assessment when used with children having special needs.

The process of formative assessment involves gathering evidence, making inferences about students' learning, giving feedback, and making instructional decisions about learning opportunities. Undergirding that process, however, is an important foundation that can powerfully affect students' learning: the assumptions that teachers make about their students' capabilities. If a teacher believes and expects that every student can and will succeed, that belief has a very different effect from feedback based on comparison with peers, with the associated assumption that some students are not as able as others and so cannot expect full success (Black & Wiliam, 1998).

Bias. Research in human cognition has taught us that learning with understanding occurs when students can connect new ideas with prior knowledge (Hiebert & Carpenter, 1992). This observation implies that teachers should take into account students' prior knowledge when planning and carrying out instruction and assessment. Prior knowledge, however, is also embedded in a student's personal experiences. A review by Rismark (1996) shows that students are frequently marginalized and their work undervalued if they use frames of reference from personal experiences outside school. Filer (1993) found that children learning handwriting and spelling in English primary school classrooms were constrained by the teacher to develop these skills in standard contexts, so that their personal experiences were "blocked out." These studies point to the importance of incorporating students' personal experiences into assessment and instructional decisions so that a teacher's self-fulfilling prophecy does not come into play (Filer, 1995).

Evidence shows that teachers' and schools' lower expectations for students from racial- and ethnic-minority and poverty backgrounds can "lower the ceiling" of their achievement (Oakes, 1990; Ladson-Billings, 1997; Secada, 1991). Conversely, the classrooms of Escalante (1990) and the Algebra Project of Moses (Moses, Kamii, Swap, & Howard, 1989) demonstrate that higher expectations can have a powerful effect on students' achievement. In fact, some argue that "all U.S. students can do enormously better than they ordinarily do in primary school mathematics" (Fuson, Smith, & Lo Cicero, 1997)— meaning that higher expectations may be called for on a broader basis than merely bringing underachievers up to grade level.

Differences in assessment outcomes are often explained by using coarse categories for students, such as gender, race, or social class. This explanation tends to treat all those within a category as homogeneous. Such studies as that by Dart and Clarke (1989) show that the experiences of students in the same classroom can be radically different, suggesting that other models need to be used to look at outcome differences. Bourdieu (1985) suggests the use of *habitus* to account for differences between individuals in the same categories. Habitus is a way of describing the orientations, experiences, and positions adopted by individuals. Essentially, it emphasizes the importance of attending to the individual members of a group rather than assuming that all members share the same experiences and thus will perform the same on assessments.

Openness: Assessment Should Be an Open Process

Portfolios as a vehicle for open assessment. Portfolios have become an increasingly popular strategy for assessment that can enable the process to be open for students. To accomplish such open assessment, the portfolio needs to be more than just a collection of student work (a file folder that contains every paper produced). Instead, the portfolio can be a collection of the student's work that meets certain criteria, such as best work or exemplars of good problem solving or conceptual understanding. When criteria are set for including work in the portfolio, students can participate in choosing which work samples to include. The student thereby is made aware of what constitutes quality work of a certain kind and is given a chance to reflect on what he or she has done. Used in this way, portfolios can allow students to actively monitor their progress and growth (Columba & Dolgos, 1995).

Student self-assessment. For assessment to be an open process, students must have opportunities to become familiar with the assessment process, as well as with the purposes, performance criteria, and consequences of the assessment. One method for providing such opportunities is through student self-assessment. Although many strategies can help students assess their own work, strong evidence from research indicates that when teachers are trained in self-assessment methods, and when students self-assess on a regular (mainly daily) basis, significant improvements can be attained in students' achievement (Fernandes & Fontana, 1996). The self-assessment methods involved teaching students to understand the learning objectives and the assessment criteria, giving students the opportunity to choose learning tasks, and using tasks that allow students to assess their own learning outcomes.

In another study that involved both self- and peer assessment in middle school science classes, the groups that were given time to engage in reflective assessment (peer assessment of presentations to the class and self-assessment) outscored the control group on three measures: a mean score on projects throughout the course, a score on two individual projects, and a conceptual test. Lower-achieving students showed greater achievement gains than higher-achieving students. Also, the students who showed the best understanding of the assessment process achieved the highest scores (Frederiksen & White, 1997). Several other studies have shown similar results from combinations of self-assessment and peer assessment (Kock & Shulamith, 1991; Higgins, Harris, & Kuehn, 1994; Stefani, 1994).

Research into assessment practices in group work (Webb, 1995) gives a picture of features that are most likely to allow all members of the group to produce their best work. First, it is vital for group members to be trained in group processes and for clear goals and achievement criteria to be set and understood by all members of the group. Members of the group then have to make a choice between having as their goal the best performance of the group as a whole and having a goal of improving individuals' performances through group collaboration. The group's composition is also an important issue. If a group's goal is to optimize group performance, established high achievers are the most productive members on well-defined tasks, whereas for more open tasks, a range of high and low achievers is an advantage. However, if the priority is individual performance, high achievers are little affected by the mix, but low achievers benefit more from a mixed group, provided that the group training emphasizes methods to draw out, rather than overwhelm, their contribution.

Inferences: Assessment Should Promote Valid Inferences About Mathematics Learning

How do teachers make valid inferences from assessment results? An inference about learning is a conclusion about a student's cognitive processes that cannot be directly observed; instead, the conclusion is based on the student's performance. For the inference to be valid, the evidence must be adequate and relevant. In addition, a study by Lorsbach, Tobin, Briscoe, and LeMaster (1992) found it is crucial that students understand the tasks they were given to do and that teachers can construct meaning for the students' responses. They also found that teachers used assessment results as if the results gave information on what students knew, whereas, in fact, the results were better indicators of motivation and task completion. Therefore, the inferences that educators make are only as valid as the evidence is relevant and of high quality.

A review of two studies of teachers' current practices in formative assessment shows several common key weaknesses:

- Classroom evaluation practices generally encourage superficial and rote learning, concentrating on recall of isolated details, usually items of knowledge which pupils soon forget.

- Teachers do not generally review the assessment questions that they use and do not discuss them critically with peers, so there is little reflection on what is being assessed.

- The grading function is overemphasized and the learning function underemphasized.

- There is a tendency to use a normative rather than a criterion approach, which emphasizes competition between pupils rather than personal improvement of each.

- The evidence is that with such practices the effect of feedback is to teach the weaker pupils that they lack ability, so that they are demotivated and lose confidence in their own capacity to learn. (Black & William, 1998, p. 10)

These results suggest that teachers' inferences, in the form of summative grades, often do not reflect the practices of making valid and reliable judgments about what students know and can do. In both the formative and summative assessments that are a daily part of classroom instruction, it is important to base inferences on good evidence.

Consistency in scoring written work. One of the key factors affecting consistency in scoring written work is the purposes of the assessment. For example, student portfolios may be constructed for accountability purposes (such as those used in Vermont at the state level), or they may be used by the teacher for instructional purposes. Benoit and Yang (1996) found a tension in teachers' scoring of portfolios between the use of scores for accountability purposes (summative) and for instructional decision making (formative). That tension affects the selection and the scoring of tasks to be included in the portfolios.

Using multiple tools and forms of evidence. Making use of a number of different sources of evidence about students' learning can increase the validity of the inferences made, because no one form of assessment can adequately depict the depth and range of students' achievement. Weaknesses in one form of assessment can be compensated for by strengths in another. For example, a multiple-choice test can be used to assess a broad range of content, whereas open-ended assessments can probe more deeply into students' strategies and understanding. Long-term projects allow students to explore mathematical ideas in depth, and portfolios may help them to reflect on their best work. Taken together, these multiple strategies can give students, teachers, and the broader public a more complete picture of the student's mathematical knowledge, skills, and understanding.

Multiple sources of evidence ensure greater validity, allowing the teacher to look for convergence of common conclusions about the student's mathematical progress. Paper-and-pencil tests can give one kind of evidence about what students know and understand, but other kinds of assessment can broaden the base of evidence and increase the validity of inferences. Portfolios, for example, have been shown to assess different constructs than standardized tests (Reckase, 1997). If only one source of data is used, the resulting inferences will not be as valid as when several different types of assessment strategies are used. In this way, teachers become researchers of sorts as they look for convergence of evidence or try to explain conflicting evidence from a variety of sources.

However, the use of multiple assessment forms raises issues. For example, what are the costs and benefits of different forms of assessment? How do various forms of assessment affect what students learn? And how often should more common forms of assessment, such as classroom tests, be used?

The actual context of the assessment can influence what students believe is required. In a study of a fifth-grade geometry class (Hall, Knudsen, & Greeno, 1995), performance was assessed two ways: using a multiple-choice test and using an assignment in which student had to design a HyperCard geometry tutorial. In the multiple-choice test, students focused on the grades awarded, whereas in the tutorial task, students engaged in more qualitative discussions of their work. Perhaps most significantly, discussion among students about the different tutorials focused more directly on the subject matter (i.e., geometry) than the intense comparison of grades on the multiple-choice test did.

Slater, Ryan, and Samson (1997) describe an experiment, in an introductory algebra course for college students, that produced no significant difference in achievement between a group engaged in portfolio production and a control group. However, the achievement test was a 24-item multiple-choice test, which might not have reflected the advantages of the portfolio approach; at the same time, the teacher reported that the portfolio group ended up asking more questions about real-world applications and had been led to discuss more complex and interesting phenomena than the control group did.

Bangert-Drowns, Kulik, and Kulik (1991) reviewed the evidence of the effects of frequent class testing. Their meta-analysis of 40 relevant studies showed that performance improved with frequent testing and increased with greater frequency up to a certain level—but beyond that (i.e., beyond one and two tests per week), it could decline again. The evidence also indicated that several short tests were more effective than fewer long ones.

Quality of feedback. A crucial aspect of making valid inferences is the feedback that the teacher gives the student after the inferences are made. Several research studies give clues about the nature of effective feedback. One study (Bangert-Drowns, Kulik, Kulik, & Morgan, 1991) revealed that feedback is most effective when it is designed to correct errors but in a thoughtful way that is related to the task, not in an artificial way. In other words, teachers should give specific feedback on errors and strategies, with suggestions on how to improve, but should keep the focus on deep understanding rather than on superficial learning of procedures (Bangert-Drowns et al., 1991; Elawar & Corno, 1985).

Too often, the most common feedback that students experience in schools is in the form of grades, which are usually normative in nature (i.e., comparing students with students). Butler (1988) found that even when feedback comments are operationally helpful for a student's work, their effect can be undermined by the negative motivational effects of grades. These results are consistent with other studies indicating that when evaluation focuses on the task rather than on the student, it is much more effective—to the extent that even the giving of praise can have a negative effect with low achievers. As many teachers know, when students are preoccupied with grades, the quality of their work suffers, especially for tasks that require higher-order and divergent thinking.

Feedback can be particularly effective when teachers work from a sound model of students' progression in learning the subject matter (Fennema et al., 1996; Fuson, Smith, & LoCicero, 1997). In one model of children's solution strategies to addition and subtraction problems, known as Cognitively Guided Instruction (CGI), teachers are trained in using the model in their teaching, relying heavily on formative assessment. That is, teachers in CGI classrooms attend closely to the strategies that students are using and the types of problems they are capable of solving, giving students specific feedback to encourage them to talk about their mathematical thinking and to use more and more sophisticated strategies. This type of teaching has been shown to improve student achievement in concepts and problem solving (Carpenter, Fennema, & Franke, 1996). When teachers use such models as their foundation for feedback, they need effective assessment instruments, as well as methods for interpreting and responding to results of assessments in a formative way (Fuchs, Fuchs, Hamlett, & Stecker, 1991).

Coherence: Assessment Should Be a Coherent Process

Students experience a coherent assessment process in three different ways. First, the phases of assessment work together as a coherent whole. That is, the teacher plans assessment events that match the purposes of the assessment, gathers evidence from multiple sources, makes inferences about students' learning based on that evidence, and then uses those inferences as the basis for decision making. A second form of coherence is evident when the design of an assessment matches its purposes. Assessment intended to give information about the mathematical knowledge of an entire population (e.g., a state or province's eighth graders) should look quite different from an assessment designed to measure an individual student's problem-solving abilities.

A third type of coherence deals with the alignment of curriculum, instruction, and assessment. Coherence here has to do with the extent to which a match is evident between what students are taught, how they are taught, and the assessments given to them. Alignment can be studied at many different levels (e.g., within the classroom or at the district or state level). When done in a thorough way, an alignment analysis is complex, time-consuming, and labor-intensive. In a recent joint project of the Council of Chief State School Officers (CCSSO) and the National Institute for Science Education, researchers developed a procedure for examining assessments, such as a state test or NAEP, and comparing it with the mathematics that classroom teachers are teaching (CCSSO, 2001). See the following section on large-scale assessment and coherence for a further look at recent work in alignment.

Issues in Large-Scale Assessment

In a recent article, Popham (1999) contends that in the current educational and political climate, "'large scale assessment' could be more accurately described as 'larger and larger scale assessment'" (p. 13). That is, over the last decade, students in nearly all grade levels have experienced an increase in the number of externally mandated assessments they take. These tests include standardized norm-referenced tests, district-level tests, and state-level tests for students in all states except Iowa. Particularly hard hit by this barrage of tests are students in the middle elementary years (Grades 3 and 4), the final years of middle school (Grades 7 and 8), and the last two years of secondary school (Grades 11 and 12). Additionally, national and international assessments such as NAEP and TIMSS depend on the voluntary participation of students in the key grades just mentioned. Thus, in the near future a typical eighth-grade student could possibly take multiple large-scale mathematics assessments within an academic year: for example, a mandated norm-referenced standardized achievement test; a mandated state-level test; and NAEP and TIMSS, if selection for both samples is possible. This battery is a prime example of Popham's description of "larger and larger" scale assessment.

Not only are the numbers of large-scale assessment programs increasing, but also the stakes for some assessments are rising for students, teachers, individual schools, and school districts. Forty-nine of the U.S. states have instituted performance standards and outcomes as measured by a test. Some states depend primarily on norm-referenced standardized tests to measure students' progress toward the objectives, whereas other states have instituted their own testing programs and developed their own grade-level assessments (National Research Council, 1999c). Performance on these mandated tests often has consequences that range from placing students into "skill improvement programs," to giving teachers monetary rewards if their students perform at high levels, to schools and school districts being taken over by the state if no performance improvement occurs over a set period of time.

Given the current emphasis on the importance and power of large-scale assessment, we might profitably think about how the six mathematics assessment standards used to organize the preceding classroom-assessment discussion in this chapter would appear when viewed through a different lens. In particular, if we focus the six assessment standards on large-scale assessment, we would arrive at statements like these:

- Large-scale assessments should reflect the mathematics that all students need to know and be able to do.
- Large-scale assessments should enhance mathematics learning.
- Large-scale assessments should promote equity.
- Large-scale assessment should be an open process.
- Large-scale assessments should promote valid inferences about mathematics learning.
- Large-scale assessment should be part of a coherent process.

To some, the preceding statements represent good common sense; these individuals believe that large-scale assessments should conform to the NCTM's Assessment Standards (1995). Others, however, would argue that because of particular aspects of large-scale assessments (e.g., their purpose, reporting methods, etc.), some of the Assessment Standards are neither reasonable nor applicable. In the remainder of this chapter, we comment on the Assessment Standards in light of large-scale assessments and, where possible, summarize relevant findings from research that apply to each. This review of research is not intended to be comprehensive but rather to highlight important findings that have implications for what happens in mathematics classrooms.

Mathematics

NCTM's *Curriculum and Evaluation Standards* (1989) presents a vision of the kind of mathematics that all students need to know and be able to do. Since the release of that document, many large-scale assessment programs claim that the NCTM Standards form the basis of the framework and item development. But how closely do the standardized tests, such large-scale survey instruments as NAEP and TIMSS, and state- and district-level assessments match the NCTM Standards?

Research studies on the match between large-scale assessment and the NCTM Standards are most prevalent for the large-scale survey instruments and some state-level tests. One component of studies of NAEP mathematics assessments (Silver & Kenney 1993, 1994; Silver, Kenney & Salmon-Cox, 1992) involved asking a panel of mathematics educators to evaluate items according to the four cross-cutting themes of the NCTM *Curriculum Standards* document—problem solving, reasoning, communication, and mathematical connections. Results from those studies suggested that about half of the items were judged not to match any of the four Standards themes. The items that matched a theme tended to cluster in particular content and process categories used by NAEP. In particular, a majority of the items that matched the communications theme were constructed-response questions; most of the items that matched the reasoning theme were classified by NAEP in the categories of geometry or data analysis, statistics, and probability; and nearly all of the items classified by NAEP as problem solving matched the Standards' themes of problem solving or reasoning.

Studies of state-level assessment and their congruence with important themes in the NCTM *Standards* had results similar to those from the NAEP mathematics studies just described. That is, for both the North Carolina End-of-Grade test at Grade 8 and the Maryland School Performance Assessment Program (MSPAP) Grade-8 test, the match between the items and the NCTM *Standards*' themes was uneven and appeared to be linked to the content assessed and the format of the item. For example, the North Carolina study revealed that the high-cognitive-demand items tended to assess topics in geometry and in data analysis, statistics, and probability, whereas the low-cognitive-demand items were classified as assessing computational skills (Sanford & Fabrizio, 1998). For the Maryland study, the results revealed that more than 80 percent of a subset of MSPAP tasks were judged to have a high cognitive demand (Schafer, 1998). This high demand spanned all content categories and was attributed primarily to the constructed-response format of the items. Results from other research studies (e.g., Nichols & Sugrue, 1999) confirm that it is difficult for large-scale assessments such as NAEP to adequately assess cognitively complex constructs such as problem solving and reasoning, especially through the use of multiple-choice items.

Learning

The assessment standard on learning states that assessment should be an integral part of mathematics instruction and, as such, should contribute significantly to students' learning. Given the proximity between classroom assessment and instruction, that idea makes sense. Is it also true, then, that large-scale assessment has a similar significant effect on student learning? And is that significant effect beneficial or harmful?

The study by Madaus and his colleagues (Madaus, West, Harmon, Lomax, & Viator, 1992) on the influence of testing on teaching mathematics and science in Grades 4–12 remains a seminal work on how tests can influence teaching and learning. The results revealed that mathematics and science standardized tests and textbook tests, with copyrights ranging from 1985 to 1991, fell short of the current standards recommended by experts in mathematics and science curriculum. Tests in particular emphasized and reinforced low-level thinking and knowledge, and they were found to have a pervasive influence on instruction, as reported by teachers.

Another important source of information is Hancock and Kilpatrick's (1993) review of literature about the effect of mandated testing on instruction. These authors analyzed findings from about thirty research studies and other documents that focused on the possible effects that such testing could have on the curriculum, teaching practice, teachers, and students. On the basis of their analyses, Hancock and Kilpatrick concluded that the effects of large-scale testing on instruction "have not been well studied and are not clear" (p. 169); furthermore, "the available research does not lead to the unqualified conclusion that mandated testing is having harmful effects on mathematics instruction. The picture is both more mixed and indistinct" (p. 168). The WYTIWIG phenomena—"what you test is what you get" (Burkhardt, et al., 1990)—is an example of how external testing can have both good and bad effects on instruction. A test based on important mathematical content topics, with an emphasis on problem solving, could be the lever that encourages teachers to ensure that students experience these topics and become familiar with problem-solving strategies. However, if the large-scale test is a multiple-choice, minimal competency test, then instruction could become narrowly focused on basic skills.

Both the Madaus et al. study and Hancock and Kilpatrick's analytical review of the literature identify important points concerning the influence of large-scale testing on instruction. And their points are even more relevant ten years later in the current climate of accountability in education; that is, school districts, school, and even individual teachers should be held accountable for students' learning, with testing as an important—and in some cases, the only—measure of such learning. We need to continue to study the effects that large-scale assessment may have on students' learning in the hope that a clearer, more distinct picture might result.

Equity

For an assessment to be equitable, it must provide each student with an opportunity to demonstrate his or her level of mathematical power. This vision allows for differences among students that go beyond the opportunity to learn the mathematics to be assessed to include differences in students' cultural, ethnic, and social backgrounds; physical condition; and gender. In the classroom, teachers have the advantage, for they can deal with students on an individual basis. However, in large-scale assessment such differences are not always accounted for, and students are not usually thought of as "individuals." Instead, the conditions of the administration must be standardized (i.e., the same for everyone) so that the results are comparable across individuals or groups. Accommodations for students (e.g., Braille or bilingual test versions, orally dictated questions for hearing-impaired students, or recorders who write answers for students with motor-skill problems) were viewed as threats to the comparability of scores, and in some cases such as NAEP such students were excluded from the sample.

A series of reports have focused on the need to include special-needs students in large-scale assessments and to assess them fairly and equitably, but these reports have focused on issues and policy and are not research reports per se. For example, the report *High Stakes: Testing for Tracking, Promotion, and Graduation (High Stakes)* (National Research Council, 1999b) recommends increasing students' participation in large-scale programs and making appropriate accommodations for the effect of a disability or limited English proficiency. Another NRC report, *Grading the Nation's Report Card* (1999a), focuses on inclusion and equity in NAEP and cites a study by Olson and Goldstein (1997) that recommends ways to expand the NAEP samples to include students with disabilities and limited English proficiency.

These and other reports raise important issues about equity in large-scale testing, especially the high-stakes testing of students who are not native English speakers. *High Stakes* (NRC, 1999b) notes the following:

> The central dilemma regarding participation of English-language learners in large-scale assessment programs in that, when students are not proficient in the language of the assessment (English), their scores on a test given in English will not accurately reflect their knowledge of the subject being assessed (except for a test that measures only English proficiency). (p. 214)

However, only a small body of literature is available on language accommodations for non-native speakers of English, with most of it based in policy instead of research and nearly all of it done with NAEP, a relatively low-stakes test (e.g., Abedie, Lord & Plummer, 1995; Anderson, Jenkins, & Miller, 1996; Olson & Goldstein, 1997). Perhaps the most important research report to date on large-scale assessment of language-minority children was jointly produced by the National Research Council and the Institute of Medicine (1997). The report focused more generally on schooling issues for these children but included testing for English-language proficiency in the children's realms of interest. One important conclusion was that "most measures used [for assessing English language proficiency] not only have been characterized by the measurement of decontextualized skills but also have set fairly low standards for language proficiency" (p. 118).

With respect to gender issues, a 1997 report by Cole summarized results from the Educational Testing Service Gender Study, a four-year effort with data from more than 400 tests and 1500 data sets involving millions of students. The findings from this massive study were that no dominant picture emerged of males excelling over females, or vice versa; the familiar mathematics and science advantage for males was quite small—significantly smaller than 30 years earlier. However, the language differences in reading and writing favored females and remained unchanged over that time period. As Cole notes, the term *bias* implies a systematic error in measuring knowledge and skill, yet the results from the ETS Gender Study indicated that the observed differences between males and females were not error based but instead were a correct reflection of differences that occur on many kinds of tests and in different kinds of students. Important for our consideration of mathematics assessment is Cole's conclusion that "the content on tests must be guided by how educationally important the content is, not what differences it produces" (1997, p. 4) and that attention must be paid to students of both genders to have the opportunity to learn important content in a variety of school subjects.

Despite these comprehensive findings on gender and testing, questions remain about the differential effects of testing on males and females. Such organizations as the American Association of University Women have monitored gender issues in large-scale assessment and, in particular, on high-stakes tests used for college admission and placement, such as the AP tests.

As noted above, much of the existing information on equity in large-scale assessment is from position papers and not from research studies. Thus, studies involving the testing of language-minority children, males and females, and children with disabilities are needed.

Openness

According to the NCTM *Assessment Standards* (NCTM, 1995), the standard of openness has three facets: (1) students, teachers, and other interested parties have timely information about how the information will be gathered and used; (2) teachers are involved in all phases of assessment development; and (3) the assessment process is open to scrutiny and modification. Evidence shows that some large-scale assessment programs satisfy at least some aspects of the openness standard. For example, the companies that produce standardized norm-referenced tests publish descriptions of the tests for teachers and parents. These descriptions include assessed topics and, in most cases, sample questions. Additional information about standardized tests can be found in such sources as the *Mental Measurements Yearbook* (Plake & Impara, 1999).

In state and district level assessments used for accountability, teachers are often included as members of framework- and item-development committees (see, e.g., North Carolina Department of Public Instruction, 1996). Also, in states that use open-ended questions (e.g., Delaware, Maryland), teachers serve as raters of students' responses. Large-scale surveys, such as NAEP, include teachers and mathematics educators on their committees, and the frameworks undergo a state-level review involving curriculum committees on which teachers serve (National Assessment Governing Board, 1994). This active participation in the assessment process gives teachers important information and allows them to consider these external tests in light of their own instructional practices.

Not all teachers have the same level of involvement in the assessment process, however. Further, the kind of information that students receive about a particular test is unknown. In many cases, large-scale assessment is an unknown quantity because of the need for secrecy of the test items. That is, to establish a level "playing field," the test questions are drawn from item banks that are not publicly available, and the scoring standards for open-ended items are not released beforehand. Critics of large-scale assessments claim that such secrecy "denies teachers, administrators, policymakers, parents, subject matter specialists, . . . students, and the public the opportunity to ensure that the goals defined by the assessment instruments match their own" (Viator, 1990, p. 2). Although it is possible to examine such public documents as frameworks and statements of objectives (i.e., the "test as specified"), it is generally not possible to match the items (i.e., the "test as developed") with the objectives so as to establish the degree of alignment between the test and the framework.

Instances have arisen, however, in which groups of mathematics educators have had the opportunity to examine secure test items and determine the degree of fidelity between the items and the frameworks—whether the items assessed important mathematics and whether the scoring standard captured the cognitive complexity of the items. For example, summaries of the results of particular efforts involving the NAEP framework and items can be found in Silver, Kenney, and Salmon-Cox (1992) and Silver and Kenney (1994). As part of an NAGB-sponsored project, the frameworks and items used in two state assessments (Maryland and North Carolina) at Grade 8 were compared to those from the 1996 NAEP mathematics assessment to determine the degree of congruity between the Maryland test and NAEP and between the North Carolina test and NAEP (Kenney & Silver, 1998; Sanford & Fabrizio, 1998; Schafer, 1998).

In sum, although some degree of openness is present in large-scale assessment (e.g., in the availability of descriptions of tests, frameworks, and sample items), such programs represent a closed system. In particular, teacher involvement in the test-development process is far from common, and the processes involved in creating large-scale assessments are not open to scrutiny by teachers, parents, or students.

Inferences

Any assessment should lead to valid inferences about a student's cognitive processes that cannot be observed directly, and that conclusion must be based on performance on multiple measures. Ascertaining the validity of inferences has long been a concern of large-scale assessment programs aimed at placing students in particular courses; projecting students' success in college or graduate school; or certifying students' competence in skills associated with a particular profession, such as nursing. Research studies of the validity of inferences abound in the educational measurement literature, and we refer interested readers to *Educational Measurement, Third Edition* (Linn, 1989) for summaries of important studies as well as to recent issues of journals such as *Applied Measurement in Education* and the *Journal of Educational Measurement.*

Most testing experts agree that inferences affecting important milestones in a student's educational life, such as promotion to the next grade level or graduation from high school, should not be based on a *single* instrument, such as a standardized norm-referenced test or a district-, state-, or province-level exit examination. Yet, as part of the testing-as-accountability movement in the United States today, a number of states have instituted high-stakes exit examinations that students must pass before earning a high school diploma. Despite the existence of multiple opportunities to pass the test, some students continue to fail, and that is especially true for minority students. Yet, what proof exists that the inference associated with these exit examinations is valid? Does failure to attain a particular cut score mean that the student has not learned what he or she was supposed to learn in school? And does failure to pass the test lead to the conclusion that the student will not be successful either in a career or in future schooling? We can find no research on the validity of inferences made from performance on such high-stakes large-scale assessments. Such studies are indeed warranted.

Coherence

Three types of agreement are necessary to satisfy the coherence standard: (1) the phases of the assessment fit together; (2) the assessment matches the purpose for which it is being done; and (3) the assessment is aligned with the curriculum and the instruction. Large-scale assessments should be considered when one thinks about coherence between them and what teachers are doing every day in mathematics classrooms. Research studies have been conducted on the alignment between the framework and items within a particular testing program (e.g., Silver & Kenney, 1994) and between two different testing programs (e.g., the Maryland/NAEP study and the North Carolina/NAEP study; see Kenney & Silver, 1998). To date and to our knowledge, however, no formal research studies have focused on the larger topic of coherence among assessment systems, ranging from classroom assessment to external assessments at the district, state, province, and higher levels. As we reported above in the section on coherence in classroom assessment, studies have been conducted on the alignment between assessments and instruction.

Teachers must understand the kinds of large-scale assessments that their students take and must think about the extent to which those tests can affect classroom instruction and assessment activities. For some high-stakes assessments, especially those that determine placement of students into particular courses or programs, teachers may want to pay attention to the content of those assessments and ascertain that their students are adequately prepared.

It is also important to be informed about how well assessments are aligned with standards for mathematics. Alignment between standards and assessments is currently being examined at the state level by researchers, for example, Webb (1999), who studied the alignment of four states' standards and assessments in mathematics and science. The analysis was carried out by 16 people, including state content specialists, state assessment consultants, content experts, and researchers, over 4 days. They analyzed state standards documents and, item by item, the state assessment instruments according to four criteria: categorical concurrence, depth-of-knowledge consistency, range-of-knowledge correspondence, and balance of representation. Other criteria, such as articulation across grades and ages, equity and fairness, and pedagogical implications, were given less emphasis.

A detailed report produced for each state and across the four states showed that the alignment of standards and assessments varied among the four states and among the content areas and grade levels within each state. One state, judged to be the most fully aligned in mathematics, exhibited the following degrees of alignment: fully aligned for all three grades on categorical concurrence, highly aligned for one of the three grades on depth-of-knowledge consistency, acceptably or highly aligned for two grades on range-of-knowledge correspondence, and acceptably or highly aligned on balance of representation for all three grades levels on balance of representation. This study is the most comprehensive one to date on alignment at the state level between standards and assessments. Other projects, such as one initiated at Project 2061 (American Association for the Advancement of Science, 1998), propose to develop alignment procedures for assessment tasks and specific learning goals.

Although alignment of all aspects of an assessment system is important, as is alignment among assessments, standards, and instruction, meeting this part of the standard should not be confused with ensuring that the assessments, or the curriculum or instruction, are of high quality. Alignment studies are essentially analyses of the degree to which a match is evident in what students are expected to know and be able to do, but it does not speak to the worthiness of those expectations. Good alignment is possible between an assessment system, or even an assessment task, and a standard or set of standards, for example, but both may be of poor quality. The alignment analysis is merely a judgment on how well the assessment and the standards are designed to meet the same goal. That goal, however, might not be a worthwhile goal, mathematically speaking.

Conclusion

A growing body of research evidence supports the ideas set forth in NCTM's *Assessment Standards*. More important, researchers and teachers are learning more about the most effective ways to meet those standards. Formative assessment in the classroom, which depends on teachers' gathering evidence from multiple sources and then acting on that evidence, can be one of the most powerful forces for learning. Listening to students, asking them good questions, and giving them the opportunity to show what they know in a variety of ways are all affirmed by research to be important ways to increase student learning.

Large-scale assessments can also meet the NCTM Standards, although in different ways than classroom assessments. A clear message from current research is the importance of the role of classroom teachers as assessors. As Paul E. Barton (1999), an Educational Testing Service researcher, writes,

> We are in danger of focusing too much on highly structured [assessment] systems . . . and not on the teacher as a professional. . . . It is, and will remain, the teacher who delivers the 'content,' who aligns his or her assessment methods to the content, and who judges performance. (p. 31)

All parties involved in large-scale assessment programs would benefit from more teacher input to the process so that the expertise that good teachers develop in their classrooms can have some influence on such programs.

REFERENCES

Abedi, J., Lord, C., & Plummer, J. (1995). *Language background as a variable in NAEP mathematics performance.* Los Angeles: National Center for Research on Evaluation, Standards, and Student Testing, University of California at Los Angeles.

Ames, C. (1992). Classrooms: Goals, structures, and student motivation. *Journal of Educational Psychology, 84,* 261–271.

American Association for the Advancement of Science. (1998). *Project 2061: Science literacy for a changing future.* Washington, DC: Author.

Anderson, N. E., Jenkins, F. F., & Miller, K. F. (1996). *NAEP inclusion criteria and testing accommodations: Findings from the NAEP 1995 field test in mathematics.* Washington, DC: Educational Testing Service.

Bangert-Drowns, R. L., Kulik, J. A., & Kulik, C-L. C. (1991). Effects of frequent classroom testing. *Journal of Educational Research, 85,* 89–99.

Bangert-Drowns, R. L., Kulik, C-L. C., Kulik, J. A., & Morgan, M. T. (1991). The instructional effect of feedback in test-like events. *Review of Educational Research, 61,* 213–238.

Barton, P. E. (1999). *Too much testing of the wrong kind; Too little of the right kind in K–12 education.* Princeton, NJ: Educational Testing Service.

Beaton, A. E., Mullis, I. V. S., Martin, M. O., Gonzalez, E. J., Kelly, D. L., & Smith, T. A. (1996). *Mathematics achievement in the middle school years.* Boston: Boston College.

Benoit, J., & Yang, H. (1996). A redefinition of portfolio assessment based on purpose: Findings and implications from a large scale program. *Journal of Research and Development in Education, 29,* 181–191.

Black, P., & Wiliam, D. (1998). Assessment and classroom learning. *Assessment in Education, 5,* 7–74.

Blumenfeld, P. C. (1992). Classroom learning and motivation: Clarifying and expanding goal theory. *Journal of Educational Psychology, 84,* 272–281.

Bourdieu, P. (1985). The genesis of the concepts of "habitus" and "field." *Sociocriticism, 2,* 11–24.

Bromme, R., & Steinberg, H. (1994). Interactive development of subject matter in mathematics classrooms. *Educational Studies in Mathematics, 27,* 217–248.

Burkhardt, H., Fraser, R., & Ridgeway, J. (1990). The dynamics of curriculum change. In I. Wirszup & R. Streit (Eds.), *Developments in school mathematics around the world: Proceedings of the Second UCSMP International Conference on Mathematics Education, 7–10 April 1988* (pp. 3–30). Reston, VA: National Council of Teachers of Mathematics.

Butler, R. (1988). Enhancing and undermining intrinsic motivation: The effects of task-involving and ego-involving evaluation on interest and performance. *British Journal of Educational Psychology, 58,* 1–14.

Carpenter, T. P., Fennema, E., & Franke, M. L. (1996). Cognitively guided instruction: A knowledge base for reform in primary mathematics instruction. *Elementary School Journal, 97*(1), 3–20.

Clarke, J. (1988). Classroom dialogue and science achievement. *Research in Science Education, 18,* 83–94.

Cole, N. S. (1997) *The ETS gender study: How females and males perform in educational settings.* Princeton, NJ: Educational Testing Service.

Columba, L., & Dolgos, K. (1995). Portfolio assessment in mathematics. *Reading Improvement, 32,* 174–176.

Cooney, T., Badger, E., & Wilson, M. (1993). Assessment, understanding mathematics, and distinguishing visions from mirages. In N. Webb (Ed.), *Assessment in the mathematics classroom* (1993 Yearbook, pp. 239–247). Reston, VA: National Council of Teachers of Mathematics.

Council of Chief State School Officers. (2001). *New tools for analyzing teaching, curriculum, and standards in mathematics and science* [Online]. Retrieved 17 October 2002 from http://www.ccsso.org

Dart, B. C., & Clarke, J. A. (1989). Target students in year 8 science classrooms: A comparison with and extension of existing research. *Research in Science Education, 19,* 67–75.

Dumas-Carre, A., & Larcher, C. (1987). The stepping stones of learning and evaluation. *International Journal of Science Education, 9,* 93–104.

Elawar, M. C., & Corno, L. (1985). A factorial experiment in teachers' written feedback on student homework: Changing teacher behavior a little rather than a lot. *Journal of Educational Psychology,* 77, 162–173.

Escalante, J. (1990). The Jamie Escalante math program. *Journal of Negro Education, 59,* 407–423.

Fennema, E., Carpenter, T. P., Franke, M. L., Levi, L., Jacobs, V. R., & Empson, S.B. (1996). A longitudinal study of learning to use children's thinking in mathematics instruction. *Journal for Research in Mathematics Education, 27,* 403–434.

Fernandes, M., & Fontana, D. (1996). Changes in control beliefs in Portuguese primary school pupils as a consequence of the employment of self-assessment strategies. *British Journal of Educational Psychology, 66,* 301–313.

Filer, A. (1993). Contexts of assessment in a primary classroom. *British Educational Research Journal, 19,* 95–107.

Filer, A. (1995). Teacher assessment: social process and social product. *Assessment in Education, 2,* 23–38.

Foos, P. W., Mora, J. J., & Tkacz, S. (1994). Student study techniques and the generation effect. *Journal of Educational Psychology, 86,* 567–576.

Frederiksen, J. R., & White, B. J. (1997). *Reflective assessment of students' research within an inquiry-based middle school science curriculum*. Paper presented at the meeting of the American Educational Research Association, Chicago.

Fuchs, L. S., & Fuchs, D. (1986). Effects of systematic formative evaluation: A meta-analysis. *Exceptional Children, 53*, 199–208.

Fuchs, L. S., Fuchs, D., Hamlett, C. L., & Stecker, P. M. (1991). Effects of curriculum-based measurement and consultation on teacher planning and student achievement in mathematics operations. *American Educational Research Journal, 28*, 617–641.

Fuson, K. C., Smith, S. T., & Lo Cicero, A. M. (1997). Supporting Latino first graders' ten-structured thinking in urban classrooms. *Journal for Research in Mathematics Education, 28*, 738–766.

Hall, R. P., Knudsen, J., & Greeno, J. G. (1995). A case study of systemic aspects of assessment technologies. *Educational Assessment, 3*, 315–361.

Hancock, L., & Kilpatrick, J. (1993). Effects of mandated testing on instruction. In Mathematical Sciences Education Board, *Measuring what counts: A conceptual guide for mathematics assessment* (pp. 149–174). Washington, DC: National Academy Press.

Hiebert, J., & Carpenter, T. P. (1992). Learning and teaching with understanding. In D. Grouws (Ed.), *Handbook of research on mathematics teaching and learning* (pp. 65–100). Reston, VA: National Council of Teachers of Mathematics.

Higgins, K. M., Harris, N. A., & Kuehn, L. L. (1994). Placing assessment into the hands of young children: A study of student-generated criteria and self-assessment. *Educational Assessment, 2*, 309–324.

Johnson, D. W., & Johnson, R. T. (1990). Co-operative learning and achievement. In S. Sharan (Ed.), *Co-operative learning: Theory and research* (pp. 23–27). New York: Praeger.

Kenney, P. A., & Silver, E. A. (1998). *Design features for the content analysis of a state assessment and NAEP*. Paper presented at the meeting of the American Educational Research Association, Montreal, Quebec, Canada.

King, A. (1991). Effects of training in strategic questioning on children's problem-solving performance. *Journal of Educational Psychology, 83*, 307–317.

King, A. (1992). Facilitating elaborative learning through guided student-generated questioning. *Educational Psychologist, 27*, 111–126.

King, A. (1994). Autonomy and question asking—The role of personal control in guided student-generated questioning. *Learning and Individual Differences, 6*, 163–185.

King, A., & Rosenshine, B. (1993). Effects of guided cooperative questioning on children's knowledge construction. *Journal of Experimental Education, 61*, 127–148.

Koch, A., & Shulamith, G. E. (1991). Improvement of reading comprehension of physics texts by students' question formulation. *International Journal of Science Education, 13*, 473–485.

Ladson-Billings, G. (1997). It doesn't add up: African American students' mathematics achievement. *Journal for Research in Mathematics Education, 28*, 697–708.

Linn, R. L. (Ed.) (1989). *Educational measurement* (3rd ed.). New York: American Council on Education and Macmillan.

Lorsbach, A. Q., Tobin, K., Briscoe, C., & LaMaster, S. U. (1992). An interpretation of assessment methods in middle school science. *International Journal of Science Education, 14*, 305–317.

Madaus, George F., West, M. M., Harmon, M. C., Lomax, R. G., & Viator, K. A. (1992). *The influence of testing on teaching math and science in grades 4–12: Executive summary*. Chestnut Hill, MA: Boston College Center for the Study of Testing, Evaluation, and Educational Policy.

Mathematical Sciences Education Board (1993a). *Measuring what counts: A conceptual guide for mathematics assessment*. Washington, DC: National Academy Press.

Mathematical Sciences Education Board (1993b). *Measuring up: Prototypes for mathematics assessment*. Washington, DC: National Academy Press.

Moses, R. P., Kamii, M., Swap, S. M., & Howard, J. (1989). The Algebra project: Organizing in the spirit of Ella. *Harvard Educational Review, 59*, 423–443.

National Assessment Governing Board. (1994). *Mathematics framework for the 1996 National Assessment of Educational Progress*. Washington, DC: Author.

National Assessment Governing Board. (1999, June). *The Voluntary National Test: Purpose, intended use, definition of voluntary, and reporting* [Online]. Available at www.nabg.org/vnt/purpose

National Council of Teachers of Mathematics. (1989). *Curriculum and evaluation standards for school mathematics*. Reston, VA: Author.

National Council of Teachers of Mathematics. (1995). *Assessment standards for school mathematics*. Reston, VA: Author.

North Carolina Department of Public Instruction. (1996). *North Carolina End-of-Grade Tests, Technical Report No. 1*. Raleigh, NC: Author.

National Research Council. (1999a). *Grading the nation's report card: Evaluating NAEP and transforming the assessment of educational progress*. Washington, DC: National Academy Press.

National Research Council. (1999b). *High stakes: Testing for tracking, promotion, and graduation*. Washington, DC: National Academy Press.

National Research Council. (1999c). *Uncommon measures: Equivalence and linkage among educational tests*. Washington, DC: National Academy Press.

National Research Council & The Institute of Medicine. (1997). *Improving schooling for language-minority children*. Washington, DC: National Academy Press.

Nichols, P., & Sugrue, B. (1999). The lack of fidelity between cognitively complex constructs and conventional test development practice. *Educational Measurement: Issues and Practice, 18*, 18–29.

Olson, J. F., & Goldstein, A. A. (1997). *The inclusion of students with disabilities and limited English proficient students in large-scale assessments: A summary of recent progress.* Washington, DC: U.S. Department of Education.

Plake, B. S., & Impara, J. C. (1998). *The thirteenth mental measurements yearbook*. Lincoln, NE: Buros Institute.

Popham, W. J. (1999). Where large scale educational assessment is heading and why it shouldn't. *Educational Measurement: Issues and Practice, 18*, 13–17.

Reckase, M. (1997). *Constructs assessed by portfolios: How do they differ from those assessed by other educational tests?* Paper presented at the meeting of the American Educational Research Association, Chicago.

Rismark, M. (1996). The likelihood of success during classroom discourse. *Scandinavian Journal of Educational Research, 40*, 57–68.

Rosenshine, B., Meister, C., & Chapman, S. (1996). Teaching students to generate questions: A review of the intervention studies. *Review of Educational Research, 66*, 181–221.

Sanford, E. E., & Fabrizio, L. M. (1998). *Results from the North Carolina–NAEP Comparison and what they mean to the End-of-Grade Testing Program.* Paper presented at the meeting of the American Educational Research Association, Montreal, Quebec, Canada.

Schafer, W. D. (1998) *The NAEP-Maryland comparison: Thoughts on process and results.* Paper presented at the Annual Meeting of the American Educational Research Association, Montreal, Quebec, Canada.

Schoenfeld, A. H. (1985). *Mathematical problem solving*. New York: Academic Press.

Silver, E. A., & Kenney, P. A. (1994). The content and curricular validity of the 1992 NAEP Trial State Assessment (TSA) in mathematics. In National Academy of Education, *The Trial State Assessment: Prospects and realities*: Background studies (pp. 231–284). Stanford, CA: The Academy.

Silver, E. A., & Kenney, P. A. (1993). An examination of the relationship between the 1990 NAEP mathematics items for grade 8 and selected themes from the NCTM *Standards. Journal for Research in Mathematics Education, 24*, 159–167.

Silver, E. A., Kenney, P. A., & Salmon-Cox, L. (1992). The content and curricular validity of the 1990 NAEP mathematics items: A retrospective analysis. In National Academy of Education, *Assessing student achievement in the states: Background studies*. Stanford, CA: The Academy.

Slater, T. F., Ryan, J. M., & Samson, S. L. (1997). Impact and dynamics of portfolio assessment and traditional assessment in a college physics course. *Journal of Research in Science Teaching, 34*, 255–271.

Sleeter, C. E. (1997). Mathematics, multicultural education, and professional development. *Journal for Research in Mathematics Education, 28*, 680–696.

Stefani, L. A. (1994). Peer, self and tutor assessment: Relative reliabilities. *Studies in Higher Education, 19*, 69–75.

Stein, M. K., Grover, B. W., & Henningsen, M. (1996). Building student capacity for mathematical thinking and reasoning: An analysis of mathematical tasks used in reform classrooms. *American Education Research Journal, 33*, 455–488.

Stiggins, R J., Griswold, M. M., & Wikelund, K. R. (1989). Measuring thinking skills through classroom assessment. *Journal of Educational Measurement, 26*, 233–246.

Viator, K. A. (1990). Introduction: The social, intellectual, and psychological prices of secrecy. In J. L. Schwartz and K. A. Viator (Eds.), *The prices of secrecy: The social, intellectual, and psychological costs of current assessment practice* (pp. 1–6). Cambridge, MA: Harvard University, Graduate School of Education.

Von Glasersfeld, E. (1987). Learning as a constructive activity. In C. Janvier (Ed.). *Problems of representation in the teaching and learning of mathematics* (pp. 3–17). Hillsdale, NJ: Erlbaum.

Webb, N. L. (1999). *Alignment of science and mathematics standards and assessments in four states.* Madison, WI, & Washington, DC: National Institute for Science Education & Council of Chief State School Officers.

Webb, N. M. (1995). Group collaboration in assessment: Multiple objectives, processes, and outcomes. *Educational Evaluation and Policy Analysis, 17*, 239–261.

Wilson, L. D., & Kenney, P. (1997). Assessing student work: The teacher knowledge demands of open-ended tasks. In J. Dossey, J. Swafford, M. Parmantie, & A. Dossey (Eds.), *Proceedings of the Nineteenth Annual Meeting of the North American Chapter of the International Group for the Psychology of Mathematics Education* (Vol. 1, pp. 131–137). Columbus, OH: ERIC Clearinghouse for Science, Mathematics, and Environmental Education.

Chapter 6

Developing Mathematical Power in Whole Number Operations

Karen C. Fuson, Northwestern University

Although much of the recent research reviewed in this chapter is empirical, much is also conceptual and analytic. Important advances have been made in describing and categorizing real-world situations in each domain; in analyzing attributes of, and potential problems with, the mathematical notations and words in a domain; and in designing conceptual supports to facilitate learners' understanding in a given domain. Some research involves a mixture of empirical and analytical approaches, such as analyzing advantages and disadvantages of particular algorithms—partly from seeing children use them—or identifying children's errors and the reasoning behind them.

Real-World Situations, Problem Solving, and Computation

Continual Intertwining for Sense Making and Computational Fluency

Traditionally in the United States and Canada, students have first learned how to compute with whole numbers and then have applied that kind of computation. This approach presents several problems. First, less-advanced students sometimes never reach the application phase, so their learning is greatly limited. Second, word problems usually appear at the end of each section or chapter on computation, so sensible students do not read the problems carefully: They simply perform the operation that they have just practiced on the numbers in the problem. This practice, plus the emphasis on teaching students to focus on key words in problems rather than to build a complete mental model of the problem situation, leads to poor problem solving because students never learn to read and model the problems themselves. Third, seeing problem situations only after learning the mathematical operations keeps students from linking those operations with aspects of the problem situations. This isolation limits the meaningfulness of the operations and the ability of children to use the operations in a variety of situations.

Research has indicated that beginning with problem situations yields greater problem-solving competence and equal or better computational competence. Children who start with problem situations directly model solutions to these problems. They later move to more advanced mathematical approaches as they progress through levels of solutions and problem difficulty. Thus, their development of computational fluency and their acquisition of problem-solving skills are intertwined as both develop with understanding.

For many years, researchers have contrasted conceptual and procedural aspects of learning mathematics, debating which aspect should come first. Recent research, however, portrays a more complex relationship between these conceptual and procedural aspects, concluding that they are continually intertwined and potentially facilitate each other. As a child comes to understand more, the child's problem-solving method becomes more integrated internally and in relation to other methods. As a method becomes more automatic, reflection about some aspect may become more possible, leading to new understanding. These conceptual and procedural interconnections are forged in individual ways, and attempts to distinguish between them may not even be useful, because doing and understanding are always intertwined in complex ways. Furthermore, different researchers may refer to the same method as a procedure or as a concept, depending on whether the focus is on carrying out the method or on its conceptual underpinnings. And, in a given classroom at a given time, some students may be using what looks like the same method, but they may well have different degrees of understanding of that method at that time. Classroom teaching can help students relate their methods to their knowledge in ways that give them fluency and flexibility.

What types of real-world situations have been identified for addition and subtraction and for multiplication and division? Such situations, in the form of word problems and real situations that students bring to the classroom, supply contexts that allow addition, subtraction, multiplication, and division operations to take on their complete range of fundamental mathematical meanings. The first part of this chapter focuses on these real-world situations, and the second part explores developmental progressions in the methods that students can use for these operations on single-digit and multidigit whole numbers.

Types of Real-World Addition, Subtraction, Multiplication, and Division Situations

Researchers from around the world have reached consensus in identifying the types of real-world situations that involve addition or subtraction, although minor variations in terminology are found. One classification of such situations is given in Table 6.1. Note that for each type of problem, each of the three quantities involved can be the unknown. The language of problems that involve comparing quantities is difficult for children at first, partly because the structure in English lumps together two kinds of information: who has more and how much more. In general, problem statements, syntax, or sentence orders that do not follow the action in a situation are more difficult than those that do. Even kindergarten children can solve many of these problems if they use objects to directly model the situation. Textbooks, however, typically include only the simplest variation of each problem type. In contrast, in the textbooks of the Soviet Union, problems were given equally across the various types and unknowns, and 40% of the problems in the first-grade books and 60% of the problems in the second-grade books were two-step problems (Stigler, Fuson, Ham, & Kim, 1986).

An important distinction should be made between a situation representation (i.e., an equation or a drawing) and a solution representation. The most powerful problem-solving approach is to understand the situation deeply—that is, to be able to draw it or otherwise represent it to oneself. This method is used naturally by young students. But textbooks and teachers influenced by textbooks push students to write solution representations that are not consistent with their view of the situation. Students will write $8 + A = 14$ for a problem like "Erica had $8. She babysat last night and now has $14. How much did she earn babysitting?" Textbooks often push students to write $14 - 8$, but many students do not represent or solve the problem in that way. Allowing students to represent the situation in their own way communicates that the goal of problem solving is to understand the problem deeply. Once they see the goal, students can experience success and move on to more difficult problems.

Less consensus is found about whole-number multiplication and division situations (see the reviews by Greer, 1992, and Nesher, 1992). These situations are usually addressed in the literature on rational numbers, which focuses on fractions as well as on whole numbers. Whole-number situations that almost all researchers have identified include grouping situations; multiplicative comparing situations; and cross-product, or combination, situations. Grouping situations involve some number of equal groups, such as three packages containing five apples each. For the two corresponding division situations, the solver may either (a) know the product (15) and the number of packages (3) and need to find out how many apples are in a package or (b) know the product (15) and the size of the packages (5 apples per package) and need to find out how many packages are required. The multiplicative comparing situations use the language "x times as many as" or the reverse fractional language "$1/x$th as many," as in "Maria had 15 books. She had five times as many as Saul. How many books did Saul have?" As with the additive comparing situations, the comparing sentence can be said in two ways; here, it can also be said as "Saul had one fifth as many books as Maria." The cross-product, or combination, situations are those in which everything in one group is combined with everything in a second group. Arrays are a good way to show these situations. Familiar examples are problems involving clothes (e.g., with three shirts and two pairs of pants, how many outfits are possible?) or sundaes (e.g., with four kinds of ice cream and two kinds of topping, how many kinds of sundaes are possible?). Area is one kind of cross-product situation.

Experience with these addition, subtraction, multiplication, and division situations, and with the language involved in them, allows students to build a mathematically adequate understanding of the operations and notations. The symbol – means more than take away, and the symbol × means more than repeated addition. Solving and

TABLE 6.1: Types of Addition and Subtraction Situations Given as Word Problems: Change-Add-To/Take-From, Put Together/Take Apart and No Action, and Compare

Change-Add-To and Change-Take-From

Change-add-to, unknown result	Change-add-to, unknown change	Change-add-to, unknown start
Miguel had 4 dollars. Giovanni paid Miguel 3 dollars for a milk carton. How many dollars does Miguel have now?	Eliany had 5 packets of ten candies and 7 loose ones and went to the store and bought some more candy. Now she has 8 packets of ten candies and 6 loose ones. How much candy did Eliany buy?	Pablo had some pencils and bought 9 more. Now Pablo has 16 pencils. How many pencils did Pablo have to start with?
Change-take-from, unknown result	**Change-take-from, unknown change**	**Change-take-from, unknown start**
Doridalia had 32 dollars. She went to the the store and paid 13 dollars for some crayons. How many dollars does Doridalia have now?	Aunt Pat had 11 ears of corn. Then the children ate some of them. Now Aunt Pat has 6 ears of corn. How many ears of corn did the children eat?	Mitzi went to Roberto's store and bought 2 packets of ten peanuts and 8 loose ones from him. Now Roberto has 4 packets of ten peanuts and 7 loose ones. How many peanuts were there in Roberto's store before Mitzi bought her peanuts?

Combine: Put Together/Take Apart and No Action

Put together, unknown total	Take apart, unknown part	Put together, unknown part
Mario bought 3 packets of ten colored pencils and 5 loose ones. Edwin bought 2 packets of ten colored pencils and 9 loose ones. How many pencils are in their bags altogether?	Rachna picked 42 apples at the tree farm. She put them in a bag for her grandmother and a bag for her mother. There were 28 in the bag for her mother. How many were in the bag for her grandmother?	Farmer Brown has 6 sheep in one field and some sheep in another field. When he puts them together he has 14 sheep. How many sheep are in the other field?
No action, unknown total	**No action, unknown part**	**No action, unknown part**
Dad baked some big cookies. 7 were red, and 9 were blue. How many cookies did he bake?	Ed has 15 kittens and puppies. 7 of them are puppies. How many of them are kittens?	Isabel has a flower shop. In her shop there are 18 roses and some carnations Altogether there are 37 flowers in Isabel's shop. How many carnations does Isabel have in her shop?

Compare *More* and Compare *Less/Fewer*

Compare, unknown difference	Compare, consistent (with *more*) Compare, inconsistent (with *less*)	Compare, inconsistent
Tom has 8 stamps. Sue has 13 stamps. How many *more* stamps does Sue have than Tom?	My friend and I went to the store to buy notebooks. My friend paid \$.64 *more* than I did. If I paid \$1.68, how much did my friend pay?	Rodrigo has 16 books. Rodrigo has 7 *more* books than Aki has. How many books does Aki have?
How many *fewer* stamps does Tom have than Sue?	I paid \$.64 *less* than my friend.	Aki has 7 *fewer* books than Rodrigo.

Note: Compare situations have an additive or subtractive character depending on the language in the comparing sentence. "More" suggests addition, whereas "less" suggests subtraction. These suggestions are even stronger in languages such as Spanish, where the same words are used for "more" and "add" and for "less/fewer" and "subtract." The difficult compare problems have inconsistent language: the word in the question suggests the operation opposite to that required to solve the problem. The comparing sentence can always be said in two ways, one using *more* and one using *fewer/less* (these are italicized above). Thus one can change a difficult inconsistent problem into a simpler consistent problem by changing the question. Also, language can be used to make the comparison: "How many books does Aki have to get to have as many as Rodrigo?" or "If Rodrigo and Aki match their books, how many extra will there be?"

posing problems from a wide range of real-world whole-number situations enable students to understand alternative meanings.

Building Fluency with Computational Methods: General Issues

Fluency with computational methods is the heart of what many people in the United States and Canada consider to be the elementary mathematics curriculum. Learning and practicing computational methods are central to many memories of learning in the 20th century. However, 20th-century mathematics teaching and learning were driven by goals and by theories of learning that are not sufficient for the 21st century, in which inexpensive machine calculators are widely available, computers increasingly appear in schools and libraries, the World Wide Web gives access to a huge variety of information, and supercomputers create demands for new kinds of machine algorithms (e.g., general multistep methods). The information age creates for all citizens the need for lifelong learning and for flexible approaches to solving problems. Everyone needs the ability to use calculating machines with understanding.

Clearly, the 21st century requires a greater focus on a wider range of problem-solving experiences and a reduced focus on learning and practicing by rote a large body of standard calculation methods. How to use the scarce hours of mathematics learning time in schools is a central issue. This decision requires in part a value judgment as to which needs are most important. But new research can also influence our choices. Educators and the public are still attempting to reach consensus on the kinds and amounts of computational fluency that are necessary today. Computational fluency is one vital component of developing mathematical power; other components include understanding the uses and methods of computation. Given that mathematics learning time is a scarce resource, educators need to know roughly the amount of time various children require to reach various levels of computational fluency. Only with such knowledge can we make sensible decisions about how to allocate scarce learning time for reaching, among all the worthwhile goals of mathematics learning, computational fluency.

The goal of computational fluency for all has been an elusive goal since the 1950s. The United States and Canada have not had a successful computational curriculum that is now at risk of being overthrown by "math reform." Research studies, national reports, and international comparisons have for decades identified many aspects of computation in which children's performance was disappointing. These results have sometimes been overshadowed by even worse results for problem solving or applications of calculations, making calculation seem less of an issue than it has consistently been. Many of the calls for school mathematics reform have at least partially focused on teaching for understanding as a way to eliminate computational errors and thus increase computational performance. For example, on standardized tests, the U.S. Grade 2 norm for two-digit subtraction requiring regrouping (e.g., 62 – 48) is 38% correct. Many children subtract the smaller from the larger number in each column to get 26 as the answer to 62 – 48. This top-from-bottom error is largely eliminated when children learn to subtract with understanding (e.g., Fuson & Briars, 1990; Fuson et al., 1997; Hiebert et al., 1997). Building on a foundation of understanding can help all students achieve computational fluency.

Several themes characterize much of the research on computational methods over the past 30 years. These themes apply across computational domains (e.g., single-digit addition and subtraction, multidigit multiplication and division). Within each computational domain, individual learners move through progressions of methods from initial, transparent, problem-modeling, concretely represented methods to less transparent, more-problem-independent, mathematically sophisticated, symbolic methods. At a given moment, each learner knows and uses a range of methods that may differ according to the numbers in the problem, the problem situation, or other individual and classroom variables. A learner may use different methods even on very similar problems, and because any new method competes for a long time with older methods, the learner may not use it consistently. Typical errors can be identified for each domain and for many methods (e.g., Ashlock, 1998), and researchers have designed and studied ways to help students overcome these errors. A detailed understanding of methods in each domain enables us to identify prerequisite competencies that all learners can develop to access those methods.

The constant cycles of mathematical doing and knowing in a given domain lead to learners' construction of representational tools that they use mentally to find solutions in that domain. For example, children initially use the counting-word list to find the number of objects in a given group—that is, for counting, adding, and subtracting. Gradually, the list itself becomes a mathematical object. The words themselves become objects that are counted, added, and subtracted; other objects are not necessary.

For students who have opportunities to learn with understanding, the written place-value notation can become a representational tool for multidigit calculations as the digits in various positions are decomposed or composed, and proportional statements can become a representational tool for solving a range of problems involving ratio and proportion.

Learners invent varying methods regardless of whether their teachers have focused on teaching for understanding or on rote memorizing of a particular method. In classrooms in which the focus is teaching for understanding, however, students develop a wider range of effective methods. In classrooms in which rote learning methods are used, students' inventiveness often generates many different kinds of errors, most of which are partially correct methods created by a particular misunderstanding. Thus, even in traditional classrooms emphasizing standard computational methods, learners are not passive absorbers of knowledge. They build and use their own meaning and doing, and they generalize and reorganize this meaning and doing.

Multidigit addition, subtraction, multiplication, and division solution methods are called *algorithms* (see Gravemeijer & van Galen, this volume). An algorithm is a general multistep procedure that will produce an answer for a given class of problems. Computers use many different algorithms to solve different kinds of problems, and inventing new algorithms is an increasingly important area of applied mathematics. Around the world, many different algorithms have been invented and taught for multidigit addition, subtraction, multiplication, and division. Students in U.S. and Canadian schools have learned different algorithms at different times. Each algorithm has advantages and disadvantages. Therefore, the decisions about computational fluency concern in part the algorithms that might be supported in classrooms and the bases for selecting those algorithms.

One goal of the following sections is to underscore the possibility of understanding various computational methods. Because such understanding has ordinarily not been a goal of school mathematics, most educational decision makers have not had an opportunity to understand the standard algorithms or to appreciate the wide variety of possible algorithms. Most teachers also have not had that opportunity, and most textbooks do not sufficiently help develop such understanding. Research indicates that some algorithms are more accessible to understanding than others and that understanding can be increased by quantity supports (e.g., manipulatives, drawings) to help children understand the meanings of the numbers, notations, and steps in the algorithms. This understanding does not conflict with developing computational fluency but rather is foundational to it. Children need scaffolded practice with whatever methods they are using to become more fluent in orchestrating the steps in any algorithm. Understanding can serve as a continual directive toward correct steps and as a constraint on the many creative calculating errors invented by students who are taught algorithms by rote.

Because children cannot understand all algorithms equally (e.g., many algorithms sacrifice comprehensibility to save space in writing), I describe at least one algorithm that has been demonstrated to be accessible to a wide range of students. My criteria for such accessible algorithms are that they scaffold the understanding of principal steps in the domain, generalize readily to large numbers, have variations that provide for individual differences in thinking, and are procedurally simple to carry out (i.e., they require the minimum of computational subskills so that valuable learning time is not required to bring unnecessary subskills to the needed level of accuracy).

Single-Digit Computation: Much More Than "Learning the Facts"

Learning single-digit addition, subtraction, multiplication, and division has for much of a century been characterized in the United States and Canada as "learning math facts." The predominant learning theory was of rote paired-associate learning in which each pair of numbers was a stimulus (e.g., 7 + 6) having an answer (e.g., 13) that students needed to memorize as the response to that stimulus. "Memorizing the math facts" has been a central focus of the mathematics curriculum, and many pages of textbooks presented these stimuli, eliciting from children their memorized responses.

This view of how children learn basic single-digit computation was invalidated by one line of research earlier in the century (Brownell, 1956/1987) and by much research from all over the world during the last 30 years. We now have very robust knowledge of how children in many countries actually learn single-digit addition and subtraction. In the following section, I explore the research for addition in some depth both because it is the largest body of research and because it sets the scene for understanding calculation in other areas.

Single-Digit Addition and Subtraction

Single-Digit Addition

Research throughout the century (e.g., Brownell, 1956/1987) has presented a complex view of children using a variety of mathematical learning methods. Substantial research now indicates that children move through an experiential progression of single-digit addition methods (e.g., see reviews in Fuson, 1992a, 1992b; Siegler, this volume). These methods are not ordinarily taught in the United States, Canada, and many other countries. Adults whose cultures are stimulating new demands for calculating also invent the methods (Saxe, 1982). Analyses of these methods reveal that learners build later methods from earlier methods by chunking, recognizing, and eliminating redundancies; using parts instead of entire methods; and using their knowledge of specific numbers. Thus, the methods seem accessible to almost all learners by natural and general learning processes. When these more advanced methods are not supported in the classroom, however, several years separate the earliest and latest users of advanced methods. In contrast, helping children progress through methods can lead all first graders to methods that are efficient enough to use for all later multidigit calculation.

Children's tools for an initial understanding of addition are the counting-word list ("one, two, three, four," etc.); the ability to count objects; an indicating act (e.g., pointing, moving objects) that ties the words said to the objects counted as each is indicated; and the count-cardinal knowledge that the last count word said tells the total number of objects. Many but not all children in the United States and Canada learn these tools in the preschool years. With these tools, young learners can do addition orally using concrete situations; they count out objects for the first addend, then for the second addend, and then for all the objects (*count all*). Children abbreviate, internalize, chunk, and abstract this general counting-all method as they become more experienced with it.

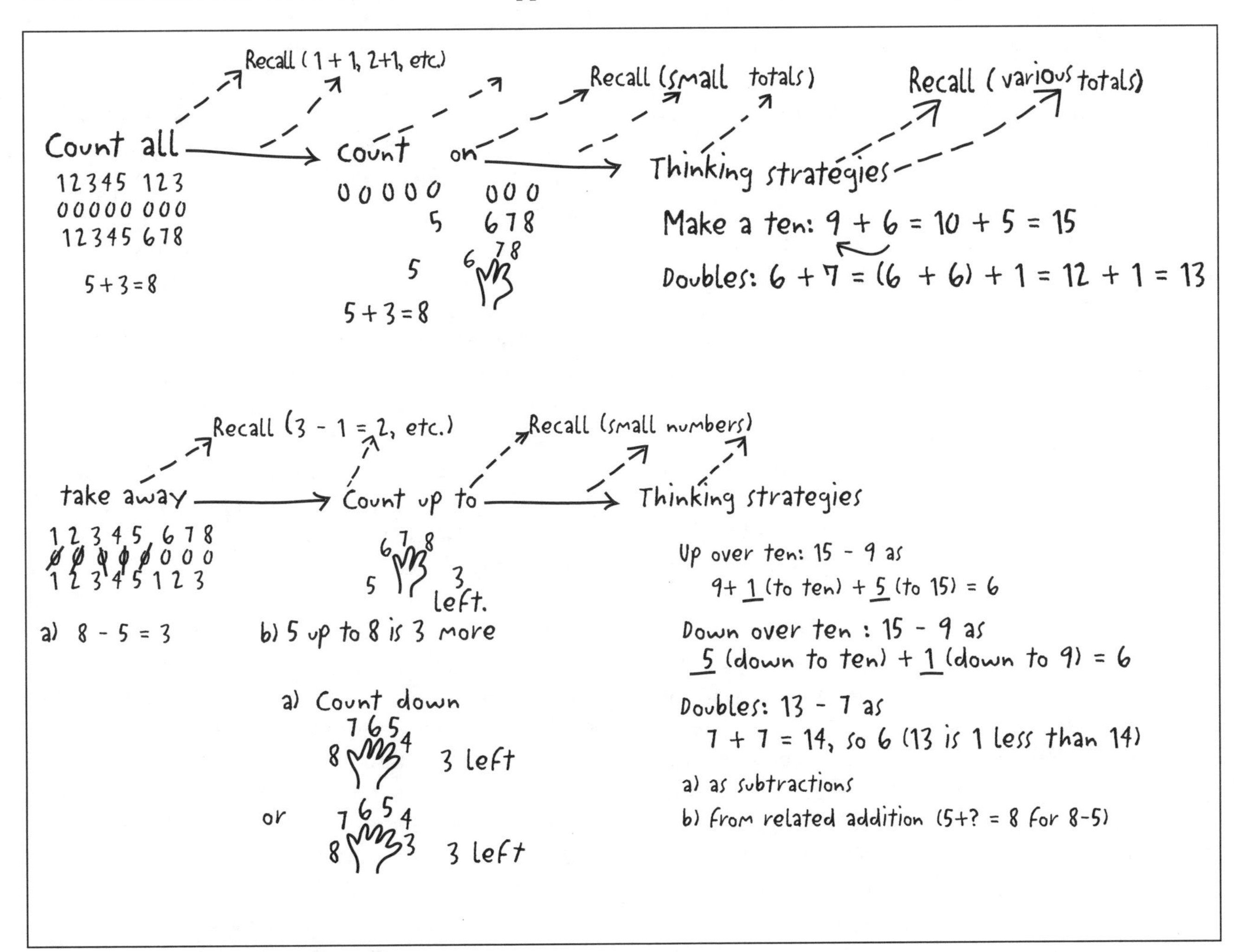

Figure 6.1. Learning progression for single-digit addition and subtraction.

The major steps in this worldwide learning progression are shown in Figure 6.1. Children notice that they do not have to count the objects for the first addend but that they can start with the number in one addend and count on the objects in the other addend; that is, they can *count on*. Children count on with objects. They then begin to use the counting words themselves as countable objects and keep track of the number of words that are counted on by using fingers or auditory patterns; in this way, the counting list has become a representational tool. With time, children chunk smaller numbers into larger numbers and use thinking strategies in which they convert an addition fact that they do not know into an addition fact that they do know. In the United States and Canada, they usually make this conversion by using a *doubles* addition (2 + 2, 3 + 3, etc.), and they learn these doubles very quickly. As noted in Figure 6.1, throughout this learning progression particular addition combinations move into the category of being rapidly recalled rather than solved in some way. Most children readily recall doubles, sums that involve adding 1, and combinations of small numbers; individuals vary in the other sums they recall readily.

In many other parts of the world, children are taught a general thinking strategy: Make a ten by giving some quantity from one addend to the other addend. Children in the United States or Canada who speak English seldom invent this "make a ten" method, nor do they learn it in textbooks. Yet the method is taught in first grade in China, Japan, Korea, and Taiwan (Fuson & Kwon, 1992; Fuson, Stigler, & Bartsch, 1988). This method is facilitated by the number words in these countries: "ten, ten one, ten two, ten three," and so forth. Many children in these countries also learn numbers and addition using a *ten-frame*: an arrangement of small circles into two rows of five. This pattern emphasizes the numbers 6, 7, 8, and 9 as 5 + 1, 5 + 2, 5 + 3, and 5 + 4. Work with this visual pattern enables many children to "see" these small sums under 10 as using a five-pattern. For example, some Japanese adults report adding 6 + 3 by thinking or visualizing [(5 + 1) plus 3] = 5 + 4 = 9. Adults do this reduction of 6 + 3 to 5 + 1 + 3 = 5 + 4 = 9 very rapidly and without effort, as automatically as recalling the sum. The ten-frame is also used to teach the make-a-ten method. For example, 8 has 2 missing in the ten-frame, so 8 + 6 requires 2 from the 6 to fill the ten-frame, leaving 4 to make 10 + 4 = 14. By the end of first grade, most children in these Asian countries rapidly use these patterns to add single-digit numbers mentally.

The make-a-ten method is also taught in some European countries. Children must learn three prerequisites to use the make-a-ten method effectively. They must know what number makes a ten with each number up to 10 (e.g., 10 = 8 + 2 or 6 + 4 or 7 + 3), be able to break apart a number into any of its two addends (so that ten can be made with any other number), and know the sum of 10 + n (e.g., 10 + 5 = 15). In countries that teach the make-a-ten method, children develop these prerequisites before teachers introduce the method. Many first and even second graders in the United States have not consolidated these prerequisites, and textbooks rarely develop them sufficiently. Counting on is not sufficient to prompt U.S. or Canadian children to move on to the make-a-ten method, for two reasons. English counting words do not signal a change at ten as the East Asian words "ten, ten one, ten two" do. And, the fingers do not use or show a ten because most children use their fingers to keep track of the words counted on ("8, 9, 10, 11, 12, 13, 14"). This use is in contrast with that of Korean or some Latino children, who make the first number on their fingers and then count on ("1, 2, 3, 4, 5, 6 is 14") as they use 6 more fingers: 2 more to fill 10 fingers and 4 more greater than ten. This counting-on method supports the make-a-ten method.

Textbooks in the United States typically have shown little understanding of children's progression of methods. Many have moved directly from counting all (e.g., 4 + 3 shows four objects and then three objects) to using numerals only. Children are then expected to begin to "memorize their facts"—which they cannot do because they have no answers available. Children respond by following the experiential trajectory of methods discussed earlier and summarized in Figure 6.1: They use their fingers or make little marks to count all, eventually invent counting on, and may go on to invent thinking strategies, especially using doubles. As indicated in Figure 6.1, particular sums all along the way become recalled sums; the answer is produced rapidly and automatically and without knowledge of a reportable solution method. Most textbooks do not support children's learning of these strategies or move them through this progression of methods. Nor are children typically given visual supports, such as a ten-frame, for adding by patterns of numbers; if these alternatives are supplied, they are often not orchestrated with sufficient practice of subskills or methods to reach mastery by all.

This lack of fit between what textbooks contain and how children think is exacerbated by other features of the textbook treatment of addition. Compared with other countries, the United States has had a very delayed placement of topics in the elementary school curriculum (Fuson, 1992a; Fuson, Stigler, & Bartsch, 1988). Children spend almost all of first grade learning addition and subtraction below 10. In second grade, teachers emphasize and review such simple problems, resulting in children's focus on many more of the easier sums and relatively fewer of the more difficult sums (Hamann & Ashcraft, 1986). Thus, in contrast with East Asian children, who in first grade learn effective methods for solving the difficult sums over 10 in visually and conceptually supported ways, many U.S. children have had little opportunity to solve such problems in first grade and do not have the support of any effective methods for doing so.

Intervention studies indicate that teaching counting on in a conceptual way makes all single-digit additions accessible to U.S. first graders, including learning-disabled students and limited-English-proficient students (Fuson & Fuson, 1992; Fuson & Secada, 1986). These results have been replicated in urban and suburban classrooms in English and in Spanish (see reviews in Fuson, 1992a, 1992b, as well as Fuson, Perry, & Kwon, 1994; Fuson, Perry, & Ron, 1996). Children in the United States show numbers on their fingers in various ways, some starting with the index finger, some with the thumb, and some with the little finger. Any of these arrangements are effective ways to keep track of the second addend while counting on. With practice, children can perform counting on rapidly and accurately enough to use this method in multidigit calculations of all kinds. Conversations with many adults reveal that adults use counting on in situations where accuracy counts. Counting on is a powerful, general, and sufficiently rapid method for most purposes.

To date, very little research has been conducted with U.S. or Canadian children on using spatial patterns for learning addition with small numbers. This approach might be very powerful, especially for some children who have difficulty with sequential information or who have strong spatial competence. A few studies have been conducted on using ten-frames or other visual supports for the make-a-ten method (Thornton, Jones, & Toohey, 1983), but none on developing the three prerequisites for that method. Some Latino children use finger methods that support the make-a-ten method (Fuson, Perry, & Kwon, 1994; Fuson, Perry, & Ron, 1994) and go on to invent and use it. How accessible the method is to all children is not clear, given the irregularities in the English number words. Given that it is taught in France and some other European countries with similar irregularities in the words used for the teens, it is worth trying in North America to see how rapid and automatic it could become. It is very useful in multidigit addition because it gives the answer already prepared for regrouping (i.e., carrying, trading) as, for example, 1 ten and 4 ones.

The prerequisites for the make-a-ten method are important learning goals themselves, so research would help educators understand how to help students learn them rapidly. More conceptual work on the meaning of "teens" words and written numbers as tens and ones would be especially valuable. Several studies using different methods indicate that for many U.S. children, what you see in a teen number is what you get: They see 15 as a one and a five, not as a ten and a five (see the reviews in Fuson, 1992a, 1992b). Many children speaking European languages have similar difficulties. Clearly, however, U.S. children can learn the meanings of teens as tens if they are supported in doing so (Fuson, Smith, & Lo Cicero, 1997).

Single-Digit Subtraction

Subtraction follows a progression that is similar to that for addition in its major categories (see Figure 6.1). Most children in the United States, however, invent counting-down methods that model the taking away of numbers from the total. Counting down is difficult (e.g., try generating an alphabet list backwards from K to C), and some children require a long time to learn it. Counting down is also subject to more errors because the cardinal correspondence of the objects in the addends is not clear. Children use two distinct counting-down methods: One starts with the total, and the other starts with the word before the total (see Figure 6.1). Many children create combinations of the two that give an answer that is either one too many or one too few. Many other children make errors when counting backward or are very slow at it.

In contrast, in Latin America and many countries in Europe, children learn to subtract by counting up from the known addend to the known total. For example, 14 – 8 is solved by counting up from 8 to 14: I have eight things, so 9, 10, 11, 12, 13, 14, that's 6 more I counted to get to 14. Counting up is even easier than counting on because you only have to listen for the total; you do not have to know and monitor a finger pattern for an addend as you count. So counting up makes subtraction easier than addition.

Intervention studies with U.S. first graders that help them see subtraction situations as taking away the first x objects enable them easily to learn a counting-up-to process for subtraction. All children, including learning-disabled and special education students, were able to learn all single-digit subtraction combinations in first grade. This result was revolutionary for first-grade teachers, who typically see children having much more difficulty with subtraction than with addition. U.S. children invent counting on for situations in which an unknown quantity is added to a known quantity. Many children later use counting on in taking-away subtraction situations because counting on is easier than counting down (Carpenter & Moser, 1984). However, that invention is delayed for many children until second grade, and many never subtract by counting up. Discussing counting up to for taking-away situations in first grade makes it accessible to all children then.

Less research is available about thinking strategies in subtraction than about those in addition. In East Asia, two methods using 10 are taught, and different children use each of these methods. One is just fast counting on with chunking at 10: 15 – 9 is 9 + 1 (up to 10) + 5 (up to 15), so 9 + 6 = 15. The other goes down over 10: 15 – 9 is 5 down to 10, and 1 more down to 9 is 6. These methods may also not have a strong directional component.

Summary

The unitary progression of methods used worldwide by children stems from the sequential nature of the list of number words. This list is first used as a counting tool, and then it becomes a representational tool in which the number words themselves are the objects that are counted (Bergeron & Herscovics, 1990; Fuson, 1986b; Steffe, Cobb, & von Glasersfeld, 1988). Counting becomes abbreviated and rapid. Some (or in some settings, many) children then chunk numbers using thinking strategies. These chunking actions convert sums that children do not know into sums that they do know, often using doubles. During this progression—which may last into third or even into fourth or fifth grade for some children who are not helped through the progression—individual children use a range of different methods on different problems. Learning-disabled children and others having difficulty with mathematics do not use methods that differ from this progression; they are just slower than others in the progression (Geary, 1994; Ginsburg & Allardice, 1984; Goldman et al., 1988; Kerkman & Siegler, 1993). Teachers can make counting on accessible to first graders; counting on makes possible rapid and accurate addition of all single-digit numbers. For U.S. children, single-digit subtraction is usually more difficult than addition is. Learning to think of subtraction as counting up to the known total, as is done in many other countries, makes subtraction as easy as, or easier than, addition. But at present, the counting-up-to method rarely appears in textbooks. Research is needed on using spatial patterns to make sums and differences less than 10 accessible visually; on supporting thinking strategies and their prerequisites for all children; and on effective organizations of practice and support that will enable all children to progress to rapid and accurate methods of single-digit sums and differences by the end of first grade (i.e., at least counting on and counting up to).

Single-Digit Multiplication and Division

Less research has been conducted on single-digit multiplication and division than on addition and subtraction. U.S. students go through an experiential progression of multiplication methods that is somewhat similar to that for addition (e.g., Anghileri, 1989; Baek, 1998; Mulligan & Mitchelmore, 1997; Steffe, 1994). Eventually they make equal groups and count them all. They learn count-on lists for different products (e.g., counting by 4s gives 4, 8, 12, 16, 20, etc.). They then count up and down these lists, using their fingers to keep track and to find different products. They may use a combination method in which they enter a list at a point they know and then count on by ones to get to the product (e.g., to find 5 × 6, 5 × 5 is 25, and 26, 27, 28, 29, 30 is 5 more). They invent thinking strategies in which they derive related products from products they know.

As with addition and subtraction, many of these methods are developed as individual inventions by children who are not supported by textbooks or instruction. Some very recent textbooks support children's pattern finding and use of count-by lists in multiplication. Yet most older books use the 19th-century model of learning multiplication by memorizing isolated facts using rote associations.

To see the problems with this limited view of rote "memorizing of multiplication facts," consider Figure 6.2. The figure shows a table of multiplication products using an alphabet analogy: Suppose you were to learn the multiplication combinations for a new counting list (C, D, E, F, G, H, I, J, K). We pose this task so that you will not already know the answers. Look at the table, and see all the interference involved in learning each of these facts separately as a rote response to two stimuli. The task is formidable because the "numbers" all look very similar (they are similar, just as 1, 2, 3, 4, etc., are for young children). Look at the table for a while, and think how you might go about this formidable task in another way. See all the patterns you can find. Do you see a pattern for multiplying by C? (Look at the top row or left column). Do you see a pattern for multiplying by CL? (Look at the last column or the bottom row). Look at the interesting pattern for G. Can you find a pattern for D? The patterns for E and F are more subtle. K has a wonderful pattern; can you explain it?

Some patterns in the table are as follows:

- C just copies the number being multiplied to give the same number as the product.
- Multiplying by CL just adds an L, so D becomes DL, E becomes EL, and so forth.

- G products alternate between G and L in the ones place; the tens place has two Cs, two Ds, two Es, and so forth.
- In the K pattern, the tens place increases by one and the ones place decreases by one; the total of the tens and ones numbers is K. This pattern occurs because any quantity ($n \times 9$) is just $[(n \times 10) - (n \times 1)]$. In English number words and numbers, 2 nines are 2 tens – 2 = 20 – 2 = 18, 3 nines are 3 tens – 3 = 30 – 3 = 27, and so forth.

Finding and using patterns greatly simplifies the task of learning multiplication combinations. Moreover, it is one of the very essences of mathematics. Thus, approaching multiplication learning as pattern finding greatly simplifies the task and constitutes a core mathematical approach.

After they have identified patterns, children still need much experience to produce count-by lists and individual products rapidly. However, little research describes how to accomplish this fluency—or how to link the patterns in the numbers with the underlying groupings spatially or conceptually (i.e., how to relate all the different groupings to the ten-groupings in the base-ten multidigit numbers). Likewise, not much research describes how children learn products in other countries. Informal inquiries of researchers outside the United States suggest some methods that might be pursued. In France, where children learn multiplication effectively (Lemaire & Siegler, 1995), educators may use an elaborate yearlong social organization involving an extra mathematics period and extra help for those who need it. In Japan, oral rhythmic chanting of the tables is used; the structure of the Japanese number words facilitates this approach. In China, educators emphasize commutativity ($4 \times 7 = 7 \times 4$), reducing the number of products by half (i.e., notice the line of symmetry in Figure 6.2 moving from top left to bottom right). The use of this approach is reflected in the reaction-time patterns of Chinese adults, which differ from those of U.S. adults (LeFevre & Liu, 1997).

Division combinations can be approached in terms of the related products. For example, 72/9 = ? can be thought of as $9 \times ? = 72$. Again, research has little to offer as yet. Not clear are whether children can be introduced to the division-multiplication relationship very early, thereby learning and practicing quotients at the same time as products, or whether they should learn products first. How to help children learn and easily use all the symbols for division (e.g., 15/3, $15 \div 3$, $\frac{15}{5}$, and the reversed $5\overline{)15}$) is also not clear.

The general methods of counting on for addition and counting up for subtraction are readily learned. However, no similar rapid general methods are available for single-digit multiplication and division. Rather, we need to

	C	D	E	F	G	H	I	J	K	CL
C	C	D	E	F	G	H	I	J	K	CL
D	D	F	H	J	CL	CD	CF	CH	CJ	DL
E	E	H	K	CD	CG	CJ	DC	DF	DI	EL
F	F	J	CD	CH	DL	DF	DJ	ED	EH	FL
G	G	CL	CG	DL	DG	EL	EG	FL	FG	GL
H	H	CD	FJ	DF	EL	EH	FD	FJ	GF	HL
I	I	CF	DC	DJ	EG	FD	FK	GH	HE	IL
J	J	CH	DF	ED	FL	FJ	GH	HF	ID	JL
K	K	CJ	DI	EH	FG	GF	HE	ID	JC	KL
CL	CL	DL	EL	FL	GL	HL	IL	JL	KL	CLL

Figure 6.2. A multiplication table for numbers C through CL.

orchestrate much specific pattern-based knowledge into accessible and rapid multiplication and division methods—and we need research into ways to support such pattern finding. Then, we must organize the necessary follow-up learning conceptually, motivationally, and socially in classrooms if U.S. children are to learn this gatekeeper knowledge by the end of Grade 3, as is accomplished in other countries where students' mathematical performance is high.

Traditional learning of addition and multiplication facts creates interference between these two operations (LeFevre, Kulak, & Bisantz, 1991; Lemaire, Barrett, Fayol, & Abdi, 1994; Miller & Paredes, 1990). Thus, when children begin learning multiplication combinations, their addition performance decreases. This phenomenon is a strong reason to encourage learning of general addition and subtraction methods; these methods do not interfere with multiplication or division. Interference between addition and multiplication for combinations involving 0 and 1 is particularly great. Children readily learn the patterns involved in $7 + 0$, $7 + 1$, 7×0, and 7×1, but they tend to confuse them. These patterns are complex across operations because although $7 + 0 = 7 \times 1$, these facts look maximally dissimilar.

Timed tests have been a controversial part of single-digit computational practice. No definitive evidence about their use exists. In some situations, students enjoy timed tests as a challenge, and individuals can watch their own progress. However, they can be counterproductive and feed math anxiety if educators use them—

- before students have acquired conceptual knowledge in a domain that enables them to generate solutions;
- in a competitive fashion, so that some students "lose" rather than focus on monitoring and improving their own individual progress;
- or in a nonsupportive environment, causing students to feel isolated or hopeless.

Multidigit Addition and Subtraction

Considerable research attests to the ways that children learn multidigit addition and subtraction methods, although this research is not nearly as copious as that on single-digit addition and subtraction. In single-digit addition and subtraction, the same learning progression occurs in many different countries despite the fact that educators do not explicitly teach methods for learning it. Multidigit addition and subtraction depend much more on what is taught, and different children within the same class may follow different learning progressions and use different methods. Multidigit addition and subtraction knowledge seems to consist of different pieces that children put together in different orders and in different ways (e.g., Hiebert & Wearne, 1986).

Difficulties with Words and Numbers

As with the "teens" words, the English number words between 20 and 100 complicate the teaching and learning processes for multidigit addition and subtraction. English names the hundreds and thousands regularly but does not do so for the tens. For example, 3333 is said "3 thousand 3 hundred thirty 3" not "3 thousand 3 hundred 3 ten 3." English-speaking children must learn a special sequence of decade words for 20, 30, 40, and so forth. This sequence, like the teens, has irregularities. Furthermore, teens words and decade words sound alike: In a classroom, children often have difficulty hearing the difference between "eighteen" and "eighty." The same numbers, 1 through 9, are reused to indicate how many tens, hundreds, thousands, and so forth, are designated. Whether a 3 signifies 3 tens or 3 hundreds or 3 thousands is shown by the relative position of the 3: How many places to the left of the rightmost number is the 3? Relative position is a complex concept. French is even more complex, with its use of twenty as a base in some number words.

The written place-value system is a very efficient system that lets people write very large numbers. Yet it is very abstract and can be misleading: The digits in every place look the same. To understand the meaning of the digits in the various places, children need experience with some kind of *size-quantity supports* (e.g., objects or drawings) that show tens to be collections of 10 ones and show hundreds to be simultaneously 10 tens and 100 ones, and so on. Educators have designed and used various kinds of supports in teaching the written system of place value. Some techniques support students' understanding of the sizes involved in place values, and some can support their understanding of patterns in the numbers (e.g., a hundreds grid). Classrooms, however, rarely have enough such supports for children themselves to use them—especially size-quantity supports—and many classrooms do not use anything. Such supports are rarely used in multidigit addition and subtraction; when used, they may be used alone at first to get answers rather than to link the written method with the manipulative method to facilitate understanding.

As a result, many studies indicate that U.S. children do not possess or are unable to use a quantity understanding of multidigit numbers (see reviews in Fuson, 1990, 1992a, 1992b). Instead, children view numbers as single digits side by side: 827 is functionally "eight two seven" and not 8 groups of one hundred, 2 groups of ten, and 7 single ones. Children make many different errors in adding and subtracting multidigit numbers, and many students who add or subtract correctly cannot explain how they got their answers.

Teaching for Understanding and Fluency

In contrast, research on instructional programs in the United States, Europe, and South Africa indicates that focusing on understanding multidigit addition and subtraction methods results in much higher levels of correct multidigit methods and produces children who can explain how they got their answers using quantity language (Beishuizen, 1993; Beishuizen, Gravemeijer, & van Lieshout, 1997; Carpenter, Franke, Jacobs, & Fennema, 1998; McClain, Cobb, & Bowers, 1998; Fuson & Briars, 1990; Fuson & Burghardt, 1997, in press; Fuson, Smith, & LoCicero, 1997; Fuson et al., 1997).With these approaches, students used some kind of visual quantity support to learn meanings of hundreds, tens, and ones, and these meanings were related to the oral and written numerical methods developed in the classrooms. Students developed many different addition and subtraction methods, often in the same classrooms (see Fuson et al., 1997, and Fuson & Burghardt, in press, for summaries of many methods). In most of these studies, children invented methods and described them to each other, but in some studies, researchers used conceptual supports to help students give meaning to a chosen algorithm. Many studies were intensive studies of children's thinking in one or a few classrooms, but some involved 10 or more classrooms, including one study of all second-grade classrooms in a large urban school district (Fuson & Briars, 1990). All studies placed a strong emphasis on children's understanding and explaining their method using quantity terms (e.g., hundreds, tens, ones, or the names of the object supports being used).

Students can use roughly three classes of effective methods for multidigit addition and subtraction, although some methods are mixtures. *Counting list methods* are extensions of the single-digit counting methods. Children initially may count large numbers by ones, but these unitary methods are highly inaccurate and are not effective. All children need to be helped as rapidly as possible to develop prerequisites for methods using tens. In counting-list methods using tens, children count on or count up by tens and by ones. These methods generalize readily to counting on or up by hundreds but become unwieldy for larger numbers. In *decomposing methods*, children decompose numbers so that they can add or subtract the like units (e.g., add tens to tens, ones to ones, hundreds to hundreds, etc.). These methods generalize easily to very large numbers. *Recomposing methods* are like the make-a-ten or doubles methods. The solver changes both numbers by giving some amount of one number to another number (i.e., in adding) or by changing both numbers equivalently to maintain the same difference (i.e., in subtracting). These methods are highly useful in such special cases as 398 + 276: the 276 gives 2 to change the 398 into 400, so 400 + 274 is 674. But they do not generalize easily to all numbers, and the addition and subtraction methods may interfere with each other if students or teachers do not understand them well enough.

Educators have successfully used different kinds of conceptual supports, such as manipulatives, in classroom research. Each type of support has its advantages and disadvantages, and each supports some methods more clearly than others (see Fuson & Smith, 1997, for an analysis). Number lines and hundreds grids of numbers (10 rows or columns showing 1 to 10, 11 to 20, etc., up to 100) support counting-list methods the most effectively. However, they do not generalize easily to numbers greater than 100. Children may use the hundreds grid in particular by rote to get answers without really seeing tens on it. Decomposition methods are facilitated by supports that enable children to physically add and subtract the different quantity units. For example, base-ten blocks show ones, tens (10 attached centimeter cubes), hundreds (a 10 cm by 10 cm flat block), and thousands (a 10 cm by 10 cm by 10 cm large cube). Educators have successfully used these blocks with children inventing their own methods and understanding chosen methods.

Because of the expense and management problems posed by objects that act as conceptual supports, some studies have also introduced some system of drawing ones, tens, and hundreds (e.g., circles or small dashes for ones, vertical sticks for tens, and squares for hundreds) or of recording on an open number line. Such drawings leave records for a teacher to see after class, and children can draw figures on the board to explain their method. The drawings are also easy to link with the written numbers so that the numbers begin to take on quantity meanings for children.

The function of size-quantity supports is to suggest meanings that children can attach to the written numerals and to the steps in the solution method with numbers. Therefore, methods of relating the size-quantity supports and the written method through linked actions and verbal descriptions of the numerical method are crucial. However, in the classroom, educators often use supports without recording anything except the answer at the end, leading students to use written methods without linking them with the steps taken using the supports. Thus, the written numerals do not necessarily take on the meaning of tens, hundreds, and so forth, and students may think of the steps in the numeral method as involving only single digits rather than their actual quantity meanings. This development leaves students vulnerable to the many errors they create without the meanings to direct or constrain them. Even for students who initially learn a meaningful method, the appearance of a multidigit number as a collection of single digits may cause errors to creep in. An important step in maintaining the meaning of the steps is to have students occasionally explain their method, using the names for their quantity support (e.g., big cubes, money, etc.).

Solution Methods and Accessible Algorithms

In different countries, children invent and use many different methods of multidigit addition and subtraction. This chapter discusses two addition methods and one subtraction method that are especially clear conceptually, easy for even less proficient students to carry out, and less prone to errors than many other methods are. It also presents the most frequently taught addition and subtraction algorithms in textbooks in the United States and Canada.

In Figure 6.3, the algorithm on the top left is the addition method currently appearing in most U.S. textbooks. It starts at the right, in contrast with reading, which starts at the left. Most methods that children invent start at the left, perhaps because they are used to reading from the left and because number words in English are read from left to right. The current addition algorithm has two major problems. One is that many children object initially to putting the little 1s above the top number. They say that you are changing the problem. And in fact, this algorithm does change the numbers it is adding, because it proceeds by adding these carries to the digits in the top number. The second method in the top row of Figure 6.3 does not change the top number: The new 1 ten is written down in the space for the total on this line (children using base-ten blocks [Fuson & Burghardt, 1993; Fuson &Burghardt, in press] invented this method so that they did not change the answer as they went). The method lets children more easily see the total 14 ones, because the 1 is written in close proximity to the 4. The second problem with the present U.S. algorithm is that it makes single-digit addition difficult. The solver must add in the 1 to the top number, remember it even though it is not written, and add that remembered number to the bottom number. If, instead, the solver adds the two numbers seen, he or she may forget to go up to add on the 1 ten (or 1 hundred). The second method remedies this difficulty: The solver just adds the two numbers seen and then increases that total by one. This method makes the adding much easier for children with less mathematics proficiency.

Both these methods require that children understand two aspects of multidigit numbers: (1) they must add like units to each other, and (2) when they get 10 or more of anything, they must give 1 group of ten of those things to the next left place and record the remaining things. The second understanding has been called "carrying" or "regrouping" or "trading." This grouping is done after the adding of each kind of unit. The make-a-ten method of single-digit addition described earlier is clearly helpful for such grouping because it converts a number into 1 ten and some ones. Multidigit addition is a useful place to use this make-a-ten method. Unless the structure of teen numbers as 1 ten and some ones is strongly experienced in the classroom, however, children may have trouble knowing how to break a teen number for regrouping. Again, the teen words in English obfuscate the tens, and all calculation is carried out by the solver using number words (even though these words may be said only internally). In one study with base-ten blocks, some first and second graders who were successfully adding four-digit numbers and explaining their methods still had trouble with the grouping step when they did not use blocks. They knew that each teen word represented a ten and some ones; they just did not know how many ones were in a given teen word. Instead, they used their knowledge of written numbers to write their total off to the side: For example, they said, "eight plus six is fourteen" and wrote 14, which they then read as 1 ten and 4 ones. Work on teens as 1 ten and x ones would have been helpful to these children.

Method B in Figure 6.3 separates the two major steps in multidigit adding. The total for adding each kind of multiunit is written on a new line, emphasizing that the solver is adding each kind of multiunit. Students do the carrying-grouping-trading as part of the addition of each kind of multiunit: The new 1 ten of the next larger multiunit is simply written in the next-left column. Students then do the final step of multidigit adding: Add all the partial additions to find the total. Students can do Method B in either direction (Figure 6.3 shows the left-to-right version). Because they write the whole value of

Typical U.S. Algorithms	Accessible Generalizable Methods Method A: New Groups Below	 Method B: See Place Values	Drawings to Show Quantities
1 1 568 +876 1444	568 + 876 → 4 (1) 568 + 876 → 44 (1 1) 568 + 876 → 1444 (1 1)	568 + 876 1300 130 14 1444	1 thousand, 1 hundred, 1 ten 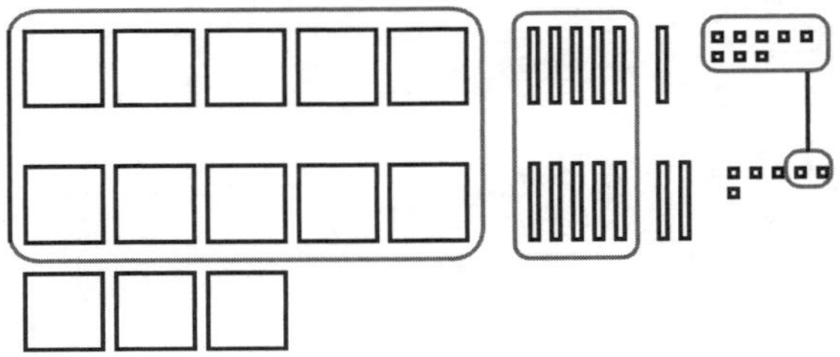
Move right to left. Add ones, carry 1 to above left; add tens, carry 1 to above left. Usually add carry to top number, remember that number while adding it to bottom number.	Move right to left. 1 new group goes below in answer space of next left column, keeping total together. Add 2 numbers you see, then increase that number by 1 to add the new group.	Can be done in either direction. Add each kind of unit first, then add those totals.	Stage 1: Sustained linking of quantities to written algorithm to build understanding of quantity meanings. Stage 2: Only do numerical algorithm, but occasionally explain using quantity words (*thousands*, *hundreds*, *tens*, *ones*).
13 13 3 3 14 1 4 4 4 − 5 6 8 8 7 6 Move right to left. Alternate ungrouping and subtracting.	Ungroup Everything First (As Necessary), Then Subtract Everywhere 13 13 14 14 14 1 4 4 4 − 5 6 8 8 7 6 left-to-right ungrouping 13 13 3 3 14 1 4 4 4 − 5 6 8 8 7 6 right-to-left ungrouping	Do all ungrouping, in any order, until every top number is larger than the bottom number. Then subtract each kind of multiunit, in any order.	

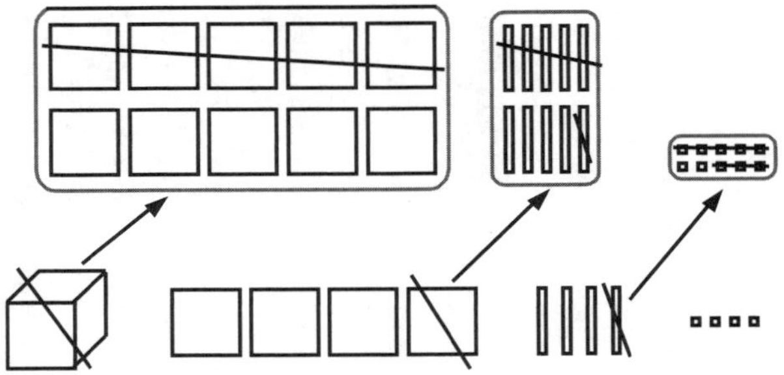

Figure 6.3. Multidigit addition and subtraction.

each addition (e.g., 500 + 800 = 1300), children think about and explain how and what they are adding more easily with this method.

The drawings at the far right can accompany any of the three Figure 6.3 methods to support understanding of the methods' major components. The different sizes of the ones, tens, and hundreds in the drawings support children's adding of those like quantities to each other. Ten of a given unit can be encircled to make 1 of the next higher unit (10 ones = 1 ten, 10 tens = 1 hundred, 10 hundreds = 1 thousand). The issue for each algorithm, then, is how to record the adding of each kind of unit, the composing of each new larger unit from 10 of the smaller units, and the adding of the partial additions to make the total. Circling the new ten units can also support the general make-a-ten single-digit methods.

The two vital elements of using drawings or objects to support understanding of addition methods are summarized under the drawing in Figure 6.3. First is a long Stage 1 in which the objects or drawings are linked with the steps in the algorithms to give meanings to those algorithms' numerical notations. A second but crucial Stage 2 then lasts an even longer time, over years, in which students carry out only the numerical algorithm but occasionally explain it with words describing quantity objects or drawings so that meanings stay attached to the steps of the algorithm. Stage 2 is vital because of the single-digit appearance of the written numerals. Numerals neither facilitate correct methods nor inhibit incorrect methods the way the objects and drawings do; errors can creep into already understood methods, especially as children learn other solution methods in other domains.

Figure 6.3 shows two subtraction methods. The method on the left is the most widely used current U.S. algorithm. It moves from right to left, alternating between the two major subtraction steps: Step 1 is ungrouping ("borrowing," trading) to get 10 more of a given unit so that unit can be subtracted (necessary when the top unit is less than the bottom unit); Step 2 is subtracting after the top number has been ungrouped. The ungrouping may be written in different ways (e.g., as a little 1 beside the 4 instead of crossing out the 4 and writing 14 above it).

Alternating between the two major subtracting steps presents three kinds of difficulties to students. One is initially learning this alternation. The second is then remembering to alternate the steps. The third is that the alternation renders students susceptible to the pervasive subtracting error: subtracting a smaller top number from a larger bottom number (e.g., doing 72 – 15 as 63). When moving left using the current method, a solver sees two numbers in a column while primed to subtract. For example, after ungrouping in 1444 – 568 to get 14 in the rightmost column and subtracting 14 – 6 to get 8, the solver sees 3 at the top and 6 at the bottom of the next column. Automatically the solver produces the answer 3 (6 – 3 = 3). This answer must be inhibited while the solver thinks about the direction of subtracting and asks whether the top number is larger than the bottom (i.e., asks whether ungrouping, or "borrowing," is necessary).

The accessible subtraction method shown in the bottom middle of Figure 6.3 separates the two steps used in the current method. First, a student asks the ungrouping ("borrowing") question for every column, in any direction. The goal is to rewrite the whole top number so that every top digit is larger than the bottom digit. This rewriting makes the conceptual goal clearer: The solver is rearranging the units in the top number to make them available for subtracting like units. Rewriting also prevents the ubiquitous top-from-bottom error because the solver fixes everything, ungrouping if necessary, before doing any subtracting. Doing the fixing in any direction allows children to think in their own way. The second major step is then to subtract the digits in every column, which the solver can do in any order.

The drawing at the bottom right of Figure 6.3 shows how a size drawing or size objects can support the two aspects of multidigit subtracting. Not enough ones, or tens, or hundreds are available to do the needed subtracting, so one larger unit is opened up to make 10 of the needed units. Students can do the subtraction from this 10, facilitating the "take from ten" single-digit subtraction method, or they can count up to find the difference in the written number problem.

The irregular structure of the English words between twenty and ninety-nine continues to present difficulties in multidigit problems, given that we perform all single-digit and multidigit calculation using the words as oral intermediaries for the written numbers but that these words do not show the tens in the numbers. Using English forms of the regular East Asian words ("1 ten 4 ones" for 14) along with the ordinary English number words has been reported to be helpful (Fuson, Smith, & Lo Cicero, 1997). This approach permits children to generalize single-digit methods meaningfully. For example, for 48 + 36, students can use their single-digit knowledge and think, "4 tens + 3 tens is 7 tens" rather than need to know the answer to "40 + 30 is ?"

Textbook and Curricular Issues

U.S. textbooks have several problematic features that complicate children's learning of multidigit addition and subtraction methods. The grade placement of topics is delayed compared with that of other countries (Fuson, Stigler, & Bartsch, 1988), and problems have one more digit each year so that this topic continues into Grade 5 or even Grade 6. In contrast, children complete multidigit addition and subtraction for large numbers in some countries by Grade 3. In the first grade in the United States, two-digit addition and subtraction problems with no regrouping ("carrying" or "borrowing") appear in textbooks, but problems requiring regrouping may not appear until almost a year later, in second grade. Problems with no regrouping lead children to make the most common errors, especially subtracting the smaller digit from the larger even when the larger digit is on the bottom (e.g., 72 – 38 = 46). This error is one major reason that on standardized tests only 38% of U.S. second graders give accurate answers to such problems as 72 – 38. Accessibility studies indicate that first graders can solve two-digit addition problems with trading if they can use drawings or quantity supports (Fuson, Smith, & Lo Cicero, 1997; Carpenter et al., 1998). Because knowing when to make 1 new ten is an excellent use of place-value knowledge, such problems can consolidate place-value ideas. Giving children from the beginning subtraction problems that require regrouping would help them understand the general nature of two-digit subtraction. Although educators might well delay this topic until second grade because children find two-digit subtraction much more difficult than addition, second graders learning with quantity supports and with a focus on understanding their methods can have high levels of success.

Textbooks or approaches characterized as "reform" may have different shortcomings than those of traditional textbooks. No study using a reform approach or focused on teaching for understanding has reported children's doing worse on multidigit or single-digit computation than children using traditional textbooks. Yet studies have reported that some children use unitary count-all multidigit strategies as late as Grades 3 and 4, suggesting insufficient attention to helping all children learn prerequisite counting and quantity understanding for effective methods using tens. A five-year longitudinal study following 20 classes of children using a reform textbook, *Everyday Mathematics* (Everyday Learning Corporation, 1988–1994), suggests other issues that educators must consider if U.S. children's multidigit performance is to improve above that of standard textbooks. Overall, achievement results were very positive (Carroll, Fuson, & Drueck, 2000; Carroll & Fuson, 1999): At every grade level, children who used *Everyday Mathematics* outperformed comparison groups using traditional U.S. textbooks on a wide range of topics. The only exception was in single-digit and multidigit addition and subtraction problems, in which children using *Everyday Mathematics* performed the same as comparison children using standard textbooks.

A focus group of teachers and researchers identified several attributes of the *Everyday Mathematics* curriculum, or its use in classrooms, that seemed responsible for the lower-than-desired multidigit performance. I summarize these attributes here because they indicate issues that educators may need to be address as classrooms move to teaching for understanding. They are relatively easy to avoid, are being addressed in *Everyday Mathematics* revisions, and can be addressed by how teachers use reform materials. Although most EM classrooms emphasized using alternative methods and explaining them, children were not using quantity supports except for the hundreds grid. The grid was usually used as a counting tool without the tens and ones explicitly indicated (i.e., the first addend was identified as the square containing 38, instead of the 38 being identified as the three rows of 10 squares and the 8 squares in the fourth row); and counting was done by rote with the vertical ten-jump rarely justified or explained. Most explanations of methods were solely verbal, so that children with less mathematics proficiency had difficulty following them. A few teachers wrote numbers on the board as children explained their thinking, but writing numbers for all problems and using quantity referents of some kinds (e.g., drawings on the board) would have made the explanations more accessible to all children.

No meaningful treatment of the standard algorithms was included in the lessons nor presented in most classrooms. Some students inevitably brought the standard algorithms from home, and teachers did not know how to help children explain them meaningfully. Furthermore, because of test pressures, some teachers taught standard algorithms right before standardized tests but without giving meaning to the algorithms.

Many of the multidigit lessons used contexts (e.g., temperature) to give real-world meanings to the numbers. Yet many lessons heavily focused on the context and insufficiently focused on the multidigit processes. Likewise, educators did not adequately emphasize having all children learn prerequisites for effective methods. Everyday Mathematics children did better on multidigit combinations in word-problem situations than in vertical columns. Their success on word problems is noteworthy, but the errors on column-form combinations indicate insufficient strength of the meanings of place-value quantities in the face of the single-digit appearance of the numbers. However, *Everyday Mathematics* did introduce the addition method on the right in Figure 6.3, and many children in some classrooms used and explained it effectively.

This review suggests central features for effective reform and traditional textbooks. Any algorithms included must be accessible to children and to teachers, and educators must support children in learning the algorithms with understanding. The research-based accessible methods in Figure 6.3 were included here to indicate algorithms that are more accessible than those presently appearing in most U.S. textbooks. Further, children need to use quantity supports in initial experiences with multidigit solving and multidigit algorithms so that they can learn these methods with meaning. Students and teachers need to use referents when discussing methods so that everyone can follow the discussion. Drawing quantities can be helpful in such discussions. Steps in algorithms need to be linked with the quantity supports.

Conclusion

Recent research clearly indicates that nontraditional approaches can help children carry out, understand, and explain methods of multidigit addition and subtraction rather than merely carry out a method. Children can accomplish this higher level of performance at earlier grades than those at which they presently must supply only answers. Features of classrooms engendering this higher level of performance are as follows:

- An emphasis on understanding and explaining methods
- Initial use by children of quantity supports or drawings showing the different sizes of ones, tens, and hundreds to give meanings to methods with numbers
- Sufficient time and support for children to develop meanings for numerical methods and for prerequisite understandings, which may be developed alongside the development of methods; and to negotiate and become more skilled with the complexities of multistep, multidigit methods

The research is not yet clear about which quantity supports, multidigit methods, or details of classroom functioning can maximize learning for all. The most effective approach at present seems to be to make learning algorithms more mathematical by considering it an important arena of mathematical pattern finding and invention that uses and contributes to robust understandings of the place-value system of written numeration. Meaningful discussion of various standard algorithms that children bring into the classroom from home (e.g., the subtraction algorithm widely used in Latin America and Europe; see Ron, 1998) has an important role. Seeking to discover why each algorithm works prompts excellent mathematical investigations. Also, educators must share accessible methods with children having less mathematics proficiency so that they acquire a comprehensible method that they can use. However, educators should focus instruction on children's understanding and explaining of mathematical concepts—not just on rote use. The three accessible methods in Figure 6.3 were invented by children but have also been shared with, and learned meaningfully by, many children, and other methods not yet discovered—or rediscovered—may prove even more powerful. Comparing methods to see how they address the domain's crucial issues facilitates everyone's reflection on the underlying conceptual and notational issues of that domain. This focus seems much more appropriate than others in the 21st century, in which new machine algorithms will be needed and new technology will require many people to learn complex multistep algorithmic processes. If the focus is accompanied by a continual focus on testing and teaching accessible methods as well as on fostering invention, all children should be able to learn and explain a multidigit addition and subtraction method as well as carry it out accurately.

Multidigit Multiplication and Division

Much less research is available on children's understandings of multidigit multiplication and division than on the operations already discussed. Educators have published sample teaching lessons (e.g., Lampert, 1986, 1992) and have explored alternative methods for accomplishing these operations (e.g., Carroll & Porter, 1998). Researchers have reported a preliminary learning progression of multidigit methods for third to fifth-grade classrooms in which teachers fostered children's invention of algorithms (Baek, 1998). These methods moved from (a) direct modeling with objects or drawings (i.e., by ones and by tens and ones), to (b) written methods involving repeatedly adding—sometimes by repeated doubling, a surprisingly effective method used historically, to (c) partitioning methods. The partitioning methods were partitioning with various partitions using numbers other than 10, partitioning one number into tens and ones, and partitioning both numbers into tens and ones.

Current and Accessible Methods

The multiplication and division algorithms currently most prevalent are complex embedded methods that are not easy to understand or to carry out (see the leftmost methods in Figure 6.4). They demand high levels of skill in multiplying a multidigit number by a single-digit number within complex embedded formats in which multiplying and adding alternate. In these algorithms, the meaning and scaffolding of substeps have been sacrificed to using a small amount of paper. The multiplication and division algorithms use aligning methods that keep the steps organized by correct place value without requiring any understanding of what is actually happening with the ones, tens, and hundreds.

Figure 6.4 presents modifications of these methods that clarify the meaning and purpose of each step. The separation of steps in each of these accessible methods also facilitates the linking of each step with the quantities involved. An array drawing shows the quantities; arrays are powerful models of multiplication and division. The accessible methods and drawings demonstrate central features in multidigit multiplication and division that students must come to understand and do.

Accessible Multiplication Methods

For multiplication, teachers may use an array-size model. Such a model provides initial support for the crucial understandings of the effects of multiplying by 1, 10, and 100. It also shows clearly how each of the tens and ones numbers in 46 and 68 are multiplied by each other and are then added after students have completed all multiplication operations. The sizes of the resulting squares or rectangles indicate the sizes of these various products and thus support understanding. As one looks across each row in the array, one can see in the top row 10×46 as 10×40 (four squares of 100) plus 10×6 (six columns of 10 each). Multiplying by 60 creates six such rows of 10 products, so multiplying by 60 is multiplying by 10 and then multiplying by 6. Then one sees eight rows of 1×46 as 1×40 and 1×6 (eight rows of each).

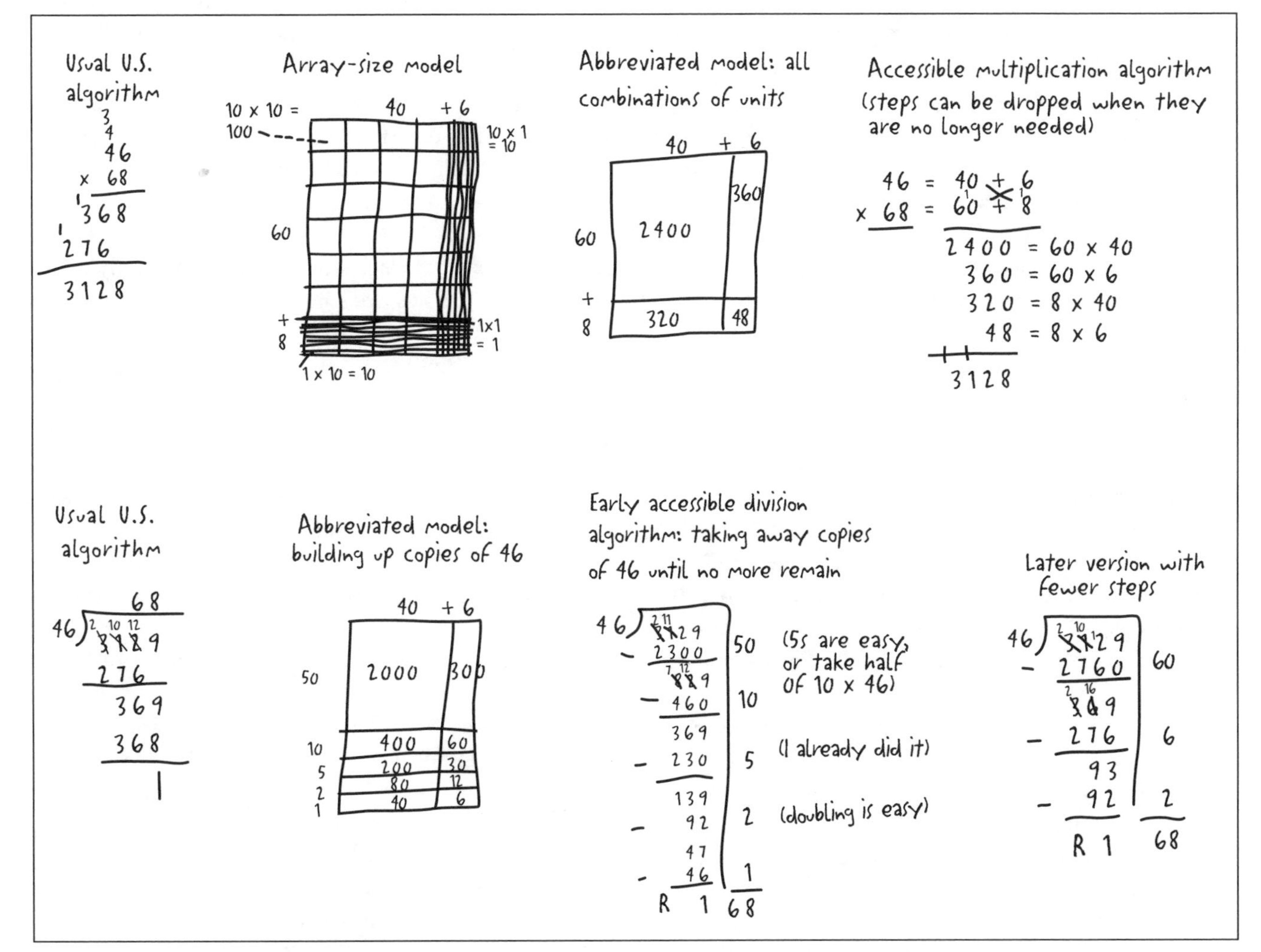

Figure 6.4. Multidigit multiplication and division.

Teachers can draw the abbreviated model shown in Figure 6.4 to summarize steps in multidigit multiplication. Its separation into tens and ones facilitates the necessary multiplication operations.

The accessible multiplication algorithm shown in the top right of Figure 6.4 is the fullest form with all possible supports. As students come to understand each aspect of multiplication, they can drop each of the supports, resulting in a streamlined version that is a simple expanded form of the usual U.S. method. Variations of the accessible algorithm have been widely used in research classrooms and in some innovative textbooks. Its main feature is a clear record of each of the four pairs of numbers (40×60, 40×8, 6×60, 6×8) that students need to multiply. The vertical and diagonal marks are a way for students to record as they go which numbers they have already multiplied. Unlike the current U.S. algorithm, which starts at the right and multiplies units first, the accessible algorithm begins at the left, as students prefer to do. This approach also has the advantage that the first product written is the largest, permitting all the smaller products to align easily under it in their correct places. Writing the factors at the side of each product emphasizes what students are actually doing in each step and permits an easy check. Writing out the separate products for 40×60 and 40×8 is much easier for students than doing the usual procedure: multiply 40×8, write part of the answer below and part above the problem, multiply 40×60, and then add the number written above the problem. The complex alternation of multiplying and adding in the usual algorithm is not necessary, is a source of errors, and obfuscates what students are actually doing in multidigit multiplying: multiplying each combination of units and adding all of them (see the abbreviated model). Students who understand and wish to drop steps in the accessible algorithm do so readily, with a result looking like the usual U.S. method except that it has four, instead of two, products to be add. These four can even be folded into two, if students wish. Therefore, the accessible model permits students to function at their own level of scaffolded understanding and helps them explain what they are doing.

The accessible algorithm also generalizes more readily to algebraic polynomial multiplication than the current U.S. algorithm does. The abbreviated drawing can show, for example, $2x + 3$ across the top and $y + 4$ along the bottom. The model, and students' previous experience with multidigit multiplication, then clarify that students just multiply each kind of unit in one number with each kind of unit in the other to find the product, or $2xy + 8x + 3y + 12$.

Multiplying by three-digit numbers is a simple extension of the two-digit version. After a conceptual development of the results of multiplying by 100 (i.e., numbers get two places larger, so they move left two places), abbreviated drawings can demonstrate the nine combinations of products that students need to find and add. Students can easily carry out the accessible algorithm for these larger numbers because it scaffolds the necessary steps. Given the accessibility of calculators, the amount of valuable school learning time that teachers should devote to such large multiplication problems is unclear. But teachers could easily introduce them in a conceptual fashion that then relates to estimating the product, especially when the largest product is found first, as in the accessible method shown in Figure 6.4.

Current and Accessible Division Methods

The usual U.S. division algorithm has two aspects that create difficulties for students. First, it requires them to determine exactly the maximum copies of the divisor that they can take from the dividend. This feature is a source of anxiety because students often have difficulty estimating exactly how many will fit. Students commonly multiply trial products off to the side until they find the exact one. Second, the current algorithm creates no sense of the size of the answers that students are writing; in fact, they are always multiplying by single digits. In the example in Figure 6.4, they just write a 6 above the line; they have no sense of 60 because they are literally only multiplying 46 by 6. Thus, students have difficulty gaining experience with estimating the correct order of magnitude of answers in division when they are using the current U.S. algorithm.

The accessible division method shown in Figure 6.4 facilitates safe underestimating. It builds students' experience with estimating and, later, their accurate assessment of calculator answers, because they multiply by the correct number (e.g., 60, not 6). The procedure is easy for those students who are still gaining mastery of single-digit multiplication because it permits the use of readily known products. For those who can manage it, the method can be abbreviated to be as brief as the current algorithm. Educators have used this accessible division algorithm in innovative materials since at least the 1960s.

The example of the accessible method given first in Figure 6.4 shows a solution that a student might do very early in division learning. Conceptually, the drawing and the written algorithm work together to show the meaning of long division: It is like a puzzle in which the solver takes away copies of the divisor (here, 46) until no further copies can be taken away. He or she is solving the equation "46 × ? = 3129" using the notion of division as the inverse of multiplication. The drawing shows these copies being added to make the total 3128 as 46 × 68 (remainder of 1), and the written algorithm subtracts each large copy as the solver keeps track of how many more still must be taken away. The drawing can scaffold the one-digit-by-two-digit multiplication necessary at each step: 50 × 46 is split into 50 × 40 = 2000 and 50 × 6 = 300 to make 2300. The scaffolding is important because this combination of multiplying and adding is complex for some students. The example shows that the student elected to multiply by 50 because fives facts are learned easily and accurately; one example is the nice repeating pattern for the 5s—the Gs—in the Figure 6.2 alphabet multiplication table. The student then sees that he or she can simply take away another 10 copies of 46. The student next cleverly uses a product that he or she has already found (50 × 46) to take away 5 copies of 46. Doubling is also easy, although many students would probably have multiplied by 3 at that point. Successive doubling represents the basis of the multiplication and division algorithms used historically in Europe. The right side of Figure 6.4 gives a version of the same problem that the same student might complete with more experience. At this point, the student may not need the drawing to scaffold the steps, meanings, or multiplication operations.

The algorithms for multiplication and division depend heavily on fluency with multiplication and addition, and in division, with multidigit subtraction. The difficulties that many students have in subtraction noticeably affect division, so understanding and fluency in multidigit subtraction are very important. Because students typically range substantially in their multiplication learning rate, many of them may not have achieved full fluency by the time their class is discussing multidigit multiplication and division. An advisable tactic is to give such students a multiplication table that they can use to check their multiplications as they go. This aid will permit them to keep up with the class and learn an algorithm. Furthermore, each verification of, or search for, a product in the table creates another learning trial for basic multiplication. Of course, presenting separate learning opportunities for multiplication combinations with which the student is not yet fluent would also be helpful.

How Much Consolidation Time?

How much valuable school mathematics time should be spent on multidigit multiplication and division is a question whose answer will probably need continual revision during the21st century. New goals will arise to compete with these domains, as they have already done. At present, time is well spent on conceptual and accessible approaches that facilitate students' understanding of how to build multidigit multiplication and division from the central concepts of place value and basic multiplication combinations. During that time, students could also bring those combinations to mastery. Drilling for long periods on problems involving large numbers seems a goal more appropriate to the 20th than to the 21st century.

General Issues in Achieving Computational Fluency

Curricular Issues

The U.S. curriculum has been characterized as "underachieving" and was recently characterized in the TIMSS international study as "a mile wide and an inch deep" (McKnight, Crosswhite, Dossey, Kifer, Swafford, Travers, & Cooney, 1989; McKnight & Schmidt, 1998; Peak, 1996). Countries whose students score high in international studies select vital grade-level topics and devote enough time for students to gain initial understanding and mastery. In the United States, no teacher and no grade level are responsible for a given topic. Educators distribute such topics as multidigit computations over several years, doing problems one digit larger each year. They devote large amounts of time at the beginning of each year and each new topic to teach concepts and operations that many students have not learned or learned incorrectly the year before.

Helping students build initial correct computational methods, however, is much easier than correcting errors. For example, second graders using base-ten blocks for initial learning of multidigit addition and subtraction explained answers and achieved high levels of accuracy that were maintained over time (Fuson, 1986a; Fuson & Briars, 1990). Older students who had been making subtraction errors for years seemingly did learn in one session with base-ten blocks to correct their errors, but many later regressed to their old errors (Resnick & Omanson, 1987). Carefully designed practice and help during learning are among the important aspects for computational fluency. But the most severe problem at this point is helping students learn in a timely fashion any correct general method that they understand. Such initial learning must be deep and accurate. Only with understanding can students reduce interference from later similar notations and methods.

Helpful Instructional Phases

What features of classrooms contribute to computational fluency? A recent review of the literature contrasts the many studies that showed an experimental instructional method to be superior to a traditional control method (Dixon et al., 1998). The less effective traditional methods involved two phases: the teacher's presentation of a topic, with students passively observing, followed by students' independent practice of that topic, with or without teacher monitoring, feedback, and so forth. In contrast, students achieved superior learning through effective methods that had three phases. First, teachers involved students in the introduction of the topic through explanations, questions, and discussion; students were active learners whose initial knowledge about a domain was elicited. Second, a long period transpired in which students were helped to move from teacher-regulated to self-regulated solution processes. Teachers structured a significant period of help that was gradually phased out. This help was given in different ways: by scaffolded problems and visual or other supports, by peers, and by the teacher or aides. During this sustained helping period, students received feedback on their performance; got corrective help so that they did not practice errors; and received, and often gave, explanations. The third phase of effective instruction was a brief assessment of students' ability to apply knowledge to untaught problems when working independently. Such independent work might then be distributed over time. Other relevant results from studies that Dixon and fellow researchers reviewed were that various kinds of strategy instruction were superior to not giving such instruction; working fewer problems in depth was more effective than working more problems quickly; writing as well as solving problems was helpful; and solving concept examples sequenced for generalization and discrimination was beneficial.

The implications for computational fluency of these results are that all students had sustained, supported time to learn a given domain deeply and accurately. Such deep, sustained, accurate learning over time is necessary for complex domains requiring multistep solution methods. Students need to learn the central principles of a domain (e.g., in multidigit addition and subtraction, that one adds or subtracts like multiunits), the overall shape of a given method, and the detailed steps of the method; and they need to interweave this developing knowledge so that it operates fluidly and accurately. These learning prerequisites are true whether students invent the method or learn it from other students or from the teacher. Practice is important, but effective practice is supported by monitoring and help that are focused on doing and understanding. In contrast, *drill and practice* frequently carries the connotation of rote practice; has little sense of monitoring or feedback; and carries no connotation of helping or of visual, conceptual, psychological, or motivational support for learning throughout the practicing phase.

At present, we do not know enough about effective ways to orchestrate the helping period to deliver feedback and help to students as they need it, especially in classrooms where students are using different methods. Given the heterogeneity of most classrooms, such orchestration is difficult. Designing and testing effective helping methods and effective ways to give feedback on answers are vital areas for further research, as are ascertaining how to facilitate peer helping, given that peer helping or cooperative learning frequently but not always results in better learning. Some methods used in other school subjects, such as jigsaw methods in which each group member is given different knowledge to contribute, have been difficult to use in mathematics.

A textbook issue that at present interferes with the more effective three-phase method—and even with effective teacher presentation of topics in the less effective two-phase traditional approach—is the common misuse of art (e.g., photographs, drawings, cartoons, etc.) in U.S. mathematics textbooks. In many other countries, textbook art is designed to support conceptual thinking. In the United States, textbook art frequently distracts from conceptual understanding because it is irrelevant or overwhelmingly busy.

Helping Diverse Learners

A related review of literature concerning school success of diverse learners (Kameenui & Carnine, 1998) identified six crucial aspects of teaching and of learning materials: (a) structuring around big ideas, (b) teaching conspicuous strategies, (c) priming background knowledge, (d) using mediated scaffolding (e.g., peer tutoring, feedback about thinking, visual supports with cues for correct methods), (e) using strategic integration (i.e., integration into complex applications to provide distributed practice in more complex situations), and (f) designing judicious review. Diverse learners are those who may experience learning difficulties because of low-income backgrounds, language deficiencies, or other reasons.

In the first crucial aspect of helping diverse learners—structuring around big ideas—an absolutely necessary component is sufficient time for students to deeply learn the core concepts of that grade level. This requirement is the antithesis of the present "mile wide and inch deep"

U.S. curriculum. This issue must be resolved for diverse learners to attain computational fluency. The next three aspects specify components of the initial active learning phase and the helping phase in the three-phase effective teaching model outlined previously.

Using strategic integration and designing judicious review are aspects of computational fluency that follow deep and effective initial learning in a domain. Strategic integration of various computational methods into moderately complex problems increases problem-solving competence by increasing the range of situations in which students use that computational method. It also affords distributed practice of the method, one of the most effective kinds of practice.

Judicious review is defined as being plentiful, distributed, cumulative, and appropriately varied. It follows initial deep learning. Distributed and monitored practice requires working one or two examples occasionally, with immediate help for students getting wrong answers. This practice is important, even after successful meaningful learning, because the nonsupportive or misleading mathematical words or notations in many domains continually suggest wrong methods (e.g., *adding* the top and the bottom numbers when adding fractions). Furthermore, many computational domains are similar, and learning new domains creates interference with old domains (e.g., one does *multiply* the tops and bottoms of fractions). Therefore, after deep and successful initial learning, distributed practice of a few problems of a given kind reveals whether errors are creeping in. Frequently, helping students correct their methods is as simple as suggesting that they remember original supports. For example, as some errors crept into multidigit methods learned with base-ten blocks, asking students to "think about the blocks" was sufficient for them to correct their own errors in subtracting with zeroes in the top number (Fuson, 1986a).

The research of Knapp and associates (Knapp, 1995; Zucker, 1995) on attributes of successful techniques in high-poverty classrooms underscores these results. They found that a balance between conceptual understanding and skills practice resulted in higher computational and problem-solving performance by lower achieving and higher achieving students. Successful teachers supported conceptual understanding by focusing students' attention on alternative solution methods, not just answers; elicited thinking and discussion about solution methods; used multiple representations and real-life situations to facilitate meaning-making; and modeled ways to probe the meaning of mathematical problems or methods. These teachers also supplied a "healthy dose" of skills practice.

Individual Differences

As in other school subjects, substantial differences in students' mathematics achievement are related to social class and ethnicity (e.g., Ginsburg & Russell, 1981; Secada, 1992). Kerkman and Siegler (1993) found that low-income children had less practice in solving problems and that they executed strategies less well. Strategy instruction and monitored practice were therefore recommended for such students.

Individual differences as early as first grade cut across gender and income levels to differentiate children into what Siegler (1988) has termed good students, not-so-good students, and perfectionists. Roughly half the not-so-good students were then identified as having mathematical disabilities by fourth grade, versus none in the other groups. On single-digit addition tasks, these students used more primitive methods and produced more errors on problems on which they could have used, but did not use, more accurate but effortful strategies (e.g., counting with fingers). Thus, these students produced incorrect answers more often, thereby creating responses that competed with their experiences of correct answers. Siegler's model for learning single-digit addition emphasizes the importance of avoiding error generation because it interferes with students' efforts to remember the correct answer. Thus, these studies underscore the importance of feedback and immediate help. Perfectionists and good students had similar long-term outcomes, but the perfectionists were much more likely to use slower and effortful methods even on simpler problems than the good students were. This finding emphasizes that methods of practice should facilitate students' understanding of their own growth and progress rather than lead to comparing individuals. Educators should also vary practice so that sometimes speed is important but at other times, using a method in a complex situation is important. An overemphasis on either could lead to rigidity rather than computational fluency.

Less research has been done on mathematics disability than on reading disability, especially with younger children. Researchers have identified different kinds of mathematics disability, and Geary's (1994) review identifies four types and recommends different kinds of learning supports for each kind. Students with semantic memory disabilities have difficulty with verbal, and especially phonetic, memory, but many have normal visuospatial skills. These students have great difficulty memorizing basic computations because these computations rely on a phonetic code. Therefore, instructional supports that use visual rather than phonetic cues and practice with strategies for basic calculations are recommended for these students. Students with procedural deficits use less advanced methods than their peers. Although many eventually catch up, this long period of using primitive methods may be detrimental. Such children do not seem to invent more advanced methods as readily as their peers do. Therefore, conceptually based strategy instruction that helps these children use and understand more advanced strategies, such as counting on, can be helpful. Students with visuospatial disabilities have difficulties with concepts that use spatial representations, such as place value. Research is not clear about the developmental prognosis of such children, but suggested methods of remediation involve using extra cues to support students' visual processing. Because directionality is a special problem with such students, the accessible methods that can be carried out in either direction might be especially helpful for them. Difficulties with mathematical problem solving that go beyond arithmetic deficits also characterize some students. Such problem-solving supports as drawing the problem situation, discussed previously in this chapter, are potentially useful for these students. Technology may also generate complex problem-solving situations that are nevertheless accessible to students with disabilities in mathematics (Goldman, Hasselman, & the Cognition and Technology Group at Vanderbilt, 1997).

Although one might think that students identified as learning disabled in mathematics might need special learning situations, the recommendations for all types of disabilities in mathematics summarized by Geary (1994, p. 285), one of the most prominent researchers in that field, sound like a summary of the results in this chapter. Thus, the kinds of teaching recommended by research for helping all students progress to general computational fluency may especially help students with mathematical disabilities. Such students tend to lag behind and become discouraged and therefore could greatly benefit from accessible methods that could help them learn more quickly and easily.

Teaching to Prepare Students for Rational Numbers

Teaching and learning with whole numbers can lay an adequate foundation for later work with rational numbers, including decimal and ordinary fractions, or it can make such work more difficult. Presently, students make many errors in decimal and ordinary fractions because they incorrectly generalize concepts from whole numbers (e.g., Hiebert & Wearne, 1986; Resnick, Nesher, Leonard, Magone, Omanson, & Peled, 1989). Approaching all domains with a focus on the meanings of the notations, and with explicit consideration of which results do and do not generalize, could improve students' competence in these advanced domains. Deep comprehension of place value and of the regular ten-for-one trades to the left as numbers get larger can help students understand decimal fractions as regular one-for-ten trades to the right as quantities get smaller. Understanding multidigit addition and subtraction as adding or subtracting like quantities (e.g., ones to ones, tens to tens, etc.) can facilitate the related understanding that adding or subtracting decimal fractions or ordinary fractions must also involve adding or subtracting like quantities (e.g., for decimal fractions, tenths to tenths or hundredths to hundredths; and for fractions, fourths to fourths or thirds to thirds). Deeply understanding quantities in fractions and decimal fractions is necessary to overcome meanings suggested by whole-number notation (e.g., that $0.25 > 0.3$ because $25 > 3$). Similarly, students must reassess the whole-number notions that "multiplication makes larger" and "division makes smaller" when they multiply and divide fractions. If whole-number knowing and doing has been a sense-making process intertwined with problem solving and explaining one's thinking, students can more easily make the necessary extensions and adjustments to their whole-number knowledge as they enter these more advanced domains.

Achieving Mathematical Power in Whole-Number Operations

The reform approach as outlined in the first round of the NCTM *Standards* documents (1989, 1991) stimulated much action in the United States and Canada and contributed to a broader view of mathematics learning and teaching. As is inevitable with such documents, however, the approach was sometimes misunderstood and distorted in ways that are counterproductive to good mathematics teaching and learning. Indeed, some individuals have mistakenly characterized "reform math teaching" as extensive unfocused, meandering discussion; mathematical content restricted to children's current knowledge and interests; real-world contexts or activities in which the mathematical content is not clear or is so complex that little mathematical learning occurs; the absence of teacher input of information of any kind, including standard mathematical vocabulary or notation; and prolonged periods of "invention" of solution methods in which children struggling with mathematics use primitive methods rather than build prerequisite knowledge for more advanced methods or be helped to learn such methods.

NCTM's *Principles and Standards for School Mathematics* (2000) attempts to clarify and correct such misunderstanding. This document instead recommends, and the research literature supports, ambitious mathematical goals; teacher-led and monitored discussion that focuses on central mathematical ideas; teachers' explaining and clarifying as well as children's explaining and clarifying; using and building on children's knowledge but extending that knowledge in mathematically important ways; and using carefully chosen real-world contexts as well as carefully designed pedagogical learning supports (e.g., selected manipulatives or drawings) to facilitate all children's meaning-building. Teachers play vital roles in helping children build initial understanding and in supporting their achievement of computational fluency and mathematical power. Researchers have produced considerably less work on such productive teacher roles than on student understanding, errors, and methods (but see, e.g., Fraivillig, Murphy, & Fuson, 1999; Fuson & Burghardt, in press; Hiebert et al., 1997; Sherin, 2002; Simon, 1995; Stipek, Salmon, Givvin, Kazemi, Saxe, & MacGyvers, 1998). As this body of research grows, teachers will have more detailed guides to developing mathematical power in all their students.

Meanwhile, the existing research gives substantial direction for improving students' mathematical capabilities. All students require a constant intertwining of understanding and doing—of building meaning, of problem solving, and of computing. Research indicates that in single-digit addition and subtraction, children around the world progress from simple methods with objects through a series of more rapid methods. Educators can help children progress through these methods to powerful and rapid general methods. Multiplication and division involve different patterns for different numbers, and students progress through a learning path of more rapid methods. Many different algorithms (i.e., general methods) exist for solving multidigit addition, subtraction, multiplication, and division problems. Research and analysis have identified some methods that are easy both to understand and to carry out; they relate to the methods commonly taught in the United States and Canada but are conceptually more powerful or easier to use. Students in the United States and Canada can learn to understand and explain computational methods if these methods are approached as sense-making endeavors. Practice is important, as is learning prerequisite knowledge that facilitates acquiring more advanced methods. Likewise, educators can call on problem solving from the beginning to give meaning to computations; then they can continually intertwine the two as methods and problem solving become consolidated.

The new research-based view of achieving computational fluency is a more complex and connected view than the past linear view consisting of counting, memorizing facts, solving problems, learning algorithms, and then solving problems with those algorithms. However, a new, more complex view is necessary to achieve the new, more complex goals of mathematics learning and teaching necessary for the 21st century. A new kind of computational fluency is needed for the challenges and changes that individuals in the United States and Canada will face in the years to come.

ACKNOWLEDGMENTS

The following reviews and summaries of the literature are used extensively in this paper: Baroody and Coslick (1998); Baroody and Ginsburg (1986); Bergeron and Herscovics (1990); Brophy (1997); Carpenter and Moser (1984); Cotton (1995); Davis (1984); Dixon, Carnine, Kameenui, Simmons, Lee, Wallin, and Chard (1998); Fuson (1992a, 1992b); Geary (1994); Ginsburg (1984); Greer (1992); Hiebert (1986, 1992); Hiebert and Carpenter (1992); Lampert (1992); Nesher (1992); Resnick (1992); and Siegler (this volume). These works include reviews carried out by experts in mathematics education, cognitive science, learning disabilities, special education, educational psychology, and developmental psychology. The reviews in Leinhardt, Putnam, and Hattrup (1992) were written especially for teachers and other educational leaders; they also include analyses of textbook approaches to teaching. To avoid excessive citations, results that are strong, salient, and clear in these reviews and summaries have not been cited separately. More specialized results are cited.

REFERENCES

Anghileri, J. (1989). An investigation of young children's understanding of multiplication. *Educational Studies in Mathematics, 20,* 367–385.

Ashlock, R. B. (1998). *Error patterns in computation.* Upper Saddle River, NJ: Prentice-Hall.

Baek, J.-M. (1998). Children's invented algorithms for multidigit multiplication problems. In L. J. Morrow & M. J. Kenney (Eds.), *The teaching and learning of algorithms in school mathematics* (pp.151–160). Reston, VA: National Council of Teachers of Mathematics.

Baroody, A. J., & Coslick, R. T. (1998). *Fostering children's mathematical power: An investigative approach to K–8 mathematics instruction.* Mahwah, NJ: Erlbaum.

Baroody, A. J., & Ginsburg, H. P. (1986). The relationship between initial meaningful and mechanical knowledge of arithmetic. In J. Hiebert (Ed.), *Conceptual and procedural knowledge: The case of mathematics* (pp. 75–112). Hillsdale, NJ: Erlbaum.

Beishuizen, M. (1993). Mental strategies and materials or models for addition and subtraction up to 100 in Dutch second grades. *Journal for Research in Mathematics Education, 24,* 294–323.

Beishuizen, M., Gravemeijer, K. P. E., & van Lieshout, E. C. D. M. (Eds.). (1997). *The role of contexts and models in the development of mathematical strategies and procedures* (pp. 163–198). Utretcht: CD-B Press/Freudenthal Institute.

Bergeron, J. C., & Herscovics, N. (1990). Psychological aspects of learning early arithmetic. In P. Nesher & J. Kilpatrick (Eds.), *Mathematics and cognition: A research synthesis by the International Group for the Psychology of Mathematics Education.* Cambridge: Cambridge University Press.

Brophy, J. (1997). Effective instruction. In H. J. Walberg & G. D. Haertel (Eds.), *Psychology and educational practice* (pp. 212–232). Berkeley, CA: McCutchan.

Brownell, W. A. (1987). *AT* Classic: Meaning and skill—Maintaining the balance. *Arithmetic Teacher, 34*(8), 18–25. (Original work published 1956)

Carpenter, T. P., Franke, M. L., Jacobs, V., & Fennema, E. (1998). A longitudinal study of invention and understanding in children's multidigit addition and subtraction. *Journal for Research in Mathematics Education, 29,* 3–20.

Carpenter, T. P., & Moser, J. M. (1984). The acquisition of addition and subtraction concepts in grades one through three. *Journal for Research in Mathematics Education, 15,* 179–202.

Carroll, W. M., & Fuson, K. C. (1999). *Achievement results for fourth graders using the Standards-based curriculum Everyday Mathematics.* Unpublished manuscript, Northwestern University.

Carroll, W. M., Fuson, K. C., & Drueck, J. V. (2000). Achievement results for second and third graders using the Standards-based curriculum *Everyday Mathematics. Journal for Research in Mathematics Education, 31,* 277–295.

Carroll, W. M., & Porter, D. (1998). Alternative algorithms for whole-number operations. In L. J. Morrow & M. J. Kenney (Eds.), *The teaching and learning of algorithms in school mathematics* (pp. 106–114). Reston, VA: National Council of Teachers of Mathematics.

Cotton, K. (1995). *Effective schooling practices: A research synthesis.* Portland, OR: Northwest Regional Lab.

Davis, R. B. (1984). *Learning mathematics: The cognitive science approach to mathematics education.* Norwood, NJ: Ablex.

Dixon, R. C., Carnine, S. W., Kameenui, E. J., Simmons, D. C., Lee, D-S., Wallin, J., & Chard, D. (1998). *Executive summary: Report to the California State Board of Education: Review of high quality experimental research.* Eugene, OR: National Center to Improve the Tools of Educators.

Everyday Learning Corporation. (1988–1994). *Everyday Mathematics* (1st ed.). Chicago: Author.

Fraivillig, J. L., Murphy, L. A., & Fuson, K. C. (1999). Advancing children's mathematical thinking in *Everyday Mathematics* reform classrooms. *Journal for Research in Mathematics Education, 30,* 148–170.

Fuson, K. C. (1986a). Roles of representation and verbalization in the teaching of multi-digit addition and subtraction. *European Journal of Psychology of Education, 1,* 35–56.

Fuson, K. C. (1986b). Teaching children to subtract by counting up. *Journal for Research in Mathematics Education, 17,* 172–189.

Fuson, K. C. (1990). Conceptual structures for multiunit numbers: Implications for learning and teaching multidigit addition, subtraction, and place value. *Cognition and Instruction,* 7, 343–403.

Fuson, K. C. (1992a). Research on learning and teaching addition and subtraction of whole numbers. In G. Leinhardt, R. T. Putnam, & R. A. Hattrup (Eds.), *The analysis of arithmetic for mathematics teaching* (pp. 53–187). Hillsdale, NJ: Erlbaum.

Fuson, K. C. (1992b). Research on whole number addition and subtraction. In D. Grouws (Ed.), *Handbook of research on mathematics teaching and learning* (pp. 243–275). New York: Macmillan.

Fuson, K. C., & Briars, D. J. (1990). Base-ten blocks as a first- and second-grade learning/teaching approach for multidigit addition and subtraction and place-value concepts. *Journal for Research in Mathematics Education, 21,* 180–206.

Fuson, K. C., & Burghardt, B. H. (1993). Group case studies of second graders inventing multidigit addition procedures for base-ten blocks and written marks. In J. R. Becker & B. J. Pence (Eds.), *Proceedings of the 15th Annual Meeting of the North American Chapter of the International Group for the Psychology of Mathematics Education* (pp. 240–246). San Jose, CA: San Jose State University, Center for Mathematics and Computer Science Education.

Fuson, K. C., & Burghardt, B. H. (1997). Group case studies of second graders inventing multidigit subtraction methods. In J. A. Dossey, J. O. Swafford, M. Parmantie, & A.E. Dossey (Eds.), *Proceedings of the 19th Annual Meeting of the North American Chapter of the International Group for the Psychology of Mathematics Education* (Vol. 1, pp. 291–298). Columbus, OH: ERIC Clearinghouse for Science, Mathematics, and Environmental Education.

Fuson, K. C., & Burghardt, B. H. (in press). Multi-digit addition and subtraction methods invented in small groups and teacher support of problem solving and reflection. In A. Baroody & A. Dowker (Eds.), *The development of arithmetic concepts and skills: Constructing adaptive expertise.* Hillsdale, NJ: Erlbaum.

Fuson, K. C. & Fuson, A. M. (1992). Instruction to support children's counting on for addition and counting up for subtraction. *Journal for Research in Mathematics Education, 23,* 72–78.

Fuson, K. C. & Kwon, Y. (1992). Korean children's understanding of multidigit addition and subtraction. *Child Development, 63*, 491–506.

Fuson, K. C., Perry, T., & Kwon, Y. (1994). Latino, Anglo, and Korean children's finger addition methods. In J. E. H. van Luit (Ed.), *Research on learning and instruction of mathematics in kindergarten and primary school* (pp. 220–228). Doetinchem, Netherlands, & Rapallo, Italy: Graviant.

Fuson, K. C., Perry, T., & Ron, P. (1996). Developmental levels in culturally different finger methods: Anglo and Latino children's finger methods of addition. In E. Jakubowski, D. Watkins, & H. Biske (Eds.), *Proceedings of the 18th Annual Meeting of the North American Chapter for the Psychology of Mathematics Education* (Vol. 2, pp. 347–352). Columbus, OH: ERIC Clearinghouse for Science, Mathematics, and Environmental Education.

Fuson, K. C., & Secada, W. G. (1986). Teaching children to add by counting on with finger patterns. *Cognition and Instruction, 3*, 229–260.

Fuson, K. C., & Smith, S. T. (1997). Supporting multiple 2-digit conceptual structures and calculation methods in the classroom: Issues of conceptual supports, instructional design, and language. In M. Beishuizen, K. P. E. Gravemeijer, & E. C. D. M. van Lieshout (Eds.), *The role of contexts and models in the development of mathematical strategies and procedures* (pp. 163–198). Utretcht, The Netherlands: CD-B Press/Freudenthal Institute.

Fuson, K. C., Smith, S. T., & Lo Cicero, A. (1997). Supporting Latino first graders' ten-structured thinking in urban classrooms. *Journal for Research in Mathematics Education, 28*, 738–760.

Fuson, K. C., Stigler, J., & Bartsch, K. (1988). Grade placement of addition and subtraction topics in Japan, mainland China, the Soviet Union, Taiwan, and the United States. *Journal for Research in Mathematics Education, 19*, 449–456.

Fuson, K. C., Wearne, D., Hiebert, J., Murray, H., Human, P., Olivier, A., Carpenter, T., & Fennema, E. (1997). Children's conceptual structures for multidigit numbers and methods of multidigit addition and subtraction. *Journal for Research in Mathematics Education, 28*, 130–162.

Geary, D. C. (1994). *Children's mathematical development: Research and practical applications*. Washington, DC: American Psychological Association.

Ginsburg, H. P. (1984). *Children's arithmetic: The learning process*. New York: Van Nostrand.

Ginsburg, H. P., & Allardice, B. S. (1984). Children's difficulties with school mathematics. In B. Rogoff & J. Lave (Eds.), *Everyday cognition: Its development in social contexts* (pp. 194–219). Cambridge, MA: Harvard University Press.

Ginsburg, H. P., & Russell, R. L. (1981). Social class and racial influences on early mathematical thinking. *Monographs of the Society for Research in Child Development, 44*(6), Serial No. 193.

Goldman, S. R., Hasselbring, T. S., & the Cognition and Technology Group at Vanderbilt. (1997, March/April). Achieving meaningful mathematics literacy for students with learning disabilities. *Journal of Learning Disabilities, 30*(2), 198–208. Reprinted in D. P. Rivera (Ed.) (1998), *Mathematics Education for Students with Learning Disabilities* (pp. 237–254). Austin, TX: Pro-Ed.

Goldman, S. R., Pellegrino, J. W., & Mertz, D. L. (1988). Extended practice of basic addition facts: Strategy changes in learning disabled students. *Cognition and Instruction, 5*, 223–265.

Greer, B. (1992). Multiplication and division as models of situation. In D. Grouws (Ed.), *Handbook of research on mathematics teaching and learning* (pp. 276–295). New York: Macmillan.

Hamann, M. S., & Ashcraft, M. H. (1986). Textbook presentations of the basic addition facts. *Cognition and Instruction, 3*, 173–192.

Hiebert, J. (Ed.). (1986). *Conceptual and procedural knowledge: The case of mathematics*. Hillsdale, NJ: Erlbaum.

Hiebert, J. (1992). Mathematical, cognitive, and instructional analyses of decimal fractions. In G. Leinhardt, R. T. Putnam, & R. A. Hattrup (Eds.), *The analysis of arithmetic for mathematics teaching* (pp. 283–322). Hillsdale, NJ: Erlbaum.

Hiebert, J., & Carpenter, T. P. (1992). Learning and teaching with understanding. In D. Grouws (Ed.), *Handbook of research on mathematics teaching and learning* (pp. 65–97). New York: Macmillan.

Hiebert, J., Carpenter, T., Fennema, E., Fuson, K. C., Wearne, D., Murray, H., Olivier, A., & Human, P. (1997). *Making sense: Teaching and learning mathematics with understanding*. Portsmouth, NH: Heinemann.

Hiebert, J., & Wearne, D. (1986). Procedures over concepts: The acquisition of decimal number knowledge. In J. Hiebert (Ed.), *Conceptual and procedural knowledge: The case of mathematics* (pp. 199–223). Hillsdale, NJ: Erlbaum.

Kameenui, E. J., & Carnine, D. W. (Eds.). (1998). *Effective teaching strategies that accommodate diverse learners*. Upper Saddle River, NJ: Prentice-Hall.

Kerkman, D. D., & Siegler, R. S. (1993). Individual differences and adaptive flexibility in lower-income children's strategy choices. *Learning and Individual Differences, 5*, 113–136.

Knapp, M. S. (1995). *Teaching for meaning in high-poverty classrooms*. New York: Teachers College Press.

Lampert, M. (1986). Knowing, doing, and teaching multiplication. *Cognition and Instruction, 3*, 305–342.

Lampert, M. (1992). Teaching and learning long division for understanding in school. In G. Leinhardt, R. T. Putnam, & R. A. Hattrup (Eds.), *The analysis of arithmetic for mathematics teaching* (pp. 221–282). Hillsdale, NJ: Erlbaum.

LeFevre, J., Kulak, A. G., & Bisantz, J. (1991). Individual differences and developmental change in the associative relations among numbers. *Journal of Experimental Child Psychology, 52*, 256–274.

LeFevre, J., & Liu, J. (1997). The role of experience in numerical skill: Multiplication performance in adults from Canada and China. *Mathematical Cognition, 3*, 31–62.

Leinhardt, G., Putnam, R. T., & Hattrup, R. A. (Eds.). (1992). *The analysis of arithmetic for mathematics teaching*. Hillsdale, NJ: Erlbaum.

Lemaire, P., Barrett, S. E., Fayol, M., & Abdi, H. (1994). Automatic activation of addition and multiplication facts in elementary school children. *Journal of Experimental Child Psychology, 57*, 224–258.

Lemaire, P., & Siegler, R. S. (1995). Four aspects of strategic change: Contributions to children's learning of multiplication. *Journal of Experimental Psychology, 124*(1), 83–97.

McClain, K., Cobb, P., & Bowers, J. (1998). A contextual investigation of three-digit addition and subtraction. In L. J. Morrow & M. J. Kenney (Eds.), *The teaching and learning of algorithms in school mathematics* (pp.141–150). Reston, VA: National Council of Teachers of Mathematics.

McKnight, C. C., Crosswhite, F. J., Dossey, J. A., Kifer, E., Swafford, J. O., Travers, K. T., & Cooney, T. J. (1989). *The underachieving curriculum: Assessing U. S. school mathematics from an international perspective.* Champaign, IL: Stipes.

McKnight, C. C., & Schmidt, W. H. (1998). Facing facts in U.S. science and mathematics education: Where we stand, where we want to go. *Journal of Science Education and Technology,* 7(1), 57–76.

Miller, K. F., & Paredes, D. R. (1990). Starting to add worse: Effects of learning to multiply on children's addition. *Cognition, 37*, 213–242.

Mulligan, J., & Mitchelmore, M. (1997). Young children's intuitive models of multiplication and division. *Journal for Research in Mathematics Education, 28*, 309–330.

National Council of Teachers of Mathematics. (1989). *Curriculum and evaluation standards for school mathematics.* Reston, VA: Author.

National Council of Teachers of Mathematics. (1991). *Professional standards for teaching mathematics.* Reston, VA: Author.

National Council of Teachers of Mathematics. (2000). *Principles and starndards for school mathematics.* Reston, VA: Author.

Nesher, P. (1992). Solving multiplication word problems. In G. Leinhardt, R. T. Putnam, & R. A. Hattrup (Eds.), *The analysis of arithmetic for mathematics teaching* (pp. 189–220). Hillsdale, NJ: Erlbaum.

Peak, L. (1996). *Pursuing excellence: A study of the U.S. eighth-grade mathematics and science teaching, learning, curriculum, and achievement in an international context.* Washington, D.C.: National Center for Educational Statistics.

Resnick, L. (1992). From protoquantities to operators: Building mathematical competence on a foundation of everyday knowledge. In G. Leinhardt, R. T. Putnam, & R. A. Hattrup (Eds.), *The analysis of arithmetic for mathematics teaching* (pp. 373–429). Hillsdale, NJ: Erlbaum.

Resnick, L. B., Nesher, P., Leonard, F., Magone, M., Omanson, S., & Peled, I. (1989). Conceptual bases of arithmetic errors: The case of decimal fractions. *Journal for Research in Mathematics Education, 20*, 8–27.

Resnick, L. B., & Omanson, S. F. (1987). Learning to understand arithmetic. In R. Glaser (Ed.), *Advances in instructional psychology* (Vol. 3, pp. 41–95). Hillsdale, NJ: Erlbaum.

Ron, P. (1998). My family taught me this way. In L. J. Morrow & M. J. Kenney (Eds.), *The teaching and learning of algorithms in school mathematics* (pp. 115–119). Reston, VA: National Council of Teachers of Mathematics.

Saxe, G. B. (1982). Culture and the development of numerical cognition: Studies among the Oksapmin of Papua New Guinea. In C. J. Brainerd (Ed.), *Progress in cognitive development research: Vol. 1. Children's logical and mathematical cognition* (pp. 157–176). New York: Springer-Verlag.

Secada, W. G. (1992). Race, ethnicity, social class, language, and achievement in mathematics. In D. Grouws (Ed.), *Handbook of research on mathematics teaching and learning* (pp. 623–660). New York: Macmillan.

Sherin, M. G. (2002). A balancing act: Developing a discourse community in a mathematics classroom. *Journal of Mathematics Teacher Education, 5*, 205–233.

Siegler, R. S. (1988). Individual differences in strategy choices: Good students, not-so-good students, and perfectionists. *Child Development, 59*, 833–851.

Simon, M.A. (1995). Reconstructing mathematics pedagogy from a constructivist perspective. *Journal for Research in Mathematics Education, 26*, 114–145.

Steffe, L. (1994). Children's multiplying schemes. In G. Harel & J. Confrey (Eds.), *The development of multiplicative reasoning in the learning of mathematics* (pp. 3–39). Albany: State University of New York Press.

Steffe, L. P., Cobb, P., & von Glasersfeld, E. (1988). *Construction of arithmetical meanings and strategies.* New York: Springer-Verlag.

Stigler, J. W., Fuson, K. C., Ham, M., & Kim, M. S. (1986). An analysis of addition and subtraction word problems in American and Soviet elementary mathematics textbooks. *Cognition and Instruction, 3*, 153–171.

Stipek, D., Salmon, J. M, Givvin, K. B., Kazemi, E., Saxe, G., & MacGyvers, V. L. (1998). The value (and convergence) of practices suggested by motivation research and promoted by mathematics education reformers. *Journal for Research in Mathematics Education, 29*, 465–488.

Thornton, C. A., Jones, G. A., & Toohey, M. A. (1983). A multisensory approach to thinking strategies for remedial instruction in basic addition facts. *Journal for Research in Mathematics Education, 14*, 198–203.

Zucker, A. A. (1995). Emphasizing conceptual understanding and breadth of study in mathematics instruction. In M. S. Knapp (Ed.), *Teaching for meaning in high-poverty classrooms* (pp. 47–63). New York: Teachers College Press.

Chapter 7

Fractions and Multiplicative Reasoning

Patrick W. Thompson, Vanderbilt University
Luis A. Saldanha, Vanderbilt University

In this chapter, we begin with a relatively simple observation and follow its implications to end with an analysis of what it means to understand fractions well. In doing so, we touch on related issues of curriculum, instruction, and convention that sometimes impede effective teaching and learning. We make these connections with the aim of bringing out aspects of knowing fractions that are important when considering the design of fraction curricula and instruction over short and long terms. We hope readers see our attempt to clarify learning goals for fractions as a helpful contribution of research to improving mathematics curricula and teaching.

Our observation is that how students understand a concept has important implications for what they subsequently can do and learn.[1] Although this observation is neither new nor breathtaking, it is rarely taken seriously. To take it seriously means to ground the design of curricula and teaching on careful analyses of what we expect students to learn and what students do learn from instruction.

Careful analyses of what students learn means more than creating a catalog of their behaviors or of the strategies we hope they employ. Careful analyses also entail tracing the implications that various understandings have for related or future learning. For example, many students understand "*a*/*b*" as denoting a part-whole relationship, for example, that "3/7" means "three out of seven" (Brown, 1993). This understanding is unproblematic until they attempt to interpret "7/3." Students often will think, if not say aloud, "7/3 sort of doesn't make any sense. You can't have 7 out of 3" (Mack, 1993, p. 91; 1995). Even further, students who understand "*a*/*b*" as meaning "*a* things out of *b* things" cannot interpret "8/(3/7)." It would have to mean something like "8 things out of (3 out of 7 things)," which does not make sense to them or to us. We see a strong possibility that nonintroductory lessons about fractions are largely meaningless to many students participating in them.

Students also can understand inscriptions as commands to engage in a sequence of actions. When students read " $7\overline{)3}$ " as command to act, they anticipate writing things below and above it until they satisfy some criterion for stopping, such as a remainder of 0 or a remainder that has appeared before. But anyone thinking of " $7\overline{)3}$ " as a command to act will find it difficult to interpret the expression "$[\,7\overline{)3}\,]\overline{)8}$ " It simply doesn't make sense from an action perspective.

To reiterate our point, the way students understand an idea can have strong implications for how, or whether, they understand other ideas. This observation is important for thinking about what students have learned or actually understand, and it has implications for how instructional and curricular designers think about what they intend for students to understand. Designers always intend some understanding, whether or not they make it available for public scrutiny. We contend that mathematics education profits from efforts to both publicize and scrutinize those intentions. Such efforts increase the likelihood that the meanings we intend students to develop actually have the potential of being consistent with, and supportive of, the meanings, understandings, and ideas we hope they develop from them.

[1] We use phrases like "understand a concept" and "concept of *x*" reluctantly. To say "understand a concept" suggests we are comparing a person's concepts and something that constitutes a correct understanding. We do not mean this at all. Rather, by "concept of *x*," we mean "conceptual structures that express themselves in ways people would conventionally associate with what they understand as *x*." But saying that is too cumbersome, so we continue to use "concept of *x*" and "understanding of *x*."

Pedagogical Contexts

Although our intent is to describe understandings that might support sophisticated fraction reasoning, we cannot ignore contexts in which learning and teaching occur. What students learn through instruction at any moment is not just a function of the instruction; it is influenced by what they already know (including beliefs they have about mathematics, doing it, and learning it) and by instruction in which they have participated. Reciprocally, a teacher's instructional actions at any moment are not simply a matter of executing a plan. They are influenced both by what the teacher understands about what he or she is teaching, and by what he or she discerns about what students know and about how students might build productively upon that knowledge. We examine each consideration briefly in regard to fractions and multiplicative reasoning.

Pedagogical Context of Present Mathematics Learning

A variety of sources suggest a problem with the nature of and coherence of mathematics instruction in the United States. The TIMSS report of eighth-grade mathematics instruction in the United States, Germany, and Japan states this deficiency clearly:

> Finally, as part of the video study, an independent group of U.S. college mathematics teachers evaluated the quality of mathematical content in a sample of the video lessons. They based their judgments on a detailed written description of the content that was altered for each lesson to disguise the country of origin (deleting, for example, references to currency). They completed a number of in-depth analyses, the simplest of which involved making global judgments of the quality of each lesson's content on a three-point scale (Low, Medium, High). Quality was judged according to several criteria, including the coherence of the mathematical concepts across different parts of the lesson, and the degree to which deductive reasoning was included. Whereas 39 percent of the Japanese lessons and 28 percent of the German ones received the highest rating, none of the U.S. lessons received the highest rating. Eighty-nine percent of U.S. lessons received the lowest rating, compared with 11 percent of Japanese lessons. (Stigler, Gonzales, Kawanaka, Knoll, & Serrano, 1999, p. iv)

The TIMSS sampling technique was to draw nationally representative samples from each of its participating countries (Stigler et al., 1999), so we can expect its results to be fairly representative. That no U.S. lesson's content received the highest quality rating from these mathematicians and that 89% of the U.S. lessons' content received the lowest quality rating suggests a general lack of attention among teachers to the ideas students develop. Instead, U.S. lessons tended to focus on having students do things and remember what they have done. Little emphasis was placed on having students develop robust ideas that could be generalized. The emergence of conversations about goals of instruction—understandings we intend that students develop—is an important catalyst for changing the present situation.

One major source of personal reform is the realization by teachers that what they are teaching does not support what students should be learning. Thus, discussions that entail descriptions of understandings we intend for students to develop also need to address how various instructional practices might support or impede this development.

Post, Harel, Behr, and Lesh (1991) and Ma (1999) shed additional light on present contexts in which students engage ideas of fractions. Post et al. (1991) gave several versions of a ratio and rate test to 218 intermediate (grades 4–6) mathematics teachers in Illinois and Minnesota. The test reflected concepts covered in the mathematics curriculum they taught. Teachers scored between the13-year-old and the17-year-old National Assessment of Educational Progress (NAEP) average on items drawn from the 1979 NAEP; overall performance among teachers varied widely across test versions, but average performance scores ranged from 60% to 69% across test versions, and more than 20% of the teachers scored less than 50%.

Post et al. (1991) found that teachers had significant difficulty with problems like "Melissa bought 0.46 of a pound of wheat flour for which she paid $0.83. How many pounds of flour could she buy for one dollar?" (p. 193). Forty-five percent of the teachers answered this question correctly, whereas 28% left the page blank or wrote, "I don't know." Teachers were also asked to explain their solutions as if to a student in their class. In the case of the Melissa problem (given here), only 10% *of those who answered correctly* could give a sensible explanation of their solution. The authors came to the following conclusion:

> Our results indicate that a multilevel problem exists. The first and primary one is the fact that many teachers simply do not know enough mathematics. The second is that only a minority of those teachers who are able to solve these

problems correctly were able to explain their solutions in a pedagogically acceptable manner. (Post et al., 1991, p. 195)

Ma (1999) compared elementary schoolteachers' understandings of mathematical topics they commonly taught. She found that Chinese teachers were far more likely than U.S. teachers to exhibit richly connected and pedagogically powerful understandings of what they expected students to learn, despite the U.S. teachers' more extensive educational backgrounds.[2] Thus, both the Post et al. (1991) and Ma studies point to the distinct possibility that phrases like "teach for deep understanding" and "teach with meaning" will not convey, for many teachers, a personally meaningful message without further professional development.

We hope no one interprets our remarks as attacking teachers or the important role they play in students' mathematical development. We are mathematics teacher educators as well as researchers; we work daily with teachers and prospective teachers. We are mathematics teachers ourselves. However, we have a "public health" perspective on problems of mathematics teaching. Communities resolve a problem most effectively when they discuss its scope, severity, and sources openly and objectively.

Learning Context of Present Mathematics Teaching

Whereas U.S. students are often asked to understand fractions in pedagogical contexts that provide little support, many U.S. teachers who are capable of engaging in appropriate instruction find themselves with students who are poorly prepared to participate in it. For example, the 1996 NAEP (Reese, Miller, Mazzeo, & Dossey, 1997) gave these items to 8th and 12th graders:

1. Luis mixed 6 ounces of cherry syrup with 53 ounces of water to make a cherry-flavored drink. Martin mixed 5 ounces of the same cherry syrup with 42 ounces of water. Who made the drink with the stronger cherry flavor? Give mathematical evidence to justify your answer. (NAEP M070401)

	1980 Population	1990 Population
Town A		
Town B		
	1000 people	1000 people

2. In 1980 the populations of Towns A and B were 5000 and 6000, respectively. In 1990 the populations of Towns A and B were 8000 and 9000, respectively.

   Brian claims that from 1980 to 1990 the two towns' populations grew by the same amount. Use mathematics to explain how Brian might have justified his answer.

   Darlene claims that from 1980 to 1990 the population of Town A had grown more. Use mathematics to explain how Darlene might have justified her answer. (NAEP M069601)

Problem 1's intent was to ascertain the extent to which students could quantify intensity of flavor (higher ratio of cherry juice to water means more intense cherry taste). Problem 2's intent was to see whether students could compare quantities additively (by difference) as well as multiplicatively (by ratio). Compared additively, both towns grew the same amount (1000 people). Compared multiplicatively, Town A's 1990 population was 8/5 (160%) as large as its 1980 population, whereas Town B's 1990 population was 9/6 (150%) as large as its 1980 population.

Although these problems might seem straightforward, they challenged 8th- and 12th-grade students who took part in the1996 NAEP (Reese et al., 1997). Less than half the 12th graders gave even partially acceptable answers to Problem 1; about one fifth of 8th graders and one fourth of 12th graders gave partially correct responses to Problem 2 (see Table 7.1). Students' performance in the 1996 NAEP was consistent with findings

TABLE 7.1. Student Performance in 1996 NAEP on Problems 1 and 2

Grade	Problem 1		Problem 2	
8	Correct:	*	Correct:	1%
	Partially correct	*	Partially correct	21%
	Incorrect	*	Incorrect	60%
	Omitted	*	Omitted	16%
	Off task	*	Off task	2%
12	Correct:	23%	Correct:	3%
	Partially correct	26%	Partially correct	24%
	Incorrect	42%	Incorrect	56%
	Omitted	9%	Omitted	16%
	Off task	1%	Off task	1%

Note. * = data not released.

[2] Chinese elementary school teachers enter normal school after ninth grade, graduating two to three years later.

from a long line of studies that examined students' abilities to reason about ratios and relative amounts (Harel, Behr, Lesh, & Post, 1994; Harel, Behr, Post, & Lesh, 1992; Hart, 1978; Karplus, Pulos, & Stage, 1979, 1983; Noelting, 1980a, 1980b; Tourniaire & Pulos, 1985). U.S. students have done poorly on such items in comparison to students in other countries (Dossey, Peak, & Nelson, 1997; McKnight et al., 1987).

Explanations of U.S. students' poor performance on questions like these are complicated. One reason for poor performance is that the understandings tapped by these questions can "go wrong" at many developmental junctures during students' schooling. Our analysis will therefore avoid focusing on any one path to understanding. Instead, we will analyze what might constitute sophisticated understandings of fractions, leaving for other discussions what instructional approaches might support their development.

Distinction Between Fractions and Rational Numbers

Before discussing what we mean by "understanding fractions," we distinguish between fractions as what Kieren (1988; 1993b) calls a "personally knowable system of ideas" and the development of what is commonly taken as the system of rational numbers. We do this for two reasons. First, we detect a tendency among textbooks to confound fractions and rational numbers. Second, we find it profitable to point out that understanding the rational number system, where "rational numbers" is used as mathematicians use it, is so far beyond the grasp of school students that curriculum and instructional designers must be clear on what they mean by "fractions" and "rational numbers" so they avoid designing for incoherent learning goals.

Mathematicians rely heavily on symbol systems to aid their reasoning. Symbol systems are tools for them. Mathematicians therefore strive to develop symbol systems (inscriptions and conventions for using them) that capture essential aspects of their intuitive understandings and means of operating, so they need not rely explicitly on conceptual imagery and operations as they move their reasoning forward or generate further insight.

In the 18th and 19th centuries, mathematicians found that their symbol systems, used according to established conventions, led to contradictory results. Many mathematicians realized that the problem was not in the symbol systems they had devised but rather in their understandings of the number and functional relationship that the symbol systems had intended to capture. The idea of "mathematical development of number systems" arose from the need to make commonly held, but tacit, meanings of number more precise and articulate. This development was also influenced by the emergence of new numeric understandings, such as the understanding that number systems could be created even if no one understood what they were.[3]

The mathematical construction of rational numbers is tremendously general and abstract for the same reasons the mathematical definition of function is general and abstract. It addresses a history of paradoxes and contradictions, and the present definition of the set of rational numbers is the result of a long line of accommodations to eliminate those paradoxes and contradictions. For example, one stimulus for the modern development of rational numbers is a spin-off from the period in which the calculus was grounded in the analysis of real-valued functions (Eves, 1976; Wilder, 1968). The original notion of a derivative as a ratio of differentials (first-order changes in two quantities' values) fits with the notion of rate of change as a relationship between two varying quantities. But contradictions that traced back to notions of derivative-as-ratio led d'Alembert, for one, to wonder whether "dy/dx" should be thought of as merely a symbol that represents one number instead of as a pair of symbols representing a ratio of two numbers (Edwards, 1979). The significance of d'Alembert's question is easily missed. He suspected that the cultural practice of considering a rate of change as being composed of two numbers was conceptually incoherent and that what was conventionally interpreted as a ratio of two numbers was indeed one number that was not the result of calculating.

Mathematicians' questions about rates-as-numbers and about continuity of functions threaded themselves into formal constructions of rational and real number systems (Eves & Newsom, 1965; Heyting, 1956; Kasner & Newman, 1940; Kneebone, 1963). To summarize that development here would be inappropriate. Rather, we wish to emphasize that the mathematical developments of rational and real number systems interconnect many issues that typically are not treated until advanced undergraduate or introductory graduate mathematics courses. Therefore, we have no idea what school mathematics textbook authors or other writers intend when they say they want middle school students to "understand" rational and irrational numbers.

[3]This was the case with complex numbers and later with hyperreal numbers (Henle, 1986; Tall & Vinner, 1981). Likewise, surreal numbers and quaternions had no immediate use except to challenge prevailing intuitions of number systems (Conway & Guy, 1997).

Conceptual Analyses of Learning Objectives

We digress momentarily to address what we mean by "to understand *x*" and to explicate our method for developing descriptions of an understanding. We do this in hopes of making our intentions precise and thereby increasing the likelihood that the reader will create meanings that we intend.

Understand has both colloquial and technical meanings. The American Heritage *Dictionary of the English Language* (4th ed.) lists eight senses of *understand*. The first six define *understand* by reference to *comprehend* and *apprehend*, and those words are defined somewhat circularly. The last two senses of *understand* operationalize it more directly:

> (7) To accept something as an agreed fact: *It is understood that the fee will be 50 dollars*, and
>
> (8) To supply or add (words or a meaning, for example) mentally. (American Heritage, 2002)

These last two senses of *understand* underpin most colloquial uses of it and match more technical meanings as well. Skemp (1979) used Piaget's notion of assimilation (Piaget, 1950, 1971a, 1971b, 1976) when he described *understanding* as "assimilating to an appropriate scheme," by which he meant attaching appropriate meanings and imagery to the utterances or inscriptions that a person interprets. Skemp's definition of understanding coincides with Carpenter's notion, as well as Hiebert and Lefevre's, of understanding as a rich set of meaningful connections by which a person acts flexibly with respect to problems he or she encounters (Carpenter, 1986; Hiebert & Lefevre, 1986).

We follow the tradition of Piaget, Skemp, Carpenter, and Hiebert and Lefevre when we speak of understanding. We choose to omit *appropriate*, however, so we can speak of a person's understanding as "assimilation to a scheme," thereby allowing us to address understandings people do have even though someone else may judge them to be inappropriate. Also, we note that to describe an understanding requires addressing two sides of the assimilation—the thing a person is attempting to understand and the scheme of operations that constitutes the person's actual understanding. We see understanding as, fundamentally, what results from a person's interpreting signs, symbols, interchanges, or conversations—assigning meanings according to a web of connections the person builds over time through interactions with his or her own interpretations of settings and through interactions with other people as they attempt to do the same.

The goal in this line of work, specifying what it might mean to understand a complicated idea like fractions, is to "consider any mental content (percepts, images, concepts, thoughts, words, etc.) as a result of operations" (Cecatto, 1947, as cited in Bettoni, 1998). That is, one must describe *consapevolezza operativa*, or conceptual operations (translated literally as "operating knowledge"),[4] to answer the question "Which mental operations do we perform in order to conceive a situation in the way we conceive it?" (Bettoni, 1998). Von Glasersfeld (1972, 1995) combined aspects of Ceccato's operational analysis and Piaget's genetic epistemology to devise a way to talk about reasoning and communicating as imagistic processes and of knowledge as an emergent aspect of them (von Glasersfeld, 1978). Doing so produced an analytic method that von Glasersfeld called *conceptual analysis*, whose aim was to describe the conceptual operations that, were people to have them, might result in their thinking the way they evidently do. Different researchers engage in varying levels of detail (Steffe, 1996; Thompson, 2000), but the aim is to describe conceptual operations in ways that are "near the surface" of the kinds of reasoning one hopes to explain.

When engaging in conceptual analysis, one may focus on understandings as they might exist at some level of sophistication or on how people might get to them. In regard to understanding fractions, we focus on what might be called "mature" understandings of fractions. This is not to say that developmental issues are unimportant. They are important, especially for designing curriculum and instruction to support students' formative understandings. We have chosen to describe "mature" understandings of fractions for the simple reason that subsequent discussions can be grounded in a common image of the overall curricular and instructional goals.

Understanding Fractions

In the spirit of being explicit about the meanings and understandings we intend for students to develop, we describe one view of what "to understand" can mean with regard to the panoply of ideas and behaviors associated with the school subject called *fractions*. Specifically, we focus on fractions and multiplicative reasoning. Our use of the conjunction *and* in "fractions and multiplicative

---

[4]Quotations translated by M. Bettoni. Phrases translated by GO Translations, http://translator.go.com./.

reasoning" points to the particular stance we take in our analysis: Coherent fractional reasoning develops by interrelating several conceptual schemes often not associated with fractions. We use the phrase *conceptual scheme* to indicate that we are talking about stable ways of thinking that entail imagining, connecting, inferring, and understanding situations in particular ways. We emphasize that we are not talking about abstract formulations that reside outside every person dealing with these situations; we are talking about ways people reason when they understand fractions in the way we are attempting to convey. The schemes we characterize here are division schemes, multiplication schemes, measurement schemes, and fraction schemes.

Our discussion in this section takes a somewhat circuitous route from examples of mildly sophisticated reasoning, to a discussion of factors in its development, to a discussion of more general characteristics of advanced multiplicative reasoning. Our discussion draws heavily from foundational research by Vergnaud (1983, 1988, 1994), Steffe and Tzur (Steffe, 1988, 1993; Tzur, 1999), Kieren (1988, 1992, 1993b), the Rational Number Project (Behr, Harel, Post, & Lesh, 1992, 1993; Lesh, Behr, & Post, 1987; Lesh, Post, & Behr, 1988; Post, Cramer, Behr, Lesh, & Harel, 1993; Post et al., 1991), and Confrey (1994; Confrey & Smith, 1995), as well as from our own work in quantitative reasoning (Saldanha & Thompson, 1998; Thompson, 1988, 1993; P. W. Thompson, 1994; Thompson, 1995, 1996).

Kieren (1988, 1993a, 1993b) and the Rational Number Project (Behr et al., 1992, 1993; Lesh et al., 1987) have given the most extensive analyses of rational number meanings. Their approach was to break the concept of rational number into subconstructs—part-whole, quotient, ratio number, operator, and measure—and then describe rational number as an integration of those subconstructs. Our belief is that their attempt to map systems of complementary meanings into the formal mathematical system of rational numbers will necessarily be unsatisfactory in regard to designing instruction for an integrative understanding of fractions. Each subconstruct is portrayed as a body of meanings, or interpretations, of the "big idea" of rational numbers. Mathematical motivations for developing the rational numbers as a mathematical system, however, did not emerge from meanings or subconstructs. Rather, they emerged from the larger endeavor of arithmetizing the calculus. So to focus on subconstructs or meanings of the mathematical system of rational numbers ultimately runs the risk of asking students to develop meanings for a big idea they do not have. Our approach is to place fraction reasoning squarely within multiplicative reasoning as a core set of conceptual operations.

As noted already, Post et al. (1991, p. 193) gave this problem to a sample of intermediate school mathematics teachers: "Melissa bought 0.46 of a pound of wheat flour for which she paid $0.83. How many pounds of flour could she buy for one dollar?" A standard solution was to set up an equation, as in

$$\frac{0.46}{0.83} = \frac{x}{1},$$

and solve for x. As Post et al. found, however, setting up this equation correctly and having a coherent understanding of it are not the same.

A solution to the Melissa problem that relies on reasoning instead of on equations might go like this: "If 0.46 lb. costs $0.83, then $0.01 (being 1/83rd of $0.83) will purchase 1/83rd of 0.46 lb. Thus $1.00 will purchase 100/83 of 0.46 lb." A more sophisticated expression of the same reasoning would be this: "$1.00 is 100/83 as large as $0.83 [100 times as large as 1/83rd of $0.83], so you can buy 100/83 of 0.46 lb. for $1.00."

What conceptual development might lead to such reasoning? A variety of sources suggest the development of a web of meanings that entails conceptualizations of measurement, multiplication, division, and fractions. We emphasize *conceptualizations* of measurement, multiplication, division, and fractions. This is not the same as measuring, multiplying, and dividing. The latter are activities. The former are images of what one makes through doing them.

Measurement Schemes

To conceive of a measured quantity is to imagine the measured attribute as segmented (Minskaya, 1975; Steffe, 1991b) or in terms of a coordination of segmented quantities (Piaget, 1970; Schwartz, 1988; P. W. Thompson, 1994).

The idea of ratio is at the heart of measurement. To conceive of an object as measured means to conceive of some attribute of it as being segmented and to realize that the segmentation is in comparison with some standard amount of that attribute. Suppose, for example, one wishes to publicize a horse racing track's distance. How one measures its distance, however, is not straightforward. Every lane has a different length, and not every horse runs in the same lane throughout. Therefore, by custom, the length of a specific lane, measured on its shortest perimeter, is taken as the "race course length."

Even when an object's attribute is clear, the matter of conceiving its measure can still remain. A mile-long racecourse, measured in yards, is 1760 yards because the length of one mile is understood to be 1760 times as long as the length called one yard. The same racecourse's length is 5,280 feet because it is 5,280 times as long as the length called one foot. "There are 5280 feet in one mile" does not mean just that a mile contains 5280 feet, in the same sense that a fielded baseball team contains nine players. The distinction we make here is between a part-whole relationship between a set and its elements and a multiplicative comparison in which the measuring unit is imagined apart from the thing measured.

The ratio nature of measurement is trivial to people who have a quantitative scheme of measurement, but it is nontrivial to students who are building one. A conceptualized measure entails an image of a ratio relationship (i.e., A is some number of times as large as B) that is invariant across changes in measurement units. For example, Figure 7.1 illustrates that if m is the measure of quantity B in units of quantity A (i.e., B is m times as large as A), then nm is the measure of quantity B in units of $1/n$th of A. Conversely, if m is the measure of quantity B in units of quantity A, then m/n is the measure of B in units of nA. Put another way, the measure of a quantity is m times as large when one dilates the unit by a factor of $1/m$, and its measure is $1/m$th as large when one dilates the unit by a factor of m.

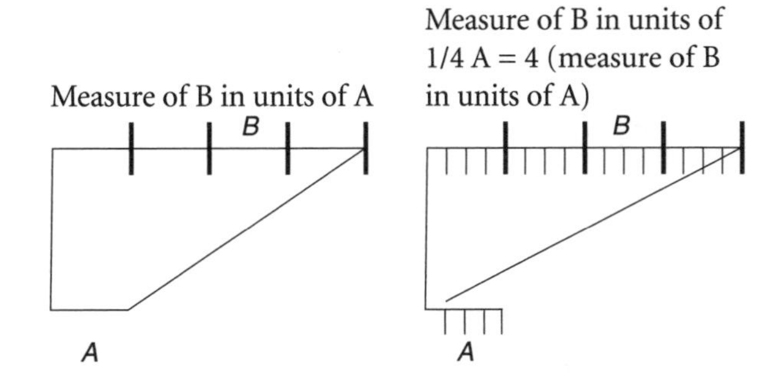

Figure 7.1. Change of measurement unit.

A conceptual breakthrough underlying students' understanding of unit substitutions is their realization that the magnitude of a quantity (i.e., its "amount") as determined in relation to a unit does not change even with a substitution of unit. Wildi (1991) emphasized this point by making two distinctions. The first was between a quantity's measure and its magnitude. A quantity's magnitude (i.e., its "amount of stuff" or its "intensity of stuff") is independent of the unit in which you measure it. If we let $m(B_U)$ denote quantity B's measure relative to unit U, then $|B|$, the magnitude of B, is $m(B_U)|U|$. A change of unit does not change the quantity's magnitude—making the unit 1/4 as large makes the measure 4 times as large, leaving the quantity's magnitude unchanged.[5]

Wildi's (1991) second distinction was between numerical equations and quantity equations. A numerical equation, such as $W = fd$, says how to calculate a particular quantity's measure. As such, the formula's result in any particular instance depends on the particular units used. Quantity equations suggest a quantity's construction. Wildi wrote the equation $[W] = [f][d]$ to say that accomplished work, as a quantity, is created by applying a force to an object and thereby moving it some distance. The equation makes no reference to measurement units. Wildi wanted the quantity formula to say that the product quantity's magnitude remains the same regardless of the units in which force or distance is measured, as long as they are measured appropriately.

Students in a fifth-grade teaching experiment on area and volume alerted us to the distinction between understanding a formula numerically and understanding it quantitatively. The first author presented the question in Figure 7.2. Portions of two students' interviews are given after the diagram.

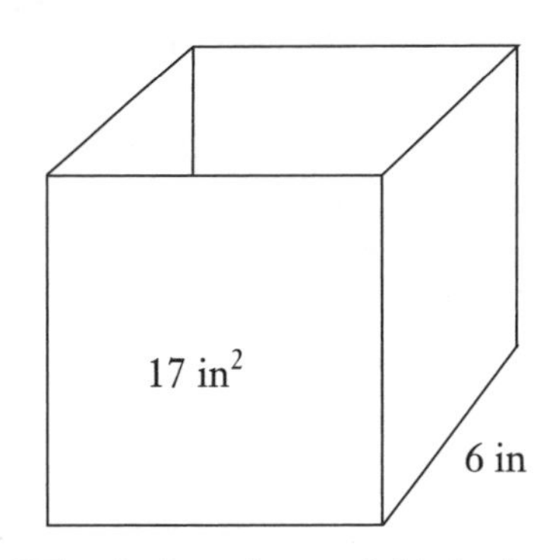

Figure 7.2. What is the volume of this box?

P.T.: (Discusses with B.J. how the diagram represents a hollow box and what about it each number in the diagram indicated.)

B.J.: (Reads question.) I don't know. There's not enough information.

P.T.: What information do you need?

B.J.: I need to know how long the other sides are.

---

[5]The first author observed a fourth-grade lesson on the metric system in which students measured their heights. He asked one boy who had measured his height with a yardstick what he got. "140 centimeters." How tall are you? "Four feet seven." Do you know how many centimeters make 4 feet 7 inches? (Pause.) "No." This child had not realized that his height, as a magnitude, was the same in both instances and therefore did not realize that 140 cm and 4 ft. 7 in. were equivalent, in that both were measures of his height.

P.T.: What would you do if you knew those numbers?

B.J.: Multiply them.

P.T.: Any idea what you would get when you multiply them?

B.J.: No. It would depend on the numbers.

P.T.: Does 17 have anything to do with these numbers?

B.J.: No. It's just the area of that face.

...

P.T.: (Discusses with J.A. how the diagram represents a hollow box and what about it each number in the diagram indicated.)

J.A.: (Reads question.) Oh. Somebody's already done part of it for us.

P.T.: What do you mean?

J.A.: All we have to do now is multiply 17 and 6.

P.T.: Some children think that you have to know the other two dimensions before you can answer this question. Do you need to know them?

J.A.: No, not really.

P.T.: What would you do if you knew them?

J.A.: I'd just multiply them.

P.T.: What would you get when you multiplied them?

J.A.: 17.

To B.J., the formula $V = LWD$ was a numerical formula. It told him what to do with numbers once he had them; however, it had no relation to evaluating quantities' magnitudes. To the second child, J.A., the formula $V = LWD$ was a quantity formula. To him, it was $V = [LW][D]$, where $[LW]$ produced an area and where $[LW][D]$ produced a volume. J.A. recognized that being provided one face's area was as if "somebody's done part of it for us," that part being the quantification of one face's area.

Proportionality and Measurement

Proportional reasoning is important in students' conceptualizing measured quantities. Vergnaud (1983, 1988) emphasized this when he placed single and multiple proportions at the foundation of what he called the multiplicative conceptual field.[6] A single proportion is a relationship between two quantities such that if the size of one quantity is increased by a factor a, then the other quantity's measure must increase by the same factor to maintain the relationship. If one is ordering food to feed guests at a party and someone announces that three times as many guests will attend, one orders three times as much food so the amount of food and the number of people remain related the way originally intended. More formally, if x is the measure of one quantity, $f(x)$ is the measure of the other, and the two are related proportionally, then $f(ax) = af(x)$.

A double proportion is a relationship among three quantities, where one is thought of as being created from the other two and in which the created quantity is proportional to each of the others. More formally, if x and y are measures of two quantities, if $f(x, y)$ is the created quantity's measure, and if the created quantity is understood as being related proportionally to the other two, then

$$\begin{aligned} f(ax, by) &= af(x, by) \\ &= bf(ax, y) \\ &= abf(x, y). \end{aligned}$$

The concept of torque—that applying a force at some distance from a resistant hinge or fulcrum creates some amount of twisting force (Figure 7.3)—provides an example of how being able to conceptualize a double proportion fosters insight into devising a measuring scheme for a new quantity.

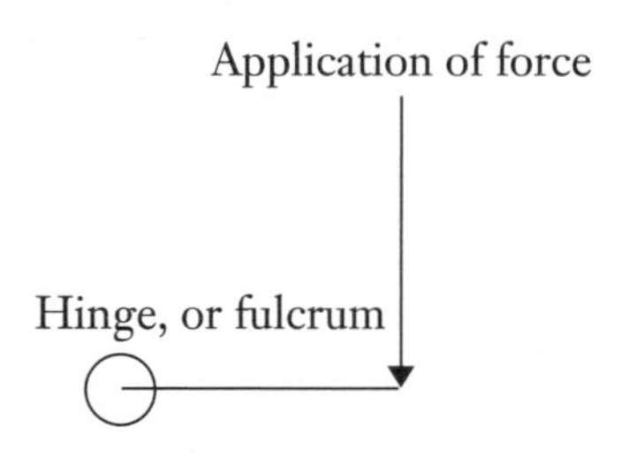

Figure 7.3. An amount of twisting force created by applying a force at a distance.

It is easy to convey an experiential understanding of torque by having students attempt to hold weights of various amounts on a broomstick at various distances from their hands. They experience personally that the same weight placed farther out is harder to hold than when placed closer in. Unfortunately, such activities do not convey an idea of how one might *measure* the amount of twisting force created by a specific weight held a spe-

[6] Confrey (1994) takes a counter position, claiming that multiplicative reasoning emanates from a primitive operation called splitting, which does not emanate from proportionality but instead underlies it. We cannot here discuss the two in contrast, as that would take us astray of our present purpose. We will say only that we agree with Steffe (1994) that when we consider the developmental origins of splitting, we see that Confrey and Vergnaud are not in opposition.

cific distance from their hands. If, however, students understand *amount of twisting force* as being related to *weight* and *distance from fulcrum* in a double proportion, then its quantification is natural:

1. Suppose that applying a force of y force-units at a distance of x distance-units from a fulcrum produces T twist-units.

2. Increasing the applied force by a factor of u while applying it at the same distance increases the amount of twist by a factor of u. That is, applying a force of uy force-units at a distance of x distance-units produces uT twist-units.

3. Increasing the distance at which the original force is applied by a factor of v increases the amount of twist by a factor of v. That is, applying a force of y force-units at a distance of vx distance-units produces vT twist-units.

Observations 1–3 in combination imply that if applying a force of y force-units at a distance from a fulcrum of x distance-units produces T twist-units, then applying a force of uy force-units at a distance of vx distance-units produces a twisting force of measure uvT twist-units. By convention, a force of 1 force-unit applied at a distance of 1 distance-unit produces 1 twist-unit. Thus, a force of $u \cdot 1$ force-units applied at a distance of $d \cdot 1$ distance-units produces $u \cdot v$ twist-units. An understanding that *amount of twisting force* is proportional to each of distance-from-fulcrum and amount-of-applied-force leads directly to quantifying torque by the formula "force times distance." That is, the quantification of torque by multiplying measures of force and distance can be grounded naturally in understanding torque as entailing a double proportion. It need not be taught as a formula to memorize.

Multiplication Schemes

Contrary to most textbooks, and contrary to Fischbein's well-known treatise (Fischbein, Deri, Nello, & Marino, 1985), multiplication is not the same as repeated addition. To be fair, we should say *conceptualized* multiplication is not the same as repeated addition.

> The conceptual foundation of multiplication of whole numbers is quite like the Biblical "multitudes"—creating many from one. That is to say, multiplication of whole numbers is the systematic creation of units of units. (Thompson, 1982, p. 316)

The difference between conceptualized multiplication and repeated addition is between envisioning the result of having multiplied and determining that result's value (Steffe, 1988). Envisioning the result of having multiplied is to anticipate a multiplicity. One may engage in repeated addition to evaluate the result of multiplying, but envisioning adding some amount repeatedly cannot support conceptualizations of multiplication.

We generalize the previous section's torque example to illustrate that the capability to conceive of multiple proportions leads generally to numerical multiplication. Suppose a double proportion relates three quantities. By convention, the standard unit of product quantity is defined as that amount made by one unit of each constituent quantity. (If we measure the three edges of a box in units of light years, centimeters, and inches, respectively, then one unit of volume is one light year-centimeter-inch.) If x and y are the constituent quantities' measures, and if $f(x,y)$ is the product quantity's measure expressed as a function of the other two, then the measure of the product quantity's standard unit is 1.

$$\begin{aligned} meas(StandardUnity) &= f(1,1) \\ &= 1 \end{aligned}$$

As such, the third quantity's measure will be the following:

$$\begin{aligned} f(x,y) &= f(x \cdot 1, y \cdot 1) \\ &= x \cdot y \cdot f(1,1) \\ &= x \cdot y \cdot meas(StandardUnity) \\ &= x \cdot y \cdot 1 \\ &= x \cdot y \end{aligned}$$

We use the statement "$f(x,y) = f(x \cdot 1, y \cdot 1)$" as a model of a particular understanding. That understanding is *not* that a student knows that x may be rewritten as $x \cdot 1$. Rather, it is that since $f(x,y)$ stands for "the measure of the object constructed from attributes having measures x and y," our rewriting x as $x \cdot 1$ and y as $y \cdot 1$ models the explicit understanding that measurement entails a ratio comparison. For example, if a student conceives a rectangle as being made by a product of its sides, where one side is 12 inches long and the other is 2 cm long, then to say $f(12, 2) = f(12 \cdot 1, 2 \cdot 1)$ reflects the move from thinking of 12 inches as a number of inches to 12 inches as a length that is 12 times as long as one inch (Figure 7.4). The rectangle's (area) measure is, then, 24 in.-cm, or 24 times the area of a rectangle of dimension 1 inch by 1 cm.

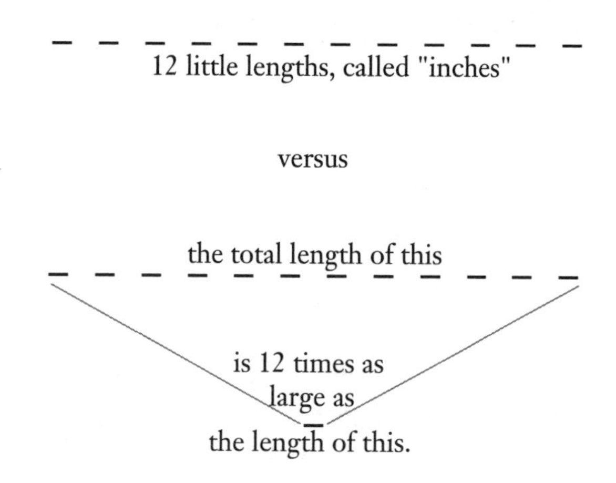

Figure 7.4. Measure as a number of things versus measure as a ratio comparison.

The fact that to understand objects as entailing a single or multiple proportion relies on understanding measures as ratio comparisons highlights the special way in which multiplication must be conceived—as entailing a multiple proportion—if it is to cohere with students' understandings of fractions. Generally, most students do not see proportionality in multiplication. In fact, a large amount of curriculum and instruction has the explicit aim that students understand multiplication as a process of adding the same number repeatedly. But an extensive research literature documents how "repeated addition" conceptions become limiting and problematic for students having them (de Corte, Verschaffel, & Van Coillie, 1988; Fischbein et al., 1985; Greer, 1988; Harel, Behr, Post, & Lesh, 1994; Luke, 1988).

As mentioned previously, Fischbein et al. (1985) and Burns (2000) propose repeated addition as the fundamental initial idea of multiplication. Because of students' prior additive, numerical, and quantitative experience, portraying multiplication as repeated addition makes multiplication easy to teach and, in this limited calculational sense, easy to learn. But thinking about multiplication as repeated addition leads to severe difficulties in later grades. Specifically, if 5 × 4 means 4 + 4 + 4 + 4 + 4, that is, add 4 five times, then it is difficult to say what 5 2/3 × 4 means. It cannot mean, "add 4 five and two-thirds times."

Introducing multiplication as repeated addition is, in principle, like introducing the quantification of work before discussing how to think of work so that it is quantifiable. Repeated addition is a quantification technique; it is not the thing being quantified. What does 4 + 4 + 4 + 4 + 4 quantify? Five fours. As described by Confrey (1994) and Steffe (1988; 1994), multiplication can be introduced to students by asking them to think about quantities and numbers in settings in which they need to envision a multiplicity of identical objects. The question "How much (many) do these make?" comes after conceiving what "these" are, and it serves to orient students toward the quantification of multiplicities. The quantification itself may or may not involve repeated addition.

This approach to conceiving multiplication as about quantifying something made of identical copies of some quantity solves another problem created by students' thinking of multiplication as repeated addition. If we start with the basic meaning of 5 × 4 as "five fours," then 5 2/3 × 4 means "(five and two-thirds) fours." The principal difference between "add 4 five times" and "five fours" is that the former tells us a calculation to perform, whereas the latter suggests something to imagine. Similarly, if students read "5 2/3 × 4 2/3" with a multiplicity meaning, they will think about some number of (four and three fifths). They will understand "5 2/3 × 4 2/3" as meaning 5 (four and three fifths) and 2/3 of (four and three fifths). When students understand what the statement refers to, they can do whatever is sensible given the surrounding context.

This example carries a larger significance: How students decode mathematical statements can have a large impact on the connections they form. If they read "5 × 4" as a command to calculate, then "5 × x" is highly problematic. But reading "5 × 4" as "five fours" supports the image that we are speaking of a product: "5 × 4" is a number that is 5 times as large as 4. So "5 × 4" becomes a noun phrase and therefore points to something. Thus, understanding "5 × x" as "a number that is 5 times as large as the number x" can be a natural extension of students' understanding of numerical multiplication, but only if their understanding of numerical multiplication is appropriately general in the first place.

We re-emphasize that when a curriculum starts with the idea that "__×__" means "some number of (or fractions of) some amount," it is not starting with the idea that "times" means to calculate. It is starting with the idea that "times" means to envision something in a particular way—to think of copies (including parts of copies) of some amount. This is not to suggest that multiplication should not be about calculating. Rather, calculating is just one thing one might do when thinking of a product. Noncalculational ways to think of products will be important in comprehending situations in which multiplicative calculations might be useful. The comprehension will enable students to decide on appropriate actions.

Although we discuss fraction schemes in a later section, it is worthwhile to point out here that for someone to understand multiplication multiplicatively, he or she must also understand fractions as entailing a proportion.

We propose that the phrase "quantity X is $1/n$ of a quantity Y" means that Y is n times as large as X. From that perspective, it does not mean that X is one of n parts of Y.[7] Thinking of $1/n$ as "one out of n parts" is to think of fractions additively—that Y is cut into two parts, one being X and the other being the rest. It is merely a way to indicate one part of a collection. When students' image of fractions is "so many out of so many," it possesses a sense of inclusion—that the first "so many" must be included in the other "so many." As a result, they will not accept the idea that we can speak of one quantity's size as being a fraction of another's size when the quantities have nothing physically in common. They will accept "The number of boys is what fraction of the number of children?" but will be puzzled by "The number of boys is what fraction of the number of girls?"

To think of multiplication as producing a product and to think at the same time of the product in relation to its factors entails proportional reasoning. To understand (5 × 4) multiplicatively, students must understand that the 4 in 5 × 4 is not just 4 ones, as in 20 = 4 + 9 + 7. Rather, the 4 is special—it is 1/5 of the product. In general, when students understand multiplication multiplicatively, they understand the product (nm) as being in multiple reciprocal relationships to n and to m:

- (nm) is n times as large as m,
- (nm) is m times as large as n,
- m is 1/n as large as (nm),
- n is $1/m$ as large as (nm).

An example using "ugly numbers" might clarify this point. In the expression 4 3/7 × 7 2/3 (read as [four and three-sevenths] seven-and-two-thirds), 7 2/3 is special. It is

$$\frac{1}{4\frac{3}{7}}$$

as large as the product. What does

"$\frac{1}{4\frac{3}{7}}$ as large as the product"

mean? Recall that the expression "a is one-nth of b" means that b is n times as large as a. So to say that 7 2/3 is

$\frac{1}{4\frac{3}{7}}$ as large as (4 3/7 × 7 2/3)

means that (4 3/7 × 7 2/3) is 4 3/7 times as large as 7 2/3 (see Figure 7.5). One shaded rectangle (7 2/3) is

$\frac{1}{4\frac{3}{7}}$ of the total

because the total is 4 3/7 times as large as 7 2/3. The other three reciprocal relationships can be elaborated similarly.

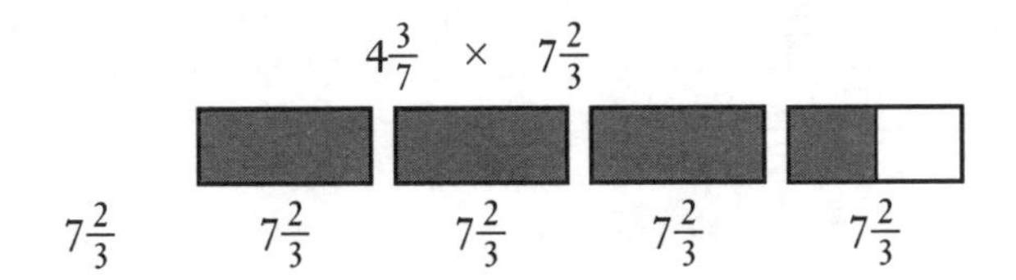

Figure 7.5. (Four and three-sevenths) seven and two-thirds.

We have attempted to convey the generality inherent in thinking of multiplication in a way that entails proportionality and multiplicity. When presenting this perspective in professional development settings, we sometimes hear the remark, "This is too complicated. Thinking of multiplication as repeated addition is much easier." We agree that repeated addition is easier than proportionality. But understanding multiplication as repeated addition keeps it divorced conceptually from measurement, proportionality, and fractions.

Division Schemes

Several research studies point to students' difficulties in conceptualizing division and its relationships with multiplication and fractions. Greer (1987) and Harel et al. (1994) noted the great frequency with which students will decide on the use of multiplication or division depending on the size or type of numbers involved, not on the underlying situation. Graeber and Tirosh (1988) found the same to be true among preservice elementary teachers. Ball (1990) found that college mathematics majors had difficulty proposing situations in which it would

[7] This poses an interesting question. To what extent do we want to respect meanings that are commonly held but that end up hurting students who adopt them? A person with a sophisticated understanding of fractions can say "one out of n," really mean it additively at the moment of saying it (like, one of those six boys has a blue shirt), and yet flip effortlessly and without awareness to "the number of boys over there is six times as large as the number of boys with blue shirts." We do not object to the use of the words "one out of n." Our objection is that many people think that "one out of n" automatically conveys the latter. We contend that it takes a consistent, systematic effort over multiple grades to ensure that the majority of students internalize the multiplicative point of view, so that it becomes "the way they see things."

be appropriate to divide by a fraction. Simon (1993) found that preservice elementary teachers' knowledge of division was tightly intertwined with their whole-number schemes, which resulted in weak connections among various meanings of division and in their inability to identify the result's unit of measurement. In short, it is unquestionable that division is a problematic area of mathematics teaching and learning.

The fact that division is a problematic area does not clarify what is problematic or what constitutes an understanding of it. In addressing this issue, we begin with a widely used distinction between two "types" of division settings, sharing and segmenting, and demonstrate that, among other things, operational understanding of division entails a conceptual isomorphism between them.

Sharing (or partitioning) is the action of distributing an amount of something among a number of recipients so that each recipient receives the same amount.[8] *Segmenting* (or measuring) is the action of putting an amount into parts of a given size. The problem "13 people will receive 37 pizzas. How many pizza's will each person receive?" is usually classified as a sharing situation because of the request to distribute the pizzas into a pre-established number of groups. The problem "You are to give 3 pizzas to each person, and you have 37 pizzas. How many people can you serve?" usually is classified as a segmenting situation because of the request to put the pizzas into pre-established group sizes. It is well known that students often see these situations as being very different (Bell, Fischbein, & Greer, 1984; Bell, Swan, & Taylor, 1981; Greer, 1987, 1988), and they are often puzzled that both can be resolved by the numerical operation of division.[9]

How is it possible that the results of sharing and segmenting are evaluated by the same numerical operation? To see why the numerical results are the same in either situation entails the development of operative imagery—the ability to envision the result of acting prior to acting—and to suppress attention to the process by which one obtains those results.

To illustrate why the same numerical operation resolves the two situations, we speak of two scenes. In Scene 1, we share an amount of chocolate (measured in bars) so that each of seven people receive the same amount of chocolate (the same number of bars) and ask how many bars each person receives. In Scene 2, we cut up the chocolate in parts the size of seven bars and ask how many parts we make. In both scenes, it is advantageous to consider the chocolate as one mass (that comprises some number of bars).

Scene 1: Share the chocolate among seven people.

- The process of sharing ends up with each recipient having the same amount of chocolate. If we put the chocolate into seven parts, each part contains 1/7 of the chocolate. So the number of candy bars in each part is 1/7 as large as the total number of candy bars that comprise the entire amount of chocolate.
- Each person receives a number of bars that is 1/7 of the total number of bars.

Scene 2: Put the chocolate into seven-bar parts

- The process of segmenting cuts up the mass of chocolate into a number of parts of a given size, perhaps with an additional part that is a fraction of the given size.
  - We cut this mass into parts, each part the size of seven candy bars, and consider any chocolate left over to make some fraction of a part (i.e., a fraction of seven candy bars).
  - We can determine the number of parts by removing one bar from each part and a proportionate part of a bar from any fraction of a part.
  - Each part (or fraction of a part) is seven times as large as what we remove from it, so the total is seven times as large as what we remove in all.
  - The mass of chocolate is seven times as large as what we remove, and what we removed counts the number of parts.
- The number of parts made by cutting up the chocolate into seven-bar-sized parts is 1/7 as large as the number of bars that make up the entire amount of chocolate.

These two analyses show that sharing (partitioning) a number of candy bars among seven people and segmenting (measuring) the number of candy bars into seven-bar-sized parts both produce a number that is 1/7 as large as the number of candy bars. The analyses also reveal that sharing (partitioning) and segmenting (measuring) are highly related even though at the level of activity they appear to be very different. When students

[8] We will speak as if all sharing situations are equi-sharing, and that all segmenting situations are equi-segmenting.

[9] It is also well known, by these same studies, that if you change the numbers so that in either situation each person receives 2/3 of a pizza, students will see these problems even differently yet. We agree with Confrey (1994) that this result is entirely an artifact of persistent experience with stereotypical problems in which products are always larger than either factor and quotients are always smaller than either the divisor or the dividend.

understand the numerical equivalence of measuring and partitioning, they understand that any measure of a quantity induces a partition of it and that any partition of a quantity induces a measure of it.

Fraction Schemes

A common approach to teaching fractions is to have students consider a collection of objects, some of which are distinct from the rest, as depicted in Figure 7.6. But, what might Figure 7.6 illustrate to students? It often is portrayed as illustrating 3/5 as if it cannot illustrate anything else. Yet a person could also see Figure 7.6 as illustrating that $1 \div 3/5 = 1\ 2/3$ —that within one whole is 1 three fifths and two thirds of another three fifths, or that $5 \div 3 = 1\ 2/3$—that within 5 is one 3 and two thirds of another 3. Finally, they could see Figure 7.6 as illustrating $5/3 \times 3/5 = 1$—that five thirds of (three fifths of 1) is 1 (Figure 7.7).

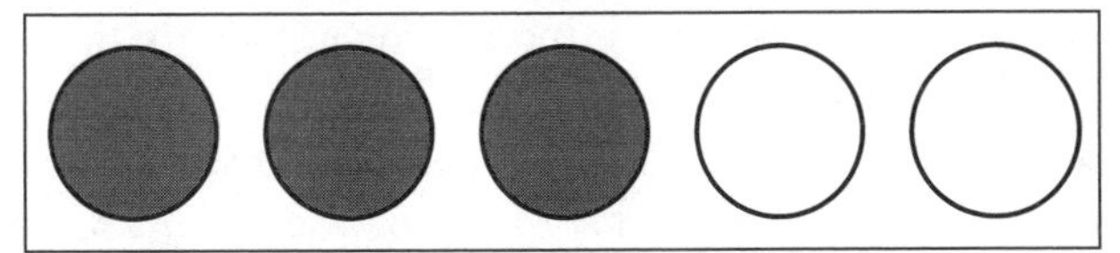

Figure 7.6. What does this collection illustrate?

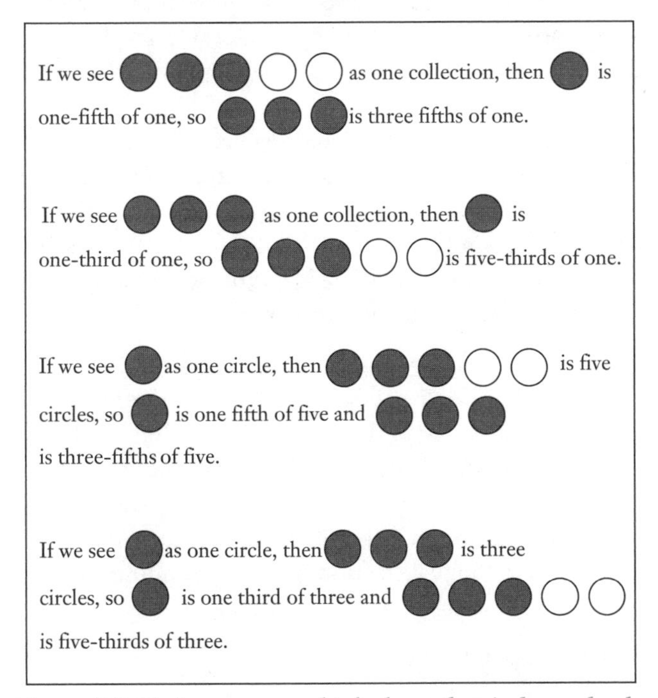

Figure 7.7. Various ways to think about the circles and collections in Figure 7.6.

We rarely observe textbooks or teachers discussing the difference between thinking of 3/5 as "three out of five" and thinking of it as "3 one fifths." How a student understands Figure 7.6 in relation to the fraction 3/5 can have important consequences. When students think of fractions as "so many out of so many," they are puzzled by fractions like 6/5. How do you take six things out of five?[10]

The system of conceptual operations comprising a fraction scheme is based on conceiving two quantities as being in a reciprocal relationship of relative size: "Amount A is $1/n$ the size of amount B" means that amount B is n times as large as amount A; "amount A is n times as large as amount B" means that amount B is $1/n$ as large as amount A. Another way to say "reciprocal relationship of relative size" is to say that the two amounts in comparison are each measured in units of the other. Saying amount B is seven times as large as amount A is saying that amount B is measured in units of A; saying amount A is 1/7 as large as amount B is saying that amount A is measured in units of B.

We worded these sentences very carefully, avoiding the phrase "$1/n$ of." This is to emphasize the slippery connection, mentioned previously, between comparison and inclusion. Fraction sophisticates can say, "A is some fraction of B" and get away with it because they imagine that A and B are separate amounts even though A might be perceived as including B (see Figure 7.4). Students are often instructed, and therefore learn, that the fractional part is contained within the whole, so "A is some fraction of B" connotes a sense of inclusion to them, that A is a subset of B. As a result, statements like "A is 6/5 of B" make no sense to them. However, when thought of multiplicatively, "A is m/n as large as B" means that A is m times as large as $1/n$ of B. So from a multiplicative perspective, "A is 6/5 as large as B" does not imply that A and B have anything in common. Rather, it means that A is 6 times as large as (1/5 of) B.

Our conceptual analysis of fraction knowledge resembles the approach taken by the Rational Number Project (Behr et al., 1992, 1993; Lesh et al., 1988). They characterized "m/n of B" as m one-$n$ths of B. The difference between their characterization and ours is that we stress the need for a direct link to students' conceptualizing reciprocal relationships of relative size. Without our insistence on students' having that link, they could appear to be speaking of fractions in ways we intend while in fact be thinking of additive inclusion—that $1/n$ of B is just one of a collection of pieces—without grounding it in an image of relative size.

Students gain considerable mathematical power by coming to understand fractions through a scheme of

[10] We often hear teachers and teacher education students say, "change 6/5 to 1 1/5 and they'll understand." This misses the point. It is problematic if a student must change 6/5 to 1 1/5, for it means that students cannot understand any situation in which they must see fractions as entailing a proportional relationship.

operations that express themselves in reciprocal relationships of relative size. For example, "related rate" problems are notoriously difficult for U.S. algebra students.

> It took Jack 20 seconds to run as far as Bill did in 17 seconds. Jack ran at about 18 ft./sec. throughout. About how fast did Bill run?

One way to reason about this problem using basic knowledge of fractions is shown below. The chain of reasoning is based on Figure 7.8. In it we see that Jack and Bill ran the same distance but in different amounts of time. It is important to note that 1 second is 1/20 of Jack's time and is 1/17 of Bill's time.

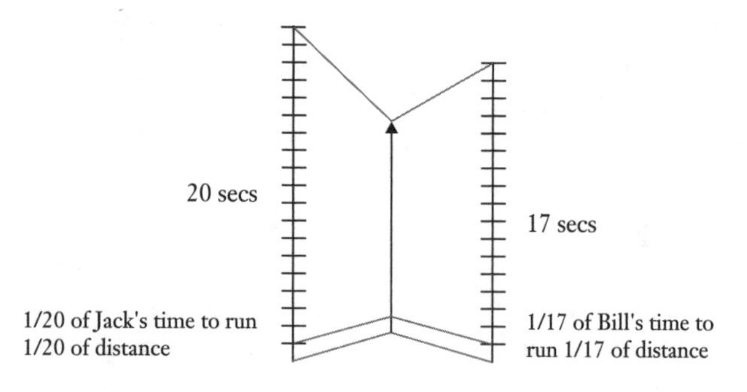

Figure 7.8. Jack and Bill took different amounts of time to run the same distance.

- Bill runs 1/17 of the distance in 1/17 of his time.
- Jack runs 1/20 of the distance in 1/20 of his time.
- 1/17 of Bill's time is as large as 1/20 of Jack's time.
- Were Bill to continue running 3 seconds, he and Jack will have run the same amount of time.
- Were Bill to continue running another 3 seconds, he will run an additional 1/17 of the distance in each second
- Bill runs 20/17 as far as Jack in the same amount of time
- Therefore, Bill runs 20/17 as fast as Jack.

To understand fractions as based in reciprocal relationships of relative size draws heavily on relationships among measure, multiplication, and division. For example, if we interpret "$a \div b$" as "the number of b's in a," then whatever number $a \div m/n$ is (i.e., whatever number is the number of m/n's in a), it is n times as large as $a \div m$ because m/n is $1/n$ as large as m (see the discussion following Figure 7.1).[11] When we recall that $(a \div m)$ is $1/m$ as large as a, we can conclude that whatever number is $a \div m/n$, it is n times as large as $1/m$ of a. Put more briefly, $a \div m/n = n/m \cdot a$. "The number of m/ns in a is n times as large as $1/m$ of a." This is a conceptual derivation of the "invert and multiply" rule for division by fractions, and its interpretation is straightforward when thinking in terms of relative sizes. Similarly, we can use fraction relationships to reason about such algebraic statements as $x = y/32$. The statement "$x = y/32$" means that x is 1/32 as large as y. That means that y is 32 times as large as x. So $y = 32x$.

The evolution of students' understandings of reciprocal relationships of relative size is still being researched, especially by Steffe and his colleagues (Hunting, Davis, & Pearn, 1996; Steffe, 1991a, 1993, 1994, in press; Tzur, 1999). The fact that this understanding happens so rarely among U.S. students makes it quite hard to research its development. But the fact that these understandings of fractions exist so rarely is a significant problem for U.S. mathematics education, for there is some evidence that it is expected more routinely elsewhere (Dossey et al., 1997; Ma, 1999; McKnight et al., 1987). It should be a dominant topic of discussion that U.S. instruction fails to support its development, and we should understand the reasons for that failure in detail. At present, we can say little more than, "It doesn't happen because few teachers and teacher educators expect it to happen." We should be able to say more.

The matter of students' and teachers' inattention to what numbers are and what it means to operate on them is reminiscent of a discussion between Samuel Kutler (1998) and John Conway (1998) in which Kutler wondered about Conway and Guy's (1996) treatment of fractions. Kutler asked the following question:

> ... they [Conway and Guy] give this example: $2/3 \times 1/4 = 4/6 \times 1/4 = 1/6$. What is the logic of this? Do [they] think that the readers have learned that the definition of the product of a/b and c/d is ac/bd and that they have forgotten it? Do they think that the reader knows or will investigate the justification for compounding the ratios of numbers, or what? (Kutler, 1998)

Conway responded this way:

> Guy and Conway DO think, unfortunately, that readers have learned that that is the definition of

[11] How you say the symbols to yourself while reading this passage can help or impede understanding it. To read "m/n" as "m slash n" or "m over n" makes it virtually impossible to make sense of the passage. To read "m/n" as "m $n$ths" helps. To read it as "m one-$n$ths" helps even more.

> the product, which it really isn't, except in formal axiomatic contexts. Suppose one stick is three-quarters as long as another, whose length is two-and-a-half inches. Then do you really think that the reason the shorter stick's length is 15/8 inches is because the DEFINITION of *a*/*b* times *c*/*d* is *ac*/*bd*? It ISN'T! It's because what actually happens is that if you cut the longer stick into 4 equal quarters and throw one of them away, the total length of what's left will actually be 15/8 inches. (Conway, 1998)

Kutler spoke from a presumed image of the result of multiplying fractions as being determined by an algorithm. Conway pointed out that the result of multiplying fractions is determined by what multiplication of fractions *means*. In his example, 3/4 × 2 1/2 refers to a length that is three times as long as is 1/4 of 2 1/2 . That is the way he turned "of 3/4 × 21/2" into something to which it referred. It happens that the result is 15/8 not by definition but by coincidence. The rule "$a/b \times c/d = ac/bd$" is a generalization that derives from the meanings of multiplication and of fractions in conjunction with a particular notational system for expressing those meanings. We agree with Conway that it is misguided to portray multiplication of fractions as being defined by a generalization, and we support the larger implication that it is misguided to portray numbers and operations as being defined by what could be (under a more enlightened pedagogy) generalizations rooted in meaning-making activities.

Although we have chosen not to discuss developmental issues of fraction understanding, we do note that Steffe (2001) describes a form of reasoning that is quite close to a conception of fraction that, while not entailing proportionality, students having it are close to understanding two magnitudes in reciprocal relationship of relative size. Steffe describes students who, when asked to make a stick that is 1/12th the length of a given stick, generated a length that, when iterated 12 times, produces a length as long as the given stick. These students did not have a conception of fractions that entailed proportionality, but they were fully aware that the longer stick was made of 12 of the shorter ones and that 12 of the shorter ones made one of the longer ones. What they were lacking was the sense of reciprocity of relative size—that the shorter stick was measured in units of the larger one as well as that the larger one was measured in units of the shorter. Steffe claims that this concept is an intermediate state between additive and multiplicative understandings of fractions and that through repeatedly generating such parts in relation to wholes, students will solidify their image into one that entails proportionality and reciprocity of relative size.

Fractions—A synthesis

The title of this section does not point to what we do in it. We do not synthesize what we have said about measurement, multiplication, division, and fraction. Rather, it points to what fractions are. Fractions *are* a synthesis. We do not make this claim in a realist sense; we mean it in a cognitive sense. Fractions become "real" when people understand them through complementary schemes of conceptual operations that are each grounded in a deep understanding of proportionality. At the same time, a focus on developing these schemes will enrich students' understanding of proportionality.

Most evidence points to the firm conclusion that fractions are rarely real to U.S. students. The extent to which other countries' cultures and educational systems support students' creation of that reality is an open question but an important one. Nevertheless, our system would be improved were it to support students' creation of fractions. We need not compare ourselves with other countries in setting goals for ourselves.

Conclusion

We have attempted to make explicit one image of what "to understand fractions" means. In doing so, we described conceptual operations that might underpin those understandings. We stress that we do not *intend* this chapter to be definitive. Rather, we offer it as a "sacrificial example" that we hope contributes to productive discussions of what we intend that students learn from instruction.

We have not discussed issues of symbolization, representation, symbolic skill, problem solving, and so on. The issues we have raised are largely independent of symbolism and symbolic procedure. Conceptual operations are about what one sees, not what one does, and about reasoning that is grounded in meaning. This is not to say that "doing" is unimportant. Rather, it expresses the commonsense adage that one should conceive of what one is trying to accomplish before trying to do it. Rules and shortcuts for operating symbolically should be generalizations from conceptual operations instead of being taught in place of them. Of course, part of building conceptual operations is the attempt to express them in symbols and diagrams. Symbolic operations can become the focus of instruction once students have developed coherent and stable meanings that they may express symbolically. But as a number of people have emphasized, it is important that students have a rich web of meanings to fall back on when they become confused or cannot recall a remembered procedure (Brownell, 1935; Cobb, Gravemeijer, Yackel, McClain, & Whitenack, 1997; Gravemeijer, Cobb, Bowers, & Whitenack, 2000; I. Thompson, 1994; Thompson, 1992; Wearne & Hiebert, 1994).

Scheme-based characterizations of learning objectives have a practical drawback. Assessment of whether students have achieved a learning objective is more complicated when expressing it in terms of schemes of conceptual operations than when expressing it in terms of behavioral skills. When learning objectives are stated in terms of skills, determining whether a student has achieved them is straightforward. When learning objectives are stated as schemes of operations, students' behavior must be interpreted to decide whether it reflects reasoning that is consistent with the objectives' achievement. This complication is unavoidable. However, coming to see reasoning in students' behavior is part of coming to understand the scheme-based approach we have used, so attempts at increasing teachers' sensitivity to students' reasoning may have the side effect of preparing them to think of learning objectives in terms of conceptual operations.

In closing, we re-emphasize our motive for writing this chapter. We believe that conditions for improved instruction entail an enduring discussion of what the community intends that students learn. When teachers and teacher educators lack clarity and conceptual coherence in what they intend students to learn, they introduce a systematic disconnection between instruction and learning. This becomes a major obstacle to creating instruction that empowers students to think mathematically. Please note that we said the problem is a lack of clarity and coherence in what teachers and teacher educators intend that students learn; we did not say that any one intention is correct. Nevertheless, not all intentions are equally coherent or equally powerful. We hope this chapter helps initiate prolonged discussions of what those intentions should be.

ACKNOWLEDGMENTS

The authors would like to acknowledge Les Steffe and Tommy Dreyfus for their very helpful suggestions on earlier drafts of this chapter.

The preparation of this chapter was supported by National Science Foundation grant number REC-9811879. All opinions expressed are those of the authors and do not reflect official positions of the National Science Foundation.

REFERENCES

American Heritage. (2002). *Dictionary of the English Language*. Retrieved 11 August 2002, from http://www.bartleby.com/61/93/U0059300.html

Ball, D. L. (1990). Prospective elementary and secondary teachers' understanding of division. *Journal for Research in Mathematics Education, 21*, 132–144.

Behr, M., Harel, G., Post, T., & Lesh, R. (1992). Rational number, ratio, and proportion. In D. Grouws (Ed.), *Handbook for research on mathematics teaching and learning* (pp. 296–333). New York: Macmillan.

Behr, M., Harel, G., Post, T., & Lesh, R. (1993). Rational numbers: Toward a semantic analysis—Emphasis on the operator construct. In T. P. Carpenter, E. Fennema, & T. A. Romberg (Eds.), *Rational numbers: An integration of research* (pp. 49–84). Hillsdale, NJ: Erlbaum.

Bell, A., Fischbein, E., & Greer, B. (1984). Choice of operation in verbal arithmetic problems: The effects of number size, problem structure and context. *Educational Studies in Mathematics, 15*, 129–147.

Bell, A., Swan, M., & Taylor, G. (1981). Choice of operations in verbal problems with decimal numbers. *Educational Studies in Mathematics, 12*, 399–420.

Bettoni, M. C. (1998, August 12). *The "Attentional Quantum" model of concepts and objects*. Retrieved 15 January 2000 from http://www.fhbb.ch/weknow/ini/front.htm

Brown, C. A. (1993). A critical analysis of teaching rational number. In T. P. Carpenter, E. Fennema, & T. A. Romberg (Eds.), *Rational numbers: An integration of research* (pp. 197–218). Hillsdale, NJ: Erlbaum.

Brownell, W. (1935). Psychological considerations in the learning and teaching of arithmetic. In W. D. Reeve (Ed.), *The teaching of arithmetic* (Tenth yearbook of the National Council of Teachers of Mathematics, pp. 1–31). New York: Bureau of Publications, Teachers College.

Burns, M. (2000). *About teaching mathematics: A K–8 resource* (2nd ed.). Sausalito, CA: Math Solutions Publications.

Carpenter, T. P. (1986). Conceptual knowledge as a foundation for procedural understanding. In J. Hiebert (Ed.), *Conceptual and procedural knowledge: The case of mathematics* (pp. 113–132). Hillsdale, NJ: Erlbaum.

Cobb, P., Gravemeijer, K. P. E., Yackel, E., McClain, K., & Whitenack, J. (1997). The emergence of chains of signification in one first-grade classroom. In D. Kirshner & J. A. Whitson (Eds.), *Situated cognition theory: Social, semiotic, and neurological perspectives* (pp. 151–233). Hillsdale, NJ: Erlbaum.

Confrey, J. (1994). Splitting, similarity, and rate of change: A new approach to multiplication and exponential functions. In G. Harel & J. Confrey (Eds.), *The development of multiplicative reasoning in the learning of mathematics* (pp. 293–330). Albany, NY: SUNY Press.

Confrey, J., & Smith, E. (1995). Splitting, covariation, and their role in the development of exponential function. *Journal for Research in Mathematics Education*, *26*(1), 66–86.

Conway, J. H. (1998, September 17). *Further fruitfulness of fractions*. Retrieved 1 December 1999 from http://forum.swarthmore.edu/epigone/historia_matematica/smanprehdol/199809180107.WAA23616@chasque.chasque.apc.org

Conway, J. H., & Guy, R. K. (1996). *The book of numbers*. New York: Springer Verlag.

De Corte, E., Verschaffel, L., & Van Coillie, V. (1988). Influence of number size, problem structure and response mode on children's solutions of multiplication word problems. *Journal of Mathematical Behavior*, 7, 197–216.

Dossey, J. A., Peak, L., & Nelson, D. (1997). *Essential skills in mathematics: A comparative analysis of American and Japanese assessments of eighth-graders* (NCES 97-885). Washington, DC: U.S. Department of Education, National Center for Education Statistics.

Edwards, C. H. (1979). *The historical development of the calculus*. New York: Springer-Verlag.

Eves, H. (1976). *An introduction to the history of mathematics* (4th ed.). New York: Holt, Rinehart, & Winston.

Eves, H., & Newsom, C. V. (1965). *An introduction to the foundations and fundamental concepts of mathematics*. New York: Holt, Rinehart, & Winston.

Fischbein, E., Deri, M., Nello, M. S., & Marino, M. S. (1985). The role of implicit models in solving verbal problems in multiplication and division. *Journal for Research in Mathematics Education*, *16*, 3–17.

Graeber, A., & Tirosh, D. (1988). Multiplication and division involving decimals: Pre-service elementary teachers' performance and beliefs. *Journal of Mathematical Behavior*, 7, 263–280.

Gravemeijer, K. P. E., Cobb, P., Bowers, J., & Whitenack, J. (2000). Symbolizing, modeling, and instructional design. In P. Cobb & K. McClain (Eds.), *Symbolizing and communicating in mathematics classrooms: Perspectives on discourse, tools, and instructional design* (pp. 225–274). Hillsdale, NJ: Erlbaum.

Greer, B. (1987). Non-conservation of multiplication and division involving decimals. *Journal for Research in Mathematics Education*, *18*, 37–45.

Greer, B. (1988). Non-conservation of multiplication and division: Analysis of a symptom. *Journal of Mathematical Behavior*, 7(3), 281–298.

Harel, G., Behr, M., Lesh, R., & Post, T. (1994). Constancy of quantity's quality: The case of the quantity of taste. *Journal for Research in Mathematics Education*, *25*(4), 324–345.

Harel, G., Behr, M., Post, T., & Lesh, R. (1992). The blocks task: Comparative analyses of the task with other proportion tasks and qualitative reasoning skills of seventh grade children in solving the task. *Cognition and Instruction*, *9*, 45–96.

Harel, G., Behr, M., Post, T., & Lesh, R. (1994). The impact of number type on the solution of multiplication and division problems: Further considerations. In G. Harel & J. Confrey (Eds.), *The development of multiplicative reasoning in the learning of mathematics* (pp. 365–386). Albany, NY: SUNY Press.

Hart, K. (1978). The understanding of ratios in secondary school. *Mathematics in School*, 7, 4–6.

Henle, J. M. (1986). *An outline of set theory*. New York: Springer Verlag.

Heyting, A. (1956). *Intuitionism: An introduction*. Amsterdam: North-Holland Publishing Co.

Hiebert, J., & Lefevre, P. (1986). Conceptual and procedural knowledge in mathematics: An introductory analysis. In J. Hiebert (Ed.), *Conceptual and procedural knowledge: The case of mathematics* (pp. 3–20). Hillsdale, NJ: Erlbaum.

Hunting, R. P., Davis, G., & Pearn, C. A. (1996). Engaging whole-number knowledge for rational number learning using a computer-based tool. *Journal for Research in Mathematics Education*, 27, 354–379.

Karplus, R., Pulos, S., & Stage, E. (1979). Proportional reasoning and the control of variables in seven countries. In J. Lochhead & J. Clement (Eds.), *Cognitive process instruction* (pp. 47–103). Philadelphia: Franklin Institute Press.

Karplus, R., Pulos, S., & Stage, E. (1983). Early adolescents' proportional reasoning on 'rate' problems. *Educational Studies in Mathematics*, *14*, 219–234.

Kasner, E., & Newman, J. R. (1940). *Mathematics and the imagination*. New York: Simon & Schuster.

Kieren, T. E. (1988). Personal knowledge of rational numbers: Its intuitive and formal development. In J. Hiebert & M. Behr (Eds.), *Number concepts and operations in the middle grades* (pp. 162–181). Reston, VA: National Council of Teachers of Mathematics.

Kieren, T. E. (1992). Rational and fractional numbers as mathematical and personal knowledge: Implications for curriculum and instruction. In G. Leinhardt, R. Putnam, & R. A. Hattrup (Eds.), *Analysis of arithmetic for mathematics teaching* (pp. 323–372). Hillsdale, NJ: Erlbaum.

Kieren, T. E. (1993a, February). *The learning of fractions: Maturing in a fraction world*. Paper presented at the Conference on Learning and Teaching Fractions, Athens, GA.

Kieren, T. E. (1993b). Rational and fractional numbers: From quotient fields to recursive understanding. In T. P. Carpenter, E. Fennema, & T. A. Romberg (Eds.), *Rational numbers: An integration of research* (pp. 49–84). Hillsdale, NJ: Erlbaum.

Kneebone, G. T. (1963). *Mathematical logic and the foundations of mathematics*. London: D. Van Nostrand.

Kutler, S. S. (1998, September 17). *Further fruitfullness of fractions*. Retrieved 1 December 1999, from http://mathforum.org/epigone/math-history-list

Lesh, R., Behr, M., & Post, T. (1987). Rational number relations and proportions. In C. Janvier (Ed.), *Problems of representation in teaching and learning mathematics* (pp. 41–58). Hillsdale, NJ: Erlbaum.

Lesh, R., Post, T., & Behr, M. (1988). Proportional reasoning. In J. Hiebert & M. Behr (Eds.), *Number concepts and operations in the middle grades* (pp. 93–118). Reston, VA: National Council of Teachers of Mathematics.

Luke, C. (1988). The repeated addition model of multiplication and children's performance on mathematical word problems. *Journal of Mathematical Behavior*, 7, 217–226.

Ma, L. (1999). *Knowing and teaching elementary mathematics: Teachers' knowledge of fundamental mathematics in China and the United States*. Mahwah, NJ: Erlbaum.

Mack, N. K. (1993). Learning rational numbers with understanding: The case of informal knowledge. In T. P. Carpenter, E. Fennema, & T. A. Romberg (Eds.), *Rational numbers: An integration of research* (pp. 85–105). Hillsdale, NJ: Erlbaum.

Mack, N. K. (1995). Confounding whole-number and fraction concepts when building on informal knowledge. *Journal for Research in Mathematics Education*, *26*(5), 422–441.

McKnight, C., Crosswhite, J., Dossey, J., Kifer, L., Swafford, J., Travers, K., et al. (1987). *The underachieving curriculum: Assessing U. S. school mathematics from an international perspective*. Urbana, IL: Stipes.

Minskaya, G. I. (1975). Developing the concept of number by means of the relationship of quantities (A. Bigelow, Trans.). In L. P. Steffe (Ed.), *Soviet studies in the psychology of learning and teaching mathematics* (Vol. 7, pp. 207–261). Palo Alto, CA, & Reston, VA: School Mathematics Study Group & National Council of Teachers of Mathematics.

Noelting, G. (1980a). The development of proportional reasoning and the ratio concept. Part 2—Problem structure at successive stages: Problem-solving strategies and the mechanism of adaptive restructuring. *Educational Studies in Mathematics*, *11*, 331–363.

Noelting, G. (1980b). The development of proportional reasoning and the ratio concept. Part 1—Differentiation of stages. *Educational Studies in Mathematics*, *11*, 217–253.

Piaget, J. (1950). *The psychology of intelligence*. London: Routledge & Kegan Paul.

Piaget, J. (1970). *The child's conception of movement and speed*. New York: Basic Books.

Piaget, J. (1971a). *Biology and knowledge: An essay on the relations between organic regulations and cognitive processes*. Chicago: University of Chicago Press.

Piaget, J. (1971b). *Genetic epistemology*. New York: Norton.

Piaget, J. (1976). *The child and reality*. New York: Penguin Books.

Post, T. R., Cramer, K. A., Behr, M., Lesh, R., & Harel, G. (1993). Curriculum implications of research on the learning, teaching, and assessing of rational number concepts. In T. P. Carpenter, E. Fennema, & T. A. Romberg (Eds.), *Rational numbers: An integration of research* (pp. 327–362). Hillsdale, NJ: Erlbaum.

Post, T. R., Harel, G., Behr, M., & Lesh, R. (1991). Intermediate teachers' knowledge of rational number concepts. In E. Fennema, T. P. Carpenter, & S. J. Lamon (Eds.), *Integrating research on teaching and learning mathematics* (pp. 177–198). Ithaca, NY: SUNY Press.

Reese, C. M., Miller, K. E., Mazzeo, J., & Dossey, J. A. (1997). *NAEP 1996 mathematics report card for the nations and states*. Washington, DC: National Center for Education Statistics.

Saldanha, L., & Thompson, P. W. (1998). *Re-thinking co-variation from a quantitative perspective: Simultaneous continuous variation*. Paper presented at the annual meeting of the North American Chapter of the International Group for the Psychology of Mathematics Education, Raleigh, NC.

Schwartz, J. (1988). Intensive quantity and referent transforming arithmetic operations. In J. Hiebert & M. Behr (Eds.), *Number concepts and operations in the middle grades* (pp. 41–52). Reston, VA: National Council of Teachers of Mathematics.

Simon, M. A. (1993). Prospective elementary teachers' knowledge of division. *Journal for Research in Mathematics Education*, *24*, 233–254.

Skemp, R. (1979). *Intelligence, learning, and action*. New York: Wiley.

Steffe, L. P. (1988). Children's construction of number sequences and multiplying schemes. In J. Hiebert & M. Behr (Eds.), *Number concepts and operations in the middle grades* (pp. 119–140). Reston, VA: National Council of Teachers of Mathematics.

Steffe, L. P. (1991a, April). *Composite units and their schemes*. Paper presented at the annual meeting of the American Educational Research Association, Chicago.

Steffe, L. P. (1991b). Operations that generate quantity. *Journal of Learning and Individual Differences*, *3*(1), 61–82.

Steffe, L. P. (1993, February). *Learning an iterative fraction scheme*. Paper presented at the Conference on Learning and Teaching Fractions, Athens, GA.

Steffe, L. P. (1994). Children's multiplying and dividing schemes: An overview. In G. Harel & J. Confrey (Eds.), *The development of multiplicative reasoning in the learning of mathematics* (pp. 3–39). Albany, NY: SUNY Press.

Steffe, L. P. (1996). Radical constructivism: A way of knowing and learning [Review of the same title, by Ernst von Glasersfeld]. *Zentralblatt für Didaktik der Mathematik*, *96*(6), 202–204.

Steffe, L. P. (2001). A new hypothesis concerning children's fractional knowledge. *Journal of Mathematical Behavior*, *20*, 267–307.

Stigler, J. W., Gonzales, P., Kawanaka, T., Knoll, S., & Serrano, A. (1999). The TIMSS *Videotape Classroom Study: Methods and findings from an exploratory research project on eighth-grade mathematics instruction in Germany, Japan, and the United States* (National Center for Education Statistics Report No. NCES 99-0974). Washington, D.C.: U. S. Government Printing Office.

Tall, D., & Vinner, S. (1981). Concept images and concept definitions in mathematics with particular reference to limits and continuity. *Educational Studies in Mathematics*, *12*, 151–169.

Thompson, I. (1994). Young children's idiosyncratic written algorithms for addition. *Educational Studies in Mathematics*, *26*, 323–346.

Thompson, P. W. (1982). *A theoretical framework for understanding young children's concepts of whole-number numera-*

tion. Unpublished doctoral dissertation, University of Georgia, Department of Mathematics Education.

Thompson, P. W. (1988). *Quantitative concepts as a foundation for algebra*. Paper presented at the annual meeting of the North American Chapter of the International Group for the Psychology of Mathematics Education, De Kalb, IL.

Thompson, P. W. (1992). Notations, conventions, and constraints: Contributions to effective uses of concrete materials in elementary mathematics. *Journal for Research in Mathematics Education, 23*, 123–147.

Thompson, P. W. (1993). Quantitative reasoning, complexity, and additive structures. *Educational Studies in Mathematics, 25*, 165–208.

Thompson, P. W. (1994). The development of the concept of speed and its relationship to concepts of rate. In G. Harel & J. Confrey (Eds.), *The development of multiplicative reasoning in the learning of mathematics* (pp. 179–234). Albany, NY: SUNY Press.

Thompson, P. W. (1995). Notation, convention, and quantity in elementary mathematics. In J. Sowder & B. Schapelle (Eds.), *Providing a foundation for teaching middle school mathematics* (pp. 199–221). Albany, NY: SUNY Press.

Thompson, P. W. (1996). Imagery and the development of mathematical reasoning. In L. P. Steffe, P. Nesher, P. Cobb, G. Goldin, & B. Greer (Eds.), *Theories of mathematical learning* (pp. 267–283). Hillsdale, NJ: Erlbaum.

Thompson, P. W. (2000). Radical constructivism: Reflections and directions. In L. P. Steffe & P. W. Thompson (Eds.), *Radical constructivism in action: Building on the pioneering work of Ernst von Glasersfeld* (pp. 412–448). London: Falmer Press.

Tourniaire, F., & Pulos, S. (1985). Proportional reasoning: A review of the literature. *Educational Studies in Mathematics, 16*, 181–204.

Tzur, R. (1999). An integrated study of children's construction of improper fractions and the teacher's role in promoting that learning. *Journal for Research in Mathematics Education, 30*, 390–416.

Vergnaud, G. (1983). Multiplicative structures. In R. Lesh & M. Landau (Eds.), *Acquisition of mathematics concepts and processes* (pp. 127–174). New York: Academic Press.

Vergnaud, G. (1988). Multiplicative structures. In J. Hiebert & M. Behr (Eds.), *Number concepts and operations in the middle grades* (pp. 141–161). Reston, VA: National Council of Teachers of Mathematics.

Vergnaud, G. (1994). Multiplicative conceptual field: What and why? In G. Harel & J. Confrey (Eds.), *The development of multiplicative reasoning in the learning of mathematics* (pp. 41–59). Albany, NY: SUNY Press.

Von Glasersfeld, E. (1972). Semantic analysis of verbs in terms of conceptual situations. *Linguistics, 94*, 90–107.

Von Glasersfeld, E. (1978). Radical constructivism and Piaget's concept of knowledge. In F. B. Murray (Ed.), *Impact of Piagetian theory* (pp. 109–122). Baltimore, MD: University Park Press.

Von Glasersfeld, E. (1995). *Radical constructivism: A way of knowing and learning*. London: Falmer Press.

Wearne, D., & Hiebert, J. (1994). Place value and addition and subtraction. *Arithmetic Teacher, 41*(5), 272–274.

Wilder, R. (1968). *Evolution of mathematical concepts: An elementary study*. New York: Wiley.

Wildi, T. (1991). *Units and conversions: A handbook for engineers and scientists*. New York: IEEE Press.

CHAPTER 8

Facts and Algorithms as Products of Students' Own Mathematical Activity

Koeno Gravemeijer, Freudenthal Institute and Vanderbilt University
Frans van Galen, Freudenthal Institute

Facts and algorithms are usually associated with drill and practice and with reproducing ideas from memory. This instrumental knowledge does not seem to fit very well with current approaches in mathematics reform that emphasize understanding, problem solving, and applications. In the history of mathematics, however, algorithms have their roots in problem solving. By formalizing, generalizing, schematizing, and developing procedures, various ways of solving applied problems have been turned into fixed algorithms, and, as we will show, facts have a similar origin. In this chapter, we elaborate the activity of what might be called "algorithmetizing" as an alternative for teaching algorithms. In such an approach, conventional algorithms might be the endpoints of an instructional sequence, but more often a sufficient aim for teachers is to develop semi-informal algorithms.

The algorithmetizing approach is an alternative for the prevailing way of teaching algorithms by trying to use concrete embodiments, such as manipulatives and visual representations, to explain them to students. Research has shown that instruction based on the use of concrete embodiments, for example, the Dienes or MAB blocks, is not very effective (see Labinowitz, 1985; Resnick & Omanson, 1987). Beyond the fact that many students are not able to show mastery of the algorithms as such, many of them fail when they must use mathematics in applied situations (Schoenfeld, 1987). A reason for these failings might lie in the fact that algorithms are often learned in isolation, leading to the effect that students know, in principle, how to do certain algorithms but do not know when those algorithms are appropriate. Hart (1981), for instance, observed this type of problem with the written algorithm for division. She found that students were more inclined to use repeated subtraction than to use the long-division algorithm when solving such "ratio division" problems as the following: "A truck drives with a speed of 75 miles per hour; how long does it take for that truck to cover a distance of 500 miles?" Apparently, students have a hard time interpreting this kind of task as a division problem.

A related problem is that students misapply rules they have learned as isolated, mathematical procedures. A well-known phenomenon is that of students' construing as a general rule the notion that multiplication makes numbers bigger and division makes them smaller. This misapplication then gets them into trouble when they must solve applied problems that require multiplying or dividing by a decimal number (Bell, Fischbein, & Greer, 1984). When students are asked to calculate the price of 0.58 kg of cheese, they divide the price of 1 kg by 0.58, reasoning that the answer must be smaller, an outcome that for them implies division.

In this chapter, we describe the learning of facts and algorithms as a process building on number sense. We argue that facts and algorithms are to be building stones of the active mathematical knowledge of students. Most of the chapter is devoted to the implications of this view for mathematics instruction. Instead of concretizing mathematical algorithms for the students, the teacher can let students develop or reinvent the algorithms themselves (Freudenthal, 1973, 1991; Kamii, Lewis, & Livingstone, 1993). Such a reinvention process starts with carefully chosen contextual problems. To solve these problems, students must model the situation to some degree. By reflecting on the solution procedures they have used, students may develop more sophisticated models and procedures that they can also use in other situations. This modeling-reinvention approach requires careful planning of instructional activities. The process is heavily dependent on the input from students, but the problems students are asked to solve and the tools they are given should be designed to offer rich opportunities for mathematics learning.

The critique of the limited value of facts and algorithms as such and of the detrimental effect of drill and practice has sometimes led educators to completely reject facts and algorithms as goals for mathematics education. We argue, however, for a mathematics education that offers students the opportunity to learn algorithms of some form. They do not necessarily have to be the canonical, standard algorithms; instead, they may be personal, semi-informal calculation procedures.

Number Sense as a Basis for Learning Facts and Algorithms

Facts and algorithms are normally considered to be quite different things, but are they? At first sight, a fact seems simply to be knowledge that, in one way or another, must be memorized. Different ways can be employed, however, to know a fact. The knowledge can be the result of being told about the fact, but it can also be the result of a self-developed insight, which as such might be the result of active engagement in mathematical problem solving. So we may ask ourselves what it means to know that $7 + 6 = 13$ or $6 \times 8 = 48$ or to know that *area = length times width*. Or, to extend this list to secondary education, what does it mean to know that the derivative of x^2 is $2x$? To extend it even further, what does it mean to know that $E = mc^2$?

However they are learned, facts do not exist without a framework of reference that gives them meaning. Most students, for example, know $6 + 7 = 13$ not as an isolated, memorized fact but as related to $6 + 6 = 12$, to $7 + 7 = 14$, or to $7 + 3 = 10$ and $6 = 3 + 3$. Knowing a fact, therefore, can hardly be separated from knowing how to make use of facts in a meaningful way. It is also hard to make a sharp distinction between facts and algorithms. The formula for the area of a rectangle, for instance, can be thought of both as a fact and as an algorithm. If the formula is known only as a phrase, it is a rather meaningless fact—as $E = mc^2$ is for many—but if one really knows how to apply the formula to calculate area, it is a procedure like an algorithm.

Over time, an alternative conception of what it might mean to learn facts and algorithms has been developed. A shift has been made from conceiving of knowledge as a commodity that can be transmitted toward conceiving of knowing as a form of activity (Sfard, 1998). In this view, facts and algorithms can be considered as an extension of number sense. We adopt here Greeno's (1991) environment metaphor, which depicts number sense as a sense of knowing one's way around in a mathematical environment. Such an environment may be conceived of as constituted by a network of number relations (Skemp, 1976; van Hiele, 1973).

An example of these number relations is the network of addition and subtraction facts for numbers under 20. Students can develop such a network in a series of activities on structuring quantities (Gravemeijer, Cobb, Bowers, & Whitenack, 2000). While doing those activities, students may learn that $2 + 2 = 4$. Basically, then, they learn that adding two items to two similar items results in four items, independently of the specific items or the situation involved. We might say that in learning this relationship, the students are generalizing over items and situations. That way, they can make the transition from thinking in terms of contextualized quantity numbers that are intrinsically linked to identifiable objects (e.g., two marbles and another two marbles make four marbles) to reasoning in terms of bare numbers and number relations (e.g., two and two equal four).

This transition can be described as construing numbers as mathematical entities that have objectlike characteristics. Here we may speak of a number existing for the student as a mathematical entity in and of itself because the numbers no longer derive their meaning from the reference to identifiable countable objects but instead from being embedded within a mathematical framework of number relations. As an example, the quantity four is no longer tied to identifiable objects, as it is in "four marbles," "four beads," "four children," and so forth, but rather is connected with number relations, such as $4 = 2 + 2$, $4 = 5 - 1$, $4 + 4 = 8$, and $4 + 6 = 10$.

The following situation is an example of an instructional activity that may foster the development of the aforementioned framework of number relations. The teacher puts a small number of chips on the overhead projector and shows this arrangement for a short time. The teacher then asks the students to tell how many chips they saw and how they saw them. The answers to the latter question will guide the teacher to start favoring solutions based on grouping over pure counting solutions (cf. McClain, 1995). In the example shown in Figure 8.1, children may report that they saw "a group of three and a group of four . . . seven chips altogether," or they may report that they saw "a group of six and one." On the basis of participating in such activities, students start to perceive number relations as being concrete.

O O O
O O O O

Figure 8.1. Counting chips on the overhead projector.

When more and more number relations are incorporated into this network, students may solve an addition problem, such as 7 + 6 = __, by using number facts, such as 6 = 3 + 3 and 7 + 3 = 10. Note that the application of a fixed procedure or algorithm is not the objective that we have in mind here. We do not presume that students are initially thinking in terms of filling out a ten, in which they decompose one number so that one of its parts can combine with the other number to make a ten. Instead, we suggest that they are initially guided by their knowledge of number relations. If in the framework of number relations that the students have constructed for themselves, such number relations as 6 = 3 + 3 and 7 + 3 = 10 are ready at hand, they can be used as building blocks to compose new relations.

Admittedly, standard practice is to teach students procedures, such as filling out a ten. Proposed here, however, is a bottom-up approach in which students develop a framework of number relations first. The idea is that students become so familiar with such relations as 3 + 3 = 6 and 7 + 3 = 10 that the solution 7 + 3 = 10, 10 + 3 = 13 presents itself as obvious. From an observer's point of view, the activity of the student may be categorized as "using the associative property and filling out a ten." From the student's point of view, however, the objective is not a question of applying some property; the student merely uses number sense and combines the number relations he or she knows in an efficient manner. In the approach we favor, only when students start to reflect on regularities in their solution methods do they realize how these solution methods can be classified as handy procedures or strategies. As a consequence, they may start to explicate strategies and routine procedures as generalizations about problem-solving methods. Subsequently, knowing these strategies and procedures may begin to guide their solution methods, and this guidance may result, eventually, in efficient routine behavior. We should note, however, that routine behavior is not always effective. Too much developing of procedures brings with it the danger of blind application—even when these procedures are well grounded in experience and reflection (van Parreren, 1981).

A second example comes from algebra in secondary education, in which students must learn rules for finding unknowns in sets of equations. We believe that instead of learning these rules as ready-made abstractions, students should be given the opportunity in concrete problem-solving situations to develop the rules themselves as generalizations of solutions. One such situation is given in Figure 8.2 (see Kindt & Abels, 1998, for similar problems). Students can find the unit costs of tacos and drinks by adding fictitious orders to the waiter's notebook. From the first order, for example, the fact that three such orders—three tacos and nine drinks—will cost \$15 follows logically. By comparing that amount to the cost of the second order, students can conclude that five drinks will cost \$5. Thus, one drink will cost \$1, and we then can see from the first order that one taco must cost \$2.

Order	Taco	Drink	Total
1	1	3	\$5
2	3	4	\$10

Figure 8.2. The waiter's notebook.

By comparing different solution methods, students learn which manipulations make sense and thereby may develop more general procedures. Initially, all manipulations are meaningful actions within the context of, for example, a Mexican restaurant. Later, the contexts may fade into the background and the manipulations with equations will acquire a meaning of their own. At that point, a more general notation may be introduced, for example, that shown in Figure 8.3.

$$\begin{cases} x+3y=5 & \vert\ 3 \\ 3x+4y=10 & \vert\ -1 \end{cases}$$

$$\begin{cases} 3x+9y=15 \\ -3x-4y=-10 \end{cases} +$$

$$5y=5$$
$$y=1$$

substitute:

$$x+3=5$$
$$x=2$$

Figure 8.3. A formal notation for the problem in figure 8.2.

So in our view, algebraic operations should be explored as operations in concrete situations that later develop into meaningful operations on more formal equations. In this process, equations become objects that can be manipulated more or less the way the quantity four in early number learning is initially bound to identifiable objects but later acquires an objectlike meaning itself.

In our opinion, therefore, the basis for learning procedures and algorithms lies in the ability to operate in a meaningful way within a mathematical environment. This environment may be quite elementary—such as the environment constituted by the network of addition and subtraction facts with results up to 20—but it also may be the environment of algebraic equations. To make possible students' ability to operate in a meaningful way, instructional sequences should start with problems that are experientially real for the students. Instead of the teacher's handing over ready-made procedures, students should develop procedures themselves by reflecting on their solutions for specific problems.

Guided Reinvention

The view of coming to know facts and algorithms described above fits with the view expressed in reform documents that emphasize the notion of mathematics as an activity. This notion of mathematics as an activity can be placed in opposition to the image of mathematics as a ready-made system of rules, procedures, and knowledge, which sometimes has led to a rejection of rules and procedures as goals for mathematics education. Educators can, however, transcend the dichotomy: Emphasizing mathematics as an activity does not imply that the subject matter of mathematics must be left out; it can be reinvented (Bednarz, Dufour-Janvier, Poirier, & Bacon, 1993; diSessa, Hammer, Sherin, & Kolpakowski, 1991; Pólya, 1963)

In making this argument, we follow Freudenthal (1971, 1973), who observed that in the history of mathematics, axioms, definitions, postulates, algorithms, and so forth, emerge as the endpoints of a long process of mathematical activity. In traditional mathematics education, these endpoints are taken as the starting points for instruction. Freudenthal sees this inversion as unwanted. Instead, he argues that the students should go through a similar process as the mathematicians go through themselves. He goes on to say that the students should be given the opportunity to reinvent mathematics. The overall objective would be, in his view, for the students to experience their mathematical knowledge as the product of their own mathematical activity.

In relation to mathematical activity, Freudenthal emphasizes organizing or *mathematizing*. Literally, mathematizing could be translated as "making more mathematical." Its implications can be construed by considering characteristics of mathematics, such as generality, certainty, exactness, and brevity. Mathematizing covers such activities as generalizing, justifying, formalizing, and curtailing; the latter also includes developing algorithms.

Mathematizing encompasses both reality-based subject matter and mathematical subject matter. Treffers (1987) distinguishes between "horizontal" mathematization and "vertical" mathematization. Horizontal mathematization involves converting a contextual problem into a mathematical problem to solve the problem mathematically. Vertical mathematization involves taking mathematical matter to a higher level. The latter outcome is possible when students mathematize their own mathematical activity. Vertical mathematizing is at the heart of mathematical progress, whereas horizontal mathematizing is essential to the goal of including a wide range of phenomena. Together, horizontal and vertical mathematizing constitute a process called "progressive mathematization.," which is in a sense the counterpart of reinvention: Reinvention from the point of view of the observer, who knows what is to be reinvented, may be progressive mathematization from the point of view of the actors, or the students.

The core idea is that the students develop mathematical concepts, notations, and procedures as organizing tools when solving problems. In such a process, informal algorithms may come to the fore as forms of well-organized routines for solving certain types of problems. With guidance from the teacher, these informal algorithms can be developed into conventional algorithms. The teacher, however, may also opt for fostering the informal algorithms as valuable end goals in and of themselves.

Guided reinvention asks for topic-specific instructional theories. To be able to organize learning processes, teachers need "local instruction theories" for each topic involved (Gravemeijer, 1998). A local instruction theory describes, with arguments, how enacting a series of instructional activities can support a specific long-term learning process. It would, for example, describe how children first learn informal ways of addition and subtraction and how they may thereby develop more standard procedures for addition and subtraction. Or it would describe how children develop calculation procedures that may or may not be similar to the standard algorithms for multiplication and division. We return to these examples in subsequent sections.

Modeling: The Case of Addition and Subtraction

Guided reinvention makes modeling a fundamental process in learning mathematics. The starting point is the commonsense solution methods that students display when solving problems that are experientially real for them. Next, the teacher models these commonsense methods. At first, acting with the model will be tightly connected with the situation that is being modeled, but gradually that acting may acquire a more independent status. This generalized model can then function as a means of support for more formal mathematical reasoning. In this approach, we more appropriately speak of "modeling" than of "model use" because the latter term suggests the use of ready-made tools from a mathematics toolbox.

This view of modeling and models contrasts with the widespread practice in mathematics instruction of using models that result from efforts to concretize expert knowledge. In that approach, the models are an attempt to make the expert knowledge accessible on the level of the student. An example is the use of base-ten blocks in what Resnick and Omanson (1987) call "mapping instruction." In this instruction, each step in the written algorithm is mapped onto an action with the blocks. From the perspective of the student, however, such rules may be completely incomprehensible.

Take, for instance, the problem of 85 – 59, where one starts with 8 tens and 5 ones. In a commonsense approach, taking away 5 tens first would be only natural. Then, knowing that 15 – 9 = 6, one might exchange 1 ten and 5 ones for 6 ones, or one might realize that taking away 9 is similar to taking away 1 ten and adding 1 one (see Figure 8.4). Solving the problem according to the standard procedure, however, the student is expected to follow a series of prescribed steps (Figure 8.5):

1. Start with the ones.
2. See if you can take away 9 from the ones.
3. That subtraction cannot be done, so exchange 1 ten for 10 ones.
4. Take away 9 ones.
5. Check whether you can take away 5 tens.
6. That subtraction can be done, so take away 5 tens.
7. Read the answer: 26.

Two problems arise with this mapping approach: The rule-governed procedure not only is rather forced but also does not prepare students adequately for the written algorithm. Whereas the students using the blocks can simply read off the answer, they will have to figure out 15 – 9 mentally when performing the written algorithm.

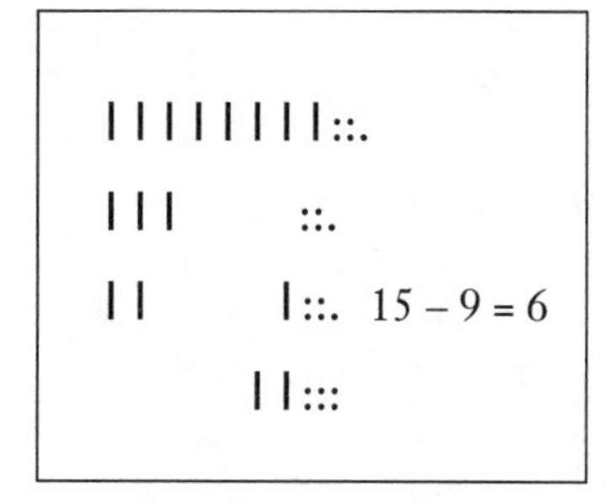

Figure 8.4. Informal procedure for the problem 85 – 59.

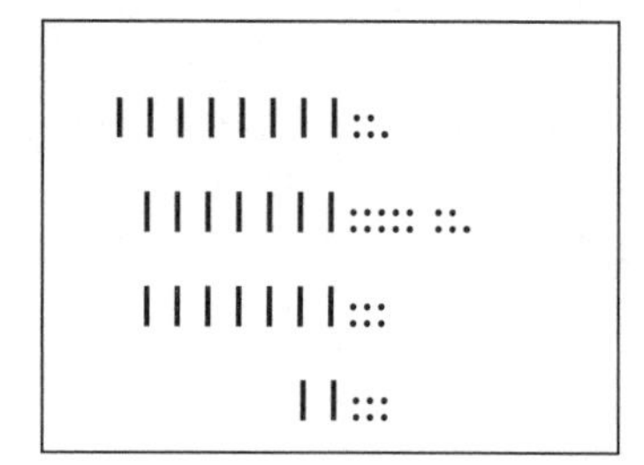

Figure 8.5. Stepwise procedure for the problem 85 – 59.

We may characterize this teaching approach as top-down because the points of reference are taken to be the ready-made column algorithm and expert knowledge of the decimal system, not the knowledge of students. The tens rods of 10, the hundreds squares, and the thousands cubes are presented as ready-made representations, but students may not yet have developed the concepts behind these representations. Also, the blocks are to be dealt with according to prescribed rules in ways that are different from what students would do if they were allowed to choose by themselves. According to research, this approach does not avoid certain consequences: a lack of understanding, a lack of proficiency, and a lack of applicability (Labinowitz, 1985; Cobb, Yackel, & Wood, 1992; Resnick & Omanson, 1987). We can picture the mapping approach as the expert stooping toward the student.

The alternative to the mapping approach is to try to help students build on and elaborate on their own informal knowledge. Toward this goal, the so-called empty number line is an appropriate tool. Instead of teaching children fixed procedures for addition and subtraction up to 100, Whitney (1985) and subsequently Treffers and de Moor (1990) proposed the use of a schematic number line on which students draw whatever number relations they see as useful for their solution. An example is given in Figure 8.6. The calculation procedures that children use in examples like these are self-developed—in contrast with ready-made algorithms—and they reflect the children's knowledge of number relations.

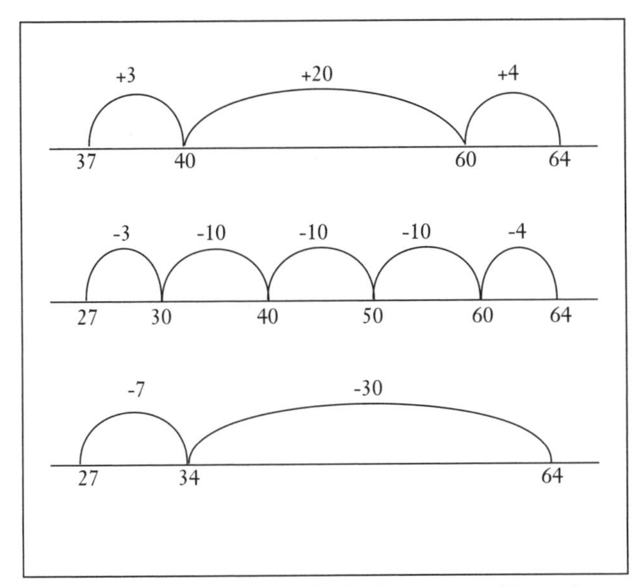

Figure 8.6. Different ways of calculating 64 – 37 on the empty number line.

Operations on the empty number line, as in the pupil's work in Figure 8.6, are already quite abstract. The tactic of introducing the empty number line as a ready-made tool just by showing its usefulness to students would therefore not be appropriate. Instead, in the necessary learning trajectory, the empty number line comes into being from reasoning in contextual problems. In the proposal of Treffers and de Moor (1990), the introduction of the number line is preceded by working with a string of 100 beads, colored in groups of 10 (see Figure 8.7). The students solve such tasks as the following: Count to 38 and add 24 more; what number do you get?

Later in this setup, the empty number line is introduced to model the activities on the bead string. So the number line first functions as a model *of* a situation, and the imagery of acting in that situation makes drawing arcs on a number line meaningful to the students. Gradually the students' attention will shift from thinking about acting on a bead string to thinking about mathematical relations among the numbers involved. As a consequence, acting on the number line can become meaningful in and of itself. Then the number line can start functioning as a model *for* mathematical reasoning.

A point of critique on the use of the bead string is its artificial character as a sort of tactile microworld. Students may come to think of the bead-string context as a world in and of itself, with its own rules that may or may not apply outside this context. These concerns prompted an attempt to design an instructional sequence that capitalizes on linear measurement as a basis for introducing the empty number line (McClain, Cobb, Gravemeijer, & Estes, 1999; Stephan, Cobb, Gravemeijer, & McClain, 1998). To describe this sequence very briefly, students measure various lengths by iterating some measurement basic unit and a larger measurement unit consisting of 10 basic units. This measuring with tens and ones is modeled with a 100-unit ruler that is structured by units of ten and one. Later, the activity of measuring is extended to incrementing, decrementing, and comparing lengths. These activities give rise to counting strategies that can be symbolized with arcs on a schematized ruler. This schematized representation is then used as a means of scaffolding and of communicating solution methods. Finally, this use of the number line is generalized for use with addition and subtraction problems in all sorts of contexts.

A feature that may cause problems is the rigid character of the ruler. On a ruler, positions must correspond to exact distances, whereas on the empty number line, the distances between the numbers do not have to correspond exactly with their numeric values. To try to strive for an exact proportional representation of all the jumps would severely hamper a flexible use of the number line. Students must therefore realize the distinction between the ruler as a measurement instrument and the empty number line as a means of expressing solution methods. This distinction is in line with the overall intent, which is *modeling an activity*, not modeling an object.

Semi-informal Algorithms

In a learning trajectory based on guided reinvention, careful consideration should be given to the characteristics of the symbolizations that are introduced. If possible, models should give students the freedom to do whatever they would have done in a concrete situation. Models should also give students opportunities to reflect on the mathematical characteristics of the solution procedures. This recommendation implies that the contextual prob-

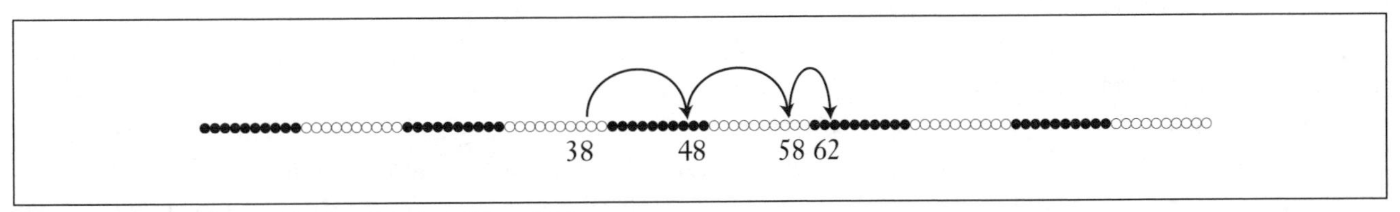

Figure 8.7. Jumping on the bead string: 38 and 24 more.

lems with which a sequence starts should be chosen in light of what is likely to follow. In the preceding section, we described this approach for addition and subtraction.

Introductory activities for algebra offer another example. In the learning sequence from which the Mexican Restaurant Problem was taken (Figure 8.2), the context of the restaurant offers a natural opportunity for introducing the waiter's notebook as a way to write down both the actual orders in the problem and fictitious orders. Adding new entries in the notebook prepares students for the procedure of transforming a given set of equations. In fact, the procedure is quite similar to the process of adding entries, but it still derives its meaning from the restaurant context. From operating in such a concrete contextual situation, students may develop more formal rules for operating with linear equations.

In time, progressive mathematization will bring students' reasoning to a higher level. In many cases, a certain degree of routine is desirable and thus suggests a certain standardizing of procedures. Designers of an instructional sequence must make choices at this point. In current mathematics education practice, for solving sets of equations, students sometimes are taught very strict rules that specify when to apply elimination or substitution and also how to do so. Another possibility, however, is to leave the solution procedure open, letting procedures function as heuristics rather than as fixed algorithms. With respect to algebra, this latter approach is more advisable, in our opinion, because the flexible use of such procedures as elimination and substitution will often result in simpler solutions. Instead of a fixed algorithmic procedure, we favor a flexible approach in which procedures become standardized to some degree and then function as semi-informal algorithms.

An example of a similar choice with respect to learning goals can be found in the multiplication algorithm. A possible approach is to develop procedures for the multiplication of integers using a ratio table (cf. van Galen & van den Heuvel-Panhuizen, 1997). A neat feature of the ratio table is its flexibility, which leaves room for a variety of solution methods. By using the ratio table, one can calculate the product 18 × 23, for instance, in various ways (see Figure 8.8).

One could stop here and accept that students use various ways to solve multiplication problems. If one wanted to, however, one could develop a standard procedure that is based on the characteristics of the decimal system and that resembles the conventional algorithm (see Figure 8.9).

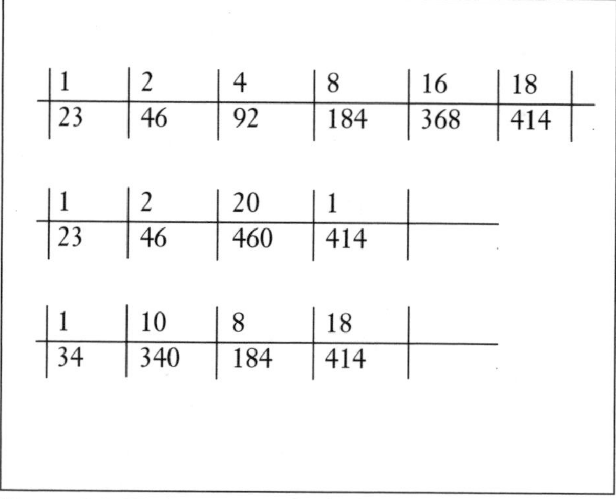

1	2	4	8	16	18
23	46	92	184	368	414

1	2	20	1
23	46	460	414

1	10	8	18
34	340	184	414

Figure 8.8. Various ways of calculating 18 × 23 with a ratio table.

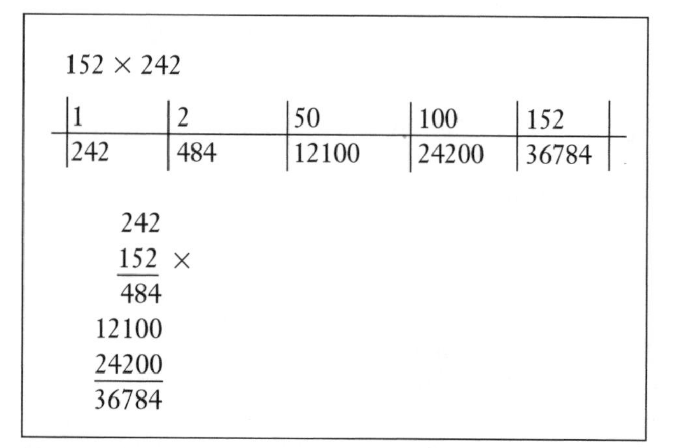

152 × 242

1	2	50	100	152
242	484	12100	24200	36784

$$\begin{array}{r} 242 \\ \underline{152} \times \\ 484 \\ 12100 \\ \underline{24200} \\ 36784 \end{array}$$

Figure 8.9. Ratio table and standard algorithm.

Conclusion

The development of algorithms is an essential component of mathematics. In the history of mathematics, methods for solving applied problems have been developed into efficient, fixed procedures. One is tempted to teach students these products of the work of the mathematicians of the past in ready-made form, especially if the goal is a more-or-less routine proficiency in mathematical procedures. And in many areas of mathematics, a certain routine is still desirable, even though now we can leave much of the work to calculators, graphing calculators, and computer algebra systems. Teaching students algorithms that they do not understand, however, has limited potential at best and, more important, leads to isolated skills that do not contribute to students' general mathematical knowledge. One reaction to this observation is to reject facts and algorithms entirely as goals of mathematics education. In this chapter, however, we have argued for an alternative approach that incorporates the learning of facts and algorithms into the core of mathematics education. Instructional sequences in which the students act like the mathematicians of the past and reinvent procedures and algorithms will contribute to their growth in mathematical understanding.

This emergent modeling-reinvention approach has been elaborated for many topics in the mathematics curriculum for grades K–12 (in, among others, the Mathematics in Context curriculum of the National Center for Research in Mathematical Sciences Education & Freudenthal Institute, 1998). The models that are part of these programs (e.g., the ratio table) can function as a basis for informal semi-algorithms. We believe this possibility implies that students do not necessarily have to reach the highest level of sophistication. In fact, we must give careful consideration to the level of automatization for which we are striving. Although a high level of automatization may be needed in some cases, such as in number relations or basic facts up to 20, the danger also exists that with too much automatization, rules and algorithms will be applied blindly. In that connection, an argument can be made for semi-informal algorithms and for keeping the sources of insight open (Freudenthal, 1973), so that students would always be able to fold back (Pirie & Kieren, 1994) to a more concrete level if the need arose.

We want to emphasize the importance of a change of perspective. The intent of a reinvention approach is to help the students generate useful and meaningful knowledge on their own level and at their own pace. This goal implies that students are to develop an attitude in which the first objective is that of understanding and coming to grips with the problem instead of looking at problems as instances of applications of some mathematical toolkit. Within this orientation, the facts and algorithms of mathematics are not the point of departure; instead, students get the opportunity to develop their own mathematical knowledge.

ACKNOWLEDGMENTS

The analysis reported in this chapter was in part supported by the National Science Foundation under grant number REC9814898 and by the Office of Educational Research and Improvement [OERI] under grant number R305A60007. The opinions expressed do not necessarily reflect the views of the Foundation or OERI. We are grateful to Karen Longhart, Paul Cobb, and Pat Tinto for comments on an earlier draft.

REFERENCES

Bell, A., Fischbein, E., & Greer, B. (1984). Choice of operation in verbal arithmetic problems: The effects of number size, problem structure and context. *Educational Studies in Mathematics, 15*, 129–148.

Bednarz, N., Dufour-Janvier, B., Poirier, L., & Bacon, L. (1993). Socio-constructivist viewpoint on the use of symbolism in mathematics education. *Alberta Journal of Educational Research, 39*, 41–58.

Cobb, P., Yackel, E., & Wood, T. (1992). A constructivist alternative to the representational view of mind in mathematics education. *Journal for Research in Mathematics Education, 23*, 2–33.

DiSessa, A. A., Hammer, D., Sherin, B., & Kolpakowski, T. (1991). Inventing graphing: Meta-representational expertise in children. *Journal of Mathematical Behavior, 10*, 117–160.

Freudenthal, H. (1971). Geometry between the devil and the deep sea. *Educational Studies in Mathematics, 3*, 413–435.

Freudenthal, H. (1973). *Mathematics as an educational task.* Dordrecht, The Netherlands: Reidel.

Freudenthal, H. (1991). *Revisiting mathematics education.* Dordrecht, The Netherlands: Kluwer.

Gravemeijer, K. (1998). Developmental research as a research method. In A. Sierpinska & J. Kilpatrick (Eds.), *Mathematics education as a research domain: A search for identity* (ICMI Study Publication 2, pp. 277–296). Dordrecht, The Netherlands: Kluwer.

Gravemeijer, K., Cobb, P., Bowers, J., & Whitenack, J. (2000). Symbolizing, modeling, and instructional design. In P. Cobb, E. Yackel, & K. McClain (Eds.), *Communicating and symbolizing in mathematics: Perspectives on discourse, tools, and instructional design* (pp. 225–273). Mahwah, NJ: Erlbaum.

Greeno, J .G. (1991). Number sense as situated knowing in a conceptual domain. *Journal for Research in Mathematics Education, 22*, 170–218.

Hart, K. M. (1981). *Children's understanding of mathematics: 11–16.* London: Murray.

Kamii, C., Lewis, B.A., & Livingstone Jones, S. (1993). Primary arithmetic: Children inventing their own procedures. *Arithmetic Teacher, 41*(4), 200–203.

Kindt, M., & Abels, M. (1998). Comparing quantities. In National Center for Research in Mathematical Sciences Education & Freudenthal Institute (Eds.), *Mathematics in context: A connected curriculum for grades 5–8* (pp. 50–77). Chicago: Encyclopedia Brittanica Educational Corporation.

Labinowitz, E. (1985). *Learning from children.* Amsterdam: Addison-Wesley.

McClain, K. (1995). *An analysis of the teacher's proactive role in supporting students' mathematical development.* Unpublished doctoral dissertation, Vanderbilt University, Nashville, TN.

McClain, K., Cobb, P., Gravemeijer, K., & Estes, B. (1999). Developing mathematical reasoning within the context of measurement. In L. Stiff (Ed.), *Developing mathematical reasoning, K–12* (1999 Yearbook of the National Council of Teachers of Mathematics, pp. 93–106). Reston, VA: NCTM.

National Center for Research in Mathematical Sciences Education & Freudenthal Institute (Eds.). (1998). *Mathematics in context: A connected curriculum for grades 5–8*. Chicago: Encyclopedia Brittanica Educational Corporation.

Pirie, S., & Kieren, T. (1994). Growth in mathematical understanding: How can we characterise it and how can we represent it? *Educational Studies in Mathematics, 26*, 165–190.

Pólya, G. (1963). *Mathematical methods in science* (Studies in Mathematics, Vol. 11). Stanford, CA: School Mathematics Study Group.

Resnick, L. B., & Omanson, S. F. (1987). Learning to understand arithmetic. In R. Glaser (Ed.), *Advances in instructional psychology* (Vol. 3, pp. 41–96). London: Erlbaum.

Schoenfeld, A. H. (Ed.). (1987). *Cognitive science and mathematics education*. Hillsdale, NJ: Erlbaum.

Sfard, A. (1998). On two metaphors for learning and the dangers of choosing just one. *Educational Researcher, 27*, 4–13.

Skemp, R. R. (1976). Relational understanding and instrumental understanding. *Mathematics Teaching, 77*, 20–26.

Stephan, M., Cobb, P., Gravemeijer, K., & McClain, K. (1998, April). *Supporting student's learning in social context*. Paper presented at the meeting of the American Educational Research Association, San Diego.

Treffers, A. (1987). *Three dimensions. A model of goal and theory description in mathematics education: The Wiskobas Project*. Dordrecht, The Netherlands: Kluwer.

Treffers, A., & de Moor, E. (1990). *Proeve van een nationaal programma voor het reken-wiskundeonderwijs, deel 2* [Proof in the national program for school arithmetic, level 2]. Tilburg, The Netherlands: Zwijsen.

Van Galen, F., & van den Heuvel-Panhuizen, M. (1997). Number tools. In National Center for Research in Mathematical Sciences Education & Freudenthal Institute (Eds.), *Mathematics in context: A connected curriculum for grades 5–8* (Vol. 1, pp. 28–55). Chicago: Encyclopedia Brittanica Educational Corporation.

Van Hiele, P. M. (1973). *Begrip en inzicht* [Understanding and insight]. Purmerend, The Netherlands: Muusses.

Van Parreren, C. F. (1981). *Onderwijsproceskunde* [Educational process theory]. Groningen, The Netherlands: Wolters-Noordhoff.

Whitney, H. (1985). *Mathematical reasoning: Early grades*. Unpublished paper, Princeton, NJ.

Chapter 9

On Appreciating the Cognitive Complexity of School Algebra: Research on Algebra Learning and Directions of Curricular Change

Daniel Chazan, University of Maryland
Michal Yerushalmy, University of Haifa

School algebra—in contrast with the abstract algebra and linear algebra of the undergraduate curriculum—might be thought of as cut and dried. Students are taught solution methods, for example, how to simplify and factor certain expressions, how to solve certain types of equations and inequalities in one unknown, and how to solve systems of two equations in two variables. Most of these methods can be described in four or five steps. The notion behind the curriculum is that if students have mastered these methods, then they will be able to apply them, as relevant, in the context of new problems.

Mathematics education research from a cognitive perspective over the past 15 years has identified ways in which this picture is too simple (see, e.g., reviews in Bednarz, Kieran, & Lee, 1996; Kieran, 1992; Leinhardt, Zaslavsky, & Stein, 1990; Wagner & Kieran, 1989; Wenger, 1987). The situation has an underlying complexity, as argued in Moschkovich, Schoenfeld, and Arcavi (1993) and in Schoenfeld, Smith, and Arcavi (1990). The strings of symbols involved in these school algebra problems elude simple classification. As a result, students may not know what particular method to choose to solve a given problem, or they may choose incorrectly (see, e.g., Davis, 1984; Matz, 1982; Kuchemann, 1978). Skilled performance involves developing a feel for symbol strings (in the sense of Usiskin, 1988, and Bell, 1995) that indicates what sorts of creatures they are and what should be done with them.[1] Developing skilled student performance requires a curriculum that does more than teach methods for solving isolated problem types. To choose appropriate methods, students must also have a feeling for what methods are appropriate with what strings of symbols and must also appreciate what those methods are meant to accomplish.

Thus the movement by the National Council of Teachers of Mathematics (NCTM) to set Standards for school mathematics with a general focus on teaching for, and learning with, understanding present unique challenges with regard to introductory school algebra. *Principles and Standards for School Mathematics* (NCTM, 2000) suggests in the Learning Principle that goals for learners should include "learning mathematics with understanding" (p. 20). As stated in the Algebra Standard for grades 9 through 12, high school students are expected to "understand the meaning of equivalent forms of expressions, equations, inequalities, and relations" (p. 296). If that understanding is to occur, teachers must figure out how to open classroom discussions to questions about equivalence—to questions about why some methods preserve equivalence and others do not. For example, the following are some questions about equivalence of equations and methods for solving equations in one variable. Our point is not that these questions are unanswerable—many teachers can respond to such questions—but that students seldom have opportunities to make such inquiries as questions occur to them.

- If one can add x to both sides of an equation in one variable—and if multiplication is repeated addition—why cannot one always multiply each side of an equation by x? In what sense is the

[1] In an essay "Second Thoughts on Paradigms," Thomas Kuhn (1977) suggests that much education in the sciences faces a similar task. Physics students learn to see the world as physicists by doing problems, by learning to see physics in the world as they learn to analyze a situation using the theories of physics.

resulting equation not necessarily an equivalent one, and why?

- When solving equations, we often undo one operation by doing its inverse. Why isn't taking the square root the inverse of squaring? Why isn't $\sqrt{x^2} = x$? And, for that matter, why does taking the square root of each of the expressions in the equation $2x - 3 = x^2 - 2x - 3$ not yield an equivalent equation?
- When I solve $10x - 45 = 5$ by operating on both sides and get $x = 5$, $x = 5$ feels quite different from $10x - 45 = 5$. When did the equation become a solution set? Or was it one all along? Is $10x = 50$ all that different from $x = 5$? If so, in what sense?
- When I solve equations in one variable, such as $3x + 7 = 2x - 4$, the solution set consists of no numbers, all numbers, or one particular number. How do I understand the equation $\sqrt{(3x-5)^2} = 3x - 5$, for which infinitely many solutions exist but for which there also exist numbers that are not part of the solution set?

For students to "understand the meaning of equivalent forms of expressions, equations, inequalities, and relations," however, teachers cannot simply entertain students' questions about the equivalence of equations or about the equivalence of expressions when those inquiries arise. Students also need opportunities to ask questions that cut across expressions and equations or across expressions and relations, questions that explore differences between types of equivalence—such questions as the following:

- When solving $3x^2 + 3x + 3 = 0$, I can think of the task as finding the zeros of the function y or $f(x) = 3x^2 + 3x + 3$. In the context of finding its zeros, I can divide the coefficients of the function by 3, but when I am working with the function y or $f(x) = 3x^2 + 3x + 3$, I cannot divide all the coefficients by 3. Why is that? How is equivalence of equations different from equivalence of functions?
- If only some equations in two variables are functions (e.g., $3x + y = 5$ is an implicit function, whereas $x^2 + y^2 = 1$ is not), and only some functions have an equation (e.g., $f(x) = 3x - 5$ has an equation, whereas $\{(0,5), (1,8), (2.3, -1.7)\}$ does not), which is a subset of which?
- From what is written on paper, how can I tell that $3(x + 2) = 3x + 6$ as an equation to solve is different from "simplifying" $3(x + 2)$ and writing $3(x + 2) = 3x + 6$? Or is the difference between the two not possible to discern simply by examining the strings themselves? Why is the equals sign used in two such different ways?
- If $25x + .05y = 100$ is a relation in two variables, how does the isolation of y, $y = (100 - 25x)\cdot 20$, turn it into a function of one variable? How can creating an equivalent relation somehow make a relation into a function? Or was $25x + .05y = 100$ a function beforehand as well? If so, could I have written $25x + .05f(x) = 100$? How do I know by looking at a relation whether it is a function? And when can I use function notation?
- How is the solution to the system $x^2 + y^2 = 1$ and $5x + y = 5$ similar to, or different from, the solution to $y = 5 - 5x$? How is the solution to this system similar to, or different from, the solution to $x^2 + (5 - 5x)^2 = 1$?

Again, these questions are not unanswerable. Most teachers can respond to such questions as these, but how often are students allowed, or even encouraged, to ask such questions?

This chapter seeks to present some dimensions of the cognitive complexity that helps define the task confronting anyone developing algebra curricula or teaching introductory school algebra. The chapter has three central parts. First we introduce distinctions, developed in the mathematics education research literature, that attempt to capture the nuances that go into developing a feel for different kinds of symbol strings. Next we concentrate on understanding "the meaning of equivalent forms of expressions, equations, inequalities, and relations" (Algebra Standard for Grades 9–12, NCTM, 2000, p. 296). In the second part, we examine the rewriting of an equation in two variables by isolating one variable, and then we examine how understandings of this sort of equivalence, as well as other sorts of equivalence, are embedded in solving simple systems of two equations in two variables by the substitution method. The discussion focuses on what attributes of equations change, as well as what attributes stay the same, as equations are manipulated. Finally we examine some recent directions of curricular change in school algebra, particularly graphing calculators and the introduction of multiple linked representations. We close by distinguishing between introduction of graphing of functions into the standard curriculum and approaches to school algebra organized around function as a central mathematical object.

Getting a Feel for the Symbols: Distinctions from the Mathematics Education Literature

Over the past 15 years, mathematics education research from a cognitive perspective (a la Davis, 1984) has identified tasks that cause learners of school algebra difficulty in systematic ways. Particular studies have focused on the equals sign (e.g., Herscovics & Kieran, 1980; Kieran, 1989), literal symbols (e.g., Kuchemann, 1978; Schoenfeld & Arcavi, 1988; Sleeman, 1984; Usiskin, 1988; Wagner, 1981), and graphing (e.g., Goldenberg, 1988; Janvier, 1987; Monk & Nemirovsky, 1994; Nemirovsky, 1994; Romberg, Fennema, & Carpenter, 1993). These studies suggest psychological dimensions of learners' experience of curriculum that may not map directly onto logical distinctions one might develop in a mathematical exposition. One way to interpret many of the studies in this tradition is that they identify complexities that learners experience but that the curriculum—with its focus on solution methods to particular problem types—does not address explicitly.

To illustrate, let us focus on equations. What is an equation? School algebra textbooks present standard answers to this question. In a standard approach, an equation is "a sentence about numbers" (Dolciani & Wooton, 1973, p. 24),[2] or "a pattern for the different statements—some true, some false—which you obtain by replacing each variable by the names for the different values of the variable" (p. 44), or "any statement of equality" (p. 583).[3] Such descriptions are true, in some sense, of all equations in an introductory algebra course. However, important differences exist between strings of symbols that are labeled as equations.

To help illustrate the psychological—as opposed to logical—complexity that mathematics education researchers see in school algebra, Zalman Usiskin (1988) presents five equations (in the sense of having an equal sign in them) involving literal symbols. Each has a different feel. (See Mason, 1989, for a related point.)

1. $A = LW$
2. $40 = 5x$
3. $\sin x = \cos x \cdot \tan x$
4. $1 = n \cdot 1/n$
5. $y = kx$

As Usiskin observes, "We usually call (1) a formula, (2) an equation (or open sentence) to solve, (3) an identity, (4) a property, and (5) an equation of a function of direct variation (not to be solved)" (p. 9).

Although readers might agree or disagree with his classification of these strings, that is part of his point. To the enculturated, each of these symbol strings reads differently, and perhaps differently enculturated readers read them differently as well. Furthermore, Usiskin (1988) argues that each of these equations has a different feel because in each case the idea of variable is put to a different use:

> In (1), A, L, and W stand for the quantities area, length, and width and have the feel of knowns. In (2), we tend to think of x as unknown. In (3), x is an argument of a function. Equation (4), unlike the others, generalizes an arithmetic pattern, and n identifies an instance of the pattern. In (5), x is again an argument of a function, y the value, and k a constant (or parameter, depending on how it is used). Only with (5) is there the feel of "variability," from which the term *variable* arose. (p. 9)

Once again, any disagreements that readers might have about Usiskin's description of the use of the idea of *variable* in any of these equations might just serve to strengthen his point: important complexities exist at the heart of school algebra. In the words of Schoenfeld and Arcavi (1988), "The mathematical meaning of a statement is determined by its context" (p. 424) and "the meaning of variable is variable" (p. 425).

In our view, the distinctions developed in this literature raise important curricular questions. If we are committed to students' developing an understanding of these various meanings of the symbol strings of algebra, how do we help students get a feel for how each of these symbol strings is different? Do we continually expose students to many different uses of variables, or should strategic decisions be made about the order in which to introduce students to particular uses of variables and when to introduce new views? We return to these questions in the fourth section of this chapter after exploring further the

---

[2]We have chosen this text because it represents some of the fruit of the efforts of the School Mathematics Study Group (SMSG). The two authors and two consultants were involved in SMSG activities, and the book identifies ways in which it explicitly builds on the work of the SMSG.

[3]A different approach would be to think of equations as simply representations of solution sets. In this view, an equation in two variables identifies a set of coordinated values that make the statement true. Function-based approaches to school algebra define equations in yet a different way, as one kind of comparison of two functions, (see, e.g., Chazan, 1993).

understanding of equivalence in introductory school algebra.

The Power of Symbols: Understanding Equivalence and Difference in School Algebra

Algebraic manipulations are powerful techniques for gaining insight into the solutions to problems of particular types; yet something is mysterious about these manipulations. By carrying out a set of manipulations and by writing equivalent symbol strings, one somehow comes to a solution that one cannot simply read from the problem description; for a schematic diagram of such mathematical modeling, see Davis and Hersh (1986, p. 59). Indeed mathematicians Philip Davis and Reuben Hersh (1986) suggest that these methods are so powerful that they led Descartes and later Leibniz to "the dream of a universal method whereby all human problems, whether of science, law, or politics, could be worked out rationally, systematically, by logical computation" (p. 7). Yet how can simply writing equivalent statements lead to such insight? How does the writing of equivalent statements support shifting one's perspective and coming to see differently? Besides whatever is being kept invariant, important change must take place as well. The purpose of this section is to appreciate the task of developing a conceptual understanding of the equivalences preserved through algebraic manipulation. We suggest that perhaps a crucial component of such an understanding is an appreciation of the differences between equivalent statements.

Differences Between Equivalent Equations in Two Variables

In this section, we extend Usiskin's (1988) and Schoenfeld and Arcavi's (1988) point about the variability of the meaning of equations and variables to the equals sign and to the use of the Cartesian coordinate system. We also question the representation of this variability in curricula by examining the point at which introductory algebra students are asked to move beyond expressions and equations in a single "variable." The question of developing a feel for different uses of variables, the equals sign, and the Cartesian coordinate system arises in a dramatic fashion when one is isolating a variable, because doing so changes the feel of an equation. With this point in mind, we focus on what attributes change and what attributes stay the same as equations in two variables are rewritten.

In standard approaches to school algebra, the focus is on what stays the same as a variable is isolated in an equation in two variables; functions of one variable are conceptualized as a special kind of relation between two variables. With such a stance, changing an equation from $Ax + By = C$ to $y = mx + b$ does not change the relation being represented, even though $Ax + By = C$ is only implicitly a function, whereas $y = mx + b$ is explicitly so (see Crowley & Tall, 1999, for students' difficulties with such notions). For example, in chapter 10 of Dolciani and Wooten (1973), students are asked to move from solving equations in one variable—the focus of chapters 4 through 9—to solving open sentences in two variables. The title of the chapter is "Functions, Relations, and Graphs." In this chapter, functions are conceptualized as a special type of relation. The chapter moves back and forth between linear functions of one variable and linear equations in two variables. Thus the method presented for solving linear equations and inequalities in two variables begins with isolating a variable. As long as one keeps in mind that a function is a particular type of relation, then the exposition in the text is clear. Any linear equation in two variables can be written as an equivalent equation that also happens to be a linear function of one variable. To create a graph of a linear equation in two variables, it may be useful to rewrite it as a linear function of one variable.

Although such an approach is mathematically sound, the distinctions introduced previously suggest that the coordinations involved here for students are quite demanding. In addition to all that has stayed the same, much has changed. To illustrate, we explore the differences between equivalent equations in two variables by starting with a situation:

> Imagine that a person is renting a car. The rental fee depends on the number of days he or she takes the car and the number of miles he or she drives (e.g., \$25/day and 5 cents/mile). The person has a coupon that provides a free rental with the maximum value of \$100. What trips can this person take that would use the maximum value of the coupon? (Adapted from *Visualizing Mathematics*, Center for Educational Technology, 1995)

When presented with this sort of situated task, a person seeking to represent this situation algebraically might think about his or her costs and the relationship of these costs to the coupon. This way of thinking about the situation can be represented by writing an equation that is an implicit function (e.g., $Ax + By = 100$). From this perspective, one has two ways of writing the costs, two ways of expressing the same number. One way to describe the total cost is by thinking of the maximum value of the coupon: $100; the other way is to consider how the total cost of trips with a value of $100 is computed: $25x + .05y$, for *x*s and *y*s that will generate $100. In such a symbol string, the *x*s and *y*s are then meant to represent the as yet unknown pairs of numbers that will satisfy the equation. Taken together, these pairs of numbers are the solution set of the equation—or the set that the equation defines, depending on one's perspective. Using the equals sign to indicate that one is equating two different ways of writing the same quantity, one gets the equation $25x + .05y = 100$.

When one graphs this equation on the Cartesian plane, three quantities are displayed: the quantities on the *x*- and *y*-axes—number of days and number of miles, respectively—and the set of points that describes trips whose total cost is $100 (Figure 9.1). In such graphing, the Cartesian plane is made up of points, and the graph consists of those points for which the equation is a true statement. For the remaining values in the plane, the statement is false. Creating the graph is a matter of finding members of the solution set. Because this equation is linear, one might find these members systematically, for example, by substituting values for one variable and solving for the other. With other sorts of equations in two variables, such as $x^2y + y^2x = 1$ (see Sfard & Linchevski, 1994, and Yerushalmy & Bohr, 1995, for discussions of student responses), in which neither variable can be isolated, figuring out how to be systematic is more complicated.

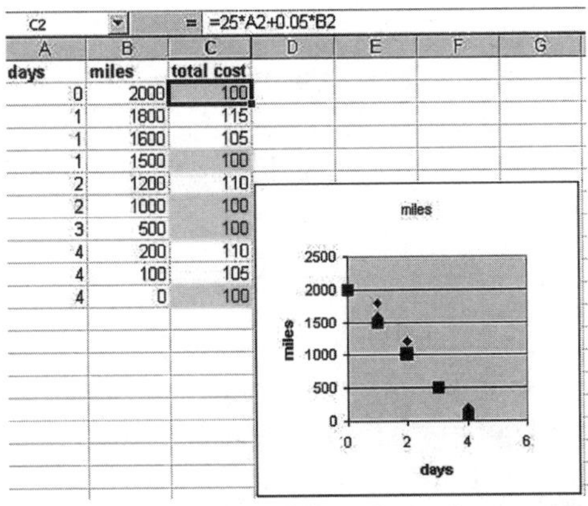

days	miles	total cost
0	2000	100
1	1800	115
1	1600	105
1	1500	100
2	1200	110
2	1000	100
3	500	100
4	200	110
4	100	105
4	0	100

Figure 9.1. Solution to a problem about the costs for renting a car, in which values for both miles and days are treated as inputs.

Alternatively, a person seeking to represent this situation might realize that if he or she is spending the full coupon, the number of miles to be traveled is a function of the number of days of rental. As the number of days gets larger, the number of miles must get smaller. That way of conceptualizing the situation can be represented with an explicit function: $y = (100 - Ax)/B$. The resulting "equation," $y = (100 - 25x)/0.5$ or $20(100 - 25x)$, being made up of numbers, an x, a y, and an equals sign, may look quite similar to the previous one, but now it represents a function. Indeed, one might even use the special notation $y = f(x) = 20(100 - 25x)$ (see Apostol, 1969, for an exposition to university students about this change of notation) to indicate that x has a different sense than y. The symbol x represents an independent variable that takes on a range of values in a systematic way. The symbol y is a dependent variable; in a certain sense y is a name for the result of the computational recipe on the other side of the equals sign. The nature of the equals sign has subtly shifted. The symbol y, or $f(x)$, is the label being assigned to the result of the process. And, finally, when one graphs this function on the Cartesian plane, although the graph appears to be identical, it has only two explicitly named quantities: the independent and dependent variables. The $100 has receded into the background with other quantities in the situation (see Figure 9.2). For any value of x, only one value of y is possible. If one rents the car for one day, one travels 1,500 miles to use up the full $100. As a result, one can create the graph by systematically substituting the values of x into the function and computing the y values.

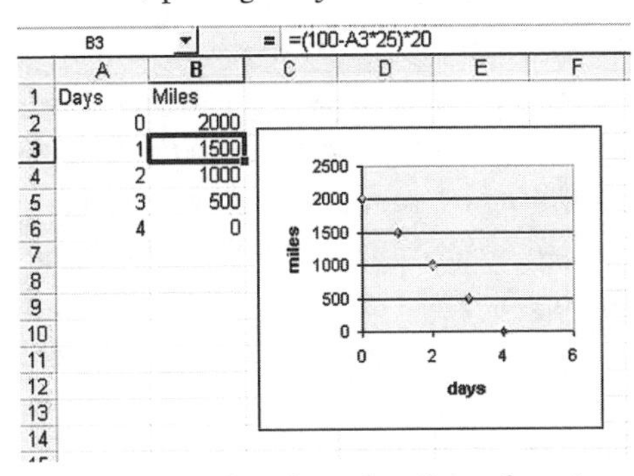

Days	Miles
0	2000
1	1500
2	1000
3	500
4	0

Figure 9.2. An expression thought of as a function of the number of miles depending on the number of days.

To summarize, these two different ways of conceptualizing the relationship between the quantities in this situation lead to algebraic representations that can be described as having a quite different feel to them (see Table 9.1).

Intriguingly, rather than write the second "equation" directly from the situation, one can manipulate the first equation to reveal the second. Doing so raises a host of questions. How does a seemingly mechanical manipulation lead to a representation that encodes such a different conception of the situation? Why does doing the same operations to both sides of an equation make for such a

Table 9.1

Comparison of the Characteristics of an Equation—an Implicit Function—and an Explicit Function

	$25x + .05y = 100$	$y = 20(100 - 25x)$
Description of the symbol string	Equation (Implicit function)	Explicit function
Letters	• x and y represent unknown numbers. • No dependence relationship between x and y	• x is an independent variable that takes on all values in a specified set. • y is a dependent variable whose values are set by those of the independent variable.
Equals sign	• Symmetric—different names for the same number	• Asymmetric—y, or $f(x)$, is the name for the result of computing the process f on the input x.
Graphical representation	• Set of points that are the solution set (pairs of numbers that make the equation a true statement) • Three quantities appear: Two on the axes and one on the graph.	• Only one point can be on any vertical line at a given x value. • One makes the graph by computing the y values from the x values, by carrying out the function recipe f on inputs x. • Only two explicit quantities and the dependence are involved.

change? In what sense are the graphs of these two "equations" the same, given the differences in how equations and functions use the same Cartesian coordinate plane? How do learners come to appreciate the fact that equations that look quite similar and may even represent the same relationship, in the sense of "ordered pairs," as in Crowley & Tall (1999), may have quite different qualities? How do learners come to appreciate the importance of isolating a variable in order to be able to graph equations in the form $Ax + By = C$?

Solving Systems of Equations in Two Variables: Equivalences That Yield Strategically Useful Differences

So far we have introduced distinctions among equations that have a different feel to them, along with related differences in uses of variables. With the example of isolating one variable in an equation in two variables, we have also suggested that writing equivalent equations can lead to equations with quite different feels to them. Implicitly, we have suggested that this process may be one source of the power of algebraic manipulation. In this section, we examine a solution method that asks students to carry out a set of manipulations in which each manipulation maintains equivalence, on the one hand, but also creates an important difference, on the other hand. We examine key manipulations involved in solving systems of two equations in two variables, manipulations that might be used to solve such a system as that shown in Figure 9.3.

Solve this system:

$$x^2 + y^2 = 1$$

and

$$5x + y = 5$$

Figure 9.3. A system of two equations in two variables.

Solving a system of equations in two variables is commonly a part of first-year algebra in North America. The sort of problem shown in figure 9.3 comes, in a standard curriculum, after students have learned to simplify expressions, solve a single equation in one unknown—for example, equations of the form $ax + b = cx + d$ and simple quadratic equations—and represent a solution set of a single equation in two variables on a Cartesian coordinate system. In this sense, the problem asks students to pull together many of the methods they have learned.

At the same time, understanding the goal of this sort of problem, while building on one's earlier understanding of solution sets, extends the notion of solution sets in new directions. Arguably such an understanding of goals lies at the heart of meaningful learning of algebraic manipulation.[4] An appreciation of the qualities of a solution to a problem type allows solvers to assess the reasonableness of their solution attempts.

If students have learned to solve equations in one unknown, they are used to solution sets whose elements are numbers rather than ordered pairs. Solution sets to equations in two variables are different; their elements are ordered pairs. For descriptions of students struggling with this difference, see Yerushalmy (1997a). But another difference exists as well. An equation in two variables indicates a relation between the values in an ordered pair that is a member of the solution set. Given partial information—that is, the value of one of the coordinates in the ordered pair—one may be able to create a solvable equation in one variable and thus ascertain the other coordinate. When searching for the solution set to a system of two equations in two variables, students must understand that each of the two equations in the system generates a solution set of this kind; if one chooses an x, one may then be able to solve each of the equations for y, or if one chooses a y, one may then solve for x.

The solution set of a system of two equations in two variables is both similar to those of single equations in one and two variables and different from them. As in an equation in two variables, the members of this solution set are ordered pairs of *x*s and *y*s, not just single numbers, such as the solution set to an equation in one variable; see Chazan, Larriva, and Sandow (1999) for a student teacher's struggles describing this situation to her students. The solution set of a system, however, is more constrained than the solution set of one equation in two variables. To compute the value of one coordinate in an ordered pair that is a member of the solution set of a system, one must first make sure that the value of the other coordinate is already a coordinate of a member of the solution set to the system. "Most" coordinates will not work. Finally, with a system of two equations, the task is to find pairs of *x*s and *y*s that are members of the solution set of each equation individually; the *and* in the problem indicates that one is operating on solution sets to the individual equations. (See Yerushalmy, 1998, or Yerushalmy & Chazan, 2002, for a discussion of students' working to generalize from solutions of equations in one variable to solutions of equations in two variables.)

Typically, students are taught to solve such systems with a variety of methods. We examine this kind of task by analyzing one of the methods usually suggested to students to solve such problems—the substitution method. Given the distinctions developed above, our focus is on what attributes change and what attributes stay the same as one manipulates the strings of symbols to approach a solution.

In solving the system $x^2 + y^2 = 1$ and $5x + y = 5$ by the substitution method to get the pairs of values $x = 1$ and $y = 0$, or $x = 12/13$ and $y = 5/13$, the following manipulations are necessary:

- $5x + y = 5$ is replaced by $y = 5 - 5x$;
- $y = 5 - 5x$ is substituted into $x^2 + y^2 = 1$ to yield $x^2 + (5 - 5x)^2 = 1$, which replaces $x^2 + y^2 = 1$;
- $x^2 + (5 - 5x)^2 = 1$ is replaced by $26x^2 - 50x + 24 = 0$.

From this point, $26x^2 - 50x + 24 = 0$ is solved and the numerical result is substituted back into the equation $y = 5 - 5x$ to ascertain the coordinated values of x and y that are the solutions to the system. Below we outline the different senses of equivalence and difference that learners coordinate in following three crucial steps of this method.

1. From an equation in two variables to a function in one

Earlier in this chapter, we outlined ways in which such equations as $5x + y = 5$ and $y = 5 - 5x$ represent the same relation, even though $5x + y = 5$ can be described as an equation in two variables and is only implicitly a function, whereas $y = 5 - 5x$ is explicitly a function of x. At this point, the difference becomes strategically important. Writing $5x + y = 5$ explicitly as a function allows it to be substituted in the next step of the method. This substitution will call on both the "equality of two different expressions" view of the equals sign and the "assignment" view. One can write $5 - 5x$ instead of y. For the points in the solution set to $5x + y = 5$, the terms y and $5 - 5x$ are both ways of writing the same number. At the same time, $5 - 5x$ tells one exactly how to compute y.

2. From an equation in two variables to an equation in one variable

This part of the solution method is different from other manipulations that students have done in introductory algebra (Kuchemann, 1981; Linchevski & Sfard,

[4] In the context of teachers' subject matter understanding, Chazan (1999) argues that being able to articulate the characteristics of the goal of a problem is an important component of a teacher's understanding of the mathematics that they teach.

1991).[5] Here, it may be easier to see differences that are created rather than equivalence that is preserved. Outside the context of the system, $x^2 + y^2 = 1$ is not equivalent to $x^2 + (5 - 5x)^2 = 1$; an equation in two variables is not usually equivalent to an equation in one. Here the notion of equivalence is at the level of the system and not at the level of the equation, where it must be justified. The equation $x^2 + y^2 = 1$ is equivalent to $x^2 + (5 - 5x)^2 = 1$ only in the context of systems whose other equation is $y = 5 - 5x$. By substituting $5 - 5x$ for y in $x^2 + y^2 = 1$, one is linking the x and y in the two equations. On the one hand, this substitution concretizes the *and* of the system, a part of this problem type that adds a new dimension to the situation. In a move analogous to constraining two expressions by linking them with an equality, one is constraining the y values in $x^2 + y^2 = 1$ to values that will also be the same as $5 - 5x$.

On the other hand, strategically speaking, this step is of course crucial. One now has $x^2 + (5 - 5x)^2 = 1$, an equation in one variable, and $y = 5 - 5x$. The solution to the first equation will be value(s) of x, not coordinated values of x and y. But this equation in one variable is not just any equation. The key point about this equation is that because the solver has constrained the ys in the two original equations to be equal, its solutions must be x values for which the y values of each of the original equations—if they exist—must be equal. Thus, the x values that solve this equation in one variable are the x values of the coordinated pairs that are solutions to the system. The coordinated y values can be generated from these x values.

3. Toward the standard form of an equation in one variable: Equivalent expressions and equivalent equations

Two key elements in an introductory algebra course involve understanding two different sorts of equivalence. Both these sorts of equivalence are present in the move from $x^2 + (5 - 5x)^2 = 1$ to $26x^2 - 50x + 24 = 0$. First, on the one hand, equivalent expressions are expressions that are superficially different; that is, they describe different computational procedures. Nonetheless, because one expression is generated by another according to an identity—a relation that holds true for all members of the replacement set with the usual default of the real numbers—these equivalent expressions return the same values for a common input. Although the manipulations involved can be justified by the properties of real numbers, which are often represented as identities, school mathematics curricula use a variety of strategies to help students appreciate such manipulations (e.g., area models and algebra tiles in Larkin, 1989). Yet such strategies do not seem to solve the problem and often cause students difficulty (Filloy & Rojano, 1984). Students develop many buggy procedures or "mal-rules" (see, e.g., Davis, 1984, and Matz, 1982, especially appendix A). In this particular case, many students might inappropriately replace the expression $(5 - 5x)^2$ with $25 + 25x^2$ by inappropriately "distributing" exponentiation over addition.

Second, on the other hand, two equivalent equations have the same solution set even though the pairs of expressions that make up the equations are different. When x takes on the value of 1, $x^2 + (5 - 5x)^2$ computes to 1, whereas $26x^2 - 50x + 24$ computes to 0. So when x takes on the value of 1, then $x^2 + (5 - 5x)^2 = 1$ is a true statement, as is $26x^2 - 50x + 24 = 0$. For both equations, 1 is a value of x for which the equations are true. In this sense, the operation on both sides of the equation has not changed the solution set even though the particular equality has changed from $1 = 1$ to $0 = 0$. Although basic properties of the real numbers suggest why adding, subtracting, multiplying, or dividing by a nonzero number will not change an equality for members of the solution set even though such an operation might have a quite different impact on values of x that are not part of the solution set, justification of this kind of manipulation has challenged teachers of many eras. Typical strategies involve recourse to such analogies as balancing scales rather than to mathematical argument (Filloy & Rojano, 1984).

Thinking about differences between the equations $x^2 + (5 - 5x)^2 = 1$ and $26x^2 - 50x + 24 = 0$ is also important, however. In the example, the equation in one variable is $x^2 + (5 - 5x)^2 = 1$, yet most solution methods for quadratic equations involve putting a quadratic equation into standard form: $Ax^2 + Bx + C = 0$. Standard form allows the solver to reconceptualize the problem of finding solutions to a quadratic equation in one variable by reintroducing a second variable. The problem of ascertaining the roots of a quadratic equation in one variable is thus the same as finding the zeros of the related quadratic function of one variable ($f(x) = 26x^2 - 50x + 24$). Thus, strategically, the solver seeks an equation for which one solution is zero and the other is in a form that supports an easy solution procedure.

Adding Representations: Graphing Technology and Equivalence

As the preceding section illustrated, "understanding the meaning of equivalent forms of expressions, equations, inequalities, and relations" is a challenging task that is

[5]It is akin to what is done in composition of functions, which is typically addressed later.

strongly connected with skilled performance on tasks of the standard introductory school algebra curriculum. Along with an awareness of the challenge of teaching introductory school algebra for understanding has come much enthusiasm for the potential of graphing technology. Arguably, technology, such as graphing calculators, that links expressions and graphs has the potential to give students visual feedback that emphasizes the various meanings of equivalence (e.g., Yerushalmy, 1991; 1999; Yerushalmy & Gafni, 1992). Supporters have described its potential to help students understand visually both the equivalence of expressions (e.g., as discussed in NCTM, 2000, p. 301, and illustrated in Figure 9.4)[6] and the equivalence of equations (as discussed in Schwartz & Yerushalmy, 1992, and illustrated in Figure 9.5).

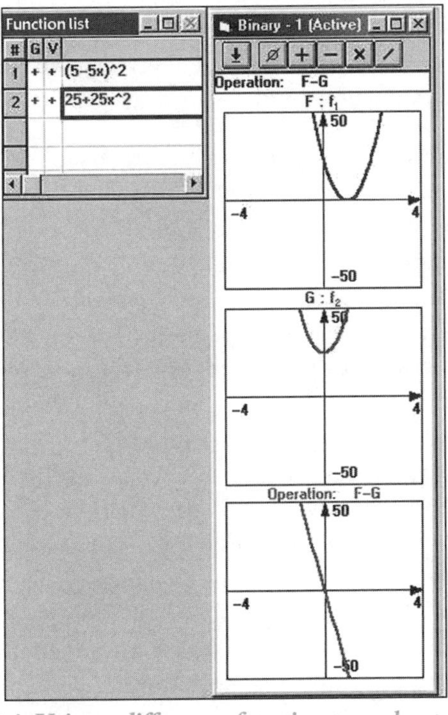

Figure 9.4. Using a difference function to understand why $(5 - 5x)^2 \neq 25 + 25x^2$.

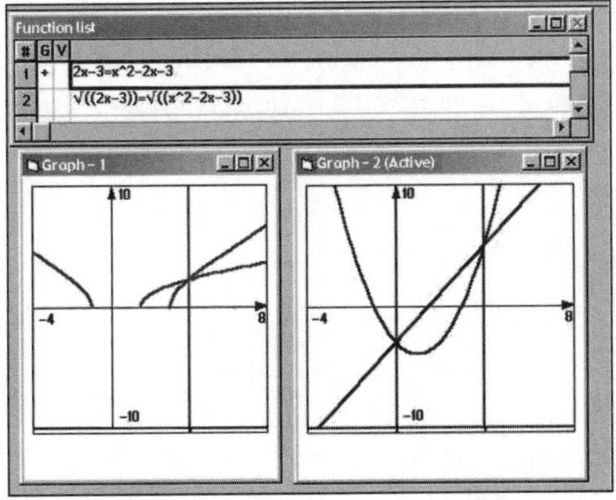

Figure 9.5. Equivalence of equations in a software environment: Understanding why $2x - 3 = x^2 - 2x - 3$ and $\sqrt{2x-3} = \sqrt{x^2-2x-3}$ are not equivalent.

However, at least three reasons can be cited for being skeptical of the power of such linked representations taken on their own without any related changes to the curriculum or to modes of instruction. First, these graphical representations are not transparent. As Goldenberg (1988) has argued (as well as others in Romberg et al., 1993), graphical representations are complex. One cannot simply expect students to be able to read these representations in the ways they are intended. The process of learning to read such representations is complex and requires teaching and learning. In particular, more needs to be made of the differences between graphing equations in which all points on the Cartesian plane can potentially be used and graphing functions in which once a value for a particular x is established, no other points in that vertical line can be used.

Second, as implemented in such technology as graphing calculators, a tension arises between the standard curriculum and the technology. With graphing calculators, the linked representations are expressions taken as the equations of functions, for example, the ever-present $y_n =$ __, and graphs of functions. Thus this technology, unlike the standard curriculum, is geared toward letters as variables, the asymmetric equals sign, and the graphing of functions. As a result, teachers using graphing calculators with the standard curriculum may experience a certain tension. For example, with the calculator, the equation $3x + 7 = 2(x + 5)$ is written into the calculator as $y_1 = 3x + 7$ and $y_2 = 2(x + 5)$. This equation in one variable results in a pair of lines on the Cartesian plane. The x-coordinate of the intersection of these two lines is 3. So the solution set to this equation in one variable is $x = 3$. In contrast, when attempting a similar task on paper as taught in standard approaches, a student would graph the system $-3x + y = 7$ and $y = 2(x + 5)$ as two intersecting lines that seem quite the same. This time, however, the solution set to the system is the coordinated values $x = 3$ and $y = 16$; see, for example, Chazan and others (1999) for a preservice teacher's ways of discussing such issues with her students. Although such questions can be addressed, this sort of tension adds yet one more wrinkle to the tasks facing learners.

[6]Critics suggest that the focus on the equivalence of these expressions in terms of the graph de-emphasizes the operations that have been used to get from one expression to the others and as a result distorts what algebra is. For such critics, the identification of a polynominal and a polynominal function creates new "epistemological obstacles" that will need to be overcome at later stages of a mathematical education.

Third, although the technology can provide students with feedback relevant to some questions, such as "Why are $(5 - 5x)^2$ and $25 + 25x^2$ not equal?" the presence of such technology does not by itself afford students opportunities to raise questions in class or to get responses to questions that explore differences between types of equivalence as raised above.

Functions-Based Approaches to Algebra

In the second section of this chapter, we illustrated the challenges involved in understanding equivalence and argued that understanding equivalence means at the same time appreciating differences among equivalent expressions, equations in one and two variables, and systems of equations. In the third section, we illustrated the potential benefits of linkages between graphs and expressions for understanding equivalence and difference. We also suggested reasons that the infusion of such representational systems into the standard course would not fully address the challenge of having students develop understandings of equivalence. Having done these things, we return to the distinctions developed in the mathematics education literature for characterizing equations having different feels to them. Does this literature simply identify potential obstacles to students' understanding? Or does the development of these distinctions have curricular ramifications? In teachers' and curriculum developers' efforts to help students develop a feel for different sorts of symbol strings and various uses of the notions of *variable*, *equals sign*, and *Cartesian coordinate system*, does it make a difference how they come to grips with these notions? Does the order of introduction matter? What are the relative merits of gradual immersion versus jumping into the deep end?

Although the research literature has only begun to explore such questions (e.g., see Kieran & Sfard, 1999, Bednarz et al., 1996, or Yerushalmy, 1997a, 1997b, 1997c), curriculum developers have already begun to explore these issues. Recently in North America, there has been much discussion of functions-based approaches to school algebra. Critics have suggested that this North American fixation is both driven primarily by a focus on the capacities of technology and not representative of algebra as a subdiscipline of mathematics (e.g., Pimm, 1995). Although functions-based approaches have been associated with technology that enables linkages among tabular, graphical, and literal representations of functions, the observations made thus far in this chapter suggest an alternative view. In our view, functions-based approaches can also be conceptualized as a series of strategic decisions about introducing students to the complexities of the tasks for which they are taught the solution methods that constitute the bulk of school algebra.

In our interpretation, functions-based approaches to school algebra initially emphasize the interpretation of—

- letters as variables rather than unknowns,
- expressions as the correspondence rules for functions,
- the Cartesian coordinate system as a space for displaying the results of calculation procedures rather than the points in a solution set, and
- the equals sign as the assignment of a name to a particular computational process ($f(x) = \ldots$) and as the indication of identity between two computational processes.

In such approaches, the notions of letters as unknowns, of the use of the Cartesian coordinate system for displaying solution sets, and of the equals sign in solving equations in one variable all appear, but as secondary notions. For example, this meaning of the equals sign is introduced as asking a question about the domain values for which the outputs of two functions will be the same; see, for example, Freudenthal's (1973) notation $(?x)\ 2x - 3 = x^2 - 2x - 3$. These interpretations are introduced systematically after students have gained reasonable control over a variety of representations of functions.

Thus in such approaches, closed-form solutions for linear and quadratic equations are postponed and graphing of functions is brought forward. Simplifications are treated as operations on a function that change the rule for moving from input to output but do not change the outputs that result for any given input. Systems of equations in two variables are no longer an introductory topic. These systems must wait for students to have access to representations for functions of two variables, comparisons of pairs of such functions, and the solution sets that result from such comparisons; see, for example, the discussion in Yerushalmy and Chazan (2002).

Conclusion

School algebra is a crucial domain in what Jim Kaput (Lacampagne, Blair, & Kaput, 1995) calls the U.S. "layer cake" mathematics curriculum. It has traditionally played a central role in filtering the educational opportunities available to college-intending students; see, for example, the argument in Moses (1994) and that in Moses, Kamii, Swap, and Howard (1989). However, as more and more districts and states institute algebra requirements for graduation from high school, the significance of algebra instruction becomes great for all students, not just the college intending.

At the same time, school algebra is a complicated curricular arena in which to describe the curriculum; for example, see Lee (1996) for an examination of the conflicting understandings of school algebra among mathematics education researchers. Although many conversations and research papers assume that school algebra is a term that refers to some shared curricular notions and to particular kinds of student experiences, the term has a host of unclarified meanings (see Lacampagne et al., 1995).

Nonetheless, school algebra is also a central subject in school reform discussion. The algebra achievement of U.S. students on the National Assessment of Educational Progress is poor; see, for example, the discussion in Moskovitch and others (1993) and that in Lindquist (1989). Handheld computer algebra systems (CASs) are available that can do all the symbolic manipulation taught in school algebra; see Podlesni (1999) for an argument about the ramifications of this development. These sorts of trends have raised a number of questions about traditional teaching of algebra (e.g., Freudenthal, 1973; NCTM Algebra Working Group, 1998; Thorpe, 1989, Yerushalmy & Schwartz, 1993).

Reformers in the mathematics education community as well as researchers in cognitive science (e.g., Gardner, 1991; Resnick, 1987a; Resnick & Klopfer, 1989) suggest that the key to improving students' performance in a subject-matter domain, such as school algebra, is not to create an ever more elaborate and fine-tuned set of procedures but rather to change the nature of instruction, resulting in recommendations like those in NCTM Algebra Working Group (1998); in Heid, Choate, Sheets, and Zbiek (1995); and in LaCampagne and others (1995). Rather than learn only well-defined tasks, students also need to work on ill-defined ones that are amenable to multiple approaches and that more closely approximate the ways in which algebraic knowledge is used in nonschool settings (Resnick, 1987b). Rather than focus in a purely mechanical way on a large number of symbol manipulations, the content should focus on a smaller set of important conceptual chunks, or "big ideas" (Prawat, 1991). Rather than ignore students' conceptions or actively try to confront misconceptions, instruction should connect with students' experience and build on the resources and strengths present in the conceptions they bring to school (Smith, diSessa, & Roschelle, 1993).

We would like this chapter to be read against the backdrop of algebra as a crucial and contested area of the curriculum in this time of educational reform. Mathematics educators are working toward the goal of having algebra students "understand the meaning of equivalent forms of expressions, equations, inequalities, and relations" (NCTM, 2000, p. 296). Cognitive research on algebra learning suggests that this goal is challenging and at the same time introduces potentially useful distinctions. Equivalent strings of symbols also have crucial differences. Students need to develop a feel for those differences. As curriculum developers and teachers attempt to reach this goal, some are drawn to the potential of functions-based approaches. One way to understand such approaches is to understand that they represent a set of decisions, decisions about what to introduce to students initially and what to emphasize. Such curricular approaches are being carefully studied—as well as challenges for teachers in implementing them (e.g., Chazan, 1999; Even, 1993; Haimes, 1996; Lloyd & Wilson, 1998; Williams, 1998; Wilson, 1994)—to ascertain whether they may lead to deeper understanding of school algebra.

ACKNOWLEDGMENTS

We are grateful to Judah Schwartz and Pat Tinto for comments on an earlier draft of this chapter.

REFERENCES

Apostol, T. (1969). *Calculus*. New York: Wiley.

Bednarz, N., Kieran, C., & Lee, L. (Eds.). (1996). *Approaches to algebra: Perspectives for research and teaching*. Dordrecht, The Netherlands: Kluwer.

Bell, A. (1995). Purpose in school algebra. *Journal of Mathematical Behavior, 14*, 41–74.

Center for Educational Technology. (1995). *Visualizing mathematics: Computer oriented inquiry curriculum for learning algebra, analysis and geometry for grades 7–12* [original in Hebrew]. Tel Aviv, Israel: Author. Information retrieved 27 March 2002, from http://www.cet.ac.il/mathinternational/first.htm.

Chazan, D. (1993). $F(x) = G(x)$?: An approach to modeling with algebra. *For the Learning of Mathematics, 13*(3), 22–26.

Chazan, D. (1999). On teachers' mathematical knowledge and student exploration: A personal story about teaching a technologically supported approach to school algebra. *International Journal of Computers for Mathematical Learning, 4*, 121–149.

Chazan, D., Larriva, C., & Sandow, D. (1999). What kind of mathematical knowledge supports teaching for "conceptual understanding"? Preservice teachers and the solving of equations. In O. Zaslavsky (Ed.), *Proceedings of the 23rd Annual Conference of the International Group for the Psychology of*

Mathematics Education (Vol. 2, pp. 197–200). Haifa, Israel: PME.

Crowley, L., & Tall, D. (1999). The roles of cognitive units, connection and procedures in achieving goals in college algebra. In O. Zaslavsky (Ed.), *Proceedings of the 23rd Annual Conference of the International Group for the Psychology of Mathematics Education* (Vol. 2, pp. 225–232). Haifa, Israel: PME.

Davis, P., & Hersh, R. (1986). *Descartes' dream: The world according to mathematics*. San Diego, CA: Harcourt, Brace, Jovanovich.

Davis, R. (1984). *Learning mathematics: The cognitive science approach to mathematics education*. Norwood, NJ: Ablex.

Dolciani, M., & Wooten, W. (1973). *Modern algebra, book one: Structure and method*. Boston: Houghton Mifflin.

Even, R. (1993). Subject matter knowledge and pedagogical content knowledge: Prospective secondary teachers and the function concept. *Journal for Research in Mathematics Education, 24*, 94–116.

Filloy, E., & Rojano, T. (1984). Solving equations: The transition from arithmetic to algebra. *For the Learning of Mathematics, 9*(2), 19–25.

Freudenthal, H. (1973). *Mathematics as an educational task*. Dordrecht, The Netherlands: Reidel.

Gardner, H. (1991). *The unschooled mind: How children think and how schools should teach*. New York: Basic Books.

Goldenberg, P. (1988). Mathematics, metaphors, and human factors: Mathematical, technical, and pedagogical challenges in the educational use of graphical representations of functions. *Journal of Mathematical Behavior, 7*, 135–173.

Haimes, D. (1996). The implementation of a "function" approach to introductory algebra: A case study of teacher cognitions, teacher actions, and the intended curriculum. *Journal for Research in Mathematics Education, 27*, 582–602.

Heid, M. K., Choate, J., Sheets, C., & Zbiek, R. M. (1995). *Algebra in a technological world*. Reston, VA: National Council of Teachers of Mathematics.

Herscovics, N., & Kieran, C. (1980). Constructing meaning for the concept of equation. *Mathematics Teacher, 73*, 572–581.

Janvier, C. (Ed.). (1987). *Problems of representation in the teaching and learning of mathematics*. Hillsdale, NJ: Erlbaum.

Kieran, C. (1989). The early learning of algebra: A structural perspective. In S. Wagner & C. Kieran (Eds.), *Research issues in the learning and teaching of algebra* (pp. 33–56). Reston, VA: National Council of Teachers of Mathematics.

Kieran, C. (1992). The learning and teaching of school algebra. In D. A. Grouws (Ed.), *Handbook of research on mathematics teaching and learning* (pp. 390–419). New York: Macmillan.

Kieran, C., & Sfard, A. (1999). Seeing through symbols: The case of equivalent expressions. *Focus on Learning Problems in Mathematics, 21*(1), 1–17.

Kuchemann, D. (1978). Children's understanding of numerical variable. *Mathematics in School, 7*(4), 23–26.

Kuchemann, D. (1981). Algebra. In K. Hart (Ed.), *Children's understanding of mathematics: 11–16* (pp. 102–119). London: Murray.

Kuhn, T. (1977). Second thoughts on paradigms. In T. Kuhn (Ed.), *The essential tension* (pp. 293–319). Chicago: University of Chicago Press.

Lacampagne, C., Blair, W., & Kaput, J. (Eds.). (1995). *The Algebra Initiative Colloquium*. Washington, DC: U.S. Department of Education.

Larkin, J. H. (1989) Robust performance in algebra: The role of problem representation. In S. Wagner & C. Kieran (Eds.), *Research issues in the learning and teaching of algebra* (pp. 120–134). Reston, VA: National Council of Teachers of Mathematics.

Lee, L. (1996). *Algebraic understanding: The search for a model in the mathematics education community*. Unpublished dissertation, Université du Québec à Montréal.

Leinhardt, G., Zaslavsky, O., & Stein, M. (1990). Functions, graphs and graphing: Tasks, learning, and teaching. *Review of Educational Research, 60*, 1–64.

Linchevski, L. & Sfard. A. (1991). Rules without reasons as processes without object—The case of equations and inequalities. In F. Furinghetti (Ed.), *Proceeding of the 15th International Conference for the Psychology of Mathematics Education* (Vol. 2, pp. 317–324). Assisi, Italy: PME.

Lindquist, M. M. (Ed.). (1989). *Results from the Fourth Mathematics Assessment of the NAEP*. Reston, VA: National Council of Teachers of Mathematics.

Lloyd, G., & Wilson, M. (May 1998). Supporting innovation: The impact of a teacher's conceptions of functions on his implementation of a reform curriculum. *Journal for Research in Mathematics Education, 29*, 248–274.

Mason, J. (1989). Mathematical abstraction as the result of a delicate shift of attention. *For the Learning of Mathematics, 9*(2), 2–8.

Matz, M. (1982). Towards a process model for high school algebra errors. In D. Sleeman & J. S. Brown (Eds.), *Intelligent tutoring systems* (pp. 26–50). New York: Academic Press.

Monk, S., & Nemirovsky, R. (1994). The case of Dan: Student construction of a functional situation through visual attributes. *CBMS Issues in Mathematics Education, 4*, 139–168.

Moschkovich, J., Schoenfeld, A. H., & Arcavi, A. (1993). Aspects of understanding: On multiple perspectives and representations of linear relations and connections among them. In T. A. Romberg, E. Fennema, & T. P. Carpenter (Eds.), *Integrating research on the graphical representation of function* (pp. 69–100). Hillsdale, NJ: Erlbaum.

Moses, R. (1994). Remarks on the struggle for citizenship and math/science literacy. *Journal of Mathematical Behavior, 13*, 107–112.

Moses, R., Kamii, M., Swap, S., & Howard, J. (1989). The Algebra Project: Organizing in the spirit of Ella. *Harvard Educational Review, 59*, 423–443.

National Council of Teachers of Mathematics (NCTM). (2000). *Principles and standards for school mathematics*. Reston, VA: Author.

National Council of Teachers of Mathematics (NCTM) Algebra Working Group. (1998). A framework for constructing a vision of algebra: A discussion document. In National Council of Teachers of Mathematics & Mathematical Sciences Education Board (Eds.), *The nature and role of algebra in the K–14 curriculum: Proceedings of a national symposium* (pp. 145–190). Washington, DC: National Academy Press.

Nemirovsky, R. (1994). On ways of symbolizing: The case of Laura and the velocity sign. *Journal of Mathematical Behavior, 13*, 389–422.

Pimm, D. (1995). *Symbols and meanings in school mathematics*. London: Routledge.

Podlesni, J. (1999). Soundoff! A new breed of calculators: Do they change the way we teach? *Mathematics Teacher, 92*, 88–89.

Prawat, R. (1991). The value of ideas: The immersion approach to the development of thinking. *Educational Researcher, 20*(2), 3–10.

Resnick, L. B. (1987a). *Education and learning to think*. Washington, DC: National Academy Press.

Resnick, L. B. (1987b). Learning in school and out. *Educational Researcher, 16*(9), 13–20.

Resnick. L. B., & Klopfer, L. E. (Eds.). (1989). *Toward the thinking curriculum: Current cognitive research*. Alexandria, VA: Association of Supervision and Curriculum Development.

Romberg, T., Fennema, E., & Carpenter, T. (Eds.). (1993). *Integrating research on the graphical representation of functions*. Hillsdale, NJ: Erlbaum.

Schoenfeld, A. H., & Arcavi, A. (1988). On the meaning of variable. *Mathematics Teacher, 81*, 420–427.

Schoenfeld, A., Smith, J., & Arcavi, A. (1990). Learning: The microgenetic analysis of one student's understanding of a complex subject matter domain. In R. Glaser (Ed.), *Advances in instructional psychology* (Vol. 4, pp. 55–175). Hillsdale, NJ: Erlbaum.

Schwartz, J., & Yerushalmy, M. (1992). Getting students to function in and with algebra. In G. Harel & E. Dubinsky (Eds.), *The concept of function: Aspects of epistemology and pedagogy* (pp. 261–289). Washington, DC: Mathematical Association of America.

Sfard, A., & Linchevski, L. (1994). The gains and the pitfalls of reification: The case of algebra. *Educational Studies in Mathematics, 26*, 191–228.

Sleeman, D. H. (1984). An attempt to understand students' understanding of basic algebra. *Cognitive Science, 8*, 387–412.

Smith, J., diSessa, A., & Roschelle, J. (1993). Misconceptions reconceived: A constructivist analysis of knowledge in transition. *Journal of the Learning Sciences, 3*(2), 115–163.

Thorpe, J. (1989). Algebra: What should we teach and how should we teach it. In S. Wagner & C. Kieran (Eds.), *Research agenda for mathematics education* (Vol. 4, pp. 11–24). Reston, VA: National Council of Teachers of Mathematics.

Usiskin, Z. (1988). Conceptions of school algebra and uses of variables. In A. F. Coxford & A. P. Schulte (Eds.), *The ideas of algebra* (1988 Yearbook, pp. 8–19). Reston, VA: National Council of Teachers of Mathematics.

Wagner, S. (1981). Conservation of equation and function under transformations of variable. *Journal for Research in Mathematics Education, 12*, 107–118.

Wagner, S., & Kieran, C. (Eds.). (1989). *Research issues in the learning and teaching of algebra*. Reston, VA: National Council of Teachers of Mathematics.

Wenger, R. (1987). Cognitive science and algebra learning. In A. Schoenfeld (Ed.), *Cognitive science and mathematics education* (pp. 217–251). Hillsdale, NJ: Erlbaum.

Williams, C. (1998). Using concept maps to access conceptual knowledge of function. *Journal for Research in Mathematics Education, 29*, 414–421.

Wilson, M. (1994). One preservice secondary teacher's understanding of function: The impact of a course integrating mathematical content and pedagogy. *Journal for Research in Mathematics Education, 25*, 346–370.

Yerushalmy, M. (1991). Effects of computerized feedback on performing and debugging algebraic transformations. *Journal of Educational Computing Research, 7*, 309–330.

Yerushalmy, M. (1997a). Designing representations: Reasoning about functions of two variables. *Journal for Research in Mathematics Education, 28*, 431–466.

Yerushalmy, M. (1997b). Mathematizing qualitative verbal descriptions of situations: A language to support modeling. *Cognition and Instruction, 15*, 207–264.

Yerushalmy, M. (1997c). Reaching the unreachable: Technology and the semantics of asymptotes. *International Journal of Computers for Mathematics Learning, 2*, 1–25.

Yerushalmy, M. (1998). Organizing the post-arithmetic curriculum around big ideas: An idea turned into practice. In T. Breiteig & G. Brekke (Eds.), *Proceedings of the Second Nordic Conference on Mathematics Education* (pp. 59–69). Kristiansand, Norway: Hogskolen i Agder.

Yerushalmy, M. (1999). Making exploration visible: On software design and school algebra curriculum. *International Journal for Computers in Mathematical Learning, 4*, 169–189.

Yerushalmy, M., & Bohr, M. (1995). Between equations and solutions: An odyssey in 3D. In L. Meira & D. Carraher (Eds.), *Proceedings of the 19th Annual Meeting of the International Group for the Psychology of Mathematics Education* (Vol. 2, pp. 218–225). Recife, Brazil: PME.

Yerushalmy, M., & Chazan D. (2002). Technologically supported curricular change, teacher knowledge, and student learning: The case of graphing and solving in school algebra. In L. D. English (Ed.), *Handbook of international research in mathematics education: Directions for the 21st century* (pp. 725–755). Mahwah, NJ: Erlbaum.

Yerushalmy, M., & Gafni, R. (1992). Syntactic manipulations and semantic interpretations in algebra: The effect of graphic representation. *Learning and Instruction, 2*, 303–319.

Yerushalmy, M., & Schwartz, J. (1993). Seizing the opportunity to make algebra mathematically and pedagogically interesting. In E. Fennema, T. Romberg, & T. Carpenter (Eds.), *Integrating research on the graphical representation of function* (pp. 41–68). Hillsdale, NJ: Erlbaum.

CHAPTER 10

Stasis and Change: Integrating Patterns, Functions, and Algebra Throughout the K–12 Curriculum

Erick Smith, Ithaca College

> In a general way, I would like to emphasize that Western schooling (and Western knowledge for that matter) looks at the world, first and foremost in terms of objects, states, situations, constant aspects of phenomena, and that dynamic aspects are introduced in a lesser and at least secondary way. (Pinxten, 1997, p. 379)

> The concept of change and state, of space and time, and of a world in which things can perdure and 'exist' while we do not focus attention on them, all these are tools the cognitive subject uses to organize and manage the flow of experience. (von Glasersfeld, 1995, p. 88)

Introduction

The relationship among patterns, functions, and algebra is not necessarily obvious. Patterns involve real or potential repetitions (audio, visual, tactile, numeric, and so on). Algebra in its traditional school form involves rules for manipulating symbols. A function, according to the definition all students must memorize, is a particular kind of mapping from one set to another. Clearly a certain amount of imagination and some historical analysis are necessary to illuminate the connections among these topics. *Curriculum and Evaluation Standards for School Mathematics* (NCTM, 1989) does make a credible attempt to bring the topics together, yet does not make a strong case for their conceptual link across Grades K–12. The Standards for Grades K–4 on patterns, for example, have little connection with later sections on patterns, algebra, and functions. Thus elementary school teachers may create rich classroom experiences around patterns, yet not have a sense of how this topic ties into the ongoing mathematical development of their students, much less into the topic of functions. Likewise, middle school teachers may create experiences that allow students to see how changes in one variable are associated with changes in another, yet see little connection with their perception of algebra or with their understanding of calculus. This chapter makes an argument for a strong connection among these topics across the curriculum for Grades K–12 and proposes the theme of "stasis and change" that could serve to unite these topics both pedagogically and conceptually.

In line with the foregoing quote by Pinxton, "stasis" refers to "objects, states, situations, (and) constant aspects of phenomena." Stasis, in looking at a pattern, would involve a description of its structure. Stasis, in looking at a function, would involve the fixed relationship between the members of two sets, for example, in an algebraic relationship of the form $y = f(x)$. Supporting Pinxton, this chapter argues that a static view (i.e., stasis) underlies much of school mathematics, particularly in the treatment of patterns, functions, and algebra. Recently, several authors have suggested that the "mathematics of change" is a missing element from the standard U.S. curriculum (Nemirovsky, Tierney, & Ogonowski, 1993; Mokros, Economopoulos, & Russell 1995).[1] From the perspective of the "mathematics of change," students are encouraged to examine patterns in relation to the ways they grow or can be extended. Functional situations, such as a trip on a bicycle, are looked at in terms of

[1] As a result of this work, the mathematics of change has been incorporated into several units of the NSF-funded exemplary elementary curriculum, "Investigations in Number, Data, and Space" (Addison-Wesley).

the change in the position (or speed) of the bicycle as time passes in equal increments. This chapter argues not only that the mathematics of change should become part of the Grades K–12 curriculum but also that the relationship between stasis and change provides a conceptual underpinning for much of the work students might do with patterns and functions across Grades K–12. This latter claim is based on two related perspectives. From a mathematical perspective, stasis and change help build the conceptual underpinnings for precalculus and calculus. From the perspective of the learner (see quote above by von Glasersfeld), stasis and change are closely connected with our fundamental ways of organizing and managing the flow of experience, that is, of our ways of learning and understanding. Thus we have reason to believe that stasis and change could unite a strong mathematical theme with a viable path for building students' understanding.

Because patterns, functions, and algebra have traditionally been treated as separate topics in the curriculum, this chapter starts with a separate review of each topic. A major theme of the chapter, however, is that these topics should be conceptually connected from early elementary through secondary school mathematics. These connections are explored more fully in subsequent sections.

Patterns

Mathematics is often called the "science of patterns," making patterns more of a defining quality of mathematics than a topic for inclusion. For this reason, we find no well-defined concept of pattern in the mathematical literature nor a specific history of patterns within the history of mathematics. Thus we might ask what makes a pattern. A fundamental idea of a pattern is that of change or repetition. We identify a pattern in that which we see repeating or imagine the possibility of its repeating. In fact, according to the *American Heritage Dictionary*, the origin of *pattern* is from the Medieval Latin *patrõnus*, "something to be imitated" (Morris, 1981, p. 962). Of course, were mathematics only about imitation in its basic sense, such terms as "math phobia" would not be a part of our vocabulary. When one imitates or repeats something, however, one is always paying attention to certain characteristics that one sees as important in the particular situation. Although the characteristics of importance are constructed in collaboration with others, the focus of each individual is ultimately his or her own construction. Thus one lens through which to look at learning mathematics is gaining the ability to construct patterns that are compatible with, or can be communicated to, others. What makes this lens both interesting and challenging is that in any situation, multiple possible patterns can always be constructed.

One (slightly modified) example from the 1989 NCTM *Curriculum and Evaluation Standards* (p. 64) is shown in Figure 10.1. The *Standards* document describes this pattern as three repeating figures—triangle, square, hexagon. One could also, however, see these figures in groups of four, where each group starts with the same figure it ends with, having two different figures in between. In many situations, the pattern that is obvious to the acculturated expert is subordinate to other patterns for the novice. In fact, one could look at many of the learning difficulties of children through this lens. Counting involves identifying a unit, creating a one-to-one correspondence between the counting sequence and each individual unit, and then associating each word number with a specific quantity of units. From an adult perspective, the pattern of interest is the accumulating quantity of units and the associated number word. Young children, however, often have learned the pattern of associating each number word with a unit but do not construct the related pattern of associating a quantity of units with a number word. In this sense, they are not constructing a pattern that is compatible with that perceived by the adult world.

Figure 10.1. Pattern of three repeating figures that could alternatively be seen as groups of four.

Because the potential for repetition is at the core of the idea of pattern, one always has two perspectives with which to view a given pattern. First, one can think of a pattern in terms of how it can be extended or repeated, in other words, in terms of change. For example, in Figure 10.1, one can focus on the process of extending the pattern, for example, adding one more group of three (or four) shapes. Likewise, one can focus on stasis, seeing the pattern as it is, for example, as a group of twelve geometric shapes that alternate. As we construct and work with patterns, we often go back and forth between these two perspectives easily and unconsciously. This distinction, however, is important in our understanding of pattern. In the pattern in Figure 10.1, the distinction between groups of three versus groups of four is largely in how one imagines extending the pattern (would the next step be to add three or four more geometric figures?).

Although many curricular activities that focus on patterns do include both stasis and change aspects of patterns, little, if any, conceptual development occurs of the importance of this relationship; rather, it is treated almost incidentally. The relationship between stasis and change, however, is a fundamental part of the approach to patterns, functions, and algebra that will be discussed subsequently.

Algebra

A major question is what we mean by *algebra*. In a brief description of the history of algebra in Western mathematics, Sfard and Linchevski (1994) provide two historic examples of what they call "rhetoric algebra," one from the Babylonians nearly 4000 years ago and one from Arabic mathematicians 3000 years later. Neither uses a well-structured symbol system; rather, as the name suggests, both are based on verbal descriptions. What makes them algebraic, from Sfard and Linchevski's perspective, is their generalization of an arithmetic process. Noting that rhetoric algebra has been around for more than 3000 years, whereas symbol-based algebra has had primacy over only the past 400 years, they state that "it is important to explicitly stress [that] the history of algebra is not the history of symbols" (p. 197). If we view algebra as generalized arithmetic, then symbols are just one linguistic tool for expressing these generalizations. As Sfard and Linchevski point out, however, algebraic symbolism is unrivaled in its power "to squeeze the operationally conceived ideas into compact chunks and thus in its potential to make the information easier to comprehend and manipulate" (p. 198). This power revolutionized algebra both epistemologically and functionally. Given this power, the relatively recent integration of symbolism with algebra is somewhat surprising. Sfard and Linchevski attribute this delay, at least partly, to the epistemological leap of treating an algebraic expression, such as $3(x + 5) + 1$, as both a process and a result.

Because the use of symbol systems has become prevalent, the word *algebra*, has become essentially synonymous with the study or use of such systems. *Algebraic thinking*, in contrast, has been used in a broader sense to indicate the kinds of generalizing that precede or accompany the use of algebra, such as with rhetoric algebra. One notable aspect of algebraic thinking lies in its general absence as a component of the traditional algebra course. This omission has had major consequences in the pedagogical treatment of algebra. Even though algebra as a generalization of arithmetic is a common perception, algebra courses give far more emphasis to learning the "linguistic rules." Essentially this prominence has meant treating algebraic manipulations as given procedures rather than as experiential constructs. This focus, along with the increasing national emphasis on algebra as a gatekeeper to higher education, has had the effect, according to James Kaput, of "alienation from mathematics for those who survive this filter and an even more tragic loss of life-opportunity for those who don't" (1999, p. 1). This emphasis is due, in no small part, to the testability of algebraic rules. The epistemological issues raised by Sfard and Linchevski (1994), however, indicate the complexity of integrating generalized arithmetic with symbolic language. Epistemology, or how one knows a topic, is intrinsically related to the language(s) one uses to create and communicate that topic (Vygotsky, 1987). Thus a symbolic language is more than just a tool that expresses generalizations; it is a tool that requires the learner to reconceptualize those generalizations.

These issues come up also in Kaput's (1999) discussions of algebra reform. He describes the following five aspects of algebra in present-day mathematics:

1. Algebra as generalizing and formalizing patterns and constraints, especially, but not exclusively, algebra as generalized arithmetic reasoning and algebra as generalized quantitative reasoning (p. 4)

2. Algebra as syntactically guided manipulation of (opaque) formalisms (p. 7)

3. Algebra as the study of structures abstracted from computations and relations (p. 7)

4. Algebra as the study of functions, relations, and joint variation (p. 8)

5. Algebra as a cluster of modeling languages and phenomena-controlling languages (p. 8)

A close look at these five aspects shows that the second one, syntactically guided manipulation of (opaque) formalisms, has been the focus of traditional algebra courses. The third aspect, the study of structures, belongs predominately to upper-level secondary and postsecondary school courses. As defined above, both these aspects focus more on the study of algebra itself than on algebraic thinking.

The first, fourth, and fifth aspects—generalization, relationships, and modeling—however, begin with algebraic thinking, allowing algebra to be a tool for expressing, manipulating, and representing these generalizations and relationships. Kaput (1999) argues that a goal of the reform movement should be the integration of these topics across the Grades K–12 curriculum, with the following result:

> Algebra disappears both as a set of isolated courses and as a set of intellectual tools, in the sense that for the carpenter, when in use the hammer becomes an extension of the arm (Polanyi, 1958). The different aspects of algebra become habits of mind, ways of seeing and acting mathematically—in particular, ways of generalizing, abstracting and formalizing across the mathematics and science curricula, including curricula leading to the world of work. (p. 16)

Although this vision may seem idealistic, the need for radical change in the way algebra is taught is evident in both the attitudes and accomplishments of our students. Although national attitudinal data are not available, many would agree with Kaput's (1999) assessment that "algebra has been transformed in the national consciousness from a joke to a catastrophe" (p. 2). In my own elementary mathematics methods course over the past five years, I have asked students to identify both their most positive and most negative mathematical experience. For most negative responses, naming a course from secondary school mathematics is, by far, the most common. Although this selection is often dependent on individual teachers, as a topic, algebra is mentioned more than any other. If we want to change the attitudes and accomplishments of our students, these future teachers hold the key. Yet algebra is the single most common topic mentioned as affecting them negatively toward mathematics.

In terms of accomplishment, the Third International Mathematics and Science Study survey ranked U.S. secondary students near the bottom in mathematics. Several other studies have indicated serious issues in students' ability to solve problems (Travers & McKnight, 1985; Brown, Carpenter, Kouba, Lindquist, & Silver, 1988; Dossey, Mullis, Lindquist, & Chambers, 1988). Although such broad-scale studies necessarily rest on debatable assumptions, they can be indicative of potential problems. The overall performance of U.S. students on standard tests intended to measure algebra competence clearly seems lower than what most educators and policymakers would like.

Of more interest are studies that have attempted to identify and analyze actual student performance in algebra, particularly in the use of variables. A classic context for such studies, first described by Clement, Lochhead, and Monk (1981), is the Students and Professors problem:

> Write an equation using the variables *S* and *P* to represent the following statement: "There are six times as many students as professors at this university." Use *S* for the number of students and *P* for the number of professors.

In several studies done in the early 1980s, between one third and one half of adults were unable to do this problem correctly (Lochhead, 1980; Clement, Lochhead, & Monk, 1981; Rosnick, 1981; Clement, 1982). Whereas the accepted answer to the problem is $S = 6P$, the most common mistake by far was to reverse the variables and answer $P = 6S$. A common conclusion is that many people treat the problem semantically as one professor being equivalent to six students, that is, *P* stands for a professor rather than the number of professors and *S* stands for students rather than the number of students. Fisher (1988) attempted to address this issue directly by modifying the variable names in the problem. Rather than *S* and *P*, she used N_S for the number of students and N_P for the number of professors. One group of students was given the original problem, and one, the modified problem. The results actually showed a decrease in performance for those receiving the modified problem.

Other studies have shown patterns of student errors in their use of variables. Tall and Thomas (1991) identify two common problems. The "lack of closure" (or "expected answer obstacle") occurs when "the child experiences discomfort attempting to handle an algebraic expression which represents a process that [he or she] cannot carry out" (p. 126). For example, in arithmetic, a final answer is always in terms of a single numeric answer, whereas an algebraic "answer" may express a process, such as 3 + 2a. A closely related issue is what Tall and Thomas call the "process-product obstacle," which is caused by the fact that an algebraic expression, such as 2 + 3a, represents both the process by which the computation is carried out and also the product of that process (p. 126). They attribute these difficulties to traditional teaching in which rules for manipulation are given primary if not sole emphasis, with the "forlorn hope" that understanding will follow.

Stacey and MacGregor (1997) gave the following problem to students in both their first year and their third year of algebra:

> David is 10 cm taller than Con. Con is *h* cm tall. What can you write for David's height? (p. 110)

Students' success was lower than the authors had expected, ranging from 50% for students in first-year algebra to 75% for more advanced students. They analyzed incorrect answers not just in terms of error types but with a focus on why different forms of the answer represented legitimate sense-making efforts by the students.

Stacey and MacGregor (1997) discuss the multiple ways in which students are asked to use letters, speculating on the difficulties in sorting them out. In particular, students must know when a letter is part of a word (or abbreviation) and when it is a mathematical variable. A common error was of the form "*Dh*," which can be read literally as "David's height." Other incorrect forms were more flexible in the meaning ascribed to *h*. Some responses with the authors' interpretations are given below:

1. $h = + 10$ (This example is an attempt to translate "the height is 10 more")

2. $h = D + 10$ (Con's height is *h;* David's height, *D*, is 10 more)

3. $Dh = h + 10$ (*Dh* is David's height, *h* is Con's height)

4. h = David − 10 (The *h* means "Con's height," but since no symbol has been given for "David's height," the word "David" is used) (p. 112)

Stacey and MacGregor conclude with a discussion of some general issues in students' interpretation of algebraic statements:

1. "operations implied in composite symbols"

For example, 5*a* means five times *a*, whereas 5 1/2 means 5 + 1/2, and 53 means ten times five, plus three.

2. "reading the equal sign as 'makes' or 'gives' and using it to link parts of a calculation"

In arithmetic, students often write multistep calculations as a continuous string, for example, $3+5 = 8 \times 7 = 56 \div 2 = 28$. Stacey and MacGregor found a similarity in students' interpretation of algebraic equations.

3. "features of natural language, such as indicators of temporal sequence, that students assume carry over into the formal language of algebra"

For example, reading $a = 28 + b$ as *a* equals 28, then add *b*. They point out how this interpretation fits more naturally with our common use of language.

Several important issues are relevant to these studies of students' problems with algebra and the use of variables. First, student errors are related to student sense-making, which, in turn, builds from prior experiences. Like so-called misconceptions in other areas, evidence shows that students do not change easily—quick fixes will not solve these problems. This conclusion concurs with those drawn by Fisher (1988) on the Students and Professors problem:

> Teachers often make the fallacious assumption, as I did here, of thinking that notation is self-explanatory and that "if I just make my explanation a little clearer, the students' problem will be solved." In fact, the students-and-professors problem and related problems continue to resist such instruction-centered modification. (p. 262)

This research elucidates the complexity of the milieu students enter as they begin to use symbol systems productively. It also helps to make somewhat clearer the epistemological jump described by Sfard and Linchevski (1994). Vygotsky (1987) and others have proposed a link between our ways of knowing and our cultural tools, including language and other representations. Each of the "error types" discussed above can be seen as instances of knowing arithmetic in a different way. This view suggests that even at a basic level, children are being asked not only to generalize arithmetic but also to reconceptualize arithmetic. Only from an adult perspective, in which the reconceptualization has already occurred, does algebra appear as generalized arithmetic.

Recognizing the complexity of the situation, two research groups have recently called for the integration of algebraic thinking (or reasoning) throughout the Grades K–12 curriculum. Erna Yackel, a member of the NCTM Algebra Working Group, describes the general agreement among the group that strict definitions of what is and what is not algebraic reasoning should not be attempted but, rather, that descriptions of the kinds of activities that would support algebraic thinking should be forthcoming (National Council of Teacher of Mathematics & Mathematical Sciences Education Board, 1998). She describes one first-grade classroom in which students focus on the possible combinations of five monkeys in two trees. She describes the point at which the class moves from naming specific combinations to speculating on whether they have found all possible combinations as an example of the kind of thinking that supports the development of algebraic thinking.

James Kaput headed the Early Algebra Group, which was supported by the Wisconsin Center for Research in Mathematics Education. Although this group did not produce a single document, an edited book is in process. The work by Kaput (1999) discussed previously is a central part of that work, and the recommendation that algebraic thinking be incorporated throughout the Grades K–12 curriculum is a central theme of the book.

Although the various authors take different perspectives, a major theme of the book is captured in the following quote from Smith and Thompson (1996):

> We believe it is possible to prepare children for different views of algebra—algebra as modeling, as pattern finding, or as the study or structure—by having them build ways of knowing and reasoning which make those mathematical practices appear as different aspects of a central and fundamental way of thinking. (n.p.)

In conjunction with these foci on early algebra, in February 1997, NCTM published a special issue dedicated to algebraic thinking in three of its member journals, *Teaching Children Mathematics*, *Mathematics Teaching in the Middle School*, and *Mathematics Teacher*. Although these three issues offer a wealth of ideas for incorporating activities to promote algebraic thinking at all grade levels, these activities tend to be offered either as isolated activities or as a set of activities appropriate at different grade levels. Although teachers at all grade levels will get many ideas for incorporating "algebraic" activities, still no conceptual basis is presented for incorporating and developing algebraic thinking across the grade levels. Given the difficulties that secondary school students have with using algebra, we cannot be at all certain that injecting such isolated activities will provide students with an adequate background for higher mathematical problem-solving activities that are part of Standards-based secondary school teaching. This issue is revisited in the last section of this chapter.

Functions

In many ways, the topic of functions should be better delineated than that of algebra. Unlike *algebra*, *function* has a definition that is essentially common to all textbooks and has been fairly consistent for some time. For example, in a 1986 textbook, Swokowski provided the following definition:

1. "A function from a set D to a set E is a correspondence that assigns to each element x of D a unique element y of E." (p. 128).

In a more recent text, Demarois, McGowen, and Whitkanack (1997) provide the following definition:

2. "A function is a process that receives input ... and returns a value called the output. ... There is exactly one output for each input." (p. 149)

The first definition focuses on a relationship between two existing sets, whereas the second gives the image of an input-output process in which the function is the process producing the output values. A commonality, however, is the focus on connecting a number in one set with a number in another set.

Like algebra, however, functions have a long history, whereas the modern definition has come into use only in the past 150 years. To understand this history, we are helped by first looking at some recent research. When we look at Table 10.1, the typical definition of function encourages us to notice a mapping from x to y: $0 \rightarrow 1$, $1 \rightarrow 4$, $2 \rightarrow 7$, and $3 \rightarrow 10$ (or one might imagine each x-value producing a y-value). In doing such a mapping, we create a correspondence between the x-values and the y-values. We also might choose to use an algebraic expression to generalize this correspondence relationship as $y = 3x + 1$.

TABLE 10.1 An x-to-y Mapping That Can Be Generalized to $y = 3x + 1$

x	y
0	1
1	4
2	7
3	10

Another way of looking at the table is by going down the column, noticing, for example, that the first value for x is 0, that the first value for y is 1, and that as x increases by 1, y increases by 3. From this perspective, one is seeing the two columns as two varying quantities and noticing the corresponding patterns of variation, that is, the *covariation* of x and y. In working with students at middle school, high school, and college levels in a course based on solving problems in contextual situations, Confrey and various associates (Borba & Confrey, 1995; Confrey & Smith, 1991, 1992, 1994, 1995) and Rizzuti (1991) found that students' initial entry into problems was typically from a covariational perspective. Using situations that would be modeled as linear, exponential, and quadratic, they found that students initially make a table of values, analyze the table from a covariational perspective, and then seek to generalize .

In a review of the historical development of concepts of function, Rizzuti (1991) distinguishes between the classical definition, often associated with Euler, and the modern definition, usually referred to as the *Dirichlet-Bourbaki* definition. An important distinction, according to Rizzuti, is that prior to the modern definition, significant importance was placed on the notion of covariation between two variables. The 10th century Arabic scholar Alhazen focused on the covariation of additive and accumulative change in his work on what we would now call quadratics, work continued by Leibniz (Dennis & Confrey, 1995). Descartes built covariation from geometric constructions (Smith, Dennis, & Confrey, 1992), and several mathematicians, from Thomas of Bradwardine to Napier, juxtaposed a variable changing additively with a variable changing multiplicatively. Ultimately this model supported Napier's development of logarithms (Smith & Confrey, 1994). Smith and Confrey conclude,

> In our current algebra curriculum, students build their understanding of analytical expressions primarily as algorithmic procedures that are decoded through a set of rules. Thus, at best, they are taught how to put a value into an expression and determine the output, very much in accord with a correspondence model of function. An alternative approach to algebra would be to see it as a means of coding the actions that one takes when expressing the variation of quantities. Although we believe that much work needs to be done in developing such a curricular approach, it offers the potential of allowing students to build their sense of covariation into their analytic expressions and finally allowing them to see this action in the expressions they and others create. (p. 336)

The final section of this chapter discusses examples of such curricular activities.

In the literature on the problems students have in learning functions, the focus on the correspondence approach is striking. MacGregor and Stacey (1993) review the various explanations that have been offered for the difficulties students have in creating an algebraic equation to represent a functional situation. They include three classifications: "syntactical translation," "misrepresentation of algebraic letters," and "interference from natural language." A striking feature about all classifications (including the explanation proposed by MacGregor and Stacey) is the focus on creating an algebraic expression as a mapping rule—that is, the focus on a correspondence approach. The researchers themselves seem to work within a paradigm wherein doing functions is a process of translating a verbal description of a situation into an algebraic representation of the mapping rule from one set to another. Given the dominance of the correspondence approach in the study of functions, this focus may not be surprising. But it seems to leave out building up relationships through an understanding of the variation of two related quantities, that is, covariation.

Slavit (1997) argues that this same focus exists in the literature on the formal properties of functions, where the emphasis on students' difficulties with such properties as domain, range, and distinguishing a functional from a nonfunctional relationship all reflect an emphasis on correspondence. Slavit argues that focusing more on a covariational approach would support a deeper student understanding of functions. The next section explores this issue more extensively, both through a theoretical approach, which attempts to integrate the topics of patterns, algebra, and functions across the Grades K–12 curriculum, and through some examples of how this approach might be instantiated at different levels.

Stasis and Change: A Framework for Integrating Patterns, Functions, and Algebra

This section develops a theme or conceptual framework for integrating patterns, functions, and algebra across the Grades K–12 curriculum in a way that incorporates many of the issues raised in the previous sections. The primary means of doing so is through the development of different examples of problems that could be appropriate at different levels of the curriculum. The theme of stasis and change, discussed in the foregoing section on patterns, becomes the inclusive framework that connects with the topics of functions and modeling from the algebra discussion and the topics of correspondence and covariation from the functions discussion and ties directly into the relationship of a function and its derivative in calculus. Thus to think of the theme of stasis and change as integrating patterns, algebra, functions, and calculus is not inaccurate.

Patterns

In the opening section, patterns were characterized as having a central component of change, repetition, or extension. Thus any pattern can be described in terms of its current state—*stasis*—or in terms of how it could be repeated or extended—*change*. Thus the pattern in Figure 10.1 can be viewed, for example, as four groups of three geometric figures—stasis—or in terms of how it could be extended by adding groups of three figures—change. For most patterns, describing change will involve choosing (or finding) a unit. Thus even young children in describing how a pattern changes can become aware of the notion of a unit of change. In Figure 10.1, the unit is a group of three shapes. A wealth of pattern activities is already available for the elementary school curriculum, and some, for middle school. Most of these pattern activities already contain elements of stasis and change. An important part of the stasis-and-change framework, however, is placing an emphasis on the distinction and the relationship between stasis and change. Thus the primary change in these pattern activities, particularly in the early grades, is, pedagogically, always to be aware of focusing on the following two questions:

- How can you describe this pattern?
- How can this pattern be repeated or extended? (What is the unit of change?)

These kinds of activities also can include number patterns. Much of the emphasis in current number activities focuses on describing a pattern in terms of the change from one number to the next—emphasizing change. Thus the two relevant questions might be these:

- What is the next number?
- How do you find it?

Once the second question is answered, activities can include finding a new number pattern by keeping the same rule for finding the next number but changing the starting number, or by asking what the rule would be to get every other number in the pattern, and so on. Such activities relate closely to the idea of change and accumulation. Tierney and Nemirovsky (1991) have described several possible kinds of relevant activities for young children's work in this area (see also Nemirovsky et al., 1993).

Patterns and Functions

We usually think of a pattern as a repeatable or extendable object. By juxtaposing the counting sequence with the units of a pattern, we create two repeatable or extendable objects, that is, a function. Such a function is represented in Figure 10.2.

Figure 10.2 A function created by juxtaposing the counting sequence with the units of a pattern.

By numbering patterns, we create a new level of kinds of questions that can frame an investigation:

- What pattern goes with 2? With 5? And so on.
- If I want to make it longer, what do I do to the top row? To the bottom row?

Functions created by juxtaposing the counting sequence with repeated pictures, patterns, activities, events, and so on, can form the basis for significant investigation in the early elementary school years. Ultimately, however, the major interest is in juxtaposing two number patterns. From Figure 10.2, for example, one might ask about keeping track of the total number of shapes in the pattern as the number of units increases. Children will invent many ways to keep track of this functional relationship, which is a juxtaposition of the counting sequence with a sequence starting at 3 when there is one unit and increasing by a constant plus 3 (3, 6, 9, 12, and so on).

At third and fourth grades, as children are learning about multiplication and division, such problems can allow such questions as "How many shapes would there be with 9 units?"

Some children may add up nine 3s, some will recognize that adding 3 nine times is the same as writing 9×3, and some may argue that there are always three times as many blocks as units and thus claim that the number of shapes going with nine units is 3×9. The first two arguments are indicative of seeing the problem from a covariational perspective, whereas the last indicates a correspondence approach. Whichever approach individual children take, each is verbally generalizing a process or pattern. By initiating discussion around how we can know that the various solution processes will have the same answer and why they will do so, children have the opportunity to generalize at a different level, focusing on their understanding of operations.

A variation on such a problem would be to introduce an initial quantity. In this example, one might place a single starter block of a different shape at the beginning of the shape pattern. Thus when the first unit is added, four shapes will be present; when the second unit is added, seven shapes will be present; and so on. This pattern is the same as that shown in Table 10.1. Although as adults, we have learned to see this relationship as a correspondence, $S = 3U + 1$ (S = #shapes, U = #units), this relationship is not evident to a novice. As mentioned previously, Confrey and her various colleagues (Borba & Confrey, 1995; Confrey & Smith, 1991, 1992, 1994, 1995; Rizzuti, 1991) found that middle and high school students have a strong tendency to analyze such problems through a covariational approach. One might expect this tendency of elementary school students also, as such an approach is closer to the actions involved in the problem. Thus, starting with one, each additional unit is equivalent to an action of adding three more shapes. For five units, I would have 1 plus 5(3s). For 9 units, I would have 1 plus 9(3s). Subsequently I discuss a similar problem in which fifth graders took just such an approach in analyzing a block pattern they were constructing.

In looking at such problems, the idea becomes apparent that each additional unit is associated with a repeated action, in this case, adding three shapes. Such an analysis underlies the argument by Confrey (1994) that in functional situations, as well as in counting situations, a unit is the repeated action of change (see also Confrey & Smith, 1994). Such a perspective on units allows the possibility of units that involve actions other than repeated addition. Confrey has argued that repeated multiplication, or "splitting," is an action (or unit type) basic to children's experience, and she has worked extensively with elementary school students on this approach (Confrey, 1995). An example of repeated multiplication would be to juxtapose the counting sequence with a series starting with a single shape and let each additional unit double the number of shapes.

Many activities that are appropriate for middle- and upper-level elementary school students involve quadratic relationships. One example related to the examples above is shown in Figure 10.3. In this situation, the action associated with the unit involves a two-step addition. First an additional square is added to the previous change unit, then the new change unit is added to the figure. One could imagine elementary school students analyzing such a pattern in terms of sums. For 7 units, they might calculate the total as, "7 triangles plus 7 circles plus (1 + 2 + 3 + 4 + 5 + 6 + 7) squares," providing a link to the topic of summing arithmetic sequences.

△□○△□□○△□□□○

# of units:	1	2	3
# of shapes:	3	7	12

Figure 3: The action associated with the unit involves a two-step addition.

One commonly used activity is working with "growing squares." In this case, one juxtaposes the counting sequence with a series of squares (see Figure 10.4). Although as adults, we tend to naturally see this relationship as a correspondence between the number in the bottom row and its square (numerically or geometrically), in an unpublished teaching interview that I conducted with a fifth grader, these figures were jointly constructed along with a table showing the values. When asked to predict the number of squares in the next figure, the student quickly responded 16. When asked how she knew, she said that the first change had been plus 3; the second change, plus 5; and therefore the third change should be plus 7. Thus what as adults may seem to be clearly a correspondence relationship may be seen quite differently by students making sense of such situations in relation to their own experiences.

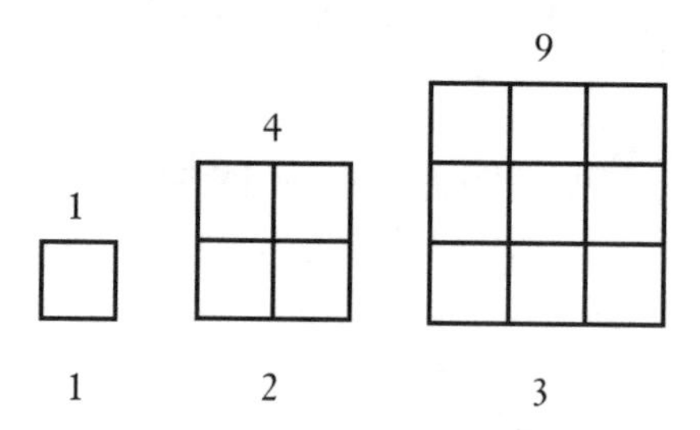

Figure 10.4. A growing-squares activity seen by one fifth grader as involving an additive change.

These examples are not intended to predict how children will see patterns and describe the changes and relationships they see. Not enough evidence is available to support such conjectures at this time. Some evidence, however, indicates that a covariational approach may be common for children working in functional situations such as these. This conclusion is based partly on the work of Confrey cited above and partly on the observation that a covariational approach is more closely related to the actual actions of making such figures, as well as creating representations for a broad range of function problems appropriate for students in the elementary grades.

The examples do hint at a broad range of investigations that all elementary school students could undertake that involve relating stasis and change (correspondence and covariation).[2] The types of situations presented are probably the most common, representing repeated addition (linear), repeated multiplication (exponential), and repeated double addition (quadratic). In addition to laying the groundwork for algebra, functions, and calculus, these activities provide a rich context for generalizing arithmetic operations. Students are called on to recognize not only that a situation calls for a certain operation (including double addition as an operation) but also that certain characteristics of the situation call for repeating the same operation. They are called on to constantly go back and forth between a context and its relationship to an operation. In addition, as they investigate the rela-

[2] Although the examples in this paper are geometric, the range of functional situations available for investigation is broad, potentially including topics across all subject matter in the curriculum.

tionship between stasis and change, they necessarily need to look at the result of applying a repeated operation.

Patterns, Functions and Algebra

Although the previous examples did not necessarily require the use of variables, they clearly involved the kind of algebraic thinking and reasoning discussed by Kaput (1999) and by Yackel. That is, students are being called on to build repeated operations from contextual situations, create ways to describe how to repeat these operations, and create ways to find and describe the results of repeating these operations. At this point, we do not know how quickly students would come to be able to verbalize these generalizations nor when they would be receptive to the use of variables. The expectation is that the use of variables would be introduced as part of ongoing classroom discussion as a shorthand way of representing some of the verbalizations that children express in class. Through introducing the use of variables in the context of ongoing linked investigations, one reasonably thinks that many of the obstacles to the use of variables discussed in the foregoing section on algebra might be overcome. Variables become part of the conversation as a means of identifying the action of doing repeated operations that are directly linked to repeated actions within a context. Moreover, they become ways to label the results of carrying out those repeated operations.

Thus the inclusion of algebra in the kinds of activities taught to elementary school children should occur naturally over a period of years as students incorporate the use of variables into their own descriptions of situations involving stasis and change. As students approach middle school, the use of variables could directly become a part of these activities, as some students will be ready to create their generalizations using variables and others will have gained the experiential background to be ready to hear or appropriate what their classmates are doing. As long as variables are discussed and used in a way that is as close as possible to the actions and descriptions the children are creating, this kind of appropriation should be a normal part of classroom activity.

Although this discussion has emphasized the covariational approach to functions, I should emphasize that this approach connects closely with the correspondence view and thus with more generalized views of functions typical of higher mathematics. In fact, one might see the covariational approach as a tool that leads into the development of a correspondence view. As will be seen in the example that follows, beginning with a covariational approach can provide the basis for developing a correspondence relationship that can then be expressed algebraically. Confrey (personal communication, 1992) has pointed out, however, that this approach to deriving an algebraic expression often leads to algebraic forms that are nonstandard. For example, the standard form of a linear function is $y = mx + b$. Students often are taught that this expression means take x m times and add b. However, this meaning may be very different from the way an individual experiences a linear situation. Developing an algebraic relationship from a covariational approach often leads to algebraic forms that are nonstandard but are built from the way the individual has constructed the relationship. One form, for example is $y = b + xm$, indicating a starting value, b, and a unit increase (decrease) of m where x indicates the number of unit changes. The covariational approach also can support the construction of algebraic formulas for other prototypic functions including exponential, quadratic, and trigonometric (Confrey & Smith, 1991, 1995). Confrey and Smith (1995) have also argued that for linear and exponential functions, students can develop ways to insert additional rational values in the domain through a process of interpolation. More generally, however, the covariational approach allows students to build an understanding of a function through the pattern (or operation) of change in each variable, to build an algebraic relationship based on that understanding, and then to use any appropriate values in the domain to create additional values for the range.

Rather than present additional activities, I present an excerpt from a summer class that I taught to elementary school teachers. This excerpt exemplifies the variety of ways in which individuals create, represent, and generalize patterns. In addition, it shows one process in which individuals working with a model that included stasis and change used their understanding of the actions involved in extending a pattern to ultimately create an algebraic representation of a numeric relationship in their pattern.

In the class, I introduced pattern blocks to the teachers and initially asked them to experiment with creating various patterns. Pattern blocks are blocks of various colors that come in certain geometric shapes: hexagon, square, triangle, rhombus, trapezoid, and so on. As might be expected, many kinds of patterns were created. After discussing the various patterns, the teachers were asked to work on creating a particular kind of pattern—one that, if repeated, would stretch in a straight band as far as one wished to expand it. As various patterns were created, people noted that what made a pattern in this context was some initial arrangement that was repeated. Thus creating such a pattern leads to the construction of a unit. As more patterns were created, I posed a question to the group: "Can you find a relationship between the number of repetitions (of the unit) and the total number of exterior sides in your pattern?"

Eventually, most groups created a table showing the number of repetitions and the number of sides. One teacher, John, put Table 10.2 on the board. He explained his reasoning in finding a pattern as follows: "I was trying to look at how much each number goes up by. I thought there might be a pattern there, but then I saw a better pattern. These numbers here (pointing at the right-hand, or ones, digits in the numbers under Sides) go down by two, and these numbers here (pointing at the left-hand, or tens, digits in the same column) go up by one. So if we use that pattern, I'm going to predict that if we had five repetitions (writes a 5 under the 4), this here would go up by one (writes a 4 under the first 3 in 33) and this here would go down by two (writes a 1 under the second 3 in 33, making the number 41 under 33), and then you get 59 (writes 59 under 41), and then you get ... (writes 67 under 59). To check it, you would have to go count all those sides, which would be kind of tedious."

TABLE 10.2. Seeking a Pattern in the Number of Sides Resulting From Repetition of the Unit

Repetitions	Sides
1	9
2	17
3	25
4	33

The first three repetitions of their pattern are shown in Figure 10.5.

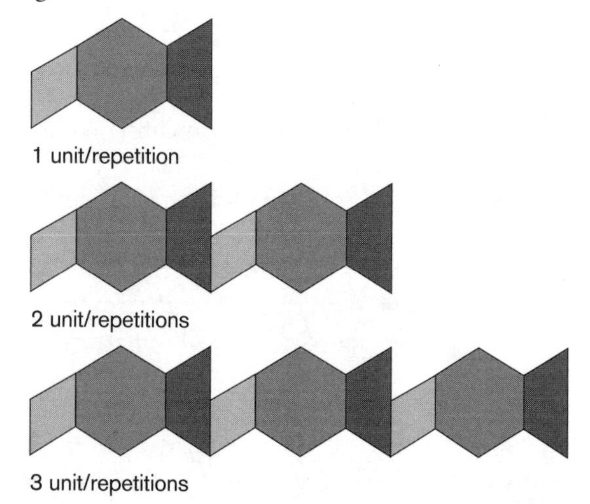

Figure 10.5. The first three repetitions of one group's pattern.

At this point, the table on the board looked like Table 10.3.

TABLE 10.3 Results of Extending Table 10.2

Repetitions	Sides
1	9
2	17
3	25
4	33
5	41
	59
	67

Another teacher suggested that each number in the original table goes up by eight. John replied, "Well actually we have, no, one was eight and the next one was nine. (Turning to his table on the board). I mean the difference between these two is eight (pointing at 9 and 17), but then the difference ..." (pauses uncertainly while pointing at 17 and 25). John had apparently miscalculated the difference between 17 and 25 as 9.

Other teachers said that the difference was eight also, then that all the differences in the original table were eight. John replied, "Are they all eight? Oh, there's a pattern there too then. All these numbers go up by eight ...(short pause). I thought that was interesting the way you could—without having to do it you can predict the way it's going to be."

John seemed to think that he had finished, and he returned to his seat. As teacher, I hypothesized (to myself) at this point that John had initially erred in calculating the differences and thus sought and found the pattern of "ones down by two, tens up by one." I was inclined to believe, however, that now knowing the existence of a constant difference of eight, he would likely adopt this model. I took his last comment in this light but posed to the class the following question: Do both of those descriptions of the pattern work?

John responded, "It seems to work as far as I have done it, eight repetitions. Forty-one plus eight is ... (pauses). That's not right, is it? (Another teacher says it should be 49.) It should be 49 if that was the pattern. So I don't think you can use that (emphasizing *that*)."

I still believed that John was rejecting his initial hypothesis, but I asked, "Which one?"

John replied, "You couldn't use that you are going to add 8 to each number to get the next."

Other teachers suggested that they needed to check it against the blocks. After a short time, John said, while pointing at the 41 on the board, "I do get 41 for the fifth repetition." John seemed to think he was verifying the pattern he suggested. He continued building another repetition. After a minute he announced, "Forty nine (turns and looks at the board). So that doesn't work."

Several teachers then pointed out that adding 8 does work and seemed satisfied. As the class continued, however, John looked alternatively between his table on the board and the blocks for several minutes.

John was reluctant to give up on his pattern even at the end of the episode when he seemed to have created a discrepancy for his pattern and at the same time created additional evidence for the constant-addition model. Eventually, however, he and the other teachers agreed that all the patterns that had been created produced tables in which the pattern of change in the number of sides involved constant addition. At this point, I posed the problem "You have found a constant difference pattern in your table. I would like you to work on finding, showing, or creating an explanation of how we might find that constant difference in the patterns you have made with the blocks."

The ensuing discussion was not taped and thus can only be summarized. After several minutes of discussion, various groups were prepared to offer an explanation. John again was the spokesman for his group. He began successively adding block units on the table while explaining what was happening. The discussion was framed in terms of the side that is "lost" in the action of appending another unit:

> Why is it we are adding eight when we know that one repetition has nine (sides), the second one only adds eight, so we are losing a side somewhere. . . . We are losing this side (pointing at the short side of the left of the unit being appended). So that's nine, but since we are losing this one side here, it's nine plus eight

John then repeated the action of appending a unit with a similar explanation. Other groups eventually offered similar explanations. In the final discussion, during which the groups were asked whether they could determine how many sides would result after 25 turns, 50 turns, or any large number of turns, their reasoning reflected this understanding. For example, John's group described how to find the number of sides after 25 repetitions by saying, "Take 25 block units of nine, which would be 25 • 9 sides." But 24 of these units would lose a side when being appended, thus their calculation was created as 25 • 9 – 24. After further class discussion, two algebraic representations of this problem were created (S = # sides; R = # repetitions):

$$S = R \cdot 9 - (R-1) \cdot 1$$

$$S = 9+(R-1) \cdot 8.$$

Each of these expressions was related to a specific way of seeing the action in the problem.

Referring to Figure 10.5, one can see that a single unit (or repetition) has nine sides. If R repetitions are conducted, $R \cdot 9$ total sides result. After the first unit, however, each additional unit loses one side when it is appended, thus $(R-1) \cdot 1$ must be subtracted. The first unit has nine sides. Each additional unit adds eight additional sides. The class concluded with an investigation of how we can know whether these two formulas are equivalent, that is, will always produce the same answer.

Two important issues are raised by this episode. First, students do not necessarily see patterns in the way the teacher expects. Like John, they are likely to be quite certain of the pattern as they see it and need both time and opportunity to explore the viability of their view. Second, the typical textbook way of writing the algebraic equation would be $S = 8R + 1$. Although concise, such a representation rarely represents the actions actually taken within a problem situation. For example, as has been observed in other situations, these teachers and students spontaneously wrote the variable as a multiplier of a constant ($R \cdot 9$ rather than $9R$), reflecting the idea that the variable represents the number of times a constant action is repeated. Students are likely to create algebraic representations that do not look standard but that do represent the actions as they see them within the problem situation. Having the opportunity to create such nonstandard representations is an essential part of making sense of variables. Also, discussing and investigating how and if different representations are equivalent provides a basis for introducing many of the rules for algebraic manipulation.

These teachers had had very little prior experience working with variables in a sense-making context, yet they were ultimately able to create the two algebraic expressions and convince themselves of their equivalence. We have almost no evidence that would allow us to predict when elementary school students might do equivalent work, especially if stasis-and-change activities had been integrated throughout their elementary school years. In a series of several teaching interviews that I conducted with fifth-grade students in a low-income inner-city elementary school, however, pairs of students were given the same task described above. After creating some extendable pattern, they were asked to investigate how the number of sides changed as they added additional units. After considerable discussion, they were finally asked whether they could find a way to predict how many sides would result with 20 units. Of four pairs of students interviewed, three were ultimately able to describe a method for doing so, all of which bore some similarity to the explanations given above, that is, they reflected the actions taken in continuously adding a unit. Although no attempt was made to introduce the use of variables, most of these students were able to quickly generalize their generalization. In other words, once they had found a way to explain how to find the number of sides for 20 units, they could apply this method if asked for another number of units.

Although this problem is typical of problems appropriate for upper elementary and middle school students, it is also representative of the kind of contextual problem that could be a basis for the secondary school curriculum. The linear, exponential, and quadratic examples in this chapter introduce more complex investigations for the secondary grades. In addition, the theme of stasis and change can also be used to investigate trigonometric, logarithmic, and higher order polynomial functions.

Discussion

Although much research clearly needs to be done on stasis and change, the argument in this chapter is that it presents a conceptually rich theme across the Grades K–12 curriculum. It has the potential not only to tie together many aspects of the curriculum, including patterns, functions, and algebra, but also ultimately to connect children's initial attempts to create contexts for repeated actions with calculus. Of the five aspects of algebraic thinking proposed by Kaput (1999), three are intrinsically connected with stasis and change. Generalization of arithmetic arises as students connect contexts with situations in which repeated operations are appropriate and in which they attempt to describe the results of repeating an operation some number of times. The distinction between functions and modeling is essentially an artificial distinction, probably related to our culture's focus on the static, as noted by Pinxten in the opening quotation. Modeling is essentially a process of simulating a process by focusing on how the variables change. Thus modeling, as it is typically practiced, is essentially a covariational (or change) approach to functions.

The topic of functions, on the one hand, as currently taught in our curriculum, focuses almost entirely on correspondence relationships (or stasis) and originates in 19th-century attempts to formalize and abstract mathematics from everyday life. Modeling, on the other hand, is the practical application of the function concept. Modelers choose a covariational approach because in most physical or biological situations, change is what is observed, what is understood, and often what is most important. Although the covariational approach could certainly be used without a context by simply having students examine tables of values and seek patterns within these tables, this technique should be done with caution. The most important aspect of functions for Grades K–12 is that students build an understanding of how contextual situations connect with patterns of change that allow the construction of a function. The study of functions provides an arena in which students can enrich and expand their understand of operations.[3]

Likewise, stasis and change merge naturally into calculus. In calculus, students learn how to calculate a derivative (change) from a correspondence functional relationship (stasis) and also how to calculate a correspondence relationship from a relationship of change. The literature on calculus reform, however, indicates that few students finish calculus with a reasonable conceptual understanding of these relationships. Instead they learn just what they have been taught—the processes for making these algebraic changes from one form to another. Just as we have no reason to compel students to wait until a formal algebra class to learn about variables, we also have no reason to make them wait until calculus to investigate the relationship between stasis and change. In the second quote opening the chapter, von Glasersfeld claims that the relationship between stasis and change is fundamental to the individual's understanding of her experiential world.

A final issue is how this theme relates to current efforts to revise NCTM's Standards. The lack of a broad research base might seemingly argue against making stasis and change, at this point, a major strand of the Grades K–12 Standards. However, the arguments in this chapter do suggest this direction for further research and future consideration. If the stasis-and-change framework becomes a major theme or strand across Grades K–12, what are currently considered independent topics in mathematics—patterns, functions, modeling, algebra, and calculus—become tools that are used in the investigation of this theme. Every year from kindergarten into college, students would be studying topics related to stasis and change, and, ideally, the various tools would be integrated as appropriate over the years. This approach

[3] It may be the case that a few students or advanced classes may want to explore functions from a more abstract perspective. In this situation, the covariational approach provides a foundation for this exploration.

would involve a major reconceptualization of the mathematics curriculum. Over the years, the formalization and dissection of the mathematics curriculum, as presented in textbooks, has essentially eliminated distinctions between mathematical tools, such as algebra, and rich conceptual themes that should play out throughout the curriculum. It seems appropriate that current work on developing standards include initiatives to identify other appropriate conceptual strands as a means of conceptually restructuring the curriculum. This question is much bigger and, almost certainly, more important than simply asking "How can we learn to teach algebra with understanding?"

ACKNOWLEDGMENTS

Doug Smeltz and Steve Monk provided comments on an earlier draft of this chapter

BIBLIOGRAPHY

Borba, M., & Confrey, J. (1995). A student's construction of transformations of functions in a multiple representational environment. *Educational Studies in Mathematics, 29*, 1–20.

Brown, C. A., Carpenter, T., Kouba, V., Lindquist, M., & Silver, E., (1988). Secondary school results for the fourth NAEP mathematics assessment: Algebra, geometry, mathematical methods, and attitudes. *Mathematics Teacher, 81*, 337–347.

Clement, J. (1982). Algebra word problem solutions: Thought processes underlying a common misconception. *Journal for Research in Mathematics Education, 13*, 16–30.

Clement, J., Lochhead, J., & Monk, S. (1981). Translation difficulties in learning mathematics. *American Mathematical Monthly, 88*, 286–290.

Confrey, J. (1994). Splitting, similarity, and rate of change: New approaches to multiplication and exponential functions. In G. Harel and J. Confrey (Eds.), *The development of multiplicative reasoning in the learning of mathematics*, pp. 291–330. Albany, NY: SUNY Press.

Confrey, J. (1995). Student voice in examining "splitting" as an approach to ratio, proportions, and fractions. In L. Meira & D. Carraher (Eds.), *Proceedings of the 19th Annual Meeting of the International Group for the Psychology of Mathematics* (Vol. 1, pp. 3–29). Recife, Brazil: PME.

Confrey, J., & Smith, E. (1991). A framework for functions: Prototypes, multiple representations, and transformations. In R. G. Underhill (Ed.), *Proceedings of the 13th Annual Meeting of the North American Chapter of the International Group for the Psychology of Mathematics Education* (Vol. 1, pp. 57–63). Blacksburg, VA: PME-NA.

Confrey, J., & E. Smith (1992). Revised accounts of the function concept using multi-representational software, contextual problems and student paths. In William Geeslin & Karen Graham (Eds.), *Proceedings of the 14th Annual Meeting of the International Group for the Psychology of Mathematics Education* (Vol. 2, pp. 153-160). Durham, NH: PME

Confrey, J., & Smith, E. (1994). Exponential functions, rate of change, and the multiplicative unit. *Educational Studies in Mathematics, 26*, 111–134.

Confrey, J., & Smith, E. (1995). Splitting, covariation and their role in the development of exponential functions. *Journal for Research in Mathematics Education, 26*, 66–86.

Demarois, P., McGowen, M. A., & Whitkanack, D. (1997). *Mathematical investigations: An introduction to algebraic thinking: Concepts and processes for the college student.* Boston: Addison-Wesley Longman.

Dennis, D., & Confrey, J. (1995). Functions of a curve: Leibniz' original notion of functions and its meaning for the parabola. *College Mathematics Journal, 26*(2), 124–131.

Dossey, J., Mullis, I., Lindquist, M. M., & Chambers, D. L. (1988). *The mathematics report card: Are we measuring up?* Princeton, NJ: Educational Testing Service.

Fisher, K. (1988). The students-and-professors problem revisited. *Journal for Research in Mathematics Education, 19*, 260–262.

Kaput, J. (1995). A research base for algebra reform: Does one exist? In D. Owens, M. Reed, & G. M. Millsaps (Eds.), *Proceedings of the 17th Annual Meeting of the North American Chapter of the International Group for the Psychology of Mathematics Education* (Vol. 1, pp. 71–94). Columbus, OH: ERIC Clearinghouse for Science, Mathematics, and Environmental Education.

Kaput, J. (1999). Teaching and learning a new algebra. In E. Fennema & T. Romberg (Eds.), *Mathematics classrooms that promote understanding* (pp. 133–155). Mahwah, NJ: Erlbaum.

Lochhead, J. (1980). Faculty interpretations of simple algebraic statements. *Journal of Mathematical Behavior, 3*, 29–38.

MacGregor, M., & Stacey, K. (1993). Cognitive models underlying students' formulation of simple linear equations. *Journal for Research in Mathematics Education, 24*, 217–232.

Mokros, J., Economopoulos, K., & Russell, S. J. (1995). *Beyond arithmetic: Changing mathematics in the elementary classroom.* Palo Alto, CA: Dale Seymour.

Morris, W. (Ed.). (1981). The American heritage dictionary of the English language. Boston: Houghton Mifflin.

National Council of Teachers of Mathematics (NCTM). (1989). *Curriculum and evaluation standards for school mathematics.* Reston, VA: Author.

National Council of Teachers of Mathematics & Mathematical Sciences Education Board. (1998). *The nature and role of algebra in the K–14 curriculum: Proceedings of a national symposium.* Washington, DC: National Academy Press.

Nemirovsky, R., Tierney, C., & Ogonowski, M. (1993). *Children, additive change, and calculus* (Working Paper 2-93). Cambridge: TERC.

Pinxten, R. (1997). Applications in the teaching of mathematics and the sciences. In A. Powell & M. Frankenstein (Eds.), *Ethnomathematics* (pp. 373–402). Albany, NY: SUNY Press.

Polanyi, M. (1958). *Personal knowledge.* Chicago: University of Chicago Press.

Rizzuti, J. (1991). *Students' conceptualizations of mathematical functions: The effects of a pedagogical approach involving multiple representations.* Unpublished doctoral dissertation, Cornell University, Ithaca, New York.

Rosnick, P. (1981). Some misconceptions concerning the concept of variable. *Mathematics Teacher, 74,* 418–420.

Sfard, A., & Linchevski, L. (1994). The gains and pitfalls of reification—The case of algebra. *Educational Studies in Mathematics, 26,* 191–228.

Slavit, D. (1997). An alternative route to the reification of functions. *Educational Studies in Mathematics, 33,* 259–281.

Smith, E., & Confrey, J. (1994). Multiplicative structures and the development of logarithms: What was lost by the invention of functions? In G. Harel & J. Confrey (Eds.), *The development of multiplicative reasoning in the learning of mathematics* (pp. 333–360). Albany, NY: SUNY Press.

Smith, E., Dennis, D., & Confrey, J. (1992). *Rethinking functions: Cartesian constructions.* A paper presentation at the Second International Conference on the History and Philosophy of Science in Science Teaching, Kingston, Ontario, 11–15 May 1992

Smith, J., & Thompson, P. (1996). *Additive quantitative reasoning and the development of algebraic reasoning.* Unpublished working paper, Center for Research in Mathematics Education, San Diego State University.

Stacey, K., & MacGregor, M. (1997). Ideas about symbolism that students bring to algebra. *Mathematics Teacher, 90,* 110–113.

Swokowski, E. W. (1986). *Precalculus.* Boston: PWS Publishers.

Tall, D., & Thomas, M. (1991). Encouraging versatile thinking in algebra using the computer. *Educational Studies in Mathematics, 22,* 125–147.

Tierney, C., & Nemirovsky, R. (1991). Young children's spontaneous representations of changes in population and speed. In R. G. Underhill (Ed.), *Proceedings of the 13th Annual Meeting of the North American Chapter of the International Group for the Psychology of Mathematics Education* (Vol. 2, pp.182–188). Blacksburg, VA: PME-NA.

Travers, K. J., & McKnight, C. (1985). Mathematics achievement in U.S. schools: Preliminary findings from the second IEA mathematics study. *Phi Delta Kappan, 66,* 407–413.

Von Glasersfeld, E. (1995). *Radical constructivism.* London: Falmer Press.

Vygotsky, L. S. (1987). Thinking and speech. In R. Rieber & A. Carton (Eds.), *The collected works of L. S. Vygotsky* (pp. 39–288). New York: Plenum Press.

CHAPTER 11

Teaching and Learning Geometry

Douglas H. Clements, State University of New York at Buffalo

Geometry in the United States for grades pre-K through 12 is the study of spatial objects, relationships, and transformations; their mathematization and formalization; and the axiomatic mathematical systems that have been constructed to represent them. Traditionally, U.S. geometry includes an informal introduction to a few basic concepts in grades pre-K through 8 and axiomatic, Euclidean geometry in high school. However, research and international practice show that much more can and should be done in all grades.

Geometry in Schools Today

The paucity of geometry before high school is a major concern. The usual preschool to middle school curriculum includes little more than recognizing and naming geometric shapes (Porter, 1989). Through the grades, the curriculum tends to name more geometric objects but not require deeper levels of analysis (Fuys, Geddes, & Tischler, 1988). Compounding matters, teachers often do not teach even the barren geometry curriculum that is available to them. Fourth- and fifth-grade teachers across entire districts spend "virtually no time teaching geometry" (Porter, 1989, p. 11). Current practices in the primary grades also promote little conceptual change: First-grade students in one study were more likely than older children to differentiate one polygon from another by counting sides or vertices (Lehrer, Jenkins, & Osana, 1998). Over time, given conventional instruction of geometry in the elementary grades, children were less likely to notice these attributes. It is little wonder that there are large individual differences in concept learning. In one study, 3% of the longitudinal students in Grade 3 had already attained the formal level of the concept equilateral triangle, whereas 21% of those in Grade 12 had not. In Grade 6, 38% understood the principles pertaining to equilateral triangles, but in Grade 12, 27% did not (Klausmeier, 1992).

Such neglect evinces itself in student achievement. Students in the United States are not prepared for learning more sophisticated geometry, especially when compared with students of other nations (Carpenter, Corbitt, Kepner, Lindquist, & Reys, 1980; Fey et al., 1984; Kouba et al., 1988; Stevenson, Lee, & Stigler, 1986; Stigler, Lee, & Stevenson, 1990). In one study, fifth graders from Japan and Taiwan scored more than twice as high as U.S. students on a test of geometry (Stigler et al., 1990). The results of the more recent Third International Mathematics and Science Study (TIMSS) are not more encouraging, showing that students in the United States do not learn much geometry from grade to grade (Mullis et al., 1997).

Similar poor geometry achievement is found in U.S. high schools. Only half the students enroll in geometry; of these, only 63% correctly identify triangles at the beginning of the year (Usiskin, 1987), and only 30% can write proofs or exhibit any understanding of the meaning of proofs at the end of the course (Senk, 1985). Unsurprisingly, high school students also rank low internationally (McKnight, Travers, Crosswhite, & Swafford, 1985; McKnight, Travers, & Dossey, 1985). In the TIMSS, U.S. secondary school students scored at or near bottom in every geometry task (Beaton et al., 1996; Lappan, 1999).

As a footnote to these comparisons, recent research shows differences even among preschoolers in various countries (Starkey et al., 1999). On a geometry assessment, 4-year-olds from the United States scored 55% compared with 84% for those from China. Thus, cultural supports are lacking from the earliest years in the United States.

In summary, U.S. curriculum and teaching in the domain of geometry is generally weak, leading to unacceptably low levels of achievement. Other research results suggest some guidelines for efforts to ameliorate this situation. This chapter reviews these results under the headings of theories of geometric thinking, learning, and teaching; instructional tools; selected geometric topics; and other issues in teaching and learning geometry.

Theories of Geometric Thinking, Learning, and Teaching

Piaget and Inhelder

Piaget and Inhelder's (1967) early influential theory included two major themes. First, representations of space are constructed through the progressive organization of the student's motor and internalized actions. Thus, the representation of space is not a perceptual "reading off" of the spatial environment but is built up from prior active manipulation of that environment. Second, the progressive organization of geometric ideas follows a definite order that is more logical than historical. That is, initially, topological relations (e.g., connectedness, enclosure, and continuity) are constructed, followed by projective (rectilinearity) and Euclidean (angularity, parallelism, and distance) relations. See Figure 11.1 for examples.

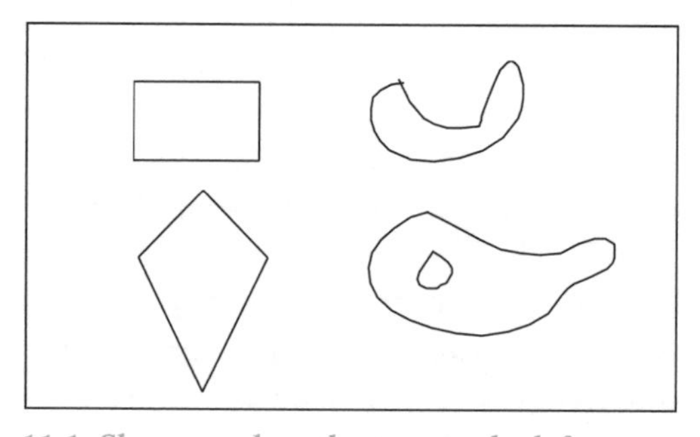

Figure 11.1. Shapes such as the two on the left were considered "Euclidean" by Piaget and Inhelder (1967) in their haptic perception experiments; the two on the right were considered to be "topological forms."

The progressive organization hypothesis has received, at best, mixed support (Clements & Battista, 1992). Mathematically, the Euclidean/topological description is not accurate, because all the forms in Figure 11.1 have both Euclidean/topological properties to an equivalent degree. Visually salient properties—such as holes, curves, and corners; simplicity; and familiarity—rather than topological versus Euclidean properties may underlie children's discrimination. Overall, research indicates that all types of geometric ideas appear to develop over time, becoming increasingly integrated and synthesized. Certainly, some Euclidean notions are present at an early age. These ideas are originally intuitions grounded in action—building, drawing, and perceiving. Children might learn actions that produce curvilinear shapes before those actions that produce rectilinear shapes. Even young children have basic competencies in establishing spatial frameworks that might be effectively built upon in the classroom.

Piaget and Inhelder's first theme—that children's representation of space is constructed from active manipulation of their spatial environment—has been supported (Clements & Battista, 1992). Children's ideas about shapes do not come from passive looking. Instead, they come as children's bodies, hands, eyes, and minds engage in active construction. In addition, children need to explore shapes extensively to fully understand them; merely seeing and naming pictures is insufficient. Finally, they have to explore the parts and attributes of shapes. The focus on attributes leads to the next influential theory.

The van Hieles

According to the theory of Pierre and Dina van Hiele, students progress through levels of thought in geometry (van Hiele, 1986; van Hiele-Geldof, 1984). Their theory is based on several assumptions. First, learning is a discontinuous process characterized by qualitatively different levels of thinking. Such levels progress from a Gestalt-like visual level through increasingly sophisticated levels of description, analysis, abstraction, and proof. Second, these levels are sequential, invariant, and hierarchical; progress is dependent upon instruction, not age. Teachers can "reduce" subject matter to a lower level, leading to rote memorization, but students cannot bypass levels and achieve understanding. Students must work through certain "phases" of instruction. Third, concepts implicitly understood at one level become explicitly understood at the next level. Fourth, each level has its own language and way of thinking; teachers unaware of this hierarchy of language and concepts can easily misinterpret students' understanding of geometric ideas.

Level 1 is the visual level, in which students can recognize shapes only as wholes and cannot form mental images of them. A given figure is a rectangle, for example, because "it looks like a door." Students do not think about the attributes, or properties, of shapes. For instance, in classifying quadrilaterals (see Figure 11.2), students at this level included imprecise visual qualities and irrelevant attributes, such as orientation, in describing the shapes while omitting relevant attributes.

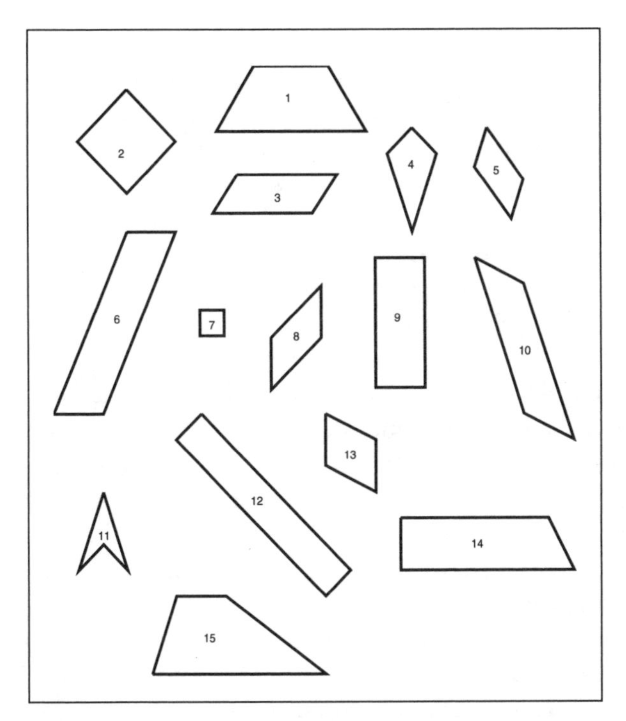

Figure 11.2. To assess the van Hiele level of geometric thinking, researchers asked students to identify and describe all the squares, rectangles, parallelograms, and rhombuses in this set of quadrilaterals (Burger & Shaughnessy, 1986).

Some researchers claim the existence of an earlier level—Level 0—at which children do not reliably distinguish circles, triangles, and squares from nonexemplars of those classes and appear unable to form reliable mental images of these shapes (Clements, Swaminathan, Hannibal, & Sarama, 1999). Other researchers prefer to categorize these children merely as beginning this first level.

At Level 2, the descriptive/analytic level, students recognize and characterize shapes by their properties. For instance, a student might think of a square as a figure that has four equal sides and four right angles. Students establish properties experimentally by observing, measuring, drawing, and model making. Students discover that some combinations of properties signal a class of figures and some do not; thus the seeds of geometric implication are planted. Students at this level do not, however, see relationships between classes of figures (e.g., a student might believe that a figure is not a rectangle because it is a square). For instance, students at this level contrasted shapes in Figure 11.2 and identified them by means of their properties. One girl said that rectangles have "two sides equal and parallel to each other. Two longer sides are equal and parallel to each other, and they connect at 90 degrees" (Burger & Shaughnessy, 1986, p. 39). Squares were not included. Many students do not reach this level until middle or even high school.

At Level 3, the abstract/relational level, students can form abstract definitions, distinguish between necessary and sufficient sets of conditions for a concept, and understand and sometimes even provide logical arguments in the geometric domain. They can classify figures hierarchically by ordering their properties and can give informal arguments to justify their classifications (e.g., a square is identified as a rhombus because it can be thought of as a "rhombus with some extra properties"). They can discover properties of classes of figures by informal deduction. For example, they might deduce that in any quadrilateral the sum of the angles must be 360° because any quadrilateral can be decomposed into two triangles each of whose angles sum to 180°. As students discover such properties, they feel a need to organize the properties. One property can signal other properties, so definitions can be seen not merely as descriptions but as ways of logically organizing properties. They see clearly why, for example, a square is a rectangle. This logical organization of ideas is the first manifestation of true deduction. The students still, however, do not grasp that logical deduction is the method for establishing geometric truths. For instance, students at this level gave minimal characterization of shapes in Figure 11.2 by arguing that a square is a parallelogram that has all the properties of a rhombus and a rectangle.

At Level 4, students can establish theorems within an axiomatic system. They recognize the difference among undefined terms, definitions, axioms, and theorems and are capable of constructing original proofs, that is, producing a sequence of statements that logically justifies a conclusion as a consequence of the "givens." For instance, one student frequently made conjectures about the shapes in Figure 11.2 and attempted to verify these conjectures by means of formal proof.

Research generally supports that the van Hiele levels are useful in describing students' geometric concept development (Burger & Shaughnessy, 1986; Clements & Battista, 1992; Fuys et al., 1988; Han, 1986). The levels are too broad for some tastes, however, and may be inaccurate in describing the geometric thinking of the youngest students. These students recognize some components and attributes of shapes and describe shapes in a variety of ways that are not necessarily "mathematical." Some researchers have proposed reconceptualizing this level as syncretic (i.e., signifying a global combination without analysis) rather than visual—implying a synthesis of verbal declarative and imagistic knowledge, each interacting with and enhancing the other (Clements et al., 1999). Other researchers prefer to eschew the notion of a "visual," or any, level because children evince a variety of ways to think about and describe geometric objects (Lehrer, Jenkins, et al., 1998). Still other researchers have proposed modifications to the theory. A synthesis of the van Hiele and SOLO (Structure of Observed Learning Outcomes) models emphasizes that children initially may be able to think of just one aspect or feature of a shape; then they can think of more than one aspect or feature, but independently of one another; and finally, they are able to relate these features. This view adds a valuable way of thinking about the levels of geometric thinking (Pegg & Davey, 1998).

Additionally, the van Hiele levels may not be discrete. Students appear to show signs of thinking at more than one level in the same or different tasks, in different contexts (Gutiérrez, Jaime, & Fortuny, 1991; Lehrer, Jenkins, et al., 1998). Movement through the levels may occur in general, but conceptualizing geometric growth as being strictly visual, then strictly descriptive/analytic, and so on, may be neither accurate nor optimal for educational theory and practice. Students instead possess and develop competencies and knowledge at several levels simultaneously, although one level of thinking may predominate. For example, de Villiers (1987) concluded that hierarchical class inclusion and deductive thinking develop independently and that hierarchical thinking depends more on teaching strategy than on van Hiele level.

In summary, although the van Hiele theory lacks detailed descriptions of students' thinking and researchers remain uncertain how well the theory reflects children's mental representations of geometric concepts, it does provide a general framework for curriculum and teaching. The theory also provides a reminder that students think about geometry in quite different ways and serves as a framework that helps us understand students' varied notions. The finding that most U.S. students are not developing through the levels at all, but that such development is possible given better curriculum and teaching, cannot be ignored. For example, most textbooks do not require students to develop higher levels of thinking through the grades (Fuys et al., 1988). Given that even young children refer to components and attributes of shapes, there is every reason to enrich multiple types of geometric thinking in students of all ages.

The van Hiele theory also includes a model of teaching that progresses through five phases in moving students from one level of thinking to the next. In Phase 1, Information, the teacher places ideas at students' disposal. In Phase 2, Guided Orientation, the teacher engages students actively in exploring objects, such as folding and measuring, so as to encounter the principal connections of the network of conceptual relations to be formed. In Phase 3, Explicitation, the teacher guides students to become explicitly aware of their geometric conceptualizations, describe these conceptualizations in their own language, and learn traditional mathematical language. In Phase 4, Free Orientation, students solve problems whose solutions require synthesizing and using those concepts and relations. The teacher's role includes selecting appropriate materials and geometric problems with multiple solution paths; giving instructions to permit various performances and encouraging students to reflect and elaborate on these problems and their solutions; and introducing terms, concepts, and relevant problem-solving processes as needed. In Phase 5, Integration, teachers encourage students to reflect on and consolidate their geometric knowledge, with an increased emphasis on the use of mathematical structures as a framework for consolidation, and eventually place these consolidated ideas in the structural organization of formal mathematics. At the completion of Phase 5, students have attained a new level of thought for the topic. Only in the Explicitation and Integration phases does the teacher sharply direct the learner's intention.

These ideas imply that teachers should enrich the geometry learning of their students by going beyond typical curriculum materials. To build students' visual-level thinking, much is required beyond naming shapes. Students might make shadows with shapes and identify shapes in different contexts, all the while describing their experiences. Especially at the early levels, children should manipulate concrete geometric shapes and materials so they can pursue their own explorations of geometric shapes. They might combine, fold, and create shapes, or they might copy shapes on geoboards by drawing or tracing. Children who are ready to explore Level 2 can investigate the parts and attributes, or properties, of shapes. They might measure, color, fold, or cut to identify properties of figures. For example, children could fold a square to figure out equality of sides or angles or to find symmetry (mirror) lines. They might sort shapes by their attributes (e.g., place all those with a square corner here) or play "guess my shape" from attribute clues.

The van Hiele work has several other implications for teaching. Imprecise language plagues students' work in geometry (Burger & Shaughnessy, 1986; Fuys et al., 1988). Instruction should carefully draw distinctions between common usage and mathematical usage. Teachers need to remember that children's concepts that underlie language may be vastly different than the teachers think. Thus when mathematical language is used too early and when teachers do not use everyday speech as a point of reference, mathematical language is learned without concomitant mathematical understanding (van Hiele-Geldof, 1984).

Language, of course, rests on a foundation of real-world experiences, and beginning with such experiences is indicated by research. Initial presentation of concepts in real-world settings and through manipulatives is especially helpful at Levels 0 and 1. Teachers should not rely solely on textbooks. Another problem with overreliance on traditional textbooks, discussed in the following section, has to do with the type of internal, or psychological, representations students form of geometric objects.

Theories of Internal Representations

A construct that has been applied extensively to geometric thinking and learning is that of the concept image—a combination of all the mental pictures and properties that have been associated with the concept (Vinner & Hershkowitz, 1980). Students often use concept images rather than definitions of concepts in their reasoning. These concept images are adversely affected by inappropriate instruction. For example, the concept image of an obtuse angle as requiring a horizontal ray might result from the limited set of examples students see in textbooks and from a "gravitational factor" (i.e., a figure is "stable" only if it has one horizontal side, with the other side ascending). Similarly, students' concept images for a right triangle are most likely to include a right triangle with a horizontal and a vertical side, less likely to include a similar triangle rotated slightly, and least likely to include a right isosceles triangle with a horizontal hypotenuse. So, for some students, this figure (⌊) is a right angle, this figure (⌋) is a "left angle," and this figure (⟩) is not an angle at all. Students have to learn the decisive role of explicitly defining concepts to avoid errors in using the terms that signify them. They have to construct a meaningful synthesis of this definition with a range of exemplars. Employing such a synthesis of analytic and verbal processes to construct robust concepts is possible, especially for students in Grade 5 and beyond (Hershkowitz et al., 1990).

All theories discussed above support the type of geometry instruction—elaborate, meaningful, and oriented toward problem solving—proposed by the NCTM Standards. We know that the quality of geometric knowledge that students develop could have a powerful effect on their mental models and subsequent use of that knowledge (Chinnappan, 1998). The following sections review research on educational issues that affect students' geometric knowledge: instructional tools, students' ideas and learning of selected geometric topics, and other issues of teaching and learning.

Instructional Tools

Diagrams, Manipulatives, and Pictures

The construct of concept images suggests that although diagrams and pictures can support geometric reasoning, they bring their own set of problems. Research substantiates that when perceiving a diagram for a proof problem, for example, a student must focus on what is essential and dismiss what is inessential—a difficult process for many (Clements & Battista, 1992). Students often attribute characteristics of a drawing to the geometric object it represents, fail to understand that drawings do not necessarily represent all known information about the object represented, and attempt to draw figures so that they preserve both viewing perspective and the student's knowledge about the properties of the object being drawn (Parzysz, 1988). Instructional attention to diagrams, such as using multiple drawings for a proof problem and discussing diagrams explicitly, may be helpful.

Similarly, using manipulatives can facilitate the construction of sound representations of geometric concepts, but they too must be used wisely (Clements & Battista, 1992). Unfortunately, U.S. textbooks only infrequently suggest the use of manipulatives in geometry, and even when they do, the suggested uses are not aimed at developing higher levels of thinking (Fuys et al., 1988; Stigler et al., 1990). In contrast, the evidently more successful Japanese instruction and instructional materials feature greater use of manipulatives. However, manipulatives must be used thoughtfully; if not, students may merely learn rote manipulations (Clements, 1999). Teachers can use manipulatives as a way to reform their mathematics teaching without reflecting on their use of representations of mathematical ideas or on the other aspects of instruction that must be changed (Grant, Peterson, & Shojgreen-Downer, 1996). Both teachers and parents often believe that reform in mathematics education indicates that "concrete" is good and "abstract" is bad. In contrast, professional teaching standards suggest that students have a wide range of understanding and tools (Ball, 1992), and research suggests that several types of "concrete" knowledge exist, with the most sophisticated and useful types *integrating* concrete and abstract ideas (Clements, 1999).

Pictures also can be helpful; they can give students an immediate, intuitive grasp of certain geometric ideas.

However, pictures need to be varied so that students are not led to form incorrect concepts (cf. the previous discussions of concept images). In addition, research indicates that it is rare for pictures to be superior to manipulatives. In some cases, pictures may not differ in effectiveness from instruction with symbols (Sowell, 1989). The reason may lie not so much in the "nonconcrete" nature of the pictures as in their "nonmanipulability"—that is, that children cannot act on them as flexibly and extensively. This limitation suggests manipulable pictures, such as graphic computer representations, the subject to which I turn.

Computers

Computers' graphic capabilities may also facilitate the construction of geometric ideas. Computer programs vary widely in nature and content; in this section, I consider computer-assisted instruction (CAI), intelligent tutors, games, Logo, and construction and interactive geometry programs.

CAI. Students instructed in geometry with computers often score significantly higher than those having just classroom instruction, from the elementary years (Austin, 1984; Morris, 1983) to high school (Hannafin & Sullivan, 1996). Consistent with other research, Hannafin and Sullivan found that learners may not accurately gauge the amount of instruction that is optimal for them; the researchers recommend using a full computer-assisted-instruction program instead of letting students choose how much time to spend working with the program. African American college students in a computer-based developmental education course in geometry (although not in algebra) scored higher than those taught by conventional instruction and had more positive attitudes (Owens & Waxman, 1994).

Intelligent tutors. Artificial-intelligence tutors might be connected with any type of application. This section briefly discusses their role in teaching proof in a CAI context. The Geometry Tutor presents the statement to be proved at the top of the screen and the given statements at the bottom (Anderson, Boyle, & Reiser, 1985). The student adds to a developing "proof graph" by pointing to statements on the screen and by typing information. Each logical inference involves a set of premises, a reason, and a conclusion. Reasoning forward, the student points to the premises, types the reason, points to relevant geometric points in the diagram, and points to the conclusion (if already on the screen) or types it. Everything the student points to or types in is connected in the proof graph with arrows. The proof is completed when a set of logical inferences connects the given statements to the statements to be proved. The tutor makes concrete two abstract characteristics of theorem proving: logical relations among the premises and conclusions, and the search process used to find a correct proof. As the student works, the tutor infers which rule the student applied by determining which one matches the student's response. If the response is correct, the tutor is silent; otherwise, it gives instruction. All instruction is thus in the context of solving problems. Research generally reports high levels of success on achievement and attitudes, although students may have difficulties when working without the tutor's assistance (Kafai, 1989; Wertheimer, 1990). Teaching with the tutor places considerable demands on the teacher (as do most computer programs) to coordinate and manage the computer laboratory (Schofield & Verban, 1988).

Protocol analyses convinced the developers of the Geometry Tutor that good theorem provers plan in terms of conceptual "configuration schemas" rather than formal theorems. Building on this work, Angle was designed to help students learn theorem proving by emphasizing these configuration schemas (see Figure 11.3). A classroom study showed that as an add-on in classes where tutor use was not well integrated into the course, Angle neither harmed nor helped. When well integrated with the curriculum, Angle led to a full standard deviation effect over comparison classes with the same teacher (Koedinger & Anderson, 1993).

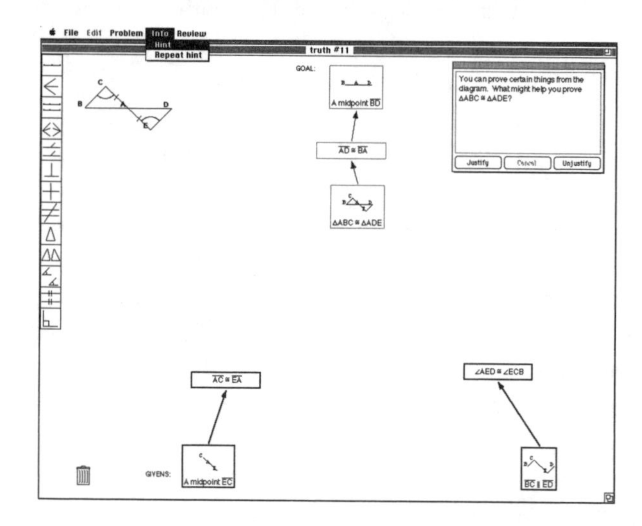

Figure 11.3. Angle, an intelligent computer program for learning geometry, emphasizes configuration schemas in theorem proving.

Games. Computer games have been found to be marginally effective at promoting learning of angle-estimation skills (Bright, 1985) and effective in facilitating achievement in coordinate geometry (Morris, 1983) and spatial abilities (Bright, Usnick, & Williams,

1992; Clements, Sarama, Battista, & Swaminathan, 1996). Game design can significantly increase the effectiveness of a program. More effective than manipulating geometric objects is manipulating representations of the specific concepts students are to learn. For example, rather than drag a shape to turn it, a "direct concept manipulation" interface might have students manipulate a representation of a turn, including turn center and amount of rotation (see Figure 11.4). Manipulating a mathematical representation of the transformation being applied to the shape in this way leads to higher achievement than manipulating the shape directly, especially when paired with appropriate use of scaffolding, such as gradually removing visual feedback aids and requiring the use of specific transformations to achieve some configurations (Sedighian & Sedighian, 1996). Thus, the effectiveness of games depends on software design, such as interface styles and scaffolding, as well as teacher and student expectations, the level of integration with other learning activities, and the pattern of use.

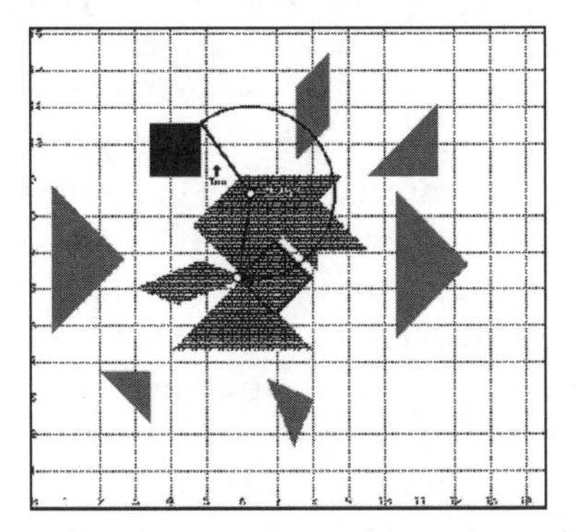

Figure 11.4. In performing a rotation in Super Tangrams, students directly manipulate representations of the turn, including the turn center, rather than the shape to be turned.

Logo. More research has been conducted on programming the Logo turtle than any other mathematics computer application. Although proportionately fewer classrooms use Logo, this research base remains significant. First, various versions of Logo, and other navigational programs, continue to be used in classrooms. Second, students' work in Logo environments has much to teach us about learning in other computer and noncomputer environments.

We have evidence that Logo activities might be used to encourage students to progress to Levels 2 (descriptive/analytic) and 3 (abstract/relational) in the van Hiele hierarchy. For instance, with the concept of rectangle, students initially are able to identify only visually presented examples, a Level 1 (visual) activity in the van Hiele hierarchy. In Logo, however, students can be asked to construct a sequence of commands, or a *procedure*, to draw a rectangle. This type of task "allows, or obliges, the child to externalize intuitive expectations. When the intuition is translated into a program it becomes more obtrusive and more accessible to reflection" (Papert, 1980, p. 145). That is, in constructing a rectangle procedure, the students must analyze the visual aspects of the rectangle and reflect on how its component parts are put together, an activity that encourages Level 2 thinking. Furthermore, if asked to design a rectangle procedure that takes the length and width as inputs, students must write a type of definition for a rectangle, one that the computer understands (see Figure 11.5). Students thus begin to build intuitive knowledge about the concept of defining a rectangle, knowledge that can later be integrated and formalized into an abstract definition—a Level 3 activity. Asking students whether their rectangle procedure can draw a square or a parallelogram, given the proper inputs, encourages students to start logically ordering figures, another Level 3 activity.

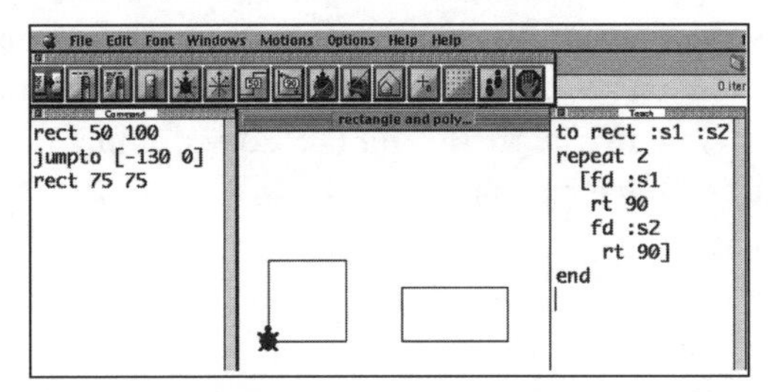

Figure 11.5. Students use Turtle Math (Clements & Meredith, 1994; Clements & Sarama, 1996) to create a Logo-based definition of a rectangle.

Research suggests that these theoretical predictions are valid. Grade 7 students' work in Logo relates closely to their level of geometric thinking (Olson, Kieran, & Ludwig, 1987). In addition, appropriate use of Logo helps elementary school students begin to make the transition from Levels 0 and 1 to Level 2 of geometric thought (Clements & Battista, 1989; Clements, Sarama, & Battista, 1998; Hughes & Macleod, 1986; Kynigos, 1993; Lehrer & Smith, 1986), possibly because Logo incorporates implicitly the types of properties that will be developed explicitly by Level 1 thinkers, something that textbooks often fail to do (Battista & Clements, 1988; Fuys et al., 1988). Logo can help students develop angle, angle-measure, and length concepts, but consistently only if properly guided by their teachers (Clements, Battista, Sarama, Swaminathan, & McMillen, 1997; Clements, Sarama, & Battista, 1996; Kieran, 1986; Olive, Lankenau, & Scally, 1986). Similar results have emerged in the area of symmetry and motion geometry (Hoyles & Healy, 1997; Olson et al., 1987). Compared with students using paper and pencil, students using Logo work with more precision and exactness (Gallou-Dumiel, 1989; Johnson-Gentile, Clements, & Battista,

1994). Thus, evidence supports the hypothesis that Logo experiences can help elementary to middle school students become cognizant of their mathematical intuitions and facilitate the transition from visual to descriptive/analytic geometric thinking in the domains of shapes, symmetry, and motions (Clements & Battista, 2001; Yusuf, 1994).

Not all research has been positive (Johnson, 1986); some has shown limited transfer (Olive et al., 1986). One problem is that students do not always think mathematically, even when the Logo environment invites such thinking. For example, some students rely excessively on visual cues and eschew analytical work (Hillel & Kieran, 1988). Students may have little reason to abandon visual approaches unless they are presented with tasks whose resolution requires an analytical approach. In addition, dialogue between teacher and students is essential for encouraging higher-level reasoning. Care must be taken to help students establish and reflect on path-command correspondences, that is, connections between geometric paths drawn by the turtle and the Logo commands that produce these paths (Battista & Clements, 1987; Clements, 1987).

In sum, studies have shown that Logo provides a context in which (a) the components and properties of geometric shapes are critical, (b) students are motivated, (c) connections between external representations of geometric objects can be maintained, (d) mathematical ideas can be expressed, and (e) abstractions that are personally meaningful and situated can be made. Further, the most positive effects involve planned sequences of carefully crafted Logo activities and teacher mediation of students' work with those activities (Clements & Battista, 2001; Kynigos, 1993; Noss & Hoyles, 1992; Sarama, 1995, 2000). Using versions of Logo (e.g., Clements & Sarama, 1995) specially designed with interface, tools, and activities supporting teaching and learning of mathematics appears more successful than using "vanilla" versions of Logo (Clements, Battista, Sarama, & Swaminathan, 1996; Clements, Battista, Sarama, Swaminathan et al., 1997; Clements, Sarama et al., 1998; Clements, Sarama, González Gómez, Swaminathan, & McMillen, in press). The potential of Logo—and, most likely, other geometric software—to develop geometric ideas will apparently be fulfilled to the extent that the computer tools, instructional materials, and teachers properly guide students' Logo experiences.

Construction and interactive geometry programs. The focus of construction programs, such as the Geometric Supposer software series (Schwartz & Yerushalmy, 1986), is to facilitate students' making and testing conjectures. The Supposer programs allow students to choose a primitive shape, such as a triangle or quadrilateral (depending on the specific program), and to perform measurement operations and geometric constructions on it. The programs record the sequence of constructions and can automatically perform it again on other triangles or quadrilaterals. In one evaluation, Supposer students performed as well as, or better than, their non-Supposer counterparts on geometry exams (Yerushalmy, Chazan, & Gordon, 1987). In addition, students' learning went beyond standard geometry content, for example, reinventing definitions, making conjectures, posing and solving significant problems, and devising original proofs. Supposer activities engendered a movement away from considering measurement evidence as proof (Chazan, 1989; Wiske & Houde, 1988), although some students still thought that counterexamples might arise to deductively proved results. Students believed that, unlike textbook theorems, Supposer-generated theorems need to be proved before they can be accepted as true, leading to ownership of the theorems. They seemed to engage in van Hiele's Phase 4 learning, orienting themselves within the network of geometric relations (Lampert, 1988). In sum, it appears that, with proper support from the teacher, students using the Supposer can come to understand the importance of formal proof as a way of establishing mathematical truth, although this outcome seems to be related more to Bell's illumination function of proof than to the verification and systematization functions (Yerushalmy, 1986).

Supposer students also made gains in understanding diagrams. They approached diagrams flexibly, treated a single diagram as a model for a class of shapes, and were aware that this model contained characteristics not representative of the class (Yerushalmy & Chazan, 1988; Yerushalmy et al., 1987). Comparison studies found that classes using the Supposer outscored classes not using it, on total scores and especially on application and higher-level problems (McCoy, 1991; Yerushalmy, 1991).

To implement successfully the Supposer's guided inquiry approach, research suggests the need for teaching strategies that connect students' inquiry with the curriculum and encourage inquiry as a way to learn successfully what needs to be known. In general, teachers in regular classrooms face multiple difficulties using the Supposer; however, such struggles seem to have the potential to change teachers' practice and thus change their beliefs about the meaning of knowing geometry and how knowledge is acquired in classrooms. Overall, although use of the Supposer was found to demand hard work from, and cause some frustration in, both teachers and students, benefits were evident (Lampert, 1988; Wiske & Houde, 1988; Yerushalmy et al., 1987).

More recent developments have led to the creation of several interactive geometry programs, such as Cabri-Geometry (Baulac, Bellemain, & Laborde, 1988) and Geometer's Sketchpad (Jackiw, 1995). (See Figure 11.6.) Interactive geometry software may change the nature of the relations between what Laborde (1996) calls "diagrams" (physical, spatial properties) and "theory" (geometrical properties). Consistent with the aforementioned discussion of diagrams, she argues that diagrams play an ambiguous role in instruction. Students believe that it is possible to abstract the properties of geometric objects from diagrams directly and thus deduce a property empirically. Interactive geometry software introduces a new type of diagram whose behavior is controlled by theory. Students' actions require construction of an interpretation in which visualization plays a crucial role but geometrical properties constrain such action. These characteristics produce a stronger connection between diagrams and theory because spatial invariants in the moving diagrams are almost certainly the representation of geometrical invariants. Work with students supports this hypothesis and shows that teachers must encourage students to move from performing construction tasks at the spatial-diagrammatic level to establishing links between that and the theoretical level. For example, students may interpret a diagram in geometric terms, give geometrical reasons for behavior of the diagram, or give geometrical reasons for the simultaneity of two diagram relations while "dragging." They should also be encouraged to move from theory to diagram by predicting what will happen to the diagram on the basis of geometric knowledge. As with Logo research, the choice of task and the teaching strategies used are important. For example, to generate motivating and useful conflict, teachers might ask students to predict properties of a diagram before allowing them to check on the computer. The facility of relaxing or modifying conditions (Cabri's "redefine an object") is one means of asking students to make predictions. Intriguing phenomena also can lead to need for proof.

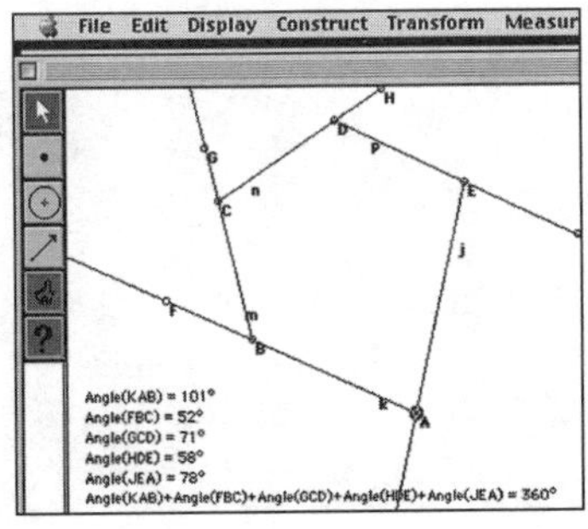

Figure 11.6. An example of dynamic geometry using The Geometer's s Sketchpad (Jackiw, 1995). Changing the vertex dynamically updates the figure and measures; of interest is what remains constant.

Students often move through phases in the strategies they use (Laborde, 1995). First, they use the primitives, or basic commands of the program, erratically. Second, they use a "by eye" strategy. Third, they make geometrical constructions without aim; the "empirical combinatorial strategy": They notice a spatial invariant but do not interpret it as *geometrical*. Fourth, their aim is to perform a geometrical construction. Features of Cabri play an important role in helping students move from one strategy to the next. For example, the drag mode disqualifies the purely visual strategies and enables students to notice spatial invariants. Because students know that the spatial phenomenon they produce by chance is the result of the use of geometrical primitives, they are convinced that using geometry can reproduce it. That is, their belief that the software "knows geometry" motivates students to search for a geometrical solution.

Even experienced teachers who were teaching with interactive geometry software for the first time experienced an uncomfortable initial loss of control in three categories: (1) management control (they believed that the new environment impaired their ability to maintain discipline), (2) personal control (they were unable to determine their own expectations of the students and to assess students' achievement), and (3) professional control (they perceived that they no longer had all the answers).[1] As the teachers learned to use the new tools, however, they gained confidence in their ability to teach effectively with the new methods and were even moved to reflect on their previous teaching practices (McDougall, 1997). Thus, mentoring should help preservice and in-service teachers accept a temporary loss of classroom control as part of the change process. In a similar vein, students may feel out of control at the beginning of a course using Sketchpad but will come to believe later in the semester that the computer is a tool or "partner" in solving geometry problems (Pokay & Tayeh, 1997).

Interactive geometry programs offer a viable way to avoid beginning with proof. According to de Villiers (1995), proof is more about explanation and discovery—and sometimes systematization and communication—than about convincing. Such software as Cabri can convince people, but can they then *understand*? Students who use interactive geometry programs in a curriculum specifically designed on the van Hiele theory do increase their levels of geometric thinking and their achievement

[1] Deborah Schifter pointed out that this may apply to other computer environments, situations of giving rich problems using manipulatives, and so forth. This is probably true; however, the empirical research cited here stemmed only from dynamic geometry environments.

(Choi-koh, 1999; Dixon, 1997), although such positive results are not guaranteed (Roberts & Stephens, 1999).

Implications. Several findings across studies using different computer environments are notable. First, researchers and teachers consistently report that in such contexts students cannot "hide" what they do not understand. That is, difficulties and misconceptions that are easily masked by traditional approaches emerge and must be dealt with when using computer environments, leading not only to some frustration on the part of both teachers and students but also to greater development of mathematics abilities (Clements & Battista, 1989; Schofield & Verban, 1988; Yerushalmy et al., 1987). Second, at least at the high school level, students can become confused regarding the purpose of different components of a course; a single location for computer work, discussion, and lecture may alleviate this confusion. A monitor or projector for group discussions is noted as important for all three environments. Third, evaluation of learning in such environments must be reconsidered (see Galindo, 1998) because traditional approaches did not assess the full spectrum of what was learned; in some cases, these approaches made little sense.

Appropriately designed software can engender higher-level interaction with geometric ideas. Certain computer environments allow the manipulation of screen objects in ways that help students view them as representatives of a class of geometric objects. Use of this feature develops students' ability to reflect on the properties of the class of objects and to think in a more general and abstract manner. Thoughtful sequences of computer activities and teacher mediation of students' work with those activities appear to be critical components of an efficacious educational environment. Perhaps even more fundamental, inquiry environments appear to have the potential to serve as catalysts in promoting teachers' and students' reconceptualization of what it means to learn and understand geometry and in promoting the growth of students' autonomy in mathematical thinking. Fundamental changes demand considerable effort on the part of teachers and call for extensive support from teacher educators or advisors, peers, and ultimately, the greater school system and culture.

Selected Geometric Topics

This section reviews research on the teaching and learning of a few selected geometric topics. (Many equally important topics have a less complete research base; space constraints eliminated others.) These topics have already been discussed in previous sections, of course; here I supplement these findings with additional specific results.

Shape, Congruence, and Similarity

Young children can identify circles quite accurately and squares fairly well; they are less accurate recognizing triangles and rectangles (Clements et al., 1999). Preschool and kindergarten children's prototypical image of a rectangle seems to be a four-sided figure with two long parallel sides and "close to" square corners. Certain mathematically irrelevant characteristics affect children's categorizations: skewness, aspect ratio, and, for certain situations, orientation (Hannibal & Clements, 2000). Most children accept triangles even if their base is not horizontal, although a few protest. Skewness, or lack of symmetry, is more important. Many, on the one hand, reject triangles because "the point on top is not in the middle." For rectangles, on the other hand, many children accept nonright parallelograms and right trapezoids. Also important is aspect ratio, the ratio of height to base. Children prefer an aspect ratio near one, that is, about the same height as width, for triangles. Children reject both triangles and rectangles that are "too skinny" or "not wide enough." In summary, although they have much to learn, they have a good start before kindergarten; on the negative side, on the same tasks, the typical school student does not perform much better right through to middle school (Clements et al., 1999). For example, Figure 11.7 displays the findings from two separate studies involving more than 1.5 thousand children, one conducted with children from 4 to 6 years old and the other with children from 6 to 12 years old. All children identified rectangles within Figure 11.2.

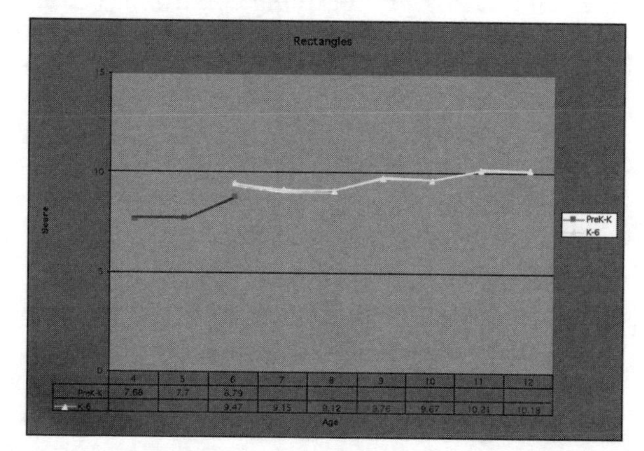

Figure 11.7. Results of two studies in which children identified rectangles within the set of shapes illustrated in Figure 11.2. The graphs represent the number of shapes correctly identified as rectangles.

Research in the primary grades supports the finding that children apply many different types of cognitive actions to shapes, from detection of features like fat or thin to comparison to prototypical forms to the action-based embodiment of pushing or pulling on one form to transform it into another (Clements & Battista, 2001; Lehrer, Jenkins et al., 1998). Again, these studies indicate that children's thinking does not change significantly over the elementary years. Indeed, first-grade children are more likely than older children to think of contrasting forms by counting the number of sides or corners. As children move through school, they are less likely to notice such attributes (Lehrer, Jenkins et al., 1998).

Young children also develop beginning ideas about congruence. Preschoolers try to judge congruence using an edge-matching strategy, although only about 50% can do it successfully (Beilin, 1984). In about first grade, they begin to use superposition. In the intermediate grades, they refer to geometric properties and motions in their justifications, although they are not always accurate (Rosser, 1994). Much remains for students to learn about congruence. One researcher classified secondary school students as either formal or concrete operational. She then asked each group to judge how closely several similarity and congruence concepts were related (McDonald, 1989). Prototypical cognitive maps could be drawn for both the formal and concrete operational students. Formal operational students' content structure was significantly more like that of subject-matter experts than that of the concrete operational students. As another example, the international average for an item from the TIMSS on properties of congruent triangles was 27% for seventh graders and 35% for eighth graders; in comparison, about two thirds of the eighth-grade students from Japan, Korea, and Singapore responded correctly (Beaton et al., 1996). Instruction can help. For example, most students taught to use transformations to learn about congruence advanced a level in the van Hiele hierarchy (Nasser, 1992).

Under the right conditions, children of all ages can apply similarity transformations to shapes. Even 4- and 5-year-olds can identify similar shapes in some circumstances (Sophian & Crosby, 1998). The coordination of height and width information to perceive the proportional shape of a rectangle (fat vs. skinny, wide, or tall) might be a basic way of accessing proportionality information and may serve as a foundation for other types of proportionality, especially fractions. Similarly, other research shows first graders can engage in, and benefit from, similarity tasks (Confrey, 1992). Older students have acquired a variety of meanings and interpretations of similarity and may not be as sensitive to similarity; it may be that to them "same shape" means "described by the same shape-name" (Vollrath, 1977). Here too, then, instruction should probably start with similarity transformations. Middle school students develop spatial abilities after completing similarity tasks (Friedlander, 1985). Computer transformations, especially those in carefully designed microworlds, can help students learn similarity and ratio concepts (Hoyles, Noss, & Sutherland, 1989).

Similar to findings regarding two-dimensional shapes, students do not perform well with solids. Most 9- and 11-year-olds have difficulty naming solids (Carpenter, Coburn, Reys, & Wilson, 1976). South African first graders have different names for solids (such as "square" for cube) but were capable of understanding and remembering features they discussed (Nieuwoudt & van Niekerk, 1997). In contrast, U.S. students' reasoning about solids was much like that about plane figures; they referred to a variety of characteristics, such as "pointyness" and comparative size or slenderness (Lehrer, Jenkins, et al., 1998). Students also treated the solid wooden figures as malleable, suggesting that the rectangular prism could be transformed into a cube by "sitting on it." The van Hiele levels have been applied to understanding solids (Gutiérrez et al., 1991): Preservice teachers with an interest in science had "complete acquisition" of Level 2 and some acquisition of Level 3, preservice teachers without such interest had complete acquisition of Level 2, and eighth-grade students had not acquired Level 1.

These studies have several educational implications. The use of plane-figure names for solids may indicate a lack of discrimination between two and three dimensions (Carpenter et al., 1976). Learning only plane figures in textbooks during the early primary grades may cause some initial difficulty in learning solids. Construction activities involving nets (foldout shapes of solids) may be valuable because they require children to switch between more-analytic 2D and synthetic 3D situations (Nieuwoudt & van Niekerk, 1997).

Secondary school students more easily interpret certain representations of solids, including verbal descriptions, perspective drawings, and prototypes, than such two-dimensional graphical representations as plans, elevations, and descriptions based on coordinates (Martin & Sweller, 1989). Perspective drawings and prototypes were more easily interpreted than verbal descriptions, but students did not find copying a prototype easier than copying a perspective. Grade 9 and 11 students were more successful than those in Grade 7 but did not differ from each other. Thus, curriculum developers and teachers should select external representations carefully. For introducing concepts, real-life representations might be preferable; two-dimensional representations are valuable for themselves and should be taught and discussed explicitly.

Researchers often recommend experience with manipulatives and computer environments. For example, students who manipulate representations of solids on the computer improve their spatial understanding and structuring and their ability to solve spatial problems both on and off the computer (Sachter, 1991). Third graders improve with experience in recognizing nets, which should include familiar and nonfamiliar solids (Bourgeois, 1986).

Symmetry and Geometric Motions

The complex topics of symmetry and geometric motions cannot be adequately treated in this chapter; this section highlights only a few findings. Children have intuitive notions of symmetry from the earliest years. However, most concepts of symmetry have not been firmly established before 12 years of age (Genkins, 1975). Vertical bilateral symmetry is easier for students to handle than horizontal symmetry (Genkins, 1975).

In the realm of geometric motions, some studies have found that second graders did learn manual procedures for producing transformation images but that they did not learn to mentally perform such transformations (Williford, 1972). Similarly, students' ability to mentally perform isometries is limited at the middle and junior high school level because it demands formal operational thought (Kidder, 1976). In contrast, other studies indicate that even young children can learn something about these motions and appear to internalize them, as indicated by increases in performance on spatial ability tests (Clements, Battista, Sarama, & Swaminathan, 1997; Del Grande, 1986). Slides appear to be the easiest motions for students, then flips and turns (Perham, 1978); however, the direction of transformation may affect the relative difficulty of turn and flip (Schultz & Austin, 1983). A transformation approach to high school geometry is possible even for average students (Usiskin, 1972).

Computer environments have been found to be particularly effective in developing conceptions of symmetry and geometric motions. The effects in one study were largely due to the greater number of Logo students drawing completely correct symmetric figures (Clements & Battista, 2001). Writing Logo commands for the creation of symmetric figures, testing symmetry by flipping figures with Logo commands, and discussing these actions apparently encouraged students to build richer and more general images of symmetric relations (possibly to the extent of overgeneralization in some cases) and to reflect on the construction of symmetric figures and possibly on the properties of symmetry. Students had to abstract and represent their actions in a more explicit and precise fashion in Logo activities than, say, in freehand drawing of symmetric figures.

A related study found no evidence that the use of the Logo-motions curriculum helped children perform tasks at Level 1 of geometric thinking; that is, such use did not appear to facilitate the ability to draw symmetric figures, draw figures given specified motions, or apply motions to problems such as sorting tasks (Johnson-Gentile et al., 1994). There was evidence that the use of Logo facilitated students' transition to Level 2 thinking. Logo students were more likely to describe the properties of geometric motions and symmetry constructs and more likely to solve problems involving these constructs at a more abstract and analytical level. Logo-group scores increased, and non-computer-group scores decreased, on tasks involving figures that had more than one line of symmetry or on parallelograms with no lines of symmetry. Logo students appeared more likely to identify horizontal and oblique lines of symmetry and less likely to be misled by the rotational symmetry of the parallelogram. This suggests that differences between the groups were on those items that required students to resist applying intuitive visual thinking and apply analytical thinking in a comprehensive manner.

Geometric motions were taught successfully by a computer game discussed previously (Sedighian & Sedighian, 1996). Secondary school students (11 to 15 years old) in France who were taught symmetry with paper and pencil or with Logo made different errors (Gallou-Dumiel, 1989). The students taught with paper and pencil included the axis of reflection as part of the figure and did not differentiate reflection from translation. Logo students tended to believe that the given figure and the image could not be intersected. Errors of orientation were similar at the beginning but decreased more rapidly in the Logo than in the paper-and-pencil environment. Also, Logo students were more likely to recognize the presence or absence of the axis of reflection.

Angle, Parallelism, and Perpendicularity

Angles are the turning points in the study of geometry and spatial relationships. Unfortunately, one does not have to look far for examples of children's difficulty with the angle concept. Many students believe that an angle must have one horizontal ray; that a right angle is an angle that points to the right; that the angle sum of a quadrilateral is the same as its area; and that two right angles in different orientations are not equal in measure (Clements & Battista, 1992).

This body of research indicates that students have many different ideas about what an angle is. These ideas include "a shape," a side of a figure, a tilted line, an orientation or heading, a corner, a turn, and a union of two lines (Clements & Battista, 1990). Students do not find angles to be salient properties of figures (Clements, Battista et al., 1996; Mitchelmore, 1989). When copying figures, students do not always attend to the angles.

Students also hold many different schemes regarding not only the angle concept but also the size of angles. They frequently relate the size of an angle to the length of the line segments that form its sides, the tilt of the top line segment, the area enclosed by the triangular region defined by the drawn sides, the length between the sides (from points sometimes, but not always, equidistant from the vertex), the proximity of the two sides, or the turn at the vertex (Clements & Battista, 1989). Some misconceptions decreased over the elementary years, such as orientation; in contrast, the effect of segment length did not change and the effects of distance between the end points of the angle's rays increased (Lehrer, Jenkins et al., 1998). Some young children can distinguish between angles on the basis of size (Lehrer, Jenkins et al., 1998). In the third grade, most children can mentally decompose figures to allow identifying and distinguishing of one or more angles. By the fifth grade, children reliably distinguish several angles, but only 9% can reliably distinguish and coordinate relationships among the four angles, 30°, 60°, 90°, and 120°.

Intermediate-grade students often possess one of two schemes for measuring angles. In the "45-90 schema," slanted lines are associated with 45° turns, and horizontal and vertical lines, with 90° turns. In the "protractor schema," inputs to turns are based on usage of a protractor in "standard" position (thus, to have a turtle at home position turn left 45°, students might use an input of 135°, which corresponds to a protractor's reading when its base is horizontal) (Kieran, Hillel, & Erlwanger, 1986).

Moreover, such schemes may be resistant to change, especially through, for example, textbook definitions and examples. When people think, they do not use definitions of concepts, but rather, concept images—a combination of all the mental pictures and properties that have been associated with the concept (Vinner & Hershkowitz, 1980). Such images can even be adversely affected by inappropriate instruction. For example, many students develop a concept image of an obtuse angle having a horizontal ray, because of the limited set of examples they see in textbooks and a "gravitational factor" (i.e., a figure is "stable" only if it has one horizontal side, with the other side ascending).

Students who not only know a correct verbal description of a concept but also have strongly associated a specific visual image, or concept image, with the concept may have difficulty applying the verbal description correctly. Instead, educators need to help them build a robust concept of angle.

One approach, researched by Mitchelmore (1993), uses multiple concrete analogies. To develop the concept of an angle, the teacher helps students progress through the following phases.

1. Practical experience in various situations leads to an understanding of angular relationships in each individual situation.

2. Angle subconcepts develop when the common features of superficially dissimilar situations are recognized. These types of situations (e.g., turns, slopes, meetings, bends, directions, corners, opening) form different angle contexts.

3. Superficial differences between contexts initially hinder children's recognition of such common features.

4. Less obvious similarities between contexts gradually become apparent, and angle subconcepts begin to emerge.

5. A full angle concept emerges when the same common features are recognized in all angle contexts.

Research on teaching activities based on these ideas revealed that most elementary age students understood physical relations. Turns, or rotations, were a difficult concept to understand in concrete physical contexts. Other research supports the importance of integration of all types. Some children have understood turns and angles in a meaningful way only after months of work (Clements, Battista, et al., 1996). Initially, they gained experience with physical rotations, especially rotations of their own bodies. During the same time, they gained limited knowledge of assigning numbers to certain turns, initially by establishing benchmarks. A synthesis of these two domains—turn-as-body-motion and turn-as-number—constituted a critical juncture in learning about turns for many elementary school students.

An implication of the Piagetian position and the emphasis on turns is that dynamic computer environments might be useful. Turns (and angles) are critical to the view of shapes as paths, and the intrinsic geometry of paths is closely related to such real-world experiences as walking.

Computer games have been found to be marginally effective at promoting learning of angle estimation skills (Bright, 1985). More extensive and promising are several research projects investigating the effects of Logo's turtle graphics experience on students' conceptualizations of angle, angle measure, and rotation. In one study, for example, responses of intermediate-grade control students were more likely to reflect little knowledge of angle or common language usage, whereas the responses of the Logo students indicated more generalized and mathematically oriented conceptualizations, including angle as rotation and as a union of two lines/segments/rays (Clements & Battista, 1989). A large group of studies have reported similar findings, although in some situations, benefits do not emerge until after a year of Logo experience (Clements & Battista, 1992). So having these experiences over several years of elementary school is recommended.

Logo experiences may foster some misconceptions of angle measure, including viewing it as the angle of rotation along the path (e.g., the exterior angle in a polygon) or the degree of rotation from the vertical (Clements & Battista, 1989; Clements, Battista, et al., 1996). In addition, such experiences do not replace previous misconceptualizations of angle measure (Davis, 1984). For example, students' misconceptions about angle measure and their difficulties coordinating the relationships between the turtle's rotation and the constructed angle may have persisted for several years during their elementary schooling, especially when not properly guided by their teachers (Clements & Battista, 1992). In general, however, Logo experience appears to facilitate understanding of angle measure. Logo children's conceptualizations of a "larger angle" are more likely to reflect mathematically correct and coherent ideas. If activities emphasize the difference between the angle of rotation and the angle formed as the turtle traces a path, misconceptions regarding the measure of rotation and the measure of the angle may be avoided. For example, students will understand that when the Logo turtle goes forward 50, right 120, forward 50, it draws an angle with a measurement of 60 degrees (see Figure 11.8).

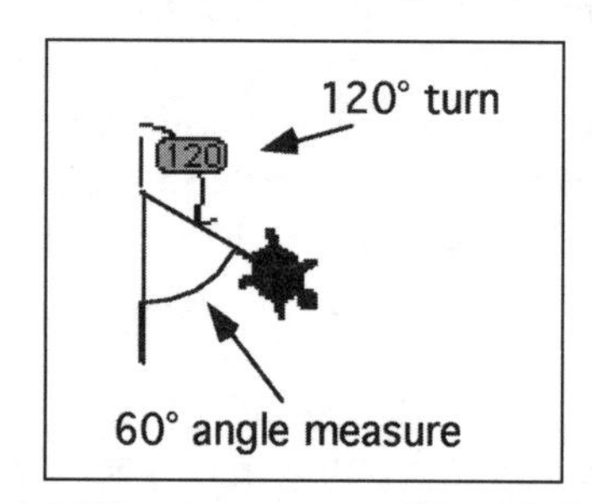

Figure 11.8. A 120° turn creates a 60° angle in a path.

To understand angles, students must understand the various aspects of the angle concept. They must overcome difficulties with orientation, discriminate angles as critical parts of geometric figures, and construct and represent the idea of turns, among others. Furthermore, they must construct a high level of integration between these aspects. This difficult task is best begun in the elementary and middle school years, as children deal with corners of figures, comparison of angle size, and turns. At the beginning of secondary school, angular regions should be integrated with angular rotations. The formulation of a single, mathematically rigorous definition of angle should follow (Mitchelmore, 1989). Throughout, students need to be more reflective about concepts of angle and angle measurement (Krainer, 1991).

Related topics include parallel and perpendicular lines. Both are difficult concepts for students in some applications. Children as young as 3 (Abravanel, 1977) and 4 years old use parallelism in alignment tasks, and 6-year-olds can name parallel and nonparallel lines, although they have difficulty locating parallels in complex figures (Mitchelmore, 1992). Further, although eighth-grade students believe that parallel lines should not intersect and should be equidistant, they also believe that parallel segments must be aligned and that curves might be parallel—beliefs that were not affected by instruction (Mansfield & Happs, 1992).

Teaching Australian students in Grade 1 about perpendicular lines was abandoned because students were unable to conceptualize perpendiculars as lines in a special angular relationship. Although all students in Grade 4 had studied right angles in school, only about two thirds seemed able to understand this concept, showing varying degrees of uncertainty in its application. Students' concept image seems to be limited to horizontal and vertical lines (Mitchelmore, 1992).

Navigation, Coordinates, and Structuring Two-dimensional Space

Spatial sense includes two main spatial abilities: spatial orientation and spatial visualization and imagery. Other important knowledge includes how to represent ideas in drawing and how and when one can use such abilities. Spatial visualization is discussed only incidentally throughout this document (for a review, see Clements & Battista, 1992). This section discusses some aspects of spatial orientation.

Spatial orientation is knowing where one is and how to get around in the world, that is, understanding and operating on relationships between different positions in space, especially with respect to one's own position. Even young children learn practical navigation early, but what do they understand and represent about spatial relationships and navigation? For example, at what age can they use and create maps and when can they build "mental maps" of their surroundings? Research has shown that even preschoolers have map-related competencies. For example, 3-year-olds can build a simple but meaningful map with landscape toys, such as houses, cars, and trees (Blaut & Stea, 1974). Not as certain is what specific abilities and strategies they are using. For example, kindergarten students may correctly make models of their classroom cluster furniture (e.g., they put the furniture for a dramatic play center together) but may not relate the clusters to one another (Siegel & Schadler, 1977). Also, it is unclear what kind of "mental maps" students possess. Some researchers believe that people first learn to navigate only by noticing landmarks, then by routes or connected series of landmarks, then by scaled routes, and finally by putting many routes and locations into a kind of "mental map." Only older preschoolers learn scaled routes for familiar paths; that is, they know about the relative distances between landmarks (Anooshian, Pascal, & McCreath, 1984). Even young children, however, can put different locations along a route into some relationship, at least in certain situations. For example, they can point to one location from another even though they never walked a path that connected the two (Uttal & Wellman, 1989).

Neither students nor adults actually have "maps in their heads"—that is, their "mental maps" are not like a mental picture of a paper map. Instead, they are filled with private knowledge and idiosyncrasies and actually consist of many kinds of ideas and processes, which may be organized into several frames of reference. The younger the child, the more loosely linked these representations are. These representations are spatial more than visual. Blind children are aware of spatial relationships by age 2, and by age 3 they begin to learn about spatial properties of certain visual language (Landau, 1988).

Even students with similar mental representations may produce quite different maps because of differences in drawing and map-building skills (Uttal & Wellman, 1989). Many third- to fifth-grade students could not structure their knowledge of their own homes and make abstractions or transformations from it so that they could portray a two-dimensional view of it on paper. They showed three levels of structuring (cognitively integrating spatial units into a unified whole) in their maps of their homes: isolated-piecemeal, with separate disconnected rooms; semi-integrated, including connected but not spatially integrated rooms, similar to route maps; and integrated, with all rooms coordinated within an objective frame of reference (Imamoglu & Imamoglu, 1983).

Most students can learn from maps. For example, 4- to 7-year-olds had to learn a route through a playhouse with six rooms (Uttal & Wellman, 1989). By the primary grades, most students are able to draw simple sketch-maps of the area around their home from memory. They also can recognize features on aerial photographs and large-scale plans of the same area (Boardman, 1990). However, these and older students still are not competent users of maps. Most school experiences fail to connect map skills with other curriculum areas, such as mathematics (Muir & Cheek, 1986).

Fundamental is the connection of primary to secondary uses of maps (Presson, 1987). Even young children form primary, direct relations to spaces on maps. They must grow in their ability to treat the spatial relations as separate from their immediate environment. These secondary meanings require people to take the perspective of an abstract frame of reference ("as if being there") that conflicts with the primary meaning. One no longer imagines oneself "inside," but rather, must see oneself at a distance, or "outside" the information. Such meanings of maps challenge people into adulthood, especially when the map is not aligned with the part of the world it represents, as when a map is held with north at the top when the map reader is heading south (Uttal & Wellman, 1989). Adults need to connect the abstract and concrete meanings of map symbols. Similarly, many of young children's difficulties do not reflect misunderstanding about space but the conflict between such concrete and abstract frames of reference. In summary, students must (a) develop abilities to build relationships among objects in space, (b) extend the size of that space, and (c) link primary and secondary meanings and uses of spatial information.

Developing students' ability to make and use mental maps is important, and so is developing geometric ideas from experiences with maps. We should go beyond teaching isolated "map skills" and geography to engaging in actual mapping, surveying, drawing, and measuring in local environments (Bishop, 1983). Such activities can begin in the early years. Students need to read and make maps meaningfully. In both these endeavors, four basic questions arise: (1) direction—which way? (2) distance—how far? (3) location—where? and (4) identification—what objects? To answer these questions, students need to develop a variety of skills.

Children must learn to deal with processes of abstraction, generalization, and symbolization. Some map symbols are icons, such as an airplane for an airport, but others are more abstract, such as circles for cities. Children might first build with objects such as model buildings, then draw pictures of the objects' arrangements, then use maps that are "miniaturizations," and then use those that employ abstract symbols (Downs, Liben, & Daggs, 1988).

As children work with model buildings or blocks, they are gaining experience with perspective. For example, they might identify block structures from various viewpoints, match views of the same structure that are portrayed from different perspectives, or try to find the viewpoint from which a photograph was taken. These experiences address such confusions of perspective as preschoolers' "seeing" windows and doors of buildings in vertical aerial photographs (Downs & Liben, 1988).

Similarly, students need to develop more sophisticated ideas about direction. They should develop navigational ideas, such as left, right, and front; and global directions, such as north, east, west, and south, from these beginnings. Such ideas, along with distance and measurement ideas, might be developed as children build and read maps of their own environments. Perspective and direction are particularly important regarding the alignment of the map with the world. Some students of any age will find it difficult to use a map that is not so aligned.

A final mathematical idea is that of location. Relative locations of objects might lead into coordinates, a topic to which I turn.

As students develop notions of two- and three-dimensional space, they must learn to construct, select, and use increasingly coordinated reference systems as frameworks for spatial organization, the foundation of spatial-geometric thinking (Clements & Battista, 1992). The Piagetian position is that such frameworks are analogous to a container made up of a network of sites or positions (Piaget & Inhelder, 1967). Objects within this container may be mobile, but the positions are stationary. From the simultaneous organization of all possible positions in three dimensions emerges the coordinate system. This conceptualization involves the gradual replacement of relations of order and distance between objects with similar relations between the positions themselves. The space is "emptied of objects." Thus, intuition of space is not an innate apprehension of the properties of objects but a system of relationships borne in actions performed on these objects.

Unlike many topics investigated by Piaget, research on this theory has been minimal. We do know that young students are more competent, and that adolescents and adults are less competent, than Piaget and Inhelder's initial work suggested. Very young children can orient a horizontal or vertical line in space (Rosser, Horan, Mattson, & Mazzeo, 1984). Similarly, 4- to 6-year-old children (a) can extrapolate lines from positions on both axes and determine where they intersect, (b) are equally successful going from point to coordinate as going from coordinate to point, and (c) extrapolate as well with or without grid lines (Somerville & Bryant, 1985). Piagetian theory seems correct in postulating that the coordination of relations develops after such early abilities. Young students fail on double-axis orientation tasks even when misleading perceptual cues are eliminated (Rosser et al., 1984). Similarly, the greatest difficulty is coordinating two extrapolations, which has its developmental origins at the 3- to 4-year-old level, with the ability to extrapolate those lines developing as much as a year earlier (Somerville, Bryant, Mazzocco, & Johnson, 1987). These results suggest an initial inability to use a conceptual coordinate system as an organizing spatial framework. Finally, not all high school seniors or college students perform successfully on tasks designed by Piaget to assess an underlying Euclidean conceptual system (Liben, 1978; Mackay, Brazendale, & Wilson, 1972; Thomas & Jamison, 1975). This is a substantial educational deficit; an understanding of grids and coordinate systems is important not only for the study of maps and two-dimensional space but also for understanding area, reading and constructing graphs, and graphing functions. Training efforts have limited success, suggesting support for Piaget and Inhelder's contention of a lack of a fully developed internal system of reference (Liben, 1978; Thomas & Jamison, 1975).

Students are capable of reading and plotting points on graphs, but they may be doing so in a rote manner. They often do not have the ability to construct axes (Leinhardt, Zaslavsky, & Stein, 1990). They may use coordinates to indicate lengths of segments on the graph. Students may interpret graph points as having density and may have difficulty with the idea that there are, for example, more points on a line than those the students graphed themselves. Students' ability to work to scale on graphs is often minimal. For example, they may believe that it is legitimate to use different scales on the positive and negative portions of the axes. In summary, dealing with a graph as an abstraction with a large amount of convention and notation to be learned is a nontrivial task (Leinhardt et al., 1990). All these areas need sustained instructional attention.

Thus young children have nascent abilities, but older students often have not developed firm conceptual grounding in grid and coordinate reference systems. To mentally structure space with such systems requires considerable conceptual work. Spatial structuring is the mental operation of constructing an organization or form for an object or set of objects in space (Battista, Clements, Arnoff, Battista, & Borrow, 1998). Structuring is a form of abstraction, the process of selecting, coordinating, unifying, and registering in memory a set of mental objects and actions. Structuring takes previously abstracted items as content and integrates them to form new structures. It creates stable patterns of mental actions that an individual uses to link sensory experiences rather than consists of the sensory input of the experiences themselves. Such spatial structuring precedes meaningful mathematical use of the structures, such as determining area or volume (Battista & Clements, 1996; Battista et al., 1998; Outhred & Mitchelmore, 1992; see the next chapter in this volume for more on these issues). Related but distinct spatial structuring precedes meaningful use of grids and coordinate systems. On the one hand, grids provided to students may aid their structuring of space; on the other hand, students face additional hurdles in understanding grid and coordinate systems. The grid itself may be viewed as a collection of square regions rather than as sets of perpendicular lines. In addition, order and distance relationships within the grid must be constructed and coordinated across the two dimensions. Labels must be related to grid lines and, in the form of ordered pairs of coordinates, to points on the grid and eventually be integrated with the grid's order and distance relationships so that they constitute a mental number and ultimately can be operated upon.

Most fourth graders still need to learn to spatially structure two-dimensional grids in this fashion (Clements et al., in press). They need to overcome conceptual hurdles of interpreting the grid structure's components as line segments or lines rather than regions; appreciating the precision of location the lines required instead of treating them as fuzzy boundaries or indicators of intervals; and learning to trace vertical or horizontal lines that are not axes. When using coordinates, students may have needed to reconstruct the levels of counting and quantification they had already constructed in the domain of counting discrete objects. Students must learn (a) to quantify what the grid labels represent, (b) to connect their counting acts with those quantities and with the labels, (c) to subsume these ideas to a part-whole scheme connected with both the grid and counting/arithmetic, and finally (d) to construct proportional relationships in this scheme. That is, students must integrate their numerical and spatial schemes to form a conceptual ruler (a mental number line connected to linear units) (Clements, Battista, Sarama, Swaminathan, et al., 1997; Steffe, 1991). They must then integrate two conceptual rulers into orthogonal number lines that define locations in that space. This integration is a distributive coordination; that is, one conceptual ruler must be taken as a mental object for input to an orthogonal conceptual ruler (Clements et al., in press). Previous research substantiated that performance on coordinate tasks progresses uniformly and continuously from Grades 1 to 6 (Carlson, 1976). Students were capable of locating points in one dimension in Grade 1, in two dimensions by Grade 4, and in three dimensions in Grade 6.

Regarding instruction, real-world contexts can be helpful initially, but mathematical goals and perspectives should be clearly articulated throughout instruction and the contexts should be faded from use as soon as students no longer need them (Clements et al., in press). Computer environments can additionally aid in developing students' ability and appreciation for the need for clear conceptions and precise work. The ethereal quality of a toggled screen grid can help scaffold students' creation of a mental construct that they project on the plane. Similarly, computer activities that remove supports for counting and require students to find numbers between others prompt students to use labels for coordinates and proportional reasoning. Curricula may introduce grid lines in multiples of 10 on paper because many students have difficulty moving to the computer initially, facing the demands of interpolation alongside those of different contextual supports. Results also imply cautions regarding some popular teaching strategies. For example, such phrases as "over and up" and "the x-axis is the bottom," which we recorded on numerous occasions, do not generalize well to a four-quadrant grid. The over-and-up strategy also hinders the integration of coordinates into a coordinate pair (Clements et al., in press). Overall, coordinates can serve as a useful vehicle to develop geometric concepts and divergent thinking (Arnold & Hale, 1971).

Proof

The process by which new mathematics is established is hidden by the deductive format in which it is recorded (Lakatos, 1976). In constructing new mathematics, problems are posed, conjectures made, counterexamples offered, and conjectures revised. A theorem results when this refinement of ideas is evaluated to have answered a substantive question. Proof has three functions in mathematics (Bell, 1976): (1) verification—concerned with establishing the truth of a proposition; (2) illumination—concerned with conveying insight into why a proposition is true; and (3) systematization—concerned with organizing propositions into a deductive system. Too many U.S. students do not appreciate or experience these functions.

Only about 30% of students in full-year geometry courses that teach proof reach a 75% mastery level in proof writing (Senk, 1985). Even high-achieving students get little meaningful mathematics out of the traditional, proof-oriented high school geometry course (Brumfiel, 1973). We need more effective ways to develop proof capabilities. I begin with an overview of theories of the development of proof. (Clements & Battista, 1992, has a more complete discussion.) The theory of the van Hieles has already been introduced. At the early levels of geometric thinking, students build "networks of relations" between shape categories. The intuitive foundation of proof "begins with a pupil's statement that belief in the truth of some assertion is connected with belief in the truth of other assertions. The notion of this connection is intuitive: The laws of such a connection can only be learned by analysis" (van Hiele, 1986, p. 124). Logic is created by analyzing and abstracting these laws, that is, by operating on the network relating statements. Deductive reasoning thus occurs at Level 3, when the network of logical relations between properties of concepts is established; students at Levels 1 or 2 do not doubt the validity of their empirical observations, so proof is meaningless to them—they see it as justifying the obvious (de Villiers, 1987). Unfortunately, more than 70% of students begin high school geometry at Levels 0 or 1, and only those students who enter at Level 2 (or higher) have a good chance of becoming competent with proof by the end of the course (Shaughnessy & Burger, 1985). On the basis of this theory, students need to attain higher levels of geometric thought before they begin a proof-oriented study of geometry. Senk (1989) reported that at the end of a course on geometric proof, students at Levels 3 and above significantly outperformed (on proof) students at Levels 2 and below; 4%, 13%, and 22% of students at Levels 0, 1, and 2, but 57%, 85%, and 100% at Levels 3, 4, and 5, respectively, were classified as having mastered proof writing.

Piaget (Piaget, 1987; Piaget, Inhelder, & Szeminska, 1960) described levels for the notions of justification and proof. In summary, at the first level, the student's thinking is nonreflective and unsystematic and, therefore, not logical. Each piece of information collected is conceived as a separate event that is not integrated with other events. The student has no plan, and his or her conclusions may be contradictory. At the second level, thought is logical but restricted to being empirical. The student's actions are more purposeful, and he or she tries to justify predictions made. Only at the third level is the student capable of logical deduction and of consciously operating within a mathematical system. The student moves through the levels on the basis of justifications and arguments with others.

Thus, van Hiele emphasizes content and Piaget emphasizes reasoning based on logical operations. Both are important. If instruction has not developed students' geometric thinking to Level 2 or higher, students may only memorize and fail to understand the purpose of proof. Teachers should have students justify their thinking at all levels and should emphasize deductive thinking from Level 3, when the network of logical relations between properties of concepts is established. Although reasoning may be less tightly tied to levels of geometric thinking than van Hiele proposed (Clements & Battista, 1992), a dynamic interaction probably occurs between the two (see McDonald, 1989). What else might we do to improve the teaching and learning of proof?

Empirical and deductive methods of finding and establishing truth should interact with and reinforce each other. For many students, however, these are two different domains (Schoenfeld, 1986). In addition, for students proof is often particular to the given figure; that is, they may indicate that a new proof would be required if a different figure were used (Martin & Harel, 1989). The "fit" between the two figures appears to be a critical issue in determining whether the same proof could be used. Few appeal to the generality of the proof. Constructions on a computer might be better for students because (a) they require more precise specification than those done with paper and pencil and (b) because the computer performs the constructions, the teacher can treat the topic less as a set of procedures to be learned and thus focus more on concept development. If used reflectively, the computer, even with its enabling of empirical strategies, appears to positively affect students' motivation and ability to learn proof.

In comparison, merely introducing informal geometry experiences first, on the basis of one interpretation of the van Hiele theory, may not be better than traditional instruction (Han, 1986). Alternative approaches, helping students expand their understanding of the nature of proof, may be more successful (Driscoll, 1983). Such approaches may include cooperative investigations in which students make conjectures and resolve conflicts by presenting arguments and evidence, prove nonobvious statements, and formulate hypotheses to prove (Fawcett, 1938; Human & Nel, 1989). This approach involves students in the crucial elements of mathematical discovery and discourse—conjecturing, careful reasoning, and the building of validating arguments that can be scrutinized by others. Computer environments, previously discussed, constitute another approach with similar aims (de Villiers, 1992). Such efforts might best be built on a cognitive model of conjecturing and argumentation (Koedinger, 1998).

Other Issues in Teaching and Learning Geometry

As discussed previously, traditional approaches dominate today's geometry instruction, and they are ineffective. In this section, we examine alternative curricula, instructional strategies, professional development, and special needs students.

Alternative Curricula

A primary cause of students' poor performance in geometry is the curriculum, both in what topics are treated and how they are treated (Jaime, Chapa, & Gutiérrez, 1992). The major focus of standard elementary and middle school curricula is on recognizing and naming geometric shapes; writing the proper symbolism for simple geometric concepts; developing skill with measurement and construction tools, such as a compass and protractor; and using formulas in geometric measurement (Porter, 1989). These curricula consist of an agglomeration of loosely connected concepts with no systematic progression to higher levels of thought. Other curricula are more effective (Idris, 1999). Although little research is available to compare such efforts, all are improvements over traditional approaches. Here I briefly describe but a small sample.

Because they observed little or no growth in geometry during the primary grades, Lehrer, Jacobson, and colleagues (1998) designed classroom environments that would promote development of student reasoning about space and geometry. They based their approach on everyday activities related to (a) perception and use of form (e.g., noticing patterns or building with blocks), leading to the mathematics of dimension, classification, transformation; (b) wayfinding (e.g., navigating in the neighborhood), leading to the mathematics of position and direction; (c) drawing (e.g., representing aspects of the world), leading to the mathematics of maps and other systems for visualizing space; and (d) measure (e.g., questions concerning "how far?" or "how big?"), leading to the mathematics of length, area, and volume measure. Students' thinking became richer and more competent, as shown by a significant increase from pretest to posttest in scores on each of the four strands.

Developers from the Netherlands similarly build their activities around sighting and projecting, locating and orientating, spatial reasoning, transforming, drawing and constructing, and measuring and calculating (de Moor, 1991). The extensive adoption of the curriculum across the country, research performed by the developers, and international comparisons all speak to the success of the approach.

Intermediate students using the reform-based University of Chicago School Mathematics Project (UCSMP) curriculum scored higher than students using more traditional materials on all geometry measures. Furthermore, fifth-grade UCSMP students outperformed comparison groups of both fifth and sixth graders (Carroll, 1998). This curriculum mixes geometry throughout the mathematics curriculum and involves students in more and more demanding geometry than traditional curricula.

The Jasper geometry adventures (Zech et al., 1998) represents another promising approach. The geometry units highlight the ubiquity of geometry in architecture, wayfinding, and measurement. Videotapes present situations that students must solve using mathematics. Assessments in Grades 6 to 8 have been positive, with dramatic improvement in students' ability to identify uses of geometry and growth in their solutions to problems of applying geometric knowledge to real-world situations. Teachers using initial versions of the activities were overwhelmed by students' many misconceptions and gave little feedback to help students learn about them. Therefore, the designers created special video "tools" to help students; initial evaluation of these tools is promising.

The Investigations in Number, Data, and Space curriculum contains geometry units and software (e.g., Battista & Clements, 1995a, 1995b; Clements, Battista, Akers, Rubin, & Woolley, 1995; Clements, Battista, Akers, Woolley, et al., 1995; Clements, Russell, Tierney, Battista, & Meredith, 1995) that were created via a scientific approach to curriculum development (Battista & Clements, 2000; Clements & Battista, 2000) and have led to multiple research reports detailing what was found about geometric learning and teaching and documenting the units' effectiveness (e.g., Battista & Clements, 1996; Battista et al., 1998; Clements, Battista, et al., 1998; Clements, Battista, et al., 1996; Clements, Battista, Sarama, & Swaminathan, 1997; Clements, Battista, Sarama, Swaminathan, et al., 1997; Clements, Sarama, & Battista, 1996; Clements, Sarama, Battista, et al., 1996). When we see research as an integral component of the design process, research as uncovering and inventing models of children's thinking and building these into a creative curriculum, then research and curriculum development can move to the vanguard in innovation and reform of education. The Investigations units are varied in their content and approach but are all based on (a) theoretically and empirically supported models of children's geometric thinking, (b) iterative field testing of activities, and (c) coherent strands of geometric topics.

Instructional Strategies

As mentioned, teachers often do not teach even the impoverished geometry curriculum that is available to them. Even when taught, geometry was the topic most frequently identified as being taught merely for "exposure," that is, given only brief, cursory coverage (Porter, 1989). Therefore, one main improvement is to teach more geometry.

Research also identifies more subtle teaching strategies. For example, high-ability students not only access a greater body of geometrical knowledge but also use that knowledge more effectively. On the more difficult problems, these students generate more information, use that generated information to access further relevant knowledge, and show more frequent management of their processing behavior (Lawson & Chinnappan, 1994). Such abilities may be teachable. Management training directed at students' attention to planning of the solution path, checking of calculations, and reviewing of the solution improves both near and far transfer performance of both high- and low-achieving high school students (Abravanel, 1977; Chinnappan & Lawson, 1992).

Research has several ramifications for teaching concepts. First, instruction that combines expository and discovery methods is more effective than structured discovery (Klausmeier, 1992). In the expository method, verbal cues aid the learner and affirmative and corrective feedback guides the learner's refinement of the concept's attributes and avoids misconceptions. Second, concepts are taught more effectively in sets of two or more related concepts than one at a time. Third, examples and nonexamples should progress from least to most typical. Fourth, students should be asked to examine each item to see whether it has each and every defining attribute.

Cultural comparisons also yield significant implications for teaching. Successful Japanese teaching employs an inductive approach, including the use of geometric representation problems in two and three dimensions and requirements to infer relationships. In comparison, in the United States, there was little room for originality and little use of physical examples; often lessons included stating definitions, using diagrams or working examples, and measuring to verify (Hafner, 1993). High-ability and high-socioeconomic-status students are often the only ones involved in the teaching practices that predict high achievement.

As in other fields, cultural strengths of U.S. students should be considered and built upon. For example, Native American, predominantly Navajo, high school students demonstrated a deep command of transformational geometry as it related to their language and cultural background (Giamati & Weiland, 1997).

"Goal free" problems (e.g., finding as many geometric relationships as possible in a diagram), compared with large numbers of conventional problems, can reduce extraneous cognitive load and promote learning (Sweller, 1993). This work also has implications for presenting problems: Cognitive load can be reduced and learning enhanced by physically integrating words and diagrams, for example, by inserting statements in an appropriate location on the accompanying diagram and so incorporating several elements into a single element. Finally, students who write more in mathematics score higher on geometry tests (DeVaney, 1996).

High school geometry classes that employed cooperative learning strategies outscored classes that did not (Nichols, 1996). Further, students' motivation was higher in the cooperative learning classes.

Taken as a whole, these results support the NCTM Standards approach to learning geometry and geometric problem solving, especially concerning higher levels of thought and more complex problems.

Professional Development

Professional development is clearly needed. Fortunately, research indicates that it can be successful. For example, one 4-week intervention program designed to enhance teachers' knowledge of geometry and their knowledge of research on student cognition in geometry (e.g., the van Hiele theory) resulted in significant positive changes in content knowledge and van Hiele level for middle-school teachers, as well as in what they taught, how they taught it, and teachers' characteristics (Swafford, Jones, & Thorton, 1997). Teachers claimed they spent more time teaching geometry and devoted more quality time to geometry instruction, emphasizing rich problems. Their perceptions and researchers' observations indicated a change in how they taught, from giving answers to generating questions and facilitating discussions. They also had higher van Hiele–based expectations for their students' thinking. They showed increased characteristics usually considered positive, such as confidence, willingness to take risks, and use of hands-on approaches and manipulatives. Teachers attributed these changes to increased geometrical content knowledge and research-based knowledge of student cognition.

Professional development and systemic reform should go together. As an example, the context and constraints within which teachers work determine the consistency between the teacher's beliefs and practices. Whereas teachers may believe that the construction of geometric understanding involves a process of socialization of students into a mathematical community of practice, the desire to cover more content hinders them from creating a classroom consistent with this belief (Frempong, 1996). Teachers require both professional development that guides reflection on practices and beliefs and systemic reform that removes constraints. Teacher education similarly needs to expand teachers' conceptual knowledge of the topics they will teach. For example, teachers score only 11.8% on a perimeter problem and 52.7% on an area problem (Reinke, 1997). Lack of such knowledge and of related pedagogical content knowledge (e.g., knowledge concerning the selection, uses, and misuses of manipulative and other instructional materials used for concrete geometry learning experiences) makes teachers vulnerable (Berenson & Blanton, 1996). Prospective and practicing teachers often struggle between a belief in teaching conceptually and the ability to act on this belief; their naive understandings of teaching and learning often result in their teaching for procedural knowledge only.

Adapting for Special Needs Students

Work with students who are deaf indicated that the teacher and students did not have substantial experience with geometry (Mason, 1995). Language did play an important role; for example, the iconic nature of the sign used for triangle is roughly equilateral or isosceles. After an eight-day geometry unit, many students spelled "triangle" instead of using signs, possibly indicating a differentiation in their minds between their new definition of the word *triangle* and what they had previously associated with the sign "triangle." In summary, low levels of achievement of students who are deaf may be the result of several factors, including lack of instruction, limited exposure to mathematical language, and the use of particular signed words in sign language. Students with special needs are quite capable of learning, provided that the material is tailored to their particular needs (Mason, 1995).

Computers can also help. Using interactive geometry software aided one high school student with cerebral palsy in understanding angles (Shaw, Durden, & Baker, 1998).

Mason (1997) also worked with mathematically talented students in grades 6 to 8. She found that about a third skipped levels in the van Hiele model. Many of them had not been exposed to or did not remember the essential defining attributes of various figures; however, they looked for similarities and differences in figures and deduced what the defining attributes might be. Their strong logical reasoning abilities may account for their ability to consistently apply their conjectured definitions and thus account for their appearing to skip levels; this facility may mask their underlying lack of basic definitions, concepts, and properties. They could do inclusion if they had the right definitions, but they do not know the "rules of the game" and thus do not know how to construct an acceptable formal proof. Thus, we need to assess talented students' knowledge of geometry. They may possess higher overall van Hiele levels than usual students and be able to reason at a high level, but they still need foundational knowledge, such as defining attributes of various classes of shapes.

Conclusions

According to a former president of NCTM, geometry is the "forgotten strand" of mathematics (Lappan, 1999). It offers us a way to interpret and reflect on our physical environment and can serve as a tool for studying other topics in mathematics and science but receives little attention in instruction. The NCTM Standards need to set the stage for a major re-emphasis on geometry education for students of all ages, grades prekindergarten through 12.

Most current curricular and teaching practices are, simply, abominable. They promote little learning or conceptual change. They often do more harm than good. They leave students unprepared for further study of geometry and the many other mathematical topics and subjects that depend on geometric knowledge. We need to do better; research provides support for the NCTM Standards and specific guidelines for teaching and learning to aid such an effort.

ACKNOWLEDGMENTS

This chapter was supported in part by the National Science Foundation under Grant Number ESI-9730804, "Building Blocks—Foundations for Mathematical Thinking, Pre-Kindergarten to Grade 2: Research-based Materials Development. " Any opinions, findings, conclusions, or recommendations expressed in this material are those of the author and do not necessarily reflect the views of the National Science Foundation. Tom Dick provided comments on an earlier draft of the chapter.

REFERENCES

Abravanel, E. (1977). The figural simplicity of parallel lines. *Child Development, 48*, 708–710.

Anderson, J. R., Boyle, C. F., & Reiser, B. J. (1985). Intelligent tutoring systems. *Science, 228*, 456–462.

Anooshian, L. J., Pascal, V. U., & McCreath, H. (1984). Problem mapping before problem solving: Young children's cognitive maps and search strategies in large-scale environments. *Child Development, 55*, 1820–1834.

Arnold, W. R., & Hale, R. (1971). An investigation of third-grade level pupils' ability in geometry. *Colorado Journal of Educational Research, 10*(4), 2–7.

Austin, R. A. (1984). Teaching concepts and properties of parallelograms by a computer assisted instruction program and a traditional classroom setting. *Dissertation Abstracts International, 44*, 2075A. (University Microfilms No. DA8324939)

Ball, D. L. (1992). Magical hopes: Manipulatives and the reform of math education. *American Educator, 16*(2), 14; 16–18; 46–47.

Battista, M. T., & Clements, D. H. (1987, June). *Logo-based geometry: Rationale and curriculum.* Paper presented at the meeting of the Learning and Teaching Geometry: Issues for Research and Practice working conference, Syracuse University, Syracuse, NY.

Battista, M. T., & Clements, D. H. (1988). A case for a Logo-based elementary school geometry curriculum. *Arithmetic Teacher, 36*, 11–17.

Battista, M. T., & Clements, D. H. (1995a). *Exploring solids and boxes.* Cambridge, MA: Dale Seymour Publications.

Battista, M. T., & Clements, D. H. (1995b). *Seeing solids and silhouettes.* Cambridge, MA: Dale Seymour Publications.

Battista, M. T., & Clements, D. H. (1996). Students' understanding of three-dimensional rectangular arrays of cubes. *Journal for Research in Mathematics Education, 27*, 258–292.

Battista, M. T., & Clements, D. H. (2000). Mathematics curriculum development as a scientific endeavor. In A. E. Kelly & R. A. Lesh (Eds.), *Handbook of research design in mathematics and science education* (pp. 737–760). Mahwah, NJ: Erlbaum.

Battista, M. T., Clements, D. H., Arnoff, J., Battista, K., & Borrow, C. V. A. (1998). Students' spatial structuring of 2D arrays of squares. *Journal for Research in Mathematics Education, 29*, 503–532.

Baulac, Y., Bellemain, F., & Laborde, J. M. (1988). *Cabri Géomètrie, un logiciel d'aide à l'apprentissage de la geometrie, logiciel et manuel d'utlizisation.* Paris: Cedic-Nathan.

Beaton, A. E., Mullis, I. V. S., Martin, M. O., Gonzalez, E. J., Kelly, D. L., & Smith, T. A. (1996, January 19, 1997). *Mathematics achievement in the middle school years: IEA's third international mathematics and science study (TIMSS).* Available: http://wwwcsteep.bc.edu/timss.

Beilin, H. (1984). Cognitive theory and mathematical cognition: Geometry and space. In B. Gholson & T. L. Rosenthanl (Eds.), *Applications of cognitive-developmental theory* (pp. 49–93). New York: Academic Press.

Bell, A. W. (1976). A study of pupils' proof-explanations in mathematical situations. *Educational Studies in Mathematics, 7*, 23–40.

Berenson, S. B., & Blanton, M. L. (1996). Preservice teachers' ideas of teaching the concept of area. In E. Jakubowski, D. Watkins, & H. Biske (Eds.), *Proceedings of the 18th Annual Meeting of the North America Chapter of the International Group for the Psychology of Mathematics Education* (Vol. 2, pp. 401–407). Columbus, OH: ERIC Clearinghouse for Science, Mathematics, and Environmental Education.

Bishop, A. J. (1983). Space and geometry. In R. Lesh & M. Landau (Eds.), *Acquisition of mathematics concepts and processes.* New York: Academic Press.

Blaut, J. M., & Stea, D. (1974). Mapping at the age of three. *Journal of Geography, 73*(7), 5–9.

Boardman, D. (1990). Graphicacy revisited: Mapping abilities and gender differences. *Educational Review, 42*, 57–64.

Bourgeois, R. D. (1986). Third graders' ability to associate foldout shapes with polyhedra. *Journal for Research in Mathematics Education, 17*, 222–230.

Bright, G. (1985). What research says: Teaching probability and estimation of length and angle measurements through microcomputer instructional games. *School Science and Mathematics, 85*, 513–522.

Bright, G., Usnick, V. E., & Williams, S. (1992, March). *Orientation of shapes in a video game.* Paper presented at the meeting of the Eastern Educational Research Association, Hilton Head, SC.

Brumfiel, C. (1973). Conventional approaches using synthetic Euclidean geometry. In K. B. Henderson (Ed.), *Geometry in the mathematics curriculum* (1973 Yearbook of the National Council of Teachers of Mathematics, pp. 95–115). Reston, VA: NCTM.

Burger, W. F., & Shaughnessy, J. M. (1986). Characterizing the van Hiele levels of development in geometry. *Journal for Research in Mathematics Education, 17*, 31–48.

Carlson, G. R. (1976). Location of a point in Euclidian space by children in grades one through six. *Journal of Research in Science Teaching, 13*, 331–336.

Carpenter, T. P., Coburn, T., Reys, R., & Wilson, J. (1976). Notes from National Assessment: Recognizing and naming solids. *Arithmetic Teacher, 23*, 62–66.

Carpenter, T. P., Corbitt, M. K., Kepner, H. S., Lindquist, M. M., & Reys, R. E. (1980). National assessment. In E. Fennema (Ed.), *Mathematics education research: Implications for the 80s* (pp. 22–38). Alexandria, VA: Association for Supervision and Curriculum Development.

Carroll, W. M. (1998). Geometric knowledge of middle school students in a reform-based mathematics curriculum. *School Science and Mathematics, 98*, 188–197.

Chazan, D. (1989). Instructional implications of a research project on students' understandings of the differences between empirical verification and mathematical proof. In D. Hergert (Ed.), *Proceedings of the First International Conference on the History and Philosophy of Science in Science Teaching* (pp. 52–60). Tallahassee: Florida State University, Science Education and Philosophy Department.

Chinnappan, M. (1998). Schemas and mental models in geometry problem solving. *Educational Studies in Mathematics, 36*(3), 201–217.

Chinnappan, M., & Lawson, M. (1992). The effects of training in use of generation and management strategies on geometry problem solving. In W. Geeslin & K. Graham (Eds.), *Proceedings of the 16th Psychology in Mathematics Education Conference* (Vol. 3, pp. 170). Durham, NH: Program Committee of the 16th Psychology in Mathematics Education Conference.

Choi-koh, S. S. (1999). A student's learning of geometry using the computer. *Journal of Educational Research, 92,* 301–311.

Clements, D. H. (1987). Longitudinal study of the effects of Logo programming on cognitive abilities and achievement. *Journal of Educational Computing Research, 3,* 73–94.

Clements, D. H. (1999). 'Concrete' manipulatives, concrete ideas. *Contemporary Issues in Early Childhood, 1*(1), 45–60.

Clements, D. H., & Battista, M. T. (1989). Learning of geometric concepts in a Logo environment. *Journal for Research in Mathematics Education, 20,* 450–467.

Clements, D. H., & Battista, M. T. (1990). The effects of Logo on children's conceptualizations of angle and polygons. *Journal for Research in Mathematics Education, 21,* 356–371.

Clements, D. H., & Battista, M. T. (1992). Geometry and spatial reasoning. In D. A. Grouws (Ed.), *Handbook of research on mathematics teaching and learning* (pp. 420–464). New York: Macmillan.

Clements, D. H., & Battista, M. T. (2000). Developing effective software. In A. E. Kelly & R. A. Lesh (Eds.), *Handbook of research design in mathematics and science education* (pp. 761–776). Mahwah, NJ: Erlbaum.

Clements, D. H., Battista, M. T., & Sarama, J. (2001). *Logo and geometry. Journal for Research in Mathematics Education* Monograph Series, 10.

Clements, D. H., Battista, M. T., Akers, J., Rubin, A., & Woolley, V. (1995). *Sunken ships and grid patterns.* Cambridge, MA: Dale Seymour Publications.

Clements, D. H., Battista, M. T., Akers, J., Woolley, V., Meredith, J. S., & McMillen, S. (1995). *Turtle paths.* Cambridge, MA: Dale Seymour Publications.

Clements, D. H., Battista, M. T., Sarama, J., & Swaminathan, S. (1996). Development of turn and turn measurement concepts in a computer-based instructional unit. *Educational Studies in Mathematics, 30,* 313–337.

Clements, D. H., Battista, M. T., Sarama, J., Swaminathan, S., & McMillen, S. (1997). Students' development of length measurement concepts in a Logo-based unit on geometric paths. *Journal for Research in Mathematics Education, 28*(1), 70–95.

Clements, D. H., & Meredith, J. S. (1994). *Turtle math.* Montreal, Quebec: Logo Computer Systems.

Clements, D. H., Russell, S. J., Tierney, C., Battista, M. T., & Meredith, J. S. (1995). *Flips, turns, and area.* Cambridge, MA: Dale Seymour Publications.

Clements, D. H., & Sarama, J. (1995). Design of a Logo environment for elementary geometry. *Journal of Mathematical Behavior, 14,* 381–398.

Clements, D. H., & Sarama, J. (1996). Turtle Math: Redesigning Logo for elementary mathematics. *Learning and Leading with Technology, 23*(7), 10–15.

Clements, D. H., Sarama, J., & Battista, M. T. (1996). Development of turn and turn measurement concepts in a computer-based instructional unit. In E. Jakubowski, D. Watkins, & H. Biske (Eds.), *Proceedings of the Eighteenth Annual Meeting of the North America Chapter of the International Group for the Psychology of Mathematics Education* (Vol. 2, pp. 547–552). Columbus, OH: ERIC Clearinghouse for Science, Mathematics, and Environmental Education.

Clements, D. H., Sarama, J., & Battista, M. T. (1998). Development of concepts of geometric figures in a specially-designed Logo computer environment. *Focus on Learning Problems in Mathematics, 20,* 47–64.

Clements, D. H., Sarama, J., Battista, M. T., & Swaminathan, S. (1996). Development of students' spatial thinking in a curriculum unit on geometric motions and area. In E. Jakubowski, D. Watkins, & H. Biske (Eds.), *Proceedings of the Eighteenth Annual Meeting of the North America Chapter of the International Group for the Psychology of Mathematics Education* (Vol. 1, pp. 217–222). Columbus, OH: ERIC Clearinghouse for Science, Mathematics, and Environmental Education.

Clements, D. H., Sarama, J., González Gómez, R. M., Swaminathan, S., & McMillen, S. (in press). Development of mathematical concepts of two-dimensional space in grid environments: An exploratory study. *Cognition and Instruction.*

Clements, D. H., Swaminathan, S., Hannibal, M. A. Z., & Sarama, J. (1999). Young children's concepts of shape. *Journal for Research in Mathematics Education, 30,* 192–212.

Confrey, J. (1992, April). *First graders' understanding of similarity.* Paper presented at the meeting of the American Educational Research Association, San Francisco.

Davis, R. B. (1984). *Learning mathematics: The cognitive science approach to mathematics education.* Norwood, NJ: Ablex.

De Moor, E. (1991). Geometry instruction in The Netherlands (ages 4–14)—The realistic approach. In L. Streefland (Ed.), *Realistic mathematics education in primary school* (pp. 119–138). Utrecht, The Netherlands: Utrecht University, Freudenthal Institute.

De Villiers, M. D. (1987, June). *Research evidence on hierarchical thinking, teaching strategies, and the van Hiele theory: Some critical comments.* Paper presented at the meeting of the Learning and Teaching Geometry: Issues for Research and Practice working conference, Syracuse University, Syracuse, NY.

De Villiers, M. D. (1992). Children's acceptance of theorems in geometry. In W. Geeslin & K. Graham (Eds.), *Proceedings of the 16th Psychology in Mathematics Education Conference* (Vol. 3, pp. 155). Durham, NH: Program Committee of the 16th Psychology in Mathematics Education Conference.

De Villiers, M. D. (1995). An alternative introduction to proof in dynamic geometry. *MicroMath, 11*(12), 14–19.

Del Grande, J. J. (1986). Can grade two children's spatial perception be improved by inserting a transformation geometry component into their mathematics program? *Dissertation Abstracts International, 47*, 3689A.

DeVaney, T. A. (1996) *The effects of instructional practices on computation and geometry achievement.* Paper presented at the meeting of the Mid-South Educational Research Association, Tuscaloosa, AL.

Dixon, J. K. (1997). Computer use and visualization in students' construction of reflection and rotation concepts. *School Science and Mathematics, 97*, 352–358.

Downs, R. M., & Liben, L. S. (1988). Through and map darkly: Understanding maps as representations. *Genetic Epistemologist, 16*, 11–18.

Downs, R. M., Liben, L. S., & Daggs, D. G. (1988). On education and geographers: The role of cognitive developmental theory in geographic education. *Annuals of the Association of American Geographers, 78*, 680–700.

Driscoll, M. J. (1983). *Research within reach: Secondary school mathematics.* St. Louis, MO: CEMREL.

Fawcett, H. P. (1938). *The nature of proof* (Thirteenth yearbook of the National Council of Teachers of Mathematics). New York: Columbia University, Teachers College, Bureau of Publications.

Fey, J., Atchison, W. F., Good, R. A., Heid, M. K., Johnson, J., Kantowski, M. G., & Rosen, L. P. (1984). *Computing and mathematics: The impact on secondary school curricula.* College Park: University of Maryland.

Frempong, G. (1996). Accessing one teacher's understanding of the teacher of geometry through stimulated recall. In E. Jakubowski, D. Watkins, & H. Biske (Eds.), *Proceedings of the 18th Annual Meeting of the North America Chapter of the International Group for the Psychology of Mathematics Education* (Vol. 2, pp. 365–369). Columbus, OH: ERIC Clearinghouse for Science, Mathematics, and Environmental Education.

Friedlander, A. (1985). Achievement in similarity tasks: Effect of instruction, and relationship with achievement in spatial visualization at the middle grades level. *Dissertation Abstracts International, 46*, 928–929A. (University Microfilms No. DA8503834)

Fuys, D., Geddes, D., & Tischler, R. (1988). The van Hiele model of thinking in geometry among adolescents. *Journal for Research in Mathematics Education Monograph 3.*

Galindo, E. (1998). Assessing justification and proof in geometry classes taught using dynamic software. *Mathematics Teacher, 91*, 76–82.

Gallou-Dumiel, E. (1989). Reflections, point symmetry and Logo. In C. A. Maher, G. A. Goldin, & R. B. Davis (Eds.), *Proceedings of the 11th Annual Meeting, North American Chapter of the International Group for the Psychology of Mathematics Education* (pp. 149–157). New Brunswick, NJ: Rutgers University.

Genkins, E. F. (1975). The concept of bilateral symmetry in young children. In M. F. Rosskopf (Ed.), *Children's mathematical concepts: Six Piagetian studies in mathematics education* (pp. 5–43). New York: Teachers College Press.

Giamati, C., & Weiland, M. (1997). An exploration of American Indian students' perceptions of patterning, symmetry and geometry. *Journal of American Indian Education, 36*(3), 27–48.

Grant, S. G., Peterson, P. L., & Shojgreen-Downer, A. (1996). Learning to teach mathematics in the context of system reform. *American Educational Research Journal, 33*, 509–541.

Gutiérrez, A., Jaime, A., & Fortuny, J. M. (1991). An alternative paradigm to evaluate the acquisition of the van Hiele levels. *Journal for Research in Mathematics Education, 22*, 237–251.

Hafner, A. L. (1993). Teaching-method scales and mathematics-class achievement: What works with different outcomes? *American Educational Research Journal, 30*, 71–94.

Han, T.-S. (1986). The effects on achievement and attitude of a standard geometry textbook and a textbook consistent with the van Hiele theory. *Dissertation Abstracts International, 47*, 3690A. (University Microfilms No. DA8628106)

Hannafin, R. D., & Sullivan, H. J. (1996). Preferences and learner control over amount of instruction. *Journal of Educational Psychology, 88*, 162–173.

Hannibal, M. A. Z., & Clements, D. H. (2000). *Young children's understanding of basic geometric shapes.* Manuscript submitted for publication.

Hershkowitz, R., Ben-Chaim, D., Hoyles, C., Lappan, G., Mitchelmore, M., & Vinner, S. (1990). Psychological aspects of learning geometry. In P. Nesher & J. Kilpatrick (Eds.), *Mathematics and cognition: A research synthesis by the International Group for the Psychology of Mathematics Education* (pp. 70–95). Cambridge: Cambridge University Press.

Hillel, J., & Kieran, C. (1988). Schemas used by 12-year-olds in solving selected Turtle Geometry tasks. *Recherches en Didactique des Mathématiques, 8*(1.2), 61–103.

Hoyles, C., & Healy, L. (1997). Unfolding meanings for reflective symmetry. *International Journal of Computers for Mathematical Learning, 2*, 27–59.

Hoyles, C., Noss, R., & Sutherland, R. (1989). Designing a Logo-based microworld for ratio and proportion. *Journal of Computer Assisted Learning, 5*, 208–222.

Hughes, M., & Macleod, H. (1986). Part 2: Using Logo with very young children. In R. Lawler, B. du Boulay, M. Hughes, & H. Macleod (Eds.), *Cognition and computers: Studies in learning* (pp. 179–219). Chichester, England: Ellis Horwood.

Human, P. G., & Nel, J. H. (1989). *Alternative teaching strategies for geometry education: A theoretical and empirical study* (RUMEUS Curriculum Materials Series No. 11). Matieland, South Africa: University of Stellenbosch.

Idris, N. (1999). Spatial visualization, field dependence/independence, van Hiele level, and achievement in geometry: The influence of selected activities for middle school students. *Dissertation Abstracts International, 59,* 2894. (University Microfilms No. DA8518842)

Imamoglu, E. O., & Imamoglu, V. (1983). Children's plan-drawings of their houses. In D. R. Rogers & J. A. Sloboda (Eds.), *The acquisiton of symbolic skills* (pp. 367–379). New York: Plenum Press.

Jackiw, N. (1995). The Geometer's Sketchpad [Computer software]. Berkeley, CA: Key Curriculum Press.

Jaime, A. P., Chapa, F. A., & Gutiérrez, A. R. (1992). Definiciones de triangulos y cuadrilateros: Errores e inconsistencias en libros de texto de E.G.B. *Epsilon, 23,* 49–62.

Johnson, P. A. (1986). *Effects of computer-assisted instruction compared to teacher-directed instruction on comprehension of abstract concepts by the deaf.* Unpublished doctoral dissertation, Northern Illinois University.

Johnson-Gentile, K., Clements, D. H., & Battista, M. T. (1994). The effects of computer and noncomputer environments on students' conceptualizations of geometric motions. *Journal of Educational Computing Research, 11*(2), 121–140.

Kafai, Y. (1989). What happens if you introduce an intelligent tutoring system in the classroom: A case study of the Geometry Tutor. In W. C. Ryan (Ed.), *Proceedings of the National Educational Computing Conference* (pp. 46–51). Eugene, OR: International Council on Computers for Education.

Kidder, F. R. (1976). Elementary and middle school children's comprehension of Euclidean transformations. *Journal for Research in Mathematics Education,* 7, 40–52.

Kieran, C. (1986). Logo and the notion of angle among fourth and sixth grade children, *Proceedings of the 10th Annual Conference of the International Group for the Psychology of Mathematics Education* (pp. 99–104). London: University of London, Institute of Education.

Kieran, C., Hillel, J., & Erlwanger, S. (1986). Perceptual and analytical schemas in solving structured Turtle-Geometry tasks. In C. Hoyles & R. Noss & R. Sutherland (Eds.), *Proceedings of the Second Logo and Mathematics Educators Conference* (pp. 154–161). London: University of London.

Klausmeier, H. J. (1992). Concept learning and concept teaching. *Educational Psychologist, 27,* 267–286.

Koedinger, K. R. (1998). Conjecturing and argumentation in high-school geometry students. In R. Lehrer & D. Chazan (Eds.), *Designing learning environments for developing understanding of geometry and space* (pp. 319–347). Mahwah, NJ: Erlbaum.

Koedinger, K. R., & Anderson, J. R. (1993). Reifying implicit planning in geometry: Guidelines for model-based intelligent tutoring system design. In S. Lajoie & S. Derry (Eds.), *Computers as cognitive tools* (pp. 241–248). Hillsdale, NJ: Erlbaum.

Kouba, V. L., Brown, C. A., Carpenter, T. P., Lindquist, M. M., Silver, E. A., & Swafford, J. O. (1988). Results of the fourth NAEP assessment of mathematics: Measurement, geometry, data interpretation, attitudes, and other topics. *Arithmetic Teacher, 35*(9), 10–16.

Krainer, K. (1991). Consequences of a low level of acting and reflecting in geometry learning—Findings of interviews on the concept of angle. In F. Furinghetti (Ed.), *Proceedings of the 15th Annual Conference of the International Group for the Psychology of Mathematics Education* (Vol. 2, pp. 254–261). Assisi, Italy: Program Committee, 15th Psychology in Mathematics Education Conference.

Kynigos, C. (1993). Children's inductive thinking during intrinsic and Euclidean geometrical activities in a computer programming environment. *Educational Studies in Mathematics, 24,* 177–197.

Laborde, C. (1995). Designing tasks for learning geometry in a computer-based environment. In L. Burton & B. Jaworski (Eds.), *Technology in mathematics teaching.* Bromley, England: Chartwell-Bratt.

Laborde, C. (1996). A new generation of diagrams in dynamic geometry software. In E. Jakubowski & D. Watkins & H. Biske (Eds.), *Proceedings of the 18th Annual Meeting of the North America Chapter of the International Group for the Psychology of Mathematics Education.* Columbus, OH: ERIC Clearinghouse for Science, Mathematics, and Environmental Education.

Lakatos, I. (1976). *Proofs and refutations: The logic of mathematical discovery.* New York: Cambridge University Press.

Lampert, M. (1988). *Teachers' thinking about students' thinking about geometry: The effects of new teaching tools. Technical Report.* Cambridge, MA: Harvard Graduate School of Education, Educational Technology Center.

Landau, B. (1988). The construction and use of spatial knowledge in blind and sighted children. In J. Stiles-Davis, M. Kritchevsky, & U. Bellugi (Eds.), *Spatial cognition: Brain bases and development* (pp. 343–371). Hillsdale, NJ: Erlbaum.

Lappan, G. (1999). Geometry: The forgotten strand. *NCTM News Bulletin, 36*(5), 3.

Lawson, M.J., & Chinnappan, M. (1994). Generative activity during geometry problem solving: Comparison of the performance of high-achieving and low-achieving students. *Cognition and Instruction, 12*(1), 61–93.

Lehrer, R., Jacobson, C., Thoyre, G., Kemeny, V., Strom, D., Horvarth, J., Gance, S., & Koehler, M. (1998). Developing understanding of geometry and space in the primary grades. In R. Lehrer & D. Chazan (Eds.), *Designing learning environments for developing understanding of geometry and space* (pp. 169–200). Mahwah, NJ: Erlbaum.

Lehrer, R., Jacobson, C., Thoyre, G., Kemeny, V., Strom, D., Horvarth, J., Gance, S., & Koehler, M. (1998). Developing understanding of geometry and space in the primary grades. In R. Lehrer & D. Chazan (Eds.), *Designing learning environments for developing understanding of geometry and space* (pp. 169-200). Mahwah, NJ: Erlbaum

Lehrer, R., Jenkins, M., & Osana, H. (1998). Longitudinal study of children's reasoning about space and geometry. In R. Lehrer & D. Chazan (Eds.), *Designing learning environments for developing understanding of geometry and space* (pp. 137–167). Mahwah, NJ: Erlbaum.

Lehrer, R., & Smith, P. C. (1986, April). *Logo learning: Are two heads better than one?* Paper presented at the meeting of the American Educational Research Association, San Francisco.

Leinhardt, G., Zaslavsky, O., & Stein, M. K. (1990). Functions, graphs, and graphing: Tasks, learning, and teaching. *Review of Educational Research, 60*, 1–64.

Liben, L. S. (1978). Performance on Piagetian spatial tasks as a function of sex, field dependence, and training. *Merrill-Palmer Quarterly, 24*, 97–110.

Mackay, C. K., Brazendale, A. H., & Wilson, L. F. (1972). Concepts of horizontal and vertical: A methodological note. *Developmental Psychology*, 7, 232–237.

Mansfield, H. M., & Happs, J. C. (1992). Using grade eight students' existing knowledge to teach about parallel lines. *School Science and Mathematics, 92*, 450–454.

Martin, C., & Sweller, J. (1989). Secondary school students' representations of solids. *Journal for Research in Mathematics Education, 20*, 202–212.

Martin, W. G., & Harel, G. (1989). The role of the figure in students' concepts of geometric proof. In G. Vergnaud, J. Rogalski, & M. Artique (Eds.), *Proceedings of the 13th Conference of the International Group for the Psychology of Mathematics Education* (pp. 266–273). Paris: University of Paris.

Mason, M. M. (1995). Geometric knowledge in a deaf classroom: An exploratory study. *Focus on Learning Problems in Mathematics, 17*(3), 57–69.

Mason, M. M. (1997). The van Hiele model of geometric understanding and gifted students. *Journal for the Education of the Gifted, 21*, 39–53.

McCoy, L. P. (1991). The effect of geometry tool software on high school geometry achievement. *Journal of Computers in Mathematics and Science Teaching, 10*, 51–57.

McDonald, J. L. (1989). Cognitive development and the structuring of geometric knowledge. *Journal for Research in Mathematics Education, 20*, 76–94.

McDougall, D. E. (1997). *Mathematics teachers' needs in dynamic geometric computer environments: In search of control.* Unpublished doctoral dissertation, University of Toronto.

McKnight, C. C., Travers, K. J., Crosswhite, F. J., & Swafford, J. O. (1985). Eighth-grade mathematics in U.S. schools: A report from the Secondary International Mathematics Study. *Arithmetic Teacher, 32*(8), 20–26.

McKnight, C. C., Travers, K. J., & Dossey, J. A. (1985). Twelfth-grade mathematics in U.S. high schools: A report from the Secondary International Mathematics Study. *Mathematics Teacher, 78*(4), 292–300.

Mitchelmore, M. (1992). Children's concepts of perpendiculars. In W. Geeslin & K. Graham (Eds.), *Proceedings of the 16th Psychology in Mathematics Education Conference* (Vol. 2) (pp. 120–127). Durham, NH: Program Committee of the 16th Psychology in Mathematics Education Conference.

Mitchelmore, M. C. (1989). The development of children's concepts of angle. In G. Vergnaud, J. Rogalski, & M. Artique (Eds.), *Proceedings of the 13th Conference of the International Group for the Psychology of Mathematics Education* (pp. 304–311). Paris: University of Paris.

Mitchelmore, M. C. (1993). The development of pre-angle concepts. In A. R. Baturo & L. J. Harris (Eds.), *New directions in research on geometry* (pp. 87–93). Brisbane, Australia: Queensland University of Technology, Centre for Mathematics and Science Education.

Morris, J. P. (1983). Microcomputers in a sixth-grade classroom. *Arithmetic Teacher, 31*(2), 22–24.

Muir, S. P., & Cheek, H. N. (1986). Mathematics and the map skill curriculum. *School Science and Mathematics, 86*, 284–291.

Mullis, I. V. S., Martin, M. O., Beaton, A. E., Gonzalez, E. J., Kelly, D. L., & Smith, T. A. (1997). *Mathematics achievement in the primary school years: IEA's third international mathematics and science study (TIMSS).* Chestnut Hill, MA: Boston College, Center for the Study of Testing, Evaluation, and Educational Policy.

Nasser, L. (1992, August). *A van Hiele–based experiment on the teaching of congruence.* Paper presented at the 16th annual meeting of the International Group for the Psychology of Mathematics Education, Durham, NH.

Nichols, J. D. (1996). The effects of cooperative learning on student achievement and motivation in a high school geometry class. *Contemporary Educational Psychology, 21*, 467–476.

Nieuwoudt, H. D., & van Niekerk, R. (1997, March). *The spatial competence of young children through the development of solids.* Paper presented at the meeting of the American Educational Research Association, Chicago.

Noss, R., & Hoyles, C. (1992). Afterword: Looking back and looking forward. In C. Hoyles & R. Noss (Eds.), *Learning mathematics and Logo* (pp. 427–468). Cambridge, MA: MIT Press.

Olive, J., Lankenau, C. A., & Scally, S. P. (1986). *Teaching and understanding geometric relationships through Logo: Phase 2.* (Interim Report, Atlanta-Emory Logo Project). Atlanta: Emory University.

Olson, A. T., Kieran, T. E., & Ludwig, S. (1987). Linking Logo, levels, and language in mathematics. *Educational Studies in Mathematics, 18*, 359–370.

Outhred, L. N., & Mitchelmore, M. (1992). Representation of area: A pictorial perspective. In W. Geeslin & K. Graham (Eds.), *Proceedings of the 16th Psychology in Mathematics Education Conference* (Vol. 2, pp. 194–201). Durham, NH: Program Committee of the 16th Psychology in Mathematics Education Conference.

Owens, E. W., & Waxman, H. C. (1994). Comparing the effectiveness of computer-assisted instruction and conventional instruction in mathematics for African-American postsecondary students. *International Journal of Instructional Media, 21*, 327–336.

Papert, S. (1980). *Mindstorms: Children, computers, and powerful ideas.* New York: Basic Books.

Parzysz, B. (1988). Knowing vs. seeing. Problems of the plane representation of space geometry figures. *Educational Studies in Mathematics, 19*, 79–92.

Pegg, J., & Davey, G. (1998). Interpreting study understanding in geometry: A synthesis of two models. In R. Lehrer & D. Chazan (Eds.), *Designing learning environments for developing understanding of geometry and space* (pp. 109–135). Mahwah, NJ: Erlbaum

Perham, F. (1978). An investigation into the effect of instruction on the acquisition of transformation geometry concepts in first grade children and subsequent transfer to general spatial ability. In R. Lesh & D. Mierkiewicz (Eds.), *Concerning the development of spatial and geometric concepts* (pp. 229–241). Columbus, OH: ERIC Clearinghouse for Science, Mathematics, and Environmental Education.

Piaget, J. (1987). *Possibility and necessity: Vol. 2, The role of necessity in cognitive development.* Minneapolis: University of Minnesota Press.

Piaget, J., & Inhelder, B. (1967). *The child's conception of space* (F. J. Langdon & J. L. Lunzer, Trans.). New York: W. W. Norton.

Piaget, J., Inhelder, B., & Szeminska, A. (1960). *The child's conception of geometry.* London: Routledge & Kegan Paul.

Pokay, P. A., & Tayeh, C. (1997). Integrating technology in a geometry classroom: Issues for teaching. *Computers in the Schools, 13*, 117–123.

Porter, A. (1989). A curriculum out of balance: The case of elementary school mathematics. *Educational Researcher, 18*, 9–15.

Presson, C. C. (1987). The development of spatial cognition: Secondary uses of spatial information. In N. Eisenberg (Ed.), *Contemporary topics in developmental psychology* (pp. 77–112). New York: Wiley.

Reinke, K. S. (1997). Area and perimeter: Preservice teachers' confusion. *School Science and Mathematics, 97*, 75–77.

Roberts, D. L., & Stephens, L. J. (1999). The effect of the frequency of usage of computer software in high school geometry. *Journal of Computers in Mathematics and Science Teaching, 18*, 23–30.

Rosser, R. A. (1994). Children's solution strategies and mental rotation problems: The differential salience of stimulus components. *Child Study Journal, 24*, 153–168.

Rosser, R. A., Horan, P. F., Mattson, S. L., & Mazzeo, J. (1984). Comprehension of Euclidean space in young children: The early emergence of understanding and its limits. *Genetic Psychology Monographs, 110*, 21–41.

Sachter, J. E. (1991). Different styles of exploration and construction of 3-D spatial knowledge in a 3-D computer graphics microworld. In I. Harel & S. Papert (Eds.), *Constructionism* (pp. 335–364). Norwood, NJ: Ablex.

Sarama, J. (1995). *Redesigning Logo: The Turtle metaphor in mathematics education.* Unpublished doctoral dissertation, State University of New York at Buffalo.

Sarama, J. (2000). Toward more powerful computer environments: Developing mathematics software on research-based principles. *Focus on Learning Problems in Mathematics, 22*(3&4), 125–147.

Schoenfeld, A. H. (1986). On having and using geometric knowledge. In J. Hiebert (Ed.), *Conceptual and procedural knowledge: The case of mathematics* (pp. 225–264). Hillsdale, NJ: Erlbaum.

Schofield, J. W., & Verban, D. (1988). Computer usage in the teaching of mathematics: Issues that need answers. In D. A. Grouws, T. J. Cooney, & D. Jones (Eds.), *Perspectives on research on effective mathematics teaching* (Vol. 1, pp. 169–193). Hillsdale, NJ: Erlbaum.

Schultz, K. A., & Austin, J. D. (1983). Directional effects in transformational tasks. *Journal for Research in Mathematics Education, 14*, 95–101.

Schwartz, J. L., & Yerushalmy, M. (1986). The Geometric Supposer series [Computer-based courseware]. Pleasantville, NY: Sunburst Communications.

Sedighian, K., & Sedighian, A. (1996). Can educational computer games help educators learn about the psychology of learning mathematics in children? In E. Jakubowski, D. Watkins, & H. Biske (Eds.), *Proceedings of the 18th Annual Meeting of the North America Chapter of the International Group for the Psychology of Mathematics Education* (Vol. 2, pp. 573–578). Columbus, OH: ERIC Clearinghouse for Science, Mathematics, and Environmental Education.

Senk, S. L. (1985). How well do students write geometry proofs? *Mathematics Teacher, 78*, 448–456.

Senk, S. L. (1989). Van Hiele levels and achievement in writing geometry proofs. *Journal for Research in Mathematics Education, 20*, 309–321.

Shaughnessy, J. M., & Burger, W. F. (1985). Spadework prior to deduction in geometry, *Mathematics Teacher, 78*, 419–428.

Shaw, K. L., Durden, P., & Baker, A. (1998). Learning how Amanda, a high school cerebral palsy student, understands angles. *School Science and Mathematics, 98*, 198–204.

Siegel, A. W., & Schadler, M. (1977). The development of young children's spatial representations of their classrooms. *Child Development, 48*, 388–394.

Somerville, S. C., & Bryant, P. E. (1985). Young children's use of spatial coordinates. *Child Development, 56*, 604–613.

Somerville, S. C., Bryant, P. E., Mazzocco, M. M. M., & Johnson, S. P. (1987, April). *The early development of children's use of spatial coordinates.* Paper presented at the meeting of the Society for Research in Child Development, Baltimore.

Sophian, C., & Crosby, M. E. (1998, August). *Ratios that even young children understand: The case of spatial proportions.* Paper presented at the meeting of the Cognitive Science Society of Ireland, Dublin.

Sowell, E. J. (1989). Effects of manipulative materials in mathematics instruction. *Journal for Research in Mathematics Education, 20*, 498–505.

Starkey, P., Klein, A., Chang, I., Qi, D., Lijuan, P., & Yang, Z. (1999, April). *Environmental supports for young children's mathematical development in China and the United States.* Paper presented at the meeting of the Society for Research in Child Development, Albuquerque, NM.

Steffe, L. P. (1991). Operations that generate quantity. *Learning and Individual Differences, 3*, 61–82.

Stevenson, H. W., Lee, S.-Y., & Stigler, J. W. (1986). Mathematics achievement of Chinese, Japanese, and American children. *Science, 231*, 693–699.

Stigler, J. W., Lee, S.-Y., & Stevenson, H. W. (1990). *Mathematical knowledge of Japanese, Chinese, and American elementary school children*. Reston, VA: National Council of Teachers of Mathematics.

Swafford, J., Jones, G. A., & Thorton, C. A. (1997). Increased knowledge in geometry and instructional practice. *Journal for Research in Mathematics Education, 28*, 467–483.

Sweller, J. (1993). Some cognitive processes and their consequences for the organization and presentation of information. *Australian Journal of Psychology, 45*, 1–8.

Thomas, H., & Jamison, W. (1975). On the acquisition of understanding that still water is horizontal. *Merrill-Palmer Quarterly, 21*, 31–44.

Usiskin, Z. (1987). Resolving the continuing dilemmas in school geometry. In M. M. Lindquist & A. P. Shulte (Eds.), *Learning and teaching geometry, K–12* (1987 Yearbook of the National Council of Teachers of Mathematics, pp. 17–31). Reston, VA: NCTM.

Usiskin, Z. P. (1972). The effects of teaching Euclidean geometry via transformations on student achievement and attitudes in tenth-grade geometry. *Journal for Research in Mathematics Education, 3*, 249–259.

Uttal, D. H., & Wellman, H. M. (1989). Young children's representation of spatial information acquired from maps. *Developmental Psychology, 25*, 128–138.

Van Hiele, P. M. (1986). *Structure and insight: A theory of mathematics education*. Orlando, FL: Academic Press.

Van Hiele-Geldof, D. (1984). The didactics of geometry in the lowest class of secondary school. In D. Fuys, D. Geddes, & R. Tischler (Eds.), *English translation of selected writings of Dina van Hiele-Geldof and Pierre M. van Hiele* (pp. 1–214). Brooklyn, NY: Brooklyn College, School of Education. (ERIC Document Reproduction Service No. 289 697)

Vinner, S., & Hershkowitz, R. (1980). Concept images and common cognitive paths in the development of some simple geometrical concepts. In R. Karplus (Ed.), *Proceedings of the Fourth International Conference for the Psychology of Mathematics Education* (pp. 177–184). Berkeley: University of California, Lawrence Hall of Science.

Vollrath, H. J. (1977). The understanding of similarity and shape in classifying tasks. *Educational Studies in Mathematics, 8*, 211–224.

Wertheimer, R. (1990). The geometry proof tutor: An "intelligent" computer-based tutor in the classroom. *Mathematics Teacher, 83*, 308–317.

Williford, H. J. (1972). A study of transformational geometry instruction in the primary grades. *Journal for Research in Mathematics Education, 3*, 260–271.

Wiske, M. S., & Houde, R. (1988). *From recitation to construction: Teaching change with new technologies* (Technical Report). Cambridge, MA: Harvard Graduate School of Education, Educational Technology Center.

Yerushalmy, M. (1986). Induction and generalization: An experiment in teaching and learning high school geometry. *Dissertation Abstracts International, 47*, 1641A. (University Microfilms No. DA8616766)

Yerushalmy, M. (1991). Enhancing acquisition of basic geometrical concepts with the use of the Geometric Supposer. *Journal of Educational Computing Research*, 7, 407–420.

Yerushalmy, M., & Chazan, D. (1988). *Overcoming visual obstacles with the aid of the Supposer*. Cambridge, MA: Harvard Graduate School of Education, Educational Technology Center.

Yerushalmy, M., Chazan, D., & Gordon, M. (1987). *Guided inquiry and technology: A year long study of children and teaching using the Geometric Supposer* (ETC Final Report). Newton, MA: Educational Development Center.

Yusuf, M. M. (1994). Cognition of fundamental concepts in geometry. *Journal of Educational Computing Research, 10*, 349–371.

Zech, L., Vye, N. J., Bransford, J. D., Goldman, S. R., Barron, B. J., Schwartze, D. L., Kisst-Hackett, R., Mayfield-Stewart, C., & the Cognition and Technology Group. (1998). An introduction to geometry through anchored instruction. In R. Lehrer & D. Chazan (Eds.), *Designing learning environments for developing understanding of geometry and space* (pp. 439–463). Mahwah, NJ: Erlbaum.

Chapter 12

Developing Understanding of Measurement

Richard Lehrer, Vanderbilt University

Measurement is an enterprise that spans both mathematics and science yet has its roots in everyday experience. Most of us can probably recollect instances from childhood when we wondered why objects look smaller as we walk away from them or why their appearance changes with transitions in perspective. Later, we recast and comprehend these everyday experiences by modeling and measuring aspects of space so that measures of triangles, coupled with assumptions about light, explain the changes wrought by transitions in perspective (Gravemeijer, 1998). I recall puzzling about how the announcers of the Mercury rocket launches at Cape Canaveral knew the height of a rocket in flight and wondering how I might know the height my model rocket attained at its apogee, even without benefit of a sky tape measure. My speculations had their practical side as well. My grandfather was a carpenter who thought every young apprentice should know not only the measure on one side of his six-foot extending ruler (e.g., 1 1/2 feet) but also its counterpart on the opposite side (e.g., 4 1/2 feet). My recollections encompass measure's dual qualities of practical grasp and imaginative reach. On the one hand, to measure is to do. On the other hand, to measure is to imagine qualities of the world, such as length and time.

This synthetic character of measure is readily apparent in its history. For example, Eratosthenes estimated the circumference of the earth nearly 2,200 years ago by both imagining and doing. He imagined or assumed that the earth was spherical, that the rays of the sun could be considered to be parallel, and that the form of a circle approximated the surface of the earth. As a matter of practicality, he knew that in a well located at Syene, sunlight penetrated all the way to the bottom. He also knew that this absence of shadow meant that light followed a line like that depicted in Figure 12.1. By practical measure, he knew that Alexandria, located approximately due north of Syene, was 5,000 stades away. A stick at Alexandria cast a shadow with an angle measure of 1/50 of the arc of a circle (about 7 1/2 degrees). So this meant that the circumference of the earth must be 250,000 stades, or about 25,000 miles, not too far from the modern estimate. Eratosthenes also appreciated the potential error in his measure, adjusting it because Alexandria is not quite due north from Syene.

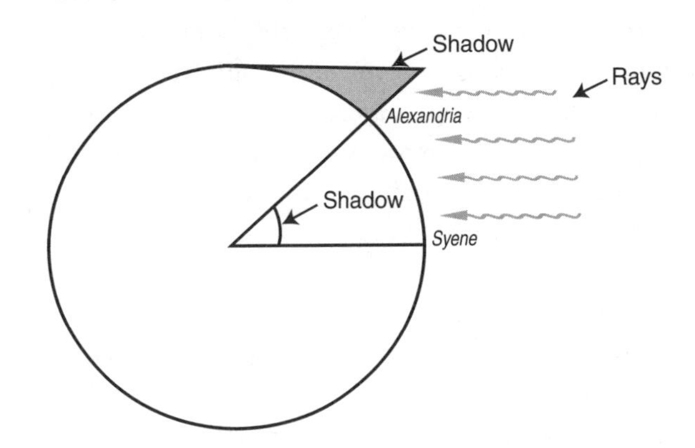

Figure 12.1. Eratosthenes approximated the circumference of the earth by using shadows.

The investigation conducted by Eratosthenes illustrates a deep connection between modeling space and exploring its extent. Measuring space requires the construction of a model to represent it and also tools that embody and extend the model. Measurement is inherently imprecise. Sources of imprecision include model-world mismatches, measurement-device qualities, and observer qualities (Kerr & Lester, 1976).

Ironically, because of Euclid's emphasis on straightedge and compass constructions, geometry is often treated separately from measurement. This approach has some virtues, especially when it leads to generalizations that go beyond the natural world. Yet measurement has traditionally served as one route for developing mathematical knowledge of space (Kline, 1959). A pedagogical implication of this history is that learning to measure qualities of space, such as length, area, volume, and angle, provides a practical means for developing fundamental understandings of its structure. For these reasons, substantial pedagogical value is gained in realigning measure and space. Moreover, conceptions involved in measuring space readily extend to other phenomena, such as mass or time, as well as to relational measures, such as rate. These extensions often prove crucial to the development of scientific reasoning (Lehrer & Schauble, 2000).

Because most of the research to date focuses on children's conceptions of spatial measure, the review reflects this research concentration. The first section of this chapter describes components of a mathematics of measure from the perspective of child development. This developmental perspective is influenced by Piaget's studies of important transitions in children's conceptions of spatial measure (Piaget, Inhelder, & Szeminska, 1960) and also by related research guided by other traditions (e.g., Davydov, 1975; Miller, 1984). In the sections that follow, I employ frameworks from this developmental literature to organize a summary of children's evolving understandings of length, area, volume, and angle measure. These progressions are based primarily on two kinds of evidence: clinical interviews conducted with children at various ages or grades and longitudinal study of transitions in children's reasoning during the course of traditional schooling.

More recent work examines the acquisition of measure concepts in classrooms that emphasize guided reinvention of the underpinnings of measure rather than simple procedural competence. This contemporary work suggests the need to revise previous accounts of development to include consideration of the mediational means (e.g., Wertsch, 1998), including forms of mathematical notation and argument that are employed by teachers and students during the course of instruction about measurement. For example, as I subsequently describe more fully, children's sense of length measure as a paced distance evolves when they shift from the plane of activity to representing their paces as a ruler that uses their footsteps as the units of measure. In the final section of this chapter, I summarize research that extends the investigation of conceptions of measure in other directions, including student thinking about the nature and sources of error of measurement.

Understanding Measure: A Developmental Perspective

Much of the research in the field draws on the seminal contributions of Piaget (Piaget & Inhelder, 1948/1956; Piaget et al., 1960), which continue to be a wellspring for contemporary research. Piaget's analysis suggested that conceptions of spatial measure were not unitary but instead consisted of a web of related constructs leading to the eventual construction and coordination of standard units. Piaget was careful to distinguish between activity, such as using a ruler, and reflective abstraction on activity, such as understanding the role played by the identical units in the ruler. In Piaget's view, understanding of measure entailed a successive mental restructuring of space, so that conceptions of measure increasingly encompassed subdivisions of space and translations of these subdivisions to comprise a measurement.

Piaget and his colleagues further suggested that conceptual change was tightly coupled with the overall development of reasoning. Accordingly, conservation (recognition of invariance under transformation, undergirded by such mental operations as reversibility, a form of mental undoing) of length, area, volume, and angle was a hallmark of, and constraint on, development in each domain of spatial measure studied by Piaget. Measurement was also considered to be tightly coupled with quantity, so that measurement was an outgrowth of counting. However, studies conducted in the past two decades generally fail to support this tight coupling of the development of understanding of spatial measure with quantity or even with general capacities for mental logic. For example, Hiebert (1981a, 1981b) conducted an instructional study in which first-grade children, some of whom conserved length and some of whom did not, were taught important underpinnings of length measure, such as iterating units (accumulating units by counting) to measure lengths. Hiebert found that acquiring ideas like iteration was generally unrelated to a child's status as a conserver. The sole component related to conservation was recognition of the inverse relationship between the length of a unit of measure and the resulting count (e.g., smaller units produce larger counts). In a like vein, Miller (1984) noted that even preschool children generally employed systematic procedures to ensure equal distributions of snacks (measured via lengths and areas) when solving problems requiring spontaneous measurement procedures. The general lack of relationship between conservation and understanding of measure is also characteristic of other domains of spatial measure (Carpenter, 1975).

Studies generally fail to support Piaget's claims that general forms of logic strongly constrain children's ideas about measure. These findings imply that little value is gained by delaying or withholding instruction until a child is mentally "ready" to learn about measurement. In contrast, it has proved useful to consider conceptual change about measure as change in a network or web of ideas related to unit. Some of the most prominent conceptual foundations include the following:

(1) Unit-attribute relations. Correspondence between units and the attribute being measured must be established. Although this relationship may seem transparent, especially in light of the ubiquity of tools like rulers and protractors, children in fact often misappropriate units of length measure for the measurement of other spatial extent, such as area, volume, and, perhaps most notoriously, angle. The suitability of a particular unit of measure usually involves a trade-off between models of the space being measured and the tools that are practical for the purposes at hand. For example, estimating the area of a playground might involve decisions about a model of it, such as a parallelogram, and the purposes for which the measure is intended. Should the measure be made to the nearest square inch? Square foot? In a similar vein, consider the distance between two cities. Can units of time (e.g., it's 2 1/2 hours between Madison and Chicago) be used instead of units of length to measure distance? If so, what assumptions need to be made?

(2) Iteration. Units can be reused; this understanding is based on subdivision (to establish congruent parts) and translation. For example, to iterate a unit of length, a child must come to understand length as a distance that can be subdivided. Moreover, these subdivisions can be accumulated and, if necessary, rearranged to measure a length. Hence, given a fixed length eight units long and a single unit, one measures the length by subdividing it into eight congruent partitions. This task is accomplished by translating the unit successively from the start point to the end point.

(3) Tiling. Units fill lines, planes, volumes, and angles. For example, to measure a length, one needs to arrange units in succession. Young children often find it useful to lay units in succession but sometimes are unaware of the consequences of leaving "cracks." If so, using counts of units as representations of length is problematic. Tiling (space-filling) is implied by subdivision of lengths, areas, volumes, and angles, but this implication is not transparent to all children.

(4) Identical units. If the units are identical, a count will represent the measure. Mixtures of units should be explicitly marked, for example, as "5 yards and 3 inches," not "8."

(5) Standardization. Conventions about units facilitate communication. Though arbitrary, standard units often have interesting histories. For example, Nickerson (1999) suggests that the relation of 12 inches to a foot likely arose from the confluence of several related developments. First, the ratio of the spans of thumb and foot is usually about 12, an important ratio for measurement centered on readily available parts of the body. Such a ratio would have had particular advantages before the advent of mass-produced tools. Second, many fractions of 12, like one half or one third, are integers, making subdivision relatively convenient. Third, 12 is the sum of 3, 4, and 5 and thus can be used to make perpendicular joints with beams: Mark 3 units along one beam from the end at which the joint will be made, mark 4 along the other, and then adjust the angle so that the 5-unit beam touches the ends of both lengths simultaneously.

(6) Proportionality. Measurements with different-sized units imply that different quantities can represent the same measure. These quantities will be inversely proportional to the size of the units. Consequently, a foot-long strip has a measure of 12 inches, or 24 half-inches.

(7) Additivity. Units of Euclidean space can be decomposed and recomposed, so that, for example, the total distance between two points is equivalent to the sum of the distances of any arbitrary set of segments that subdivide the line segment. For example, if B is any point on the segment AC, then $AB + BC = AC$. This recognition is implicit in studies of conservation, in which, for example, the length of an object is not affected by its translation to a new point. Similarly, the lengths of two paths may be different even if they begin and end at the same points on the plane, because the sum of the parts of one path exceeds the sum of the parts of another. The recognition of this component of spatial measure is often a significant intellectual milestone.

(8) Origin (zero-point). Measurement often involves the development of a scale. Although scale properties vary with different systems of measure, measures of Euclidean space conform to ratios, so that, for example, the distance between 0 and 10 is the same as that between 30 and 40. This conformity implies that any location on the scale can serve as the origin. Other common forms of measurement may not have these properties. For example, when judging how much one likes chocolate ice cream on a five-point scale, it is difficult to

know whether a judgment of 4 indicates twice as much affection as a judgment of 2.

Collective coordination among these eight components constitutes an informal theory of measure. Studies conducted in the past two decades suggest that children's developing sense of measurement is marked by gradual coordination and consolidation of these components. Research has been oriented toward tracking transitions across a profile and range of understandings rather than a unitary construct of measurement. For example, a child might be able to subdivide a line, yet fail to appreciate the function of identical subdivisions. Children's theories, of course, have a limited scope and precision when contrasted with those developed by mathematicians and scientists. Nevertheless, understanding these constituents of measure and their relations establishes a firm ground for future exploration of the mathematics of measure, and their acquisition also implies coming to understand the (Euclidean) structure of space.

In the following sections, I review developmental progressions of children's understandings of length, area, volume, and angle measure with an eye toward documenting important milestones in children's notions of the components of measure noted previously. The review is intended to be representative, not exhaustive. Two kinds of studies are included: cognitive development studies and classroom studies. Studies of cognitive development typically engage groups of children in activities designed to reveal how they think or understand an issue. Children at different ages (cross-sectional) are compared, or the same children are followed for a period of time, (longitudinal) to observe transitions in thinking. These studies provide glimpses of children's thinking under conditions of activity and learning that are typically found in the culture. In contrast, classroom studies modify instruction and then investigate the effects of these modifications on children's thinking or understanding. The resulting portraits of student reasoning are not always in close agreement with those obtained from studies of cognitive development. One reason may be that the cultural experiences of measure are less frequent and perhaps less thought-provoking than those deliberately created in the design of instruction.

Length Measure

Studies of Cognitive Development

Length measure builds on preschoolers' understanding that lengths span distances (e.g., Miller & Baillargeon, 1990). Measure of distances requires restructuring space so that one "sees" counts of units as representing an iteration of successive distances. Iteration refers to accumulating units of measure to obtain a quantity, such as 12 inches. It rests on a foundation of subdividing length and ordering the subdivisions (Piaget et al., 1960). Thus, a count of *n* units represents a distance of *n* units. Studies of children's development suggest that acquisition of this understanding involves the coordination of multiple constructs, especially those of unit and zero-point. As noted previously, the construction of unit involves a web of foundational ideas including procedures of iteration, recognition of the need for identical units, understanding of the inverse relationship between magnitude of each unit and the resulting length measure, and understanding of partitions of unit. Understanding zero-point involves the mental coordination of the origin and endpoint of the scale used to measure length, so that the length from the 10 cm mark on the scale to the 20 cm mark is considered equivalent to that between 2 cm and 12 cm. Developmental studies indicate that these constructs are not acquired in an all-or-none manner, nor are they necessarily tightly linked. Most studies suggest that these understandings of units of length are acquired over the course of the elementary grades, although significant variations in developmental trajectories occur when different forms of instruction are employed.

Children's first understandings of length measure often involve direct comparison of objects (Lindquist, 1989; Piaget et al., 1960). Congruent objects have equal lengths, and congruency is readily tested when objects can be superimposed or juxtaposed. Yet young children (first grade) also typically understand that they can compare the length of two objects by representing the objects with a string or paper strip (Hiebert, 1981a, 1981b). This use of representational means likely draws on experiences of objects "standing for" others in early childhood, such as in pretend play in which a banana can represent a telephone yet retain its identity as a fruit (Leslie, 1987) or in the ability to distinguish, yet coordinate, models or pictures and their referents (deLoache, 1989). Moreover, first graders can use given units to find the length of different objects, and they associate higher counts with longer objects (Hiebert, 1981a, 1981b, 1984). Most young children (first and second graders) even understand that counts of smaller units will be larger than counts of larger units, given the same length (Carpenter & Lewis, 1976; Lehrer, Jenkins, & Osana, 1998b).

These understandings of the practical use of units are probably grounded in childhood experiences in which children observe others use rulers and related measurement devices and incorporate the resulting lessons learned into their play. However, facility with counting need not imply understanding of length measure as a metric distance. Because early measure understanding emerges as a collection of developing concepts, children may understand qualities of measure, such as the inverse relation between counts and size of units, yet not fully appreciate other constituents of length measure, such as the function of identical units or the operation of iteration of unit (Lehrer et al., 1998b). These concepts are much more problematic for primary-grade children (i.e., ages 6 to 8). Children often have difficulty creating units of equal size (Miller, 1984), and even when provided equal units, first and second graders often do not understand their purposes, so they freely mix, for example, inches and centimeters, counting all to "measure" a length (Lehrer et al., 1998b). For these students, measure is not significantly differentiated from counting (Hatano & Ito, 1965). For example, younger students in the study by Lehrer , Jenkins, and Osana (1998b) often imposed their thumbs, pencil erasers, or other invented units on a length, counting each but failing to attend to inconsistencies among these invented units (and often mixing their inventions with other units). Even given identical units, significant minorities of young children fail to spontaneously iterate units of measure when they "run out" of units, despite demonstrating procedural competence with rulers (Hatano & Ito, 1965; Lehrer et al., 1998b). For example, given 8 units and a 12-unit length, some of these children lay the 8 units end to end and then decide that they cannot proceed further. They cannot conceive of how one could reuse any of the 8 units, perhaps because they have not mentally subdivided the remaining space into unit partitions. Moreover, young children (e.g., first grade or kindergarten) may coordinate some of the components of iteration, such as use of units of constant size and repeated application, yet not others, such as tiling. For example, first and second graders may leave "spaces" between identical units even as they repeatedly use a single unit to "measure" a length (Horvath & Lehrer, 2000; Koehler & Lehrer, 1999).

Children's understanding of zero-point is particularly tenuous. Only a minority of young children understand that any point on a scale can serve as the starting point, and even a significant minority of older children (e.g., fifth grade) respond to nonzero origins by simply reading off whatever number on a ruler aligns with the end of the object (Lehrer et al., 1998a). Many children throughout schooling begin measuring with one rather than zero (Ellis, Siegler, & Van Voorhis, 2001). Parts of units create additional complexities. For example, Lehrer, Jacobson, Kemeny, & Strom (1999) noted that some second-grade children (7 to 8 years old) measured a 2 and 1/2-units strip of paper by counting "1, 2, [pause], 3 [pause], 3 and a half." They explained that the 3 referred to the third unit counted, but "there's only a half," so in effect the last unit was represented twice, first as a count of unit and then as a partition of a unit. Yet these same children could readily coordinate different starting and ending points for integers (e.g., starting at 3 and ending at 7 yielded the same measure as starting at 1 and ending at 5).

Classroom Studies

Recent work has focused on establishing developmental trajectories for understanding of linear measure in classrooms that promote representation and communication. These studies suggest that important gains are realized in understanding when children's learning is mediated by systems of inscription (e.g., what children write) and notation (Greeno & Hall, 1997). For example, Clements, Battista, and Sarama (1998) reported that using computer tools that mediated children's experience of unit and iteration helped children mentally restructure lengths into units. Other recent studies place a premium on making transitions from embodied forms of length measure, such as pacing, to inscribing and symbolizing these forms as *foot strips* and other kinds of measurement tools (Lehrer et al., 1999; McClain, Cobb, Gravemeijer, & Estes, 1999). Inscriptions like foot strips help children reason about the mathematically important components of activity (e.g., the lengths spanned while pacing) so that paces are transformed into units of measure. By constructing tools, children have the opportunity to discover the measurement principles that guide the design of these tools. Although constructing and using tools have a long tradition in teaching practice, recent studies of mediated activity provide important details about how these practices contribute to conceptual change (e.g., Wertsch, 1998). Generally, asking children to represent their experiences tends to help them select and make visible mathematically important components of activity. For example, when pacing, mathematically fruitful components include "lifting out" paces as units that can be iterated to obtain a length measure and deciding what is meant by "walking straight." Other elements of the activity, such as maintaining one's balance while pacing, are placed in the (mathematical) background.

Further work in classrooms suggests the importance of providing opportunities for children to repeatedly "split" (Confrey, 1995; Confrey & Smith, 1995), or partition, lengths to come to understand unit partitions. For example, in second-grade classrooms where students had the opportunity to design rulers, students were motivated by their previous experience with rulers to add "marks" that would help them measure lengths that were parts of units, for instance, 3 1/2. To develop these unit partitions, children folded their unit (represented as a length of paper strip) in half, then repeated this process to create fourths, eighths, and even sixty-fourths (Lehrer et al., 1999). In addition to developing procedures for partitioning, children were able to examine the resulting partitions (inscribed as fold lines) by unfolding the resulting strips so that they could design their own ruler marks. Eventually, these actions helped children develop their first understandings of operator conceptions of fractions (e.g., half of half of half, etc.).

Classrooms are excellent forums for understanding the importance of conventional units. For example, if children measure the length of objects with different units, such questions as "Which object is longest?" may direct attention to the communicative functions of standard units. Classroom studies also point to creative ways of melding measure and the study of form in the elementary grades in ways that recall their historical co-development. For example, children in Elizabeth Penner's first- and second-grade class searched for forms (e.g., lines, triangles, squares) that would model the configuration of players in a fair game of tag (Penner & Lehrer, 2000). Attempts to model the shape of fairness initiated cycles of exploration involving length measure and properties related to length in each form (e.g., distance from the corners of a square to the center). Eventually, children decided that circles were the fairest of all forms because the locus of points defining a circle was equidistant from its center. This insight was achieved by developing understanding of linear measure and by employing this understanding to explore the properties of shape and form. Such tight coupling between space and measure is reminiscent of Piaget's investigations but helps one to see these linkages as objects of instructional design rather than as pre-existing qualities of mind.

Area Measure

Cognitive Developmental Studies

In many ways, studies of children's conceptions of area measure parallel those discussed for length. I focus here on a longitudinal investigation of 37 children in grades 1 through 3 who were followed for 3 years, because the results of this study are representative of much of the literature (Lehrer et al., 1998b). Young children (e.g., in first and second grades) often treat length measure as a surrogate for area measure. For example, Lehrer et al. (1998b) found that some young children measured the area of a square by measuring the length of a side, then moving the ruler over a bit and measuring the length between the sides again, and so on, treating length as a space-filling attribute. When provided manipulatives (i.e., squares, right triangles, circles, and rectangles) for use in finding the area measure of a variety of forms, most children in grades 1 through 3 freely mixed units and then reported the total count of the units. The two most commonly observed strategies with use of manipulatives were *boundedness* and *resemblance*. That is, children deployed units in ways that would not violate the boundaries of closed figures, and they often used units that resembled the figure being measured (e.g., triangles for triangles). Young children were also liable to ignore the space-filling properties of units, preferring instead to honor the boundaries of the forms, so when presented with a choice between "leaving cracks" and overlapping a boundary, they invariably chose the former. Figure 12.2 displays two second-grade students' approaches to measuring the "space covered" by their hands (Lehrer et al., 1998a). The solution labeled "a" uses beans as units of measure, and the one labeled "b" uses spaghetti. Both units were chosen because they "looked like" (resembled) the contour of the hand. Both solutions also ignore the "cracks" (space-filling). A third solution, proposed by their teacher and labeled "c," consisted of an overlay of square units of measure. The class initially rejected this solution, both because the squares "went over" the outline of the hand (boundedness) and because the squares "looked wrong" (did not resemble the contours of the hand).

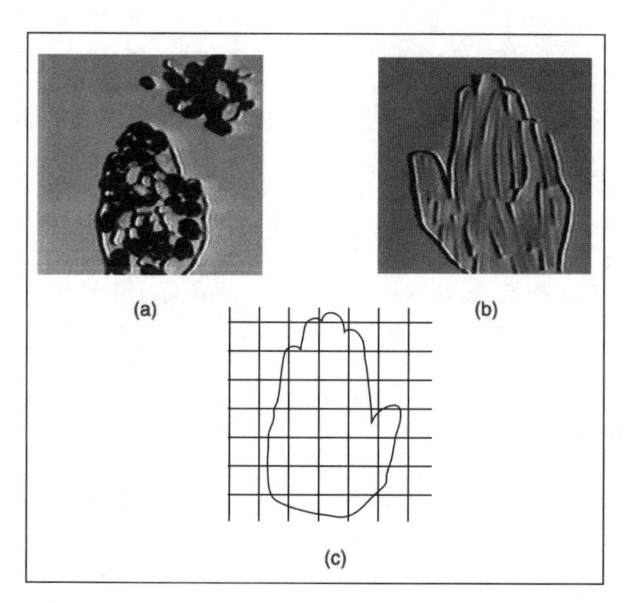

Figure 12.2. Students proposed measures using (a) beans, (b) spaghetti, and (c) an overlay of square units to measure the space covered by their hands.

Over the elementary grades, area measure becomes differentiated from length measure, and the space-filling (tiling) property of the unit becomes more apparent to most children (Lehrer et al., 1998a). However, other aspects of area measure remain problematic, even though students can recall standard formulas for finding the areas of squares and rectangles. Fewer than 20% of the students in the study by Lehrer et al. (1998b) believed that area measure required identical units, and fewer than half could reconfigure a series of planar figures so that known area measures could be used to find the measures of the areas of unknown figures. Just as linear measure requires restructuring a length into a succession of distances, area measure requires restructuring of the plane. Consequently, students found it very difficult to decompose and then recompose the areas of forms to see one form as a composition of others.

Similarly, Battista, Clements, Arnoff, Battista, and Can Auken Borrow (1998) reported that students in the primary grades often cannot structure a rectangle as an array of units. In a more extensive exploration of children's strategies for structuring rectangular arrays across grades 1 through 4, Outhred and Mitchelmore (2000) found a wide range of conceptions of array. Students were asked, for example, to find the number of 1-cm squares needed to cover a 6-cm by 5-cm rectangle. Many first and second graders either incompletely covered the rectangle or drew coverings that varied the size of the square unit. Third graders were more successful in using concrete units to cover the rectangle, but most did not use the structure of the rows of the array to accomplish this task, relying instead on perception. Only in the fourth grade did the majority of students use the dimensional structure of the array to measure it. These students used one dimension to find the number of units in each row and used the other to find the number of rows. Nevertheless, a significant minority (about 30%) of these older students simply tried to count squares using the more primitive strategies. These findings are especially troublesome in light of the widespread use of area models of fractions and the use of array models for multiplication, which apparently assume knowledge that may not be in place.

In sum, conceptual development in area measure lagged behind that of length measure. Understanding core conceptual notions, such as identical units and tiling, was typical of students by the end of the elementary grades for length measure but not for area measure. Younger children often employed resemblance as the prime criterion for selecting a unit of area measure, suggesting the need for attention to the qualities of unit that make it suitable for area measure. Other studies focus on students' conceptions of the area measure of rectangles. Most often, rectangular area is treated in schooling as a simple matter of multiplying lengths, but the research suggests that many students in the elementary grades do not "see" this product as a measurement. Many fail to structure even a simple form like a rectangle as an array that could conceivably be measured with unit squares. Current practices of giving students squares as units may lead to apparent procedural competence but fail to challenge students' preconceptions about what makes a unit suitable. Moreover, many students understand square units as things to be counted rather than as subdivisions of the plane.

Classroom Studies

As with length measure, studies of developmental trajectories of area measure in classrooms that emphasize representation and communication reveal significant departures from patterns typically described in the literature. For example, Lehrer et al. (1998a) found that an effective way for second-grade students to begin working on ideas about area was to solve problems involving partitioning and reallotment of areas without measuring. In the course of this partitioning and rearranging, students came to regard one of the partitions as a unit so that counts of this unit afforded ready comparison among areas. Later, children explored the suitability of different units (e.g., beans) for finding the areas of irregular forms, such as handprints, and found that units like squares had the desirable properties of space-filling and identity. By the end of the school year, these children had little difficulty creating two-dimensional arrays of units for rectangles. For example, one student's solution to the question of whether the areas of 5×8 and 4×10 rectangles (with unlabeled dimensions) were the same or different is displayed in Figure 12.3. This student first used length measure to partition each rectangle into unit squares and then demonstrated that skip-counting by columns or by rows of either resulted in the same count (40). He also concluded that if he rotated the rectangle, he could readily "see" commutativity because "$5 \times 8 = 8 \times 5$."

These second graders often spontaneously imposed these kinds of arrays on nonpolygonal forms to find approximate solutions to area. For example, one student's solution to finding the space covered by her hand is displayed in Figure 12.4. She used color regions to represent approximate partitioning of units, such as one fourth or one third. She collected these parts of units and added them to the units completely enclosed by the figure

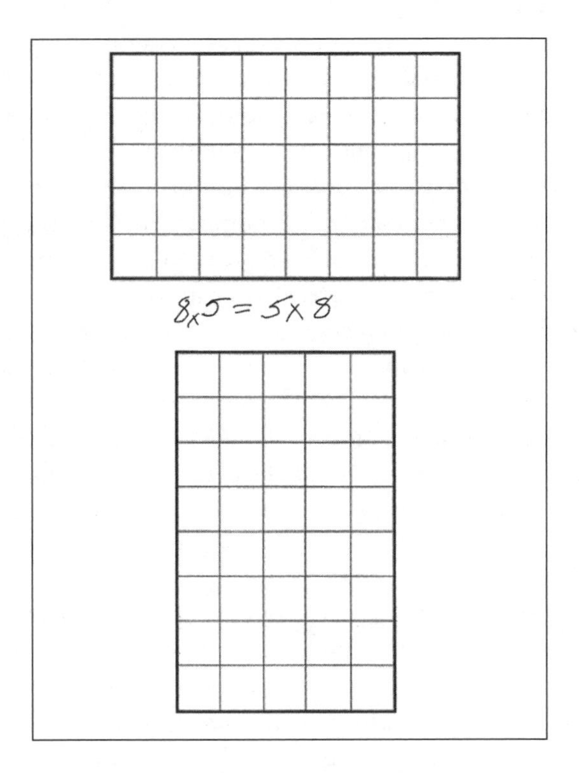

Figure 12.3. A student's visual demonstration of commutativity of multiplication.

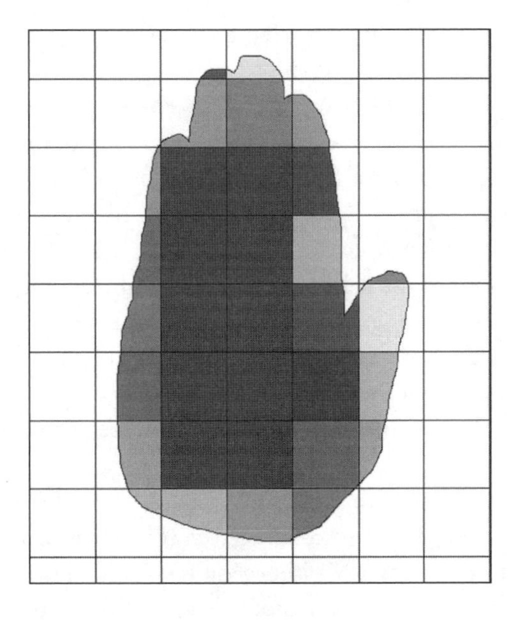

Figure 12.4. A second grader represents parts of units with different colors and combines like-colored regions to approximate whole units.

to arrive at an estimate of the area of her hand. These findings suggest that mental restructuring of units of arrays is assisted by classroom emphasis on representation and argument. The developmental patterns noted so often in the literature probably reflect the shortcomings of typical experiences, including instruction.

Volume Measure

The measure of volume presents some additional complexities for reasoning about the structure of space, primarily because units of measure must be defined and coordinated in three dimensions. The research conducted often blends classroom study with description of individual change, so this section and the one that follows on angle measure reflect this synthesis.

An emerging body of work addresses the strategies that students employ to structure a volume, given a unit. For example, Battista and Clements (1998) noted a range of strategies employed by students in the third and fifth grades to mentally structure a three-dimensional array of cubes. Many students, especially the younger ones, could count only the faces of the cubes, resulting in frequent instances of multiple counts of a single cubic unit and a failure to count any cubic units in the interior of the cube. The majority of fifth-grade students, but only about 20% of third-grade students, structured the array as a series of layers. Layering enabled students to count the number of units in one layer and then multiply or skip-count to obtain the total number of cubic units in the cube. These findings suggest that, as with area and length, students' models of spatial structure influence their conceptions of its measure.

Classroom studies again suggest that forms of representation heavily influence how students conceive of structuring volume. For example, third-grade students with a wide range of experiences and representations of volume measure structured space as three-dimensional arrays. Unlike the younger students described in the Battista and Clements (1998) study, all could structure cubes as three-dimensional arrays. Most even came to conceive of volume as a product of area and height (Lehrer, Strom, & Confrey, 2002). For example, one third grader's solution for finding the volume of a cylinder is displayed in Figure 12.5. The solution draws on the method described in Figure 12.4 of finding parts to compose whole units but refines the method to describe part-whole relations as fractional pieces, such as one fourth. Fractional pieces were then composed to estimate area units, for example, $1/4 + 1/2 + 1/4 = 1$. After estimating the area of the circle in this manner, this student proposed finding volume by multiplying the estimated area by the height of the cylinder "to draw it [the area of the base] through how tall it is."

Battista (1999) followed the activity of three pairs of fifth-grade students as they predicted the number of cubes that fit in graphically depicted boxes. He found that student learning was affected both by individual activity and by socially constituted practices like collective reflection. Thus, traditional notions about trajectories of development may need to be revised in light of more careful attention to classroom talk and related means of representing volume. Although the solution of the third-grade student depicted in Figure 12.5 is unusual in the literature, it was commonplace in a third-grade classroom where students had prolonged opportunities to explore the mathematics of space and measure.

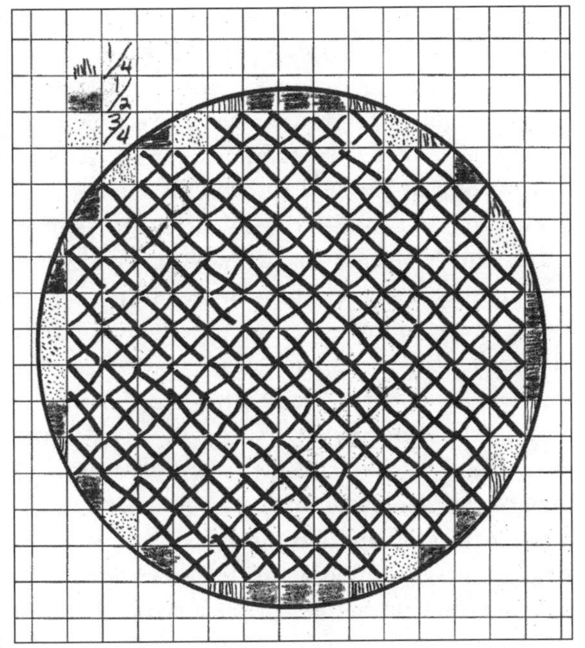

Figure 12.5. A third grader first uses colors to explicitly mark fractional pieces of units to approximate the area of the base of a cylinder and then finds volume as the product of area and height.

Angle Measure

Freudenthal (1973) suggested that multiple mathematical conceptions of angle should be entertained during the course of schooling. Henderson (1996) suggests three conceptions: (a) angle as movement, as in rotation or sweep; (b) angle as a geometric shape, a delineation of space by two intersecting lines; and (c) angle as a measure, a perspective that coordinates the other two. Mitchelmore (1997, 1998) and Lehrer et al. (1998b) found that students in the elementary grades develop separate mental models of angle as movement and angle as shape. In Mitchelmore's (1998) study, students in Grades 2, 4, and 6 increasingly perceived how different types of turning situations might be alike (e.g., those involving unlimited turning, like a fan, and those involving limited turning, like a hinge), but they rarely related these to situations involving "bends" or other aspects of intersecting lines. Lehrer et. al (1998b) asked children to find ways of measuring the bending in a hinge (with a sweep demonstrated from one position to another) and the bending in a bent pipe cleaner. Like the students in Mitchelmore's studies, students in Grades 1 through 5 rarely saw a relationship between these situations, but their measurement actions were very similar. Children most often chose to measure the distance between the jaws of the hinge and the ends of the pipe cleaners. In these static contexts ("bends"), students typically thought that angle measures were influenced by the lengths of the intersecting lines or by their orientation in space. The latter conception decreased with age, but the former was robust at every age (Lehrer et al., 1998b).

As noted for length, area, and volume measure, the tools employed or invented by students significantly affected their developing conceptions of angle measure. Studies of student learning with the Logo computer program generally confirm the existence of distinct models of angle as static or dynamic, respectively. Logo's Turtle geometry affords the notion of angle as a rotation, although students often confuse the interior and exterior (turtle) angles of figures traced by the Turtle. Nevertheless, with well-crafted instruction, tools like Logo mediate the development of angle measures as rotations (Lehrer, Randle, & Sancilio, 1989). However, students rarely bridge these rotations to models of the space in the interior of figures traced by the Turtle (e.g., Clements, Battista, Sarama, & Swaminathan, 1996). Simple modifications to Logo help students perceive the relationship between turns and traces (the path made by Logo's Turtle), and in these conditions students can use turns to measure static intersections of lines (Lehrer et al., 1989).

It remains a major challenge to design pedagogy to help students develop understanding of angle and its measure. Unlike the spatial structuring of linear dimensions (length, area, and volume), developing understanding of angle requires novel forms of representation that are perhaps not as prevalent in the culture (e.g., developing notions of turn, tracing a locus of a turning movement, relating turning movements to traces in environments like Logo). In addition, understanding angle involves the coordination of several potential models and integration of these models in a theory of their measure (Mitchelmore & White, 1998). Common admonitions to teach angles as turns usually fail because students rarely spontaneously relate situations involving rotations to those involving shape and form. As I have stated often in this tour of measure, the form of mediation (e.g., the tools, what students write, the models they explore) matters as much as the problems posed.

Expanding the Scope of Measure

Most research has been inspired by the Piagetian tradition of tight coupling between study of measure and study of space, and research findings indeed indicate that coming to understand measure is a productive means for learning about the structure of space. Nevertheless, the scope of measure often must extend beyond spatial extent, especially to support scientific reasoning. Although research here is sparser, it suggests that children's conceptions of measure can be extended to nonspatial realms.

Understanding the Natural World

Measurement is essential for developing an understanding of the natural world (Crosby, 1997). By quantifying and otherwise *mathematizing* nature (Kline, 1980), students can model the natural world, even at an early age. Although studies in science education often refer to the importance of measure, they generally do not "unpack" its conceptual foundations, so measurement is often viewed more as a matter of procedure than as a matter of conception. Yet a small number of studies suggest that developing concepts of measure can support better understanding of natural phenomena. In short, when children understand measure, its application to natural phenomena yields enhanced understanding in a manner reminiscent of the tight linkages between spatial extent and spatial structure noted previously. For example, third- and fifth-grade students who had the opportunity to develop understanding of spatial measure (e.g., length, area, volume) readily extended these understandings to the measure of mass. That is, with appropriate reminders from their teachers, they decided that units of mass should be identical, conventional, iterable, and so on. These understandings proved crucial for subsequent modeling material kind as a ratio of mass and volume—that is, density (Lehrer, Schauble, Strom, & Pligge, 2001). Similarly, third-grade children who had histories of learning about measure readily extended their ideas about unit to encompass rate, a ratio measure, to support reasoning about the growth (e.g., change in height per day) of plants (Lehrer, Schauble, Carpenter, & Penner, 2000). The challenges of ratio measures like these veer into more general considerations of children's understanding of rational number and multiplicative structure (e.g., Harel, Behr, Post, & Lesh, 1992). Unfortunately, little research addresses these potential relationships between measure and multiplicative structure.

Precision and Error

Much of the research about measurement explores precision and error of measure in relation to mental estimation (Hildreth, 1983; Joram, Subrahmanyam, & Gelman, 1998). To estimate a length, students at all ages typically employ the strategy of mentally iterating standard units (e.g., imagining lining up a ruler with an object). In their review of a number of instructional studies, Joram et al. (1998) suggest that students often develop brittle strategies closely tied to the original context of estimation. They suggest that instruction focus on children's development of reference points (e.g., landmarks) and on helping children establish reference points and units along a mental number line. Mental estimation would also likely be improved with more attention to the nature of unit, as suggested by many of the classroom studies reviewed previously.

Although mental estimation is one potential source of imprecision in measure, error is a fundamental quality of measure, a recognition that historically was quite troubling to scientists (Porter, 1986). Acts of measuring yield a range of estimates and, to the extent that errors are random, often a Gaussian ("normal") distribution. Hence, understanding of error is tied to conceptions of distribution. Conceptions of error are also central to scientific experimentation (Mayo, 1996). Schauble (1996), for example, found that fifth- and sixth-grade students who conducted experiments often confounded error and the variation due to a small, but reliable, effect of a variable. Perhaps they would not do so if they had opportunities to consider likely sources of error. Kerr and Lester (1976) suggest that instruction in measure should routinely encompass considerations of sources of error, especially (a) the assumptions (e.g., the model) about the object to be measured, (b) the choice of measuring instrument, and (c) the way the instrument is used (e.g., method variation). Variation among individuals is also commonly considered, especially in work in the social sciences. Recent work in classrooms explores children's ideas about some of these sources of error.

Varelas (1997) examined how third- and fourth-grade students made sense of the variability of repeated trials. Many children apparently did not conceptualize the differences among repeated observations as error, and children often suggested that fewer trials might be preferable to more. These conceptions seemed bound with relatively diffuse conceptions of representative values of a set of repeated trials. In a related study, Lehrer et al. (2000) found that with explicit attention to ways of ordering and structuring trial-to-trial variability, second-grade children made sense of trial-to-trial variation by suggesting representative (typical) values of sets of trials. Choices of typical values included "middle numbers" (i.e., medians) and modes, with a distinct preference for the latter.

In follow-up work with fourth-grade students, Petrosino, Lehrer, & Schauble (in press) further investigated children's ideas about sources and representations of measurement error. In one portion of the Petrosino et al. (in press) classroom teaching experiment, fourth-grade students measured the length of a pencil and the height of a flagpole. They represented each distribution of measurements, an accomplishment that involved ordering and putting like values in "bins." Students noticed differences in the comparative spread of each distribution (see Figure 12.6, in which the right panel depicts the pencil measures and the left panel, the flagpole measures) and readily attributed these differences to the relative precision of measure of the instruments available to conduct each measure (rulers vs. "height-o-meters"). The distribution obtained in each case was attributed primarily to individual differences in use of each instrument.

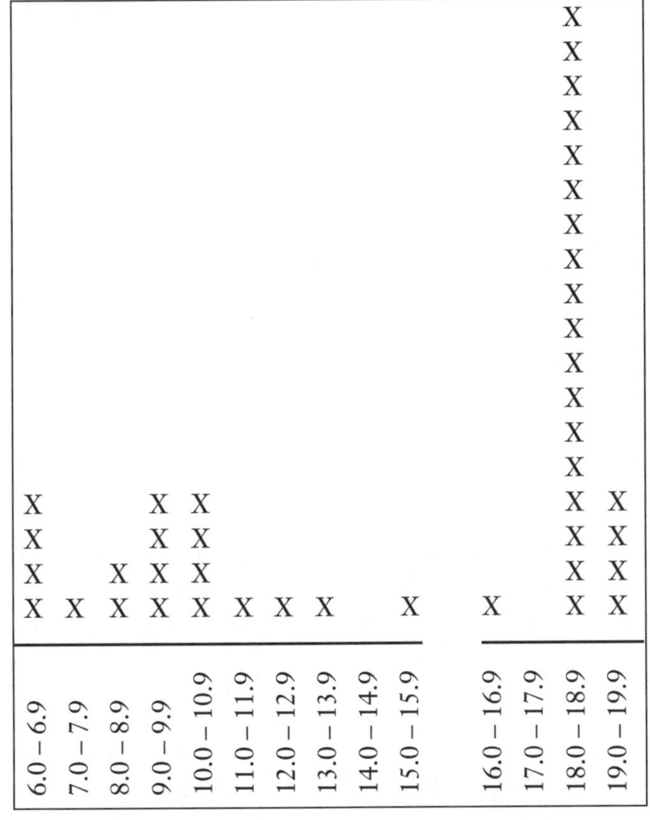

Figure 12.6. Distributions of fourth-grade students' measurements of the height of a flagpole (in m, left) and the length of a pencil (in cm, right).

When pressed to describe the differences in the spread that they perceived, students, with the assistance of their teacher, developed the notion of a "spread number" to quantify variation. One of the procedures invented by the class consisted of finding the median of the differences between observed measures and the typical measure (in this class, a median). Armed with these understandings, students went on to investigate a variety of sources of error and their consequences for a distribution of values. For example, Figure 12.7 displays the distributions of differences of observed values from the median for flagpoles measured with two different instruments (the left panel depicts errors with a handmade instrument, and the right panel depicts errors with a machined instrument). These distributions sparked examination of the comparative reliability of each instrument. This research suggests how relationships between variation and sources of error might be explored and elaborated in later grades. Such understandings are important foundations to the conduct of experiment and related forms of scientific explanation (Lehrer, Schauble, & Petrosino, 2001).

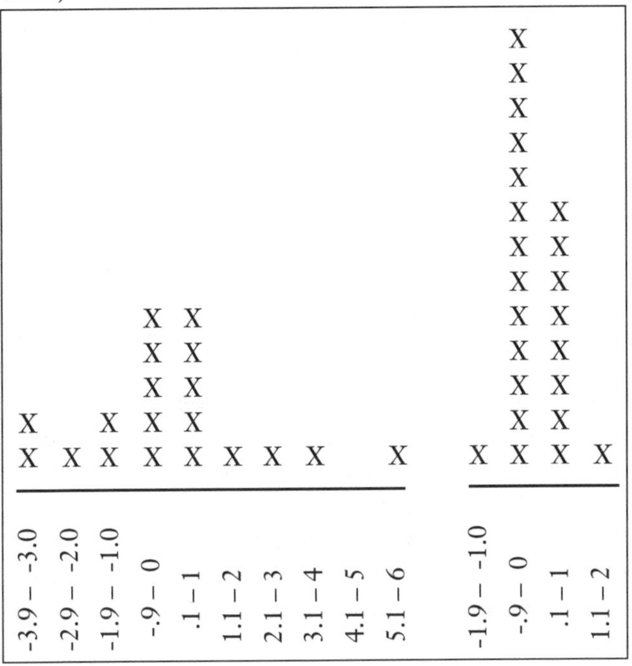

Figure 12.7. Distributions of differences of observed values from the median value of the height of a flagpole for measurements made with homemade (left) and machined (right) instruments.

Concluding Comments

Children are tacit measurers of nearly everything. Early and repeated experiences with cultural artifacts like rulers, and with the general epistemology of quantification that is characteristic of many contemporary societies, provides a fertile ground for developing mathematical understanding of measure. Developmental research suggests that children's conceptions of measure reflect a collection of emerging concepts whose coordination gradually unfolds as a network of relations, not a unitary concept of measure. Understanding some of these relationships spans the course of schooling. Classroom research points to the importance of helping children go beyond procedural competence to learn about the mathematical underpinnings of measure so that procedures and concepts are mutually bootstrapped. Measuring always involves doing, and such activity is always conducted in light of some model of the attribute being measured. Consequently, measuring a length, area, volume, or angle affords opportunities for developing ideas about the structure of space, such as its dimension, array, and curvature. Understanding structures like these, in turn, provides a platform for increased comprehension of number, as exemplified in spatial models like the number line or area models of rational number.

No clear-cut "best" sequence of instruction seems to exist in any domain of measure nor any reasonable list of prescriptions or proscriptions other than the need to avoid exclusive reliance on the development of procedural competence. As in other domains of mathematics, procedural competence (e.g., measuring with a ruler) can bootstrap conceptual development (e.g., inferring more general principles on the basis of the design of the ruler). Developing knowledge of effective procedures is a form of conceptual development, and developing concepts is aided and abetted by constructing and reflecting on ways to measure. Classroom studies emphasize the importance of helping children understand the rationale of familiar tools, such as rulers, and of finding productive ways for engaging children in the guided reinvention of such crucial concepts of measure as unit-attribute relations (e.g., length is not the measure of all things) and the very idea of unit. Teachers who understand the growth and development of student reasoning about measure are our collective "best bet" for generating productive learning about measurement.

ACKNOWLEDGMENT

Mary Linquist provided comments on an earlier draft of this chapter.

REFERENCES

Battista, M. T. (1999). Fifth graders' enumeration of cubes in 3D arrays: Conceptual progress in a inquiry classroom. *Journal for Research in Mathematics Education, 30,* 417–448.

Battista, M. T., & Clements, D. H. (1998). Students' understanding of three-dimensional cube arrays: Findings from a research and curriculum development project. In R. Lehrer & D. Chazan (Eds.), *Designing learning environments for developing understanding of geometry and space* (pp. 227–248). Mahwah, NJ: Erlbaum.

Battista, M. T., Clements, D. H., Arnoff, J., Battista, K., & Can Auken Borrow, C. (1998). Students' spatial structuring of 2D arrays of squares. *Journal for Research in Mathematics Education, 29,* 503–532.

Carpenter, T. P. (1975). Measurement concepts of first- and second-grade students. *Journal for Research in Mathematics Education, 6,* 3–13.

Carpenter, T. P., & Lewis, R. (1976). The development of the concept of a standard unit of measure in young children. *Journal for Research in Mathematics Education,* 7, 53–64.

Clements, D. H., Battista, M. T., & Sarama, J. (1998). Development of geometric and measurement ideas. In R. Lehrer & D. Chazan (Eds.), *Designing learning environments for developing understanding of geometry and space* (pp. 201–225). Mahwah, NJ: Erlbaum.

Clements, D. H., Battista, M. T., Sarama, J., & Swaminathan, S. (1996). Development of turn and turn measurement concepts in a computer-based instructional unit. *Educational Studies in Mathematics, 30,* 313–337.

Confrey, J. (1995). *Student voice in examining "splitting" as an approach to ratio, proportions, and fractions.* Paper presented at the International Conference for the Psychology of Mathematics Education, Universidade Federal de Permanbuco, Recife, Brazil.

Confrey, J., & Smith, E. (1995). Splitting, covariation and their role in the development of exponential functions. *Journal for Research in Mathematics Education, 26,* 66–86.

Crosby, A. W. (1997). *The measure of reality.* Cambridge: Cambridge University Press.

Davydov, V. V. (1975). The psychological characteristics of the "prenumerical" period of mathematics instruction. In L. P. Steffe (Ed.), *Soviet studies in the psychology of learning mathematics* (Vol. 7, pp. 109–205). Chicago: University of Chicago Press.

DeLoache, J. S. (1989). The development of representation in young children. In H. W. Resse (Ed.), *Advances in child development and behavior* (Vol. 22, pp. 1–39). New York: Academic Press.

Ellis, S., Siegler, R. S., & Van Voorhis, F. E. (2001). *Developmental changes in children's understanding of measurement procedures and principles.* Unpublished paper.

Freudenthal, H. (1973). *Mathematics as an educational task.* Dordrecht, The Netherlands: Reidel.

Gravemeijer, K. P. (1998). From a different perspective: Building on students' informal knowledge. In R. Lehrer & D. Chazan (Eds.), *Designing learning environments for developing understanding of geometry and space* (pp. 45–66). Mahwah, NJ: Erlbaum.

Greeno, J. G., & Hall, R. (1997). Practicing representation: Learning with and about representational forms. *Phi Delta Kappan, 78,* 1–24.

Harel, G., Behr, M., Post, T., & Lesh, R. (1992). The blocks task: Comparative analysis of the task with other proportion tasks and qualitative reasoning skills of seventh-grade children in solving the task. *Cognition and Instruction, 9,* 45–96.

Hatano, G., & Ito, Y. (1965). Development of length measuring behavior. *Japanese Journal of Psychology, 36,* 184–196.

Henderson, D. W. (1996). *Experiencing geometry on plane and sphere.* Upper Saddle River, NJ: Prentice Hall.

Hiebert, J. (1981a). Cognitive development and learning linear measurement. *Journal for Research in Mathematics Education, 12,* 197–211.

Hiebert, J. (1981b). Units of measure: Results and implications from National Assessment. *Arithmetic Teacher, 28,* 38–43.

Hiebert, J. (1984). Why do some children have trouble learning measurement concepts? *Arithmetic Teacher, 31,* 19–24.

Hildreth, D. J. (1983). The use of strategies in estimating measurements. *Arithmetic Teacher, 30,* 50–54.

Horvath, J., & Lehrer, R. (2000). The design of a case-based hypermedia teaching tool. *International Journal of Computers for Mathematical Learning, 5,* 115-141.

Joram, E., Subrahmanyam, K., & Gelman, R. (1998). Measurement estimation: Learning to map the route from number to quantity and back. *Review of Educational Research, 68,* 413–449.

Kerr, D. R., & Lester, F. K. (1976). An error analysis model for measurement. In D. Nelson & R. E. Reys (Eds.), *Measurement in school mathematics* (pp. 105–122). Reston, VA: National Council of Teachers of Mathematics.

Kline, M. (1959). *Mathematics and the physical world.* New York: Thomas Y. Crowell.

Kline, M. (1980). *Mathematics: The loss of certainty.* Oxford: Oxford University Press.

Koehler, M., & Lehrer, R. (1999). Understanding children's reasoning about measurement: A case-based hypermedia tool for professional development (Technical Report and functioning software). Madison: Wisconsin Center for Education Research.

Lehrer, R., Jacobson, C., Kemeny, V., & Strom, D. (1999). Building on children's intuitions to develop mathematical understanding of space. In E. Fennema & T. Romberg (Eds.), *Mathematics classrooms that promote understanding* (pp. 63–87). Mahwah, NJ: Erlbaum.

Lehrer, R., Jacobson, C., Thoyre, G., Kemeny, V., Strom, D., Horvath, J., Gance, S., & Koehler, M. (1998a). Developing understanding of space and geometry in the primary grades. In R. Lehrer & D. Chazan (Eds.), *Designing learning environments for developing understanding of geometry and space* (pp. 169–200). Mahwah, NJ: Erlbaum.

Lehrer, R., Jenkins, M., & Osana, H. (1998b). Longitudinal study of children's reasoning about space and geometry. In R. Lehrer & D. Chazan (Eds.), *Designing learning environments for developing understanding of geometry and space* (pp. 137–167). Mahwah, NJ: Erlbaum.

Lehrer, R., Randle, L., & Sancilio, L. (1989). Learning preproof geometry with Logo. *Cognition and Instruction, 6,* 159–184.

Lehrer, R., & Romberg, T. (1996). Exploring children's data modeling. *Cognition and Instruction, 14,* 69–108.

Lehrer, R., & Schauble, L. (2000). Modeling in mathematics and science. In R. Glaser (Ed.), *Advances in instructional psychology* (Vol. 5, pp. 101–105). Mahwah, NJ: Erlbaum.

Lehrer, R., Schauble, L., Carpenter, S., & Penner, D. E. (2000). The inter-related development of inscriptions and conceptual understanding. In P. Cobb, E. Yackel, & K. McClain (Eds.), *Symbolizing and communicating in mathematics classrooms: Perspectives on discourse, tools, and instructional design* (pp. 325–360). Mahwah, NJ: Erlbaum.

Lehrer, R., Schauble, L., & Petrosino, A. (2001). Reconsidering the role of experiment in science education. In K. Crowley, C. Schunn, & T. Okada (Eds.), *Designing for science: Implications from everyday, classroom, and professional settings.* Mahwah, NJ: Erlbaum.

Lehrer, R., Schauble, L., Strom, D., & Pligge, M. (2001). Similarity of form and substance: Modeling material kind. In S. M. Carver & D. Klahr (Eds.), *Cognition and instruction: 25 years of progress* (pp. 39–74). Mahwah, NJ: Erlbaum.

Lehrer, R., Strom, D., & Confrey, J. (2002). Grounding metaphors and inscriptional resonance: Children's emerging understanding of mathematical similarity. *Cognition and Instruction 20,* 359–398.

Leslie, A. M. (1987). Pretense and representation: The origins of "theory of mind." *Psychological Review, 94,* 412–426.

Lindquist, M. (1989). The measurement standards. *Arithmetic Teacher, 37*(1), 22–26.

Mayo, D. (1996). *Error and the growth of experimental knowledge.* Chicago: University of Chicago Press.

McClain, K., Cobb, P., Gravemeijer, K., & Estes, B. (1999). Developing mathematical reasoning within the context of measurement. In L. V. Stiff & F. R. Curcio (Eds.), *Developing mathematical reasoning in grades K–12* (pp. 93–106). Reston, VA: National Council of Teachers of Mathematics.

Miller, K. F. (1984). Child as measurer of all things: Measurement procedures and the development of quantitative concepts. In C. Sophian (Ed.), *Origins of cognitive skills* (pp. 193–228). Hillsdale, NJ: Erlbaum.

Miller, K. F., & Baillargeon, R. (1990). Length and distance: Do preschoolers think that occlusion bring things together? *Developmental Psychology, 26,* 103–114.

Mitchelmore, M. C. (1997). Children's informal knowledge of physical angle situations. *Learning and Instruction,* 7, 1–19.

Mitchelmore, M. C. (1998). Young students' concepts of turning and angle. *Cognition and Instruction, 16,* 265–284.

Mitchelmore, M. C., & White, P. (1998). Development of angle concepts: A framework for research. *Mathematics Education Research Journal, 10,* 4–27.

Nickerson, R. S. (1999). Why are there twelve inches in a foot? *International Journal of Cognitive Technology, 4,* 26–38.

Outhred, L. N., & Mitchelmore, M. C. (2000). Young children's intuitive understanding of rectangular area measurement. *Journal for Research in Mathematics Education, 2,* 144–167.

Penner, E., & Lehrer, R. (2000). The shape of fairness. *Teaching Children Mathematics,* 7(4), 210–214.

Petrosino, A., Lehrer, R., & Schauble, L. (in press). Structuring error and experimental variation as distribution in the fourth grade. *Mathematical Thinking and Learning*.

Piaget, J., & Inhelder, B. (1948/1956). *The child's conception of space*. London: Routledge & Kegan Paul.

Piaget, J., Inhelder, B., & Szeminska, A. (1960). *The child's conception of geometry*. New York: Basic Books.

Porter, T. M. (1986). *The rise of statistical thinking 1820–1900*. Princeton, NJ: Princeton University Press.

Schauble, L. (1996). The development of scientific reasoning in knowledge-rich contexts. *Developmental Psychology, 32*, 102–119.

Varelas, M. (1997). Third and fourth graders' conceptions of repeated trials and best representatives in science experiments. *Journal of Research in Science Teaching, 34*, 853–872.

Wertsch, J. V. (1998). *Mind as action*. New York: Oxford University Press.

Chapter 13

Reasoning About Data

Clifford Konold, Scientific Reasoning Research Institute, University of Massachusetts Amherst
Traci L. Higgins, TERC, Cambridge, Massachusetts

Statistics is a rising star in the Grades K–12 mathematics curriculum. Ten years ago precollege students rarely learned about statistics. Now, following recommendations made in *Curriculum and Evaluation Standards for School Mathematics* (National Council of Teachers of Mathematics [NCTM], 1989), statistics is featured prominently in all current mathematics curricula and is popping up in the Grades K–12 science curricula as well.

There are several reasons the field has come to view statistics as important even for kindergartners. Educationally, statistics provides links to other areas of study—science, geography, and history, for example—in which students can apply mathematical ideas to model and reason about real-world situations. The American Statistical Association (1997) has suggested that more needs to be done to develop the "synergy between teaching statistics and teaching basic mathematics concepts" (Section 2, para. 4). At the practical level, knowledge of statistics is a fundamental tool in many careers, and without an understanding of how samples are taken and how data are analyzed and communicated, one cannot effectively participate in most of today's important political debates about the environment, health care, quality of education, and equity. For those who have traditionally been left out of the political process, probably no skill is more important to acquire in the battle for equity than statistical literacy.

Unfortunately, developers of recent curricula have not had years of experience teaching these topics at the pre-college level to draw on in designing their curricula. Furthermore, much of the research describing how students think about and learn these topics is not quite appropriate for their purposes. Because until recently students in North America encountered statistics for the first time at the university, most of the research on statistical reasoning has been conducted with adults and has focused on ideas central to formal statistical inference. Garfield and Ahlgren (1988) and Shaughnessy (1992) provide reviews of this research from the point of view of statistics education. Whereas these ideas are relevant to objectives at the high school level (see Scheaffer, Watkins, & Landwehr, 1998), they do not figure centrally in the curricula for Grades K–8.

In this chapter, we focus primarily on what we have learned more recently from research about how younger students reason about data, concentrating on ideas that begin developing in early elementary school. We therefore do not review the literature related to statistical inference. One reason for not reviewing that literature here is that a reasonable treatment would require us to also review the development of probabilistic thinking (see the next chapter in this volume). But more importantly, core ideas in reasoning about data tend to get shoved to the wings as soon as statistical inference takes the stage. The issues we discuss here, though basic, are still essential to statistical reasoning in the upper grades.

To clarify our point, let us make a distinction between *statistical inference* and *data analysis*. We can think of statistical inference as playing the role of a jury—deciding what we can conclude about a population on the basis of evidence from a sample. Until recently, it had been common practice among social scientists and others who use statistics to collect data and submit them to some test of significance (*t* test, ANOVA, chi square) without ever examining the data in more detail and without looking at the distributions of values for patterns or trends that might lead to additional insight into the question of interest. Once they had conducted the statistical tests, the analysis was essentially over. Indeed, researchers had been warned of the danger of conducting post hoc analyses on questions they had not formulated before collecting the data (Hand, 1998).

Tukey (1977) regarded this exclusive focus on testing hypotheses as shortsighted. He advocated that statisticians could and should do more with data than simply confirm the theories they had formulated prior to collecting data. He encouraged statisticians to take on as well the role of detective, to search among data for interesting and unexpected results. This he described as "exploratory" data analysis (EDA). To assist the researcher in detecting trends and patterns, Tukey developed a variety of methods for displaying data graphically, including the box plot and stem-and-leaf plot. Though he designed many of his techniques for making quick displays with pencil and paper, fast computing and quality computer graphics have actually been a boon to his general approach, making it possible to quickly manipulate and redisplay data in numerous ways. The computer continues to change the way statisticians work (Cobb, 1997).

Although these new developments in data analysis may be a blessing for statisticians, they can be a curse for the educator who is left to figure out how best to teach a field as it is undergoing rapid change. To complicate matters, currently we must assume that students at every grade have had little prior experience with statistics. Therefore, when current research indicates that students at a certain age have difficulty with a particular concept, we do not know what students this age could accomplish if they had years of prior instruction. Thus we find ourselves stitching together a Grades K–12 instructional sequence in statistics and data analysis that, if effective, will probably need overhauling in a few years.

It is all the more important, given this state of flux, that our efforts to teach these topics be guided by an understanding of the core ideas in data analysis. Otherwise we will likely find our students familiar with an array of graphing and data-collection skills but with no ability to reason intelligently about data.

As mentioned previously, many countries, including the United States, have only recently begun to teach statistics and data analysis prior to college; thus research in these areas has not been a priority. We draw heavily in this chapter on the work of Russell, Schifter and Bastable (2002a, 2002b, 2002c), who have compiled as part of course materials for preservice and in-service mathematics teachers a set of cases written by elementary school teachers about their attempts at teaching data analysis. In our opinion, the reflections of these teachers and their descriptions of students' thinking are the richest source to date on children's reasoning about data and on how children's thinking evolves during instruction.

Overview: Data Analysis as an Iterative Process

We can think of data analysis as a four-stage process: (1) ask a question, (2) collect data, (3) analyze data, and (4) form and communicate conclusions (Friel & Bright, 1998). Listing the stages in this order makes sense because it is more or less the sequence in which a research investigation proceeds. It would be hard, after all, to analyze data before collecting them or to form conclusions before doing any analyses. That these are meaningful stages is underscored by the fact that scientists in many fields compose research reports with major sections that correspond to these ordered stages: hypotheses, methods, analyses, results, and conclusions. Similarly, we have organized this chapter into major sections treating, in turn, issues related to data collection, organization, and interpretation.

Real research, however, seldom proceeds in this orderly fashion. One reason is that conscientious researchers often find themselves backtracking. While writing their report, they think of another analysis to do or decide they need to return to the study site to collect more data. But research does not proceed linearly for a more profound reason, which is that these research phases are interdependent (Wild & Pfannkuck, 1999).

Experienced researchers look forward from the beginning. Although they do not analyze data before collecting them, they imagine doing so and make guesses, or hypotheses, about what they will find. They develop and refine their questions and decide what data to collect by thinking ahead to which statistical methods they can use and to the audience they want to convince. Experienced researchers also look backward. When the time comes to analyze the data, they do so from the perspective of their original question, testing the intuitions they started with against what the data reveal. And their questions often evolve and change as they discover unanticipated results in the data (Moore, 1990).

In these respects, data analysis is like a give-and-take conversation between the hunches researchers have about some phenomenon and what the data have to say about those hunches. What researchers find in the data changes their initial understanding, which changes how they look at the data, which changes their understanding, and so forth.

We need to keep this more complex picture of data analysis in mind as we consider what the research literature tells us about students' statistical thinking. Simplistic views can lead to the use of recipe approaches to data analysis and to the treatment of data as numbers only, stripped of context and practical importance. Conversely, staying grounded in the data and attentive to what they have to say keeps the tools of data analysis—the collecting, graphing, and averaging—in their appropriate, subservient role. As one third-grade teacher observed, "Analyzing data is more than just the sum of using data analysis techniques. It's important not to lose sight of what the data themselves have to tell us" (Russell, Schifter, & Bastable, 2002a, case 19).

Turning Observations Into Data

A data investigation usually begins with a question about the real world, for example, "Are children more active than adults?" One of the first challenges is to transform that general question into a statistical one that we can answer with data, for example, "If we put odometers on the feet of a sample of adults and a sample of fourth graders, which group will travel further in a day?" Among other things, the statistical question allows us to develop measurement instruments and data-collection procedures. By analyzing the data, we answer our statistical question, which ideally, but not always, tells us something about the real question we started with.

In learning how to formulate questions and to collect and analyze data to answer them, students must learn to walk two fine lines. First, they must figure out how to make a statistical question specific enough so they can collect relevant data yet make sure that in the process they do not trivialize their question. Second, they must learn to see the data they have created as separate in many ways from the real-world event they observed yet not fall prey to treating data as numbers only. They must maintain a view of data as "numbers in context" (Moore, 1992) and at the same time abstract the data from that context.

"Creating data" may seem an odd phrasing. However, data are not lying around like melons on the ground to gather up and cart off to the table. Turning observations into data involves an explicit process of abstraction, a process more like impressionist painting than snapshot photography (Hancock, Kaput, & Goldsmith, 1992).

Forming a Statistical Question

We collect data to answer a question or solve a real problem. For example, students in the upper grades at a Grades K–8 school believed that the water from the fountains on the third floor, where their classrooms were located, was better than the water from the fountains on the floors below (Rosebery & Warren, 1992). A combined class of seventh and eight graders decided to determine whether the water on different floors of the school building really was different.

Their first task was to reformulate their question as a statistical one: "In a blind taste test of water samples from each of the three floors, which sample will most students prefer?" On the basis of this question, they could develop a plan for data collection. Rosebery and Warren (1992) do not provide the details of this process, but we can presume that students made a number of decisions before collecting data: Whom would they use as tasters? How many tasters should they test? Should tasters drink directly from fountains or from cups? Should the same students taste all three water samples? How should tasters indicate their preference? Such decisions are part of the process of making a general question a statistical one.

Elementary school students can learn a lot about data as they grapple with issues that arise in formulating statistical questions, especially when they anticipate conducting surveys with the questions they design. By thinking about how they would answer a proposed question, students quickly discover not only the range of different responses but also that multiple interpretations of a question are often possible and that the wording of the question matters. In the words of a second grader, "Everyone has to understand your question. If they don't understand your question, everyone will be answering any old way" (Russell et al., 2002a, Case 6).

In Russell et al. (2002a, case 5), a class of fifth-grade students worked in groups to develop questions to guide a data investigation. One group wanted to find out the number of languages spoken by classmates. They suggested asking the following question in a survey: "Do you speak more than one language?" The teacher pushed them to think more carefully about their question:

Teacher: How do we know when someone speaks another language? For example, is knowing how to say, "Where is the bathroom?" in French speaking French?

Student: No. We mean speaking fluently. (Case 5)

This, of course, raises the issue of what is meant by "speaking fluently." The students discussed ways to further refine their question. They were learning that one needs to formulate questions so that respondents interpret the question in the same way.

Often in trying to make a question more precise, students lose track of what they want to know in the first place. For example, second graders Nadia and Keith were interested in finding out from classmates, "How many states have you visited?" (Russell et al., 2002a, Case 6). However, they soon discovered that students they surveyed interpreted "visited" in many different ways. Nadia offered further criteria (described here by her teacher) for defining a visit:

> A visit could only count if you were going to that state for a specific purpose other than simply driving through to reach another destination. Airports could not count. If you stayed with a friend from out of state it could only count if you really, really wanted to see them and you stayed with them for more than a day. (Case 6)

Nevertheless, in their final survey, the question was phrased as follows: "How many states have you ever set foot in?" This wording had apparently been adopted at Keith's prompting because students could answer it with little ambiguity in meaning. However, Nadia was not satisfied. She thought the phrase "set foot in" missed the point. She wanted to know whether students had traveled to, rather than through, a state. In transforming a general question to a statistical one, the problem is not only in wording it so that people will interpret it consistently but also in making sure the question gives you the information you "really, really" want.

Sampling

Part of formulating a statistical question is deciding what population you want to study. Some of the questions we investigate involve collecting information on the entire population of interest (e.g., everyone in a classroom). However, many questions involve populations that would be impossible or impractical to study exhaustively. In these cases, we collect data from only a part of the population—*a sample*. The quality of our study depends heavily on how we obtain this sample. We will probably get a very poor idea of the outcome of a national election by polling, for example, our friends and neighbors—not only because the sample is small but because it is likely biased, that is, not representative of the voting populace. Our friends and neighbors, after all, tend to hold opinions similar to ours. This problem was dramatically demonstrated in the 1936 presidential election (see Gallup, 1972). On the basis of a sample of more than 2 million people, the *Literary Digest* predicted that the Republican candidate, Alfred Landon, would soundly defeat the Democratic candidate, Franklin Roosevelt. From much smaller samples, both Gallup and Roper correctly predicted a Roosevelt victory. The *Literary Digest* had sent questionnaires to addresses they got from lists of owners of telephones and automobiles. Although this was an easy way to get addresses, it tended to exclude poorer segments of the population who at the time did not own phones or cars but did tend to vote the Democratic ticket.

Thus, it is more important that our sample include a representative cross section of the population than that it be large. And the best way to guarantee a representative sample is to draw it randomly. In a random sample, every member of the population has an equal chance of being included. To draw a random sample of 60 students from a large school, we could put the names of all the students in a container, mix them up very well, and blindly draw out 60.

According to research by Jacobs (1999), some students reject the idea of sampling in general "because they drastically underestimate the difficulties of asking everyone in larger surveys, such as surveys of entire states" (p. 245). Others struggle with how information from a sample could give useful information about the whole population.

Understanding the relationship between sample and population requires grasping that X can represent Y without being Y. The majority of the elementary school students studied by Metz (1999) were unwilling to generalize from a sample of crickets they had studied to the larger population of crickets. Among the arguments students gave for not generalizing from their sample were that (a) you can only know about the cases you observe; (b) to characterize a group, you must test every member of that group; and (c) sampling does not work because of variability in the population.

Jacobs (1999) asked fifth graders to evaluate various sampling methods. She used a variety of contexts, including estimating the number of students likely to buy raffle tickets, identifying favorite lunchroom items, and determining which animal students would prefer as a classroom pet. When faced with conflicting results from samples collected in different ways, nearly a third of the students did not differentiate between results produced by biased versus unbiased sampling methods. Rather, they based conclusions on personal experience or said they would ask a trusted authority, such as a teacher or principal, to estimate the outcome. Another 12% evaluated sampling methods on the basis of whether the results fit with their expectations or were decisive. The latter students favored a survey in which, for example, 100% of those sampled said they would buy raffle tickets over a survey that showed a split opinion, because "50-50's not going to decide it for you" (p. 245).

Students who do accept the idea of sampling often favor methods that are biased. Many elementary school students, for example, prefer self-selected samples because they minimize hurt feelings that might result from being excluded from a sample. One upper elementary school student said,

> The people will choose [to participate] if they want to.... Like if they wanted to do the survey, they will, but if they would not want to, they don't have to—so they're not pressuring anybody. (Jacobs, 1999, p. 244)

Schwartz, Goldman, Vye, and Barron (1998) described various biased sampling techniques that sixth graders used to estimate attendance at a particular booth at an upcoming fun fair. Some students suggested they would survey their friends or those they thought likely to visit the booth. They wanted to sample students likely to attend the booth rather than to estimate the distribution of positive and negative responses for the entire school. Other students favored sampling techniques that directly influenced the outcome of the survey. For example, asked to choose a sample to estimate the gender makeup of their school, a quarter of the sixth graders suggested a "fair split" schema, sampling 25 boys and 25 girls. Watson and Moritz (2000) reported similar findings with third- and sixth-grade students.

Some upper elementary school students do select and evaluate sampling methods on the basis of the potential for bias. Roughly one third of the fifth graders Jacobs (1999) studied used expertlike reasoning to evaluate whether various samples would adequately represent the population. The methods they endorsed included randomization and stratified-randomization techniques. For example, one student favored sampling from a population of school students by randomly selecting five girls and five boys from each grade, because

> that way he has a mixture of boys and girls and who are different ages...because sometimes girls and boys can have different opinions on things and also one age might really like something, but an older age might think that was a terrible idea. (p. 244)

Likewise, 40% of sixth graders in a study by Schwartz et al. (1998) proposed sampling methods that avoided obvious bias. A follow-up study indicated that fifth and sixth graders overwhelmingly preferred a stratified or stratified-random sample to a biased sample. However, the students remained somewhat skeptical about using truly random sampling methods. For example, to select a sample of schoolchildren, roughly 60% of the students indicated a preference for selecting the first 60 children in a line over drawing 60 student names from a hat. Indeed, some students worried that randomization might produce a biased sample: "She might pull out [of the hat] all first-grade names" (p. 255). Students may reject randomization procedures precisely because "the selection of population characteristics [is left] up to chance" (p. 256). Thus, when they can, students purposefully select individuals to represent the crucial population characteristics.

It is interesting that students accept randomization in games of chance and see merit in stratification techniques when sampling opinions. What they often struggle with is how these two ideas connect. Schwartz et al. (1998) observe,

> Even though the children could grasp stratifying in a survey setup and randomness in a chance setup, they did not seem to have a grasp of the rationale for taking a random sample in a survey setup. They had not realized that one takes a random sample precisely because one cannot identify and stratify all the population traits that might covary with different opinions. (p. 257)

Differentiating Between the Observed Event and the Data

As we mentioned previously, the process of turning observations of events into data involves simplification and abstraction. Sampling is but one of several stages in this process. To reason intelligently about the data, however, we must keep in mind that the data refer to those more complex, real-world events. If we forget that fact, we begin treating data as numbers only, making no connections back to the context or real-world question.

Other problems arise if we treat the data as if they were the events we observed, failing to distinguish "between the world and a representation of that world" (Lehrer & Romberg, 1996, p. 70). For example, fifth graders in a study by Hancock and his colleagues (1992) collected data to determine which of three cafeteria meals students liked most. The student researchers conducted surveys in the cafeteria on different days, asking students whether they had "bought," "brought," or skipped lunch on that day ("none"). Their plan was to determine menu preferences by comparing the number of students buying versus bringing lunch on a particular day. Their rationale was that because daily menus were published in advance, many students would decide whether to bring a lunch on a particular day depending on whether they liked what was offered. As they began analyzing their data, the student researchers discovered that they had failed to record what meal was served. On the day they recorded the data, a mark under a column on the survey had a clear meaning, a meaning that was gone once they had forgotten the menu of the day.

In Russell et al. (2002a), a kindergarten teacher gave each of her students a bag of M&M's to count. The class created a frequency bar graph using stick-on notes on which each student had recorded the number of M&M's in his or her package. The teacher asked, "What can you tell from this graph?"

Farub: We eat M&M's.

Rocky: Joy has the most.

Desmond: We know how many I ate.

Tammy: Andrea's is the most...because hers is a bigger number. (Case 16)

The students associated names with values even though the graph they were interpreting did not show who had counted each bag. The students were basing their interpretations on their memories of counting and eating M&M's rather than on the data they had abstracted from that event.

For many students, data serve merely as pointers to the more complex event (Konold, Higgins, & Russell, 2000). In forming conclusions, they draw without awareness on their memories of the event as well as on the objectified data. As a way to help her students begin to distinguish the information in the coded data from what they knew from observing the events, the teacher in the foregoing episode suggested to her students that they "pretend that the principal walks into our room and looks at this chart, what would he know from this chart?" (Russell et al., 2002a, Case 16.)

In keeping with their view of data as pointers, we see many examples in Russell et al. (2002a) of children producing elaborate drawings that show as much about the events as possible. Their iconic representations, often called *pictographs*, can be difficult and time-consuming to draw and typically include detail that would seem to convey no useful information, what Tufte (1983) referred to as "chartjunk." However, whereas to our eye pictographs may be distracting and unnecessary, for students they may help establish explicit links between the data and the event, possibly helping them reason about the data in the appropriate context. Furthermore, the level of abstraction appropriate for a particular representation depends on the questions students have asked. Given that younger students are drawn to questions about who has the most and where they personally fall within a range of values, it makes sense that their representations would make it easy for them to read off individual values and to identify to whom these values belong.

One indication that students are separating data from events is the emerging ability to reason about data in ways they had not anticipated before collecting the data. In their classroom study of fifth graders, Lehrer and Romberg (1996) worked with students to design a survey of student interests. Among other things, the survey asked students to list their favorite school subject, list their favorite winter sport, and estimate the hours they spent watching TV. After collecting the surveys, the instructors asked the student researchers to come up with questions they could "ask about the data." To these fifth graders, this request was ridiculous. Questions, they countered, could be posed to people, but certainly not to data. The instructors prompted the students with various examples, such as "Which is the least-favorite school subject?" The students successfully used these examples to generate a list of similar questions. But they needed further assistance to see that they could answer many of these questions by analyzing the data they already had. Their initial impulse was to conduct another survey using these new questions.

As Hancock and his colleagues (1992) pointed out, once recorded, data become objects in their own right, objects that we can manipulate and query quite independently of the real-world events they model. We can organize data by stacking, grouping, and ordering cases, operations that would be difficult or impossible to perform on the real-world events themselves.

Because students can reorganize the data in a number of different ways, they can pose and answer questions that may not have occurred to them before collecting the data. For example, kindergarten and first-grade students worked with data gathered from a school lunch count (Russell et al., 2002a, Case 1). While gathering data, students became concerned that although a total of 18 students were in their class, they had recorded only 15 data values. The three missing students turned out to have been absent that day. One student was intrigued that they had gotten an attendance count from data they had gathered about lunches. She wondered how this was possible:

> Can the clothespins [markers used to record answers to the survey questions] tell only one thing? If the clothespins tell us how many school lunches and how many home lunches there are, can they tell us how many are at school—I mean at the same time? (Case 1)

This student understood that the sum of the "yes" and "no" counts would equal the number of students in the class that day. But she struggled with the idea that "fifteen could stand for how many in school as well as how many students were getting lunch" (Russell et al., 2002, Case 1). We see her struggle as the beginning of the discovery that data, once recorded, have a life of their own and that in examining them, new questions may arise that the data can answer.

Organizing and Displaying Data

In this section, we examine issues that arise as students organize data into tables and plots. We continue to emphasize the important role that questions play in determining the representations students create and the information they cull from those representations.

Creating Useful Representations of Data

Data are usually somewhat useless in the form we first collect them. A stack of completed questionnaires is like a messy room in need of a good cleaning. To find what we want, we must organize the information, and how we organize the information, or data, depends on what we want to know (Biehler, 1989; Cleveland, 1993).

There are no fixed criteria for judging one data display as superior to another. Rather, the relative value of a plot depends on its intended purpose. Thus, a plot with labeled axes is not necessarily better than one without labeled axes. On the one hand, suppose students wanted to quickly make a frequency bar graph to help them determine where the data were centered. It would be unnecessary in this case to label the axes. In fact, taking the time to do so or to add other fine detail to the graph would be squandering time that could be spent thinking about their question. On the other hand, if these same students made a graph to communicate their findings to the whole class, then labeling the axes and taking care to make the graph easily readable would probably be crucial to achieving their goal.

Likewise, one *type* of representation is not inherently superior to another. Graphs are not better than tables; bar graphs are not better than pictographs. Each of these representations is good for some purposes and not as good for others. Sconiers (1999) described a project undertaken by a kindergarten class that frequently asked their teacher for help tying shoes. The class reasoned that if they all knew which of them could tie shoes, those who did not know how could get help from those who did. After conducting a survey, the students posted a list of names of classmates who could tie shoes. Had the class not been trying to solve that specific problem, they might have made a graph showing how many students could and could not tie shoes, but that plot would have been useless for their purposes. The list worked.

Although no paticular type of representation is inherently better than another, some representations are harder than others to learn to interpret (Bright & Friel, 1998). Roth and Bowen (1994) argued that as we represent numerical data with maps, lists, graphs, and equations, we move from concrete to increasingly abstract statistical representations. As we move along this progression, information about individual data values becomes obscured, disappearing into larger aggregates. According to Feldman, Konold, and Coulter (2000), increasing the level of aggregation allows one to

> perceive ever more general features of the data at the expense of being able to identify individual data values. One can easily forget, however, the learning required to interpret the more abstract statistical plots. As a result, [educators] often encourage students to use plots and summaries before they sufficiently understand them and, by doing so, effectively pull the rug from beneath them. (p. 119)

As Bright and Friel (1998) pointed out, graphs also use axes and other display elements in a variety of ways. We can see some of these differences in the two plots in Figures 13.1 and 13.2 that fifth-grade students located in a textbook (Eicholz, 1991) while searching for graphs they could use to compare two data sets (Russell et al., 2003c). Both are described as "double bar graphs." On quick inspection, they seem to represent data in exactly the same way. However, a closer inspection reveals an important difference. Figure 13.1 uses a bar to represent the cost of each individual item (e.g., hat, shoes). The length of each bar is proportional to that item's cost. We prefer to call this representation a *case-value* plot. This term refers to the fact that each case (e.g., a hat, a cape) is represented by a separate bar indicating that case's value (i.e., price in dollars). Thus the bar's length shows the magnitude or value for that case—a "case value."

The graph in Figure 13.2 shows the number of respondents who chose various personalities as people they would most like to have known. In this type of graph, called a *frequency* bar graph, a bar's height is not the value of an individual case but rather the number (frequency) of cases (respondents) that all have a particular value. In this instance, the leftmost bar shows that 16 respondents (cases) selected Susan B. Anthony. Of course, we could use a stack of Xs instead of bars to represent the number who selected each famous person. Some refer to plots made with Xs as *line* plots. We prefer to call them stacked dot plots.

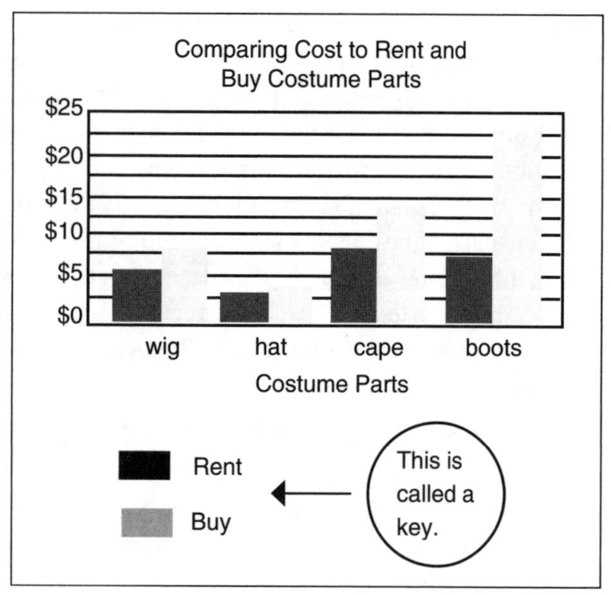

Figure 13.1. A case-value plot of the cost of renting versus buying various costume parts; from Eicholz (1991, p. 96).

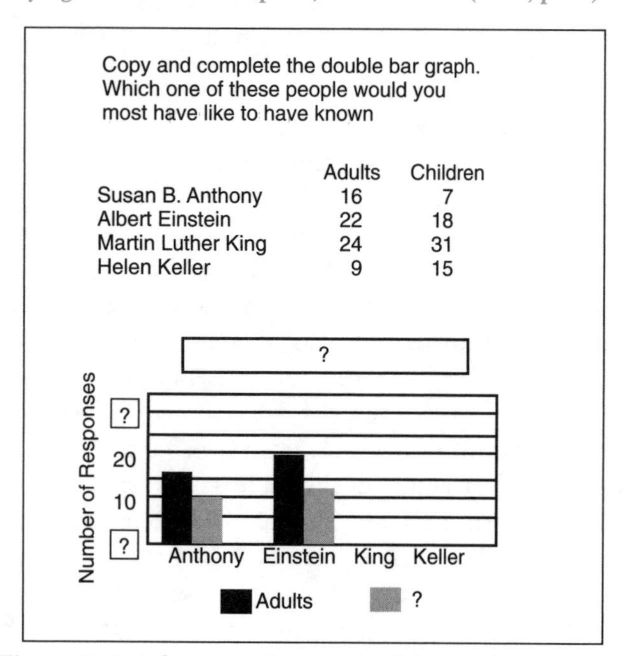
Copy and complete the double bar graph.
Which one of these people would you most have like to have known

	Adults	Children
Susan B. Anthony	16	7
Albert Einstein	22	18
Martin Luther King	24	31
Helen Keller	9	15

Figure 13.2. A frequency bar graph of the number of adults and children who named various personalities as people they would like to have known; adapted from Eicholz (1991, p. 97).

Anticipating, as Bright and Friel (1998) reported, that students would have more difficulty working with frequency bar graphs and stacked dot plots than with case-value plots, Cobb and his associates had students in their seventh-grade teaching experiment spend the first few weeks working solely with case-value plots (Cobb, 1999). The materials and software "mini" tool they used displayed these bars horizontally to facilitate the later transition to stacked dot plots. Furthermore, they started with problems in which it was relatively natural to display individual data values as length (e.g., length of lizards, braking distance of cars, hours of service from a battery).

After a few weeks working with case-value plots, the researchers introduced students to the stacked dot plot by demonstrating how it could be created from a case-value plot. This introduction involved a progression of intermediate stages: first the bars were erased and only their endpoints plotted; and finally, these "floating" points were allowed to collapse down onto the horizontal axis such that cases with the same values would stack on top of one another.

We demonstrate this progression using weight (in pounds) of the backpacks carried by 55 elementary school students. Figure 13.3 is a case-value plot of the data, with the cases ordered from lightest backpacks (bottom) to the heaviest (top).

In Figure 13.4, each bar in Figure 13.3 has been replaced by a small circle located at what was the right-hand end of the bar. The value of each case is indicated by a circle's position over the x-axis.

Finally, in Figure 13.5, the circles of Figure 13.4 have collapsed onto the x-axis to form stacks showing the number of students with backpacks of a given weight.

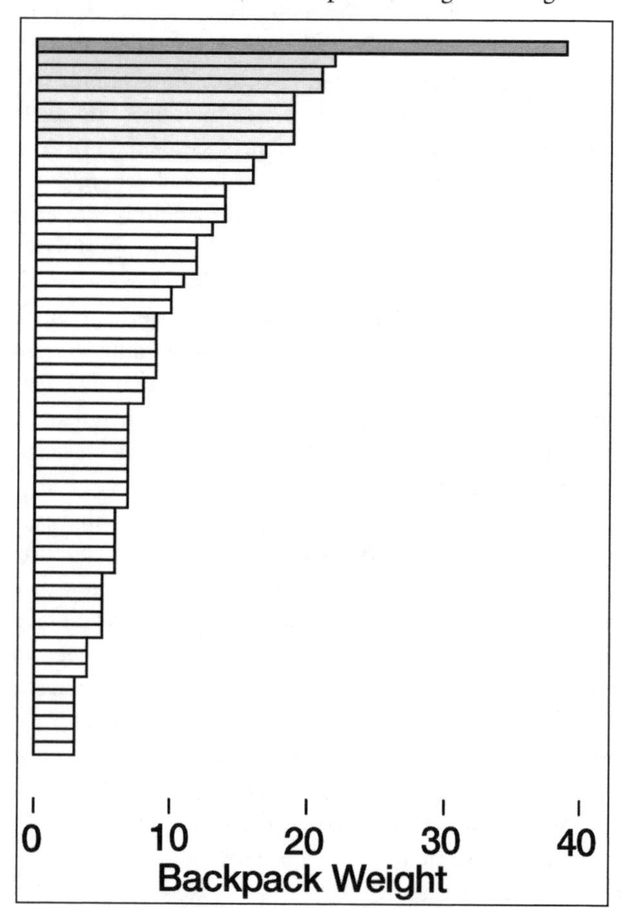

Figure 13.3. Case-value plots showing weight (in pounds) of the backpacks of 55 elementary school children, plotted with Tinkerplots (Konold & Miller, 2002).

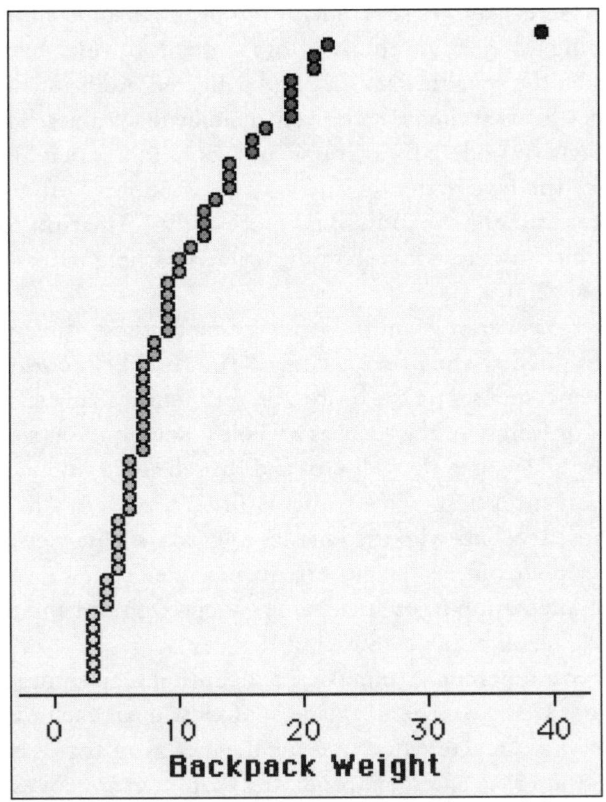

Figure 13.4. Intermediate step in the animation of the case-value plot of Figure 13.3 to the frequency graph of Figure 13.5.

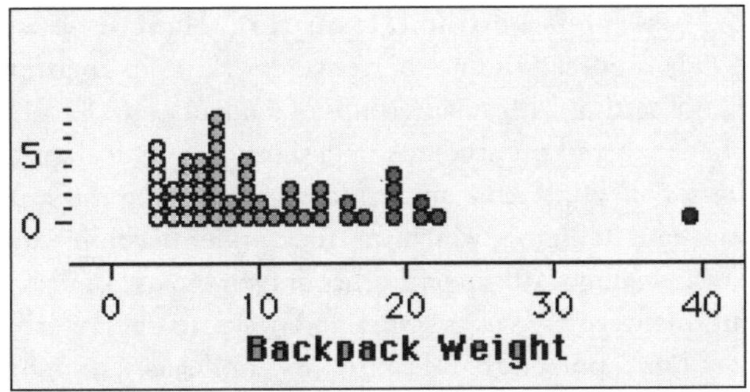

Figure 13.5. Final step in the animation of the case-value plot in Figure 13.3 to a stacked dot plot.

Cobb and his associates believe that by grounding the stacked dot plot in the case-value plot, students were able to view the Xs (or circles) in stacked dot plots as magnitudes (Cobb, 1999). As we discuss next, many of the difficulties students have in representing and interpreting data with stacked dot plots (e.g., putting a value of 12 next to 10 when there is no value of 11) stem from their failure to view the Xs as measures of magnitude.

Deciding About Scale

Deciding on plot scales and on what data should be included in their plots poses a number of interesting challenges to students. Some students described in Russell et al. (2002a) thought that plot scales should not extend beyond the range of observed values, whereas others argued that the scales should extend to include values that could have occurred or far enough at least to make a pleasant boundary.

Working with categorical data, many students argue that they should not include categories with frequencies of zero. Frequencies of zero also raise important scaling issues when students are working with numerical data. Students may be unsure of how or whether to plot nonoccurring data values. For example, some fourth graders (Russell et al., 2002a) used the scale "50, 52, 54, 55, 56, 57, 58, 59, 60, 62, 64" to represent data about classmates' heights on a stacked dot plot, omitting from the axis any value they had not observed. Without gaps along the scale representing nonoccurring heights, it is difficult to see clumping in the data or to judge the magnitude of the difference between various heights. Some of the difficulties students have may result from the fact that they often construct their scales at the same time they plot the data. This practice may make it harder for them to distinguish between the numbers along their scale and the actual data they are plotting.

In making graphs, some students are driven more by conventions they have learned than by the information they want to convey. As Roth and McGinn (1997) pointed out, "In schools...students make graphs for the purpose of making graphs" (p. 95). Students are well practiced, therefore, in setting aside their own intentions and purposes and getting down to the business of producing "good" graphs. This orientation is evident in much of the dialogue among students in the study by Russell et al. (2002a) as they discussed not what they *could* do in plotting their data but what they *ought* to do. Scaling decisions represent ideas about how to frame the data. Thus, they should be tailored to the questions one wants to answer.

Imagine drawing a face on an inflated balloon and then popping it. Next imagine stretching that popped balloon over various objects—a book, a doorknob, a basketball. The face would look quite different stretched over these various objects. The issues in selecting scales for plotting data are exactly the same as those in deciding between doorknobs and basketballs in the balloon example. What will be the minimum and maximum values on the axes, what will be the interval sizes between numbers, and what ratio should we use for the relative sizes of the x- and y-axes (e.g., do we want tall, skinny bars or short, fat ones)? Each of these choices affects how the data appear. In making these decisions, therefore, we cannot afford to adhere to strict conventions but must be mindful of our particular purposes. And we have no reason to puzzle long over these questions. Unlike the balloon example, we have no ideal scale that will make the data appear as they "really" are. We can try out several alternative plots and scales and learn what we can from each. When it comes time to summarize our results for others, we can then pick those representations that do the best job of telling the story sharply and fairly.

Describing and Interpreting Data

David Moore (1990) proposed five core ideas of statistics. Topping his list is the awareness that variation is everywhere. "Individuals are variable; repeated measurements on the same individual are variable" (p. 135). The idea that individuals vary is apparent to even young students. Variability is literally everywhere around them. Their classmates come in a variety of heights, hair colors, and temperaments. The weather they encounter when they step outside varies not only from season to season but from day to day, and sometimes from one minute to the next. If students know nothing else when they begin collecting data, they know that they will get a variety of values.

Anticipating and observing variability within a group are not difficult. What is difficult is figuring out how to quantify variability and perceive and characterize the group as a whole when individuals in that group differ from one another. In this section, we explore how students who tend first to describe and reason only about individual cases begin to describe group characteristics, to summarize and compare groups despite the fact that individuals in each group vary.

Describing Group Features

In early experiences with data, students tend to focus on describing individual data points, or clusters of similar individuals. Russell et al. (2002a, Case 7) described a kindergarten episode in which students reported their favorite color. As the teacher recorded the information on the board, students spontaneously commented on which color was ahead—the modal value. However, the next day when the teacher asked what the graph showed, none of the students mentioned the mode. Instead they made such comments as these:

- We know what everyone's favorite color is.
- My favorite color is red.
- My shirt is blue.
- We learned English and Chinese colors. (Case 7)

Here again we see students not distinguishing all the things they knew about colors and color preferences from the specific data they had collected. Additionally, they focused almost exclusively on individual values. The teacher wondered why it was so obvious to her that blue was the favorite color and why her students "did not seem to pull the individual pieces of information together to share ideas about the data as a whole" (Russell et al., 2002a, Case 7).

However, it is unlikely that the mode these students attended to the previous day as the teacher recorded their responses on the board was, for them, a characteristic or feature of the "data as a whole." Rather, it was simply "the winner." As Mokros and Russell (1995) found in their interviews with students in Grades 4, 6, and 8, those students who used modes to describe data viewed the mode only as the most frequent value. They showed no inclination to use that value to represent or summarize the other values as well.

Students need to make a conceptual leap to move from seeing data as an amalgam of individuals each with its own characteristics to seeing the data as an aggregate, a group with emergent properties that often are not evident in any individual member. "To be able to think about the aggregate," claimed Hancock and colleagues (1992), "the aggregate must be 'constructed.'" (p. 355).

This leap is a difficult transition. Hancock et al. (1992) reported their experiences working in an after-school setting with small groups of students ages 8 to 15. They involved the students in designing statistical questions, collecting data, and exploring data using the software tool Tabletop, which was then under development. The teaching staff explicitly encouraged the use of distributional terms, such as *cluster* and *range*, to characterize data but reported that despite this emphasis, "students often focused on individual cases and sometimes had difficulty looking beyond the particulars of a single case to a generalized picture of the group" (p. 354).

The researchers characterized this individual-based analysis as resulting in blow-by-blow descriptions of results: "This person said 'yes' to Question 1 and to Question 2, but this person said 'yes' for Question 1 but she didn't say 'yes' for Question 2…" (Hancock et al., 1992, p. 354). We see similar responses throughout the cases in Russell et al. (2002a), especially in the earlier grades. Asked what they learned from their survey about who liked their vacation, one kindergartner responded: "11 people said yes, 2 people said no, and 8 people said something else. That makes 21" (Case 9). As a kindergarten teacher observes, this individualistic orientation seems to preclude students from seeing group features:

> In the end, they seemed to attend to names on the chart and the information that was recorded about each person. They did not seem to pull the individual pieces of information together to share ideas about the data as a whole. (Russell et al., 2002a, Case 7)

Hancock and his colleagues (1992) reported similar observations. For example, Tabletop allowed students to display labels of individual data points. When they could, students kept a label showing the identity of each data point (e.g., the names of individual respondents). Worried that students were focusing primarily on individual data points, the teaching staff encouraged students to remove these labels, hoping that doing so would make the features of the whole distribution more salient. However, students could usually remember which data point belonged to which individual and therefore continued to draw on this information in interpreting the data.

With the individuals as the foci, it is difficult to see the forest for the trees. If the data values students are considering vary, however, why should they regard or think about those values as a whole? Furthermore, the answers to many of the questions that interested students—for instance, Who is tallest? Who has the most? Who else is like me?—require locating individuals, especially themselves, within the group. We should not expect students to begin focusing on group characteristics until they have a reason to do so, until they have a question whose answer requires describing features of the distribution. As we discuss subsequently, asking if two groups differ may be one such question.

Types of averages. Only one type of average—the mode—is useful in describing qualitative data, such as favorite color. With numeric data, on the other hand, there are many ways we can characterize the center of distributions. Statistics instruction has traditionally focused on the mean (arithmetic average), median, and mode, with special emphasis on the mean. There are, in fact, many other types of averages. The geometric mean, for example, is often used in economics to find average growth rates and is computed by multiplying all *n* values and then finding the *n*th root of the product.

By second or third grade most children have heard of grade averages and average temperatures. Although familiar with the term, their ideas about average are based on everyday meanings that draw on qualitative, rather than quantitative, notions of typicality. In the study by Russell et al. (2002a), we see students using averages of all sorts—means, medians, and especially modes. In addition, we see some using the midpoint between two extreme values (the midrange) and various other intuitive notions, such as the "modal clump" (Konold et al., 2002).

The ideal average that young students appear to have in mind is an actual value in the data set that is also the most frequently occurring value (the mode), positioned midway between the two extremes both in terms of value (the midrange) and order (the median), and that is not too far from all the other cases. In symmetric, mound-shaped distributions with lots of data, one can often find an average value that has all, or nearly all, these properties. But with many of the small data sets students explore, most of these conditions cannot be met. And when students have to start giving up items on their "average" wish list, most of them hang tenaciously onto the mode. Thus, teachers in the Russell et al. (2002a) study describe their students as heading "straight for the mode" and considering it "the end-all way to describe what's typical in a set of data" (Case 12, Case 19).

Mokros and Russell (1995) found in their interviews that most of the fourth graders, and even a few older students at Grades 6 and 8, used the mode in situations in which its use was inappropriate. For example, they asked children to price nine bags of potato chips such that the "typical or usual or average" price of the nine bags would be $1.38 without any of the individual prices' being $1.38. Students who conceived of averages only in terms of the most common value were unable to solve the task without using $1.38 as one of the nine prices.

The mean is noticeably absent from the foregoing list of students' ideal average. We get some insight as to why from the research of Strauss and Bichler (1988). As part of their study of student understanding of the mean, they described seven fundamental properties of the mean. Of those properties, the only one that is clearly among those students regard as important is that the mean is located between (though not necessarily midway between) the extreme values. Two of the seven properties—(1) that the mean does not necessarily equal one of the values in the data set and (2) that it does not necessarily have any counterpart in physical reality—are, in fact, two of the reasons students give for dismissing the mean as a useful average. For example, in Case 25 of Russell et al. (2002a), a third grader objected to using the mean of his two attempts at blowing a Styrofoam cylinder as far as he could because "I didn't get that as one of my distances. It wouldn't be true. It's a lie!"

Traditionally, students in U.S. schools have been introduced to the mean beginning at Grades 4 or 5. In their survey of 122 children in Philadelphia schools, Gal, Rothschild, and Wagner (1990) found that 2% of the third graders, 85% of the sixth graders, and 90% of the ninth graders referred to the add-and-divide algorithm when asked how weather forecasters determined average temperatures. Most researchers who have explored students' use and understanding of means are now recommending that we place much less emphasis in the elementary grades on teaching the mean. As Mokros and Russell (1995) found, those students who do know the standard algorithm for computing the mean generally misunderstand and misapply it. They speculated that "premature introduction of the algorithm ... may cause a short circuit in the reasoning of some children" (p. 37).

Additional research involving students from middle school through college suggests that few students know much more about means than how to compute them (Cai, 1998; Pollatsek, Lima, & Well, 1987). Thus, even children who can compute the mean often do not understand why the computation works or what the result represents. Although the add-and-divide algorithm is relatively simple to execute, developing a conceptual underpinning that allows one to use the mean sensibly is surprisingly difficult. Based on observations of teachers in the study by Russell et al. (2002a), the same difficulties appear in trying to teach students to use the median.

Students frequently use the midrange as an average. Although it may seem a crude average at best, it can be a useful index with large data sets. Eisenhart (1971) pointed out that long before astronomers used means of multiple observations to estimate the actual position of stars, they used the midrange as their estimate. This and other evidence suggests that, historically speaking, the midrange was the precursor to the mean.

A type of average that Russell et al. (2002a) observed many students using is what one third grader called the "middle clump" (Case 21). A middle clump is a cluster of values in the heart of the distribution. For example, a fourth-grade teacher had her students make a stacked dot plot of the number of people in their families. Using stars for emphasis, the students wrote the following statements below their plot (Russell et al., 2002a, Case 3):

- One person has 18 in her family.
- The range of the data 4–18.
- Most typical number of people in the family is 5 or 6.

This bullet summary includes descriptions of spread, center, and a value of special interest.

Additional research by Konold et al. (2002) suggested that students often select a "modal clump" so that it includes all, or most, of the ideal features of averages listed above. The clump of 5 to 6 in the distribution of family sizes in Russell et al. (2002a, Case 3) included the mode and the median and was near most of the data: two thirds of the cases were in the interval 4 to 7. In describing a distribution, statisticians often specify values for both center and spread. They might summarize this distribution by saying that its median is 6 and that the middle 50% of the data (the interquartile range) lies between 5 and 9. Konold et al. (2002) suggested that the modal clump potentially serves a somewhat similar purpose for students, letting them express at the same time both what is average and how variable the data are.

For example, in Case 12 of Russell et al. (2002a), third- and fourth-grade students worked with data indicating the number of years students' families had lived in town (see Figure 13.6).

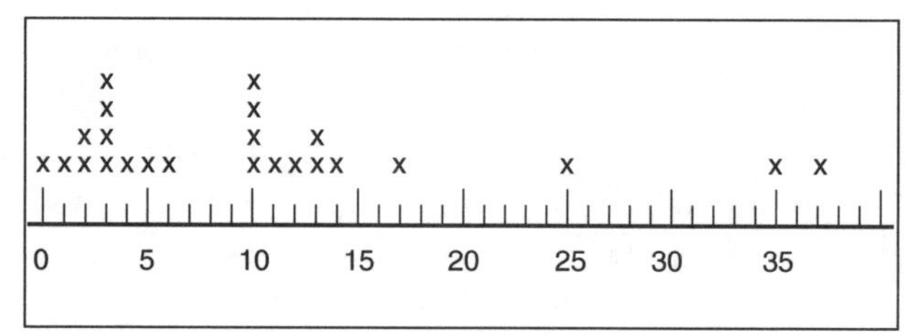

Figure 13.6. Number of years students' families had lived in town.

Excerpts from group reports and class discussion suggest that students were using the idea of modal clumps to make sense of the data as a whole:

Zia & Tuyet: Not a lot of people have lived here for a long time. A lot of people have been here for three years. There are two big clumps and a lot of gaps.

Anna, Amber, & Janis: At the beginning we can plainly see that there is a very big clump of 0, 1, 2, 3, 4, and 5.

Kevin: Most of the Xs are between 0 and 6. It's the biggest clump.

Anna: There's eleven [in the clump]. That's almost half.

Reflecting on how her students had summarized the data, the teacher noted:

> That first big clump clearly needed to be part of it [the summary] in the kids' eyes, and the fact that it also contained that mode at 3 didn't hurt either.... "Almost half" was good enough for them. It seemed to carry some weight of significance, and as I thought about it, I realized that it did for me also. This was a meaningful statement to make about our data.... (Russell et al., 2002a, Case 12).

Cobb (1999) described how the seventh graders in his teaching experiment began reasoning about data sets as wholes once they were able to isolate and converse about what they called the "hills" in the data. Many researchers have emphasized that we should encourage students to use less formal methods like these, and that in many situations what students come up with as descriptors of average are as good as, or better than, the mean or median in summarizing the data (Bakker, 2001; Cobb, 1999; Mokros & Russell, 1995). Not only are such averages as modal clumps (or hills) often good enough for the task at hand, but such ideas can provide the bases for later constructing meaningful interpretations of more traditional averages, such as means and medians.

Interpretations of averages. Of course, as with any average, not all students who describe a modal clump are thinking of it as a description of the group as a whole. Just as they often do with the mode, some students think of the modal clump simply as the "winner." Indeed, they may be using the clump as an extended mode, as a description of the most common values.

To explore further how students think about and use averages, we need to distinguish more carefully between the types of averages they use (modes, medians, midranges, modal clumps, etc.) and the meanings they give those averages. A fourth-grade teacher in the study by Russell et al. (2002a, Case 26) made this distinction as she tried to explore what significance her students gave to the mean they had just computed. After Robert answered that they had "just used a calculator" to get the average, the teacher made this clarification:

Teacher: I'm actually not asking you *how* you got the answer. I'm asking what your answer means. Do you have a sense of that?

Robert: Not really.

We might argue that Robert interpreted the mean simply as a result of a computation, which would explain why he answered the way he did. Earlier in the episode, others had questioned why Robert and his group had used only four values to get an average to decide how many fourth graders they would need to build a 100-foot tidal wave. To others in the class, this sample seemed too small. Robert replied, "You don't need a whole lot of people to get an average," by which he may have meant that to get an answer by adding and dividing, four will do. Of the 20 students Mokros and Russell (1995) interviewed, 3 seemed to view averages just as Robert did, simply as a procedure. The researchers described this approach as"algorithmic" and described it along with four other approaches that they observed students using.

Konold and Pollatsek (2002) have suggested a somewhat different set of "interpretations," which include among them ones that more experienced data analysts employ. Konold and Pollatsek (2002) characterized an interpretation as "the goal a person has in mind when he or she computes or uses an average. It is the answer a person might give to the question 'Why did you compute the average of those values?'" (p. 267). The interpretations they describe include average as a data reducer, as a fair share, and as a typical score.

Viewed as a data reducer, an average is a way to boil down a set of numbers into one value. A high school student interviewed by Konold, Pollatsek, Well, and Gagnon (1997) gave the following rationale for why she would use a mean or median to describe the number of hours worked by students at her school:

> We could look at the mean of the hours they worked, or the median... It would go through a lot to see what every, each person works. I mean, that's kind of a lot, but you could look at the mean. (Konold and Pollatsek, 2002, p. 268)

The data need to be reduced because of their complexity, in particular because of the difficulty of holding in memory the individual values. The student quoted above went on to say, "You could just go through every one... [but] you're not going to remember all that."

An average is interpreted as a fair share when we imagine redistributing a quantity among individuals so that in the end each has the same amount. The computation for the mean is probably first encountered in elementary school in the context of fair-share problems, with no reference to the result's being a mean or average. Most of the tasks Strauss and Bichler (1988) used in their research were problems that seemed to suggest a fair-share interpretation, such as in the following example:

> Ruth brought 5 pieces of candy, Yael brought 10 pieces, Nadav brought 20, and Ami brought 25. The children who brought many gave some to those who brought few until everyone had the same number of candies. How many candies did each girl end up with? (Adapted from Strauss & Bichler, 1988)

An average interpreted as a typical score, Konold and Pollatsek (2002) suggested, includes ideas related to the majority, mode, median, and midrange. Teachers in the Russell et al. (2002a) casebook often posed questions to students hoping to elicit this interpretation of a "typical" or "representative" value. The teacher posing a question such as "How tall is a typical fourth grader?" is presumably thinking of a value that is representative of the entire group.

However, many of the students' responses suggested they believed a typical value described a characteristic of a particular case, or set of cases, in the distribution. A third-grade teacher (Russell et al., 2002a, Case 22) asked her class, "What would you say is the average height of kids in our room?" Brita volunteered, "It's me. I think I am average." She seemed not to be focusing on an average height but rather on a characteristic of a person: "I'm average." After students lined up by height, other students used a similar notion in describing average, viewing average as a characteristic of a person rather than the group.

Phoebe: I think I'm taller than average because I notice that on the playground.

Brita: I was right. Sam is average, and I'm average too. We are the same.

Tiffany: I'm average too.

Katie: I'm not average. I'm shorter.

To claim that Sam is of "typical" or "average" height is to characterize him and not necessarily the group as a whole. We are unsure whether these students would consider Sam's height to be a good characterization of the whole group any more than they would consider Katie's diminutive height to be a useful summary. This use of averages to describe particular individuals rather than the group is supported by common usage in which we frequently speak of the "average" or "typical" student.

Mokros and Russell (1995) suggested that students who are beginning to make sense of averages as representative values may draw on intuitive ideas about what seems "reasonable." Students using this "reasonable" approach drew on their experiences with the phenomenon to explain, for example, why all the values would not be the same. This approach was used by 20% of the students they interviewed. Here is how a fourth grader described the distribution of allowances she had created by placing individual tiles on a stacked dot plot such that the average would be $1.50:

> [It depends] on how old they are.... If there are some kids that were like 15 and 16 years old and there are other kids that were 10 years old.... It depends on how rich their parents are sometimes.... If the typical [allowance] is $1.50, you're not going to really go above $5.00 for any kid. If I got $5.00, it would be good.... And you know that when you run around with a lot of kids, most of them are like $1.50 or $1.75 or $1.25 or $1.00, something like that. (p. 30)

Students using this approach also relied on intuitions that averages are roughly in the center, thinking of an average not as a precise location but as an around-about sort of thing.

Interviewer: Tell me how you're thinking about this one.

Suzanne: Well, just as they get higher, sometimes they should get lower. And you said the typical allowance is about $1.50, so some kids can get $1.50. And if it were $1.75 that would be pretty close and so would [$1.25], because that's around it.... Parents don't like to waste their money on kids. (p. 30)

Comparing Groups

If we consider statistics instruction as a staircase beginning in early elementary school and continuing up to Grade 12, the ability to compare two groups should be seen as a major landing midway up the stairs. We might think of it as the place where instruction in the early years is headed and as the foundation from which further statistics will rise. Making such comparisons is the heart of statistics. Most of the important issues and questions argued with data amount to comparing two groups, for example, treatment and control groups in medicine, before-and-after groups in various interventions and educational studies, and females versus males in gender equity studies. Furthermore, students will not understand the rationale of statistical inference until they first are comfortable summarizing a difference by comparing two groups using some measure of center. Nor are the skills required in comparing groups separate from those that allow us to perceive trends and patterns in the first place. When we go snooping about in data, we are searching for just these sorts of group differences, and the question is, How do we spot them?

Research has demonstrated that students do not initially know how to approach a group-comparison problem. This is true even of students who appear to know quite a bit about averages. Gal, Rothschild and Wagner (1990) interviewed students in Grades 3, 6, and 9 to determine, among other things, their understanding of how means were computed and used. They also gave the students nine pairs of stacked dot plots. In one version of their materials, graphs showed the results of a frog-leaping contest between two teams, with Xs on the graphs representing the distances jumped by individual frogs of each team. The students' task was to decide from the data which team won the contest. Only half of the sixth- and ninth-grade students who knew how to compute means went on to use means to compare the two groups, even when the groups were of unequal size.

A number of studies have observed students who appeared to use averages to describe a single group but did not use them to compare two groups (Hancock et al., 1992; Watson & Moritz, 1999; Jones et al., 1999; Konold et al., 2002). Bright and Friel (1998), for example, questioned eighth-grade students about a stem-and-leaf plot that showed the heights of 28 students who did not play basketball. They then showed them a stem-and-leaf plot that included these data along with the heights of 23 basketball players. The heights of basketball players were indicated, as they are in Figure 13.7, in bold type.

10	
11	
12	
13	8 8 8 9
14	1 2 4 7 7 7
15	0 0 1 1 1 1 2 2 2 2 3 3 5 6 6 7 8
16	
17	1
18	**0 3 5**
19	**0 2 5 7 8 8**
20	**0 0 2 3 5 5 5 5 7**
21	**0 0 0 5**
22	**0**
23	

Figure 13.7. Heights of students and basketball players (bold); adapted from Bright and Friel (1998, p. 81). The row headed by 13 (the stem) contains four cases (leaves), three students each of 138 centimeters, and a fourth student of 139 centimeters.

Asked about the "typical height" in the distribution of the heights of nonplayers, two of four interviewed students specified a modal clump (e.g., 147–151 cm). But shown the plot with both distributions, these students could not generalize their method to determine "How much taller are basketball players than students?" The students who did make comparisons compared selected individuals from each group (e.g., pointed out that the tallest student was shorter than the shortest basketball player). In the words of Bright and Friel (1998), some of these students could "describe a 'typical' student or basketball player, but they did not make the inference that the 'typical difference' in heights could be represented by the 'difference in typicals'" (p. 80).

Konold et al. (1997) found similar difficulties among high school seniors who had just completed a yearlong course in probability and statistics. On many occasions during the course, the students had used both medians and means to compare groups. But in the classroom they were supported by the curricula, software, and instructor and thus to a large extent were not choosing for themselves the methods they would use. During a postcourse interview, where they were less constrained in their choice of methods, they seldom used medians, means, or percentages when comparing two groups, though they did use them when summarizing single groups.

Using averages to compare two groups requires viewing averages as a way to represent or describe the entire group and not just a part of it. For this reason, Konold et al. (1997) argued that students' reluctance to use averages to compare two groups suggests that they have not developed a sense of average as a measure of a group characteristic that can be used to represent the group. Thus, students may use averages as part of a description of a single group and yet not accept them as representative of that group. The students interviewed by Bright and Friel (1998) may have been thinking of typicality as a characteristic of the heights of just those people in the center of a distribution. If so, then it is understandable that they would not consider using that modal clump as a characteristic or measure of the whole group for the purpose of comparing it to another group.

The cases in Russell et al. (2002a) demonstrate the power of comparison tasks to engage student interest. Even more important, they appear to prompt students to shift their focus from individuals in a data set to the group as a whole. However, as a teacher notes in Case 27, comparison problems can be "a little puzzling to kids at this age [Grades 3 and 4]—how can you talk about the group, after all, as something somewhat separate from the individuals in the group?" Indeed, the power of comparison questions is that they provide a clear problem that must be resolved—the problem of what to use as group measures that can be compared.

The teaching experiment by Cobb (1999) does suggest that students can appropriate the idea of modal clumps for the purpose of comparing groups. As the researchers were designing the teaching experiment, they had planned on developing the median as the primary indicator of center. However, though the median was a frequent topic of class discussion, students rarely used it to compare groups. Students made most of their decisions about group difference by comparing the numbers of individuals in each group within narrow slices of the range. Konold et al. (1997) reported the same tendency to compare slices among high school students , even when the two groups were of radically different sizes.

Cobb (1999) described a critical episode during the seventh-grade teaching experiment when a student, Janice, used the idea of modal clumps, or "hills" as she called them, to compare two groups. The class was investigating speeds of two groups of cars sampled before and after a police speed trap. During a class group discussion, Janice suggested:

> If you look at the graphs and look at them like hills, then for the before group, the speeds are spread out and more than 55, and if you look at the after graph, then more people are bunched up close to the speed limit [50 mph], which means that the majority of the people slowed down close to the speed limit. (p. 19)

Cobb (1999) reported that this was the first time in a whole-group discussion that a student had "described a data set in global, qualitative terms by referring to its shape" (p. 19). This language was soon taken up by other students in the class and became a standard way for them to describe and compare groups. As they progressed to comparing data sets of different sizes, they began talking not just about the location of two hills but about the number of cases in a hill relative to the number of cases in its group.

Judgments About Covariation

In analyzing data, we frequently want to know whether and how two things are related. Are increases in air-borne pollutants related to warmer-than-average temperatures? Does the death penalty act as a deterrent to serious crime? Is classroom size related to student achievement? In the previous section, we focused on group-comparison questions in which one variable was numeric (such as speed of car) and the other was categorical (such as timing of observation). We can think about comparing such groups as trying to decide whether two variables are related—whether their values covary. Concluding from data that cars on a certain stretch of road tend to go slower after a speed trap than before implies that a car's speed varies with the time of observation (before versus after the speed trap).

In this section, we review research related to judgments about two other types of covariation: determining whether a relationship exists between (1) two categorical variables (e.g., Is having a curfew related to gender?) and (2) two numeric variables (e.g., Does blood pressure increase with age?). As in the previous section, we concern ourselves here with how students learn to examine data in an attempt to detect such relationships. We do not review the research on how students decide whether a trend is strong enough (i.e., statistically significant) to justify concluding that the relationship exists.

Comparing two categorical variables. To make judgments about the relation between two categorical variables, we often display data in a contingency table. Instructors of introductory college courses are often heard to say that contingency tables are the hardest displays their students encounter. The difficulty of reasoning about such data has been borne out in numerous studies.

For example, Batanero, Estepa, Godino and Green (1996) gave data like those in Table 13.1 to more than 200 Spanish students in their final year of secondary school. Their task was to decide whether the data indicated a relationship between the two variables. Most of the students had had instruction in elementary statistics and probability, but they had not yet been introduced to contingency tables.

TABLE 13.1. Incidents of bronchial disease among smokers and nonsmokers; adapted from Batanero et al. (1996).

	Bronchial disease	No bronchial disease	Total
Smoke	90	60	150
Not smoke	60	40	100
Total	150	100	250

According to the data in Table 13.1, smoking is not associated with bronchial disease. One way to see this lack of dependency is by comparing the percentage of the two groups who have the disease. Of the 150 smokers, 90 (or 60%) have bronchial disease, but 60% of the nonsmokers also have it. We would therefore conclude on the basis of these data that smoking has no apparent effect on bronchitis (but, of course, these data are fictional).

Students in the Batanero et al. (1996) study solved about 30% of such problems correctly, a high success rate compared to results from similar studies. The majority of incorrect responses were, however, well predicted by previous research. Many students (13% in the Batanero et al. study) seemed to attend only to the number 90 in deciding that smoking is associated with bronchitis. Even more common was for students to attend to the fact that 90 of the 150 smokers had bronchial disease. Because more of the smokers had the disease than did not, many students believed that the data from the smokers alone was sufficient to establish a connection between smoking and the disease.

The high school seniors interviewed after a statistics course by Konold et al. (1997) used a similar strategy in evaluating data collected from students at their school. Two students, M and J, who were analyzing these data using statistical software, posed the question of whether more female students held jobs than male students. They generated the contingency table shown in Table 13.2. The interview excerpt below suggests that these students attended only to the information about those who held jobs and that they viewed the data about students who did not hold jobs as irrelevant to their question.

TABLE 13.2. Incidents of jobs among male and female high school students; adapted from Konold et al. (1997).

	Job		
Sex	No	Yes	Total
F	23 (0.29)	57 (0.71)	80
M	16 (0.22)	57 (0.78)	73
Total	39 (0.25)	114 (0.75)	153

M: So, oh no, it's pretty even.

I: So tell me how—So, you're, you're looking at the percentage of males and—

M: Yeah.

J: Yeah, the difference.

M: Yeah, of females and males who have jobs.

J: Or who don't.

M: And that don't. And for the amount of—that do have jobs, that females and males are pretty even.

I: So what—tell me the numbers. I can't read them.

M: 57 males and 57 females. (p. 158)

In correcting J's *or* to *and*, M seemed to be stressing that the question about males and females who hold jobs was, in her mind, separate from the question about those would did not hold jobs. In her subsequent statements, M made it clear that she was looking at those who held jobs in forming her conclusion. But because the numbers of males and females are different, we cannot simply compare the numbers of males and females who have jobs. We must compare the rate of jobs among males to that among females to answer the question.

We frequently see this kind of reasoning in news reports or advertising where we are given only half the information we need. "Seventy percent of headache sufferers who take AcheAway feel relief within an hour." The implication is that AcheAway did the trick. But what percent of suffers who take nothing feel relief within an hour? For all we know, it is 80%, and AcheAway tends to make things worse.

Follow-up research by Batanero, Estepa, and Godino (1997) suggests just how difficult it is for students to learn to reason about relationships between two categorical variables. They worked with 19 high school students for approximately 30 hours in an intervention that included instruction on group differences and detection of covariation in numeric and categorical data. Before instruction, the students could correctly solve about 20% of the problems involving contingency tables. After instruction, this value went up to only about 30%. In contrast, before instruction, students could successfully evaluate relationships in scatterplots 35% of the time; this value increased to 60% after instruction.

Comparing two numeric variables

To explore relationships between two numeric variables, data are traditionally displayed in two dimensions, with each case represented by a point. The horizontal position of a case is determined from its value on one numeric variable, and its vertical position is determined by its value on the other numeric variable. Data that vary over time (time series) are often displayed this way. Because time series are true functions—only one y value is associated with every x value—we frequently connect the points with a line, which can help us see general trends in the data.

Figure 13.8 shows the time series of gold-medal times for the men's 100-meter dash for all the years since the Olympics began. Ben-Zvi and Arcavi (2001) gave these data to 13-year-old Israeli students who were participating in a teaching experiment. The students' first task was a general one—to learn what they could from these data. Though this plot could be rescaled to make it easier to perceive, we can still detect an overall downward trend, with running times generally improving (i.e., getting smaller) over the years. To see this trend, we must focus on the data as a whole. As we saw with the task of group comparisons, many students attend to particular data values, or pairs of values, and therefore do not perceive such trends.

The students in this teaching experiment, A and D, were working with this plot and a corresponding table of values. They initially focused only on single values.

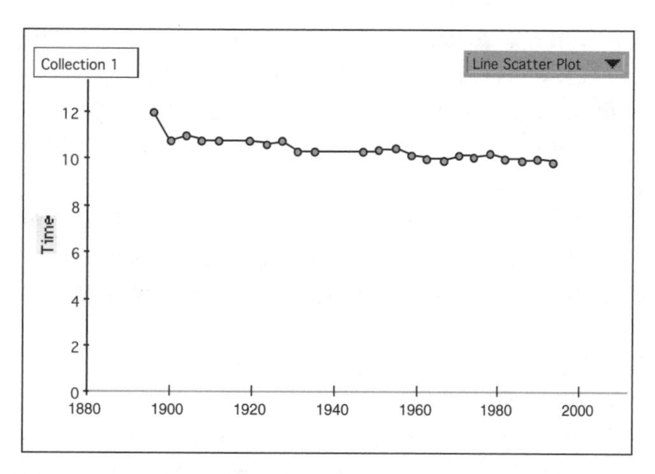

Figure 13.8. Olympic gold medal times for the men's 100-meter dash. Reproduced using Fathom; adapted from Figure 13.2 in Ben-Zvi and Arcavi (2001).

A: [Reading the question] What do you learn from this graph? We learn from the graph, in which year there was which running time.

D: What running time was achieved in what year.

Their teacher intervened in an effort to help them move from these localized perceptions to more global ones.

T: Does it decrease all the time?

A&D: No.

T: No. Does it increase all the time?

A&D: No.

T: No. So, what does it do after all? ...

D: It generally changes from Olympiad to Olympiad. Generally, not always. (p. 53)

With additional prompting from their teacher, the students began to notice also that the differences between adjacent years were not constant. The researchers speculated that the students may have been drawing on earlier experience they had had using spreadsheets to explore linear functions in which the differences between adjacent values were constant. This experience may have provided the background against which the fluctuating differences in the Olympic data stood out and became the focal point for the students.

With continued prompting, the students began to take note of the overall trend, concluding in their written report,

> The overall direction is increase in the records, yet there were occasionally lower (slower) results, than the ones achieved in previous Olympiads. (p. 54)

Scatterplot displays are similar to the time-series plot except that for every value of x, multiple values of y may occur. Thus, with scatterplots, not only is the trend not constant, but a scattering of values further masks that trend. The scatterplot in Figure 13.9 shows the relationship between students' weights and the weights of their backpacks for 55 elementary school students. Heavier students tend to carry heavier backpacks, though we can see plenty of exceptions to this trend. This, by the way, is a good example of the importance of the warning that "correlation does not prove causation." It is probably not the case that students carry more weight in their backpacks because they are heavier. Rather, the students are getting heavier as they age and move into the higher grades. And as any sixth grader will tell you, they tend to get more homework assignments than do first graders.

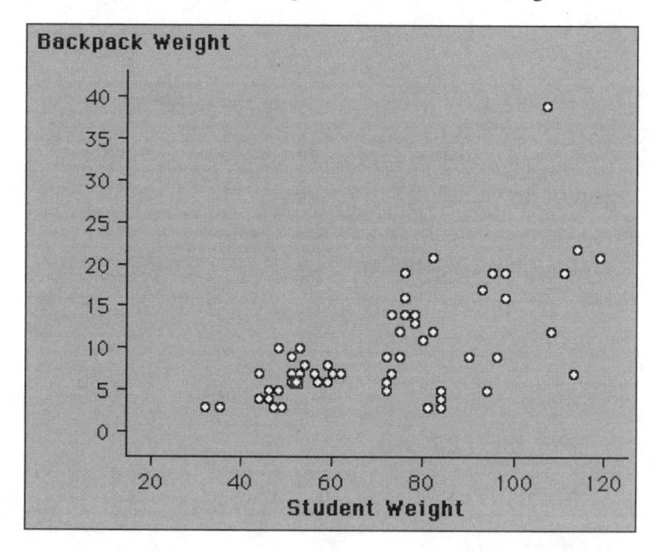

Figure 13.9. The weights (in pounds) of students and their backpacks, plotted using Tinkerplots.

In research on reasoning in the workplace, Noss, Pozzi, and Hoyles (1999) studied the statistical reasoning of practicing nurses. As part of a follow-up teaching experiment, nurses analyzed a health database of British adults. Using statistical software, the nurses quickly generated a scatterplot to answer the question of whether a relationship existed between age and blood pressure. The nurses knew from experience that with increasing age, blood pressure tends to rise. However, they were unable to see this relationship in the scatterplot, the trend apparently masked by the variability in the data. The researchers prompted the nurses to make several vertical slices in the scatterplot, essentially recomposing the continuous age variable into several discrete categories (e.g., ages 18–29, 30–44, 45–59, 60–75). By computing and comparing the average blood pressure in each of these groupings, the nurses then could perceive the trend they had expected.

Cobb and his associates have used this same "slicing" technique in their teaching experiment with middle school students, providing them computer "minitools" that allow students to form such vertical partitions easily (Cobb, McClain, & Gravemeijer, in press). Their intention is to build systematically on understandings that students first develop from comparing two sets of numeric data. By seeing each vertical slice of data in a scatterplot as a distribution of a discrete group, their hope is that students can visually locate the centers (or "hills") of each slice to detect a general trend.

However, scatterplots may not be the best representation for novices to begin using as they explore covariation in two numeric variables. Emerging evidence suggests that it may be easier for students to perceive relationships in two-column tables or case-value plots in which the values of one variable have been ordered and displayed next to the corresponding values of the other variable (see Konold, 2002). Figure 13.10 shows such a paired case-value plot for the same backpack data displayed previously. Knowing that the cases have been ordered according to student weight, one can see the increasing trend of backpack weight in the plot on the right by tracking the changing length of the bars while moving one's eye up the plot.

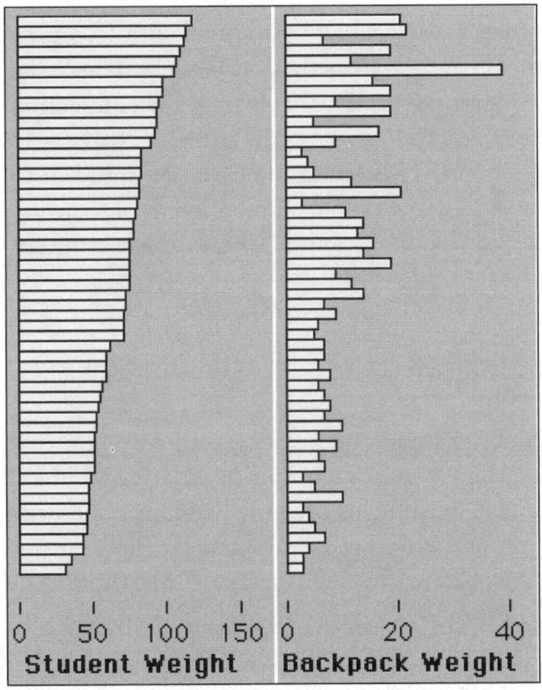

Figure 13.10. The weights (in pounds) of students and their backpacks, plotted using Tinkerplots. The two bars immediately across from each other belong to the same student.

Cobb and his colleagues (in press) mention that before they had introduced their students to scatterplots, they gave them data on carbon dioxide produced by cars going various speeds. They asked students to generate a plot that would allow them to make recommendations about highway speed limits. None of the students made scatterplots; nearly half of them drew a paired double-bar graph like the one shown in Figure 13.10. At the end of the eighth-grade teaching experiment, Cobb and colleagues interviewed 11 of the students. The researchers gave students data from a similar context and asked them to choose from among several representations the one that would best help them judge whether there was a relationship between the two variables. Even though the scatterplot was the only representation they had worked with during the 14-week teaching experiment, only four of the students chose a scatterplot display to make that judgment. The other students chose either a paired double-bar graph or the corresponding table of paired values (J. Cortina, personal communication, 21 December 2000).

Relating Data to the Observed Event

We have discussed some of the issues that arise as students generate, represent, and interpret authentic or meaningful data. Previously we emphasized that in making sense of data, students need to develop an understanding of data as being related to, but not identical with, real events—as models of those events. After reviewing some of the complexities students face as they construct the idea of data as an aggregate and begin to work through problems of data analysis, we find that we have come full circle. The further students become immersed in the tools of data analysis, the more important it becomes that they maintain the connection between the data, the events they represent, and the question that motivated their construction. Too often students treat data as numbers only, forgetting that those numbers have a context and that the reason for analyzing them is to learn more about that context. Students seem particularly vulnerable to treating data as numbers only when they work with data they themselves have not collected.

Cobb (1999) reported that before instruction, the seventh-grade students with whom he and his colleagues worked tended to view data analysis as "doing something with the numbers" (p. 12). In summarizing student responses, Cobb concluded that it was "doubtful whether most of the students were actually analyzing data, in that the numbers they manipulated did not appear to signify measures of attributes of a situation about which a decision was to be made" (p. 13). Early in the subsequent teaching experiment, the researchers saw the same tendency as students began reasoning about the data in Figure 13.11, which showed hours of use of "Always Ready" and "Tough Cell" batteries displayed as a case-value plot.

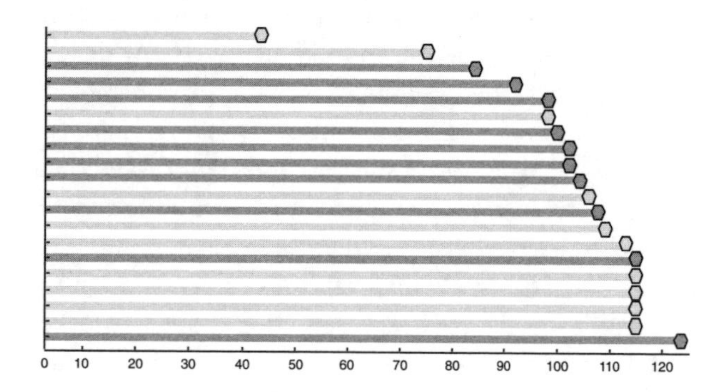

Figure 13.11. Case-value plot of hours of use of "Always Ready" (light gray) and "Tough Cell" (dark gray) batteries; adapted from Cobb (1999).

The two brands of batteries appeared as green and pink bars on the computer screen the students were viewing. In Figure 13.11, as well as in the class dialogue below, we have changed these colors to light and dark gray, respectively. During the first day working with this display, the students referred mostly to numbers and colors. Noticing this tendency, the teacher began to encourage them to talk instead about *batteries*.

Casey: And I was saying, see like there's 7 [light gray] that last longer.

Teacher: OK, the [light gray] are the Always Ready, so let's make sure we keep up with which is which. OK?

Casey: OK, the Always Ready are more consistent with the 7 right there, and then 7 of the Tough ones are like further back, I was just saying 'cause like 7 out of 10 of the [light gray] were the longest, and like...

Ken: Good point.

Janice: I understand.

Teacher: You understand? OK, Janice, I'm not sure I do, so could you say it for me?

Janice: She's saying that out of 10 of the batteries that lasted the longest, 7 of them are [light gray], and that's the most number, so the Always Ready batteries are better because more of those batteries lasted longer. (pp. 14–15)

Cobb (1999) concluded that a an essential step in these students' learning to reason about data was coming to expect that statements and claims about various data displays should extend beyond mere numbers by making reference to a specific real-world situation.

In most of the activities described in the cases compiled by Russell et al. (2002a), students collected their own data. Yet we see clear examples of students' losing the connection between the data and the real-world situation. The failed connection often occurs as students begin learning how to describe general features of the data. For example, in Case 23, third-grade students wrote summaries describing a stacked dot plot of daily temperatures they had collected in February. One student described the data this way: "At first very spread out. Then it gets more bunched up."

Concerning the student who wrote this summary, the teacher wondered,

> Did he know it wasn't just a clump of Xs, but a representation of a real thing, which was indicating a predominance of a certain temperature on the high side of the range of temperatures for the month? ... I wondered how to help him see that what he noticed about how the data *looked* implied something significant about what the temperature was like in February. (p. 134)

Feldman et al. (2000) cited several examples from data-intensive science projects of what happens when we give students data about phenomena far removed from their experience and with no clear questions to answer. They concluded that

> nearly every problem associated with...keeping them engaged in analysis ultimately stems from students not making, or losing, the connection between the data they have and a real-world question. This being the case, the solution to most of the problems can be found in focusing on how to make and maintain these connections. (p. 127)

Conclusion

The interdependency of the stages of data analysis places high demands on young students and their teachers. The students are only beginning to learn how to turn observations into data and are not yet aware of how they can probe data to answer questions. Unlike the expert, novices have little relevant experience they can use to plan ahead. Helping students raise questions that interest them and that they can productively pursue is a challenge for the teacher. Left on their own, students are often overwhelmed or wander off track; given too much structure and assistance, they can lose sight of the big picture and their motivation for looking at data.

The challenge for teachers is to "help students keep hold of the big picture as they explore the parts" (Russell et al., 2002a, Case 21). Doing so requires finding ways to manage complexity so that students can think about the questions they are pursuing, focusing not on "features of the graph" per se, but on "its implications as a representation of something real."

The research and cases we have presented call attention to the need for students to work with real data throughout the elementary and middle grades. Understanding data representation and analysis involves many complex issues, from sorting through what different numbers on a graph mean, to choosing appropriate measures to summarize and compare groups, to identifying relationships between variables. Through multiple experiences with a variety of data sets, students begin to develop the tools and concepts they need to use data themselves and to interpret the data they will encounter throughout life.

ACKNOWLEDGMENTS

Portions from this chapter appear in an article titled "Highlights of Related Research," by Clifford Konold and Traci L. Higgins, in *Working With Data*, edited by Susan Jo Russell, Deborah Schifter, and Virginia Bastable, pages 165–201 (Parsippany, N. J.: Dale Seymour Publications, 2002). These portions are included here with permission.

The writing of this chapter was supported in part by the National Science Foundation under Grant Nos. REC-9725228, ESI-9730683, ESI-9731064, and ESI-9818946. Any opinions, findings, conclusions, or recommendations expressed here are those of the authors and do not necessarily reflect the views of the National Science Foundation.

Joan Garfield and John Carter provided comments on an earlier version of the chapter.

REFERENCES

American Statistical Association. (1997). *Report from the American Statistical Associations' Advisory Review Group.* Alexandria, VA: Author. Available: http://www.stat.ncsu.edu /stated/standards.html

Bakker, A. (2001). Historical and didactical phenomenology of the average values. In P. Radelet-de Grave (Ed.), *Proceedings of the Conference on History and Epistemology in Mathematics Education* (Vol. 1, pp. 91–106). Louvain-la-Neuve & Leuven, Belgium: Catholic Universities of Louvain-la-Neuve & Leuven.

Batanero, C., Estepa, A., & Godino, J. D. (1997). Evolution of students' understanding of statistical association in a computer-based teaching environment. In J. B. Garfield & G. Burrill (Eds.), *Research on the role of technology in teaching and learning statistics: Proceedings of the 1996 IASE Round Table Conference* (pp. 191–205). Voorburg, The Netherlands: International Statistical Institute.

Batanero, C., Estepa, A., Godino, J. D., & Green, D. R. (1996). Intuitive strategies and preconceptions about association in contingency tables. *Journal for Research in Mathematics Education, 27*, 151–169.

Biehler, R. (1989). Educational perspectives on exploratory data analysis. In R. Morris (Ed.), *Studies in mathematics education, Vol. 7: The teaching of statistics* (pp. 185–201). Paris: UNESCO.

Ben-Zvi, D., & Arcavi, A. (2001). Junior high school students' construction of global views of data and data representations. *Educational Studies in Mathematics, 45*, 35–65.

Bright, G. W., & Friel, S. N. (1998). Graphical representations: Helping students interpret data. In S. P. Lajoie (Ed.), *Reflections on statistics: Learning, teaching, and assessment in grades K–12* (pp. 63–88). Mahwah, NJ: Erlbaum.

Cai, J. (1998). Exploring students' conceptual understanding of the averaging algorithm. *School Science and Mathematics, 98*, 93–98.

Cleveland, W. S. (1993). *Visualizing data.* Summit, NJ: Hobart Press.

Cobb, G. W. (1997). Mere literacy is not enough. In L. A. Steen (Ed.), *Why numbers count: Quantitative literacy for tomorrow's America* (pp. 75–90). New York: College Entrance Examination Board.

Cobb, P. (1999). Individual and collective mathematical development: The case of statistical data analysis. *Mathematical Thinking and Learning, 1*(1), 5–43.

Cobb, P., McClain, K., & Gravemeijer, K. (in press). Learning about statistical covariation. *Cognition and Instruction.*

Eicholz, R. (1991). *Addison-Wesley Mathematics—Grade 5.* Reading, MA: Addison-Wesley.

Eisenhart, C. (1971). *The development of the concept of the best mean of a set of measurements from antiquity to the present day.* Unpublished notes from the presidential address at the 131st meeting of the American Statistical Association, Fort Collins, Colorado.

Feldman, A., Konold, C., & Coulter, R. (2000). *Network science, a decade later: The Internet and classroom learning.* Mahwah, NJ: Erlbaum.

Friel, S. N., & Bright, G. W. (1998). Teach-Stat: A model for professional development in data analysis and statistics for teachers K–6. In S. P. Lajoie (Ed.), *Reflections on statistics: Learning, teaching, and assessment in grades K–12* (pp. 89–117). Mahwah, NJ: Erlbaum.

Gal, I., Rothschild, K., & Wagner, D. A. (1990). *Statistical concepts and statistical reasoning in school children: Convergence or divergence?* Paper presented at the meeting of the American Educational Research Association, Boston.

Gallup, G. (1972). Opinion polling in a democracy. In J. M. Tanur, F. Mosteller, W. H. Kruskal, R. F. Link, R. S. Pieters, & G. R. Rising (Eds.), *Statistics: A guide to the unknown* (pp. 146–152). San Francisco: Holden-Day.

Garfield, J., & Ahlgren, A. (1988). Difficulties in learning basic concepts in probability and statistics: Implications for research. *Journal for Research in Mathematics Education, 19*, 44–63.

Hancock, C., Kaput, J. J., & Goldsmith, L. T. (1992). Authentic inquiry with data: Critical barriers to classroom implementation. *Educational Psychologist 27*, 337–364.

Hand, D. J. (1998). Data mining: Statistics and more? *The American Statistician, 52*, 112–118.

Jacobs, V. R. (1999). How do students think about statistical sampling before instruction? *Mathematics Teaching in the Middle School, 5*(4), 240–246, 263.

Jones, G. A., Thornton, C. A., Langrall, C. W., Mooney, E. S., Perry, B., & Putt, I. J. (1999). *A framework for assessing students' statistical thinking.* Paper presented at the meeting of the Research Presession of the National Council of Teachers of Mathematics, San Francisco.

Konold, C. (2002). Teaching concepts rather than conventions. *New England Journal of Mathematics*, 34(2), 69–81.

Konold, C., Higgins, T., & Russell, S. J. (2000). Developing statistical perspectives in the elementary grades. In M. L. Fernandez (Ed.), *Proceedings of the 22nd Annual Meeting of the North American Chapter of the International Group for the Psychology of Mathematics Education* (p. 329). Tucson, Arizona.

Konold, C., & Miller, C. M. (2002). Tinkerplots [Computer software, in development]. Amherst, MA: University of Massachusetts, SRRI.

Konold, C., & Pollatsek, A. (2002). Data analysis as the search for signals in noisy processess. *Journal for Research in Mathematics Education, 33*, 259–289.

Konold, C., Pollatsek, A., Well, A., & Gagnon, A. (1997). Students analyzing data: Research of critical barriers. In J. B. Garfield & G. Burrill (Eds.), *Research on the role of technology in teaching and learning statistics: 1996 Proceedings of the 1996 IASE Round Table Conference* (pp. 151–167). Voorburg, The Netherlands: International Statistical Institute.

Konold, C., Robinson, A., Khalil, K., Pollatsek, A., Well, A., Wing, R., & Mayr, S. (2002). *Students' use of modal clumps to summarize data.* Paper presented at the Sixth International Conference on Teaching Statistics, Cape Town, South Africa.

Lehrer, R., & Romberg, T. (1996). Exploring children's data modeling. *Cognition and Instruction 14* (1), 69–108.

Metz, K. E. (1999). Why sampling works or why it can't: Ideas of young children engaged in research of their own design. In F. Hitt & M. Santos (Eds.), *Proceedings of the 21st Annual Meeting of the North American Chapter of the International Group for the Psychology of Mathematics Education* (pp. 492–498), Morelos, Mexico.

Mokros, J., & Russell, S. (1995). Children's concepts of average and representativeness. *Journal for Research in Mathematics Education, 26*, 20–39.

Moore, D. S. (1990). Uncertainty. In L. A. Steen (Ed.), *On the shoulders of giants* (pp. 95–137). Washington, DC: National Academy Press.

Moore, D. S. (1992). Teaching statistics as a respectable subject. In F. S. Gordon & S. P. Gordon (Eds.), *Statistics for the twenty-first century* (MAA Notes no. 26, pp. 14–25). Washington, DC: Mathematical Association of America.

National Council of Teachers of Mathematics. (1989). *Curriculum and evaluation standards for school mathematics.* Reston, VA: Author.

Noss, R., Pozzi, S., & Hoyles, C. (1999). Touching epistemologies: Meanings of average and variation in nursing practice. *Educational Studies in Mathematics, 40*, 25–51.

Pollatsek, A., Lima, S., & Well A. D. (1987). Concept of computation: Students' understanding of the mean. *Educational Studies in Mathematics, 12*, 191–204.

Rosebery, A. S., & Warren, B. (1992). Appropriating scientific discourse: Findings from language minority classrooms. *Journal of the Learning Sciences, 2*(1), 61–94.

Roth, W.-M., & Bowen, G. M. (1994). Mathematization of experience in a grade 8 open-inquiry environment: An introduction to the representational practices of science. *Journal of Research in Science Teaching, 31*, 293–318.

Roth, W.-M., & McGinn, M. K. (1997). Graphing: Cognitive ability or practice? *Science Education, 81*(1), 91–106.

Russell, S. J., Schifter, D., & Bastable, V. (2002a). *Developing mathematical ideas: Working with data.* Parsippany, NJ: Dale Seymour Publications.

Russell, S. J., Schifter, D., & Bastable, V. (2002b). [Fourth-graders representing data of height on stacked dot plot.] Unpublished raw data.

Russell, S. J., Schifter, D., & Bastable, V. (2002c). [Fifth-grade students comparing graphs.] Unpublished raw data.

Scheaffer, R. L., Watkins, A. E., & Landwehr, J. M. (1998). What every high-school graduate should know about statistics. In S. P. Lajoie (Ed.), *Reflections on statistics: Learning, teaching, and assessment in grades K–12* (pp. 3–31). Mahwah, NJ: Erlbaum.

Schwartz, D. L., Goldman, S. R., Vye, N. J., & Barron, B. J. (1998). Aligning everyday and mathematical reasoning: The case of sampling assumptions. In S. P. Lajoie (Ed.), *Reflections on statistics: Learning, teaching, and assessment in grades K–12* (pp. 233–273). Mahwah, NJ: Erlbaum.

Sconiers, S. (Ed.) (1999). *Bridges to classroom mathematics: Mathematics handbook.* Lexington, MA: COMAP.

Shaughnessy, J. M. (1992). Research in probability and statistics: Reflections and directions. In D. Grouws (Ed.), *Handbook of research on the teaching and learning of mathematics* (pp. 465–494). New York: Macmillan.

Strauss, S., & Bichler, E. (1988). The development of children's concepts of the arithmetic average. *Journal for Research in Mathematics Education, 19*, 64–80.

Tufte, E. R. (1983). *The visual display of quantitative information.* Cheshire, Connecticut: Graphics Press.

Tukey, J. W. (1977). *Exploratory data analysis.* Reading, MA: Addison-Wesley.

Watson, J. M., & Moritz, J.B. (1999). The beginning of statistical inference: Comparing two data sets. *Educational Studies in Mathematics, 37*, 145–168.

Watson, J. M., & Moritz, J.B. (2000). Developing concepts of sampling. *Journal for Research in Mathematics Education, 31*, 44–70.

Wild, C. J., & Pfannkuck, M. (1999). Statistical thinking in empirical enquiry. *International Statistical Review, 67*, 223–265.

CHAPTER 14

Research on Students' Understandings of Probability

J. Michael Shaughnessy, Portland State University

In the early 1970s, both mathematics educators and psychologists began to undertake systematic investigations of students' understandings and beliefs about chance events. Research in this area has continued to grow, and over the past decade investigations into students' understandings of probability and statistics have mushroomed to the point at which it has become difficult to stay current with all the national and international activity. Research on the teaching and learning of probability and statistics is regularly presented at the research presessions of the National Council of Teachers of Mathematics (NCTM), the International Group of Psychology and Mathematics Education (PME), and the Mathematics Education Research Group of Australasia (MERGA), as well as at sessions of the International Congress of Mathematics Education (ICME) and the International Conference on the Teaching of Statistics (ICOTS). The International Study Group for Research in Probability and Statistics publishes a quarterly electronic *Statistical Education Research Newsletter* that contains summaries of recent publications on data and chance and that supplies information about research to be presented at upcoming meetings. It is quite impossible to discuss all the aspects of the research into students' conceptions of probability in one short chapter.

In this chapter, I provide a discussion of selected discoveries that have been made about students' conceptions of probability. I also include a discussion of some trends in student performance on probability items on the 1996 National Assessment of Educational Progress (NAEP). In light of what is known about student thinking and performance in probability, I present some suggestions for the teaching of probability. Although this chapter focuses on probability, I point out that a separation of research discussions of probability and statistics is artificial, just as artificial as the separation of data and chance when teaching. *Principles and Standards for School Mathematics* (NCTM, 2000) aptly places probability and statistics under one shared heading. I believe the most interesting research questions for the future reside in the joint realm of the areas of probability and statistics, just as the most interesting teaching challenges for the future lie in making interconnections between these two areas.

Conceptions and Beliefs About Chance That Students Bring to the Classroom

Several summary analyses of research have provided a comprehensive and detailed picture of the complexities of secondary and tertiary students' thinking in probability (Borovcnik & Perd, 1996; Garfield & Ahlgren, 1988; Kapadia & Borovcnik, 1991; Shaughnessy, 1992). Metz (1998) has provided a comprehensive discussion of the emergence of ideas of chance among primary students. The litany of conceptions and beliefs about probability that has been identified in the research literature can appear to be a daunting list of potential roadblocks to students' understanding of probability. However, a more positive view of the research in probability is that it has helped to identify numerous teaching opportunities, opportunities for teachers to assess students' thinking about chance and to challenge their beliefs.

Researchers have investigated students' conceptions involving the ratio concept, belief in equiprobability, the randomness concept, the conjunction fallacy, the outcome phenomena, troubles with conditionals, the representativeness heuristic, and the availability heuristic. In addition, investigations have been made into young children's understandings of probability and children's growth in understanding under instruction. This is not an exhaustive list of all the research in understanding probability. However, research on each of these topics has repeatedly identified and documented students' probabilistic thinking at various levels, from primary students to tertiary students.

The Ratio Concept

Because probabilities are often expressed as ratios, an understanding of ratios provides the cornerstone for an understanding of probability. If young children were asked the item below, we might expect many of them to choose Bag B because it has more black counters.

> Two bags have black and white counters.
>
> Bag A: 3 black and 1 white
> Bag B: 6 black and 2 white
>
> Which bag gives the better chance of picking a black counter?
>
> A) Same chance
> B) Bag A
> C) Bag B
> D) Don't know
>
> Why?

What's interesting is that Green (1983) found that many older students still believe that Bag B gives a better chance of pulling a black. More than 50% of the several thousand 11- to 16-year-old students Green questioned chose Bag B. Furthermore, 39% of all Green's students gave as their reason, "because there are more blacks in Bag B." It is the *relative* size of a particular outcome in a probability experiment, not the absolute size, that students must focus on when comparing the likelihood of events. Metz (1998) notes that an understanding of part-whole relationships is fundamental to understanding probability. Part-whole relationships are the key for the bag problem. Metz reviewed research that found that 5-year-olds already possess important intuitions of part-whole relationships when they come to school. Part-whole relationships are the building blocks for the ratio concept, and a solid understanding of ratio and proportion is critical for understanding relative frequency, as noted by Fischbein, Pampu, and Manzit (1970).

Young Children's Understanding of Chance

A few years ago Kuzmak and Gelman (1986) conducted a study that gave strong support to the notion that young children are able to distinguish between random outcomes (balls in a gum jar) and predetermined outcomes (visible balls lined up ready to appear through a hole one at a time). Their study avoided much of the criticism of some previous work, such as that of Piaget and Inhelder (1975), in which the framing of tasks might have led young children to respond in particular ways. Piaget claimed that the probability concept does not develop until the stage of formal operations, in part because the ratio concept has not yet fully developed in young children. Contrary to Piaget, however, Kuzmak and Gelman found evidence in support of the findings of Fischbein and his colleagues (Fischbein, 1975; Fischbein et al., 1970; Fischbein & Gazit, 1984), who claimed that children have some primary intuitions about probability that will support further learning. Although children do not have a complete understanding of ratio, they do have some notions of chance or randomness.

Recent work by Jones, Langrall, Thorton, and Mogill (1999) appears to confirm some of the earlier work of Fischbein. Jones and his colleagues found that young (third-grade) children have notions of probability that can provide the basis for developing an instructional framework for teaching probability. They theorized that children exhibit four levels of thinking about probability situations: subjective, transitional, informal quantitative, and numerical. For example, when asked what color a spinner would come up, some students responded, "Blue, it's my favorite color" (subjective) or "Blue because there's more blue on the spinner" (transitional). Other students used quantitative expressions in their reasoning: "Blue, there's three blue pieces on the spinner and one green piece" (informal quantitative) and "The chance of blue is three out of four and green is one out of four" (numerical). Jones and his colleagues found significant growth in these levels of probabilistic thinking among third graders during instruction in probability.

Other research by J. Truran (personal communication, April 1999) investigated children's (ages 8 to 10) intuitive understandings of variance in a sampling task. In order to recognize variance in sampling, students need some intuitive notion of what outcomes are the likely ones in a sampling experiment. Although Truran's work was admittedly preliminary in nature, it is worth noting because most of Truran's 32 interviewed subjects were aware that extreme expectations in the sampling would be "very surprising." Some of Truran's subjects gave reasonable ranges for the expected frequency of a particular color drawn in repeated samples from a known (two blue, one green) distribution. Truran concluded that the children with the best interpretations of variability were those who looked for structure and that their understanding of variance seemed to depend on their computational skills.

In summary, research seems to suggest that (1) young children do indeed have some intuitions about probability prior to instruction, and (2) young children can learn more about probability in the context of particular instructional settings, and in some cases, can even change their thinking from their prior intuitions. The work of Jones et al. (1999) indicates that about third grade may be a good place to start teaching probability to children in a systematic way.

Equiprobability

Suppose a bag is presented to a young child who is told that there are two red marbles and one black marble in the bag. The child is then asked, "Which is more likely, that we pull out a red marble or that we pull out a black marble? Or, do you think that these two things have the same chance of happening?" LeCoutre (1992) and her colleagues have documented what they call the "equiprobability" bias, in which students may believe that all outcomes from a probability experiment have the same chance of happening. In addition, Jacobs (1997, 1999) identified the issue of "fairness" for children. Children believe that to be "fair," everything should have an equal chance of happening in a sampling experiment. It is interesting that LeCoutre found that the equiprobability bias persisted across students with no, a little, or even substantial exposure to probability and statistics concepts. For some of the students' tasks, LeCoutre found very little difference in the results across various ages and levels of probability background. LeCoutre claimed that students believe that random events are equiprobable "by nature." Equiprobability may even characterize some students' concept images of "randomness." Students sometimes respond to a probability task by saying, "Well, it could be anything, anything can happen," suggesting that all outcomes have the same chance of happening. That type of response suggests an equiprobability bias but is also characteristic of what Konold (1989) calls the *outcome approach*.

The Outcome Approach

Konold (1989, 1991) first introduced the terminology *outcome approach* to describe some students' responses on probability tasks. These students might not possess any process model for chance experiments, because they do not envision the results of a single trial of an experiment as just one of many possible outcomes that will vary across a sample space if the experiment is repeated. Konold reported that some subjects perceive each single trial of an experiment as a separate, individual phenomenon and believe that their task is to correctly predict the outcome of a probability experiment rather than to recognize what is likely to occur, or what would occur more often if the experiment was repeated.

For example, in a recent experiment (Shaughnessy, Watson, Moritz, & Reading, 1999), middle school and secondary school students were presented with a jar containing 100 colored chips: 50 red, 30 blue, 20 yellow. Researchers first asked students how many reds they thought would be pulled out in a handful of 10 chips. Then they asked students what would happen if the experiment were repeated six times, each time pulling out 10 chips. What would the numbers of reds be? Because half the chips in the jar were red, students who took into account the distribution of the chips tended to predict an interval of reds that clustered around 5. However, the researchers also found that about 17% of the students (n = 324) predicted an abnormally wide range for the number of reds, such as from 1 to 10, or wrote lists like "1, 3, 5, 7, 9, 10" for the numbers of reds that would be pulled in repeated samples. In their written explanations or during follow-up interviews, these students said, "It could be any number of reds," or "Anything could happen." This type of thinking is indicative of Konold's outcome approach to probability experiments.

The Representativeness and Availability Heuristics

Psychologists Daniel Kahneman and Amos Tversky and many of their students and colleagues stimulated an increased research interest in students' ideas of chance and data nearly three decades ago. They pursued a line of research in human judgment and decision-making under uncertainty that hypothesized that humans rely on certain judgmental heuristics, such as *representativeness*, *availability*, or *anchoring*, when estimating the likelihood of events or when comparing probabilities of outcomes. Detailed accounts and reviews of these heuristics have been presented by a number of researchers (e.g., Kahneman & Tversky, 1972, 1973; Konold et al., 1993; Shaughnessy, 1977, 1992). I discuss representativeness and availability briefly below and include some recent research that challenged the original claims that people reason by heuristics when estimating probabilities (Gigerenzer, 1994, 1996).

Representativeness

According to the representativeness heuristic, people estimate the likelihood of an event on the basis of how well it is "representative" of the parent population from which it is drawn or on how well it "represents" the process that generates it. One oft-used task that evokes representativeness asks subjects to compare likelihoods of sequences of outcomes that have been generated by a binomial process.

1. Which of the following sequences is most likely to result from flipping a fair coin six times?

 (a) HHHTTT
 (b) HTHTTH
 (c) HHHHTH
 (d) HTHTHT
 (e) All four sequences are equally likely

Give a reason for your answer.

According to Kahneman and Tversky (1972), subjects who reasoned by representativeness chose (b) because it is representative of the 50-50 expected ratio of heads to tails in the parent distribution and of a random process. Some subjects said that (a) was not representative of the random process of flipping a coin. Others said that (d) is "too regular" to be a likely outcome, or that it is not random enough (Shaughnessy, 1977). Kahneman and Tversky did not give the choice of (e), nor did they ask subjects to give a reason for their answers in their early studies. They forced their subjects to make a "most likely" choice among one of these equally likely binomial sequences. Later researchers allowed the choice of "All four sequences are equally likely." Even with the inclusion of the correct answer among the choices, reasoning by representativeness still accounts for a high number of responses to this type of task.

Interestingly, representativeness is not the only type of intuitive reasoning that occurs in tasks like these. Slight changes in the task can evoke other kinds of reasoning by subjects. Konold et al. (1993) used a version of the task that evoked an "outcome approach" response under one wording but then may have evoked representativeness from some subjects when the wording of the task was changed.

A. Which of the following sequences is most likely to result from flipping a fair coin five times? (Circle one)

 (a) HHHTT
 (b) THHTH
 (c) THTTT
 (d) HTHTH
 (e) All four sequences are equally likely

Justify your answer.

B. Which of the following sequences is least likely to result from flipping a fair coin five times? (Circle one)

 (a) HHHTT
 (b) THHTH
 (c) THTTT
 (d) HTHTH
 (e) All four sequences are equally likely

Justify your answer.

Konold and colleagues (1993) found some students who picked ""All four sequences are equally likely"" in the *most likely* version of the task and then switched, picking a particular sequence as *least likely* in the least likely version. Konold and colleagues claimed that these students might be reasoning from an outcome approach in the most likely version, because each one of those four sequences "could" happen on a single trial. Evidently for some students, this reasoning did not necessarily imply that each of these four sequences "couldn't" happen with equal weighting. They switched to a representativeness argument in the least likely version when they claimed that HTHTH was "just too regular" (the most unrepresentative, and thus the *least likely*). One of the important points highlighted in this piece of research by Konold and his colleagues is that people use different, sometimes competing, intuitive and personal theories when reasoning about probability tasks. The world of intuitive probabilistic reasoning is quite murky!

The framing or wording of the task influenced the type of reasoning evoked for some of the students in Konold and colleagues' (1993) study. Gigerenzer (1994) has claimed that probability tasks can actually be framed in a way to reduce, or even eliminate, reliance on reasoning by representativeness. Gigerenzer posed tasks from a frequency perspective, using whole numbers as input data, as opposed to ratios, percentages, or decimals, which frequently appear in the wording of probability problems. For example, consider these two framings of the same task:

A. Painting records for a large, old subdivision of a city that is undergoing renovation indicate that about 20% of the houses in that area had portions of their interiors painted with a lead paint. There is a chemical agent available that provides a quick test to analyze paint for lead contents; however, this chemical agent correctly identifies the presence of lead in only 90% of trials on old brands of paint. If the test indicates that your house has lead paint, what is the probability that you really do have lead paint in your house?

B. Painting records for a large, old subdivision of a city that is undergoing renovation indicate that about 20 out of every 100 houses in that area had portions of their interiors painted with a lead paint. There is a chemical agent available that provides a quick test to analyze paint for lead contents; however, this chemical agent correctly identifies the presence of lead in only about 90 of every 100 trials on old paint. If 1,000 houses in the subdivision are tested, about how many of them would you expect to test positive for lead? Of the houses that are identified as containing lead, how many of them would you expect to actually be lead free?

Researchers in human decision-making have long used these types of bivariate tasks in their attempt to study what information people actually use when estimating likelihoods. In the first framing, A, responses like "There is a 90% chance that the house has lead" may indicate that the responder attended only to the 90% accuracy of the test. People who respond "90%" may be using the test accuracy as a representative estimate for the *conditional* probability that a house actually has lead if the test said it did. This type of reasoning disregards the base-rate information, the known 20% incidence of lead in houses in that area. Gigerenzer (1994, 1996) recommended presenting such tasks via a frequency framing (B). He claimed people are more likely to incorporate base-rate information into their solution process if the information is given in terms of frequencies, because frequencies are easier for people to reason

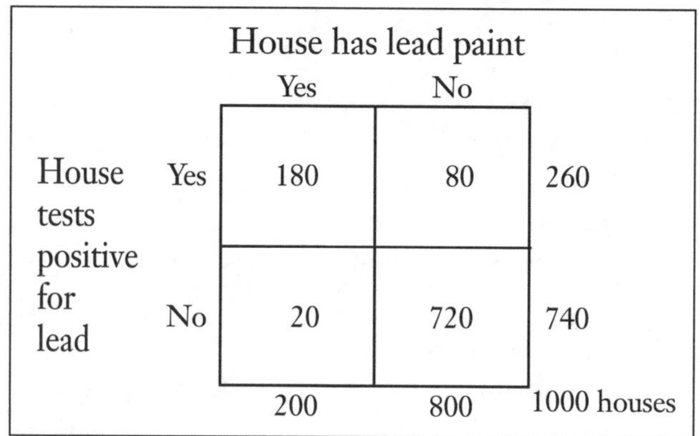

		House has lead paint		
		Yes	No	
House tests positive for lead	Yes	180	80	260
	No	20	720	740
		200	800	1000 houses

Figure 14.1. Frequencies used in a 2×2 contingency table to model a bivariate task.

with than percentages. Gigerenzer thus agreed with earlier recommendations (Shaughnessy, 1992) that frequencies should be used in a 2×2 contingency table to model bivariate tasks (see Figure 14.1).

The margin frequency totals in Figure 14.1 can help to decode the problem. From a sample of 1,000 houses, about 200 are expected to have lead paint problems and 800 are expected to have no problems. Of the 200 with lead problems, the test correctly identified 180 (90 of every 100) as containing lead. But the test also incorrectly indicated the presence of lead in about 80 of the 800 lead-free houses because the test is only 90% accurate. Row 1 in the table shows that the test identified a total of 260 houses that contained lead. Of the 260 houses identified as having lead—the margin total of row 1—only 180 of them actually do have lead, which are those in the "Yes-Yes" box. Thus, the chance is only 180/260, or about 69%, that a house does have lead if it tested positive for lead. Frequencies may provide a more intuitive approach to contingency tables and bivariate data than, say, Bayes' theorem. Some recent curriculum materials in statistics do introduce probability concepts through data and use this 2×2 representation of bivariate data through frequencies (Hopfsensberger, Kranendonk, & Scheaffer, 1999).

Availability

When people estimate the likelihood of events on the basis of on how easy it is for them to call to mind particular instances of the event, they use the availability heuristic (Kahneman & Tversky, 1973). This judgmental heuristic can induce significant bias because of one's own narrow experience and personal perspective. For example, if a person has driven through a town and been hit by a car running a stop sign, that person is more likely to overestimate the frequency of accidents in the town than someone who has driven accident-free in the town for years. In fact, both of these individual perspectives may be far from the truth. The accident danger in the town is probably less than the person who was in the accident feels, but more than the accident-free person feels. Both persons based their estimate only on their available personal experience, and in doing so also relied on a very small sample size.

People have egocentric impressions, based on their own experiences, of the frequency of events. Often these impressions are biased because even a single occurrence of an event can take on inflated significance when it happens to us. People do not perceive events that happen to them as just one more tally in an enormous, uncaring, objective frequency distribution.

Evaluating Events as Random

A number of researchers have used tasks in which string outcomes are presented to students and then the students are asked to choose which one was generated by a random process (Green, 1983; Falk, 1981; Batanero & Serrano, 1999). These studies have shown that humans are poor judges when identifying or constructing random events. Falk gave secondary school students sequences of 21 green and yellow cards and asked them to pick the more random sequence. For example, how random does each of these sequences appear?

A) Y Y Y G G Y Y G G Y Y Y G G G G G G Y Y Y
B) G Y G Y G Y G Y Y G Y G Y Y G Y G G Y G G

Students tended to pick the sequences in which more "switches" occur, such as in B. In fact, sequences with longer runs of one color are actually more likely to be the random ones. The second sequence appeared "more mixed" to students because of the frequent switches, thus more random in their eyes. Falk (1981) obtained similar results when she asked students to generate their own sequences of 21 cards. Students included more switches, and shorter runs of the same outcome, than are predicted by theory. According to Falk, humans are not very reliable either at recognizing random outcomes or at generating them. Falk concludes that we tend to see patterns in random events when none exist, and thus we reject the notion that certain outcomes are random. Also, we tend to infer randomness when it is not really present. The message from Falk's work is quite clear: Human beings should never be responsible for trying to generate random choices without using random devices.

Conjunctions and Conditionals

Both psychological and mathematical conceptions interfere when students are asked questions that involve compound probability. Compound probability problems can involve conjunctions in which the events are independent of one another or in which the events are dependent on one another. For example, if two chips are pulled from our jar of 50 red, 30 blue, and 20 yellow chips and replaced after each draw, the chance of getting a red on the second pull is not affected by the knowledge that we got a red on the first pull; it is still 1/2. In that case, our probability experiment involves repeated trials that are independent events. However, if we do not replace the chip after each draw, the chance that the second one is red is 49/99 if red was drawn on the first try, and 50/99 if red was not drawn on the first try. In this latter instance, the repeated trials are statistically dependent. Students have difficulty just sorting out the mathematics of whether events are statistically dependent or independent in probability problems. However, Tversky and Kahneman (1983) also identified psychological difficulties that can confound students' reasoning on conjunctions. Consider this situation:

> On a given day, all the people who commute to work in a large city are monitored. Which is more likely to occur?
>
> (A) A person has an automobile accident.
> (B) A person has an automobile accident and is under 21 years of age.

Many people pick choice B in this problem. However, on any day the set of people who have an accident and are under 21 (B) is a subset of the set of all people who have an accident (A). So $P(B) \leq P(A)$ because B is included in A. The psychological interference that leads people to pick B as more likely comes from the high incidence rate of accidents among young drivers. People are therefore more likely to attribute accidents to young drivers. However, on a given day, the number of people who have an accident is larger than the number of people who have an accident and in addition are under 21.

People who pick B may also be confusing this conjunction with the conditional statement, ""Given that a person is under 21, what is the chance that the person had an accident?" That's an altogether different question than the conjunction. Tversky and Kahneman (1983) have posed a number of scenarios involving conjunctions to tertiary level students. They found that the bias toward the conjunction fallacy—thinking that compound events are more likely to occur than simple events—is very robust across many different scenarios. Gigerenzer (1994) disagreed again with Tversky and Kahneman, providing evidence that the effects of the "conjunction fallacy" can be diminished if the question is posed in a frequency format.

Would the "conjunction fallacy" disappear if the scenario did not involve a potentially misleading or entrapping context, such as the automobile accident scenario? What happens if students are given a conjunction problem that is more decontextualized? In an item released from the 1996 National Assessment of Educational Progress (NAEP) items on probability and statistics, 12th-grade students were shown two identical circular spinners, each shaded 1/2 black and 1/2 white (Figure 14.2). The students were asked whether they agreed with the following statement and then asked to defend their answer: A student claims that if we spin both spinners simultaneously, there is a 50% chance that both of them will end up on black. Do you agree or disagree? Explain.

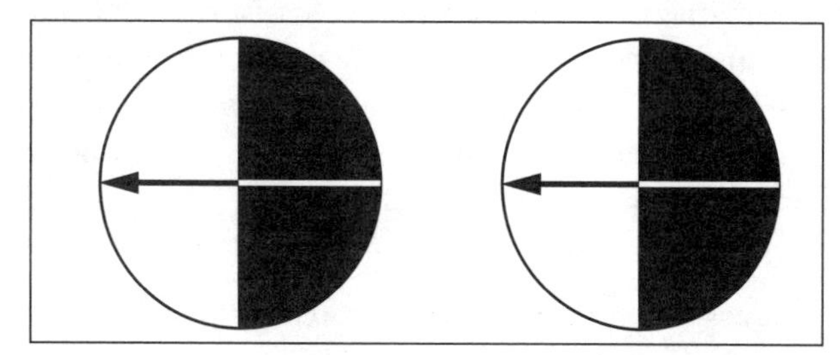

Figure 14.2. Identical spinners shown to students on 1996 NAEP examination.

In an analysis of this item, Zawojewski and Shaughnessy (2000) analyzed the responses of more than 1,000 12th-grade students who were given this problem. Only 8% disagreed with this statement and correctly reasoned that the chance of both spinners landing on black was only 1/4. An additional 20% of the students disagreed with the statement and said they somehow thought it "should be less than 50% for both to land on black." Zawojewski and Shaughnessy obtained a convenience sample of student work for 306 of these 1996 NAEP 12th graders (Shaughnessy & Zawojewski, 1999). In that sample, more than half the students agreed the chance was 50% that both spinners would land black, and 104 of the students in the sample gave one of the following reasons: "There is a 50% chance that both will land black because each spinner is half black," "They are cut in half," or "There is an equal distribution of colors." The NAEP data on this item indicate that students have trouble with conjunctions even in a decontextualized situation. The analysis of the NAEP made by Zawojewski & Shaughnessy (2000) suggests that students' difficulties with conjunctions are more than just psychological. Students lacked the mathematical skills to analyze or list the outcomes for the spinner problem discussed above. Students need more experience building sample spaces and more experience listing the set of all possible outcomes in probability experiments.

National Assessment and Students' Understanding of Probability

Over the past decade, the NAEP has gradually been increasing the coverage of topics in probability and statistics. The NAEP is normally administered once every four years to obtain a snapshot of how school children and adolescents are progressing in subject matter areas in the United States. Information about students in Grades 4, 8, and 12, as well as about their performance on mathematical tasks, is gathered from a random sample of participating schools from across the nation. In the 1996 administration of the NAEP, the percentage of items classified as probability and statistics had risen from 6% in 1986 to 20% in 1996 for the Grade 12 students (Shaughnessy & Zawojewski, 1999). Similar increases have occurred at the other grade levels, with coverage at eighth grade growing to 15% and more items being included at fourth grade as well.

On the one hand, Grade 12 NAEP questionnaires give some indication of an increase in the number of students who say they have studied data analysis, statistics, and probability at some time during their school experience (from about 12% in 1990 to about 20% in 1996). On the other hand, nearly 80% of graduating secondary school students in the NAEP sample had no experience in probability or statistics in their schooling. In light of the persistent calls during the past 25 years for increased attention to probability and statistics in the schools (Conference Board of the Mathematical Sciences, 1975; NCTM, 1980, 1989, 2000), the NAEP student survey data indicate that an enormous amount of work still remains to be done to mobilize teachers, counselors, and parents to encourage and provide opportunities for students to investigate data and chance.

The 1996 NAEP data on student performance provide mixed reviews on how our schoolchildren are doing on probability. For example, Grade 12 students performed quite poorly on the spinner task that was discussed above, with only 8% indicating the four possible outcomes for spinning the two spinners. It should be noted that the NAEP sample was drawn from all 12th graders, not just from those actually taking a mathematics course in their senior year. I have obtained much better results in Grade 12 mathematics classes—precalculus, calculus, and AP statistics—with 60% to 75% of the students in those classes correctly listing the sample space for the spinner problem. Nonetheless, the poor performance on the spinner task at Grade 12 is indicative of a chronic problem throughout school mathematics identified by Zawojewski and Shaughnessy (2000) in their analysis of the NAEP data on probability and statistics. Our students are weak in dealing with sample spaces (set of all outcomes) for probability experiments. Even if students do correctly list or identify the sample space for a problem in chance, NAEP found that students were weak in using that sample space to answer further questions or to make predictions. Students' interpretive skills in probability were quite poor on the 1996 NAEP.

Analysis of NAEP data also indicates that Grade 4 and Grade 8 students have difficulty with probability problems in which the probability of an event is a ratio of the form m/n, or "m chances out of n." In contrast, problems where the probability of an event is $1/n$ were quite easy for children in Grades 4 and 8. The difficulty with m/n probabilities foreshadows the problems that older students have with sample space. Whether the difficulty with m/n type problems stems from a lack of understanding of ratios, from errors in listing or creating the sample space, or both, or perhaps even from other phenomena is not well known. Most of the NAEP items that investigated simple probability were either multiple choice or short response, with no attempt to request student thinking or reasoning. Thus, further research is needed in this area to get at the roots of students' conceptions.

In summary, although the 1996 NAEP data give some credence to the position that increased attention is beginning to be paid to probability and statistics in schools, student performance on some probability items was remarkably low. Furthermore, only about 20% of our Grade 12 students are graduating from secondary schools with some study of probability and statistics. One cannot expect a very high performance on the NAEP items on probability and statistics if students do not study it during their secondary years. We still have a very long way to go in teaching probability and statistics in the schools.

Some Thoughts on the Teaching of Probability

Probability is just one piece of the study of *stochastics*, which generally includes the overall study of data, chance, and statistics. There are ever-increasing applications and uses of probability and statistics in business and industry. Informed citizens need to be numerate in data and chance and need to know how to decipher and make sense out of information that is presented in newspapers, medical reports, consumer reports, and environmental studies. Results of the 1996 NAEP suggest that many graduating secondary students are not yet numerate enough in data and chance to make decisions about data that are influenced by models of chance. A recent medical study (Schwartz, Woloshin, Black, & Welch, 1997) found that regardless of the format (percentages, frequencies, etc.) in which women were given information about their risk of breast cancer and the effectiveness of mammograms, their ability to judge the benefits of mammography was related to their performance on several numeracy tasks.

People will be required to make more and more decisions in their lives under conditions of uncertainty. An understanding of chance and risk, an ability to read and interpret graphs, and an ability to question situations involving data and chance are all essential skills for making stochastic decisions. Davis and Hersh (1991) make an excellent case for a world view that treats knowledge as uncertain in nature. In their essay "The Stochasticized World," they describe an entire epistemology based on chance, data, and uncertainty. They conclude that "bottom line probability" will always be among us. Bottom line probability is the "hunch" variety of probability, the kind of decision-making driven by such heuristics as representativeness or the outcome approach described above. Teachers, mathematics educators, parents, and administrators, must provide their children and their students with alternative ways of approaching data and chance—alternatives to a subjective "bottom line hunches and beliefs" approach.

Whereas much of the research above sights reasons students have some difficulties with probability concepts, there is also a potential silver lining to tap in this "bottom line probability" of Davis and Hersh (1991). On the one hand, research suggests that students do have lots of hunches and beliefs about probability problems, what Fischbein (1987) calls "primary intuitions." On the other hand, students also enjoy testing out their hunches and beliefs through experiments or simulations. "Bottom line probability" provides a fertile learning opportunity to build on, and research suggests some strategies for teaching probability:

1. Begin to teach probability concepts at a young age, and continue them throughout the school-age years.

According to Jones et al. (1999), opportunities arise to begin to systematically teach probability concepts and the language of chance to young students as early as third grade. Metz (1988), Fischbein, Pampu, and Manzit (1970), and Fischbein (1975) also noted that young children have acquired some of the concepts and skills necessary to begin constructing an understanding of chance at an early age.

2. Emphasize the value and importance of building the sample space for a probability experiment.

The NAEP data indicate that students are weak on the concept of sample space. Students need more opportunities to determine and discuss the set of all possible outcomes for a probability experiment or a sampling activity. Disagreement often arises as to what the outcomes of a probability experiment are or on how to assign probabilities to those outcomes. For example, students may argue that there are 11 outcomes (the sums), 15 outcomes (the number pairs without considering order), or 36 outcomes (the ordered number pairs) possible when tossing two six-sided dice. Each of these sample spaces can be used to model the outcomes for tossing two dice, but one must assign probabilities in different ways for each model.

3. Make connections between probability and statistics. In particular, connect the notion of the sample space in probability with the concept of variation in statistics.

Which are the likely outcomes in a probability experiment? Are some more likely to occur than others? Why? Which outcomes are extreme or unlikely to occur? Suppose the experiment were repeated several times, or many times? What would the data—the results of the repeated experiments—look like? What is the range of likely outcomes if the experiment is repeated? These questions are both statistical and probabilistic in nature. They make connections between two big stochastical ideas: the sample space in probability and the nature of variation in statistics (Shaughnessy, 1997). Outcomes can vary, but they occur predominantly in a "likely range" that is governed by the sample space for a probability experiment. For example, in a gumball machine with 50 red, 30 blue, and 20 yellow balls, one would expect to obtain about half reds in a sample, but one would also expect a "range of likelihood" around 50% for the number of reds in repeated samples. One would not get exactly 50% reds in every sample! Connections exist between the probability of an event, which is given as a point value, and the range of likely outcomes, which is a confidence interval around that point value. Confidence intervals model the variability of the likely point values for a repeated probability experiment. The concepts of sample space and variability are closely connected.

4. Introduce probability *through* data; start with statistics to get to probability.

Previous recommendations suggested that students start with probability experiments, conduct simulations, and gather actual data to test probability conjectures (Garfield & Ahlgren, 1988; NCTM, 1980, 1989, 2000; Shaughnessy, 1977). This approach starts from probability and introduces a data environment through a chance situation. Probability precedes the collection and analysis of data. For example, in the infamous problem of the three doors, a prize is hidden behind one of the doors and students are asked to pick a door at random. Then they are shown one of the doors that has no prize behind. Which strategy gives them a better chance of winning: sticking with their first choice or switching to the remaining door? Or does it make no difference? Simulation is a powerful tool to identify the sample space for this innocent-sounding problem (Shaughnessy & Dick, 1991).

Recent work suggests that perhaps it should be the other way around and that statistics should motivate probability questions. This approach recommends that the teaching of probability should actually start with data, and that probability questions should be raised from data sets (Hopfsenberger et al., 1999). For example, using the data in Figure 14.1 from lead paint in houses, consider the following probability questions: What is the probability that a house will test positive for lead (740/1000)? What is the probability that a house is clean of lead paint (800/1000)? What is the probability that a house does have lead paint even though the test says it does not (20/740)? The importance of probability questions in the context of real data supersedes past approaches to probability that started either with counting problems or with games of chance. Doctors, lawyers, consumers, statisticians, politicians, and citizens must be able to glean actual probabilities from real data and to understand, and perhaps challenge, claims made about data. As entertaining as games of chance are, it may be more important for students to be first introduced to probability through the lens of data analysis.

5. Adopt a problem-solving approach to probability. Give students opportunities to investigate probability problems or chance situations on their own and to conduct their own stochastics projects.

No amount of teaching probability theory in a direct way can ever supplant the need for students to deal with the actual data gathered in a probability experiment or data obtained through a statistical sampling procedure. "Bottom line approaches"—beliefs and hunches about chance—can best be challenged by the students themselves. However, in order to challenge beliefs, students need to get their hands on actual data and make their own conjectures about chance. They need to test their conjectures, find ways to represent the data they gathered, and communicate their findings to their fellow students. Only then can probability begin to become "the ultimate problem-solving proving ground" in our classrooms as Bergman suggested (Shaughnessy & Bergman, 1993).

ACKNOWLEDGMENTS

Joan Garfield and Dave Robathan provided comments on an earlier draft of this chapter.

REFERENCES

Batanero C., & Serrano L. (1999). The meaning of randomness for secondary school students. *Journal for Research in Mathematics Education, 30*, 558–567.

Borovcnik, M., & Peard, R. (1996). Probability. In A. Bishop, K. Clements, C. Keitel, J. Kilpatrick, & C. Laborde (Eds.), *International handbook of mathematics education* (pp. 239–287). Dordrecht, The Netherlands: Kluwer.

Conference Board of the Mathematical Sciences. *Report of the National Advisory Committee on Mathematical Education: Overview and analysis of school mathematics grades K–12.* Washington, D.C.: Author.

Davis, P., & Hersh, R. (1991). *Descartes' dream.* Boston: Houghton Mifflin.

Falk, R. (1981). The perception of randomness In C. Comiti & G. Vergnaud (Eds.), *Proceedings of the Fifth International Conference for the Psychology of Mathematics Education* (pp. 222–229). Grenoble, France: Laboratoire I. M. A. G.

Fischbein, E. (1975). *The intuitive sources of probabilistic thinking in young children.* Dordrecht, The Netherlands: Reidel.

Fischbein, E. (1987). *Intuition in science and mathematics.* Dordrecht, The Netherlands: Reidel.

Fischbein, E., & Gazit, A. (1984). Does the teaching of probability improve probabilistic intuitions? *Educational Studies in Mathematics, 15*, 1–24.

Fischbein, E., Pampu, I., & Manzit, I. (1970). Effects of age and instruction on combinatory ability of children. *British Journal of Educational Psychology, 40*, 261–270.

Garfield, J., & Ahlgren, A. (1988). Difficulties in learning basic concepts in probability and statistics: Implications for research. *Journal for Research in Mathematics Education, 19*, 44–63.

Gigerenzer, G. (1994). Why the distinction between single-event probabilities and frequencies is important for psychology (and vice versa). In G. Wright & P. Ayton (Eds.), *Subjective probability* (pp. 129–161). Chichester, England: Wiley.

Gigerenzer, G. (1996). On narrow norms and vague heuristics: A rebuttal to Kahneman and Tversky. *Psychological Review, 103*, 592–596.

Green, D. R. (1983). A survey of probability concepts in 3000 pupils aged 11–16 years. In D. R. Grey, P. Holmes, V. Barnett, & G. M. Constable (Eds.), *Proceedings of the First International Conference on Teaching Statistics* (pp. 766–783). Sheffield, England: Teaching Statistics Trust.

Hopfsenberger, P., Kranendonk, H., & Scheaffer, R. (1999). *Data driven mathematics: Probability through data.* Palo Alto, CA: Dale Seymour Publications.

Jacobs, V. R. (1997). *Children's understanding of sampling in surveys.* Paper presented at the meeting of the American Educational Research Association, Chicago.

Jacobs, V. R. (1999). How do students think about statistical sampling before instruction? *Mathematics Teaching in the Middle School, 5*, 240–246, 263.

Jones, G., Langrall, C., Thorton, C., & Mogill, T. (1999). Students' probabilistic thinking in instruction. *Journal for Research in Mathematics Education, 30*, 487–519.

Kahneman, D., & Tversky, A. (1972). Subjective probability: A judgment of representativeness. *Cognitive Psychology, 3*, 430–454.

Kahneman, D., & Tversky, A. (1973). Availability: A heuristic for judging frequency and probability. *Cognitive Psychology, 5*, 207–232.

Kapadia, R., & Borovcnik, M. (Eds.) (1991). *Chance encounters: Probability in education.* Dordrecht, The Netherlands: Kluwer.

Konold, C. (1989). Informal conceptions of probability. *Cognition and Instruction, 6*, 59–98.

Konold, C. (1991) Understanding students' beliefs about probability. In E. von Glasersfeld (Ed.), *Constructivism in mathematics education* (pp. 139–156). Dordrecht, The Netherlands: Kluwer.

Konold, C., Pollatsek, A., Well, A., Lohmeier, J., & Lipson, A. (1993). Inconsistencies in students' reasoning about probability. *Journal for Research in Mathematics Education, 24*, 392–414.

Kuzmak, S., & Gelman, R. (1986). Young children's understanding of random phenomena. *Child Development, 57*, 559–566.

Metz, K. (1998). Emergent ideas of chance and probability in primary-grade children. In S. Lajoie (Ed.), *Reflections on statistics: Learning, teaching, and assessment in grades K–12.* Mahwah, NJ: Erlbaum.

LeCoutre, V. P. (1992). Cognitive models and problem spaces in "purely random" situations. *Educational Studies in Mathematics, 23*, 557–568.

National Council of Teachers of Mathematics. (1980). *An agenda for action: Recommendations for school mathematics of the 1980s.* Reston, VA: Author.

National Council of Teachers of Mathematics. (1989). *Curriculum and evaluation standards for school mathematics.* Reston, VA: Author.

National Council of Teachers of Mathematics. (2000). *Principles and standards for school mathematics.* Reston, VA: Author.

Piaget, J., & Inhelder, B. (1975). *The origin of the idea of chance in children.* London: Routledge & Kegan Paul.

Schwartz, L., Woloshin, S., Black, W. C., & Welch, H. G. (1997). The role of numeracy in understanding the benefit of screening mammography. *Annals of Internal Medicine, 127*, 966–972.

Shaughnessy, J. M. (1977). Misconceptions of probability: An experiment with a small-group, activity-based, model building approach to introductory probability at the college level. *Educational Studies in Mathematics, 8*, 285–316.

Shaughnessy, J. M. (1992). Research in probability and statistics: Reflections and directions. In D. Grouws (Ed.), *Handbook of research on mathematics teaching and learning* (pp. 465–494). New York: Macmillan.

Shaughnessy, J. M. (1997). Missed opportunities in research on the teaching and learning of data and chance. In F. Biddulph & K. Carr (Eds.), *People in mathematics education* (Vol. 1, pp. 6–22). Waikato, New Zealand: Mathematics Education Research Group of Australasia.

Shaughnessy, J. M., & Bergman, B. (1993). Thinking about uncertainty: Probability and statistics. In P. Wilson (Ed.), *Research ideas for the classroom: High school mathematics* (pp. 177–197). Reston, VA: National Council of Teachers of Mathematics.

Shaughnessy, J. M., & Dick, T. P. (1991). Monty's dilemma: Should you stick or switch? *Mathematics Teacher, 84*, 252–256.

Shaughnessy, J. M., Watson, J., Moritz, J., & Reading, C. (1999). *School mathematics students' acknowledgment of statistical variation.* Paper presented at the meeting of the National Council of Teachers of Mathematics, San Francisco.

Shaughnessy, J. M., & Zawojewski, J. S. (1999). Secondary students' performance on data and chance in the 1996 NAEP. *Mathematics Teacher, 92*, 713–718.

Tversky, A., & Kahneman, D. (1983). Extensional versus intuitive reasoning: The conjunction fallacy in probability judgment. *Psychological Review, 90*, 293–315.

Zawojewski, J. S., & Shaughnessy, J. M. (2000). Data and chance. In E. A. Silver & P. A. Kenney (Eds.), *Results from the Seventh Mathematics Assessment of the National Assessment of Educational Progress* (pp. 235–268). Reston, VA: National Council of Teachers of Mathematics.

Chapter 15

Reasoning and Proof

Erna Yackel, Purdue University Calument
Gila Hanna, Ontario Institute for Studies in Education of the University of Toronto

The increasing awareness that reasoning is central to mathematics and mathematics learning is clearly demonstrated in the recently published *Principles and Standards for School Mathematics* (*Principles and Standards*) (National Council of Teachers of Mathematics [NCTM], 2000). In addition to asserting that "systematic reasoning is a defining feature of mathematics" (p. 57), *Principles and Standards* calls for reasoning and proof to "be a consistent part of students' mathematical experience in prekindergarten through grade 12" (p. 56).

An emphasis on reasoning and proof has not always been reflected in all parts of Grades K–12 mathematics instruction. Some areas of school mathematics, such as geometry and certain higher-level topics, have traditionally been thought of in terms of logical reasoning and proof. Others, however, notably arithmetic and school algebra, have typically been more rule-oriented, with skill development and procedural proficiency as the primary goals. Consequently, NCTM, in *Principles and Standards*, is taking a bold position by setting forth *reasoning and proof* as one of the five process standards for all grade levels.

The purpose of this chapter is to review some recent research in reasoning and proof in school mathematics, pointing out the views that underpin the research as well as its potential implications for the curriculum. We first outline the importance of reasoning and proof from the perspectives of educational theory and of mathematics itself. Next, we discuss perspectives that can enlighten our conceptions of reasoning and proof. Finally, we discuss selected research studies that shed light on the possibilities and limitations students encounter when learning to engage in the activities of reasoning and proving.

The Importance of Reasoning and Proof

The Importance of Reasoning from the Perspective of Educational Theory

The current emphasis on reasoning, especially at the early grade levels, reflects a shift from the predominantly behaviorist theory of learning that until quite recently dominated much of educational thinking and practice in the United States. Today the educational landscape is dominated by theories emphasizing the processes by which students come to know (Steffe & Kieren, 1994). It is important to be aware that shifts such as these are the result of many forces, prominent among them the desire of researchers and educators to explain phenomena that are less easily explained by other theoretical perspectives. (The issues that can be successfully explored and the explanations that can be developed are clearly constrained by the theoretical perspectives that are adopted.) In this sense the emphasis on reasoning as a central aspect in all areas and at all levels of mathematics instruction is a deliberate choice that mathematics educators have made as a result of a better understanding of how individuals come to know.

One prominent supporter of this view is Ernst von Glasersfeld, as Smith (1997) explains: "He [von Glasersfeld] is very much in the tradition of seeing reason[ing] as the process by which a knower comes to know" (p. 108). According to this position, which posits that knowledge is built up by the cognizing individual (von Glasersfeld, 1990), it follows that reasoning is the process through which someone learns. Thompson (1996) expresses the same point of view when he says, "I am unable to separate matters of learning from matters of reasoning" (p. 267).

The Importance of Proof in Mathematics and Mathematics Learning

An emphasis on reasoning at all levels of mathematics education calls attention to mathematical argumentation and justification. For mathematicians, the end result of argumentation is often a formal mathematical proof. Thus the notion of proof is viewed as a central construct in mathematical thinking, and learning to understand and develop formal proofs is seen as an important aspect of students' mathematical learning.

The increased attention now being given to mathematical proof in mathematics education is a reversal of the trend of recent years. Over the past 30 years or so, proof had been relegated to a less prominent role in the school mathematics curriculum. Concerned with this state of affairs, Greeno (1994) expressed alarm at "what appears to be a trend toward making proofs disappear from precollege mathematics education" and laid the blame squarely on misconceptions as to the nature of proof. He suggested that this situation might be redressed "by a more adequate theoretical account of the epistemological significance of proof in mathematics" (pp. 270–271).

Proof has many functions within mathematics, including verification, explanation, systematization, discovery, communication, construction of empirical theory, exploration of definition and of the consequences of assumptions, and incorporation of a well-known fact into a new framework (Bell, 1976; de Villiers, 1990; Hanna & Jahnke, 1996). In the past, students have often encountered proof only as the ultimate method of verification. But the functions of proof that may have the most promise for mathematics education are those of explanation and communication.

An emphasis on the explanatory function of proof is consistent with the views of many mathematicians. Formal proof is sometimes thought of only as chains of logical argument that follow agreed-on rules of deduction and is often characterized by the use of formal notation, syntax, and rules of manipulation. But for mathematicians, proof is much more than a sequence of logical steps; it is also a sequence of ideas and insights (Jaffe, 1997; Kleiner, 1991; Manin, 1998; Rota, 1997). Thus, proof for mathematicians involves interpretation, understanding, reasoning, and sense-making.

From this perspective, chains of logical argument do not function as satisfactory proofs unless they serve explanatory and communicative functions for an interpreting individual. Thurston (1994) expressed this view as follows: "How do mathematicians advance human understanding of mathematics? ... We [mathematicians] are not trying to meet some abstract production quota of definitions, theorems and proofs. The measure of our success is whether what we do enables people to understand and think more clearly and effectively about mathematics" (p. 163). Manin (1977) made the same point somewhat differently, stating that mathematicians would like good proofs to make them wiser. In this view, a good proof is one that also helps one understand the meaning of what is being proved: to see not only that it is true but also why it is true. In addition to leading to further understanding, such a proof also enhances communication.

Research on Reasoning and Proof

In the past 15 years, but especially since 1989, when *Curriculum and Evaluation Standards for School Mathematics* (NCTM) was published, numerous mathematics educators and teachers have taken up the cause of fostering mathematical reasoning in classroom instruction. Concurrently, several researchers have conducted research in an attempt to explain what reasoning means and how students learn to reason mathematically. These studies have included, but have not been limited to, investigations of explanation, argumentation, discourse, norms for justification, and proof.

The Meaning of Reasoning

Writing about reasoning in mathematics is complicated by the fact that the term *reasoning*, like understanding, is widely used with the implicit assumption that there is universal agreement on its meaning. On this assumption, in fact, most mathematicians and mathematics educators use this term without any clarification or elaboration. One attempt to be explicit has been made by Thompson (1996), who describes mathematical reasoning as "purposeful inference, deduction, induction, and association in the areas of quantity and structure" (p. 267). In this view he is following Piaget (1970), for whom scientific and mathematical reasoning are activities orienting the individual's understanding of quantity and structure.

In contrast, mathematics educators such as Bauersfeld (1980), Krummheuer (1995), Cobb (1994), and Saxe (1991) take a social perspective as well, focusing on the communal aspects of reasoning and other mathematical activity. From this perspective, mathematical reasoning is a communal activity in which learners participate as they interact with one another to solve (resolve) mathematical problems.

Instruction That Emphasizes Reasoning

The view of mathematics as reasoning can be contrasted with the view of mathematics as rule-oriented activity. In his seminal article "Relational Understanding and Instrumental Understanding," Skemp (1978) referred to the first of these views as *relational mathematics* and to the second as *instrumental mathematics*. The thrust of his argument is that the school subject called mathematics does not have an immutable identity but will in effect assume different identities, depending on which of these two views predominates in the classroom. The instrumental view could be taken to imply that school mathematics should have skill development and procedural proficiency as its primary goals.

Skemp (1978) also applied the labels *instrumental* and *relational* to understanding itself. For him, instrumental understanding is an understanding of rules and procedures, whereas relational understanding is an understanding of underlying relationships. Relational understanding is more encompassing because it permits one to develop multiple ways to navigate within the mathematical environment. It is analogous to knowing one's way around a community without having to follow a map step by step. In contrast, instrumental understanding is limited to knowing how to use a few specific procedures. Skemp compares the latter to having one or two ways to move from one location to another by following specific directions.

The importance of Skemp's distinction here is that it is analogous to the difference between emphasizing mathematics as reasoning and emphasizing mathematics as rules and procedures. That is not to say that students do not engage in any form of reasoning in instrumental mathematics. As Skemp noted, one might argue that in instrumental mathematics, students are developing some form of understanding. The nature of that understanding is the issue; it differs qualitatively from the nature of the understanding that students develop when the emphasis is on relational mathematics.

These distinctions have been elaborated extensively by Cobb, Wood, Yackel, and McNeal (1992) in a detailed analysis of two elementary school mathematics classes—a second-grade class and a third-grade class—that follow different classroom traditions. Following Richards (1991), Cobb et al. (1992) call these the *school mathematics tradition* and the *inquiry mathematics tradition*. As Cobb and his colleagues show, mathematics in the third-grade class was constituted as following rules and procedures (school mathematics tradition), whereas in the second-grade class it was constituted as a meaning-making activity (inquiry mathematics tradition).

An interactional analysis of episodes from lessons in each of the classes shows that in the third-grade class a solution or answer was acceptable if the child had followed the mathematical instructions correctly. Explanations and justifications were not only unnecessary but also redundant because the instructions children followed were fixed rules. In contrast, in the second-grade class, children developed personally meaningful solutions. Here a solution or an answer was acceptable only if the student's (mathematical) activity, including his or her creation of, and action on, mathematical objects, was acceptable. In this class, students were obliged to describe their activity so that others could agree that it was legitimate. Such descriptions necessarily required students to explain and justify their actions. Thus, the classroom that followed the inquiry mathematics tradition, like classrooms that foster relational understanding, can be described as offering mathematics instruction that emphasizes reasoning.

Explanation and Justification as Interactional Accomplishments

As the discussion above shows, explanation and justification are key aspects of students' mathematical activity in classrooms in which mathematics is constituted as reasoning. A question that comes to mind immediately is this: How do students learn to give mathematical explanations and justifications? Here, research that takes a socioconstructivist perspective is useful. From this perspective, explanations and justifications are considered to be aspects of communication, and as such they are taken as interactively constituted by the participants (Bauersfeld, 1980). For example, when students or the teacher give explanations in a mathematics classroom, they are attempting to develop a basis for communication by elaborating aspects of their mathematical activity, including their mathematical thinking, that they think are not readily apparent to others (Cobb et al., 1992). Similarly, justifications are given in an attempt to communicate the legitimacy of one's mathematical activity. In either situation the individual giving the explanation or justification makes the assumption that others are genuinely attempting to participate in the communication as well. For example, an explanation or justification may be a response to a question or a challenge. Further, those who take this perspective would not say that genuine communication was involved unless someone attempted to interpret the explanation or justification.

In this view, the adequacy of explanations and justifications as communal acts is determined not only by the speaker but by the listeners as well. Yackel and Cobb (1996) refer to what counts as an acceptable mathematical explanation and justification as a sociomathematical norm. They argue that the teacher and students negotiate such norms as they interact in the classroom. For example, they describe an episode in a second-grade classroom in which several students challenged an explanation for the sum of 12 and 13 because it was strictly procedural in nature. The explanation given was that "one plus one is two, and three plus two is five" (p. 470). The numbers were treated as four individual digits, without reference to the quantities that the digits represented. Several students challenged this explanation with "That's 20 ... That's taking a 10 right here. This [is] 10 and 10. That's 20. ... And this is five more and it's 25." In response, the teacher said, "That's right. It's 25" (p. 470).

Yackel and Cobb (1996) argue that the teacher, by legitimizing the challenge, contributed to the ongoing negotiation of what is acceptable as an explanation in this classroom as one that describes actions on mathematical objects. In this way the teacher promoted the evolution of an inquiry mathematics classroom (Voigt, 1995), that is, a classroom that fosters mathematics as reasoning.

A Discussion of Selected Research Studies

A number of research studies have investigated students' development of notions of reasoning, explanation, justification, and proof in classroom settings, from the elementary to the university level. Through these studies we are able to gain some insight into what types of reasoning students are capable of at various age and grade levels, how their notions of reasoning and proof develop over time, and what limitations in reasoning they exhibit. In the remainder of this paper we discuss several of these studies, turning first to those in which the emphasis was on reasoning, explaining, and justifying and then to those in which the emphasis was on proof and the activity of proving.

Reasoning, explaining, and justifying. Lampert (1990), Ball (1991), and Yackel and Cobb (1994) have all conducted year-long investigations in elementary school classrooms in which reasoning, explanation, and justification were central aspects of students' mathematical activity. Lampert's purpose was "to examine whether and how it might be possible to bring the practice of knowing mathematics in school closer to what it means to know mathematics within the discipline" (p. 29). For Lampert, it was a kind of existence proof of what is possible.

Using Lakatos' (1976) work on proofs and refutations and Pólya's (1954) work on intellectual courage and honesty in problem solving to guide her thinking, Lampert (1990) set about developing a form of classroom instruction in a fifth-grade class that engaged students in making and testing mathematical hypotheses. To do so required carefully selecting problem tasks, creating a classroom environment in which it was psychologically safe for students to share their emerging and tentative ideas, and establishing classroom norms that supported reasoning and mathematical argumentation as the primary source of legitimacy for ideas and assertions. Facilitating these activities, in turn, required that she initiate and support classroom social interactions appropriate to making mathematical arguments in response to students' conjectures.

Lampert (1990) presents an analysis of episodes from a class session in which the students' task was to figure out the last digit in each of 5^4, 6^4, and 7^4 without doing the multiplication and to be able to provide a justification for their assertions. The analysis shows that students moved back and forth between inductive and deductive arguments in the course of the lesson. By the end of the lesson, each student had made an assertion about a pattern, provided a proof that a pattern could continue, or given an interpretation of another student's assertion. This is compelling evidence that young children can and do engage in sophisticated mathematical reasoning. Further, as Lampert points out, the students learned not only about the mathematical content, in this example about the laws of exponents, but also about how to justify them within the domain of mathematics. In doing so, they were learning about the process of justification itself. That is, they were learning how truth is established in mathematics.[1]

It is important to note that in the Lampert (1990) study, as in those by Ball (1991) and by Yackel and Cobb (1994), students had to learn about justification. Normative understandings of what constituted acceptability had to be negotiated. For example, both Lampert and Ball report that early in the year students suggested voting as one means to resolve differences. They also report situations in which students used social reasons, such as agreeing with good students, as a basis for argumentation.

Similarly, Yackel and Cobb (1994) report that initially some students in their project classrooms used their interpretation of implicit cues from the teacher to make decisions about the correctness of their responses. In one instance, a student changed her answer simply because the teacher repeated the question after the student responded. The student assumed that, consistent with her

[1]It is important to note that this lesson did not occur in isolation but was one in a yearlong sequence in which the same type of reasoning and mathematical thinking was fostered throughout.

previous schooling, the teacher repeats a question only when the response given is incorrect. In the classroom Yackel and Cobb studied, the teacher used the occasion to initiate a discussion about the reasons the children used for their responses. Similarly, in the Ball (1991) study, the issue of whether voting is appropriate as a basis for determining mathematical correctness became an explicit topic for discussion. These studies show that students do not come to the classroom with an understanding of what is taken by the larger mathematical community as acceptable means of argumentation. That is, in classrooms that emphasize reasoning, students not only engage in reasoning but also, concurrently, learn what constitutes acceptable mathematical reasoning.

A longitudinal case study conducted by Maher and Martino (1996) complements the studies described in the preceding paragraphs by providing information about the development of one child's mathematical reasoning and argumentation over a 5-year span. The purpose of the study was to trace the development of Stephanie's use of mathematical justification over time as she worked on combinatorics tasks (beginning in first grade and continuing through fifth grade). Of particular interest is the Towers Problem, in which the task is to figure out how many different towers four (or five) cubes tall can be made selecting from red and blue cubes. Stephanie and her classmates worked on this problem, first for four cubes and later for five, at various times in third grade through fifth grade.

Maher and Martino's (1996) analysis shows a progression in the means Stephanie used to generate towers and to justify that she had exhausted all possibilities. Initially Stephanie used trial-and-error and guess-and-check strategies to create new towers and search for duplicates. When asked to show that she had found all possible towers, she explained that she had continued until she could not find any that were different from those she already had. By the middle of her fourth-grade year, in a discussion of towers five cubes tall, Stephanie introduced an indirect method of proof to account for all possibilities. By the spring of fourth grade, she explored a version of proof by cases for towers four cubes tall in an individual interview setting, and several days later she confidently presented a proof by cases to a group of her classmates to demonstrate that she had found all possible towers three cubes tall. Early in her fifth-grade year, in a written assessment task, she produced what Maher and Martino refer to as an "elegant" written version of the proof by cases to justify her solution to the towers problem.

Maher and Martino (1996) are quick to note that the classroom environment that prevailed during mathematics instruction made possible the type of progress Stephanie made. The learning environment was one that nurtured multiple opportunities for children to explore ideas, by providing flexibility of content and extended periods of time for exploration, and also one that valued student perseverance and initiative.

This research study is important for several reasons. First is its longitudinal nature, documenting a student's development of notions of justification and proof over time. Second, the study is significant in that it suggests how notions of formal proof might develop naturally out of an emphasis on justification and explanation. In this sense, it provides a link between a treatment of mathematics as reasoning and the interest of mathematicians in formal proof. Finally, the study documents that a high level of sophistication in understanding justification and proof can be attained as early as the beginning of fifth grade in supportive classroom conditions.

Proof and the activity of proving. The research studies discussed above show that as early as the elementary grades, students can engage in mathematical reasoning, including developing and testing conjectures using both inductive and deductive reasoning. Yet studies at the high school level and above have shown that developing an understanding of mathematical proof and proving remains a challenge for many students. By its very nature, mathematical proof is highly sophisticated and seems to be much more challenging intellectually than many other parts of the school mathematics curriculum. To a large extent this is so because the kind of reasoning required in mathematical proof is very different from that required in everyday life. Reasoning in everyday life does not require the rigor of mathematical proof nor the careful attention to process demanded by mathematical proof, which seeks, for example, to make clear distinctions among assumptions, theorems, and rules of inference.

Fishbein (1999), in discussing intuitions and schemata in mathematical reasoning, has identified five situations in which everyday reasoning may differ from mathematical reasoning, in process or in result. These situations are those in which (1) a statement is accepted intuitively and no proof is requested; (2) a statement is accepted intuitively, but in mathematics it is also formally proved (coincidence between intuitive acceptance and a logically based conclusion); (3) a statement is not intuitive or self-evident and may be accepted only on the basis of a formal proof; (4) a conflict appears between the intuitive interpretation (solution) of a statement and the formally based response; and (5) two conflicting intuitions may appear.

In some situations, a mathematically correct result may be accepted by a layperson as intuitively obvious even though from the point of view of mathematicians it still has to be proved (for example, the statement that in an isosceles triangle the angles adjacent to the base are equal). In other situations, intuition may provide no guidance as to the truth or falsehood of a result, although the result may be decided as true or false through the use of formal methods (for example, the statement that the sum of the angles in a triangle in the plane is always equal to two right angles). In yet other situations, everyday reasoning may actually conflict with mathematical reasoning: a result that seems intuitively obvious may be shown to be false using formal methods (for example, it seems intuitively clear that the set of natural numbers and the set of even numbers are not equivalent, but mathematical reasoning proves that they are equivalent). When everyday reasoning can differ in so many ways from mathematical reasoning, it is understandable that formal methods, and proof in particular, become challenging for students to learn.

In one study, Chazan (1993) interviewed high school geometry students from two different schools on their views of empirical evidence and mathematical proof. The students had all completed both a traditional instructional unit in geometry about deductive proofs and how to construct them, and an instructional unit, also in geometry, specifically designed to highlight differences between empirical evidence and deductive proof. Chazan's analysis showed that despite specific instruction designed to foster students' understanding of the strengths and limitations of empirical evidence versus deductive proof, their justifications for their views about empirical evidence and mathematical proof spanned a broad spectrum. Chazan found several kinds of students' misunderstandings about deductive proof and empirical evidence. For example, some students viewed empirical evidence as legitimate mathematical proof; others viewed deductive proof as no more than empirical evidence applying only to one single diagram. One student put it this way: "The [deductive proof] could be true for this triangle [in the associated diagram or in diagrams of that type] but up here it [the statement] says in any triangle. I would have to think of all the other types of triangles it would be true for" (p. 372). In general, students were still formulating their views of what constitutes a justification that a statement is always true.

Chazan's (1993) analysis showed that students were concerned with the persuasiveness and explanatory power of arguments, in particular for their own understanding. They were interested in deciding for themselves whether they believed a statement to be true, without reference to the mathematical legitimacy of the argument involved. Assuming that the arguments these students thought were persuasive were also valid, one could say that the students were interested primarily in what Hanna (1989) has called proofs that explain, whereas the researcher was more interested in probing students' understanding of proofs that prove. The students in this study, like the elementary school students in the studies discussed previously, apparently were interested first and foremost in mathematics as a sense-making activity. The adequacy of explanations and justifications was based on the extent to which they served communicative functions for the group of individuals involved and not on some external criteria.

In a British study, Coe and Ruthven (1994) examined the proof practices and constructs of a group of advanced-level students who had followed an inquiry-based secondary school mathematics curriculum. The new courses, for students ages 16 to 19, were designed to engage students more fully in the self-generation of mathematical knowledge, with a particular emphasis on proof. The students studied were age 17 and had reached the end of their first year under the new curriculum. They were interviewed with the aim of exploring their general conceptions of proof, their views of the functions of proof, and their views on insight and understanding. Although most of them stated that the function of proof is to ensure the certainty of what is proved, few students explained why rules or patterns occur, and their proof strategies were predominantly empirical as opposed to deductive.

In a study of post–secondary school students, Moore (1994) examined the cognitive difficulties that university students experience in learning to develop formal proofs. All students in the study were majoring in mathematics or mathematics education. On the basis of his analysis, Moore identified three major sources for the students' difficulties with the transition to formal proof: (1) conceptual understanding, (2) mathematical language and notation, and (3) getting started on a proof. Moore stated that most post–secondary school students have difficulty with formal proof because they "begin their upper-level mathematics courses having written proofs only in high school geometry and having seen no general perspective of proof or methods of proof" (p. 249). He concluded that although the students had learned to do proofs in the course and had acquired some notion of their purpose, "they probably were not ready to use proof and deductive reasoning as tools for solving mathematical problems and developing mathematical knowledge" (p. 264).

Balacheff (1991) noted that when the emphasis in instruction is on the written form of proof rather than on students' reasoning, students do not gain an appreciation for the role of proof as a tool that allows a mathematician to establish the validity of a statement and to convey that validity to others. His investigations also show, however, that involving students in situations in which they must arrive at an agreed-on solution to a problem is insufficient as well, because in such situations social behaviors influence the mathematical processes. In the adversarial environment that often develops in such circumstances, students often do not want to lose face by admitting their errors in front of their peers or are reluctant to acknowledge the correctness of an opposing group's solution.

Balacheff (1991) also noted that students engage in behaviors and activities that lead them to act as practical persons rather than as theoreticians. They tend to frame their solutions in a format they expect will be acceptable to peers and teacher. Their aim is then to produce a solution, not to produce knowledge. Balacheff concluded that "in some circumstances social interaction might become an obstacle, when students are eager to succeed, or when they are not able to coordinate their different points of view, or when they are not able to overcome their conflict on a scientific basis" (p. 188).

Taken together, these studies support Hoyles's (1997) conclusion that we are a long way from understanding how students acquire proof. Hoyles went on to suggest that to influence classroom practice it is not enough to focus on student and teacher and that it is essential to take into account the wider influences of curriculum organization. The challenge is to design situations that help construct a coherent and connected conception of proof. Blum and Kirsch (1991) and Dreyfus and Hadas (1996) have given some recommendations in this regard. Blum and Kirsch have argued for use of "preformal" proofs, in which students develop a chain of correct, but not formally represented, conclusions starting from valid, nonformal premises. Their recommendation is that classroom instruction make frequent use of preformal proofs, translating back and forth between formal and preformal arguments, and that it include explicit discussions with students about proof and proving.

For their part, Dreyfus and Hadas (1996) have proposed choosing situations and constructing activities "in such a manner as to effectively lead students up to the surprise and to the feeling that there is something to explain or prove" (p. 11). Their recommendation is based on the position that one of the cognitive difficulties posed by the nature of proof is the failure of students "to see a need for proof, as well as their failure to see the explanatory and convincing roles of proof" (p. 1). They point to interactive geometry computer programs as one way to increase the willingness and ability of students to investigate, generalize, and conjecture. In an observation consistent with Chazan's (1993) research, however, they noted that such computer programs do not necessarily strengthen the understanding of proof and can lead students to difficulties in distinguishing between empirical evidence and proof. Nevertheless, Dreyfus and Hadas outline "how the empirical approach can be used as a basis for creating didactic situations in which students actually require proofs" (p. 2).

In a university-level mathematics course for prospective elementary school teachers, Simon and Blume (1996) conducted a study of a somewhat different nature on how students develop an understanding of justification and proof. Their goal was to investigate the range of student responses to situations demanding justification, as well as the means by which their understanding of what constitutes justification might be furthered. The instructor was committed to developing the class as "a mathematical community in which mathematical knowledge is developed and validated by the community" (p. 5) while being fully aware that the previous school experiences of this group of students had been in traditional classrooms. Using a classroom teaching experiment, with one of the researchers as instructor, Simon and Blume set out to promote a classroom environment in which "mathematical validation and understanding would become important foci and to endeavor to make mathematical ideas problematic in ways that students are likely to see a need for deductive proof and 'proofs that explain'" (p. 9).

Simon and Blume's (1996) emphasis on the interactive constitution of justification by the class is particularly useful for highlighting the impact of students' conceptual understanding on the process of justification. Their data provide strong support for their conclusion that the conceptual understanding of the community members affects what they can accept as valid justification. For example, in one class session, students were asked to decide, without doing any calculations, which juice has a stronger strawberry flavor between one made from four strawberry cubes and three blueberry cubes or one made from three strawberry cubes and two blueberry cubes. The task was for the students to demonstrate their thinking using actual cubes they were given. The solutions students developed included some that relied on additive reasoning (there is one more strawberry cube than blueberry in each case, so they will taste the same) and some that relied on multiplicative reasoning (the second is a

3-to-2 mixture, whereas the first can be thought of as a 3-to-2 mixture to which a 1-to-1 mixture has been added, resulting in a dilution).

Simon and Blume (1996) point out that the justifications based on multiplicative reasoning were not persuasive to many of the students, even though they were consistent with the norms for validation that had been established for the class. The students who were not persuaded were those who had not yet constructed the requisite cognitive structures to reason multiplicatively about comparisons.

A second important point we can take from the study by Simon and Blume (1996) is that explanations and justifications must be built on what is taken-as-shared by the community involved. At the same time, what is taken-as-shared in a classroom evolves as the school term progresses. As mathematical practices new to the classroom become taken-as-shared, they no longer require justification, as we have noted elsewhere (Yackel, 1997), and hence what students use (and require) as support for their arguments, in the form of warrants and backing,[2] will also evolve. Thus there is a reflexive relationship between what becomes taken-as-shared in the classroom and the nature of the arguments that students give.

As pointed out, the rationales that students give as data, warrants, and backing for their explanations and justifications contribute to the development of what is taken as shared by the classroom community, that is, to the mathematical practices of the community. Consequently, a strong link exists between the nature of the mathematical explanations and justifications that students give in a classroom and the mathematical learning that takes place.

Summary of Research Lessons

These research studies provide ample evidence that students as early as the primary grades of elementary school, given a classroom environment constituted to support mathematics as reasoning, can and do engage in making and refuting claims, use both inductive and deductive modes of reasoning, and generally treat mathematics as a sense-making activity—that is, they treat mathematics as reasoning. However, these studies also demonstrate clearly that creating a classroom atmosphere that fosters this view of mathematics is a highly complex undertaking that requires explicit effort on the part of the teacher. The activity of initiating social and sociomathematical norms that support a view of mathematics as reasoning has itself been the object of study in several of the investigations reported here, notably the research of Lampert (1990) and that of Simon and Blume (1996).

The studies also provide ample evidence that not all students reason with the same levels of sophistication. The studies of the understanding of proof and the activity of proving on the part of high school and post-secondary school students in particular show that many of these students had not to that point developed an understanding that mathematical justification is the means by which assertions are legitimized (i.e., by which assertions become mathematical truths for a community). We hasten to add that in each of these studies, many of the students were participating in their first mathematical experience that had expectations of mathematical justification and explanation. They may have been disadvantaged by a number of years of prior mathematics instruction that fostered a rule-oriented view dominated by procedures and emphases on correct answers. An important task for mathematics educators is to develop meaningful ways to help students make the transition to formal proof from their early experiences with reasoning, explaining, and justifying.

Another important issue raised by these studies is that of feasibility on a large scale. In several of the studies discussed here, the researcher or a member of the research team was also the teacher. In others, for example that by Cobb et al. (1992), the regular classroom teacher was responsible for the instruction. In the latter studies considerable time and effort was spent working with the teacher as he or she developed a form of practice that made possible the type of learning that took place. A challenge for mathematics educators is to design means to support teachers in developing forms of classroom mathematics practice that foster mathematics as reasoning and that can be carried out successfully on a large scale. The current vision for mathematics education will be realized only when this challenge is met.

ACKNOWLEDGMENT

Frank Lester provided comments on an earlier version of this chapter.

REFERENCES

Balacheff, N. (1991). The benefits and limits of social interaction: The case of mathematical proof. In A. J. Bishop, E. Mellin-Olsen, & J. van Dormolen (Eds.), *Mathematical knowledge: Its growth through teaching* (pp. 175–192). Dordrecht, The Netherlands: Kluwer.

[2]See Krummheuer (1995) for a detailed exposition of argumentation in the mathematics classroom. Krummheuer uses the work of Toulmin to outline a framework for argumentation based on analyzing conclusions, data, warrants and backing.

Ball, D. L. (1991). What's all this talk about "discourse"? *Arithmetic Teacher, 39*(3), 44–48.

Bauersfeld, H. (1980). Hidden dimensions in the so-called reality of a mathematics classroom. *Educational Studies in Mathematics, 11*, 23–41.

Bell, A. (1976). A study of pupils' proof-explanations in mathematical situations. *Educational Studies in Mathematics, 7*, 23–40.

Blum, W., & Kirsch, A. (1991). Preformal proving: Examples and reflections. *Educational Studies in Mathematics, 22*, 183–203.

Chazan, D. (1993). High school geometry students' justification for their views of empirical evidence and mathematical proof. *Educational Studies in Mathematics, 24*, 359–387.

Cobb, P. (1994). Where is the mind? Constructivist and sociocultural perspectives on mathematical development. *Educational Researcher, 23*(7), 13–20.

Cobb, P., Wood, T., Yackel, E., & McNeal, B. (1992). Characteristics of classroom mathematics traditions: An interactional analysis. *American Educational Research Journal, 29*, 573–604.

Coe, R., & Ruthven, K. (1994). Proof practices and constructs of advanced mathematics students. *British Educational Research Journal, 2*, 41–53.

De Villiers, M. (1990). The role and function of proof in mathematics. *Pythagoras, 24*, 17–24.

Dreyfus, T., & Hadas, N. (1996). Proof as an answer to the question why. *Zentralblatt für Didaktik der Mathematik (International Reviews on Mathematical Education), 96*(1), 1–5.

Fishbein, E. (1999). Intuitions and schemata in mathematical reasoning. *Educational Studies in Mathematics, 38* (1–3), 11–50.

Greeno, J. (1994). Comments on Susanna Epp's chapter. In A. Schoenfeld (Ed.), *Mathematical thinking and problem solving* (pp. 270–278). Hillsdale, NJ: Erlbaum.

Hanna, G. (1989). Proofs that prove and proofs that explain. In G. Vergnaud, J. Rogalski, & M. Artigue (Eds.), *Proceedings of the 13th Conference of the International Group for the Psychology of Mathematics Education* (Vol. 2, pp. 45–51). Paris: PME.

Hanna, G., & Jahnke, H. N. (1996). Proof and proving. In A. Bishop, K. Clements, C. Keitel, J. Kilpatrick, & C. Laborde (Eds.), *International handbook of mathematics education* (pp. 877–908). Dordrecht, The Netherlands: Kluwer.

Hoyles, C. (1997). The curricular shaping of students' approaches to proof. *For the Learning of Mathematics, 17*(1), 7–16.

Jaffe, A. (1997). Proof and the evolution of mathematics. *Synthese, 3*(2), 133–146.

Kleiner, I. (1991). Rigor and proof in mathematics: A historical perspective. *Mathematics Magazine, 64*, 291–314.

Krummheuer, G. (1995). The ethnography of argumentation. In P. Cobb & H. Bauersfeld (Eds.), *The emergence of mathematical meaning: Interaction in classroom cultures* (pp. 229–269). Hillsdale, NJ: Erlbaum.

Lakatos, I. (1976). *Proofs and refutations.* Cambridge: Cambridge University Press.

Lampert, M. (1990). When the problem is not the question and the solution is not the answer: Mathematical knowing and teaching. *American Educational Research Journal, 27*, 29–63

Maher, C. A., & Martino, A. M. (1996). The development of the idea of a mathematical proof: A 5-year case study. *Journal for Research in Mathematics Education, 27*, 194–214.

Manin, Y. (1977). *A course in mathematical logic.* New York: Springer-Verlag.

Manin, Y. (1998). Truth, rigour, and common sense. In H. G. Dales & G. Oliveri (Eds.), *Truth in mathematics* (pp. 147–159). Oxford: Oxford University Press.

Moore, R. C. (1994). Making the transition to formal proof. *Educational Studies in Mathematics, 27*, 249–266.

National Council of Teachers of Mathematics. (1989). *Curriculum and evaluation standards for school mathematics.* Reston, VA: Author.

National Council of Teachers of Mathematics. (2000). *Principles and standards for school mathematics.* Reston, VA: Author.

Piaget, J. (1970). *Genetic epistemology.* New York: Columbia University Press.

Pólya, G. (1954). *Induction and analogy in mathematics.* Princeton, NJ: Princeton University Press.

Richards, J. (1991). Mathematical discussions. In E. von Glasersfeld (Ed.), *Radical constructivism in mathematics education* (pp. 13–52). Dordrecht, The Netherlands: Kluwer.

Rota, G-C. (1997). The phenomenology of mathematical proof. *Synthese, 3*(2), 183–197.

Saxe, G. B. (1991). *Cultural and cognitive development: Studies in mathematical understanding.* Hillsdale, NJ: Erlbaum.

Simon, M. A., & Blume, G. W. (1996). Justification in the mathematics classroom: A study of prospective elementary teachers. *Journal of Mathematical Behavior, 15*, 3–31.

Skemp, R. R. (1978). Relational understanding and instrumental understanding. *Arithmetic Teacher, 26*(3), 9–15.

Smith, E. (1997). Constructing the individual knower—A review of *Radical Constructivism. Journal for Research in Mathematics Education, 28*, 106–111.

Steffe, L. P., & Kieren, T. (1994). Radical constructivism and mathematics education. *Journal for Research in Mathematics Education, 25*, 711–733.

Thompson, P. W. (1996). Imagery and the development of mathematical reasoning. In L. P. Steffe, P. Nesher, P. Cobb, G. A. Goldin, & B. Greer (Eds.), *Theories of mathematical learning* (pp. 267–283). Mahwah, NJ: Erlbaum.

Thurston, W. P. (1994). On proof and progress in mathematics. *Bulletin of the American Mathematical Society, 30*(2), 161–177.

Voigt, J. (1995). Thematic patterns of interaction and sociomathematical norms. In P. Cobb & H. Bauersfeld (Eds.), *The emergence of mathematical meaning: Interaction in classroom cultures* (pp. 163–201). Hillsdale, NJ: Erlbaum.

Von Glasersfeld, E. (1990). An exposition of constructivism: Why some like it radical. In R. B. Davis, C. A. Maher, & N. Noddings (Eds.), *Constructivist views on the teaching and learning of mathematics* (pp. 19–29). Reston VA: National Council of Teachers of Mathematics.

Yackel, E. (1997, April). *Explanation as an interactive accomplishment: A case study of one second-grade classroom.* Paper presented at the meeting of the American Educational Research Association, Chicago.

Yackel, E. & Cobb, P. (1994, April). *The development of young children's understanding of mathematical argumentation.* Paper presented at the meeting of the American Educational Research Association, New Orleans.

Yackel, E., & Cobb, P. (1996). Sociomathematical norms, argumentation, and autonomy in mathematics. *Journal for Research in Mathematics Education, 27*, 458–477.

Chapter 16

Communication and Language

Magdalene Lampert, University of Michigan
Paul Cobb, Vanderbilt University

Since 1989, when the National Council of Teachers of Mathematics (NCTM) published its *Curriculum and Evaluation Standards for School Mathematics*, mathematics education researchers have identified several new areas of study, growing out of a single, overarching goal: That all students will learn to *do* mathematics. If school lessons are to involve learners doing mathematical work, classrooms will not be silent places where each learner is privately engaged with ideas. If students are to engage in mathematical argumentation and produce mathematical evidence, they will need to talk or write in ways that expose their reasoning to one another and to their teacher. These activities are about communication and the use of language.

Like other aspects of mathematics, communication and language need to be taught and learned in school classrooms. But they are also a primary means by which mathematics is taught and learned. Thus, more than any other topic, they bring to the fore the tension between *acquisition* and *participation* as metaphors for learning (Sfard, 1998). These metaphors are not mutually exclusive, and, as Sfard argues, any viable approach to the learning and teaching of mathematics has to cope with the tension between them. Nonetheless, it is useful for analytic purposes to distinguish between approaches that characterize learning primarily as the acquisition of mathematical knowledge, on the one hand, and those that treat it as a process of coming to participate in established mathematical practices, on the other hand.

When we think of mathematics as something students are to acquire, we imagine that some form of instruction is given and that students come away from the learning situation with knowledge and skills they did not have before. Whether we believe this knowledge is acquired by active construction or by passive listening, this perspective orients us to view activities in the mathematics classroom as a *treatment* and the learning outcomes as measurable increases in students' mathematical knowledge. For communication and language, this causal relationship can be viewed in two different ways: Either we can think of communication as part of the treatment with mathematical understanding as the outcome, or we can think about the instructional intervention as the treatment and learning to communicate as the outcome.

This choice is somewhat less cut-and-dried if we see learning as increasingly competent participation in mathematical practices that have been developed over a period of centuries and that constitute students' intellectual inheritance. In this view, communication is one aspect of participation in the activities of a mathematical community. Consequently, learning to communicate as a goal of instruction cannot be cleanly separated from communication as a means by which students develop mathematical understandings. In any particular area of mathematics, the types of communication in which students can engage are constrained by their current mathematical understandings, and conversely students develop more sophisticated mathematical understandings as they attempt to communicate their reasoning (Simon & Blume, 1996).

The differences between the world views of researchers who emphasize the acquisition metaphor and those who emphasize the participation metaphor make it difficult to simply say "what is known" about students' learning in the area of communication and language as it develops across Grades pre-K to 12. As we discuss these two bodies of research, it will become clear that we generally find the participation metaphor to be more useful when addressing issues of mathematics learning and teaching in the current era of reform. Further, in line with this metaphor, we will not always draw a hard-and-fast distinction between learning to communicate and communicating to learn.

Acquiring Knowledge of Mathematical Communication

Work guided by the acquisition metaphor is linked with the process-product tradition in educational research. What is "known" within this tradition about students' learning as it relates to language is based largely on a loosely defined concept of classroom communication. In one line of research, instruction in communication is generally equated with teaching students to work in small groups; it focuses on interaction among peers (as opposed to interaction between the teacher and a single student or between either the teacher or a student and the whole class), and it does not consider mathematics as having any special interactive requirements. Summarizing this work, O'Connor (1998) cites a cumulative study conducted by Webb in 1991 and based on 17 research projects across grade levels on many different mathematical topics. According to O'Connor, Webb found that

> (a) low achievement correlates with receiving nonresponsive feedback (e.g., only being told the correct answer by one's peers, with no further information) and (b) high achievement correlates with the behavior of giving "elaborate explanations" to one's teammates. (pp. 17–18)

She concluded that instruction aimed toward high achievement would involve teaching students "to give one another elaborate explanations" (p. 18). O'Connor continues with the following:

> However, [Webb] cites a study that [involved teaching students to give one another elaborate explanations] and found no effects on achievement (Swing & Peterson, 1982). The results of just one study that involved only a two-session training program with follow-up support of course do not invalidate Webb's proposal, but they do underline just how little is known about why such interactions might correlate with achievement. (p. 18)

As O'Connor observes, both Webb and Swing and Peterson seem to assume that giving elaborate explanations enhances the explainer's achievement in a direct causal fashion. A number of studies, some of which are based on the acquisition metaphor, call into question the assumption that particular features of the classroom have an impact on achievement in a direct, unmediated manner. Also, in focusing on achievement outcomes, Webb's work does not help us understand either the learning opportunities that arise for students in the course of small-group interactions or the process of their learning. It could be, for example, that the ability to give elaborate explanations is itself an indicator of high achievement. Webb's findings do not clarify whether giving elaborate explanations might further enhance either such students' mathematical understandings or their ability to communicate. As O'Connor makes clear, we would need to have a better idea of what small-group interactions consist of by analyzing them directly in order to investigate these issues.

A second line of research guided by the acquisition metaphor has focused on the possible benefits of whole-class discussions. Although several studies in the process-product tradition have found positive effects on a variety of intellectual and cognitive outcomes, the overall picture is unclear because of the lack of a clear definition of what constitutes *group discussion* (Gray, 1993). Further, the features of discourse measured in these studies are typically generic and do not assess characteristics of communication that are specific to mathematics. Hiebert and Wearne (1993) conducted the most definitive study of this type. In their investigation, Hiebert and Wearne gathered data from four traditional and two alternative second-grade classrooms to relate classroom tasks and discourse to students' achievement. They analyzed the classroom discourse by coding the questions the teachers asked. They used four categories that ranged from low-level recall questions to higher-level explain-and-analyze questions. Their findings indicate that higher-level questioning was associated with increased student achievement. They caution, however, that "given the complexity of classrooms and the undoubted interaction of many classroom features, it is impossible to isolate specific features and connect them to specific learning outcomes" (p. 420). The difficulty in drawing clear-cut conclusions from studies of this type again points to the value of analyzing both classroom discourse and the process of students' learning as they participate in it more directly.

Participating in Mathematical Communication to Learn

Several lines of research that emphasize the participation metaphor have focused on the process of learning to communicate mathematically. One set of analyses conducted by Cobb, Yackel, and Wood (1989) describes how a second-grade teacher initiated and guided the development of a classroom culture in which mathematical inquiry was valued. Making a distinction between *talking about* mathematics and *talking about talking about* mathematics proved important. As the teacher and students talked about the mathematics they were working on at the beginning of the school year, incidents frequently occurred that conflicted with the teacher's inquiry-oriented agenda. The teacher framed these incidents, as well as others that fit with her agenda, as occasions to talk about her expectations with the students. What she was doing was talking about how they should talk about mathematics. In the course of these discussions of specific cases, she and the students developed a common language in which to talk about new roles and responsibilities in the classroom. For example, they developed a shared understanding of what it meant to give a mathematical explanation, to understand another's explanation, and to collaborate to learn mathematics. This initial grounding in specific incidents in turn made it possible for them to talk directly about their roles and responsibilities as the school year progressed.

A related analysis of this same classroom focused on the opportunities of mathematical learning that arose as students worked together in small groups (Cobb, 1995). In this analysis, it was important to see the students' small-group work as part of a broader system of activities in the classroom. For example, the students' realization that they would subsequently be expected to explain and justify their reasoning in a whole-class discussion influenced their small-group work. The analysis identified two necessary conditions for productive small-group relationships. First and most obvious, it was essential that the students develop an adequate basis for mathematical communication. In one group, for example, the students made every effort to share their mathematical reasoning with one another but nonetheless were unable to understand one another's thinking. Second, students had to establish a relatively symmetrical relationship in which none of the students was viewed as a mathematical authority by the others. In cases in which one student did become a mathematical authority in the group, the other students typically deferred to this student's judgments without either challenging his or her explanations or asking clarifying questions. The analysis indicates that interactions of this type were productive neither for the student giving explanations nor for those listening. This finding bears on Webb's (1991) study in that its suggestion that learning opportunities do not necessarily arise for students who give elaborate explanations to their peers. It could be that students merely verbalize what they already know unless they anticipate that their explanations might be questioned and critiqued by others in the group. It is therefore important to consider the social context of students' small-group relationships when assessing whether particular types of activity, such as explaining, are productive for mathematical learning.

Research by Lampert, Rittenhouse, and Crumbaugh (1998) on the mathematical character of students' talk in small groups further complicates the relationship between participation in mathematical activity and learning in school. One of the primary activities that engage students in this setting is making assertions about how to solve a problem and what solutions make sense. We know that when students disagree, they have discussions that can lead to improvements in their mathematical understanding as they muster evidence to convince their peers of different points of view (Hatano & Inagaki, 1991). But what we are only beginning to understand is how and what students need to learn in order to disagree in ways that are both mathematically productive and socially acceptable. The study by Lampert et al. (1998) examined this problem at the upper elementary level and provides an interesting developmental contrast with the Cobb et al. (1989) findings in lower grades, suggesting that participation in mathematical communication, both as a strategy for mathematical learning and as a learning outcome, might be different at different ages.

In contrast with these case studies of individual classrooms, Forman (1996) compared traditional middle school level mathematics classrooms with classrooms at the same level in which students worked together on mathematical problems. Forman views mathematical learning as an apprenticeship into the discourse and reasoning practices of mathematically literate adults. She argues that the mathematical discourse that students are to master is a specialized type or genre of speech that she calls the *mathematics register*. In the course of their apprenticeship, students participate in this discourse in increasingly substantial ways as they come to understand the skills, norms, values, and ideas that are shared by mathematically literate adults. Forman's findings reveal that more opportunities arose for students to learn to use the mathematics register and thus communicate mathematically in the reformed classrooms than in the traditional classrooms. Her analysis also indicates that the mathematics register is not learned as a separate language but instead is intertwined with everyday speech in effective classroom discussions.

Taken together, the investigations we have discussed go a long way toward addressing the problems that O'Connor raised when commenting on research conducted in the process-product tradition. This work differs from studies guided by the acquisition metaphor, such as those reported by Webb (1991) and by Hiebert and Wearne (1993), on several counts:

- It attends to the specifically *mathematical* elements of classroom discourse.
- It attends strongly to the *teacher's role* in supporting the development of mathematically productive discourse.
- It looks at communication in small-group and whole-class settings and considers the ways in which the teacher is a participant as well as a director of classroom conversations.
- Although it considers the achievement of mathematical understanding to be a valued outcome of mathematics instruction, it attends to the improvement of social processes, such as small-group relationships and mathematical argumentation, as well as to individual outcomes.

Borrowing concepts from both cognitive and social psychology, these studies see talking and writing to be aspects of doing mathematics and regard the classroom as a community of learners, led by the teacher, in which learners are socialized to accept new norms of interaction and learn new meanings for mathematical words and symbols as they work together on problems.

From this point of view, the task of assessing whether participating in certain kinds of classroom talk result in more desirable mathematical understandings involves investigating the kinds of curriculum and instruction that will support that talk. Further, the task of examining whether various instructional activities support better mathematical communication involves both investigating what kinds of mathematical talk and writing are possible in classroom settings and clarifying what counts as an improvement in students' capacities to communicate.

What Is Taught and Learned in Mathematical Communication?

A number of studies reported during the past 10 years have attempted to understand these matters more clearly from the perspective of classroom practice in Grades pre-K to 12. The authors of these studies consistently argue that we need a rich vision of what talk can be and a complex concept of learning that is sensitive to social and cultural context so as to relate teaching and curriculum with learning. To develop this vision, researchers have looked carefully at a few classrooms in which mathematical talk is primary among the activities that are structured to give students opportunities to learn.

Mathematizing

One focus of this work has been to analyze the gradual emergence of mathematical understanding from initial instructional activities that focus on material objects, actions, and events. This process of coming to see concrete situations in mathematical terms is called *mathematizing*. Although it is often assumed that mathematical understanding emerges directly from the manipulation of physical objects, these analyses indicate that a series of subtle shifts occurs and that both talk and writing play a crucial role in this mathematization process. In her influential book *The Mastery of Reason*, Valerie Walkerdine (1988) presents two detailed analyses of teacher-student interactions that focus on the emergence of elementary addition and of place-value numeration. In the case of elementary addition, the shifts that Walkerdine describes include the following:

- Talking of "putting them [two collections of blocks] altogether" while actually carrying out the action (concrete)
- Redescribing the action as "three and four make ..." (concrete/iconic)
- Using a drawing of three circles to organize the action such that blocks are first placed in two of the circles and then moved together into the third circle (iconic)
- Replacing the blocks with drawings of blocks in two of the circles (iconic/abstract)

Superficially, the interaction between the teacher and students could be interpreted in terms of a gradual transition from the concrete to the iconic and eventually to the abstract. As Walkerdine's analysis reveals, however, the actions on the blocks and on the written symbols do not make sense apart from the talk that describes them. Her analysis is paradigmatic in that it reveals that mathematization involves the development of a chain of symbols, of ways of acting on those symbols, and of ways of talking about those actions. What happens in the process of mathematizing as we have described it here is a developmental process that initially involves learning how to use the words that go with objects or sets of objects and learning how to use the words that go with actions on those objects (putting together, sharing equally, etc.). Both O'Connor (1994) and Sfard (2000) have pointed out a complication of this abstracting process that is central to mathematical learning. As this kind of talk moves beyond its initial phases, although experiences and actions occur, no objects correspond to abstract mathematical concepts that can be pointed to directly. In Walkerdine's example, although the teacher and students could point to, and talk about, collections of blocks, they could not point directly to either numbers as abstract mathematical concepts or actions on them, such as adding. Sfard, in fact, describes the world of mathematical concepts as a virtual reality to distinguish it from the concrete reality of physical objects and actions. Walkerdine's work illustrates that this virtual reality does not spring directly from physical reality. Instead, the development of abstract mathematical concepts at even the most elementary level requires careful guidance in which talk and writing play a crucial role.

Cobb and his colleagues (Cobb, Boufi, McClain, & Whitenack, 1997) have studied this gradual process of moving from the concrete through the iconic to the abstract in the classroom using the notion of *reflective discourse*. This type of classroom discourse is characterized by repeated shifts such that what the teacher and students say and do subsequently itself becomes an explicit object of discussion. For example, one sample episode began with a class of first graders generating the different ways in which six monkeys could be distributed between two trees (four monkeys in one tree and two in the other, etc.). Later, the results of this activity became the topic of discussion when the students attempted to check whether they had found all the possibilities by searching for patterns in the combinations they had generated (e.g., there could be four in one tree and two in the other, or vice versa, but no other combinations involving these numbers). Discourse of this type might in fact be called *mathematizing discourse* because a strong parallel exists between its structure and psychological accounts of mathematical development in which actions or processes are transformed into objects of mental mathematical objects (Dubinsky, 1991; Freudenthal, 1983; Gray & Tall, 1994; Pirie & Kieren, 1994; Sfard, 1991). We can reasonably conjecture that talking about mathematics in this way might give rise to opportunities for students to learn by reflecting on and objectifying prior activity. Further, we have some indication that as a consequence of participating in this kind of communication, students develop what might be termed a *mathematizing attitude* that involves organizing the results of prior mathematical reasoning by searching for patterns.

Symbolic records in which students write or draw representations of their reasoning play a crucial role in supporting reflective shifts in classroom discourse. In the case of the sample episode involving the monkeys in the two trees, for example, the teacher made a table to record the possibilities the students generated. The students then began to refer to the table entries as they checked whether they had found all the possibilities. Thus, the shift in discourse was accompanied by a change in the function of the table such that it was no longer a mere record. Instead, the students used it to organize the combinations they had generated. In a very real sense, they might be said to have reasoned with the table in that they would not have been able to reflect on their prior activity without it. The central role of written symbols in supporting reflective shifts in discourse is highly compatible with Gravemeijer's (1997) more general analysis of the role of symbols in mediating between concrete, informal mathematical activity and abstract mathematical reasoning. In addition, it is consistent with Dörfler's (2000) discussion of the role of symbolic records, which he calls *protocols of action*, in supporting the development of abstract mathematical reasoning that involves understanding. The interplay between shifts in discourse and changes in the function of symbols is in fact so close that the choice of which to highlight is primarily a matter of convenience.

It is important to note that a one-sided push for mathematization can result in some of the participating students' reasoning becoming decoupled from concrete situations and events. In extreme cases, a teacher's attempts to initiate reflective shifts in discourse can degenerate into a social guessing game in which students try to infer what the teacher wants them to say. In light of this possibility, the teacher's role might be thought of as probing to assess whether students can step back and reflect on what they are currently doing. A related notion that also helps temper a one-sided drive for abstraction is that of the folding back of discourse described by McClain and Cobb (1998). The process of folding back occurs when the mathematical relationships under discussion are redescribed in terms of specific situations or events. This notion was derived from Pirie and Kieren's (1994) theory of the growth of mathematical understanding and reflects the conjecture that the folding back of discourse helps students ground their increasingly abstract mathematical reasoning in situation-specific imagery. In addition, the folding back of discourse can play a vital role in enabling students to communicate their reasoning effectively to one another. In particular, students who talk past one another when they explain their reasoning in relatively abstract terms can often communicate effectively when they redescribe their thinking in terms of a concrete situation.

The opposite problem—students' becoming mired in mathematically irrelevant details of a concrete problem situation—is perhaps equally common as work in classrooms moves increasingly toward work on real-life problems. McNair (1998) analyzed three classrooms (Grades 2, 5, and 10) in which students solved problems of this type and their teachers attempted to move from talk about specific problem situations to talk about mathematical relationships. McNair describes a range of roles that the teachers and students played, not all of which gave rise to opportunities for the students to learn to do mathematics. In an investigation that also focused on the mathematization process, Stevens and Hall (1998) examined the mathematical communication between a high school student studying graphing functions and his tutor. They found that the way in which the tutor talked about the graphs the student was making made a significant difference in whether the student could understand mathematical ideas that went beyond the concrete details of the specific example they were working on together.

Negotiated Defining and Genre Instruction

As students increase their participation in talk about mathematical ideas, they mix informal definitions with formal ones, challenging the conventional meanings of familiar terms. One the one hand, this participation means that they are using terms in ways that relate them closely to the ideas they are intended to signify. On the other hand, it suggests a potential problem in communicating about those ideas with others, outside the local classroom community.

As O'Connor (1998) observes,

> Dictionary definitions have long been a staple of elementary school math and science, and precise use of such definitions allows students to refer to objects and processes in an unambiguous and technically correct fashion. They are as important as ever, but in reform classrooms, other kinds of defining must take place. In traditional classrooms the creation of one's own definitions was rarely required. But in classrooms that attempt to support "real" inquiry in math and science, the need to agree on the precise meaning of an expression within the local context arises frequently. Within open-ended projects, it is sometimes necessary to develop a working definition of some phenomenon or process, a definition that will change as understanding increases. In such classrooms, the development of one's own symbols, measures, and terminology is often required, with complex and poorly understood difficulties often arising. (p. 43)

The teacher thus has a new role in managing this dilemma. In the context of a case in which a third-grade student asserts that the number 6 "can be both odd and even," Deborah Ball (1993) examines her own conflicts as a teacher facing contradictory goals. Other examples of this problem are analyzed in regard to the use of the terms *length* and *width* (O'Connor, 1992), *points* and *corners* (Russell & Corwin, 1993), and *minus one half* (Lampert, 1992a). In these studies, the teacher takes on the role of mediator between students' informal and local uses of these terms and the conventional mathematical use of them.

Managing the dilemma of teaching mathematical word meanings is a small part of coping with the more general tension between academic genres of speech and writing and the genres that are familiar in students' lives outside school. The genre movement (as it has become known among literacy educators) has been a direct effort to reform teaching and learning around the ideas that specific genres structure academic disciplines and that domain-specific teaching must include explicit genre instruction (see Cope & Kalantzis, 1993; Reid, 1987). Sociolinguist and literacy educator Deborah Hicks (1998) comments,

> Such instruction might, in a mathematics educational setting, include things like teaching students how to construct a mathematical narrative (Longo, 1994) or explanation. To some extent, such deliberate "genre instruction" is part of what happens at least implicitly in reform mathematics classrooms. Teachers deliberately call students' attention to the forms of communication appropriate for group discussions. . . . The genre instruction movement in Australia and New Zealand goes even further than this in weaving explicit grammar instruction (that is, the *text grammars* of academic discourses) into the teaching of subjects like social studies and science. What we in this country would refer to as language arts instruction is woven into discipline-specific teaching. In the context of learning about the greenhouse effect, for instance, students receive grammar instruction (e.g., modeling, structured practice sessions) on how to write a scientific explanation (see Callaghan, Knapp, & Noble, 1993). (p. 248)

The teacher studied by Cobb et al. (1989) who is described above as using "talk about talk" to teach mathematical communication provides an example of this approach. As a pedagogy of "inclusion and access"

(Cope & Kalantzis, 1993), genre instruction has the potential to afford students from a wide range of cultural and economic backgrounds access to mathematical communication, both within and beyond the classroom. As with negotiated defining, the teacher manages the dual goal of accepting the ways of talking that students bring with them into the classroom and making them aware of other alternatives.

Participation Structure for Doing and Learning Mathematics

The investigations we have discussed thus far indicate that giving simultaneous attention to intellectual content and social processes is necessary when identifying elements of teaching and learning that occur at the intersection of subject matter-transactions and classroom dynamics (Erickson, 1982). This synthesis is highlighted in Lampert's research on pedagogical design (1990, 1992b, 2001). Drawing on Pólya's (1954) and Lakatos's (1976) analyses of the specifically mathematical aspects of mathematical discourse, Lampert began to organize whole-class discussions around students' different answers to a problem and to call their answers "conjectures" until the class as a whole discussed the legitimacy of various solution strategies. In these discussions, the mathematical importance of conditions, assumptions, and interpretations were made more explicit, and students were pushed toward revising definitions using their own terms of reference. Because this activity was conducted in a classroom, it required additional work from the teacher and the students to deflect the agenda from the simple production and assessment of the answer so that students could learn the mathematical processes of refining language and verifying assertions. Instruction had to be slowed down and the discourse opened up to enable students to participate in evaluating their own thinking.

Three studies have examined particular instructional processes that were used in Lampert's classroom to move students away from the typical pattern of interaction (i.e., teacher demonstrates a procedure, teacher or book poses a question or problem, students produce answers, teacher judges answer, and teacher moves on or reteaches). Examining what the teacher said about what was expected in small-group work, Blunk (1998) found that supporting a way of talking and working that put the responsibility on students for investigating whether what they were doing made sense in mathematical terms was a yearlong element of the teachers' role. Rather than teach students how to talk about mathematics at the beginning of the year and then leave them on their own to do it, the teacher needed to reassert mathematical norms and values at several points throughout the entire course of instruction. In working on mathematics with students, Lampert was constantly "stepping in and stepping out" of the discourse. Like the teacher studied by Cobb and his colleagues, she both talked about mathematics with her students and talked about what appropriate mathematical talk was supposed to sound like (Rittenhouse, 1998). A close linguistic analysis of teacher-student talk in this classroom suggests that the teacher can also play a significant role in helping students distinguish between ordinary politeness in which they expressed respect for one another and mathematical politeness in which they expressed respect for one another's ideas (Weingrad, 1998).

A related line of work has focused on norms for classroom action and interaction that are specific to mathematics. This work has been influenced by Lampert's (1990) observation that classroom communities negotiate both what counts as a solution and what it means to understand. Examples of these so-called sociomathematical norms discussed by Voigt (1995) and by Yackel and Cobb (1996) include what counts as a different mathematical solution, a sophisticated mathematical solution, an efficient mathematical solution, and an acceptable mathematical explanation. These norms can differ significantly from one classroom to another, thereby influencing the learning opportunities that arise for students and, indeed, for teachers. Further, in consciously guiding the negotiation of sociomathematical norms that fit with their agendas, teachers can support students' development of what is termed a *mathematical disposition* in *Professional Standards for Teaching Mathematics* (NCTM, 1991).

Conceptual Discourse and Big Mathematical Ideas

In their research on teaching and teacher thinking, Alba and Pat Thompson made a distinction between conceptual and calculational orientations in teaching (Thompson, Philipp, Thompson, & Boyd, 1994). This distinction can usefully be extended by talking about conceptual and calculational discourse (Sfard, Nesher, Streefland, Cobb, & Mason, 1998). An important point to clarify that calculational discourse does not refer to conversations that focus on the procedural manipulation of conventional symbols that do not necessary signify anything for students. Instead, calculational discourse refers to discussions in which the primary topic of conversation is any type of calculational process. As an example, in whole-class discussions of elementary word problems, first graders typically report a variety of counting and thinking strategy solutions. Discussions that focus on the students' methods are calculational in nature in that they are concerned with the calculational processes the students used to arrive at answers. As this example illustrates, calculational discourse can involve insight and understanding. Many of the illustrations given in the literature of discourse that is compatible with the various Standards documents (NCTM, 1989, 1991, 1995) are in fact calculational.

Calculational discourse can be contrasted with conceptual discourse in which the reasons for calculating in particular ways also become explicit topics of conversation. In this latter case, conversations encompass both students' calculational processes and the task interpretations that underlie those ways of calculating. In the instance of elementary word problems, the teacher might help students explain how they related the quantities described in the problem statement (e.g., in solving a problem corresponding to 4 + _ = 11, some students might have included the 4 in the 11, whereas others might have treated them as separate quantities and added them to arrive at an answer of 15). Practical experience in a number of teaching experiments indicates that discussions in which the teacher judiciously supports students' attempts to articulate their task interpretations can be extremely productive settings for mathematical learning. As the parenthetical example illustrates, students' participation in discussions of this type increases the likelihood that they might come to understand one another's mathematical reasoning. Had the discussion remained purely calculational, the students who added 4 and 11 could have understood other students' explanations only by creating entirely on their own a task interpretation that lay behind the calculational method. In contrast, their participation in conceptual discourse provides them with resources that might enable them to reorganize their initial interpretation of the task. An important outcome to note is that a significant mathematical idea, that of numerical part-whole relations, became an explicit topic of conversation in the example. This idea was not imported into the conversation from the outside by the teacher but instead emerged in what might be termed a natural way from students' problem-solving efforts. At least in the early grades, within a few weeks most students can learn to give conceptual explanations and to ask others clarifying questions that bear directly on their underlying task interpretations.

How Is Mathematics Taught and Learned in Classroom Communication?

An important focus in the study of classroom communication is the ways in which students and teachers learn to take on different roles in the conversation. Studies in this area have attended primarily to the development of a mathematical *participation structure*, that is, to the ways in which teachers teach students to talk as they do mathematics in the classroom. Equally important, but much less understood, is the teachers' role in assuring that classroom talk in mathematics lessons is both about appropriate mathematical topics and productive of mathematical understanding.

Even though current reform efforts focus on increasing the amount of mathematical communication that occurs among students, discourse is not a goal in and of itself (cf. Ball, 1991). Specific discussions must be justified in terms of what students might be learning as they participate in them, whether it be learning to communicate mathematically or communicating to learn mathematics. This requirement calls for planning that involves capitalizing on what students do and directing their activity toward important mathematical issues—a process that O'Connor calls "managing the intermental" (Lampert, 1992b, 2001; O'Connor, 1996). As part of this planning process, the teacher monitors the students' reasoning as they work individually or in small groups to develop conjectures about mathematically significant issues that might emerge as topics of conversation in the subsequent discussion (Simon, 1995). Further, at the beginning of a discussion, the teacher might call on specific students selected in advance because he or she anticipates that a comparison of their solutions might lead to substantive mathematical conversations that advance the pedagogical agenda.

In one of the few case studies of classroom communication that focuses on a mathematical idea rather than on the quality of student interaction or the development of mathematical language, Hall and Rubin (1998) traced the ways in which ideas about rate moved from teacher to students and among students as they worked on time-speed-distance problems. Hall and Rubin describe how a teacher can construct topical teaching in response to students' work on problems. The work of Thompson and Thompson (1996) also involved time-speed-distance problems and focused on communicating to learn mathematics. Although the interactions they describe are between a teacher and a single student, their analysis does highlight the teacher's role in managing the discourse so that the significant mathematical ideas could become the explicit focus of attention. Taking a longer-term view, Gravemeijer, Cobb, and colleagues describe the gradual emergence of mathematical ideas during a series of classroom teaching experiments that each lasted several weeks. These ideas include part-whole reasoning in elementary arithmetic (Gravemeijer, Cobb, Bowers, & Whitenack, 2000), place-value numeration (Bowers, Cobb, & McClain, 1999), and distribution as a big idea in statistics (Cobb, 1999).

The work of O'Connor and Michaels (1993, 1996) provides a further example of research that attends to the subtleties of classroom interaction while taking mathematical content issues seriously. They analyzed whole-class discussion to understand how teachers move their pedagogical agendas forward by helping students take on particular roles (e.g., the proposer of a conjecture) in relation to both one another and the mathematical ideas under discussion. A type of intervention that proved to be important is *revoicing*, in which the teacher reformulates a student's contribution verbally or in writing. O'Connor and Michaels found that teachers revoice students' explanations for a variety of reasons, including clarifying ideas, introducing new terms for familiar ideas, directing the discussion in a new and potentially productive direction, and helping students explicate their reasoning. For the last purpose, for example, the teacher might reformulate a student's conjecture so that the grounds for it are expressed in a clear and coherent manner. In addition, teachers revoice students' explanations to bring them into relation with one another. That use might involve recasting two contributions apparently made independently as conflicting theories, thereby the placing the students in opposition to each other with respect to the ideas being discussed. Alternatively, the teacher might revoice a sequence of responses so that they become a string of alternative ways of solving the same task.

O'Connor and Michaels (1996) extended their analysis of revoicing by comparing the discourse in an innovative United States classroom with that in a Japanese classroom, reported by Wertsch and Toma (1995), in which similar instructional activities were used. They found striking differences, the most significant being that the U.S. teacher revoiced 10 times more frequently than her Japanese counterpart. In the U.S. classroom, the students tended to take separate turns without explicitly attempting to link their explanations to others' contributions. A primary function of the U.S.teacher's revoicing was therefore to establish relationships between these independent contributions. In contrast, the Japanese students frequently cited one another when developing explanations. For example, a Japanese student might explain that she is using a previous students' argument as support for her argument against a third students' conjecture. Wertsch and Toma call discourse of this type, in which students view one another's contributions as devices to think with, *multivocal discourse*. They report that Japanese students are explicitly taught in the early elementary grades to refer to one another's ideas and to use them when constructing their own ideas. This observation gives some indication of the capabilities that U.S. students might develop if their teachers were to consistently revoice their contributions, as described by O'Connor and Michaels.

Teaching Mathematical Communication to All Students

An emphasis on communication and language in the mathematics classroom leads naturally to a consideration of diversity and of equality of opportunities to learn mathematics. Although these issues have thus far received only limited attention in mathematics education research, they have been the explicit focus of research in a number of other fields, particularly language and literacy. Part of the challenge is to learn from this work while developing research agendas that take account of the specific characteristics of communication about mathematical ideas.

Warren and Rosebery's (1995) work in science education illustrates an approach that focuses on students' out-of-school discourse while taking content issues seriously. They note that the details of reform are typically worked out in mainstream settings on the assumption that they can be imported without problems into nonmainstream settings. Warren and Rosebery question this assumption and cite evidence indicating that the ways of knowing, talking to, interacting with, and valuing low-income, African American, and linguistic minority communities differ from those typically established in mainstream science (and mathematics) classrooms. They therefore argue that attempts to develop classroom sense-making communities, whether in science or in mathematics, must explicitly attend to, and build from, the discourse practices of students' home communities. The issue that Warren and Rosebery have investigated in their work with linguistic minority Haitian Creole students and their teachers is how to make science an activity in which all students can participate successfully. In the approach they take, much of the science emerges from the questions students pose, the dilemmas they meet, the observations they make, the experiments they design, and the theories they articulate.

To assess what a class of seventh- and eighth-grade students learned over the course of a school year, interviews were conducted in September and then the following

June. Warren and Rosebery (1995) report that in addition to learning about the topics they had investigated (e.g., water pollution, aquatic ecosystems), the students learned to use hypotheses to give direction to their inquires by linking conjecture and experimentation. In addition, they learned to conceptualize evidence not simply as information that was already known but also as the product of experiments they could undertake to confirm or disconfirm a given hypothesis. Beyond these specific results, Warren and Rosebery's work is relevant to mathematics educators in that it exemplifies a methodology that attends to students' out-of-school discourse practices when establishing classroom communities that, in many respects, parallel science as it is practiced in professional communities. One can imagine similar studies in mathematics education that investigate how students from nonmainstream communities might master specifically mathematical discourse that constitutes what Forman (1996) called the *mathematics register*.

A second, more speculative, approach draws on the notions of silencing and marginalization that have been used by scholars whose primary concern is with equity (see Secada, 1995). Silencing is a broad term that applies to the process by which ideas are devalued and people are marginalized. Fine (1987) reports a series of analyses that focus on silencing as it occurs in classroom discourse. She illustrates how certain groups of students have no voice in the classroom in that their opinions are not heard and the issues they raise are not discussed. The relevance of silencing to our concerns as mathematics educators becomes apparent when we note that the teacher is an instructional leader in the classroom. In proactively supporting students' mathematical learning, the teacher necessarily has to treat students' contributions to classroom discussion differentially. The decisions the teacher makes might be entirely justifiable from a mathematical point of view. However, in differentiating between students' contributions, the teacher implicitly communicates to students that certain opinions and ways of reasoning are particularly valued and others are less valued. The danger therefore exists that, against the background of the diverse out-of-school types of discourse in which students participate, some might interpret their voices as being silenced and become marginalized in the classroom despite the teacher's best intentions. This possibility alerts us to the need to complement a mathematical point of view with what might be termed an equity perspective that focuses on the relationship between classroom discourse and the out-of-school discourse practices of students' home communities.

Conclusion: Caveats and a Reason to Be Hopeful

Many of the teaching and learning environments studied by the researchers cited in this chapter are unusual, not only because of their current scarcity but also because they have been created with uncommon access to many kinds of resources. They are what psychologist Ann Brown (1992) calls *design experiments*. Similar to prototypical work in fields such as medicine and industry, they are created to be studied. The practices that are described have been constructed through a process of design and practical experiment influenced by research on teaching, research on understanding, and research on the nature of mathematics. Classrooms in which mathematical communication and language development are focal are not typical. But we do have evidence that with appropriate support and structures in place, teachers can improve the quality of their mathematics instruction to build the capacity of students to think, reason, solve complex problems, and communicate mathematically (Brown, Stein & Forman, 1996).

The QUASAR (Quantitative Understanding Amplifying Student Achievement and Reasoning) project, working for 5 years with diverse student populations in six urban school districts, has shown that this sort of improvement is both feasible and responsible. Silver, Smith, and Nelson (1995) provide vignettes of QUASAR classrooms in which students examine one another's reasoning and learn to express their mathematical ideas effectively. Stein, Grover, and Henningsen (1996) analyzed a representative sample of lessons and found that 75% of instructional episodes involved mathematical tasks intended to provoke students to engage in conceptual understanding, reasoning, or problem solving. Schools with an overwhelming majority of poor children, serving large subgroups whose first language is not English, have produced gains in students' proficiency with respect to mathematical understanding, problem solving, and communication (Silver & Stein, 1996).

As with all the case studies cited above, the QUASAR project points to the complexity of separating treatments and outcomes when mathematical communication is the concern. Mathematical communication can be reduced neither to a matter of curriculum nor to a matter of instructional processes—it is both. It is not a treatment or an outcome in relation to classroom instruction and learning—it is both.

ACKNOWLEDGMENTS

Support for the preparation of this paper was provided by the National Science Foundation under Grant No. REC9814898 and by the Office of Educational Research and Improvement under Grant No. R305A60007. We are grateful to Clea Fernandez and Steve Leinwand for comments on an earlier draft.

REFERENCES

Ball, D. L. (1991). What's all this talk about "discourse"? *Arithmetic Teacher, 39*(3), 44–48.

Ball, D. L. (1993). With an eye on the mathematical horizon: Dilemmas of teaching elementary school mathematics. *Elementary School Journal, 93*, 373–397.

Blunk, M. (1998). Communication about small groups in one classroom. In M. Lampert & M. Blunk (Eds.), *Talking mathematics in school: Studies of teaching and learning* (pp. 190–212). Cambridge: Cambridge University Press.

Bowers, J. S., Cobb, P., & McClain, K. (1999). The evolution of mathematical practices: A case study. *Cognition and Instruction, 17*, 25–64.

Brown, A. (1992). Design experiments: Theoretical and methodological challenges in creating complex interventions in classroom settings. *Journal of the Learning Sciences, 2*(2), 141–78.

Brown, C. A., Stein, M. K., & Forman, E. A. (1996). Assisting teachers and students to reform the mathematics classroom. *Educational Studies in Mathematics, 31*, 63–93.

Cobb, P. (1995). Mathematics learning and small group interactions: Four case studies. In P. Cobb & H. Bauersfeld (Eds.), *Emergence of mathematical meaning: Interaction in classroom cultures* (pp. 25–129). Hillsdale, NJ: Erlbaum.

Cobb, P. (1999). Individual and collective mathematical learning: The case of statistical data analysis. *Mathematical Thinking and Learning, 1*, 5–43.

Cobb, P., Boufi, A., McClain, K., & Whitenack, J. (1997). Reflective discourse and collective reflection. *Journal for Research in Mathematics Education, 28*, 258–277.

Cobb, P., Yackel, E., & Wood T. (1989). Young children's emotional acts while doing mathematical problem solving. In D. McLeod & V. M. Adams (Eds.), *Affect and mathematical problem solving: A new perspective* (pp. 117–148). New York: Springer-Verlag.

Cope, B., & Kalantzis, M. (Eds.). (1993). *The powers of literacy: A genre approach to teaching writing*. London: Falmer.

Dörfler, W. (2000). Means for meaning. In P. Cobb, E. Yackel, & K. McClain (Eds.), *Communicating and symbolizing in mathematics classrooms: Perspectives on discourse, tools, and instructional design* (pp. 99–132). Mahwah, NJ: Erlbaum.

Dubinsky, E. (1991). Reflective abstraction in advanced mathematical thinking. In D. Tall (Ed.), *Advanced mathematical thinking* (pp. 95–123). Dordrecht, The Netherlands: Kluwer.

Erickson, F. (1982). Classroom discourse as improvisation: Relationships between academic task structure and social participation structure. In L. C. Wilkerson (Ed.), *Communicating in the classroom* (pp. 153–181). New York: Academic Press.

Fine, M. (1987). Silencing in public schools. *Language Arts, 64*(2), 157–175.

Forman, E. (1996). Forms of participation in classroom practice: Implications for learning mathematics. In P. Nesher, L. Steffe, P. Cobb, G. Goldin, & B. Greer (Eds.), *Theories of mathematical learning* (pp. 115–130). Hillsdale, NJ: Erlbaum.

Freudenthal, H. (1983). *Didactical phenomenology of mathematical structures*. Dordrecht, The Netherlands: Reidel.

Gravemeijer, K. (1997). Mediating between concrete and abstract. In T. Nunes & P. Bryant (Eds.), *Learning and teaching mathematics: An international perspective*. Mahwah, NJ: Erlbaum.

Gravemeijer, K., Cobb, P., Bowers, J., & Whitenack, J. (2000). Symbolizing, modeling, and instructional design. In P. Cobb, E. Yackel, & K. McClain (Eds.), *Communicating and symbolizing in mathematics classrooms: Perspectives on discourse, tools, and instructional design* (pp. 225–273). Mahwah, NJ: Erlbaum.

Gray, L. (1993). *Large group discussion in a 3rd/4th grade classroom: A sociolinguistic case study*. Unpublished doctoral dissertation, Boston University.

Gray, E. M., & Tall, D. O. (1994). Duality, ambiguity, and flexibility: A "proceptual" view of simple arithmetic. *Journal for Research in Mathematics Education, 25*, 116–146.

Hall, R., & Rubin, A. (1998). ...there's five little notches in here: Dilemmas in teaching and learning the conventional structure of rate. In J. G. Greeno & S. V. Goldman (Eds.), *Thinking practices* (pp. 189–235). Hillsdale, NJ: Erlbaum.

Hatano, G., & Inagaki, K. (1991). Sharing cognition through collective comprehension activity. In L. B. Resnick, J. M. Levine, & S. D. Teasley (Eds.), *Perspectives on socially shared cognition* (pp. 331–348). Washington, D.C.: American Psychological Association.

Hicks, D. (1998). Closing reflections on mathematical talk and mathematics teaching. In M. Lampert & M. L. Blunk (Eds.), *Talking mathematics in school: Studies of teaching and learning* (pp. 241–252). Cambridge: Cambridge University Press.

Hiebert, J., & Wearne, D. (1993). Instructional tasks, classroom discourse, and students' learning in second-grade arithmetic. *American Educational Research Journal, 30*, 393–425.

Lakatos, I. (1976). *Proof and refutations: The logic of mathematical discovery* (J. Worrall & E. Zahar, Eds.). Cambridge: Cambridge University Press.

Lampert, M. (1990). When the problem is not the question and the solution is not the answer: Mathematical knowing and teaching. *American Educational Research Journal, 27*, 29–64.

Lampert, M. (1992a). Practices and problems in teaching authentic mathematics in school. In F. Oser, A. Dick, & J.-L. Patry (Eds.), *Effective and responsible teaching: The new synthesis* (pp. 295–314). New York: Jossey-Bass.

Lampert, M. (1992b, August). *The collaborative construction of the mathematics curriculum using teacher constructed problems.* Paper presented at the Seventh International Congress on Mathematics Education, Quebec, Canada.

Lampert, M. (2001). *Teaching problems and the problems of teaching.* New Haven, CT: Yale University Press.

Lampert, M., Rittenhouse, P., & Crumbaugh, C. (1998). Agreeing to disagree: Developing sociable mathematical discourse. In D. R. Olson & N. Torrance (Eds.), *Handbook of Education and Human Development* (pp. 731–764). Oxford: Blackwell.

Longo, P. (1994, February). *The dialogical process of becoming a problem solver: A sociolinguistic analysis of implementing the NCTM standards.* Paper presented at the Ethnography in Education Research Forum, University of Pennsylvania, Philadelphia.

McClain, K., & Cobb, P. (1998). The role of imagery and discourse in supporting students' mathematical development. In M. Lampert & M. Blunk (Eds.), *Talking mathematics in school: Studies of teaching and learning* (pp. 56–81). Cambridge: Cambridge University Press.

McNair, R. (1998). Building a context for mathematical discussion. In M. Lampert & M. Blunk (Eds.), *Talking mathematics in school: Studies of teaching and learning* (pp. 82–106). Cambridge: Cambridge University Press.

National Council of Teachers of Mathematics. (1989). *Curriculum and evaluation standards for school mathematics.* Reston, VA: Author.

National Council of Teachers of Mathematics. (1991). *Professional standards for teaching mathematics.* Reston, VA: Author.

National Council of Teachers of Mathematics. (1995). *Assessment standards for school mathematics.* Reston, VA: Author.

O'Connor, M. C. (1992). *Negotiated defining: The case of length and width.* Unpublished manuscript, Boston University.

O'Connor, M. C. (1994, October). *Communication in classrooms that promote understanding.* Symposium at the National Center for Research in Mathematics and Science Education, University of Wisconsin—Madison.

O'Connor, M. C. (1996). Managing the intermental: Classroom group discussion and the social context of learning. In D. Slobin, J. Gerhardt, A. Kyratzis, & J. Guo (Eds.), *Social interaction, social context, and language* (pp. 495–509). Hillsdale, NJ: Erlbaum.

O'Connor, M. C. (1998). Language socialization in the mathematics classroom: Discourse practices and mathematical thinking. In M. Lampert & M. Blunk (Eds.), *Talking mathematics in school: Studies of teaching and learning* (pp. 17–55). Cambridge: Cambridge University Press.

O'Connor, M. C., & Michaels, S. (1993). Aligning academic task and participation status through revoicing: Analysis of a classroom discourse strategy. *Anthropology and Education Quarterly, 24,* 318–335.

O'Connor, M. C., & Michaels, S. (1996). Shifting participant frameworks: Orchestrating thinking practices in group discussions. In D. Hicks (Ed.), *Discourse, learning, and schooling* (pp. 63–103). Cambridge: Cambridge University Press.

Pirie, S., & Kieren, T. (1994). Growth in mathematical understanding: How can we characterize it and how can we represent it? *Educational Studies in Mathematics, 26,* 61–86.

Pólya, G. (1954). *Induction and analogy in mathematics* (2 vols.). Princeton, NJ: Princeton University Press.

Reid, I. (Ed.). (1987). *The place of genre in learning: Current debates.* Deakin, Australia: Deakin University, Centre for Studies in Literary Education.

Rittenhouse, P. (1998). The teacher's role in mathematical conversation: Stepping in and stepping out. In M. Lampert & M. Blunk (Eds.), *Talking mathematics in school: Studies of teaching and learning* (pp. 163–189). Cambridge: Cambridge University Press.

Russell, S. J., & Corwin, R. B. (1993, March). Talking mathematics: "Going slow" and "letting go." *Phi Delta Kappan, 74,* 555–558.

Secada, W. G. (1995). Social and critical dimensions for equity in mathematics education. In W. G. Secada, E. Fennema, & L. B. Adajian (Eds.), *New directions for equity in mathematics education* (pp. 146–164). Cambridge: Cambridge University Press.

Sfard, A. (1991). On the dual nature of mathematical conceptions: Reflections on processes and objects as different sides of the same coin. *Educational Studies in Mathematics, 22,* 1–36.

Sfard, A. (1998). On two metaphors for learning and the dangers of choosing just one. *Educational Researcher, 27*(2), 4–13.

Sfard, A. (2000). Symbolizing mathematical reality into being: How mathematical discourse and mathematical objects create each other. In P. Cobb, E. Yackel, & K. McClain (Eds.), *Communicating and symbolizing in mathematics classrooms: Perspectives on discourse, tools, and instructional design* (pp. 37–98). Mahwah, NJ: Erlbaum.

Sfard, A., Nesher, P., Streefland, L., Cobb, P., & Mason, J. (1998). Learning mathematics through conversation: Is it as good as they say? *For the Learning of Mathematics, 18*(1), 41–51.

Silver, E. A., Smith, M. S., & Nelson, B. S. (1995). The QUASAR Project: Equity concerns meet mathematics education reform in the middle school. In W. G. Secada, E. Fennema, & L. B. Adajian (Eds.), *New directions in equity in mathematics education* (pp. 9–56). Cambridge: Cambridge University Press.

Silver, E. A., & Stein, M. K. (1996). The QUASAR project: The "revolution of the possible" in mathematics instructional reform in urban middle schools. *Urban Education, 30,* 476–521.

Simon, M. A. (1995). Reconstructing mathematics pedagogy from a constructivist perspective. *Journal for Research in Mathematics Education, 26,* 114–145.

Simon, M. A., & Blume, G. W. (1996). Justification in the mathematics classroom: A study of prospective elementary teachers. *Journal of Mathematical Behavior, 15,* 3–31.

Stein, M. K., Grover, B. W., & Henningsen, M. A. (1996). Building student capacity for mathematical thinking and reasoning: An analysis of mathematical tasks in reform classrooms. *American Educational Research Journal, 33*, 455–488.

Stevens, R., & Hall, R. (1998). Disciplined perception: Learning to see. In M. Lampert & M. Blunk (Eds.), *Talking mathematics in school: Studies of teaching and learning* (pp. 107–152). Cambridge: Cambridge University Press.

Swing, S. R., & Peterson, P. L. (1982). The relationship of student ability and small-group interaction to student achievement. *American Educational Research Journal, 19*, 259–274.

Thompson, A. G., Philipp, R. A., Thompson, P. W., & Boyd, B. (1994). Calculational and conceptual orientations in teaching mathematics. In D. B. Aichele & A. F. Coxford (Eds.), *Professional development for teachers of mathematics* (1994 Yearbook of the National Council of Teachers of Mathematics, pp. 79–92). Reston, VA: NCTM.

Thompson, A., & Thompson, P. W. (1996). Talking conceptually about rates, Part 2: Mathematical knowledge for teaching. *Journal for Research in Mathematics Education, 27*, 2–24.

Voigt, J. (1995). Thematic patterns of interaction and sociomathematical norms. In P. Cobb & H. Bauersfeld (Eds.), *Emergence of mathematical meaning: Interaction in classroom cultures* (pp. 163–201). Hillsdale, NJ: Erlbaum.

Walkerdine, V. (1988). The *mastery of reason: Cognitive development and the production of rationality*. London: Routledge.

Warren, B., & Rosebery, A. S. (1995). Equity in the future tense: Redefining relationships among teachers, students, and science in linguistic minority classrooms. In W. Secada, E. Fennema, & L. Adajian (Eds.), *New directions in equity for mathematics education* (pp. 298–328). Cambridge: Cambridge University Press.

Webb, N. M. (1991). Task-related verbal interaction and mathematics learning in small groups. *Journal for Research in Mathematics Education, 22*, 366–389.

Weingrad, P. (1998). Teaching and learning politeness during mathematical argument in school. In M. Lampert & M. Blunk (Eds.), *Talking mathematics in school: Studies of teaching and learning* (pp. 213–240). Cambridge: Cambridge University Press.

Wertsch, J., & Toma, C. (1995). Discourse and learning in the classroom: A sociocultural approach. In L. P. Steffe & J. Gale (Eds.), *Constructivism in education* (pp. 159–174). Hillsdale, NJ: Erlbaum.

Yackel, E., & Cobb, P. (1996). Sociomathematical norms, argumentation, and autonomy in mathematics. *Journal for Research in Mathematics Education, 27*, 458–477.

CHAPTER 17

Representation in School Mathematics: Learning to Graph and Graphing to Learn

Stephen Monk, University of Washington, Seattle, and TERC, Cambridge, Massachusetts

A new view of mathematical representations in general and graphing in particular has slowly emerged in the past decade. Instead of being isolated curricular items to be taught and tested as ends in themselves, graphs, along with diagrams, charts, number sentences, and formulas, are increasingly seen as "useful tools for building understanding and for communicating both information and understanding" (National Council of Teachers of Mathematics [NCTM], 2000). As such, graphs and other representations have come to play an increasingly important role in mathematical activities in school. This change results from the work of overlapping communities within mathematics education, of researchers who have described in detail the resources students bring to the symbolization process and the difficulties they have with it, and of curriculum developers who, in collaboration with teachers, have designed new modes of classroom activity to help students become fluent users of representations. A number of other recent shifts in education have supported the emergence of this new view of representation. Among these shifts is a heightened awareness of the complexities of representation as a cognitive and social process and of how it is inextricably linked with the knowledge people have of the situation being represented. In addition, new classroom configurations in which students work together, sharing and elaborating their mathematical thinking, have brought these processes directly into the foreground.

In this chapter, I try to convey to a reader who is not necessarily a specialist in mathematics education the fullest sense possible of what it means to use a graph as a tool for building and communicating understanding, and how such use plays out for students and teachers. This is done primarily through examples of students' work with graphs and other visual displays that come from the research literature, published work by teacher-researchers, curriculum materials, and my own experience working with students in Grades K–12 and beyond. These examples are from two fields—phenomena of change in time (e.g., of growth and the motion of objects) and data analysis—that are sufficiently disparate to convey those qualities essential to graphs and graphing. The examples were chosen for their power to illustrate the main ideas and themes of the chapter; they do not represent the full range of the known work in graphing.

Two Uses of Graphs

Students' use of graphs was first discussed in the research literature of the 1970s primarily in terms of their understanding of the central mathematical concept of a function. Although students' work with graphs of such specific real situations as moving cars and growing plants was reported there, contexts like these were seen primarily as a concrete means for students to approach this abstract mathematical concept. Although some articles describe students' success in using graphs, the literature consists primarily of reports of the difficulties they have with what were then regarded as relatively straightforward questions on graphs related to familiar situations. The cumulative effect of this literature was to bring fully into view the complexity of the processes of constructing and reading graphs of quantities in real situations. This complexity is reflected in the 1990 survey article on students' understanding of functions and graphs (Leinhardt, Zaslovsky, & Stein, 1990), which gives a classification of graphing tasks and describes both the difficulties students have with them and several sources of these difficulties.

The complexity that teachers and researchers now see in function graphs flows from the fact that a graph has many potential meanings and can be interpreted in many ways. This complexity exists even in a graph as simple as the one in Figure 17.1, which shows the velocity of a car as a function of time. In addition to the height of the graph at any particular time, which gives the velocity of the car at that time, several aspects of the graph can be related to the car's motion, each in several ways. People who can comfortably interpret the graph tend to screen most of these possibilities out, since they do not fit with the way they see it. Among these are the following:

1. The graph consists of two straight-line segments. One is oblique and the other is horizontal. Many people respond to this contrast by taking the oblique segment to indicate a change and the horizontal to indicate no change. This interpretation leads to their describing the car as increasing its speed and then standing still, which conflicts with the fact that the velocity of the car is 30 mph throughout the second time interval. Although the assumption that the vertical axis label is of *position* and not *velocity* would support this interpretation, studies indicate that students can give this primarily visual interpretation even while demonstrating that they know that the vertical axis label is velocity (Monk, 1987, 1994; Monk & Nemirovsky, 1994).
2. The straightness of the two segments is a salient feature of this graph for many, but the response it evokes varies. For some, the straightness of a line is in contrast with its being "wiggly" or "wavy," whereas for others, the contrasting case is a graph that bends gradually. Thus straightness can signify either steadiness or a lack of overall change.
3. When people read a graph related to a moving car, they want to be able to tell a story of what happens from the beginning of the time to the end. For many, the story begins at time $t = 0$ minutes, with the scene in a state of rest. The action of the story commences with the events described by the points of the graph beyond $t = 0$ minutes. But at $t = 0$ minutes the graph reads 10 mph, so that the car is already moving when the story begins. Many people find this situation genuinely difficult to envision.
4. The "corner" in the graph at $t = 2$ minutes, where the two straight-line segments meet, seems to most people sufficiently prominent to somehow correspond to an important aspect of the car's motion. But what would the corresponding pattern of motion be? The car's velocity does not vary in any unusual way. Perhaps this variation is a theoretical possibility that is practically impossible. Or perhaps one's sense of change in motion must be refined in order to detect how the corner reflects an aspect of the way the car moves.

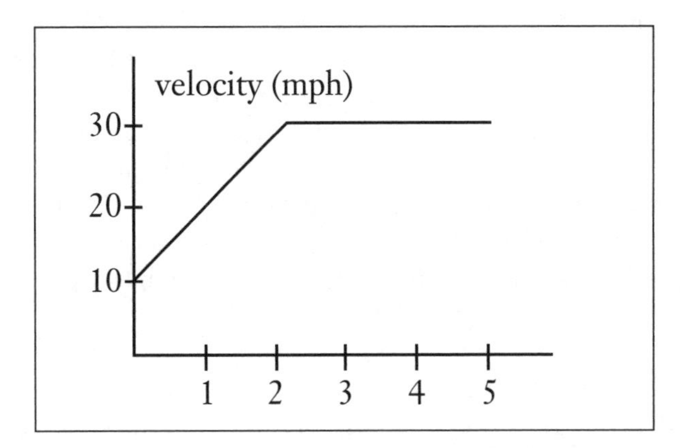

Figure 17.1. Velocity of a car as a function of time.

The complexity of a graph as apparently simple as the one in Figure 17.1 helps explain why the research literature on graphing contains so many reports of student errors and difficulties. One of the most important uses of a graph is as a tool for communicating information and understanding. Getting the message right is crucial to such use. Since those who are fluent graph users can easily forget what the real difficulties in communication are, documenting the sources of error and how robustly they operate is important. In the past decade, however, this same complexity has led a number of mathematics educators to widen their attention from an exclusive focus on the information contained in a graph to the process by which an individual constructs that information and, more deeply, makes meaning of all kinds through the use of a graph. This is a process of exploration, one in which errors and missteps are likely but in which the graph user can learn something about graphing and motion along the way. This approach to graphing is captured in the expression "a graph is a tool for making meaning." Whereas a graph had earlier been seen exclusively as a *conduit*, a carrier of information, for example, about the motion of a car, it can now also be seen as a *lens* through which to explore that motion.

Although these two uses of a graph, as a tool for communication and as a tool for generating meaning, are very different, the distinction between them is not made to set them *against* each other—to create a polarity that must then be resolved. Both communicating and making meaning play important roles in graphing; each is appropriate in certain situations, depending on the underlying setting, the task, and the experience and knowledge of the graph user. Many examples of graph use reflect both ways of using a graph, although one use is likely to dominate at any given time. Educators must keep this distinction in mind, however, because activities designed to support one use are likely to be different from activities designed to support the other, and the criteria used to determine whether an activity or any particular student's work is going well will also differ. One use calls for preciseness and clarity, whereas the other calls for creativity and an ability to handle ambiguity and open-ended tasks. The magnitude of the change in the view of graphing that has taken place over the past decade is strongly evident when one realizes that graphing involves both these uses, and that each is beyond a simple application of skills and procedures.

In the next section, the notion of using a graph as a tool for generating understanding is fleshed out by a description of a number of different meaning-making processes that can take place within such use. These processes are illustrated by a wide variety of examples of students' use of graphs. The examples are followed by a discussion of the particular possibilities and difficulties inherent in the visual nature of graphs and then by a description of recent work by researchers and curriculum designers addressing some of the difficulties students have with graphing.

Ways of Making Meaning with a Graph

In this section, I make more specific and substantial the notion of using a graph to make meaning. I describe six different ways in which meaning can be made, each illustrated by examples of student work from various grade levels. Many of these examples also involve the use of a graph as a tool for communication of meaning and understanding. This catalog of ways of making meaning is intended to show how widespread and diverse the phenomenon of meaning-making through graphs is. The catalog is not exhaustive, and the items in it are not entirely separable, since one way of making meaning can often lead to, or overlap with, another.

Many of the examples are from the *Working with Data Casebook* of Developing Mathematical Ideas (DMI) (Russell, Schifter, & Bastable, 2002). As is true of earlier modules in this professional development material, each case is written by a classroom teacher, identified by a first-name pseudonym along with her grade level, and is based on actual conversations with students.

1. *By using graphs, students can explore aspects of a context that are not otherwise apparent.*

The graphs shown in Figure 17.2 were made by Amelia, a fourth-grade student, from a record she kept of daily measurements of the heights of three plants she grew from seed (Nemirovsky & Tierney, 2001). Although graphs such as these might seem to be merely a record of plant height, people often make graphs such as these with the hope that information that is not otherwise apparent will emerge from them. For instance, Amelia's graphs show similarities in the patterns of the plants' growth that are not likely to be noticed by watching the plants themselves or even by examining a table of values.

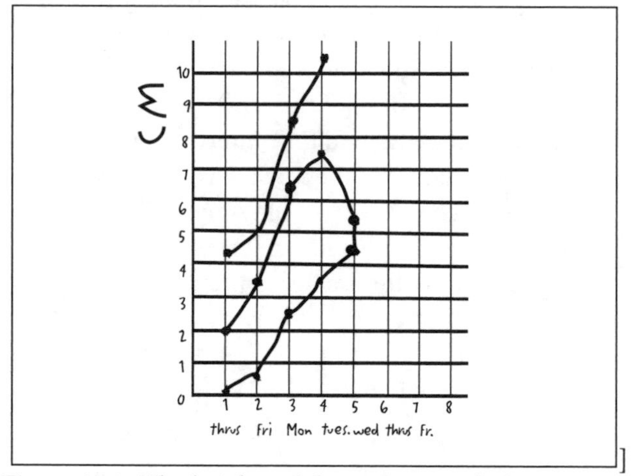

Figure 17.2. The heights of three plants each day.

This same reason for making graphs motivated Maura, a teacher-author of the DMI *Working with Data Casebook*, to ask her third- and fourth-grade students to make a bar graph (Figure 17.3) to represent their data after they surveyed their own class with the question "How many years has your family lived in our town?" Making this graph led to a discussion of its clear visual pattern of having two "clumps," one indicating those families who had lived in town for 10 years or more and the other indicating those families who had lived there for 6 years or fewer. These two distinct groups, which would not have been evident without making the graph, became the basis for a conversation about the children's diverse experiences of living in their town. The clumps were so compelling to the children that Maura was unable to shift their attention to the traditional single-value measures of typicality—mean, median, and mode—which, in this case, would not have yielded as much information about the class as the clumps did.

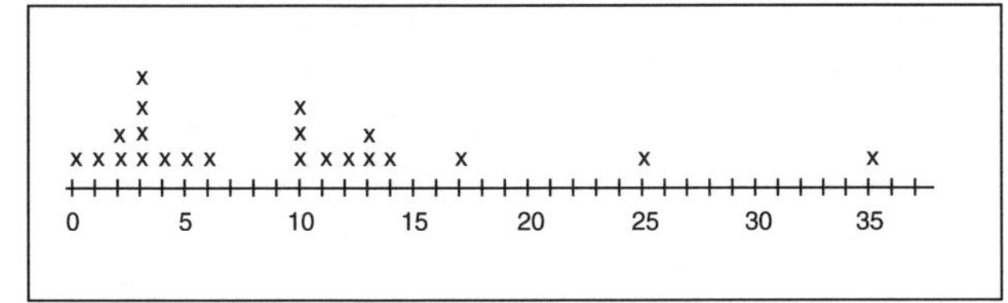

Figure 17.3. Bar graph of years of family residence.

2. *The process of representing a context can lead to questions about the context itself.*

In several of the cases reported by teachers in the DMI *Working with Data Casebook*, young children made graphs by first collecting data from their own class using questions they invented. They asked such questions as, "Are you wearing jeans?" (Sally, Grade 1), "How many states have you visited?" (Andrea, Grade 2), and "How many people are in your family?" (Olivia, Grade 4). But often, the process of collecting data raised questions about both the data and the context: Is a denim skirt a form of jeans? What does it mean to "visit" a state? What is a family? Does it include your uncle who lives with you? What about your married sister who lives with her husband, or stepparents and foster parents? The question of what a family is came into sharper focus in Olivia's fourth-grade class when the students had to choose a scale for their graph and were led to address the questions "Can a family have just one member? How about two?"

Even the task of keeping a record of the growth of their plants (as in Amelia's graphs) can raise questions for students. The task of measuring the height of a young plant is simple enough, but once the plant begins to grow more leaves and they become more complex, the "top" of the plant, from which to measure its height, becomes less clearly marked. New problems arise if the plant begins to lean over, as plants often do. Should the height be measured along the stem, an endeavor that presents mechanical difficulties, or vertically down in a straight line from the top of the plant to the dirt below? Later, when the plants are fully mature, children tend to question whether height was the best way to describe how "big" a plant is, since a short bushy plant can seem as big as a tall scraggly one.

Children's experience of comparing the growth of their plants through graphs and asking questions about how to measure the size of a plant can then be a jumping-off place for new studies of plant growth using graphs. Such a process is reported in a classroom study of a third-grade class in which the children investigated the relationship between plant width and height (Lehrer, Schauble, Carpenter, & Penner, 2000). The graphs the children made raised new questions about the relationship between changes in the height and the width of a plant as it grows. These questions then led the children to draw three-dimensional models of the "growth space" of a plant. The process of making these new representations then served as an occasion for new questions and conjectures about the life of a plant.

3. *Using a graph to analyze a well-understood context can deepen a student's understanding of a graph and graphing.*

In my course, Hands-On Calculus, taught to in-service and preservice high school mathematics teachers, I give the teachers the graph and table shown in Figure 17.4. It shows the step sizes of two people, Amanda and Joe, who are standing next to each other and then begin to walk along parallel lines. The sizes of their steps are given by the graphs and table. For instance, the table tellsus that

Step no.	A's step size (ft.)	J's step size (ft.)
1	3.00	0.0
2	2.75	0.5
3	2.50	1.0
4	2.25	1.5
5	2.00	2.0
6	1.75	2.5
7	1.50	3.0
8	1.25	3.5
9	1.00	4.0
10	0.75	4.5

(a)

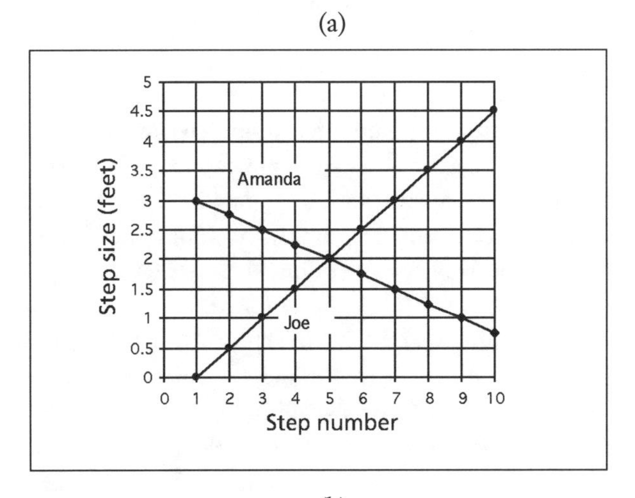

(b)

Figure 17.4. Graph of step size at each step.

Amanda takes a step that is 3 feet long, then a step that is 2.75 feet long. Their task is to predict what Amanda and Joe's trip will look like: Who will be ahead at various times? Will one person be catching up or falling behind at any given time?

The teachers in the course are struck by the graphs' shapes and have difficulty using the graphs to predict how the two walkers will move in relation to one another. I ask them to go out into the hallway in pairs to act out Amanda and Joe's walk, using the values on the table. They often take more than 30 minutes to do so. This experience of enacting a table of values through the motion of their own bodies leads the teachers to a much fuller sense of both the distinction and the connections between their step sizes and their relative positions. That distinction then serves as the basis for making sense of the graphs. Amanda's graph is headed downward because she takes smaller and smaller steps. She pulls away from Joe for the first five steps because these steps of hers are all bigger than Joe's. When the graphs cross at the fifth step, Joe and Amanda's steps are the same size. Then, starting with the sixth step, Joe takes bigger steps than Amanda and so begins to catch up. Therefore, Amanda and Joe are furthest apart when the graphs cross.

Many of these teachers report, with considerable surprise, that they "knew what the crossing of the graphs meant," but never realized before how much information can be included in this feature of graphs. This example is an instance of deeper understanding of graphing resulting from understanding of its context, not the other way around.

4. *Students can construct new entities and concepts in a context by beginning with important features of a graph.*

Mathematics teachers have long used graphs to make present and real for students phenomena and situations that can serve as the basis for the construction of new ideas and concepts. By gathering data into bar graphs of frequencies of quantities, such as how long the children have lived in their town (Figure 17.3), the teacher-authors in the DMI *Working with Data Casebook* create worlds in which such entities as "holes" and "clumps" in graphs come to have a reality and a meaning that is then attributed to actual groups and individuals in the class, such as "the lower group" (meaning the more recent arrivals) and "outliers." Similarly, the teacher in the study by Lehrer et al. (2000) uses the many graphs of plant growth made by the children in her class as the material from which to build a shared notion of the *typical* growth pattern of the kinds of plants they work with. The typical growth pattern then becomes sufficiently real to be compared with the growth pattern of a particular plant.

Latour (1990) points out that representations such as graphs can be used to stabilize fluid and dynamic situations, such as that of a moving car, so that they can be pointed to and reasoned about. In this way, motion, a transitory phenomenon, is made the subject of a conversation. Then, as the conversation becomes more focused on the graph and less on the moving object, a language and a set of concepts based on the graph and its geometric qualities begin to develop and stand for the corresponding physical quantities. For example, in the graph of position versus time in Figure 17.5, the vertical height of any point on the graph corresponds to the position of the object at that time. Consequently, the *difference* between the heights of the two points shown, indicated by the double-headed arrow, corresponds to the distance covered by the object in the time that has elapsed. This lexicon can then be extended to include other constructs in the graph (such as the slope of the secant line L) and numerical quantities related to motion (such as the average velocity over a time interval). These entities in a graph and the physical quantities they stand for can be connected so closely for a fluent graph user that they are often referred to as if they were the same. Thus, a person talking about the graph in Figure 17.5 might point to the straight line L and refer to it as the average velocity.

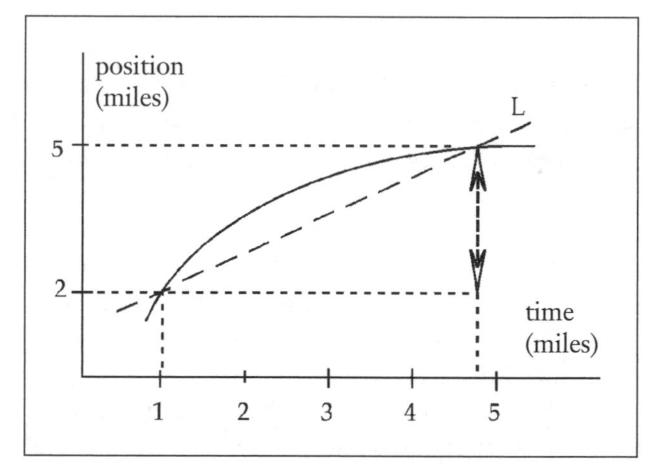

Figure 17.5. Graph of position versus time.

5. *Students can elaborate their understanding of both a graph and its context through an iterative and interactive process of exploring both.*

The examples given thus far are one-directional in that knowledge or understanding flows from the graph to the context, or vice versa. An interactive process is possible, however, in which understanding of both the context and the graph builds at the same time. This growth in understanding is likely to occur for anyone who works with the velocity versus time graph in Figure 17.1. In a study of how children use their visual response to graphs and their kinesthetic sense of the movement of their own bodies as essential resources in learning to graph, Nemirovsky, Tierney, & Wright. (1998) describe how a 10-year-old girl named Eleanor learned about graphs of distance versus time through the use of a computer-based motion detector. Eleanor held a small metal reflector, called a "button," in her hand while the computer recorded the distance between it and a sonic detector (the "tower"). As she moved back and forth, the computer screen displayed, in real time, the changing distance between the button and the tower in the form of a graph of distance versus time. After experimenting with the relationship between her movement and the graphs on the screen, Eleanor gave increasingly articulate descriptions of her movement and made increasingly precise observations about the graphs that appeared on the screen in response to her movement. For example, she said, "Let me move it farther away. The line goes up. Maybe this is the farthest it can go. The closer to the tower it gets, the lower, I think." Slowly, her predictions and explanations about the relationship between patterns in the graphs on the screen and her movement stabilized and gave clear evidence that she was learning about graphing. She was not simply learning about graphing from knowledge of her bodily motion that she already possessed, since aspects of the knowledge she used came from seeing her own bodily motion through the lens of the motion detector. She used the feedback of the device to articulate her knowledge of her bodily motion and of the graphs on the screen—both at the same time.

In another paper, Nemirovsky (1994) describes the case of a high school student named Laura who, using the motion detector and a toy car, came to understand the concept of negative velocity as the speed of an object that is moving backward. The process by which Laura came to that understanding is remarkably similar to Eleanor's process of making sense of graphs. Laura iteratively and interactively built her understanding of subtle features of both the car's motion and the shapes of the graphs on the screen, as well as the relationships between them.

Monk (1994) describes a process of building understanding in both a graph and its context as exhibited by a college student named Carl in a series of clinical interviews. Carl was working with a device that displayed the graphs of velocity versus time (Figure 17.6) for two cars on a computer screen and could then run the two cars along a 9-foot track according to these graphs. Carl's task was to predict and explain the motion of the cars in terms of these graphs, particularly in the time period between 8 and 14 seconds. Carl understood that the fact that the graphs get closer and closer does *not* mean that the cars get closer and closer. In fact, he reasoned that since the speed of the red car is always greater than the speed of the blue car in this time interval, the red car must be pulling away from the blue car throughout the interval. The fact that the graphs get closer in this time interval must mean that their speeds are getting closer and hence that the red car is pulling away from the blue car at a decreasing rate.

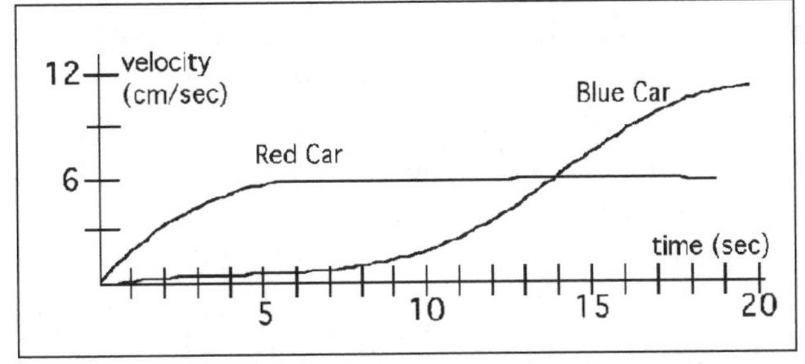

Figure 17.6. Graph of velocity versus time for two cars.

This work with the graphs, however, raised the following question about motion for Carl: "Doesn't moving apart at a slower and slower rate mean the same as getting closer?" After much struggle and repeatedly changing his mind on the question of the two cars' motion, even as he watched the two cars run, Carl recalled an experience he had had with moving cars. When one enters a freeway, one goes faster and faster and one's speed gets closer to the speed of the cars already on the freeway. "But they are still pulling away from you." Throughout this series of interviews, Carl called upon his knowledge of graphs and his knowledge of the motion of cars, using each to elaborate and illuminate the other, in an interactive process through which he arrived at a more adequate understanding of both.

6. *A group can build shared understanding through joint reference to the graph of phenomena in a context.*

In a well-known paper, Eleanor Ochs and her colleagues analyze conversations among a group of solid-state physicists to show how professionals use visual representations to create a shared world of understanding (Ochs, Jacoby, & Gonzales, 1994; see also Ochs, Gonzales, & Jacoby, 1996). Ochs and her colleagues describe the graphlike visual display at the center of the conversations as a "stage on which scientists dramatize understandings of their own and others' works" and describe the scientists as using this stage in such a way that "when narrators are speaking, gesturing, and drawing, they are thus asking co-participants not only to look through the graph to some represented world but also at the graph as a referential object in and of itself" (Ochs et al., 1994, p. 10).

This use of a graph as a window into a phenomenon and as a world of meanings in its own right is also found among students working with graphs of data and motion. One example of such collective meaning-making around a graph is in the DMI *Working with Data Casebook* in a second case by Maura, the third- and fourth-grade teacher-author. After collecting data and making separate bar graphs (Figure 17.7) of the heights of the third graders and fourth graders in the class, the students were asked in a homework assignment to look at the graphs and write about what the data told them about the question: "Is one class taller than the other?" The classroom conversation the next day was remarkably similar to the one among the research physicists in the ways in which the speakers inhabited the world of their graphs. In such statements as "I see that with the fourth graders, the Xs are really scattered around" and "There is a big clump of fourth graders at the beginning," these children spoke as if the graphs and the groups of students whose heights were represented were the same. Through this productive immersion in a shared space, which Nemirovsky, Tierney, and Wright (1998) refer to as "fusion" of symbol and referent, these students considerably refined and developed their understanding of the subtle relations between the two distributions of height.

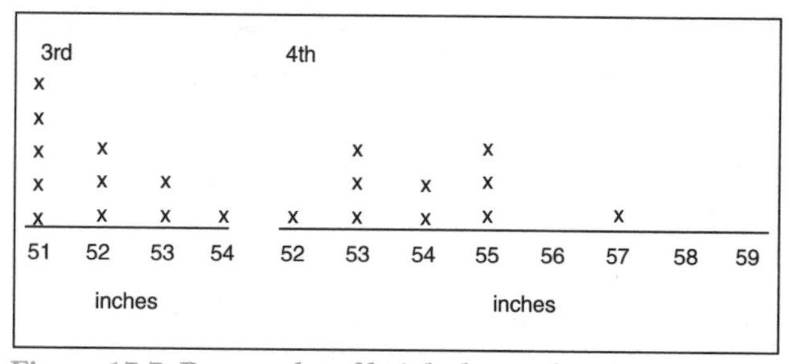

Figure 17.7. Bar graphs of height by grade.

Using graphs as a shared symbolizing space to create meaning is the first of the modes of meaning-making described here that explicitly refers to the use of a graph by a group. This reference is in contrast with the earlier modes, which can take place, at least in principle, when a person is working alone. Clearly, however, the effect of any of these modes of making meaning is greatly enriched by conversations among several people, where there is a greater possibility of new points of view emerging, new questions being raised, and iterative and interactive meaning-making taking place.

The many examples in this section demonstrate the sheer diversity of ways of making meaning with graphs, of grade levels in which this process can take place, and of kinds of mathematical issues that can arise in the process. The common theme beneath this diversity is evident, however, when one considers the list of modes of making meaning separate from the examples:

1. By using graphs, students can explore aspects of a context that are not otherwise apparent.
2. The process of representing a context can lead to questions about the context itself.
3. Using a graph to analyze a well-understood context can deepen a student's understanding of a graph and graphing.
4. Students can construct new entities and concepts in a context by beginning with important features of a graph.
5. Students can elaborate their understanding of both a graph and its context through an iterative and interactive process of exploring both.
6. A group can build shared understanding through joint reference to the graph of phenomena in a context.

Set out in this way, these modes of making meaning with graphs can be seen as particular versions of more general processes of intellectual engagement. The task of mathematics educators is to ensure that these processes play as important a role in students' mathematical activities as the more traditional processes of problem solving, using algorithms, and communicating information accurately.

Graphs—The Visual Medium

The distinguishing quality of graphs that sets them apart from other standard mathematical representations is that they are a visual medium. Although the visual aspect—the visuality—of graphs makes them an extremely rich and powerful medium for generating meaning, it is also the source of many of the incorrect responses students have to graphs. In this section, I discuss the role of visuality in graphing—how it supports meaning-making and communication and the particular way in which it is a source of students' difficulties. This discussion naturally leads to a consideration of how students can be helped to productively and efficiently use this resource—of how they can learn to see graphs in the way experienced graph users do.

The numerous and diverse ways in which the visual attributes of graphs play an important role in graphing is apparent in the many examples given in the previous sections. In the graph in Figure 17.1, such geometric properties as straightness, obliqueness, horizontality, and "corner" are all potential sources of meaning about the motion of the car. Examining the graphs of plant heights in Figure 17.2, one notes that they start at different heights and are parallel to one another, and one wonders what this configuration means about the plants' growth. The "holes" and "clumps" in bar graphs indicating groupings and absences of children along a range of possibilities, such as in Figure 17.3, are just two of several aspects of what is known in the *Investigations in Number, Data, and Space* material as the *shape* of the data (Russell, Tierney, Mokros, & Economopoulos, 1998). The complexity of the information in a graph is evident in the two velocity- versus-time graphs that Carl struggles with (Figure 17.6), since they contain sufficient information to give a very detailed and accurate "play-by-play description" of the cars' relative position and relative speed, as well as the ways in which these factors change. Learning how to use the visual information in these graphs and coordinating it with other kinds of information to give such a description go directly to the heart of understanding some of the central ideas in calculus.

In a paper in which they describe a series of interviews with a high school student named Dan, Monk & Nemirovsky (1994) show that the visual information in a graph can be a rich resource for making sense of phenomena. In these interviews, Dan works with a computer-controlled device that monitors and regulates the flow of air in and out of an air reservoir in order to explore the relationship between the graphs of *volume* of air in the reservoir and *rate* of airflow in and out. Using his visual sense of the shapes of these graphs, together with his kinesthetic sense of the motion of the flow of air in and out of the reservoir, Dan comes to understand something of the complicated relationship between these two graphs, a relationship that reflects one of the main ideas of calculus. The linkage among our kinesthetic, narrative, and visual capacities and our understanding of mathematics supported by graphs is also shown in the work of Nemirovsky and his colleagues. (Nemirovsky, 1994, Nemirovsky et al., 1998), described previously, in which Eleanor and Laura use the motion detector to learn about graphs of position versus time and of negative velocity, respectively.

The particular strength of graphs for studying motion is that they enable the user to explore what is called *across-time* behavior—how the position or velocity of the object changes as time passes. In a 1987 study, Monk shows that students find across-time questions far more difficult to answer correctly than pointwise questions, which ask about the object's position or velocity at a particular point in time. A corresponding distinction can also be made in graphs used for data analysis in which the shape of graphs is essential in thinking about a *population* instead of *individuals*, a distinction that is known to cause difficulties for children (Hancock, Kaput, & Goldsmith, 1992). The two DMI cases reported by Maura demonstrate the importance of shape in helping these students begin to move beyond their focus on individuals to be able to compare populations and subgroups within them.

Among the most widely agreed upon conclusions of research on graphing in the past 25 years is that visuality is a key source of difficulties for students using graphs. "Iconic interpretation," that is, interpreting a graph as a literal picture, and other inappropriate responses to visual attributes of a graph are "the most frequently cited student errors with respect to interpreting and constructing graphs" (Leinhardt, Zaslavsky, & Stein, 1990, p. 39). Errors due to this kind of interpretation not only are pervasive but also seem highly resistant to change. To get a sense of the force exerted on one's thinking by the visual features of a graph, one need only interpret the graphs of step size of the two walkers in Figure 17.4. To *not say* that when the graphs come together, the two walkers get closer—*somehow*—is genuinely difficult. (It is, of course, their step sizes that get closer.) Some authors have tried to analyze the difficulties people have with such graphs in terms of the existence of two kinds of information, visual and quantitative (Kaput, 1987), but ultimately, these types cannot be disentangled. More recently, researchers have viewed the problem as one of seeing, taken in its broadest sense, so that learning to read such graphs is a matter of what Stevens and Hall (1999) refer to as "disciplined perception."

People have difficulty thinking about how to help someone else see things differently. This is because once they see a scene or an image in a certain way, then what they see just seems "there"—to be seen by anyone who simply looks at it. Seeing things in the way one does feels so natural that one can scarcely imagine how someone else could see them any other way. For instance, people routinely see three-dimensional figures on two-dimensional pieces of paper as if they were really there, and they cannot imagine how anyone else might *not* see them that way. Yet, the technique of perspective drawing for making such representations was invented in Europe during the Renaissance and has not emerged in other cultures. One's ability to make and read such images is not at all "natural" but is learned.

Stevens and Hall (1999) describe a conversation about straight-line graphs on the Cartesian coordinate plane between a high school student ("Adam") and an interviewer ("Bluma") who is mathematically well trained. Frequent breakdowns occur in the conversation as they talk about the slope and direction of these graphs, drawn on a computer screen, because Bluma sees the computer screen as a version of the conventional Cartesian graphing plane, divided into four quadrants, with points and lines located in relation to the two coordinate axes, whereas Adam tends to see the outside border of the screen as the significant visual feature for the purpose of orienting lines on the screen. Studies like these strongly suggest that attempts to show Adam, by pointing and telling, how to see the graphing plane in a different way would not be productive. Moreover, the way he does see the graph can be an important resource for him.

In trying to understand difference and change in the way people see, one must keep in mind that seeing is a constructed activity that is closely linked with one's thought and actions. In the words of philosopher of science Norwood Hanson (1958), "seeing is theory-laden." What I see in a graph is not necessarily what you see in the same graph because what each of us sees is the result of our interpretive activities, which are intertwined with the ways in which we think, talk, and act—in short, with our practices. As members of interpretive communities, such as those who know Cartesian graphing, we come to think that what we see in the graph is actually in the graph because we routinely interact with others who see such a graph in the same way we do. Similarly, the holes and clumps the children see in the graphs in Figures 17.3 and 17.7 are there because these groups of children have come to agree that they are there in the course of their shared practice of talking about and analyzing graphs together. If students do not see what we see in graphs, this is true, in part, because they have not participated in the practices of the community of graph users in the way we have.

The work of Hall and Stevens (Hall, 1996; Hall & Stevens, 1995; Stevens & Hall, 1999), which is based on close study of conversations in collaborative groups working with visual displays—both groups of students and groups of professionals—offers a strong suggestion for how students' interpretations of the visual aspects of graphs can be shaped and coordinated with the responses of others and with other features of graphs. Their suggestions are supported by evidence in the DMI *Working with Data Casebook* and by my own ongoing work in a combination third-and-fourth-grade classroom (Monk, 2000) in which children are learning to use graphs and other representations of motion. Just as being a member of a community of practice helps people experience what they see as "there" and "natural," coming to participate in a community of practice is a means of changing how one sees things. Students can learn to make meaning from graphs in ways that are consistent with those of the wider mathematical community by participating in a sustained mathematical community of their own, including other students and the teacher, whose practices are aligned with those of wider community.

Graphs and Related Representational Forms

As the shift has taken place in mathematics education away from graphs as curricular objects involving skills and procedures and toward the thinking and communication processes entailed in their use, a concomitant expansion has occurred in the range of representational forms used by students in activities around graphing. Rather than teach students to make and read the standard version of graphs, some curriculum materials and teachers engage students in broader communicative processes around purposeful shared activities using various representational forms. Some of these forms are entirely new, and others are local variants of conventional graphs. In some instances, several linked representations, each with its own purposes and strengths, are used in what is called a "cascade of inscriptions" (Latour, 1990; Hall & Stevens, 1995; Lehrer et al., 2000). In this section, I describe some of these alternative representational forms and how they are used in classrooms.

According to Greeno and Hall (1997), "When representations are used as tools for understanding and communicating, they are constructed and adapted for the purposes at hand. Often a nonstandard representation serves these purposes better than a standard form." Teachers and curriculum materials now introduce some forms of this kind because they serve as bridges between particular phenomena and graphs. For instance, in the fifth-grade unit of the Mathematics of Change strand of *Investigations in Number, Data, and Space*, called *Patterns of Change* (Tierney, Nemirovsky, & Noble, 1996), students are taught to make a record of a "trip" they take along a straight line in their classroom by dropping a beanbag on the floor at regular time intervals (e.g., every 2 seconds) as they walk. Students can then transcribe the record onto a piece of paper as in Figure 17.8, and that record can then serve as the basis for making tables and graphs of, for instance, the position versus time for a walker. In a classroom study, Monk (2000), reports on a series of seventeen 45-minute sessions in which a combination third-and-fourth-grade class used *beanbag diagrams* as a crucial representational form for analyzing and communicating about motion along a straight line. Once the children were able to use this representation of motion fluently, they extended their capacities to tables and graphs as a way of representing motion. Such a use of these representational forms would normally seem beyond the reach of children this age.

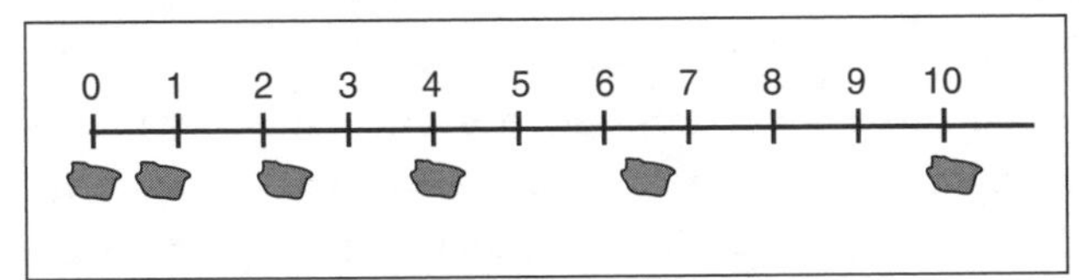

Figure 17.8. Beanbag diagram of a trip.

Several of the teacher-authors in the DMI *Working with Data Casebook* use a representation called *kidpins*, which consists of a large piece of cardboard (see Figure 17.9) with a line down the middle and the words yes and no written on each side of the line. When the teacher asks a yes-or-no question of the class ("Will you buy the school lunch today?" "Do you like to eat soup?"), each child can clip a clothespin on the cardboard according to whether his or her answer to the question is yes or no. Like beanbag diagrams, this form is visually more directly analogous to its referent than tables or graphs and is therefore easier to use. This form of representation is also found in several units in the curricular material *Investigations in Number, Data, and Space*. These include *Counting Ourselves and Others* (Economopoulos & Russell, 1998) and *Mathematical Thinking in Kindergarten* (Economopoulos & Murray, 1998). It is significant, in the light of the quote from Greeno and Hall (1997) above, that using such an immediate and accessible representation, the children "push" this form well beyond its original intentions and specifications to allow answers that are neither yes or no. One child in a class using kidpins clipped his clothespin in the middle along the bottom to indicate his uncertainty as to his answer. The children in another class used various locations on the edge of the cardboard to indicate degrees of agreement or disagreement with a yes-or-no question. These inventions support what other researchers have found: that students, even the youngest, have surprising representational competence when their activity is within a coherent community with a sustained purpose (diSessa, Hammer, Sherin, & Kolpakowski, 1991; Greeno & Hall, 1997; Lehrer & Schauble, 2002).

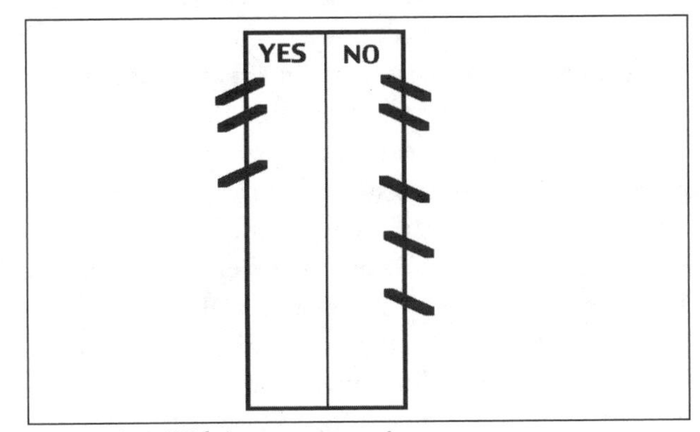

Figure 17.9. Kidpin representation.

Another alternative to conventional graphing is to encourage students to invent their own representations or to vary conventional graphs and then to work with the class in relating these "spontaneous," or "idiosyncratic," forms to the standard ones. An example of such a form is the graph made by the fourth grader, Amelia, of the heights of three plants, shown in Figure 17.2. All three of her graphs show a sharp upward move between the second and third day. But this feature does not mean that the plants suddenly shot upward then. Since Amelia only included those days on the horizontal axis on which she recorded her plants' height, the period between the second and third readings is three days long instead of one, because it spanned a weekend. Thus, the plants' *rate of change* is not as great as the steepness of the graphs suggests. Although this practice breaks conventional rules of graphing and can even lead to a misinterpretation of her graph, it was regarded as "successful" in the particular classroom situation because it followed the criterion set out by the teacher: that it tell "the story of your plant so that someone else can understand it without additional information" (Nemirovsky & Tierney, 2001). This is not to suggest, however, that "anything goes" in such a classroom. In this particular case, the teacher then asked the students to work together to create group graphs in which they would display their several graphs on a common set of axes. This strategy was effective because some of the children had included the weekend days, as the teacher had expected, thus leading to a productive conversation in a later session about scales on axes and the need for standardization. In another situation, the teacher might have elected to teach the students the conventional rules for labeling and scaling axes. Or she might have elected to do nothing until later when the issue of using a graph to tell how fast a plant is growing was made explicit. This observation points to one of many of what Nemirovsky and Tierney (2001) refer to as "transition issues," of how to address the "continuities and discontinuities between inventions and conventions."

It must be emphasized, however, that the goal is not to select one or two representational forms for students to learn and use in *all* situations but, rather, to teach students to adapt representations to a particular context and purpose and even to use several representations at the same time. This goal represents a shift from represent*ations* to represent*ing*. Those researchers who report on classroom studies in which students use alternative forms of representation all emphasize the importance of actively engaging students in an ongoing discussion of the reasons for choosing one kind of representation over another, as well as for the conventions and stipulations that underlie the specific representation's use. In a classroom study in which sixth graders "invented" graphs of motion, diSessa and his colleagues (1991) describe a series of sessions in which "the students invented, critiqued, improved, applied, and moved fluidly among a diverse collection of representational forms" (p. 147). One of their main conclusions was that students of this age possess considerable capacities for inventing representations and reflecting on their inventions—what the authors call "meta-representational skill." These authors, as well as the others cited in this section, believe that this mode of engagement can be a powerful means of promoting students' fluent and productive use of graphs.

Conclusion

In this chapter, I have tried to convey the sense and meaning of the change in the approach to graphing that has taken place over the last decade, from one in which graphs are taken to be a product, a skill, and a topic to be taught, to one in which they are taken to be a medium for making meaning, as well as for communicating information and meaning. The discussion of the changes in graphing should make evident that they are not likely to have taken place without other, more general, shifts that have occurred in mathematics education. Among these broader shifts are the following:

- Widely shared ideas of what it means to know and do school mathematics have been made more complicated, from a focus on procedures and concepts to an awareness of more general processes of meaning-making that result from interaction with one's material and social environment and with the forms and symbols of mathematics.
- Conceptualizations of thinking processes have been broadened to include the use of such diverse personal and experiential resources as students' visual, kinesthetic, and narrative capacities.
- The possible units of analysis in the study of mathematical activity have expanded, from the contents and processes of an individual's mind to a community's practices.

The relationship between the changes in the particular area of graphing and the more general aspects of mathematics education, however, is not unidirectional. Not only have the more general shifts made the change in the approach to graphing possible, but graphing has been an important site on which the work of creating these more general shifts has been carried out. Changes in the approach to graphing not only have occurred because of a widening of the view of what it means to do mathematics but also have contributed to that widening.

Central to the changes in graphing has been an increased awareness of the visual aspects of mathematics, in how the things we see and the way we see them are intimately tied to what we say and do and, therefore, how we think. This new interest in visuality in mathematics and how changes in the ways we see things can be fostered has promoted new points of view from which to address the problem of the persistent errors students make in graphing. At the same time, the growing realization of the complexities and importance of graphing has led to a more flexible and diverse approach to teaching graphs as a representational form. The focus is no longer on using a single visual representation of a given mathematical situation but on employing many visual displays, of several kinds, one of which might be a conventional graph, and having students participate in the choices and evaluation of the representations used. Cumulatively, all these changes have led to an increase in the importance of graphing in school mathematics. Graphing is no longer a topic consisting of a few skills and procedures to be taught once and for all. As a means of communication and of generating understanding, graphing must repeatedly be encountered by students as they move across the grades from one area of school mathematics to another.

ACKNOWLEDGMENTS

Much of the work leading up to this paper was supported by the following grants: National Science Foundation Grants No. MDR-9155746, and DUE-No. 9653068 to TERC, and U.S. Department of Education, Office of Educational Research and Improvement, Cooperative Agreement No. R305A60007-98 to the National Center for Improving Student Learning and Achievement in Mathematics and Science, University of Wisconsin. All opinions, findings, conclusions, and recommendations expressed herein are those of the author and do not necessarily reflect the views of the funders.

REFERENCES

DiSessa, A. A., Hammer, D., Sherin, B., & Kolpakowski, T. (1991). Inventing graphing: Meta-representational expertise in children. *Journal of Mathematical Behavior, 10*, 117–160.

Economopoulos, K., & Murray, M. (1998). *Mathematical thinking in kindergarten.* A unit of Investigations in number, data, and space. Glenview, IL: Scott Foresman (formerly published by Dale Seymour Publications).

Economopoulos, K., & Russell, S. (1998). *Counting ourselves and others.* A unit of Investigations in number, data, and space. Glenview, IL: Scott Foresman (formerly published by Dale Seymour Publications).

Greeno, J., & Hall, R. (1997, January). Practicing representations: Learning with and about representational forms. *Phi Delta Kappan*, 361–367.

Hall, R. (1996). Representation as shared activity: Situated cognition and Dewey's cartography of experience. *Journal of the Learning Sciences, 5*, 209–238.

Hall, R., & Stevens, R. (1995). Making space: A comparison of mathematical work in school and professional design practices. In S. Star (Ed.), *The culture of computing* (pp. 118–145). London: Basil Blackwell.

Hancock, C., Kaput, J. J., & Goldsmith, L. T. (1992). Authentic inquiry with data: Critical barriers to classroom implementation. *Educational Psychologist, 27*, 337–364.

Hanson, N. (1958). *Patterns of discovery*. London: Cambridge University Press.

Kaput, J. J. (1987) Representation systems and mathematics. In C. Janvier (Ed.), *Problems of representation in the teaching and learning of mathematics* (pp. 19–26). Hillsdale, NJ: Erlbaum.

Latour, B. (1990). Drawing things together. In M. Lynch & S. Woolgar (Eds.), *Representation in scientific practice* (pp. 19–68). Cambridge, MA: MIT Press.

Lehrer, R., & Schauble, L. (2002). Modeling in mathematics and science. In R. Glaser (Ed.), *Advances in instructional psychology* (Vol. 5, pp. 101–159). Mahwah, NJ: Erlbaum.

Lehrer, R., Schauble, L., Carpenter, S., & Penner, D. (2000). The inter-related development of inscriptions and conceptual understanding. In P. Cobb, E. Yackel, & K McClain (Eds.), *Symbolizing and communicating in mathematics classrooms: Perspectives on discourse, tools, and instructional design* (pp. 325–360). Mahwah, NJ: Erlbaum.

Leinhardt, G., Zaslovsky, O., & Stein, M. (1990). Functions, graphs, and graphing: Tasks, learning, and teaching. *Review of Educational Research, 60*, 1–64.

Monk, G. S. (1987) *Students' understanding of functions in calculus courses.* Unpublished manuscript, University of Washington.

Monk, G. S. (1994, January). *What does it mean to understand a mathematical concept? It depends on your point of view.* Paper presented at the Joint Mathematics Meetings, Cincinnati.

Monk, S. (2000, April). *Why would run be in speed?": Negotiating meaning within classroom representational practices.* Paper presented at the meeting of the American Educational Research Association, New Orleans.

Monk, S., & Nemirovsky, R. (1994). The case of Dan: Student construction of a functional situation through visual attributes. *Research in Collegiate Mathematics Education, 1*, 139–168.

National Council of Teachers of Mathematics. (2000). *Principles and standards for school mathematics.* Reston, VA: Author.

Nemirovsky, R. (1994). On ways of symbolizing: The case of Laura and the velocity sign. *Journal of Mathematical Behavior, 13*, 389–422.

Nemirovsky, R., & Tierney, C. (2002). Children creating ways to represent changing situations: On the development of homogeneous spaces. *Educational Studies in Mathematics, 45*, 67–102.

Nemirovsky, R., Tierney, C., & Wright, T. (1998). Body motion and graphing. *Cognition and Instruction, 16*, 119–172.

Ochs, E., Gonzales, P., & Jacoby, S. (1996). When I come down I'm in the domain state: Grammar and graphing representation in the interpretive activity of physicists. In E. Ochs, E. A. Schegloff, & S. Thompson (Eds.), *Grammar and interaction* (pp. 328–369). Cambridge: Cambridge University Press.

Ochs, E., Jacoby, S., & Gonzales, P. (1994). Interpretive journeys: How physicists talk and travel through graphic space. *Configurations, 2*(1), 151–171.

Russell, S. J., Tierney, C., Mokros, J., & Economopoulos, K. (1998). *Investigations in number, data, and space.* Glenview, IL: Scott Foresman (formerly published by Dale Seymour Publications).

Russell, S., Schifter, D., & Bastable, V. (2002). *Developing mathematical ideas: Working with data—Casebook.* Parsippany, NJ: Dale Seymour Publications.

Stevens, R., & Hall, R. (1999). Disciplined perception: Learning to see in technoscience. In M. Lampert & M. Blunk (Eds.), *Talking mathematics in school: Studies of teaching and learning* (pp. 107–149). New York: Cambridge University Press.

Tierney, C., Nemirovsky, R., & Noble, T. (1996). *Patterns of change: Tables and graphs.* A unit of investigations in number, data, and space. Glenview, IL: Scott Foresman (formerly published by Dale Seymour Publications).

Chapter 18

Representation in School Mathematics: Children's Representations of Problems

Stephen P. Smith, Northern Michigan University

My goal in this chapter is to draw on a fine-grained analysis of children engaged in problem solving to offer insight into children's idiosyncratic representations. Given pedagogy reflective of the National Council of Teachers of Mathematics' (NCTM, 2000) *Principles and Standards for School Mathematics*, idiosyncratic representations are both inevitable and necessary for learning mathematics. Yet representations that are general—that aid the solution of classes of problems—are essential to mathematics. Children must connect their idiosyncratic representations with mathematical ones if they are to progress very far within the discipline. The research I describe starts from "inside" children and explores their creation and use of idiosyncratic representations. I consider how the relationship between children's views of mathematics and the role of the problem context plays out in their representations. In the process, the research described contributes to answering one of the questions raised by Goldin in chapter 19: "How can we infer or measure students' internal representational capabilities?" The research is also relevant to his other questions. Before presenting the research, I briefly discuss the impact of reforms on representations and their significance in learning mathematics.

Impact of Reform on Representations

Principles and Standards for School Mathematics (Principles and Standards) (NCTM, 2000) provides a framework for reform of mathematics education. The authors advocate increased use of contextualized problems that ground learning mathematics in real-world situations (see, e.g., pp. 52–53, 116, 183). Thus, reform-oriented teachers have students solve problems individually, in pairs, or in small groups. The students then share solutions—including representations—that may be questioned and analyzed by peers (and the teacher). Such pedagogy results in a proliferation of representations: Students create idiosyncratic representations, draw on conventional representations, and blend idiosyncratic and conventional representations as they solve contextualized problems.

The representations students use are resources to resolve problem situations. In traditional mathematics classrooms, the emphasis has been on the use of abstract and general representations (e.g., the long-division algorithm). In reform-oriented classrooms, students are given more freedom in creating and using representations. Students in such classrooms may create idiosyncratic representations that (a) faithfully reproduce the action of a story problem; (b) strip away the context, attending only to numerical aspects of the problem; or (c) combine some of both approaches. In either traditional or reform classrooms, representations "re-present" the ideas of the problem in a form that allows for a solution.

Rather than use the general and abstract representations of mathematics—representations capable of solving a whole class of problems—children may create ad hoc representations. These representations may facilitate the solution of the problem at hand but not necessarily engage students in mathematical reasoning. After all, solving any particular problem is not the goal; students are unlikely to encounter this same problem in the real world or in a career dependent on mathematical reasoning. The goal of problem solving is for students to use problem situations as gateways to abstraction and generalization—to develop the ability to mathematize situations. Doing so involves stripping away the context and examining the mathematical underpinnings of the story. It also requires looking across problem contexts and solution methods for commonalties and differences. In this view, students must progress from idiosyncratic and ad hoc representation of particular problems to conventional, abstract, and general representations that function for classes of problems. By examining students' representations, their language as they construct the representations, and their attitude toward mathematics and about how it is learned, one can infer their representational capabilities. Further, one can explore their understanding of the role of representations in learning mathematics.

Principles and Standards (NCTM, 2000) states the following:

> Instructional programs from prekindergarten through grade 12 should enable all students to—
>
> - create and use representations to organize, record, and communicate mathematical ideas;
> - select, apply, and translate among mathematical representations to solve problems;
> - use representations to model and interpret physical, social, and mathematical phenomena. (p. 67)

The text designed to illuminate these bullets makes clear that the authors intend to encourage the use of idiosyncratic representations. They argue that teachers should "encourage students to represent their ideas in ways that make sense to them, even if their first representations are not conventional ones" (NCTM, 2000, p. 67). The authors argue that such representations "can play an important role in helping students understand and solve problems" (p. 68). In this process, the representations provide "meaningful ways to record a solution method and to describe the method to others" (p. 68). Students also need to "develop an understanding of the strengths and weaknesses of various representations" (p. 70). The representations students develop help teachers understand "students' ways of interpreting and thinking about mathematics" (p. 68). The teacher's role includes helping students build bridges from idiosyncratic to conventional representations—helping students see the similarity (from a mathematical perspective) among multiple problem contexts. Thus teachers need to understand both idiosyncratic and conventional capabilities to help students build bridges.

A representation that solves only one specific problem or a small subset of a category of problems may not be considered mathematical. The creation of mathematical representations involves abstraction and generalization. Children may have little intrinsic inclination to create or use anything other than ad hoc representations. They may have little inclination to connect their representations with those of the discipline. The very real-world contexts in which reform-oriented teachers embed problems may constrain some children. Those children who do strip away contexts, seeking to work in mathematical forms, may have varied reasons for doing so. To help students progress from idiosyncratic to discipline-valued representations, educators need to understand how children view and relate to both sorts of representations. Again, this sort of understanding on the part of teachers and researchers is important to help students foster the connection between the idiosyncratic and the abstract and general.

Definitions of Representations

Various definitions of representations can be found in the literature. For example, Brinker (1996) offered an object-oriented definition:

> *Representations* here refer to students' notations and pictures, already-made drawings such as pictures of partitioned objects, and structured materials such as fraction strips and Cuisenaire rods. *Structured* in this case refers to materials that have been designed for instruction of particular mathematical concepts. (p. 1, emphasis in original)

In this definition, agents seem curiously separated from their representations. Representations are products, that is, physical embodiments. Instructional materials (e.g., Cuisenaire rods) are often introduced by teachers who invest them with meaning for the entire class and for use across contexts. In contrast, individually designed representations ("students' notations and pictures") are invested with meaning by an individual (or a small group) for localized use, for solution of particular problems. Brinker's definition omits the intent of the creator of such idiosyncratic representations.

In Chapter 19 of this volume, Goldin defines a representation as:

> a configuration of signs, characters, icons, or objects that can somehow stand for, or "represent," something else. According to the nature of the representing relationship, the term *represent* can be interpreted in the many ways, including the following (the list is not exhaustive): correspond to, denote, depict, embody, encode, evoke, label, mean, produce, refer to, suggest, or symbolize. (p. 276, emphasis in original)

This definition also treats the representation as separate from the individual. As with the Brinker (1996) definition, it focuses on the product. If one were considering the abstract and general representations of the mathematical community, this definition would be satisfactory. To understand the idiosyncratic representations of children, however, it is not sufficient.

Pimm (1995) noted that representations "re-present something of the world using something else" (p. 119). Traditional worksheets offer ways to represent problems no less than do pictures, manipulatives, tables, graphs, and so forth. Each form of representation is an abstraction of the real world to some degree. For example, using four markers to represent adding two apples and two apples abstracts the real-world action. Drawing four tally marks is a step further away from "reality," because they are not manipulable as are the markers. Writing 2 + 2 = 4 is an even more abstract representation. Some representations offer images of the real world, whereas others have little tangible connection with the context. The issue of which of two representations is closer to the real-world action is not always straightforward, however. Which, for example, is closer to the real-world action: representing four apples with a picture of four apples or with four non-apple-shaped markers that can be physically manipulated?

Kaput (1985) offered a semiotic-oriented definition:

> In its broadest sense a representation is something that stands for something else, and so must inherently involve some kind of relationship between symbol and referent, although each may itself be a complex entity. . . . A rigorous specification of a representation should include the following entities:
>
> - The represented world.
> - The representing world.
> - What aspects of the represented world are being represented.
> - What aspects of the representing world are doing the representing.
> - The correspondence between the two worlds. (pp. 383–384)

Once again, the creator or agent is noticeably absent. This definition treats the representation as doing the work; people gain value from representations through the "correspondence between the two worlds." Yet, because representations are often differently interpreted, *multiple* correspondences clearly may result. To understand what "something else" a representation "stands for," one must divine (or invest) intention.

Lesh, Post, and Behr (1987) gave a definition that suggests a bond between a representation and its creator: "The term *representation* here is interpreted in a naive and restricted sense as external (and therefore observable) embodiments of students' internal conceptualizations—although this external/internal dichotomy is artificial" (p. 33, emphasis in original). Although the Lesh et al. text does not make the conclusion clear, one might interpret the final clause as indicating the necessity of understanding the creator's intent when analyzing a representation. Such a definition moves closer to capturing the ways of doing mathematics of the children of the study described in this chapter.

As Goldin and Kaput (1996) note, representations are embedded: "They usually belong to highly structured systems, either personal and idiosyncratic or cultural and conventional" (p. 398). In the context of the study considered here, I regard cultural and conventional representations as mathematical, but personal and idiosyncratic representations that children use to solve problems not to be mathematical, because the representations lack abstraction and cannot be generalized. (This distinction is not meant to disparage the children's abilities or efforts. I merely distinguish what is valued in the mathematical community from much of what the children did in solving problems.) If one considers representations to lie on a continuum from idiosyncratic to mathematical, two of the children described subsequently, Danielle and Nicole, tended to create representations closer to the idiosyncratic end; the other two, Teri and Bert, inclined toward the mathematical end (although generally still a significant distance from that end). I inferred each child's internal reasoning from his or her external representations and from conversations around the development of those representations.

Goldin and Kaput (1996) distinguish between imagistic and formal systems. In their view, imagistic representations "include internal imagery ... which is 'imagined' [or] visualized" (p. 414). Further, they argue that "the way in which representations carry meaning may be through analogy and even metaphor" (p. 415). All the children whose work is discussed in the following created representations that were imagistic to a great extent. Danielle's and Nicole's representations most clearly carried the meaning of the story in their pictures. Bert's and Teri's representations were focused more on the mathematical underpinnings of the story.

Each of the foregoing definitions suits its author's purposes and furthers our understanding of representations and their role in doing mathematics. For the study of children's mathematical reasoning discussed subsequently, I based my analysis on the definition of Davis, Young, and McLaughlin (1982): "A representation may be a combination of something written on paper, something existing in the form of physical objects and a carefully constructed arrangement of idea in one's mind" (p. 23). (Although having seen many children present solutions using their often painfully crafted representations, I doubt they all have a "carefully constructed arrangement" of their ideas.) This definition underscores the value of both the creator of a representation and the connection between the creator's ideas and the resultant representation. One who uses or interprets a representation may find one's own value in the representation. This definition allows for the internal imagery of one's creation and the alternative use or interpretation of the same representation by another individual

Children's Creation and Use of Representations

In this section, I briefly present and discuss four children's responses from a study that examined (in part) the creation and use of representations from the child's perspective (Smith, 1999). The children created mostly idiosyncratic representations to solve the nonroutine problems posed to them. By examining the creation and use of the representations, I make inferences about the children's capabilities and their view of the role of representations. As part of a series of individual interviews, third graders were asked to solve the Candy-Bar Problem:

> I have a sister who lives outside of Detroit. She has four children, and I always struggle deciding what to get them for Christmas because I don't have lots of money, but I want to get them something they will like. So one year I got them a box of nice candy bars, you know, expensive, fancy candy, because I know that they all like candy. I bought the four of them one box of candy bars and it had 30 bars in it. When they split them up, how much do you think they each got?

All four children described here had spent nearly two academic years together in the same reform-oriented mathematics classroom. On typical days they worked alone, with partners, or in small groups on a small number of problems. Selected children then presented their solutions, and the class analyzed them by questioning one another's assumptions, reasoning, and computations. The students in this class were more accustomed to using pictures than manipulatives to solve problems. The interviews that generated the data discussed here were intended to elicit the children's reasoning, language, and representations in contexts congruent with their classroom experiences. As a consequence, for this and most interview activities, the children had access to paper and pencils but not to manipulatives. The Candy-Bar Problem was the first task of the second interview for each child and took place before the class formally studied either division or fractions. The interviews were videotaped and audiotaped, transcribed, and analyzed for the children's reasoning and use of representations and language.

In a sense, all the children created similar representations to solve the Candy-Bar Problem; they all partitioned 30 objects among four locations. However, significant differences were observed in both the creation and use of representations that, when viewed in light of each child's attitudes, led to different inferences. Some representations involved drawing 30 objects and then distributing them; some involved objects created in the process of distribution. Some objects closely resembled candy bars, and some were mere tallies. Some distribution of candy bars was individual, and some involved grouping them first. Some groups were formed for partitioning, and some, for repeated subtraction forms of division.

All four children drew pictures to solve the problem. The extent to which these pictures were re-enactments of the story line varied considerably, however, as did the children's language in explaining as they worked.

Danielle

Danielle started by saying, "Draw a box." She drew a rectangle in which she then drew "candy bars"—rectangles arrayed closely in rows much like one would expect to see in a purchased box of candy bars (see Figure 18.1 for her completed representation). While drawing these candy bars, she asked about my sister's children—whether they were all boys. Upon learning they were two boys and two girls, she asked their names. She then drew four heads with torsos and labeled each with initials. One niece was even given long hair. Danielle "distributed" the bars by circling groups of four and labeling one bar per group for each child. This portrayal also reflects what might have happened according to the story: My sister grabs a group of four bars and gives one to each of the four children. On this problem (as in all her interviews), Danielle consistently used the names of the objects of the story; she always counted candy bars or people rather

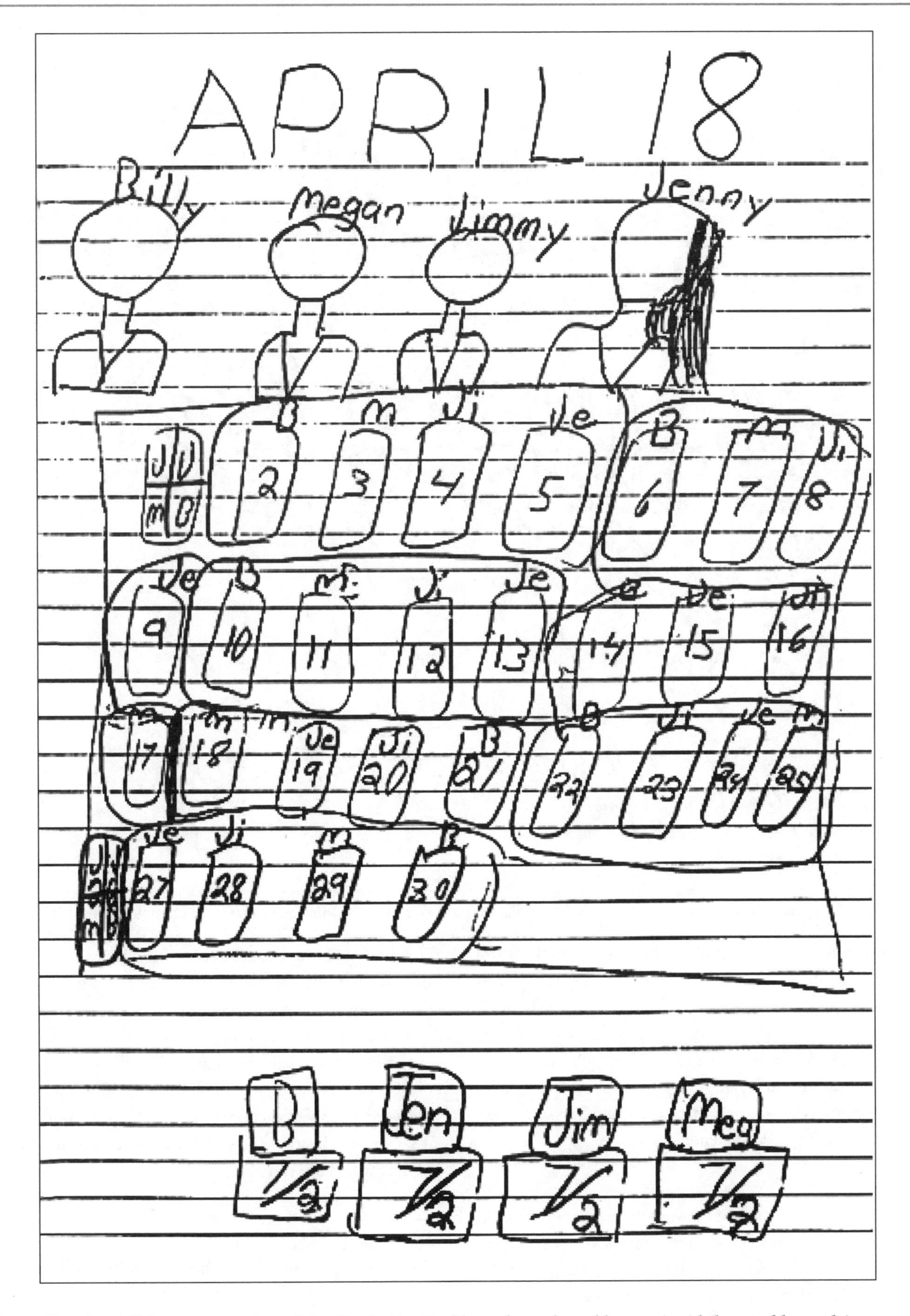

Figure 18.1. Danielle's representation of the Candy-Bar Problem; the real-world scenario aids her problem solving.

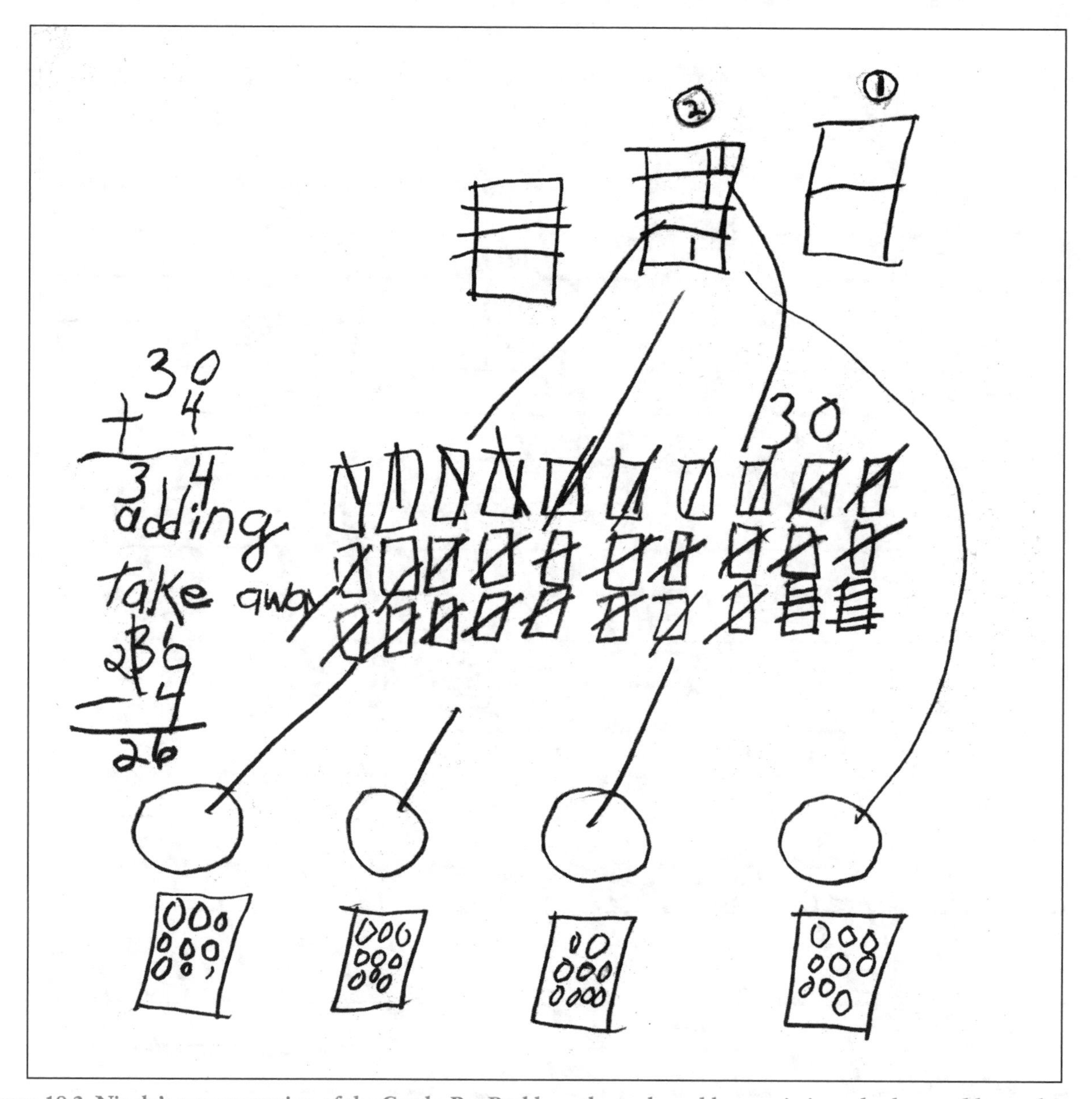

Figure 18.2. Nicole's representation of the Candy-Bar Problem; the real-world scenario impedes her problem solving.

than just numbers. For example, she would say, "Billy got seven-and-a-half candy bars." Danielle's approach contrasts with that of two of the children, who consistently dropped the objects of the story, saying things like, "They got seven-and-a-half each." Danielle used numbers as adjectives modifying the characters and objects of the story; the two other children used numbers as nouns, dropping the contextual elements.

Nicole

Nicole started by asking how many candy bars were in the "bag."[1] She then drew 30 rectangles arrayed in rows (see Figure 18.2). Next she drew four circles that she identified as the children and added a "box" (rectangle) beneath each circle. Rather than group the candy bars in fours to distribute them, Nicole put a slash through each bar and then redrew it in a child's "box." In other words, she dealt out the bars as one might a deck of cards. Like Danielle, Nicole was consistent across interviews in using numbers as adjectives, but Danielle used her close connection to the story lines as support. Throughout the interviews, Danielle's constant reference to real-world situations both provided her time to think and enabled her to find an answer that fit the situation. In contrast,

[1] Nicole often confused elements of the story. In class, the teacher generally wrote problems on the chalkboard so that students would have plenty of opportunity to refer to them; in the interviews, this and most other problems were presented orally, with details, such as quantities, repeated on request.

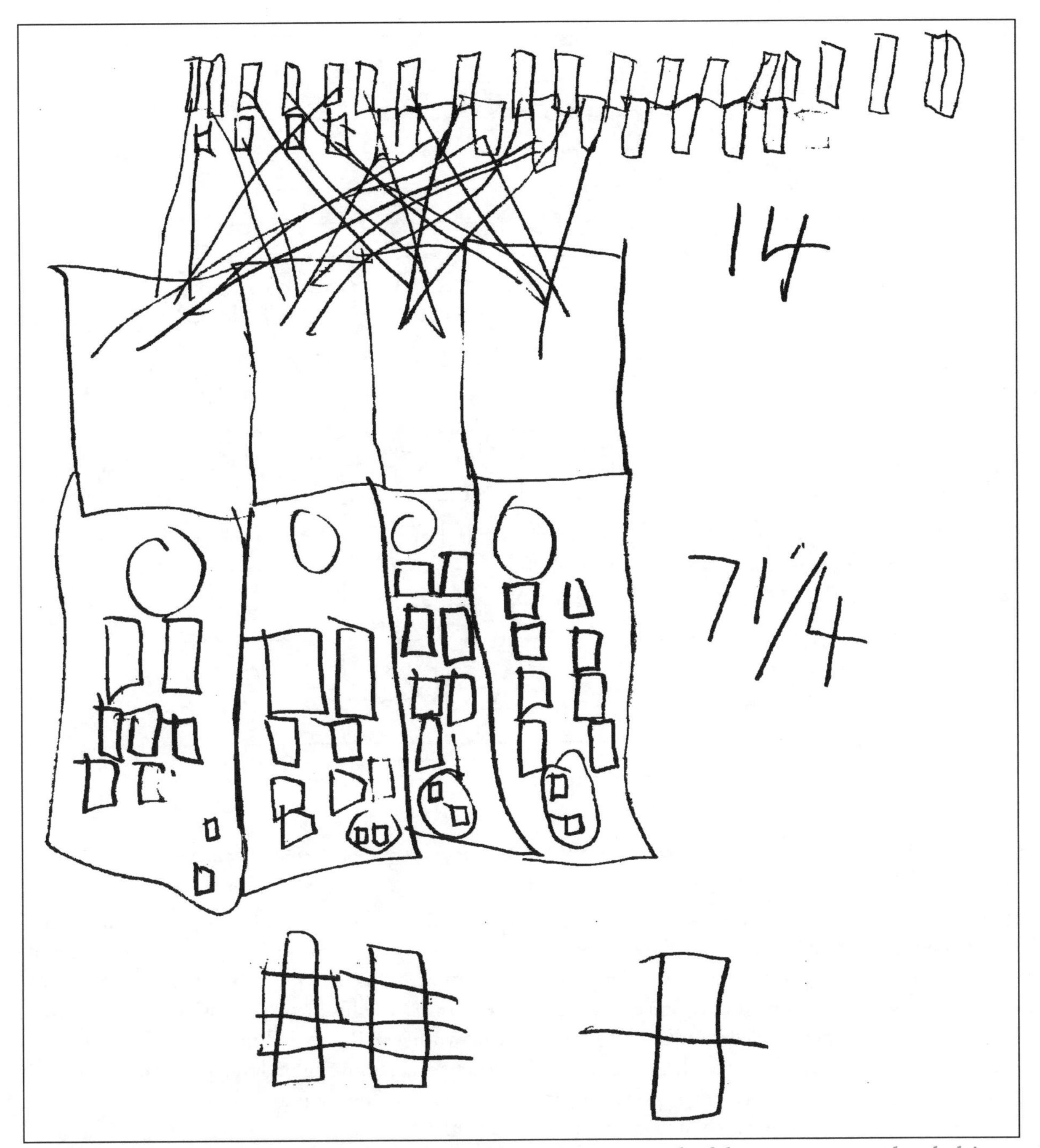

Figure 18.3. Bert's representation of the Candy-Bar Problem strips away much of the context to reveal underlying mathematical structure.

Nicole's close connection with stories often impeded her reaching a correct solution. In this problem, she did not reason through the division of the two "extra" bars. Having given each child seven bars, she stopped. When asked about the last two bars, she divided each into four pieces. Then, after adding one piece from each extra bar to each child, she concluded that the children had each received nine bars (the original seven plus one from each of the extra bars, which she counted as whole bars). After further probing, Nicole essentially argued that sharing a candy bar equally is possible for two people but that one could not physically make an equal division for more than two people. For Nicole, "reality" obviated the need to mathematize this situation—the ideal fractions of mathematics were irrelevant.

Bert

In contrast with Danielle and Nicole, Bert and Teri tended to use relatively generic representations. That is, their pictures bore only passing resemblance to children

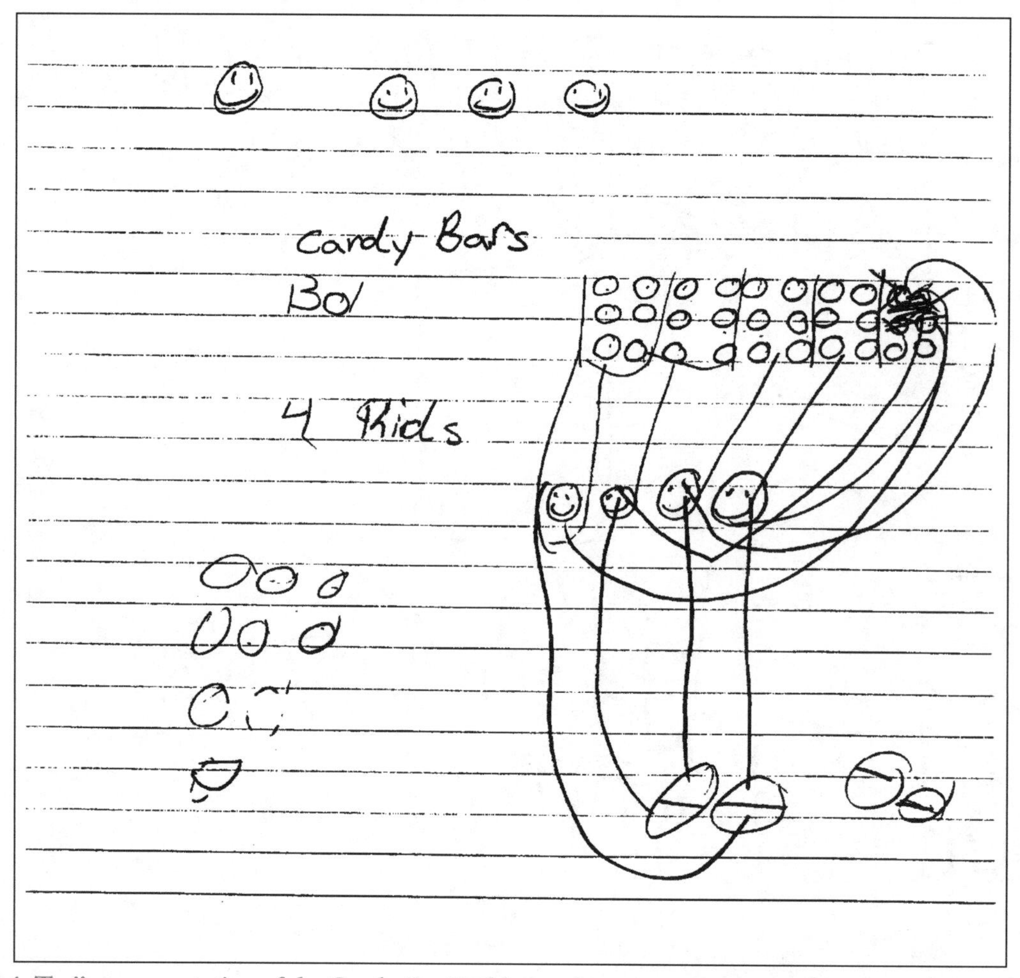

Figure 18.4. Teri's representation of the Candy-Bar Problem strips away even more context, employing a hypothetical, exemplar child to compute a solution.

and candy bars in boxes; they usually omitted the referent and treated numbers as nouns. Bert's first effort at solving the problem used rectangles for the candy bars arrayed in two unequal rows. He abandoned that representation when his attempt to connect the bars sequentially with four locations (children) resulted in a confusing mess (see top portion of Figure 18.3). His second representation did not involve first drawing all 30 bars and then distributing them; instead, he drew bars directly into four locations as he distributed them (see bottom portion of Figure 18.3). Throughout this and other problems, Bert usually dropped the nouns of the story; the numbers became nouns, and children, if mentioned, were referred to as "him" (or, often, "'em").

Teri

Teri's picture was even more generic. After she had drawn three rows of 10 circles (not the typical shape of candy bars), she drew four "smiley faces," saying, "One kid, two kid, three kid, four kid" as she did so (see Figure 18.4). She divided 24 (the left eight columns of three circles) into four groups of six each and connected each group with a smiley face. She then divided four of the remaining circles into "groups" of one by connecting each with a smiley face. Finally, she distributed the two "extras" by redrawing (enlarging) and dividing them into four halves. She then distributed the halves as she had the others.

Teri's representation was not a pictorial "re-presentation" of the story: She used circles, not rectangles, to represent the candy bars and made no mention of the box. She did draw smiley faces for the children, giggling as she did so; however, she made no distinctions among the children. She made connections from each of the three sizes of "groups" of bars (six, one, one half) to an individual child. She then used an exemplar child to compute how many bars each received. This last tactic was hypothetical: She did not count how many candy bars a particular "kid" received but, rather, gathered one set of ovoid objects (larger than her original circles) as representatives of each group in a separate location. Her creation of faces rather than mere circles seems more like something she found fun to draw than a connection with the real world that she envisioned as supporting her work. As with Bert, numbers were nouns; Teri stripped away the context of the story and dealt with the underlying numbers.

Uses of Context

Reform-oriented teachers often situate problems in real contexts familiar to students. A real context helps learners make sense of, and reason through, problems using their understanding of daily life. In the examples presented here, although all four children had been in the same class for two (of three) elementary school years, their interactions with the same tasks differed. Danielle and Nicole were inclined to use the contexts of the problems to guide the design of their representations, whereas Bert and Teri tended to strip the context away and create less idiosyncratic representations. The actions each child took—staying close to the story context or stripping it away, using representations that more or less faithfully re-enacted the story—were remarkably consistent across tasks and interviews (see Smith, 1999, for details of interview tasks and responses). As the next section shows, the children's idiosyncratic creation and use of representations reflected their views of mathematics and their prior experiences.

Superficial observation of the four children's representations suggests that the children solved the problem in similar fashion. Yet examining the children's work from "inside" the child—examining language in conjunction with the development of representations—shows that each child followed an individual path of creation and invested his or her representation with specific meaning. The next section takes this argument a step further by examining the children's work in light of their views of mathematics, problem contexts, and representations. Such examination fosters insight into the children's tendencies toward abstraction or contextualization and their relationship with representations.

Views of Mathematics, Contexts, and Representations

I next turn to a discussion of each child's views of mathematics generally and the role that specific problem contexts played in their representations. During the series of interviews, Danielle said she liked mathematics but doubted her ability to understand many concepts, such as negative numbers, fractions, and division. By regularly referring to the story to confirm her actions as reasonable and to determine next steps, she ably used contexts to help her solve the tasks posed. When tasks offered an opportunity to delve into mathematical concepts, however, she often recognized the opportunity and the concepts but declined the invitation to engage with them. For example, during one task, she posed herself the question, "Is negative one a number?" and then provided an answer that avoided engagement with the abstract ("It shouldn't be, ... [because] we don't like negative things in our school").

Danielle's use of the real-world context to ground her solutions brought her great success in problem solving; she demonstrated a powerful ability to create and use idiosyncratic representations. On the one hand, this grounding in the context, combined with her lack of self-confidence, especially about abstract concepts, suggests potential difficulty in bridging from idiosyncratic and ad hoc representations to abstract and general ones. An approach that appears powerful for the present may impede her mathematical growth. On the other hand, Danielle was constantly questioning; she questioned her peers, her teacher, her interviewer, and, most of all, herself. In many ways, she came across as a poster child for reform. She had curiosity about, and insight into, what was at issue in specific problems.

Teri likewise doubted her mathematical ability and understanding (although she was arguably the most successful at solving the interview tasks). She said she "let others decide" in class when conceptual issues arose. She seldom gave evidence of recognizing underlying mathematical concepts; she deflected probes intended to connect specific problems with abstract and general concepts of mathematics. She did so with giggles, jokes, and "I don't know." For Teri, the problem context played little role except as a delivery vehicle.

Teri's internal system of representation was powerful enough to allow her to be the most successful problem solver of the children. Despite a lack of confidence that kept her from contributing to classroom discussions, she successfully used the mathematical conclusions reached in those same discussions. She communicated her ideas well with me, and her representations were further along the continuum toward the general and abstract than those of the other children in the study. She appeared to distance herself from her peers' discussion and the real-world contexts to avoid becoming confused by them.

Bert and Nicole both expressed confidence in their mathematical reasoning ability; both believed that everyone had, in Nicole's words, "good ideas" in mathematics. Bert said he listened to others' ideas and compared them with his, modifying his own ideas or questioning the others'. In class and during the interviews, he quickly moved past the specifics of a problem to underlying mathematical concepts. In contrast with Danielle, when Bert recognized underlying mathematical concepts, he generally tried to improve his understanding of them. Unfortunately, his quick leap to ponder concepts sometimes left mathematical details behind; for example, he consistently subtracted in unconventional ways (e.g., $3 - 2 = -1$ and $2 - 3 = 1$).

Bert's work demonstrated many of the features of a powerful internal representational system. Yet his faulty conventions and frequently casual attitude when creating representations yielded external manifestations that might lead one to conclude he had a weak internal system. During our conversation (often while drawing representations), Bert repeatedly referred to the value and influence of participating in the same classroom discussions from which Teri stayed removed. Clearly, he saw mathematics as socially constructed (at least in his classroom). He described the class' explorations of multiple representations and methods of solution as helping him learn mathematics.

Nicole's belief that everyone had good ideas seemed to combine with her attachment to real-world contexts to render her the least successful on the tasks. She saw the mathematical conversations in her classroom as opportunities for each child to display his or her "good ideas." In contrast with Bert, she appeared not to use such conversations to examine her own or others' reasoning; for Nicole, mathematical conversations were opportunities for displaying ideas, not for offering them for analysis. She recognized ideal objects of mathematics (such as fractions), but the contexts in which problems were embedded could remove those objects from playing a role in determining a solution. If dividing a candy bar into five exactly equal pieces is impossible in the real world, Nicole saw no reason to invoke fractions.

Nicole's faith that everyone has "good ideas" seems to have allowed her to opt out of building an internal representational system. If all ideas are good, no one else's idea is better (or worse) than hers; she sought no contradictions and found no reason to develop a "system." Nicole had a positive attitude toward mathematics and willingly engaged with each task presented. The supportive, discussion-based mathematics classroom allowed her to develop a view of idiosyncratic representations as appropriate for real-world problems and general (mathematical) representations for decontextualized problems.

This synopsis of the children's beliefs about mathematics, reasoning, and how mathematics is learned suggests that merely looking at the external representations or the language of students can be misleading. Danielle adeptly used the context to abet her mathematical work. Teri stripped the context away to solve problems. Yet they had a similar aversion to engaging with the abstraction of mathematics. Nicole used the context to frame possible outcomes, and because her own ideas were intrinsically of value, did not examine them. She found little reason to engage with abstractions. Bert seldom used the context to examine the reasonableness of answers. He was drawn to the abstraction of mathematics.

Looking for Internal Representations

Kaput (1987b) described mathematical representations as "formally independent of the form of the external symbols used because the structure itself is treated as an abstraction or idealization" (p. 23). Although this view may hold for mathematicians, it clearly does not do so for many children (e.g., Smith, 1999). In fact, children seem just as likely to treat representations as instantiations of a problem situation. For Nicole, although her representation may have been an "abstraction or idealization" of the story, the "reality" embedded in the story controlled how the representation was to be used. Kaput argued, "*The idea of representation is continuous with mathematics itself*" (p. 25, emphasis in original). This perspective assumes that children doing mathematics are engaged in using mathematical representation systems, although Kaput also indicates the necessity of "translation between nonmathematical representations and mathematical ones" (p. 21). Thus, in Kaput's view, representation systems seem global and discipline based (e.g., the "base-ten placeholder representation system," p. 162).

In contrast, Lesh (1987) posited a more local version as well. He saw children engaged in solving individual problems as "trying to get things into a single coherent representational system so they can" solve the problem (p. 100). Of the children described in the foregoing, only Bert seemed so engaged. Danielle and Teri recognized and accepted the existence of coherent systems, although they may not have envisioned a *single* system; Nicole's responses left the question open.

Kaput (1987a) argued that by looking at "what the characteristics are of particular representations," mathematics educators could determine why those representations are "effective in some cases and ineffective in others" (p. 101). Mason (1987) portrayed such an approach as proceeding "from the outside," whereas he found it "more engaging and realistic to try to capture experience of the inner world from the inside, as it were, by using metaphor and descriptive frameworks that resonate with other people's experience" (pp. 207–208). Goldin's chapter in this volume defines internal and external systems of representations. My sense is that, when Goldin asks how we may infer students' internal representational capabilities, he is advocating a research agenda that will illuminate children's ways of understanding and learning mathematics "from the inside." In this chapter I describe research that examined children in the process of creating representations viewed through their language and their views of both mathematics and the role of problem contexts. To find internal representations, we must start by looking from inside the child.

Conclusion

In this chapter, I have explored two relationships: one between idiosyncratic and general (conventional) representations and one between contextualized and abstract reasoning. Although any teacher's goals should include having students move from idiosyncratic to general representations, some students, engaged by the real-world contexts, may see little need for doing so. Other students may tend that way of their own accord but in doing so miss the benefits the contexts are expected to provide. The insights into children's beliefs about mathematics, reasoning, and how mathematics is learned that are illuminated by the study (Smith, 1999) suggest that merely looking at the external representations or the language of students can be misleading. By examining their reasoning from "inside" these children, that is, by analyzing their creative processes through their language and beliefs, one sees more than the artifacts demonstrate. The children generally did not treat representations as part of a representational system. Although they may have seen mathematics as a systematic, coherent subject, their work did not clearly reveal that they saw representations in a similar light. Further, precious little consistency was observed across children in terms of encoding mathematical ideas. The idiosyncratic representations that children create may be ad hoc; they design representations to solve particular problems, not as tools for general use. Yet to understand and communicate mathematics, children need to bridge from idiosyncratic to mathematical representations.

Teachers and researchers need to understand how children use representations to understand mathematics. To the knowledgeable other (the teacher or researcher), multiple representations presented by different children may appear similar and seem helpful to learners. Children may view and use them in different ways; they may find multiple representations confusing. Children need to progress from solving particular problems to using the process and product of problem solving to better understand the structure of mathematics. How children understand the role of the process of sharing and analyzing representations in learning mathematics is an important aspect of learning. Children need to view analyzing one another's solutions as a tool for deepening their mathematical understanding. Research that proceeds from "inside" the child, research that attempts to understand the "internal representational systems" of children, is necessary to understand how representations enable and constrain learning.

ACKNOWLEDGMENTS

The author wishes to thank Deborah Schifter and Frank Lester for their many helpful comments on previous drafts of his chapter. He is also grateful to the anonymous reviewers for their suggestions on a preliminary version.

REFERENCES

Brinker, L. (1996). *Representations and students' rational number reasoning*. Unpublished doctoral dissertation, University of Wisconsin—Madison.

Davis, R. B., Young, S., & McLaughlin, P. (1982). *The roles of "understanding" in the learning of mathematics.* Urbana: University of Illinois, Curriculum Laboratory.

Goldin, G. A., & Kaput, J. J. (1996). A joint perspective on the idea of representation in learning and doing mathematics. In L. P. Steffe, P. Nesher, P. Cobb, G. A. Goldin, & B. Greer (Eds.), *Theories of mathematical learning* (pp. 397–430). Mahwah, NJ: Erlbaum.

Kaput, J. J. (1985). Representation and problem solving: Methodological issues related to modeling. In E. A. Silver (Ed.), *Teaching and learning mathematical problem solving: Multiple research perspectives* (pp. 381–398). Hillsdale, NJ: Erlbaum.

Kaput, J. J. (1987a). Messy context. In C. Janvier (Ed.), *Problems of representation in the teaching and learning of mathematics* (pp. 101–102). Hillsdale, NJ: Erlbaum.

Kaput, J. J. (1987b). Representation systems and mathematics. In C. Janvier (Ed.), *Problems of representation in the teaching and learning of mathematics* (pp. 19–26). Hillsdale, NJ: Erlbaum.

Lesh, R. (1987). Cluster of representations. In C. Janvier (Ed.), *Problems of representation in the teaching and learning of mathematics* (p. 100). Hillsdale, NJ: Erlbaum.

Lesh, R., Post, T., & Behr, M. (1987). Representations and translations among representations in mathematical learning and problem solving. In C. Janvier (Ed.), *Problems of representation in the teaching and learning of mathematics* (pp. 33–40). Hillsdale, NJ: Erlbaum.

Mason, J. (1987). Representing representing: Notes following the conference. In C. Janvier (Ed.), *Problems of representation in the teaching and learning of mathematics* (pp. 207–214). Hillsdale, NJ: Erlbaum.

National Council of Teachers of Mathematics. (2000). *Principles and standards for school mathematics*. Reston, VA: NCTM.

Pimm, D. (1995). *Symbols and meanings in school mathematics*. New York: Routledge.

Smith, S. P. (1999). *Children, learning theory, and mathematics: An analysis of the role of language and representations in children's mathematical reasoning*. Unpublished doctoral dissertation, Michigan State University, East Lansing.

Chapter 19

Representation in School Mathematics: A Unifying Research Perspective

Gerald A. Goldin, Rutgers University

> The term *representation* refers both to process and to product—in other words, to the act of capturing a mathematical concept or relationship in some form and to the form itself. ... Moreover, the term applies to processes and products that are observable externally as well as to those that occur "internally," in the minds of people doing mathematics. (NCTM, 2000a, p. 67)

> Some forms of representation—such as diagrams, graphical displays, and symbolic expressions—have long been part of school mathematics. Unfortunately, these representations and others have often been taught and learned as if they were ends in themselves. This approach limits the power and utility of representations as tools for learning and doing mathematics. (NCTM, 2000b, p. 14)

This chapter has two main purposes: (1) to give a brief, self-contained overview of a research-based perspective on representation in mathematical learning and problem solving that I have been involved in developing for some time, and (2) to point out how such a perspective reconciles and unifies some important ideas that, when espoused in the name of such movements as behaviorism or radical constructivism, have too often been taken as mutually exclusive.

The context of the discussion is not just theoretical. Vigorous conflicts have developed in some U.S. states around public policy in mathematics education. "Traditionalists," including some well-known mathematicians, have squared off against "constructivists," including some well-known education researchers, expressing strong, opposing views of curriculum goals, teaching methods, and ways of measuring student achievement. Huge policy swings have resulted, with classroom teachers caught in between. Convincing, research-based answers to important questions are missing, partly because the prevailing "isms" have been narrowly dismissive of one another's constructs and methods. Writing as both an active mathematical scientist and one with experience in systemic educational change, I regard the need for a unifying perspective as pressing.

To some extent, *Principles and Standards for School Mathematics* (National Council of Teachers of Mathematics, 2000a) can be seen as a valiant effort to balance conflicting curricular priorities. From a scientific standpoint, research on developing systems of representation provides some concrete findings that can improve the teaching and learning of school mathematics. But in my view this research does far more. Rejecting dismissive claims, this perspective offers a broad, sound, and inclusive framework for addressing key controversial issues, including the characterization of what conceptual and problem-solving power means in mathematics, the formulation and evaluation of mathematical objectives, and the assessment of individual students' mathematical development.

First, I describe some fundamental theoretical concepts, including aspects of a descriptive model for mathematical learning and problem solving that I am continuing to develop. Then I mention several ongoing research directions in the study of representation, including the study of affect from a representational point of view and some methodological and teaching considerations. Finally, I discuss the unification of diverse ideas from mathematics, cognitive science, and behaviorist and constructivist perspectives in mathematics education, with implications for educational practice.

Fundamental Concepts

This section addresses the meaning of representation and systems of representation; the nature of representing relationships; the distinction between representational systems internal and external to individuals; the kinds and characteristics of representational systems that pertain to mathematical learning, thinking, and problem solving; stages in the development of systems of representation; and the essential role played by ambiguity.

Representations and Systems of Representation

To begin with some definitions, a *representation* is a configuration of signs, characters, icons, or objects that can somehow stand for, or "represent" something else. According to the nature of the representing relationship, the term *represent* can be interpreted in many ways, including the following (the list is not exhaustive): correspond to, denote, depict, embody, encode, evoke, label, mean, produce, refer to, suggest, or symbolize. This type of definition (Goldin, 1987, 1998; Kaput, 1985; Palmer, 1978) has been termed by Kaput (1998, p. 266) an "abstract correspondence approach." To make the definition concrete requires subsequently specifying what kinds of entities are involved in doing the representing and being represented, and in what ways one entity stands for another.

For example, a numeral—the written symbol or the spoken word—can denote a concrete set of objects that have been counted, whether "like" objects such as oranges, or otherwise unrelated physical objects. Alternatively, a numeral can denote the more abstract *class* of sets that can be placed into one-to-one correspondence with each other and with the particular numeral through counting. The representing relationship of *denoting*, numeral to set or numeral to equivalence class, was established through a historical process leading eventually to shared social conventions, for instance, through wide adoption of the Hindu-Arabic numeration system. The encoding of the denotation relationship, together with associated competencies, exists now in billions of individual human brains. Similarly, a Cartesian graph is a representation that is established by convention and that may correspond to an algebraic equation or function, depict a set of data, express a qualitative relationship, and so forth. Numerals and graphs are examples of concrete, static, external representations—such configurations can be found in textbooks, produced and interpreted by teachers and students, pointed to, and discussed.

Representing relationships are often two-way, with the depiction or symbolization going in either direction according to the context (Goldin & Kaput, 1996; Vergnaud, 1998). Thus a graph may provide a visual representation of an algebraic equation, or the equation may describe the graph with algebraic symbols.

These examples already illustrate the idea that individual representational configurations—whether mathematical or not—rarely stand alone. Hindu-Arabic numerals belong to arithmetical and algebraic symbol systems that include base-ten multidigit configurations, representations of fractions and decimals as well as whole numbers, symbols for arithmetic and algebraic operations, equality and inequality, and so on. Cartesian graphs are formed according to general rules for associating points with pairs of numbers using perpendicular axes. Thus it is essential to consider the notion of a *representational system*, without which particular configurations are scarcely meaningful.

The building blocks of a representational system are its primitive *characters* or *signs*—terms that I shall use interchangeably—embodied in some fashion. At this level we want to be able to speak of these components without ascribing to them any further interpretation, in particular, without yet making a commitment to any representing relationship. Characters may be drawn from a fairly well defined set, such as letters of an alphabet, numerals and arithmetic signs, or standard English words, or they can be drawn from a more ambiguously defined universe, such as real-life objects and their attributes. Characters can even be abstract mathematical entities, such as numbers, vectors, and matrices, or physical entities, such as masses, velocities, accelerations, and forces.

A representational system further incorporates rules and practices, often ambiguous ones, for combining the signs into permitted *configurations*. Thus single-digit numerals and commas may be combined according to place-value rules to yield multidigit numerals, and letters, numerals, and arithmetic signs as characters may be combined in certain sequences to form algebraic formulas and equations. Other sequences do not form valid expressions.

The typical representational system also has higher *structure*. Here the possibilities include configurations of configurations; syntactic structures (e.g., as in natural languages); such relations as partial or total orderings on the family of configurations; mathematical operations of various kinds; rules for valuing configurations (e.g., truth values in symbolic logic); and, very important, ways to move from one permitted configuration or set of configurations in the system to another. In ordinary algebra, one has symbol-manipulation rules for obtaining new formulas from those already obtained, for transforming and solving equations, and so forth.

One way of giving *meaning* to the signs and configurations of a given representational system is through their relationship with one another, established by higher-level structures in the system. This way is a purely *syntactic* or *structural* interpretation of meaning, familiar in the context of formal logic and axiomatic foundations of mathematics. The more common, everyday notion of meaning derives from the *semantic* interpretation, referring to what the configuration denotes or signifies—often a configuration in a different system of representation but sometimes an external concrete object, attribute, or relation. When we think of a vector as an element of a vector space obeying certain standard axioms, our meanings are structural; when we think of a vector as denoting a velocity or a force, we refer to its semantic interpretation.

Note that the definition of a representational system is not limited per se to situations in which intelligent beings have invented and produced the representing or represented configurations, as with mathematical representations. For instance, sequences of base pairs in DNA encode biochemically (i.e., represent in a certain way, through subsequent productions) the amino acid sequences in proteins and, ultimately, the physical structures of living creatures.

Internal and External Systems

It is essential to consider and distinguish between an individual person's *internal* psychological systems of representation and systems that are *external* to that person This distinction permits exploration of the interactions between the systems.

At all grade levels, most of the stated objectives in the Representation Standard (NCTM, 2001a) focus explicitly on students' production and use of external representations and only implicitly on their internal representations. This focus is in a way reasonable because no one can directly observe another's internal representations. Rather, one continuously makes inferences about the internal states of others, based on their production of, or interaction with, representations external to them. Highly capable teachers thus form their own internal models of their students' mathematical conceptions and misconceptions on the basis of students' statements, questions, responses, and representational productions. Teachers often form these models tacitly and, at least partly, unconsciously.

Internal representations, of course, must also involve configurations built up from some set of components. I still call these components *signs*, although they are internal to the individual. These may be sensations and perceptions, visualized or otherwise imagined objects or symbols, or even emotional feelings (cf. Rogers, 1983; Zajonc, 1980). Internal systems include the natural language of individuals; their visual imagery and spatial, tactile, and kinesthetic representation; their problem-solving heuristics and strategies; their personal capabilities, including conceptions and misconceptions, in relation to conventional mathematical notations and configurations; their personal symbolization constructs and assignments of meaning to all these; and their affect (especially in relation to mathematics), as discussed below.

External representation systems include normative natural languages (e.g., "standard" English); conventional graphical, diagrammatic, and formal notational systems of mathematics; structured learning environments that may include concrete manipulative materials or computer-based microworlds; and sociocultural structures, such as those of kinship, economic relationships, political hierarchies, or school systems. The (external) formal symbol-systems of mathematical representation are structured by their underlying assumptions and conventions, which are in a strictly logical sense arbitrary, and in a practical and historical sense are the results of a complex social process. But once such a system is established, the patterns occurring in it are no longer arbitrary. They are "in the mathematics," present to be discovered by the individual.

To describe representational systems, such as our systems of numeration, of algebraic notation, or of Cartesian graphing, as being external to a particular individual is, of course, to idealize and to some degree standardize their usage across the mathematically engaged population.

Sometimes it is most valuable to regard the internal as representing the external, as when a student forms a mental image of a mathematical relationship described by the teacher, and one wants to ask if the relationship has been effectively or correctly visualized. At other times, one regards the external as representing the internal, as when the student writes a formula, draws a diagram, or seeks to communicate in words his or her mathematical ideas. This flexibility again illustrates the two-way nature of the representing relationship.

But internal representations do not only encode or represent what is external—they can also represent one another other in various ways. A student's internal, conceptual representations of the expression πr^2 may include an image of the formal notation itself; its pronunciation, pi-*r*-squared; and such related, interpretive verbal phrases as "the area of a circle" or "pi times the square of the circle's radius." In addition, the student may see visual-spatial images, such as that of a circle with its area shaded and a radius entered and labeled. Perhaps the student knows a definition of π as the circumference of the circle divided by the diameter, with accompanying visualization and kinesthetic encoding of ways to measure these lengths. The expression may evoke images of numerical or calculator-based procedures for computing the area given the radius or the radius given the area; some approximate values of π; a schema for ratio and proportion; concept images for taking the square of a quantity; and many other complex configurations. We may usefully say that the original algebraic expression, the verbal expressions, the visual images, the kinesthetically encoded procedures, and the internally accessed computational methods are internal configurations from different representational systems in the individual student that can and do internally evoke, suggest, and give meaning to—that is, represent—one another.

The creation of efficient, standard systems of representation is an important theme in the history of mathematics, and teaching students to use these systems has been central to school mathematics. But as noted in the second introductory quotation, teaching standard systems as ends in themselves may fail to develop students' mathematical power. Where, then, does the mathematical power of the individual reside? One place to look is in the translation processes among representational modes, both external and internal (Goldin, 1987; Lesh, 1981; Lesh, Landau, & Hamilton, 1983).

Internal representational systems are not carbon copies of external systems. Rather, they are complex and dynamic, involving multiple encodings and diverse processes (cf. Dreyfus, 1991). Individual power in a natural language, such as English, includes, of course, competence in standard usage. But it also includes the ability to speak and understand the language in new situations; to reason in the language; to draw analogies and similes; to create and understand metaphors; to interact meaningfully with others; and to visualize, comprehend, and express deep relationships in the language, including emotive aspects. Likewise, mathematical power requires competence in standard representations and their manipulation. But it also includes the ability to recognize and visualize structural relationships; to think spatially; to generalize and particularize; to formulate problem-solving strategies; to employ a variety of heuristic techniques and creative methods; and to experience such feelings as curiosity, bewilderment, frustration, purposefulness, elation, and satisfaction as appropriate. Mathematical concepts are learned powerfully when a variety of appropriate internal representations, with appropriate relationships among them, have been developed. An overarching goal of school mathematics, then, must be to develop in students powerful internal representational systems of different kinds and to learn how to infer these systems from their observable, external manifestations.

Thus we are led to some essential, general questions behind the research: What makes internal representational systems powerful? How can we infer or measure students' internal representational capabilities? How can we make the best use of structured, external mathematical representations in fostering the development of strong internal ones? What are the anticipated obstacles to this development?

Some Types and Characteristics of Representational Systems

Any characterization of a system of representation involves some arbitrariness. Decisions as to where one system leaves off and another begins, or whether to consider a complex system as being just one system or a combination of more elementary systems, are matters of convention and convenience of description. This observation means that the distinction between syntactic and semantic notions of meaning is also fundamentally a matter of convention.

My efforts to arrive at a unified description of mathematical problem-solving competency structures and how they develop led to a model, described at greater length elsewhere, incorporating five kinds of systems of internal representation. All five kinds seem to me to be essential for any reasonably parsimonious, psychologically adequate description of these structures (Goldin, 1987, 1992, 1998, 2000a; Goldin & Kaput, 1996, and references therein; see also Davis, 1984; Janvier, 1987a; Johnson, 1987; Kempson, 1988; Kieren & Pirie, 1991; Kosslyn, 1980; Lakoff, 1987; Paivio, 1983; Pylyshyn, 1973; Rogers, 1983; Thompson, 1996). These types of internal representation are (1) *verbal/syntactic systems* of natural language; (2) *imagistic systems* of visual-spatial, auditory-rhythmic, and tactile-kinesthetic encoding; (3) *formal notational systems* of mathematics, as internalized; (4) a *system of planning, monitoring, and executive control* that guides problem solving; and (5) an *affective system* that includes not only general, global affect but changing states of feeling, or emotion, during problem solving. As discussed, the relations of symbolization between different internal configurations within and among these systems, as well as those between internal and external representations, encode mathematical "meanings" or "understandings."

My inclusion of *affect* as an internal representational system, encoding information essential to human understanding, is a point of view that relatively few others have taken. Traditionally, emotions have been seen as interacting with cognition but not as having a powerful representational capability in their own right. However, Zajonc (1980), Rogers (1983), and others have argued for the idea of an "emotional code," suggesting that something crucial is thus "encoded." Recent related ideas include those of Picard (1997) and Damasio (1999).

External representational configurations are instantiated in media, and their representational characteristics are influenced by the characteristics of these media. Three distinctions describing features of media important to the interaction of external mathematical representations with internal representations are (1) dynamic versus static media, (2) interactive versus inert media, and (3) recording versus nonrecording media (Goldin & Kaput, 1996). Traditional external representations in mathematics are static, involving at any time fixed diagrams, graphs, equations, or configurations of such concrete objects as base-ten blocks, that the individual can look at, change, or augment. Graphing calculators, computer-based microworlds, and Internet–based interactions, however, now give us external systems in which linked representations change dynamically. Inert media such as textbooks or television sets do not respond to the individuals' actions, whereas interactive media such as calculators do. Recording media such as pencil and paper keep accessible a record of what has happened; nonrecording media such as typical hand-held calculators do not.

Stages in the Development of Systems of Representation in Mathematics

The model above is itself dynamic, incorporating three main stages in the development of representational systems in learners: (1) an *inventive/semiotic* stage, in which internal configurations are first assigned meaning with reference to previously established representations (Piaget, 1969); (2) a period of *structural development*, driven by the meanings first assigned, in which relationships within the newer system are constructed on the template of the prior one; and (3) an *autonomous* stage, in which the representational system detaches from the earlier one, functioning flexibly with new meanings and in new contexts. Such stages can also describe the historical development of mathematical ideas.

When we consider these stages in detail, we can discover possible sources of cognitive obstacles, or epistemological obstacles, occurring in students' conceptual development (e.g., Goldin & Herscovics, 1991; Goldin & Shteingold, 2001). In the semiotic stage, after a new character or configuration has been introduced, its initial meaning often serves for a long time psychologically as the "real meaning," thus the outcome of counting concrete objects can function as what a number "really is" for a child, or the operation of repeated multiplication can function as what exponentiation "really is." In each case, the new representation is modeled on a preexisting system. Many structural properties are then developed using the earlier system as the sole interpretation (e.g., seeing addition as joining sets of objects or obtaining laws of exponents from the repeated multiplication interpretation). As long as this second stage continues without other significations, the notion of the *real* meaning is confirmed. Essential mathematical aspects, however, are not supported by these representing relationships: zero, negative numbers, fractions, and complex numbers cannot be understood by counting objects, and zero and negative exponents, fractional exponents, and complex exponents cannot be understood through repeated multiplication.

Ultimately the new systems increase in power and become autonomous only when new interpretations in other representational systems are possible. These interpretations pose cognitive obstacles because they require abandonment of the initial, semiotic connections on which the system was constructed. At first, such extensions of meaning may seem metaphorical rather than literal—until new semiotic acts have been performed to provide new meanings.

Ambiguity as Fundamental

In mathematics, *ambiguity* of any kind is usually regarded as undesirable. But in describing the role of representation, both internal and external, in learning mathematics, ambiguity is inescapable. Rather than seek to eliminate ambiguity, it is useful to study its psychological functions and consequences. One finds that ambiguity is essential to the power of many representational systems, including natural language.

Ambiguities may occur in the family of signs or characters possible in a system, in the conditions pertaining to configurations in the system, in higher-level structures within a system, or in the symbolic relationship between two systems. Most representational systems are only partially characterized by means of syntactic and semantic rules, and attempts to compile exceptions become unnatural and unmanageable. The resulting ambiguities are, however, resolved in practice via the context (i.e., with reference to configurations outside the original system or systems). This possibility of the contextual resolution of ambiguity gives a representational system great flexibility and power of applicability in diverse situations.

Some Research Directions

In 1998, Claude Janvier and I were guest editors of two special issues of the *Journal of Mathematical Behavior* that reflected several years' activity of the Working Group on Representations of the International Group for the Psychology of Mathematics Education (PME). In this section, I mention some of the research reported by the working group and some developing areas of work. Contributions to the present book by Stephen Monk (chapter 17) and Stephen Smith (chapter 18), respectively, extend this discussion in two directions: (1) learning of and by graphical representation and (2) children's problem representations.

Research on representation in specific conceptual domains

For any particular mathematical concept or conceptual scheme, especially those involving standard, conventional representations, one priority is to study students' external and internal representations, how these representations develop and interact with each other, how they can be strengthened, and what the most common cognitive or epistemological obstacles are. Each major domain of school mathematics—numbers, operations with numbers, and the base-ten system of numeration; fractions, decimals, and percents; algebra and algebraic representation; geometry and diagrammatic representation; and so on—has a certain body of results.

For example, the notion of a *function* in mathematics has been one important domain for the study of representation and of students' obstacles having to do with representation (Janvier, 1987b). Thompson (1996, p. 272) discusses (internal) "concept imagery" in explaining the acceptance by students of an erroneous (external) formal notational expression. Even (1998) considers factors involved in linking representations of functions; she is able to contrast, for example, pointwise versus global approaches in students' reasoning as inferred from the graphs that they produce. Hitt (1998), making use of structured questionnaires that present (external) graphical representations, reports on the difficulties of teachers and those of students in the articulation of different representations relating to functions. Janvier (1998) discusses the notion of *chronicle*—defined as a variable that changes implicitly with time—as an epistemological barrier to the function concept; he too draws his inferences from students' graphical representations.

Representation, Problem Solving, and Imagery

Another fertile field of study has been the construction and use of representations during problem solving, an area of mathematics education with a long and successful history (Dienes & Jeeves, 1965, 1970; Goldin & McClintock, 1984; Jeeves & Greer, 1983; Schoenfeld, 1985). The work in this field of study takes account of the structure and complexity of external problem state-space representations; the role in learning of problem isomorphisms and homomorphisms; internal, imagistic representation of problem situations; and other essential ideas.

Recently Greer and Harel (1998) again considered research on isomorphic problems and their representations, addressing isomorphism within representational acts and isomorphisms as teaching aids. The study of imagery in mathematical problem solving has burgeoned (English, 1997; Presmeg, 1998). Owens and Clements (1998) considered primary school students solving spatial problems in the classroom that were intended to evoke visual imagery. They described the effects of imagistic processing, noting instances of concrete imagery, dynamic imagery, pattern imagery, action imagery, and procedural imagery. Cifarelli (1998) described how representations are developed during problem solving by first-year calculus students, inferring mental, or internal, representations from the diagrams that the students drew and modified as they represented and interpreted problem situations.

Affect as Representation

The affective domain refers to feelings that pertain to mathematics, to the experiencing of mathematics, or to oneself in relation to mathematics (Gomez-Chacon, 2000; McLeod & Adams, 1989). I wish to highlight here the idea that affect serves a *representational function* in the individual and that as a representational system, it enhances or impedes mathematical understanding in certain ways. Local, changing states of feeling are not just experienced but *utilized* by problem solvers and learners to store information, to monitor, and to evoke heuristic processes. As this utilization takes place, well-traveled affective pathways are constructed that result in long-term, global structures (Goldin, 2000a).

For example, consider the common situation of a student who experiences frustration while trying to solve a mathematical problem. The feeling of frustration signifies something about the interaction. It may, for instance, encode the fact that a laborious strategy has been tried several times without noticeable progress. In a powerful problem solver, this encoding serves as a kind of internal evaluation mechanism, stimulating the student to step back from the problem and try a new strategy in place of the one that is not working. It may influence the student toward an approach more likely to provide an immediate result, such as solving a simpler, related problem. Further, the frustration may represent that, for this student, the problem is nonroutine. This realization signals the possibility of another emotion—the anticipatory joy of having solved a challenging problem.

The widespread belief that frustration is merely a negative emotion does not take account of such a complex representational role. And the impulse that some mathematics teachers may have to alleviate the frustration by intervening with immediate guidance to the student would in this example be counterproductive to the development of powerful affect.

But emotions are highly individual. A less mathematically powerful student may respond to what seems like a similar feeling of frustration with a sense of panic or a desire to give up. The student may ask the teacher for help or may simply cease to engage. Whereas the first student anticipates the satisfaction of success, the second student's frustration has come to represent the anticipated pain of failure. In the affective system of this second student, the emotion serves a warning function and possibly evokes such defense mechanisms as avoidance and denial. These examples, although idealized and simplified, suggest that powerful and constructive affective representation, including negative as well as positive feelings, is necessary for powerful mathematics.

Partly on the basis of data from task-based interviews in a longitudinal study of elementary school children, Valerie DeBellis and I (DeBellis, 1996, 1998; DeBellis & Goldin, 1997, 1999) have proposed a tetrahedral model that considers (1) emotional states, (2) attitudes, (3) beliefs and belief structures, and (4) values, ethics, and morals in connection with mathematics. Among the important constructs related to mathematical development are the following: *meta-affect*, which can include not only affect about cognition but also affect about affect, wholly transforming the affect; *affective structures*, built up as affect repeatedly interacts with cognition in mathematical contexts; *mathematical intimacy*, the involvement and vulnerable engagement of the individual with mathematics; and *mathematical integrity* or self-acknowledgment, involving an open awareness of mathematical understanding or the absence of understanding.

Whether or not one regards affect as fundamentally representational, it seems undeniable that excellence in mathematics education requires careful attention to the affective domain. In my view, the most serious omission in NCTM's (2001a) *Principles and Standards for School Mathematics* is a separate principle or standard having to do with affect. Every student should experience curiosity and its fulfillment, personal satisfaction in doing and understanding mathematics, occasional elation at breaking through in a difficult problem, effective ways of managing impasse and frustration during mathematical problem solving, and a sense of self that includes mathematical literacy and power.

Task-Based Interview Research Methodology

Empirical research on representation in mathematics education involves observations of individual students' interaction with external representations, or their production of such representations. Inferences about students' internal cognitive representation follow. For the research to be generalizable, and thus of use to the wider educational community, its goal must be to make such observations as scientifically reliable and reproducible as possible.

One tool for such research is the planned, structured *task-based interview*. In my view, the best such interviews incorporate 10 general principles, discussed in Goldin (2000b) in detail: (1) design interviews to address well-formulated research questions posed in advance, not just to "see what occurs"; (2) choose tasks that interviewees can represent meaningfully but that offer challenge; (3) choose tasks embodying rich and varied representational structures, including mathematics subject to abstract characterization, meaningful semantic relationships, and visual imagery; (4) describe the interviews explicitly, using scripts with explicit criteria for major contingencies; (5) encourage free problem solving to the maximum extent possible; (6) maximize interaction with the external learning environment, providing a variety of external representational possibilities; (7) decide in advance what will be recorded, and record as much as possible; (8) train the interviewers, and pilot-test each interview outside the main study; (9) explicitly build into the design an alertness to new or unforeseen possibilities; and (10) be prepared to compromise appropriately when conflicts occur among principles.

Classroom Activity Involving Representation

The classroom teacher, of course, has some different priorities from those of the researcher. Nevertheless, the principles above have their counterparts, or analogs, for classroom activity addressing representation. I would paraphrase them as follows: (1) design activities to address well-formulated mathematical learning goals, posed in advance, that include internal and external representational capabilities; (2) choose tasks that the students can represent meaningfully but that offer challenges; (3) choose tasks that embody rich and varied representational structures, including contextual mathematics, abstract mathematics, and visual imagery; (4) plan for major contingencies, making use of research that identifies cognitive obstacles or common representational difficulties; (5) encourage free problem solving by students, with guiding interventions; (6) maximize students' interaction with the learning environment, encouraging a variety of external representational modes, standard and nonstandard; (7) incorporate multiple, ongoing means of assessing students' learning through their representations; (8) develop a repertoire of proven, successful activities; (9) be alert to students' novel representations, strategies, and insights; and (10) balance these considerations with one another, compromising where appropriate and valuing highly your personal teaching style.

In addition, the best classroom activity offers opportunities for rich affective representation. Problems should sometimes be easy and straightforward, so that students come to feel powerful and confident. But sometimes problems should lead to impasse, evoking puzzlement, bewilderment, and frustration, yet offer the possibility of proceeding with renewed determination and achieving the elation of sudden insight or the satisfaction of performing a difficult feat.

Implications for Theory and Practice

To return to the theme of reconciling and unifying diverse theoretical perspectives, an obstacle to progress has been the dismissal by prevailing belief systems of important constructs from other systems, on a priori (but unscientific) grounds. These dismissals, when taken seriously, have had damaging consequences for educational practice in mathematics.

For example, positivists, including behaviorists and many neobehaviorists, reject on first principles the very idea of internal mental states or mental representations. Empirically observable responses to stimulus situations and, possibly, constructs built from internal responses are taken as the only admissible entities. This point of view is not very fashionable today. The resulting "behavioral objectives" approach to school mathematics, ascendant during the 1970s and the consequences of which still endure, tends to ignore higher-level or deeper conceptual understandings in favor of easily observable measures of discrete skills.

In contrast, radical constructivists, also on a priori grounds, exclude the possibility of knowledge about the real world, rejecting all that is external to individual human "worlds of experience." Apart from the cognitions of individual knowers, mathematical structures are likewise rejected. The idea of *viability*, an entirely subjective notion, is stressed, but objective *validity* is dismissed as objectivist or absolutist. Population studies and investigative methods based on controlled experimentation are de-emphasized. Radical social constructivists see mathematical truth merely as social consensus, dismissing any possible objective sense in which reasoning could be correct or incorrect. An extreme educational consequence has been the devaluing of correct answers or analyses in some mathematics classrooms.

At one time, some cognitive researchers claimed that all internal encoding should be represented propositionally, dismissing a priori the idea of internal imagistic representation. Some computer-oriented cognitive scientists have insisted that all thinking must be information processing. Not dissimilarly, some mathematical formalists consider "real mathematics" to consist exclusively of computational and inferential procedures, theorems and proofs, and their applications. When taken seriously, these ideas downplay or eliminate teaching with powerful problem-solving strategies that either are difficult to represent verbally, logically, or formally or are hard to simulate mechanically—that is, those strategies involving imagery and visualization, analogy and metaphor.

Currently, other cognitive theorists seem to believe that all thought—and, in particular, all mathematical thought—consists exclusively of metaphors of various sorts (Lakoff & Nunez, 2000). If this view is taken seriously, we are likely to see a further devaluation and discrediting of formal systems and abstract mathematics as classroom topics, again with unfortunate consequences. Furthermore, the prevailing cognitive theories of mathematics education have placed little emphasis on affect.

In my view, behaviorism, constructivism and social constructivism, various schools of cognitive science and information-processing theory, other schools of thought I have not discussed here—such as structuralism, socioculturalism, and social practice theory—and, of course, mathematics itself have all offered essential constructs necessary to mathematics education research. The problem is how to join these constructs into an integrated whole.

The theory of representational systems lets us do so. Ideas that pertain to mathematical structure and validity as independent of the individual student can be expressed easily in terms of external representational systems.

Complex behaviors and patterns of behavior can be observed as the individual's interaction with the structured, external representational environment or in the production of new external representations. Thus both the behaviorist concern for observability and the mathematical concern for formal structures apart from individual knowers are accommodated.

The various types of internal representational systems proposed, allowing for the presence of ambiguity, permit a major role for visual imagery, image schemata, and kinesthetic encoding and metaphor, as well as affect, without in any way downplaying verbal and propositional encoding, formal logical processes, or complex problem-solving heuristics. Furthermore, we can consider interactions among the internal systems and between the internal and the external, avoiding metaphysical dualism.

The idea that knowledge is constructed by learners over time rather than achieved through direct transcription is also expressible—by research-based description of the ways in which internal systems of representations are built up, developing through identifiable stages. Furthermore, the multiple encoding and redundancy provided by several different internal representational systems can account for the persistence of schemata in long-term memory, for individuals' ability to reconstruct temporarily forgotten mathematical concepts, and for the longevity of individuals' belief systems in relation to mathematics.

Sociocultural factors can be understood (historically) as influencing the developing structure of external representational systems, including normative systems of cultural expectations. They can also be understood (in the present) as affecting the interaction of the individual with his or her human environment. This effect occurs through contingencies in representational communications, especially those that are affective in nature. Thus one is not required to diminish the importance of individual cognitions, of socially shared cognitions and expectations, or of the sociocultural contexts in which individuals learn mathematics; one can attend instead to all of them. In short, one has the basis for an inclusive approach to research.

The ideas discussed here also have consequences for practice. For example, universality of access to high-level mathematical achievement is explicit in NCTM's (2001a) Equity Principle and is embodied in many state standards documents. However, few school districts have found and put into practice methods that actually achieve breakthroughs in achievement for large numbers of children. Many people, from parents to research mathematicians, remain skeptical that such visionary goals are possible; they continue to see powerful mathematics as the province of relatively few students with innate talent.

It is indisputable that some, but relatively few, children spontaneously solve difficult mathematical problems quickly and accurately and develop flexible imagistic, heuristic, and affective representational processes without having been taught them. However, a representational perspective suggests that the other children, the large majority, are *not inherently limited* in their ability to understand mathematical ideas, including advanced concepts of algebra and geometry. Rather, when one characterizes conceptual and problem-solving ability by means of systems of representation and their interactions, it becomes possible to provide all children with the experience base needed to develop powerful systems.

In formulating mathematical learning objectives or assessing individual students' mathematical development, it becomes important to strike a good balance between the standard manipulation of formal notational systems—to which most school mathematics and standardized testing is still devoted—and the development of other representational modes: imagistic thinking, involving visualization; visual imagery, pattern recognition, and analogical reasoning; heuristic planning, involving diverse problem-solving strategies; and affective representation. Bona fide representational power in the latter systems does not stand in opposition to formal proficiency but, rather, strengthens it.

ACKNOWLEDGMENTS

I would like to thank Rutgers University and the Alexander von Humboldt Foundation for research and travel support. I am grateful to Frank Lester for comments on an earlier version of the chapter.

REFERENCES

Cifarelli, V. (1998). The development of mental representations as a problem solving activity. *Journal of Mathematical Behavior*, *17*, 239–264.

Damasio, A. (1999). *The feeling of what happens: Body and emotion in the making of consciousness.* New York: Harcourt Brace.

Davis, R. B. (1984). *Learning mathematics: The cognitive science approach to mathematics education.* Norwood, NJ: Ablex.

DeBellis, V. A. (1996). Interactions between affect and cognition during mathematical problem-solving: A two-year case study of four elementary school children. (Doctoral dissertation, Rutgers University, 1996). *Dissertation Abstracts International*, *57*(07), 2922A. (UMI No. 96-30716)

DeBellis, V. A. (1998). Mathematical intimacy: Local affect in powerful problem solvers. In S. Berenson, K. Dawkins, M. Blanton, W. Coulombe, J. Kolb, K. Norwood, & L. Stiff (Eds.), *Proceedings of the 20th Annual Meeting of PME-NA* (Vol. 2, pp. 435–440). Columbus, OH: ERIC Clearinghouse for Science, Mathematics, and Environmental Education.

DeBellis, V. A., & Goldin, G. A. (1997). The affective domain in mathematical problem solving. In E. Pehkonen (Ed.), *Proceedings of the 21st International Conference for the Psychology of Mathematics Education* (Vol. 2, pp. 209–216). Lahti, Finland: University of Helsinki, Lahti Research and Training Centre.

DeBellis, V. A., & Goldin, G. A. (1999). Aspects of affect: Mathematical intimacy, mathematical integrity. In O. Zaslavsky (Ed.), *Proceedings of the 23rd Conference of the International Group for the Psychology of Mathematics Education* (Vol. 2, pp. 249–256). Haifa, Israel: Technion Printing Center.

Dienes, Z. P., & Jeeves, M. A. (1965). *Thinking in structures.* London: Hutchinson Educational.

Dienes, Z. P., & Jeeves, M. A. (1970). *The effects of structural relations on transfer*. London: Hutchinson Educational.

Dreyfus, T. (1991). Advanced mathematical thinking processes. In Tall, D. (Ed.), *Advanced mathematical thinking* (pp. 25–41). Dordrecht, The Netherlands: Kluwer.

English, L. (Ed.). (1997). *Mathematical reasoning: Analogies, metaphors, and images.* Mahwah, NJ: Erlbaum.

Even, R. (1998). Factors involved in linking representations of functions. *Journal of Mathematical Behavior, 17*, 105–121.

Goldin, G. A. (1987). Cognitive representational systems for mathematical problem solving. In C. Janvier (Ed.), *Problems of representation in the teaching and learning of mathematics* (pp. 125–145). Hillsdale, NJ: Erlbaum.

Goldin, G. A. (1992). On developing a unified model for the psychology of mathematical learning and problem solving. In W. Geeslin & K. Graham (Eds.), *Proceedings of the 16th International Conference for the Psychology of Mathematics Education* (Vol. 3, pp. 235–261). Durham: University of New Hampshire, Department of Mathematics.

Goldin, G. A. (1998). Representational systems, learning, and problem solving in mathematics. *Journal of Mathematical Behavior, 17*, 137–165.

Goldin, G. A. (2000a). Affective pathways and representation in mathematical problem solving. *Mathematical Thinking and Learning, 2*, 209–219.

Goldin, G. A. (2000b). A scientific perspective on structured, task-based interviews in mathematics education research. In A. E. Kelly & R. A. Lesh (Eds.), *Handbook of research design in mathematics and science education* (pp. 517–545). Mahwah, NJ: Erlbaum.

Goldin, G. A., & Herscovics, N. (1991). Toward a conceptual-representational analysis of the exponential function. In F. Furinghetti (Ed.), *Proceedings of the Fifteenth International Conference for the Psychology of Mathematics Education* (Vol. 2, pp. 64-71). Genoa, Italy: University of Genoa, Department of Mathematics.

Goldin, G. A., & Kaput, J. J. (1996). A joint perspective on the idea of representation in learning and doing mathematics. In L. P. Steffe, P. Nesher, P. Cobb, G. A. Goldin, & B. Greer (Eds.), *Theories of mathematical learning* (pp. 397–430). Hillsdale, NJ: Erlbaum.

Goldin, G. A., & McClintock, C. E. (Eds.). (1984). *Task variables in mathematical problem solving.* Philadelphia: Franklin Institute Press (acquired by Erlbaum in Hillsdale, NJ).

Goldin, G. A., & Shteingold, N. (2001). Systems of representations and the development of mathematical concepts. In A. Cuoco (Ed.), *The roles of representation in school mathematics* (2001 Yearbook of the National Council of Teachers of Mathematics, pp. 1–23). Reston, VA: NCTM.

Gomez-Chacon, I. M. (2000). *Matemática emocional: Los efectos en el aprendizaje matemático* [Emotional mathematics: The effects in learning mathematics]. Madrid: Narcea, S. A. de Ediciones.

Greer, B. & Harel, G. (1998). The role of isomorphisms in mathematical cognition. *Journal of Mathematical Behavior, 17*, 5–24.

Hitt, F. (1998). Difficulties in the articulation of different representations linked to the concept of function. *Journal of Mathematical Behavior, 17*, 123–134.

Janvier, C. (Ed.). (1987a). *Problems of representation in the teaching and learning of mathematics.* Hillsdale, NJ: Erlbaum.

Janvier, C. (1987b). Representation and understanding: The notion of function as an example. In C. Janvier (Ed.), *Problems of representation in the teaching and learning of mathematics* (pp. 67–71). Hillsdale, NJ: Erlbaum.

Janvier, C. (1998). The notion of chronicle as an epistemological obstacle to the concept of function. *Journal of Mathematical Behavior, 17*(1), 79–103.

Jeeves, M. A., & Greer, B. (1983). *Analysis of structural learning.* London: Academic Press.

Johnson, M. (1987). *The body in the mind.* Chicago: University of Chicago Press.

Kaput, J. J. (1985). Representation and problem solving: Methodological issues related to modeling. In E. A. Silver (Ed.), *Teaching and learning mathematical problem solving: multiple research perspectives* (pp. 381–398). Hillsdale, NJ: Erlbaum.

Kaput, J. J. (1998). Representations, inscriptions, descriptions and learning: A kaleidoscope of windows. *Journal of Mathematical Behavior, 17*, 265–281.

Kempson, R. M. (Ed.). (1988). *Mental representations: The interface between language and reality.* Cambridge,: Cambridge University Press.

Kieren, T. E., & Pirie, S. (1991). Recursion and the mathematical experience. In L. P. Steffe (Ed.), *Epistemological foundations of mathematical experience* (pp. 78–101). New York: Springer-Verlag.

Kosslyn, S. M. (1980). *Image and mind.* Cambridge, MA: Harvard University Press.

Lakoff, G. (1987). *Women, fire, and dangerous things.* Chicago: University of Chicago Press.

Lakoff, G., & Núñez, R. E. (2000). *Where mathematics comes from: How the embodied mind brings mathematics into being.* New York: Basic Books.

Lesh, R. (1981). Applied mathematical problem solving. *Educational Studies in Mathematics 12*, 235–264.

Lesh, R., Landau, M., & Hamilton, E. (1983). Conceptual models and applied problem solving research. In R. Lesh & M. Landau (Eds.), *Acquisition of mathematics concepts and processes* (pp. 263–343). New York: Academic Press.

McLeod, D. B., & Adams, V. M. (Eds.). (1989). *Affect and mathematical problem solving: A new perspective.* New York: Springer-Verlag.

National Council of Teachers of Mathematics. (2000a). *Principles and standards for school mathematics.* Reston, VA: NCTM.

National Council of Teachers of Mathematics. (2000b). *Principles and standards for school mathematics: An overview.* Reston, VA: NCTM.

Owens, K. D., & Clements, M. A. (1998). Representations used in spatial problem solving in the classroom. *Journal of Mathematical Behavior, 17*(2), 197–218.

Palmer, S. E. (1978). Fundamental aspects of cognitive representation. In E. Rosch & B. Lloyd (Eds.), *Cognition and categorization.* Hillsdale, NJ: Erlbaum.

Paivio, A. (1983). The empirical case for dual coding. In J. Yuille (Ed.), *Imagery, memory, and cognition: Essays in honor of Allan Paivio* (pp. 307–322). Hillsdale, NJ: Erlbaum.

Piaget, J. (1969). *Science of education and the psychology of the child.* New York: Viking Press.

Picard, R. W. (1997). *Affective computing.* Cambridge, MA: MIT Press.

Presmeg, N. (1998). Metaphoric and metonymic signification in mathematics. *Journal of Mathematical Behavior, 17*, 25–32.

Pylyshyn, Z. (1973). What the mind's eye tells the mind's brain: A critique of mental imagery. *Psychological Bulletin, 80*, 1–24.

Rogers, T. B. (1983). Emotion, imagery, and verbal codes: A closer look at an increasingly complex interaction. In J. Yuille (Ed.), *Imagery, memory, and cognition: Essays in honor of Allan Paivio* (pp. 285–305). Hillsdale, NJ: Erlbaum.

Schoenfeld, A. H. (1985). *Mathematical problem solving.* New York: Academic Press.

Thompson, P. W. (1996). Imagery and the development of mathematical reasoning. In L. P. Steffe, P. Nesher, P. Cobb, G. A. Goldin, & B. Greer (Eds.), *Theories of mathematical learning* (pp. 267–283). Hillsdale, NJ: Erlbaum.

Vergnaud, G. (1998). A comprehensive theory of representation for mathematics education. *Journal of Mathematical Behavior, 17*, 167–181.

Zajonc, R. B. (1980). Feeling and thinking: Preferences need no inferences. *American Psychologist, 35*, 151–175.

SECTION 2

Perspectives on Teaching and Learning

CHAPTER 20

Implications of Cognitive Science Research for Mathematics Education

Robert S. Siegler, Carnegie Mellon University

Over the past 20 years, a great deal of cognitive science research has focused on mathematics learning. The majority of this research has examined such basic capabilities as counting, understanding numerical magnitudes, doing arithmetic—both word problems and purely numerical problems—and learning prealgebra concepts. Another, somewhat smaller, body of research has been devoted to students' understanding of, and learning about, algebra, geometry, and computer programming. This research now allows relatively firm conclusions to be drawn about a number of aspects of mathematics learning relevant to the Standards of the National Council of Teachers of Mathematics (NCTM, 1989, 1991, 1995). This chapter focuses on eight areas in which such conclusions can be drawn:

1. Pitfalls in mathematics learning
2. Cognitive variability and strategy choice
3. Individual differences
4. Discovery and insight
5. Relations between conceptual and procedural knowledge
6. Cooperative learning
7. Promoting analytic thinking and transfer

The chapter examines conclusions based on cognitive science research on each of these topics. This research informs us about how children typically learn particular skills and concepts, the stumbling blocks that many of them encounter, and instructional practices that can produce greater learning. The examples that are discussed focus on the acquisition of particular mathematical procedures and concepts rather than on broad philosophical issues about the nature of children as learners or about what mathematics should be taught. These latter issues are largely outside the purview of cognitive science approaches, or indeed any empirically based approach to mathematics learning. Further, one of the main lessons of cognitive science research is that successful teaching and learning depend on careful, detailed analysis of the particulars of individual children learning specific skills and concepts. Therefore, this chapter deliberately steers clear of statements about mathematics learning in general and instead focuses on findings concerning how children learn certain key ideas and procedures. Most of those findings concern mathematics learning by elementary school children.

Mathematical Understanding Before Children Enter School

Children's learning of mathematics in school contexts builds on a substantial base of understandings that they acquire before they begin their formal education. Understanding what children already know when they enter school is crucial both for identifying what they still need to be taught and for identifying strengths on which further instruction can be based. Some of these acquisitions are universal; others depend on environments within which children develop. Differences among social and cultural groups in the degree to which they master this latter, more variable, group of skills are clearly related to subsequent differences in learning of mathematics at school. Also striking is the fact that from early in life, children possess a relatively abstract, as well as a concrete, understanding of numbers and mathematical concepts.

Cardinality

One fundamental underpinning of understanding mathematics, the concept of cardinality, appears to be universally present from the first months out of the womb. By four months of age, perhaps earlier, infants can discriminate one object from two, and two objects from three (Antell & Keating, 1983; Starkey, Spelke, & Gelman, 1990; van Loosbroek & Smitsman, 1990). The researchers made this observation by using the following *habituation paradigm*: Infants were shown a sequence of pictures, each of which contained a small set of objects, such as three circles. The sets differed from trial to trial in the objects' size, brightness, distance apart, and other properties, but they always had the same number of objects. Once the infants habituated to displays with this number of objects, they were shown a set that was comparable in other ways to the displays they had seen but that had a different number of objects. Their renewed looking attested to their having abstracted the number of objects in the previous sets.

These nascent understandings of cardinality also enabled infants to realize the consequences of adding and subtracting small numbers of objects (Simon, Hespos, & Rochat, 1995; Wynn, 1992). In one task, 5-month-olds saw one or two objects, saw a screen come down in front of them, saw a hand place another object behind the screen, and then saw the screen rise. Sometimes the result was what would be expected by adding the one new object to the one or two that were already behind the screen; at other times, through trickery, it was not. The infants looked for a longer time when the number of objects was not what it should have been, thus suggesting that they expected the correct number of objects to be present.

Not until three or four years later, however, do children discriminate among even slightly larger numbers of objects, such as four objects versus five or six (Starkey & Cooper, 1980; Strauss & Curtis, 1984). This limitation suggests that the competencies that develop in infancy are produced by *subitizing*, a quick and effortless process of recognition that people can apply only to sets of one to three or four objects. In other words, when we see a row of from one to four objects, we feel as though we immediately know how many there are; in contrast, with larger numbers of objects, we usually feel less sure and often need to count. Adults and 5-year-olds are similar to infants in being able to very rapidly identify the cardinal value of one to three or four objects, but not of larger sets, through subitizing (Chi & Klahr, 1975).

At 3 or 4 years of age, children become proficient in another means of establishing the cardinal value of a set—*counting*. This proficiency allows them to assign numbers to larger sets than can be subitized. Gelman and Gallistel (1978) noted the rapidity with which children learn to count and identified a set of *counting principles* on which this rapid learning seems to be based. The fact that preschoolers possess such principles is particularly important because it indicates that from early in learning, children's understanding of mathematics includes abstract knowledge, as well as established procedures and factual information. Equally important, from the beginning, the abstract knowledge influences the child's learning and execution of procedures. Gelman and Gallistel identified the following five counting principles:

1. The *one-one principle*: One and only one number word is assigned to each object.
2. The *stable order principle*: The numbers are always assigned in the same order.
3. The *cardinal principle*: The last count indicates the number of objects in the set.
4. The *order irrelevance principle*: The order in which objects are counted is irrelevant.
5. The *abstraction principle*: The other principles apply to any set of objects.

Several types of evidence indicate that children understand all these principles by age 5, and some of them by age 3 (Gelman & Gallistel, 1978). Even when children err in their counting, they show knowledge of the one-one principle because they assign exactly one number word to most of the objects. For instance, they might count all but one object once, either skipping or counting twice the single miscounted object. These errors seem to be errors of execution rather than of misguided intent. Children demonstrate knowledge of the stable order principle by almost always saying the number words in a constant order. Usually this order is the conventional one, but occasionally it is an idiosyncratic order, such as "1, 3, 6, ..." The important phenomenon is that even when children use an idiosyncratic order, they use the same idiosyncratic order on each count. Preschoolers demonstrate knowledge of the cardinal principle by saying the last number with special emphasis. They show understanding of the abstraction principle by not hesitating to count sets that include different types of objects. Finally, the order-irrelevance principle seems to be the most difficult, but even for it, 5-year-olds demonstrate understanding. Many of them recognize that counting can start in the middle of a row of objects, as long as each object is eventually counted. Although few children can state the principles, their counting suggests that they know them.

Ordinality

Mastery of ordinal properties of numbers, like mastery of cardinal properties, also begins in infancy but seems to begin a little later, between 12 and 18 months.

The most basic ordinal concepts are *more* and *less*. To test when infants understand these concepts as they apply to numbers, Strauss and Curtis (1984) found that 16- to 18-month-olds have a rudimentary understanding of these concepts. Babies who had been reinforced for reaching for a square with two dots rather than one, and then for a square with three dots rather than two, subsequently selected a square with four dots rather than three, thus indicating understanding of the ordinal property "more numerous." They also succeeded when choosing the square with fewer dots was reinforced.

As with cardinality, the ability to extend these early understandings of ordinality beyond sets with a few objects takes a number of years to develop. The task most often used to examine later understandings of ordinality involves asking such questions as "Which is more: six oranges or four oranges?" Not until they are 4 or 5 years old can children from middle-class backgrounds consistently solve these ordinality problems correctly for the numbers from 1 to 9 (Siegler & Robinson, 1982). The greatest difficulty in choosing which is more occurs with numbers that are relatively large and close together (e. g., 7 and 8). Counting skills may be important in developing this ordinal knowledge as well as in doing arithmetic computations; the number that occurs later in the counting string is always the larger number, and it is easier to remember which number comes later when the numbers are far apart in the counting string.

Although most children from middle-income households know the relative magnitudes of all the single-digit numbers when they enter school, comparable understanding does not exist among children from low-income backgrounds, at least in the United States (Griffin, Case, & Siegler, 1994). Such children often have little or no sense of the relative magnitudes of single-digit numbers when they enter first grade. This lack of understanding causes particular difficulty for such children in understanding the basis of simple arithmetic operations, and that difficulty is likely related to their slow learning of the basic arithmetic facts (Jordan, Huttenlocher, & Levine, 1992). Relatively poor counting skills also appear to contribute to these children's difficulty in learning both numerical magnitudes and arithmetic.

A similar point is relevant for interpreting differences in scores on standardized achievement tests between children in the United States and those in other countries. The relatively poor performance on these international comparisons is often attributed to formal instruction's being inferior in the United States. However, substantial differences between arithmetic knowledge of children in the United States and East Asia exist before children in either country receive formal instruction in arithmetic (Geary, Fan, Bow-Thomas, & Siegler, 1993). This finding does not mean that mathematics education in U.S. schools is as effective as that in East Asian schools, but it does demonstrate that differences in mathematics achievement in different countries reflect cultural differences that influence mathematics learning outside the classroom as well as inside it.

Pitfalls in Mathematics Learning

As noted in the previous section, from the preschool years onward, children learn abstract mathematical concepts and principles in addition to procedures and facts. Fairly often, however, they either fail to grasp the concepts and principles that underlie procedures or grasp relevant concepts and principles but cannot connect them with the procedures. Either way, children who lack such understanding frequently generate flawed procedures that generate systematic patterns of errors. Depending on how one looks at it, these systematic errors can be seen as either a problem or an opportunity. On the one hand, the errors are a problem in that they indicate that children do not know what teachers have tried to teach them. On the other hand, the errors are an opportunity, in that their systematic quality points to the source of the problem and thus indicates the specific misunderstanding that needs to be overcome. Examples can be found in many areas of mathematics learning. I examine here three areas with particularly prominent systematic misconceptions: (a) the long subtraction algorithm, (b) arithmetic and magnitude comparison involving fractions, and (c) algebraic equations.

Buggy Subtraction Algorithms

Brown and Burton (1978) investigated students' acquisition of the multidigit subtraction algorithm. They used an error analysis method that involved presenting problems on which incorrect rules, or "bugs" (e.g., failing to borrow when subtracting a larger digit from a smaller one), would lead to specific errors and then examining an individual child's pattern of correct answers and errors to see whether they fit the pattern that would be produced by a buggy rule.

Many children's errors reflected such bugs. Consider the pattern in Figure 20.1. At first glance, it is difficult to draw any conclusion about this boy's performance, except that he is not very good at subtraction. With closer analysis, however, his performance becomes understandable. All three of his errors arose on problems where the *minuend*, or the top number, included a zero, suggesting that his difficulty was due to failing to understand how to subtract from zero.

307	856	606	308	835
−182	−699	−568	−287	−217
285	157	168	181	618

Fig. 20.1. Example of a substraction bug

An analysis of the problems on which the boy erred—the first, third, and fourth problems from the left—and the answers that he gave suggests the existence of two bugs that would produce these particular answers. Whenever a problem required subtraction from 0, he simply reversed the two numbers in the column with the 0. For example, in the problem 307 – 182 = __, he treated 0 – 8 as 8 – 0 and wrote "8" as the answer. The boy's second bug involved not decrementing the number to the left of the zero, in this case, not reducing the 3 to 2 in 307 – 182 = __. This lack of decrementing is not surprising, because, as indicated in the first bug, the boy did not "borrow," or regroup, anything from this column. Thus, the three wrong answers, as well as the two right ones, can be explained by assuming a basically correct subtraction procedure with two particular bugs.

Although such bugs are common among U.S. children, they are far less common among Koreans (Fuson & Kwon, 1992). A major reason appears to be that Korean children have a firmer grasp of the base-ten system and its relation to borrowing. When children have such understanding, their borrowing is more likely to maintain the value of the original number.

Fractions

When presented the problem 1/2 + 1/3 = __, many children answer 2/5. They generate such answers by adding the two numerators to form the sum's numerator and by adding the two denominators to form its denominator. The misunderstanding is far from transitory. Many adults enrolled in community college mathematics courses make the same mistake (Silver, 1983).

Much of children's difficulty in fractional arithmetic arises from their not thinking of the magnitude represented by each fraction. That omission is evident in children's errors in estimating the answer to 12/13 + 7/8 = __ (Table 20.1). On a national achievement test, fewer than one third of U. S. 13- and 17-year-olds accurately estimated the answer to this simple problem (Carpenter, Corbitt, Kepner, Lindquist, & Reys, 1981). Yet how could adding two numbers that were each close to 1 result in a sum of 1, 19, or 21?

Table 20.1. Performance on Item Involving Estimating the Sum of Two Fractions (Carpenter et al., 1981, p. 36)

Item: Estimate the answer to 12/13 + 7/8. You will not have time to solve the problem using paper and pencil.

	Percentage Choosing Answer	
Answer	Age 13	Age 17
1	7	8
2	24	37
19	28	21
21	27	15
I don't know	14	16

A similar misunderstanding of the relation of symbols to magnitudes is evident in children's attempts to deal with decimal fractions. Consider how they judge the relative size of two numbers such as 2.86 and 2.357. The most common approach of fourth and fifth graders on such problems is to say that the larger number is the one with more digits to the right of the decimal point (Resnick et al., 1989). Thus, they would judge 2.357 to be larger than 2.86. Such choices appear to be based on an analogy between decimal fractions and whole numbers. Because a whole number with more digits is always larger than one with fewer digits, some children assume that the same is true of decimal fractions.

Another group of children made the opposite responses. They consistently judged that the larger number was the one that had fewer digits to the right of the decimal point. Thus, they thought of 2.43 as being larger than 2.897. Many of these children reasoned that .897 involves thousandths, .43 involves hundredths, and hundredths are bigger than thousandths, so .43 must be bigger than .897.

The difficulty in understanding decimal fractions does not quickly disappear. Zuker (cited in Resnick et al., 1989) found that one third of a sample of seventh and ninth graders continued to make one of the two errors described above. Thus, with decimal fractions as with long subtraction, children's failure to understand the number system, or to link that understanding with specific procedures, leads to systematic and persistent errors.

Algebraic Equations

Systematic errors are also evident in children's efforts to represent concrete situations in algebraic equations. Even students who do well in algebra classes often do so by treating the equations as exercises in symbol manipulation, lacking any connection with real-world contexts.

This superficial understanding creates a situation in which mistakes often arise. Many such mistakes arise from incorrect extensions of correct rules (Matz, 1982; Sleeman, 1985). For example, because the distributive principle indicates that

$$a \times (b + c) = (a \times b) + (a \times c),$$

some students draw such superficially similar conclusions as

$$a + (b \times c) = (a + b) \times (a + c).$$

Students use a variety of procedures to determine whether transformations of algebraic equations are appropriate. Among 11- to 14-year-olds, the most frequent strategy is to insert numbers into the original and transformed equations to see whether they yield the same result (Resnick, Cauzinille-Marmeche, & Mathieu, 1987). This procedure reveals whether the transformation is allowable, although it rarely indicates why. Another common approach is to justify the transformation by citing a rule. Some students cite appropriate rules, but many others cite distorted versions of rules, such as the incorrect version of the distributive law cited above.

These problems are not quickly overcome. Even college students encounter difficulty with them. For example, more than one third of the freshman engineering students at a major state university could not write the correct equation to represent the simple statement "There are six times as many students as professors at this university" (Clement, 1982). Most wrote $6S = P$, which reflects a superficially reasonable but deeply flawed understanding of the relation between algebraic equations and the situations they represent. On a more positive note, such errors also indicate the source of the problem—in this case, that students are not analyzing in any depth the relation between what they have written and the problem they are representing. Deliberate practice in representing problem situations with equations and in analyzing why various equations do or do not accurately represent the problem situation seems likely to be helpful in overcoming this problem.

More generally, research on children's systematic errors points to a central lesson that has emerged from cognitive science research: the importance of cognitive task analysis. To promote effective learning, teachers must analyze in detail the particular procedures and concepts to be learned (Anderson, Reder, & Simon, 1997), provide students with instruction and examples that help them learn the component skills and understandings, anticipate types of misunderstandings that most often arise in the learning process, and be prepared with means for helping students move beyond these misunderstandings.

Cognitive Variability and Strategy Choice

Children's thinking has often been described as similar to a staircase on which children first use one approach to solve problems, then adopt a more advanced approach, and later adopt a yet more advanced approach. For example, researchers studying children's basic arithmetic (e.g., Ashcraft, 1987) have proposed that when children start school, they add by counting from one; sometime during first grade, they switch to adding by counting from the larger addend; and by third or fourth grade, they add by retrieving the answers to problems from memory.

More recent studies, however, have shown that children's thinking is far more variable than such staircase models suggest. Instead of adding by using the same strategy all the time, children use a variety of strategies from early in learning and continue to use both less and more advanced approaches for many years. Thus, even early in first grade, the same child given the same problem will sometimes count from one, sometimes count from the larger addend, and sometimes retrieve the answer from memory. Even when children master strategies that are both faster and more accurate, they continue to use older strategies that are slower and less accurate as well. That continued use occurs not just with young children but with preadolescents, adolescents, and even adults (Kuhn, Garcia-Mila, Zohar, & Anderson, 1995; Schauble, 1996).

This *cognitive variability*—using multiple thinking strategies when solving problems of the same type—is a spontaneous feature of children's thinking. Efforts to change it do not usually meet with much success. For example, in one study, teachers of children in Grades 1 to 3 were interviewed regarding their beliefs about their students' arithmetic strategies and their evaluation of whether their students' use of multiple strategies was a good thing (Siegler, 1984). All the teachers recognized that the children used multiple strategies, and most viewed that tendency as a bad thing. One teacher said that she was constantly discouraging her students from using such strategies as counting on their fingers. When asked how often she had done so with the pupil in her class who did it most often, she asked, "How many days have there been in the school year so far?" This teacher and others recognized that even when they explicitly told students not to use their fingers, the students did so anyway, even if they had to do it by putting their fingers in their laps, under their legs, or behind their backs.

A certain logic supports the teacher's view. Older students and those who are better at mathematics do not use their fingers, whereas younger and less-apt students do. One goal of education is to help younger and less apt students become more like older and more apt students. Therefore, children who use their fingers should be discouraged from doing so.

Children actually learn better, however, when they are allowed to choose the strategy they wish to use. Immature strategies generally drop away naturally when students have enough knowledge to answer accurately without them. Permitting basic strategies such as counting on one's fingers allows students to generate correct answers, whereas forbidding the use of such strategies would lead to many errors. Further, students who use a greater variety of different strategies for solving problems also tend to learn better subsequently (Alibali & Goldin-Meadow, 1993; Chi, de Leeuw, Chiu, & LaVancher, 1994; Siegler, 1995). This better learning occurs in part because the greater variety enables the students to cope with whatever kinds of problems they encounter rather than just with a narrow range of problems. Allowing children to use the varied strategies that they generate, as well as helping them understand why superficially different strategies converge on the right answer and why superficially reasonable strategies are incorrect, seems likely to build deeper understanding (Siegler, 2002).

Children's use of diverse strategies makes it essential that they choose appropriately among strategies. To choose appropriately, they must adjust both to situational variables and to differences among problems. Situational variables include time limits, instructions, and the importance of the task. For example, in a so-called magic-minute exercise, in which children are asked to do as many worksheet problems as possible in a minute, it is adaptive for children to state answers quickly, even when they are not absolutely sure of them. In contrast, if being correct is very important in a particular situation, then checking the correctness of answers becomes more worthwhile. At least from second grade onward, children shift their choices appropriately to adapt to such situational variations.

Adaptive choice also involves adjusting strategy use to the characteristics of particular problems. When children are faced with a simple problem, the ideal choice for them is often to use a strategy that can be executed quickly because that strategy will be sufficient to solve the problem. In contrast, when faced with a more difficult problem, children may need to adopt a more time-consuming and effortful strategy to generate the correct answer. Adaptive choice involves using quick and easy strategies when they are sufficient and reverting to increasingly effortful ones when they are necessary to being correct.

Research on strategy choice has also revealed some surprising similarities in the performance of children from different socioeconomic groups. Children from low-income backgrounds, particularly low-income African-American backgrounds, are often depicted as choosing strategies unwisely. Researchers have suggested that instruction should focus on improving these children's *metacognition*, or knowledge about their own thinking, and their strategy selection. The selection of appropriate strategies does not seem to be their main problem, however, at least in the context of arithmetic. Their strategy choices are just as systematic and just as sensitive to problem characteristics as those of children from middle-income backgrounds (Kerkman & Siegler, 1993). Instead, their problem seems to be that they do not possess adequate factual knowledge. That problem in turn seems to be due to less practice in solving problems and to poorer execution of strategies and not to any high-level deficiency in their thinking. The findings indicate that greater practice and instruction in how to execute strategies may be the most useful approach to improving such students' arithmetic skills.

Individual Differences

Substantial individual differences exist in cognitive variability and in the kinds of strategy choices that children make. These differences are in both their knowledge and the degree of confidence they need to have in an answer before they will state it. As early as first grade, children can be divided into three groups on the basis of their strategy choices in arithmetic: good students, not-so-good students, and perfectionists (Siegler, 1988). Good students and not-so-good students differ in all the ways that would be expected from the terms used to describe them. Compared with the not-so-good students, the good students are faster, are more accurate, use more advanced strategies, and perform better on standardized mathematics achievement tests.

The differences between the perfectionists and the other two groups are more interesting. The perfectionists are just as accurate as the good students. They also have equally high mathematics achievement and equally high IQ scores. Their performance in later grades also appears to be equally strong (Kerkman & Siegler, 1993). In making their strategy choices, however, the perfectionists choose a higher proportion of slow and effortful strategies. Unless they are very sure of the answer, they do not rely on memory, preferring instead to use such strategies as counting from one or from the larger addend. This behavior appears to reflect a stylistic preference. Perfectionists seem to set a very high criterion for being sure enough to state an answer without checking it via a backup strategy, such as counting on their fingers. They state remembered answers on the easiest problems, but only on such problems, whereas good students, with comparable knowledge, state remembered answers on a considerably broader range of problems.

These individual difference patterns are present among both boys and girls, and among both suburban children from middle-class households and inner-city children from low-income households. The relative frequencies of children in each group are also comparable for boys and girls and for children from low-income and middle-income households (Kerkman & Siegler, 1993; 1997). Both these findings —that, at least at this point in learning, individual difference patterns in mathematics are comparable among boys and girls and among children from low- and middle-income households—are different than stereotypes would suggest.

These classifications of children's individual difference patterns are related to traditional ones. Roughly half the not-so-good students in the Siegler (1988) study went on to be classified as having mathematical disabilities by fourth grade, versus none of the good students or perfectionists. The cognitive assessments, however, go beyond the traditional ones in showing stylistic as well as knowledge-based differences in mathematics performance.

As has often been noted (e.g., Geary, 1994), mathematical disabilities, as defined by poor performance in class and poor standardized test scores, constitute a very serious problem in the United States. Approximately 6% of children are labeled as having such disabilities. Like the not-so-good students, these children have difficulty in both executing backup strategies and retrieving correct answers from memory. As first graders, they frequently use immature counting procedures (e.g., counting from one rather than from the larger addend), execute backup strategies slowly and inaccurately, and use retrieval rarely and inaccurately. By second grade, they use somewhat more sophisticated counting procedures, such as counting from the larger addend, and their speed and accuracy improve. Still, and for years thereafter, however, they continue to have difficulty retrieving correct answers from memory (Geary, 1990; Geary & Brown, 1991; Goldman, Pellegrino, & Mertz, 1988; Jordan, Levine, & Huttenlocher, 1995). As they progress through school, these children encounter further difficulties in the many skills that build on basic arithmetic, such as multidigit arithmetic and algebra (Zawaiza & Gerber, 1993; Zentall & Ferkis, 1993).

Why do some children encounter such great difficulties with arithmetic? One reason is their limited exposure to numbers before entering school. Many children labeled "mathematically disabled" come from poor families with little formal education. By the time children from such backgrounds enter school, they already are far behind other children in counting skill, knowledge of numerical magnitudes, and knowledge of arithmetic facts. Another key difference involves working memory capacity. Learning arithmetic requires sufficient working memory capacity to hold the original problem in memory while computing the answer so that the problem and the answer can be associated. Children labeled as mathematically disabled, however, cannot hold as much numerical information in their memory as their age peers can (Geary, Bow-Thomas, & Yao, 1992; Koontz & Berch, 1996). Limited conceptual understanding of arithmetic operations and counting adds further obstacles to these children's learning of arithmetic (Geary, 1994; Hitch & McAuley, 1991). Thus, mathematical disabilities reflect a combination of limited background knowledge, limited processing capacity, and limited conceptual understanding. All these difficulties need to be addressed for such children to learn mathematics to a reasonably high level of proficiency.

Discovery and Insight

The findings described previously concerning preschoolers' understanding of counting principles showed that even before children enter school, they think abstractly about certain mathematical concepts. A surprisingly long time is needed, however, for this understanding to be expanded to other concepts, even ones that also pertain directly to understanding numbers. One such understanding that children discover in the first few years of elementary school is the *inversion principle*—the idea that adding and subtracting the same number leaves the original quantity unchanged. The understanding of this principle can be assessed by examining performance on problems of the form $a + b - b =$ __ (e.g., $25 + 8 - 8 =$ __). Children who solve such problems through applying the inversion principle would answer in the same amount of time regardless of the size of b because they would not need to add and subtract it. In contrast, children who solve the problem by adding and subtracting b would take longer when b was large than when it was small, because adding and subtracting large numbers takes longer than adding and subtracting small ones.

For children between 6 and 9 years of age, performance on $a + b - b =$ __ problems becomes faster; however, 9-year-olds, like 6-year-olds, take longer on problems in which b is large than on ones in which it is small (Bisanz & LeFevre, 1990; Stern, 1992). The improved speed on all problems appears to be due to improved procedural competence in addition and subtraction. The continuing difference between times on problems in which b is large and ones in which it is small suggests that neither 6- nor 9-year-olds have sufficient competence in understanding the inversion principle to consistently answer such problems without adding and subtracting. Not until age 11 do most children, like almost all adults, ignore the particular value of b and solve all such problems equally quickly, thus demonstrating understanding of the inversion principle.

A related concept that children also need a surprisingly long time to understand is that of *mathematical equality*. Even third and fourth graders frequently do not understand that the equals sign means that the values on each side of it must be equal. Instead, they believe that the equals sign is simply a signal to execute an arithmetic operation. On typical problems, such as $3 + 4 + 5 =$ ___, this misinterpretation does not cause any difficulty. On atypical problems, such as $3 + 4 + 5 =$ ___ $+ 5$, however, the misunderstanding leads most third and fourth graders either to add just the numbers to the left of the equal sign and answer "12" or to add all numbers on both sides of it and answer "17" (Alibali & Goldin-Meadow, 1993; Goldin-Meadow, Alibali, & Church, 1993; Perry, Church, & Goldin-Meadow, 1988; 1992).

Research on how children solve problems of mathematical equality indicates that they frequently make hand gestures that indicate knowledge that is not evident in their verbal statements. For example, given $3 + 4 + 5 =$ ___, some children who answer "12" and explain that they just added 3, 4, and 5 also motion with their hands toward the 12 in ways that indicate equality between the two. Children who on a pretest show such discrepancies between their speech and their gestures subsequently learn more from instruction in how to solve these problems than do children whose gestures and speech on the pretest reflect the same understanding (Alibali & Goldin-Meadow, 1993; Goldin-Meadow, et al., 1993; Perry et al., 1988, 1992). When asked to evaluate videotapes of children solving such problems, teachers and other children rate more highly those solutions that include advanced gestures than ones that do not, even when what the children say and write is identical (Garber, Alibali, & Goldin-Meadow, 1998). An implication of that research is that to the extent possible under classroom conditions, teachers should interpret children's nonverbal gestures as well as their verbal statements as indicators of their understanding and readiness to learn.

One lesson that has emerged from many recent studies is that children discover new strategies and concepts both when existing approaches are succeeding and when they are failing (e.g., Karmiloff-Smith, 1992; Kuhn et al., 1995; Miller & Aloise-Young, 1996; Siegler & Jenkins, 1989). We often assume that necessity is the mother of invention, and sometimes it is. However, children frequently generate new approaches on problems that they have previously solved using existing methods and when they have been succeeding on the preceding problems. One implication of this pervasive finding is that many discoveries do not require the creation of special "discovery learning" situations. In the same way that adults often generate new ideas while in the shower, while driving, or while working on unrelated or minimally related problems, so also do children.

Relations Between Conceptual and Procedural Knowledge

Throughout the 20th century, instructional reform oscillated between emphasizing mastery of facts and procedures, on the one hand, and emphasizing understanding of concepts, on the other (Hiebert & Lefevre, 1986). Few today would argue that either type of mathematical knowledge should be taught to the exclusion of the other. Much less agreement is found, however, about the balance that should be pursued between the two or about how to design instruction that will inculcate both types of knowledge.

Multidigit addition and subtraction have proved to be especially fruitful domains for studying the relations between conceptual and procedural knowledge. Children spend several years learning multidigit arithmetic. They must learn the regrouping procedure for addition, or "carrying," and that for subtraction, or "borrowing."

Understanding these procedures requires understanding the concept of place, that each position in a multidigit number represents a successively higher power of ten. It also requires understanding that a multidigit number can be represented in different ways; for example, 23 can be represented as 1 ten and 13 ones.

Many children have difficulty understanding place value and, as noted previously, frequently use buggy procedures that reflect this lack of understanding. For example, second graders often do not correctly carry when adding multidigit numbers (Fuson & Briars, 1990). Instead, they either write the two-digit sums beneath each column of single-digit addends (e. g., 568 + 778 = 121316) or ignore the carried values (e.g., 568 + 778 = 1236).

Although there are exceptions, procedural skill and conceptual understanding are usually highly correlated. One source of evidence for that view is cross-national studies. For example, comparisons of Korean and U.S. elementary school children have revealed parallel national differences in conceptual and procedural knowledge of multidigit addition and subtraction. Fuson and Kwon (1992) asked Korean second and third graders to solve two- and three-digit addition and subtraction problems that require carrying or borrowing. Then the children were presented several measures of conceptual understanding: ability to identify addition and subtraction problems that were worked correctly or incorrectly, ability to explain the basis of the correct procedure, and ability to indicate the place value of digits within a number. Almost all the children used correct procedures to solve the problems and also succeeded on all the measures of conceptual understanding. Stevenson and Stigler (1992) reported similar procedural and conceptual competence in first through fifth graders in Japan and China.

In contrast, a number of studies reviewed in Fuson (1990) indicated that U.S. children from Grades 2 to 5 frequently lack both conceptual and procedural knowledge of multidigit addition and subtraction. U.S. students' lack of conceptual understanding was evident in findings that almost half the third graders studied incorrectly identified the place value of digits within multidigit numbers (Kouba, Carpenter, & Swafford, 1989; Labinowicz, 1985), and in findings that most second through fifth graders could not demonstrate or explain ten-for-one trading with concrete representations (Ross, 1986). U.S. students' lack of procedural knowledge was evident in findings that children of these ages frequently erred while using paper and pencil to solve multidigit addition and subtraction problems (Brown & Burton, 1978; Fuson & Briars, 1990; Kouba et al., 1989; Labinowicz, 1985; Stevenson & Stigler, 1992). Taken together, these results suggest that conceptual and procedural knowledge are related; Asian children have both, and U.S. children are deficient in both.

Within the United States, conceptual and procedural competence are also highly correlated. Second and third graders who correctly execute the subtraction borrowing procedure also are more accurate in detecting conceptual flaws when they see a puppet performing subtraction procedures than are children who do not consistently execute the subtraction algorithm correctly (Cauley, 1988). Conceptual understanding of multidigit addition and subtraction and the ability to invent effective computational procedures are also positively correlated in first through fourth graders (Hiebert & Wearne, 1996).

This correlation leaves open not only the possibility that conceptual understanding might be causally related to children's inventing adequate computational procedures but also the possibility that knowing the correct procedure might be causally related to children's increased conceptual understanding resulting from being allowed to reflect on why the correct procedure is correct. One relevant source of evidence comes from an examination of the order in which individual children gain procedural and conceptual competence. A substantial percentage of children first gain conceptual understanding and then procedural competence, but another substantial percentage do the opposite (Hiebert & Wearne, 1996).

Studies aimed at improving the teaching of multidigit addition and subtraction typically emphasize linking steps in the procedures with the concepts that support them. In general, these teaching techniques successfully increase both conceptual and procedural knowledge. Although not currently conclusive, the studies suggest that instruction that emphasizes conceptual understanding as well as procedural skill is more effective in building both kinds of competence than instruction that focuses only on procedural skill is (Fuson & Briars, 1990; Hiebert & Wearne, 1996).

A question that remains is which type of knowledge should be emphasized first. Many opinions have been offered, but until recently, no directly relevant experimental evidence was available. A recent study by Rittle-Johnson and Alibali (1999), however, provides such evidence. They examined fifth graders' performance on mathematical equality problems of the form $a + b + c =$ ___ $+ c$. Some randomly selected children were given conceptually oriented instruction; other children were given procedurally oriented instruction; and still others were given neither. Then all children practiced solving problems, after which they took a posttest that assessed both conceptual and procedural knowledge. The conceptually oriented instruction produced substantial gains in both kinds of knowledge; the procedurally oriented instruction produced substantial gains in procedural knowledge and smaller gains in conceptual knowledge. To the degree that this result turns out to be general, it suggests that conceptual instruction should be undertaken before instruction aimed at teaching procedures.

Cooperative Learning

Children discover new strategies not only while solving problems on their own but also while working with others toward common goals. Some of that problem solving involves *scaffolding* situations in which a more knowledgeable person helps a less knowledgeable one to learn by providing a variety of kinds of help. Such scaffolding occurs in the context of parents helping their children, teachers helping their students, coaches helping their players, and more advanced learners helping less advanced ones (Freund, 1990; Gauvain, 1992; Rogoff, Ellis, & Gardner, 1984; Wood, Bruner, & Ross, 1976). The goal of such interactions is for the less knowledgeable learner to construct strategies that the more advanced learner already possesses.

In other situations, equally knowledgeable peers learn together. Such cooperative learning often enhances problem solving and reasoning to a greater degree than working independently would achieve (Gauvain & Rogoff, 1989; Kruger, 1992; Teasley, 1995; Webb, 1991). One particularly effective type of cooperative learning is reciprocal instruction. Reading teachers using reciprocal instruction read paragraphs with small groups of students and model such crucial metacognitive activities as summarizing, identifying ambiguities, asking questions, and predicting subsequent content. A recent review of 16 studies on reciprocal instruction (Rosenshine & Meister, 1994) indicated that results were generally positive, with students ranging from fourth graders to adults, with both low-achieving and average students, with groups ranging from 2 to 23 students, and with either experimenters or classroom teachers as the instructors.

Cooperative learning often fails to result in increased learning, however, and at times leads to worse learning than trying to solve problems on one's own (Levin & Druyan, 1993; Russell, 1982; Russell, Mills, & Reiff-Musgrove, 1990; Tudge, 1992). As noted by Ann Brown (personal communication, 4 March 1998), perhaps the greatest expert on cooperative learning, designing effective cooperative learning situations requires at least as much engineering as standard classroom instruction does. Without such careful structuring, problems of free-loading and disorganization can lead to inferior learning. Thus, fostering effective cooperative learning requires more than just assigning children to a group and telling them to work together on a problem or project.

Different types of collaborative organizations tend to have different effects not only on learning but also on instructional interactions. Damon and Phelps (1989) distinguished among three types of collaborative arrangements: peer tutoring, cooperative learning, and peer collaboration. In peer tutoring, a child who is knowledgeable about a topic instructs another child who is less knowledgeable about that topic. In cooperative learning, classrooms are divided into small groups or teams of usually three to six students of heterogeneous ability who try to solve a problem or master a task. In a common variant of cooperative learning, the jigsaw method, each child becomes the group's expert on a particular part of the task, and the task solutions require the contributions of all the experts. Finally, peer collaboration involves a pair of novices working together to solve problems that neither could solve initially on his or her own. These arrangements tend to differ in the degree to which they promote equality among participants (higher in peer collaboration and cooperative learning than in peer tutoring) and in the degree to which discussions tend to be extensive and engaging (highest in peer collaboration). Damon and Phelps argued that the collaborative arrangements that generated the most productive instructional dialogs were those that encouraged joint problem solving and that discouraged competition among students.

How can the effectiveness of collaborations be improved? One way is to examine factors that differentiate successful from unsuccessful interactions. To obtain such information, Ellis, Klahr, and Siegler (1993) examined fifth graders solving decimal fraction problems of the following form: "Which is bigger, .239 or .47?" This seemingly simple task often, as noted previously, causes children of this age considerable difficulty. In particular, the children often misapply mathematical rules acquired while learning about whole numbers or common fractions, either consistently choosing the number with more digits as larger (the whole number rule) or consistently choosing the number with fewer digits as larger (the fraction rule).

Ellis and others (1993) found that children learned more when they worked with a partner during instruction than when they worked alone; that this benefit occurred only when the children were also provided feedback by the experimenter about which answer was right; that external feedback was just as essential for partners who started with different rules as it was for those who started with the same rule; and that social aspects of the interaction, as well as external feedback, influenced learning. One particularly important factor was the enthusiasm of the partner's reactions to the child's statements. In this context, *enthusiasm* meant strong interest in, rather than agreement with, the partner's ideas. Children whose partner reacted enthusiastically during the instructional session answered correctly more often on the posttest than those whose partners showed less enthusiasm did. Among children who worked with a partner and received feedback from the experimenter, the enthusiasm of the partner's reactions was the best single predictor of learning. The example illustrates that attention to both cognitive and social variables is crucial for successful cooperative learning.

Promoting Analytic Thinking and Transfer

Analytic thinking refers to a set of processes for identifying the causes of events. Analytic thinking is among the central goals of mathematics education, in part because it is an inherently constructive process. Analysis demands that children *think actively* about the causes of events. Obtaining a general sense of the typical course of events, or the way in which things work, is possible without actively analyzing them. More active thinking is required, however, to distinguish features that usually accompany events from those that cause them to occur. Thus, distinguishing between features that typically accompany the use of a particular mathematical problem-solving technique and features that are essential for the technique to apply usually requires an analysis of why the technique is appropriate or inappropriate.

Analytic reasoning is both a cause and a consequence of a second useful quality: purposeful engagement. When children have a specific reason for wanting to learn about a topic, they are more likely to analyze the material so that they truly understand it. In that sense, analytic reasoning is a consequence of purposeful engagement. Analytic reasoning, however, also promotes purposeful engagement. Children who try from the beginning of learning about a topic to deeply understand it become more engaged in learning it than children who accept what they are told without thinking about it.

A third way in which analytic thinking is central is in promoting *transfer*. When children are actively engaged in understanding why things work the way they do, transfer follows naturally and without great effort. In contrast, when understanding stays close to the surface, and does not penetrate underneath, transfer is unlikely (Brown, 1997). Such passive learners lack ways of distinguishing the core information from the incidental details. Encouraging learners to reason analytically more often will therefore also create learners who transfer what they learn to new situations.

A variety of types of evidence attest to the importance of such explanatory activities. For both adults and children, students who ask themselves more questions about the meaning of a textbook as they are reading it learn more from their reading than children who read without asking many questions do. That result has been shown for learning of both physics and computer programming (Chi, Bassok, Lewis, Reimann, & Glasser, 1989; Pirolli & Recker, 1994). Both the quality and the quantity of explanations that children generate while reading are related to their learning. For example, when the best learners study example problems, they are especially likely to connect particular aspects of the examples with particular statements in the text (Pirolli & Bielaczyc, 1989).

How can such analytic thinking be encouraged? One effective way is to ask children to explain the correct conclusions or answers of other people. Children as young as 5 years of age benefit when they solve a difficult problem incorrectly, are told the correct answer, and then are asked, "How do you think I knew that?" (Siegler, 1995). That procedure combines advantages of discovery learning with those of didactic methods. Like discovery-learning approaches, explaining the correct conclusions of others promotes active engagement with the task because the children must generate the underlying logic for themselves. Like didactic methods, the procedure is efficient; instead of going down blind alleys, children spend their time thinking about the logic that led to desired conclusions. An added advantage of this approach is that it can be applied to a very wide variety of problems. Teachers can ask, "How do you think I knew that?" or "Why do you think I think that?" about almost any conclusion. Encouraging children to explain other people's reasoning in many contexts may lead them to internalize such an analytic stance to the point where they ask such questions reflexively, even when not prompted to do so.

Asking children to explain also why incorrect answers are incorrect may be even more effective than just asking them to explain why correct answers are correct. Such activities are featured in Japanese classrooms and are associated with high levels of mathematics achievement in that country (Stigler & Perry, 1990). These activities are also effective with U. S. children. In a recent experiment on understanding mathematical equality, children were randomly assigned to one of three conditions: (a) explain both why correct answers are correct and why incorrect answers are incorrect, (b) just explain why correct answers are correct, or (c) just try to solve mathematical equality problems and receive feedback (Siegler, 2002). Asking children to explain both why correct answers are right and why incorrect answers are wrong led to greater learning than just asking the former type of question. Especially encouraging was that it greatly increased transfer to problems that were superficially dissimilar to the originally presented ones.

Teaching children computer-programming skills has been proposed as another means of promoting transfer. In particular, advocates of providing such experience have contended that it would produce not only skill at programming but also enhanced general problem-solving ability and analytic skills. In one notable effort in this direction, Papert (1980) designed the Logo language with the goal of helping children acquire such broadly useful skills as dividing problems into their main components, identifying logical flaws in one's thinking, and generating thoughtful plans.

When learned in standard ways, Logo has proved insufficient to meet these goals. However, *mediated instruction*, in which Logo is taught with an eye toward building transferable skills, has been quite successful in producing the desired effects (Klahr & Carver, 1988; Lehrer & Littlefield, 1991; 1993; Littlefield, Delclos, Bransford, Clayton, & Franks, 1989). Like conventional instruction in computer programming, mediated instruction involves teachers' demonstrating to students how to use commands and concepts and giving students feedback on their attempts to use them. But mediated instruction also involves teachers' explicitly noting when particular commands and programs illustrate general programming concepts and drawing explicit analogies between the reasoning used to program and that used to solve problems in other contexts, including mathematics.

Such mediated instruction has produced various kinds of desirable transfer. For example, Klahr and Carver (1988) demonstrated that mediated instruction in Logo can create debugging skills that are useful outside as well as inside the Logo context. Their instructional program was based on a detailed task analysis of debugging. Within this analysis, the debugging process begins with the debugger determining the outcome that a procedure yields and observing whether and how its results deviate from what was planned, for example, by running a computer program and examining its output. In the second step, the debugger describes the discrepancy between desired and actual outcomes and hypothesizes the types of bugs that might be responsible. The third step is to identify parts of the program that could conceivably produce the observed bug. This step demands dividing the program into components, so that specific parts of the program are identified with specific functions. In the final step, the debugger first checks the relevant parts of the program to see which, if any, fail to produce the intended results, rewrites the faulty component, and then runs the debugged program to determine whether it now produces the desired output. The 8- to 11-year-olds who received this instruction took barely half as long to solve Logo debugging problems as children who did not receive it. They also improved their general problem-solving skills in areas outside of programming—in particular, in revising essays. The improvement seemed to have occurred because the children applied the skills taught in the program: analyzing the nature of the original discrepancy from the anticipated results, hypothesizing possible causes, and focusing their search on relevant parts of the instructions instead of simply checking them line by line.

Conclusions

This chapter summarizes a number of empirical findings and theoretical conclusions about children's mathematics learning. Translating these findings and conclusions into improved instructional practices, however, will take a considerable amount of work. Stigler and Hiebert's (1997) description of the Japanese emphasis on *continuous improvement* in teaching points toward a process that seems essential in U.S. classrooms as well. The process they describe involves groups of teachers working together to perfect the way in which they teach particular concepts and procedures. Neither controlled scientific experimentation nor theoretical analyses automatically translate into prescriptions for classroom instruction. Both experimentation and analyses can provide useful frameworks for thinking about teaching and learning, can indicate sources of difficulty that children encounter in learning particular skills and concepts, and can demonstrate potentially effective instructional procedures. A process of translation into the particulars of each classroom context, however, is necessary for even the most insightful frameworks and the most relevant findings to be used in ways that improve learning. Both institutional support for such continuous improvement and teacher dedication to meeting this goal are essential if research is to lead to superior instruction.

ACKNOWLEDGMENTS

I am grateful to participants in the Conference on Research Foundations for the NCTM Standards (Atlanta, March 1998) who provided comments on a preliminary draft of this chapter, and to Glen Allinger and Joan Garfield for comments on a subsequent draft. The funding provided by NIH Grant 19011 and by a grant from the Spencer Foundation also was invaluable for conducting much of the research described in this paper.

REFERENCES

Alibali, M. W., & Goldin-Meadow, S. (1993). Gesture-speech mismatch and mechanisms of learning: What the hands reveal about a child's state of mind. *Cognitive Psychology, 25*, 468–573.

Anderson, J. R., Reder, L., M., & Simon, H. A. (1997). Radical constructivism and cognitive psychology. In D. Ravitch (Ed.) *Brookings papers on education policy* (pp. 227–278). Washington, DC: Brookings Institution Press.

Antell, S. E., & Keating, D. P. (1983). Perception of numerical invariance in neonates. *Child Development, 54*, 695–701.

Ashcraft, M. H. (1987). Children's knowledge of simple arithmetic: A developmental model and simulation. In J. Bisanz, C. J. Brainerd, & R. Kail (Eds.), *Formal methods in developmental psychology: Progress in cognitive development research* (pp. 302–338). New York: Springer-Verlag.

Bisanz, J., & LeFevre, J. (1990). Strategic and nonstrategic processing in the development of mathematical cognition. In D. P. Bjorklund (Ed.), *Children's strategies: Contemporary views of cognitive development* (pp. 213–244). Hillsdale, NJ: Erlbaum.

Brown, A. L. (1997). Transforming schools into communities of thinking and learning about serious matters. *American Psychologist, 52*, 399–413.

Brown, J. S., & Burton, R. B. (1978). Diagnostic models for procedural bugs in basic mathematical skills. *Cognitive Science, 2*, 155–192.

Carpenter, T. P., Corbitt, M. K., Kepner, H. S., Lindquist, M. M., & Reys, R. E. (1981). *Results from the Second Mathematics Assessment of the National Assessment of Educational Progress*. Washington, DC: National Council of Teachers of Mathematics.

Cauley, K. M. (1988). Construction of logical knowledge: Study of borrowing in subtraction. *Journal of Educational Psychology, 80*, 202–205.

Chi, M. T. H., Bassok, M., Lewis, M., Reimann, P., & Glasser, R. (1989). Self-explanations: How students study and use examples in learning to solve problems. *Cognitive Science, 13*, 145–182.

Chi, M. T. H., de Leeuw, N., Chiu, M.-H., & LaVancher, C. (1994). Eliciting self-explanations improves understanding. *Cognitive Science, 18*, 439–477.

Chi, M. T. H., & Klahr, D. (1975). Span and rate of apprehension in children and adults. *Journal of Experimental Child Psychology, 19*, 434–439.

Clement, J. (1982). Algebra word problem solutions: Thought processes underlying a common misconception. *Journal for Research in Mathematics Education, 13*, 16–30.

Damon, W., & Phelps, E. (1989). Critical distinctions among three approaches to peer education. *International Journal of Educational Research, 13*, 9–20.

Ellis, S., Klahr, D., & Siegler, R. S. (1993, March). *Effects of feedback and collaboration on changes in children's use of mathematical rules*. Paper presented at the meeting of the Society for Research in Child Development, New Orleans.

Freund, L. S. (1990). Maternal regulation of children's problem solving behavior and its impact on children's performance. *Child Development, 61*, 113–126.

Fuson, K. C. (1990). Conceptual structures for multiunit numbers: Implications for learning and teaching multidigit addition, subtraction, and place value. *Cognition and Instruction, 7*, 343–403.

Fuson, K. C., & Briars, D. (1990). Using a base-ten blocks learning/teaching approach for first- and second-grade place-value and multidigit addition and subtraction. *Journal for Research in Mathematics Education, 21*, 180–206.

Fuson, K. C., & Kwon, Y. (1992). Korean children's understanding of multidigit addition and subtraction. *Child Development, 63*, 491–506.

Garber, P., Alibali, M. W., & Goldin-Meadow, S. (1998). Knowledge conveyed in gesture is not tied to the hands. *Child Development, 69*, 75–84.

Gauvain, M. (1992). Social influences on the development of planning in advance and during action. *International Journal of Behavioral Development, 15*, 377–398.

Gauvain, M., & Rogoff, B. (1989). Collaborative problem solving and children's planning skills. *Developmental Psychology, 25*, 139–151.

Geary, D. C. (1990). A componential analysis of an early learning deficit in mathematics. *Journal of Experimental Child Psychology, 49*, 363–383.

Geary, D. C. (1994). *Children's mathematical development: Research and practical implications*. Washington, DC: American Psychological Association.

Geary, D. C., Bow-Thomas, C. C., & Yao, Y. (1992). Counting knowledge and skill in cognitive addition: A comparison of normal and mathematically disabled children. *Journal of Experimental Child Psychology, 54*, 372–391.

Geary, D. C., & Brown, S. C. (1991). Cognitive addition: Strategy choice and speed-of-processing differences in gifted, normal, and mathematically disabled children. *Developmental Psychology, 27*, 398–406.

Geary, D. C., Fan, L., Bow-Thomas, C., & Siegler, R. S. (1993). Even before formal instruction, Chinese children outperform American children in mental addition. *Cognitive Development, 8*, 517–529.

Gelman, R., & Gallistel, C. R. (1978). *The child's understanding of number*. Cambridge, MA: Harvard University Press.

Goldin-Meadow, S., Alibali, M. W., & Church, R. B. (1993). Transitions in concept acquisition: Using the hand to read the mind. *Psychological Review, 100,* 279–297

Goldman, S. R., Pellegrino, J. W., & Mertz, D. L. (1988). Extended practice of basic addition facts: Strategy changes in learning disabled students. *Cognition and Instruction, 5,* 223–265.

Griffin, S., Case, R., & Siegler, R. S. (1994). Rightstart: Providing the central conceptual prerequisites for first formal learning of arithmetic to students at risk for school failure. In K. McGilly (Ed.), *Classroom lessons: Integrating cognitive theory and classroom practice* (pp. 25–49). Cambridge, MA: MIT Press/Bradford Books.

Hiebert, J., & Lefevre, P. (1986). Conceptual and procedural knowledge in mathematics: An introductory analysis. In J. Hiebert (Ed.), *Conceptual and procedural knowledge: The case of mathematics* (pp. 29–57). Hillsdale, NJ: Erlbaum.

Hiebert, J., & Wearne, D. (1996). Instruction, understanding and skill in multidigit addition and subtraction. *Cognition and Instruction, 14,* 251–283.

Hitch, G. J., & McAuley, E. (1991). Working memory in children with specific arithmetical learning disabilities. *British Journal of Psychology, 82,* 375–386.

Jordan, N. C., Huttenlocher, J, & Levine, S. C. (1992). Differential calculation abilities in young children from middle- and low-income families. *Developmental Psychology, 28,* 644–653.

Jordan, N. C., Levine, S. C., & Huttenlocher, J. (1995). Calculation abilities in young children with different patterns of cognitive functioning. *Journal of Learning Disabilities, 28,* 53–64.

Karmiloff-Smith, A. (1992). *Beyond modularity: A developmental perspective on cognitive science.* Cambridge, MA: MIT Press.

Kerkman, D. D., & Siegler, R. S. (1993). Individual differences and adaptive flexibility in lower-income children's strategy choices. *Learning and Individual Differences, 5,* 113–136.

Kerkman, D. D., & Siegler, R. S. (1997). Measuring individual differences in children's addition strategy choices. *Learning and Individual Differences, 9,* 1–18.

Klahr, D., & Carver, S. M. (1988). Cognitive objectives in a Logo debugging curriculum: Instruction, learning, and transfer. *Cognitive Psychology, 20,* 362–404.

Koontz, K. L., & Berch, D. B. (1996). Identifying simple numerical stimuli: Processing inefficiencies exhibited by arithmetic learning disabled children. *Mathematical Cognition, 2,* 1–23.

Kouba, V. L., Carpenter, T. P., & Swafford, J. O. (1989). Number and operations. In M. M. Lindquist (Ed.), *Results from the Fourth Mathematics Assessment of the National Assessment of Educational Progress* (pp. 64–93). Reston, VA: National Council of Teachers of Mathematics.

Kruger, A. C. (1992). The effect of peer and adult-child transactive discussions on moral reasoning. *Merrill-Palmer Quarterly, 38,* 191–211.

Kuhn, D., Garcia-Mila, M., Zohar, A., & Andersen, C. (1995). Strategies of knowledge acquisition. *Monographs of the Society for Research in Child Development, 60*(4, Serial No. 245).

Labinowicz, E. (1985). *Learning from children: New beginnings for teaching numerical thinking.* Menlo Park, CA: Addison-Wesley.

Lehrer, R., & Littlefield, J. (1991). Misconceptions and errors in Logo: The role of instruction. *Journal of Educational Psychology, 83,* 124–133.

Lehrer, R., & Littlefield, J. (1993). Relationships among cognitive components in Logo learning and transfer. *Journal of Educational Psychology, 85,* 317–330.

Levin, I., & Druyan, S. (1993). When sociocognitive transaction among peers fails: The case of misconceptions in science. *Child Development, 63,* 1571–1591.

Littlefield, J., Delclos, V. R., Bransford, J. D., Clayton, K. N., & Franks, J. J. (1989). Some prerequisites for teaching thinking: Methodological issues in the study of Logo programming. *Cognition and Instruction, 6,* 331–366.

Matz, M. (1982). Towards a process model for high school algebra errors. In D. Sleeman & J. S. Brown (Eds.), *Intelligent tutoring systems* (pp. 25–50). New York: Academic Press.

Miller, P., & Aloise-Young, P. (1996). Preschoolers' strategic behaviors and performance on a same-different task. *Journal of Experimental Child Psychology, 60,* 284–303.

National Council of Teachers of Mathematics. (1989). *Curriculum and evaluation standards for school mathematics.* Reston, VA: Author.

National Council of Teachers of Mathematics. (1991). *Professional standards for teaching mathematics.* Reston, VA: Author.

National Council of Teachers of Mathematics. (1995). *Assessment standards for school mathematics.* Reston, VA: Author.

Papert, S. (1980). *Mindstorms: Children, computers, and powerful ideas.* New York: Basic Books.

Perry, M., Church, R. B., & Goldin-Meadow, S. (1988). Transitional knowledge in the acquisition of concepts. *Cognitive Development, 3,* 359–400.

Perry, M., Church, R. B., & Goldin-Meadow, S. (1992). Is gesture/speech mismatch a general index of transitional knowledge? *Cognitive Development,* 7, 109–122.

Pirolli, P., & Bielaczyc, K. (1989). Empirical analyses of self-explanation and transfer in learning to program. In *Proceedings of the 11th Annual Conference of the Cognitive Science Society* (pp. 450–457). Hillsdale, NJ: Erlbaum.

Pirolli, P., & Recker, M. (1994). Learning strategies and transfer in the domain of programming. *Cognition and Instruction, 12,* 235–275.

Resnick, L. B., Cauzinille-Marmeche, E., & Mathieu, J. (1987). Understanding algebra. In J. A. Sloboda & D. Rogers (Eds.), *Cognitive processes in mathematics* (pp. 169–203). Oxford: Clarendon Press.

Resnick, L. B., Nesher, P., Leonard, F., Magone, M., Omanson, S., & Peled, I. (1989). Conceptual bases of arithmetic errors:

The case of decimal fractions. *Journal for Research in Mathematics Education, 20*, 8–27.

Rittle-Johnson, B., & Alibali, M.W. (1999). Conceptual and procedural knowledge of mathematics: Does one lead to the other? *Journal of Educational Psychology, 91*, 175–189.

Rogoff, B., Ellis, S., & Gardner, W. P. (1984). Adjustment of adult-child instruction according to child's age and task. *Child Development, 20*, 193–199.

Rosenshine, B., & Meister, C. (1994). Reciprocal teaching: A review of the research. *Review of Educational Research, 64*, 479–530.

Ross, S. H. (1986, April). *The development of children's place-value numeration concepts in grades two through five.* Paper presented at the meeting of the American Educational Research Association, San Francisco.

Russell, J. (1982). Cognitive conflict, transmission and justification: Conservation attainment through dyadic interaction. *Journal of Genetic Psychology, 140*, 283–297.

Russell, J., Mills, I., & Reiff-Musgrove, P. (1990). The role of symmetrical and asymmetrical social conflict in cognitive change. *Journal of Experimental Child Psychology, 49*, 58–78.

Schauble, L. (1996). The development of scientific reasoning in knowledge-rich contexts. *Developmental Psychology, 32*, 102–119.

Siegler, R. S. (1984). Research on learning. In T. Romberg & D. Stewart (Eds.), *School mathematics: Options for the 1990s* (pp. 75–80). Madison: Wisconsin Center for Education Research.

Siegler, R. S. (1988). Individual differences in strategy choices: Good students, not-so-good students, and perfectionists. *Child Development, 59*, 833–851.

Siegler, R. S. (1995). How does change occur: A microgenetic study of number conservation. *Cognitive Psychology, 28*, 255–273.

Siegler, R. S. (2002). Microgenetic studies of self-explanations. In N. Granott & J. Parziale (Eds.), *Microdevelopment: Transition processes in development and learning* (pp. 31–58). New York: Cambridge University Press.

Siegler, R. S., & Jenkins, E. A. (1989). *How children discover new strategies.* Hillsdale, NJ: Erlbaum.

Siegler, R. S., & Robinson, M. (1982). The development of numerical understandings. In H. W. Reese & L. P. Lipsitt (Eds.), *Advances in child development and behavior* (Vol. 16, pp. 241–312). New York: Academic Press.

Silver, E. A. (1983). Probing young adults' thinking about rational numbers. *Focus on Learning Problems in Mathematics, 5*, 105–117.

Simon, T. J., Hespos, S. J., & Rochat, P. (1995). Do infants understand simple arithmetic? A replication of Wynn (1992). *Cognitive Development, 10*, 253–269.

Sleeman, D. H. (1985). Basic algebra revised: A study with 14-year-olds. *International Journal of Man-Machine Studies, 22*, 127–149.

Starkey, P., & Cooper, R. S. (1980). Perception of numbers by human infants. *Science, 210*, 1033–1035.

Starkey, P., Spelke, E. S., & Gelman, R. (1990). Numerical abstraction by human infants. *Cognition, 36*, 97–128.

Stern, E. (1992). Spontaneous use of conceptual mathematical knowledge in elementary school children. *Contemporary Educational Psychology, 17*, 266–277.

Stevenson, H. W., & Stigler, J. W. (1992). *The learning gap: Why our schools are failing and what we can learn from Japanese and Chinese education.* New York: Summit Books.

Stigler, J. W., & Hiebert, J. (1997, September). Understanding and improving classroom mathematics instruction: An overview of the TIMSS Video Study. *Phi Delta Kappan, 79*, 14–21.

Stigler, J. W., & Perry, M. (1990). Mathematics learning in Japanese, Chinese, and American classrooms. In J. W. Stigler, R. A. Shweder, & G. Herdt (Eds.), *Cultural psychology: Essays on comparative human development* (pp. 328–353). New York: Cambridge University Press.

Strauss, M. S., & Curtis, L. E. (1984). Development of numerical concepts in infancy. In C. Sophian (Ed.), *The origins of cognitive skills* (pp. 131–155). Hillsdale, NJ: Erlbaum.

Teasley, S. D. (1995). The role of talk in children's peer collaborations. *Developmental Psychology, 31*, 207–220.

Tudge, J. (1992). Processes and consequences of peer collaboration: A Vygotskian analysis. *Child Development, 63*, 1364–1379.

Van Loosbroek, E., & Smitsman, A. W. (1990). Visual perception of numerosity in infancy. *Developmental Psychology, 26*, 916–922.

Webb, N. (1991). Task-related verbal interaction and mathematics learning in small groups. *Journal for Research in Mathematics Education, 22*, 366–389.

Wood, D., Bruner, J., & Ross, G. (1976). The role of tutoring in problem solving. *Journal of Child Psychology and Psychiatry, 17*, 89–100.

Wynn, K. (1992). Addition and subtraction by human infants. *Nature, 358*, 749–750.

Zawaiza, T. R., & Gerber, M. (1993). Effects of explicit instruction on math word-problem solving by community college students with learning disabilities. *Learning Disability Quarterly, 16*, 64–79.

Zentall, S. S., & Ferkis, M. A. (1993). Mathematical problem solving for youth with ADHD, with and without learning disabilities. *Learning Disability Quarterly, 16*, 6–18.

Chapter 21

Situative Research Relevant to Standards for School Mathematics

James G. Greeno, Stanford University

In this article, I review research in the situative perspective that can help educators interpret the 2000 revision of the *Standards* documents of the National Council of Teachers of Mathematics (NCTM, 1989, 1991, 1995). I first examine some issues in the purposes and uses of standards generally and in education in particular. Next, I discuss how research findings can be useful in relation to formulating and interpreting standards in education, and I discuss an alternative to the research-development-dissemination-evaluation model. Finally, I provide an overview of a situative perspective in research, exploring three main questions: (1) What kinds of learning, teaching, and assessment practices are considered in a situative perspective, and how do those practices make a difference in students' learning? (2) What are some conditions in classroom practices, especially patterns of discourse, that foster students' constructive participation in learning activities? (3) What evidence shows that students and teachers can learn to participate productively in constructivist practices of mathematics education, and what support is needed for that learning?

Alternative Ideas About Standards

Standards can be formulated in two quite different ways, with purposes and uses that are quite distinct. One kind of formulation intends to enforce uniformity of a product or activity. When the standard is enforced successfully, some aspect of a product or activity can be predicted reliably for use. Another kind of formulation intends to characterize qualities of an activity that are highly valued. When that kind of standard is adopted, practitioners and observers of the activity use it to represent their aspirations and expectations for highly valued levels of performance in the domain and to organize learning activities so that performance can progress toward those qualities. Both these kinds of standards can be useful, depending on the domain of practice in which they are applied.

Standards as Resources for Reflective Discourse and Design

Standards can be formulated as expressions of qualities that are valued in a practice. This formulation usually occurs in domains of some complexity; for example, a symphony orchestra may be said to maintain a high standard of performance, or we may say that we expect court judges to meet a high standard of integrity, or officials at the World Cup may assert that the refereeing there meets a high standard, or the National Endowment for the Arts may be required to apply a standard of decency in the awarding of grants. Romberg (1993) pointed out that the term *standard* also refers to a banner that is raised high so that people can rally around it. We use this term when we identify an exemplary individual or performance that captures qualities of excellence to which other practitioners can aspire, and we say that that individual sets a high standard for the community. When we use the term in this broader sense, it overlaps with aims—what practitioners try to achieve—and with goals.

An example of standards of this kind was reported by Heath (1991) in an analysis of participation and learning in Little League baseball. The coach's goals for the team were "(a) to have a good time, (b) to learn and practice teamwork and sportsmanship, and (c) to learn a little more about the game of baseball" (p. 104). He frequently appealed to a standard of play and behavior by asking "How would they do this in the major leagues?" (p. 105). The coach's and team's understanding of professional practice provided them with a scheme for evaluating performance and discussing ways to improve. Heath summarized the focus as follows:

> The focus was on prototypes or generic categories of behavior for major league players, catchers, batters, and "good sports." ... The models or experts to whom the boys linked their own behaviors lay beyond the coach and the vagaries of team membership; they rested in the collective knowledge of team members as they read about baseball, watched games on television, or heard them on the radio. (p. 105)

When standards are formulated to enforce uniformity, they must be accompanied by an enforcement process to ensure that practitioners comply with them. (I discuss examples in the next section.) Standards that express valued qualities of a practice generally stop short of specifying in detail exactly what should be done in the practice or exactly what should constitute its products. Thus, one response to a set of standards like that of the NCTM (1989) is that it is not very specific, nor does it include precise methods for deciding whether teaching and learning conform to them. How, then, can standards like these affect school mathematics?

Standards that express valued qualities of a practice can have important effects when they function as resources for reflective discourse and design. Practitioners can refer to the standards in their discussions in which they design, plan, evaluate, and make sense of the activities in which they engage. Such standards can serve as guidelines, in the form of design principles, in activities of developing material resources—such as textbooks or learning software—for use in the practice. They also provide a set of terms, concepts, and principles for discussions about practice in the wider society, among policymakers and other citizens.

A question for formulating standards, then, is, How can a set of standards provide effective resources for reflective discourse and design? To address this question, we need to consider the processes in which practitioners, designers, and others conduct their work, as well as the role that a set of standards can have in those processes.

A considerable body of research is developing as investigators study teachers' and schools' efforts to change their practices in line with proposals for educational reform (e.g., Fennema & Scott Nelson, 1997). Much of this research is situative in character; that is, it examines characteristics of systems in which teachers participate, interacting with other people and in the contexts of the schools in which they work.

A consistent finding of this research is that characteristics of teachers' *communities of practice* are crucial factors in whether efforts to change teaching practices are successful. It is important that the local group of teachers shares a commitment to bringing about the change (Elmore, Peterson & McCarthey, 1996; Lieberman, 1997/1998). The community of teachers needs to organize itself as a learning community, with expectations and occasions for lateral communication (i.e., between teachers) about challenges and successes in their practice (Elmore et al., 1996; Stokes, Sato, McLaughlin, & Talbert, 1997). The community needs to offer ways for new arrivals to learn changing practices through participation and interaction with more seasoned members of the community (Stein, Silver, & Smith, 1998). The learning activities need to support generative learning that allows individuals as well as the community to grow in their capabilities (Lieberman, 1997/1998). For a successful reform to scale up to the level of schoolwide change—beyond a few classrooms in which the change has been fostered by a special program—serious investments need to be made in teacher leadership and administrative support, including a commitment by the school administration and allocations of time and other resources (Stokes et al., 1997).

How can standards provide a resource for productive discourse by teachers and others promoting educational change? Here are some hypotheses about questions that are considered in such discourse:

- What are we accomplishing now?
- What could we accomplish that we would value if we changed our practices?
- Why would that accomplishment be valuable?
- What would our changed practices be like?
- What resources would we need to accomplish these changes?
- In the process, what would be lost that we would regret?

The revision of the NCTM *Standards* documents (1989, 1991, 1995) can guide discussions of such questions in its presentation of a general view of mathematics learning, its recommendations about mathematical content and teaching and learning processes for key mathematical topics, and its consideration of methods for designing and evaluating curricula, textbooks, and other resource materials. For example, the standards should support discussions of why the practices that they characterize are valuable, how such practices can be achieved and evaluated, and how different ways of carrying out the practices would be more or less effective in realizing the qualities that the community values.

Standards to Enforce Uniformity

As mentioned earlier, another way of formulating standards is to aim them at enforcing uniformity of a product or some aspect of activity. A major source of motivation for standardization, imposed by central authorities, was termed *projects of legibility* by Scott (1998). Management from the center requires systems in which performance can be measured, records can be kept, and progress can be assessed in ways that can be understood from outside of the situation. Scott, a specialist in agricultural ethnography and politics, identified a pattern in recent state enterprises that he called *high modernism*, "a strong (one might even say muscle-bound) version of the beliefs in scientific and technical progress that were associated with industrialization in Western Europe and in North America from roughly 1830 until World War I" (p. 89). High-modernist projects proceed with the apparent authority of scientific knowledge and an ambition to change nature, including human nature, to improve conditions of life.

Another aspect of debates about standardization was discussed by Porter (1995), who emphasized conflicts between efforts by governmental agencies to regulate professional activities and the professional communities who viewed the imposition of uniform imposition of standards as counterproductive interference with their work. Government agencies urge standards of practice that reduce the prerogatives of practitioners to make decisions and support their proposals on the basis of their expert judgment. The professionals resist these pressures if they can, and their ability to resist depends on their establishing and maintaining their credibility as legitimate experts in the society. Using cases involving insurance companies and corps of engineers, Porter argued that the imposition of standards—such as government agencies' standards of practice—results from societal judgments that practitioners are untrustworthy. If a professional community is trusted by a governing agency to exercise responsible and effective judgment, the agency is likely to leave important matters to the expertise of professionals. The imposition of standards involving rules that objectify judgments and assessments occurs when governing agencies are unwilling to trust professional experts.

Scott (1998) was particularly concerned about understanding cases in which well-intentioned projects led to major human disasters—his main examples were collectivization in the Soviet Union in the early 1930s and villagization in Tanzania in the mid-1970s. He concluded that these disasters can be understood as projects in which the high-modernist impulse toward rationalized reform teamed with unchecked state power to support uprooting people, relocating them, and constraining their activities in ways that satisfied the state's need for legitimacy and control but destroyed the capacity of the people for productive living and use of resources.

Scott also discussed a pattern of agricultural reform involving legitimacy requirements that were counterproductive but not as disastrous. One example was scientific forestry management, developed in Europe—especially in Prussia and Saxony—during the late eighteenth century and serving as the basis for forest management techniques in France, England, the United States, and throughout the Third World by the end of the nineteenth century. These management principles were designed to produce maximum yields of salable timber. In a scientifically managed forest, a single species of tree was grown, chosen for its rapid growth and the quality of its timber. The trees were planted in rows to reach maturity simultaneously and therefore be harvestable efficiently. This approach contrasted with forests that grew naturally, with multiple varieties of trees growing together with smaller plants in an uneven but mutually supportive ecosystem. Scott (1998) wrote that scientific management of a forest attempted

> to create, through careful seeding, planting, and cutting, a forest that was easier for state foresters to count, manipulate, measure, and assess. The fact is that forest science and geometry, backed by state power, had the capacity to transform the real, diverse, and chaotic old-growth forest into a new, more uniform forest that closely resembled the administrative grid of its techniques … . The tendency was toward regimentation, in the strict sense of the word. The forest trees were drawn up into serried, uniform ranks, as it were, to be measured, counted off, felled, and replaced by a new rank and file of lookalike conscripts. As an army, it was also designed hierarchically from above to fulfill a unique purpose and to be at the disposition of a single commander. At the limit, the forest itself would not even have to be seen; it could be "read" accurately from the tables and maps in the forester's office. (p. 19)

Scott's account is a cautionary tale, because the scientific management of forests led to their demise. Although the productivity of managed forests was high in their first cycle, it diminished in subsequent cycles, to the extent that

> a new term, *Waldsterben* (forest death) entered the German vocabulary to describe the worst cases … . Same-age, same-species forests not only created a far less diverse habitat but were also more vulnerable to massive storm-felling. The very uniformity of species and age among, say, Norway spruce also provided a favorable habitat to all the "pests" which were specialized to that species. Populations of these pests built up to epidemic proportions, inflicting losses in yields and large outlays for fertilizers, insecticides, fungicides, or rodenticides. (p. 20)

Scott also discussed food-crop agriculture from the point of view of practices that promote legibility. The practices of monocropping in production of food for humans and other animals have been successful in producing high yields of commercially valuable products, except for the periods in which low levels of rainfall caused the ploughed land to become barren and wind-blown rather than fertile. Even at its best, however, the practice requires troubling massive chemical interventions of fertilizers and pesticides. And when these practices are transported to other climes, they can be even more counterproductive.

Of course, Scott's cautionary tales of agriculture do not support wholesale rejection of standardization. In many instances, standardization has enabled significant economic and other human benefits without major negative effects. For example, "standard time" was introduced in the 1880s because railroads needed to establish schedules that could be used in different places. Previously, each locality set its time independently of others, usually to have "noon" correspond to the middle of the solar day. This was unsatisfactory for the practices of railroad transportation, in which it is necessary to have a single time scale used by all the trains. For example, it is crucial to be able to schedule the use of tracks so that two trains will not try to occupy the same space at the same time, and achieving that is too complicated if different trains have their clocks set according to the times in different localities. The solution, accomplished in 1883, was to enforce a single time within each "time zone" and a correspondence across zones so that by knowing the time in any one place, one could say just what time it was everywhere else (Cronon, 1991). Note that both the necessity and the possibility of enforcing this standard depends on technology that has not always been available. To set clocks at the same time in different localities, one needs to be able to communicate between those localities at high speeds, or at least to carry a reliable timepiece from one place to another and to use it as the standard.[1]

Another example of standards used to enforce uniformity productively is the grading of wheat. Before the 1850s, wheat that was produced on farms in the Midwest was sent in sacks to the places where it was sold to someone who would mill it. Sacks of wheat were transported to a market city such as Chicago or St. Louis, usually by wagon or boat, where they might be sold for local use but more often were transferred to a commission merchant who arranged their shipment to another, larger city such as New York or New Orleans. The ownership of each sack of wheat remained with the original shipper until it was sold to a wholesaler or distant mill, who determined the price they were willing to pay by judging a "representative sample" of the sacks in a lot. The wheat remained in the sack it was in when it left its point of origin, and records had to be kept of each sack's progress. In the 1850s, however, Chicago merchants invented the system in which sacks of wheat were graded according to standards. For example, Number 1 spring wheat is distinguished from Number 2 spring wheat by its plumpness, purity, cleanliness, and weight, and also from Number 1 winter wheat. Once a sack (or currently, a truckload) of wheat has been given a grade and type, the grower can be paid and that wheat can be combined with other wheat of the same grade and type. The need for, as well as the possibility of, these standards depended on the development of technologies for handling grain, particularly the development of steam-driven grain elevators (Cronon, 1991). Very large amounts of grain can be stored at intermediate points to allow the economies of shipping in large containers (e.g., railroad cars). It also enables the development of commodities markets, in which people buy and sell descriptions of wheat, in the form of obligations to supply actual wheat by some specified date. As Porter (1995) remarked in discussing this case,

> A successful trader of wheat no longer had to spend his time at farms, ports, and rail terminals judging the quality of each farmer's produce. By 1860 the knowledge needed to trade wheat had been separated from the wheat and the chaff. It now consisted of price data and production data, which were to be found in printed documents produced minute by minute. (p. 48)

[1] Readers familiar with the theory of special relativity will recognize that the problem of setting clocks became a deep theoretical problem in physics when it was addressed at high levels of precision.

This system has great economic consequences. In addition to the economies of scale involved in use of grain elevators and bulk shipment, commodities markets allow large profits to be made by investors in futures if they are wise and lucky, or if a group of them has conspired and successfully "cornered" the market so they can control the price of actual wheat that they were contracted to receive at a price that has been fixed (Cronon, 1991).

Another case in which standardization is beneficial involves potency of drugs and efficacy of medical treatments. Porter (1995) discussed the evolution of standardization of drugs such as digitalis, diphtheria antitoxin, and insulin, based on biological assays that determined the minimal dose that would be fatal to a laboratory animal, would counteract a sample of toxin, or would produce a certain degree of hypoglycemia in a laboratory animal of specified species and weight. The regulation of drugs, principally now in the United States by the Food and Drug Administration, requires a complex set of institutional interactions in which standards of statistical inference and experimental control are negotiated continuously.

Uses of standards to enforce uniformity require that the standardized qualities are stable as well as measurable. They also depend on a socially organized system in which measurements are used to regulate activity. We could not have the system of standard time that we use if clocks on trains ran at significantly different speeds depending on the directions that the trains were running,[2] and the socially organized system of train scheduling, assignment of tracks and gates at railway stations, publication of timetables, and handling of variations from the planned schedules make the measurement of time actually work in the system of transportation and transport.

Similarly, the standards for grading wheat function only because there is a system for regulating the conditions of storage and shipping. The standards are effective only because the grade given to a truckload of wheat when it is purchased from a grower provides information about the wheat that someone will receive at a later time and a distant location. Of course, a system must be established in which measurements or judgments of the relevant qualities are made in the same way everywhere, and a system of regulation and inspection needs to be in place to prevent judges from assigning false grades to benefit their business associates. In addition, the systems of shipment and storage must ensure that the quality of wheat is not significantly affected in those processes, so standards must be established for storing and shipping wheat, as well as for grading it, and there must be regulations and inspections to prevent storers or shippers from mixing inferior wheat with high graded wheat and selling the mixture at the price of the higher grade.[3] It also assumes that the physical properties of wheat that are responsible for its grade and that affect the utility of the product are stable over the conditions of its use. For example, if different batches of wheat that were of the same plumpness, purity, cleanliness, and weight produced flour of significantly varying quality for making bread, perhaps because of different chemical compositions arising from the use of different fertilizers, the grading standards would not be sufficient.

The difficulties of standardizing drug potency illustrate how the instability of the product can be problematic. Diphtheria antitoxin, insulin, and other drugs were standardized by having a centralized agency—the League of Nations and later the World Health Organization—maintain and distribute samples of a standard version of the drug. The standard samples slowly wore out, despite the precautions of keeping them dried, sealed, surrounded with an inert gas such as nitrogen, and kept in the dark at a constant temperature. It was concluded that the usefulness of these potency standards depended on the experience and judgment of the laboratory teams who maintained the standardized samples, tested their potency on the local populations of laboratory animals, and advised other laboratories on their use and reliability (Porter, 1995).

Scott (1998) and Porter (1995) suggested conditions in which standardization is likely to occur and be effective. Scott summarized the conditions as follows:

> An explicit set of rules will take you further when the situation is cut-and-dried. The more static and one-dimensional the stereotype, the less the need for creative interpretation and adaptation. ... One of the major purposes of state simplifications, collectivization, assembly

[2] Nor would we have the system if any number of other nonuniformities or different uniformities were operating than the one we depend on. Many wonderful possibilities were invented by Lightman (1993), any of which would wreak havoc on our system of temporal bookkeeping.

[3] Even with material like wheat and a standardized method of grading, the socially organized system of storing and marketing can be manipulated to give some participants unfair advantages. Elevator operators were able to increase their profits by mixing grain that they purchased in a higher grade with grain that they purchased in a lower grade, as long as the resulting mixture was at or above the threshold of the higher grade. Farmers complained, newspapers and elected officials sided with the farmers, and the Board of Trade and the farmers succeeded in getting laws passed that prohibited mixing wheat of different grades (Cronon, 1991).

> lines, plantations, and planned communities alike is to strip down reality to the bare bones so that the rules will in fact explain more of the situation and provide a better guide to behavior. To the extent that this simplification can be imposed, those who make the rules *can* actually supply crucial guidance and instruction. This, at any rate, is what I take to be the inner logic of social, economic, and productive de-skilling. If the environment can be simplified down to the point where the rules do explain a great deal, those who formulate the rules and techniques have also greatly expanded their power. They have, correspondingly, diminished the power of those who do not. To the degree that they do succeed, cultivators with a high degree of autonomy, skills, experience, self-confidence, and adaptability are replaced by cultivators following instructions. (p. 303)

In addition, as Porter argued, the imposition of standards that objectify professional practice represents dominance of the need for centralized regulation over the prerogatives of expert judgment and practice, which can be maintained only if the professional community is trusted in the broader society.

Standards in Mathematics Education

Some discussions of standards in mathematics education have advocated their use in ways that would attempt to enforce strong forms of uniformity by characterizing desired outcomes of mathematics learning as knowledge and skill that would be specified in detail. Setting aside the issue of whether the particular kinds of knowledge and skill that people have in mind are the most important aspects of mathematics to be learned, there are substantial reasons, based on research findings, to object to that kind of standards.

The approach taken in the 1989 NCTM *Standards* document was to express qualities that are valued in practices of mathematics learning and teaching, expressing a vision of practice that can be used as a resource in planning, designing, and evaluating resources and activities. Well-formulated standards can be a helpful resource for formulating policy; for designing curriculum materials and other learning materials; and for planning, conducting, and evaluating practices of teaching and learning. The 1989 *Standards* were written in this spirit, as was made clear in the document:

> A standard is a statement that can be used to judge the quality of a mathematics curriculum or methods of evaluation. Thus, standards are statements about what is valued. ... When a set of curricular standards is specified for school mathematics, it should be understood that the standards are value judgments based on a broad, coherent vision of schooling derived from several factors: societal goals, student goals, research on teaching and learning, and professional experience. (pp. 2, 7)

To develop standards that can facilitate mathematics education, we need to consider the kind of environment in which we want mathematics learning to occur. The agricultural systems that Scott (1998) discussed provide analogies that make some aspects of the choice helpfully concrete.

The contrast between natural-growth forests and scientifically managed forests is particularly stark. Many of us did most of our in-school learning in a physical arrangement that resembles the straight rows of trees in a managed forest, arranged to allow the teacher the best possible access to the actions of every student at all times. Like the ideal of a scientifically managed forest, the goal of learning in a behavioristically managed classroom is uniform growth of a homogeneous product.

A natural forest seems analogous to everyday learning outside of school, where people develop the ability to participate in the significant activities of their communities through observation and guidance by their more experienced compatriots. As with a natural forest, the activities that people learn in everyday environments are those that are afforded in the community's practices. There is now considerable research showing that significant learning of mathematical reasoning and other aspects of literacy occur in natural environments (e.g., Nunes, Schliemann, & Carraher, 1993). Important aspects of mathematical learning, however, are not optimized in naturally occurring activities, particularly the coherence and systematicity of mathematical concepts and methods.

The contrast of polyculture and monoculture in food-production agriculture presents an even more realistic analogy between choices of school mathematics learning environments. Quite typically, the school day is divided into plots of time, in the form of a schedule, with each plot assigned to a single topic in the curriculum—just as plots of land are laid out, each to be planted with a single crop. The topics that are given time in the curriculum are chosen for their presumed value in the marketplace of skills and knowledge that contribute to the society's economic and cultural capital. Within each assigned lesson, teachers and students engage in routines that are designed to promote steady growth by all students. Of course, not all students grow at the same rate, and that leads to separation of the students into tracks, where uniform growth can be approximated better. This procedure is strongly analogous to monocultural agriculture, where a single crop is selected, at least for a large field of land, to produce optimal value in the prevailing market. Surely parts of the field would provide a better environment for growing something else, but the main criterion is how to maximize profit when the harvested crop is sold, so those parts of the field that can be altered to conform to the overall plan will be modified. The variability that remains will be tolerated, with the understanding that although the chosen crop will not grow as well in some places as in others, those failures will be relatively inconsequential in relation to the overall efficiency of planting and harvesting a single crop with large-scale machinery.

People who believe that learning mathematics is a process that can be reduced to a set of rules will probably prefer a mathematical learning environment analogous to monocropping. But there are reasons to consider productive learning in mathematics (as in other domains) to be more complex and emergent, and if that view is taken, polycultural learning environments are probably preferable. This preference is recognized in the 1989 NCTM *Standards*' emphasis on mathematical connections. According to the *Standards*, mathematics is not learned well if the mathematical activity is treated in isolation from the rest of students' experience and understanding. Mathematical discussions can be organized so that students' informal understanding of quantities plays a salient role in the classroom conversation. Different opinions can be encouraged by the teacher and considered in relation to one another in the class's process of constructing meaning. (An example from Lampert, 1990b, is discussed below under patterns of discourse.) Discussions can also be arranged in which children's everyday experiences with quantities are important as a basis for developing formal mathematical representations and concepts. (The Algebra Project—Moses & Cobb, 2001; Moses, Kamii, Swap, & Howard, 1989; Silva, Moses, Rivers, & Johnson, 1990—discussed below under examples of successful programs, emphasizes such discussion in its curriculum.)

Likewise, when students' activities are organized as projects, their mathematical thinking and learning interact naturally with their understanding of concepts and ways of thinking in other domains, including other topics studied in school. (This interaction is illustrated by several programs discussed below, including the study conducted by Boaler, 1997; the curricula developed by Goldman, Moschkovich, and the Middle-School Mathematics through Applications Project (MMAP) Team, 1995; and the Jasper Project of Bransford, Zech, Schwartz, Barron, Vye, and the Cognition and Technology Group at Vanderbilt, 1996.) When the learning of mathematics is mixed with students' non-school experiences and with reasoning from other school domains, the learning is more complicated to organize and manage and is less predictable; students' progress is also harder to assess. People who prefer such an approach do so because they believe—with growing evidence to support their belief—that it engages most students more effectively and that the students' learning is more robust and general. The approach is analogous to polycultural agriculture, in which different crops are planted in locations where they are likely to do well. The diversity of neighboring crops is a virtue; it balances the nutrients that are needed across the field and protects them from wholesale infestations. The kinds of standards that can help in a polycultural learning environment are not rules to be followed; rather, they are guidelines—ways of conducting experiments and recognizing successful episodes—and heuristics for improving the prospects for productive growth by different students in interaction.

A common approach to formulating standards is to specify a list of mathematical things that students should know at each grade level (e.g., California State Board of Education, 1998). This approach is problematic for many reasons, not least because the question of whether a student knows a particular mathematical thing is indeterminate. Whether we think of knowledge as a class of behaviors that we can observe or as a cognitive structure that a student has acquired, there is strong evidence from research that whether students demonstrate that they have acquired that knowledge depends very strongly on the circumstances of testing. For example, Lave (1988) and Nunes, Schliemann, and Carraher (1993) found that grocery shoppers or street merchants performed mathematical inferences about prices successfully in their practices, but if they were asked to perform the paper-and-pencil calculations corresponding to those inferences, they were much less successful. The question of whether these people had knowledge of that arithmetic is simply not answerable; consequently, it is pointless to try to set standards that depend on assessing whether that knowledge has been acquired by students. Having or not having a piece of "knowledge" is not stable enough to serve as the basis for educational standards. It is as though a truckload of wheat were to be graded as No. 2 spring when it was received at the local grain elevator, but its quality were to be No. 1 spring when it was tested at a flour mill. Under those circumstances, the system for setting standards for wheat would be useless.

Another reason for uniformity in standards involves the need for students to be able to progress reasonably through their school careers in spite of moving from one school to another. This movement creates a problem that is analogous to the need for standard time to support the railway transport system. Some discussions of standards assume that the problem of transferring from school to school could be solved simply by requiring each school to cover the same content in the same time periods in the same grades and tracks. Many of these discussions, however, leave open the question of methods of teaching, so that different schools or groups of students would receive instruction by different methods. Research findings (e.g., Boaler, 1997) have shown that when learning activities are organized differently, students acquire different forms of knowledge. Students would be unlikely, then, to be able to transit smoothly in either direction between a school in which mathematics is taught didactically and one in which they are expected to participate actively in a discourse of formulating and evaluating problems and explanations, even if the same mathematical topics have been discussed in both places.

Assessing mathematics learning is not like grading wheat, where a successful grading system can be achieved by combining a relatively simple judgment with technologies, regulations, and inspections for maintaining uniformity in the conditions of storing and shipping a product. Education is a complex domain of practice. Reduction of representations of learning to a single-dimensional scale, analogous to the monetary revenue of a crop, harmfully distorts our understanding of learning and teaching.

The 1989 NCTM *Standards* rest on an assumption that "'knowing' mathematics is 'doing' mathematics. A person gathers, discovers, or creates knowledge in the course of some activity having a purpose" (p. 7). This view focuses mathematics education on the preparation of students to participate in activities rather than on their acquisition of skills and cognitive structures, and it encourages the formulation of standards in terms of kinds of mathematical activities that are important for students to participate in. Specifications of activities can be trans-situational in ways that specifications of knowledge and skill cannot. Further, the kinds of activities that are made available to students can be organized to afford meaningful participation by students who bring different kinds and amounts of relevant experience to them—so the need for uniformity across schools is not a requirement that every school be at the same place at the same time. Finally, as Porter (1995) pointed out, the legitimacy of practices and standards that allow practitioners flexibility depends on trust in the expertise of those practitioners. The concerns we have in formulating standards to influence discourses of practice need to be inherently intertwined with issues of professional development.

How Research Findings Can Shape Standards

I assume in this section that the main uses of the NCTM (1989, 1991, 1995) *Standards* documents are as resources for reflective discourse and design regarding educational practice rather than as instruments for enforcing uniformity of practices and products. Of course, the discussions that practitioners and others have about mathematics education can lead to more uniform practices. Discourse shaped by those standards can distribute a set of ideas in the community, and if practitioners find the aims, assumptions, and methods described in the standards to be valuable and become committed to them, the standards will bring about more uniformity, but the process will be quite different from the enforcement of uniformity for standards like those in grading wheat.

If the questions that practitioners and others need to discuss include those that I mentioned above (e.g., What are we accomplishing? What could we accomplish? How could we do that? What would we be giving up?), and if standards can affect that discussion by providing reasons that the practices they characterize are valuable, indicating how they can be achieved and evaluated as well as how different versions of practice would be more or less effective, then what kinds of research findings can guide the formulation of such standards? I believe there are two general answers to this question.

First, research can affect standards about the range of possible practices to be considered. Standards provide recommendations for practice, and these recommendations should be realistic—at least, as aims toward which to strive. Practitioners and the public need to understand what practices are possible, how those practices can achieve important aims in education, whether they are feasible, and what conditions may be favorable to their effectiveness. Results of research provide information about practices of teaching and learning, showing how they can be conducted and providing information about possible outcomes of the practices studied. Another general answer is that research develops ways of conceptualizing and understanding activities of learning and teaching, and these conceptualizations and understandings can be appropriated in discussions and debates about choosing between alternative practices and evaluating the achievements of schools, teachers, and students.

Models of the Research-Improvement Interaction

Discussions of relations between research and practice usually focus on the problem of translating findings from research into guides for practice. A recent discussion by Stokes (1997) argued that these prevalent discussions suffer from being based on a flawed model of the relations between research and practice. We usually treat the relation as being linear, in both a conceptual and a temporal sense. Conceptually, we believe that there is a single dimension that extends from basic research, through applied research, through design and implementation, to dissemination and evaluation. This RDDE (research, development, dissemination, evaluation) model causes us to locate any research activity somewhere on the dimension, so that if a piece of work is more applied, it necessarily is less basic. Temporally, we assume that research precedes practice—so the results of basic research are used in designing applied research, the results of applied research are used in designing resources and activities of practice, and the results of design work are used in implementing and disseminating changes of practice, which are then evaluated.

Stokes (1997) argued that a more valid model locates research and improvement in two dimensions rather than one. Some programs are initiated mainly to advance general scientific understanding, with little concern for solving a problem of general social concern. Other programs are intended mainly to develop technologies and practices that improve the functioning of an aspect of social life, with little concern for improving the concepts and principles that are available to provide fundamental understanding and explanation. Our mistake, however, has been to assume that these two kinds of activity exhaust the enterprise of research and improvement. Instead, Stokes argued that important activities also exist that contribute to general scientific understanding and to the improvement of technology and practice for an important general social problem. Representing the situation with this idea requires a two-dimensional space in which an activity may be high or low in its "quest for fundamental understanding" and high or low in its "considerations of use."

Stokes's (1997) representation, thus, has four quadrants. He used Neils Bohr's research that led to a new model of the atom to epitomize activity that is high in its quest for fundamental understanding and low in its considerations of use. He used Thomas Edison's invention of the light bulb and development of a distribution system for electrical energy to epitomize activity that is high in its considerations of use and low in its quest for fundamental understanding. And he used Louis Pasteur's research that led to a new theory of disease and fermentation as well as the development of fundamental advances in medical practice and food processing (e.g., sterilization, pasteurization, refrigeration) to epitomize activity that is high both in its quest for fundamental understanding and in its considerations of use.

The two-dimensional view provides a different way of conceptualizing different research-and-improvement activities. Instead of having to locate activities as being basic or applied, we can consider the potential contribution of various activities both with respect to developing concepts and principles to advance fundamental understanding and with respect to contributing to the improved functioning of a system of general social importance. It also provides a different sense of the temporal relations between research and improvement of practice. Rather than having R, then D, then D, and then E as the default, we can understand that the various activities of research and the changes in practice usually interact in more complex and symmetric ways, with conceptual insights affecting changes in practice and with results of efforts to change practice contributing to the formulation and evaluation of hypotheses involving fundamental explanatory concepts and principles.

Although Stokes (1997) did not discuss educational research, his proposal for a two-dimensional view is productive for educational research and improvement as well as for the domains of physical and biological science and technology. Recent reports of both the National Academy of Education (NAE, 1999) and the National Research Council (NRC, 1999) have recommended giving increased priority to research and development activities in Pasteur's quadrant. The NAE report calls this kind of research *problem-solving research and development*. Other names for it have been *design experiments* (Brown, 1992; Collins, 1992) and *interactive research and design* (Greeno & the MMAP Group, 1998). In projects of this kind, researchers, program developers, and professional educators work collaboratively with a shared commitment to make a concrete improvement in some aspect of educational practice. The improvement may be in the educational effectiveness of a local school or district, a new curriculum that affords more meaningful learning than those currently in use, or some other concrete change. Unlike traditional development activities that are expected only to use the results of past research, these collaborative projects also take on the goal of including research that develops and evaluates general explanatory concepts and principles that contribute to fundamental scientific knowledge and understanding. A defining feature of these collaborations is that researchers, developers, and professional educators all share in the accountability of the project to bring about improved practice as well as to contribute to the advancement of fundamental scientific knowledge and understanding.

The NAE (1999) report recommended that projects organized as problem-solving research and development should also be committed to developing resources for the local innovations to "travel." That is, a project's outcomes should include information and material resources that can be used at other sites where people want to adopt the innovation. The exact nature of those resources is not yet clear, but they would probably include results of analytical research that identifies conditions in the local site that helped the project succeed or hindered its success, material resources such as handbooks and implementation guides, and consultants who could assist people who want to use the project as a model.

Stokes's (1997) model of research and improvement involves complex and messy interactions among efforts to advance fundamental scientific understanding, design improved educational resources and practices, evaluate progress in educational reform, and support travel of educational innovations. The contrast of this model with the prevalent RDDE model provides another analogy to Scott's (1998) contrast between monocropping and polycropping in agriculture. We are accustomed to thinking about basic research, applied research, development, dissemination, and evaluation as separate activities, carried out by different communities of workers, who metaphorically produce their different crops in socially segregated fields of activity. By organizing research and improvement efforts mainly according to this model, we have created predictable difficulties of communication between the activities, given that we segregated them to begin with. The difficulties of translating research findings into practice, and of formulating research problems that address important educational problems, are well known and seemingly intractable, given the ways in which we have organized the practices of research and improvement in education.

The prevailing RDDE model has had advantages for managing the enterprise, as monocropping generally does. Funding announcements, university departments, and individual scientists can conveniently partition work into the categories of the model, with some researchers, projects, and academic programs identifying themselves as basic researchers (usually affiliated with scholarly or scientific disciplines); others, as applied researchers (in education); and still others, as developers, implementers, or evaluators. The kind of project that Stokes (1997) called *use-inspired basic research*, and that the NAE (1999) and NRC (1999) reports advocated, would blur those boundaries, making it difficult to account separately—whether in separate budget lines or departmental responsibilities—for the functions of advancing knowledge and understanding and of improving educational practice. But these disadvantages for management may be outweighed by making the research-and-improvement system more productive and effective for advancing knowledge and improving education.

If researchers move their activities more in the direction of polycultural research and improvement, we should not assume that all the resulting activity will focus on one kind of product. The field will still need to advance the fundamental sciences of learning, cognition, communication, and social interaction through studies of educational processes, just as it will need to develop more effective resources and practices in local settings and to create ideas and resources to support the travel of productive innovations. Participants in the projects of problem-solving research and development will have responsibilities in addition to designing and improving the local practices that they focus on as a group. Researchers will need to use materials generated in those projects to contribute to the empirical and theoretical literatures of research disciplines, developers will need to transform locally effective informational resources into products that can be used elsewhere, and educational professionals will need to incorporate the insights of the projects into their own and their colleagues' practices. In agriculture, practices in which various crops are grown in proximity do not remove the need for a variety of food products to sustain life, and practices in which various kinds of progress are made in educational research will not remove the need for a variety of products that advance scientific understanding and improve educational resources and practices.

Projects that combine fundamental and applied research with design and development generally have cycles of activity in which design and analysis alternate with trials of provisional versions of the project's product (e.g., Collins, 1992; Gravemeijer, 1995; Greeno et al., 1999). These cycles provide significant opportunities for research that sheds light on the fundamental processes of learning, cognition, and communication involved when students and teachers use different versions of the materials and practices developed at different stages of the project; on characteristics of products that make them more or less successful; and on the social and cognitive processes involved in evaluating the progress that the group is making.

Advocates of increased emphasis on collaborative projects of problem-solving research and development are careful to point out that these projects should not become the only kind of work done in the field of educational research. Such projects probably represent the domain that would be most profitable to increase in our research-and-improvement agenda, but it will continue to be important to have projects prompted mainly or entirely by concerns for advancing fundamental understanding or by concerns for improving the effectiveness of educational institutions and practices. The multiple agendas of improved theoretical and practical understanding and knowledge are best pursued in close integration, but a variety of methods will be needed to make significant progress in all aspects.

Research in Support of Polycultural Education

The most prevalent form of educational research has been located in Edison's quadrant of Stokes's (1997) model, concerned mainly with comparing and evaluating programs to judge which was more successful in some aspect of its functioning. This approach in educational research has supported practices of legibility in managed school learning, analogous to the support for monocropping in agriculture by research identifying the single best hybrid or dosage of fertilizer that then becomes the single crop or fertilizing regime prescribed for all farmers. A view of school learning that identifies well-defined tasks and analyzes requirements for their performance into components of skill and knowledge was developed in behaviorist psychology (e.g., Gagné, 1968), and a more refined version of this program has been developed in cognitive science, where performance is analyzed in terms of components of information-processing procedures and stored knowledge structures (e.g., Anderson, 1983).

Both the behaviorist and cognitive perspectives develop representations of knowledge for solving problems and answering questions that can be used as blueprints in the design of instructional sequences. When these sequences are carried out methodically, students who are sufficiently engaged in the learning activity can acquire the prescribed knowledge quite reliably. To be most effective, the regimen should be applied to each individual, keeping track of that student's progress in acquiring the components of behavioral or cognitive structure to ensure that the prerequisites for adding the next component are in place before that next step is attempted and to avoid unnecessary presentations of material that has already been learned. Although individual tutoring is generally thought to be the ideal form of instruction in these perspectives, it is not cost effective. Efforts have been exerted to develop technologies in which students could acquire the prescribed skills and knowledge by interacting with programmed textbooks (e.g., Skinner, 1958) or interactive computer programs (e.g., Anderson, Boyle, & Reiser, 1985). Schools, however, have not had the resources needed to arrange even an inanimate tutor for each student. The dominant solution, therefore, has been to attempt to create as much homogeneity in each class as possible, so that a teacher can present instruction that will be uniformly beneficial to all students in the class.

Research that has provided much of the scientific basis for monocropping in school instruction has a closer relation to the field of agriculture than mere analogy. The research technology of experimental design and statistical analysis that was developed for agricultural experiments was adopted quite literally for the empirical study of efficacy of different instructional methods, as it was for field trials in medical treatments. The strategy in much educational research, as in clinical field trials, has been to devise two instructional treatments—alike in all ways except for some variable (analogous to different hybrids or amounts of fertilizer)—to apply these treatments to samples of classrooms, to measure the learning that each student achieved with each treatment, and to draw conclusions about which treatment is more effective on average. A significant difference in favor of one of the treatments has been taken as support that the variable that distinguished the treatments makes a difference in students' learning. In principle, this research program is supposed to supply a collection of variables that should be taken into account in designing curriculum materials and in teaching, so that by conducting instruction with more favorable values of these variables, the average amount of measured learning will be optimized.

A growing sense among educational researchers is that this methodology of research and design is unproductive, but there is disagreement as to why. One view is that the paradigm is correct but that its use has not been sufficiently rigorous. This view, a version of Scott's (1998) high modernism, maintains faith in the epistemic efficacy of abstract prescriptive knowledge, acquired through identification of relevant variables from controlled experimental comparisons. It recommends an emphasis on field trials, analogous to clinical trials in medicine, to establish the most effective materials and practices, presumably leading to standards that would require those materials and practices in all schools—that is, to a stronger form of research-based monocropping in educational practice.

The other view questions this modernist paradigm, holding instead that learning and teaching must necessarily be considered as complex, situated activities. Research in this perspective is messier than research that neatly separates basic inquiry, applied inquiry, and design and development; it often takes the form of problem-solving research and development of the kind characterized by Stokes (1997) as being in Pasteur's quadrant. It also attempts to contribute to the development of standards and to educational improvement in ways that acknowledge that the responsibility for determining educational practice is fundamentally in the hands of professional educators—administrators and teachers—and that the optimal role of research is to provide these professionals with information that can help them in their challenging work, especially in discourse in which they plan, reflect on, and evaluate their ongoing efforts to continually improve their practices.

Overview of a Situative Perspective

The framing assumptions that I refer to as the *situative perspective*[4] have been developed and discussed recently by Greeno and the MMAP Group (1997, 1998), Lave (1988), Lave and Wenger (1991), Suchman (1987), Wenger (1998), and others. I use the term *situative* to refer to a broad set of theoretical programs and ideas, with roots in American pragmatism, especially Dewey (e.g., 1916/1966) and Mead (e.g., 1934), as well as in Soviet activity theory (e.g., Vygotsky, 1934/1987), and with current lines of research in sociocultural theory (e.g., Forman, this volume), ecological psychology (e.g., Reed, 1996), and ethnographic and ethnomethodological studies (e.g., Hutchins, 1995).

As I understand it, research in the situative perspective is distinguished by its analytic focus. Situative research combines concepts and methods that have been developed in two relatively separate research paths. One of these paths has been in psychology, which has focused mainly on individual behavior and cognition (Greeno, Collins, & Resnick, 1996). The other path has been in the social sciences, especially in studies that have taken an ethnographic perspective, focusing on processes of interaction, including communication, between individuals (Rogoff & Chavajay, 1995). Behaviorist and cognitive studies have provided strong analyses of learning and motivation, including detailed analyses of the organization of information that students need to learn to succeed in school tasks. In contrast, interactional studies have provided strong analyses of ways in which individuals coordinate their actions to participate in conversations and other collaborative activity.

Situative research attempts to combine these perspectives in studies that analyze both the informational aspects of cognition and behavior and the interactional aspects of socially organized activity. In the situative perspective, the most fundamental aspect of learning is *participation* in social practices and becoming more able to participate successfully, both to contribute to group functions and to achieve personal accomplishments and growth (Greeno, 1997). Learning the conceptual contents and technical methods of a subject-matter domain is important because it contributes to a person's ability to participate in the community's activities. This perspective involves a shift in the focus of research away from cognitive and behaviorist perspectives—where individual learning of concepts and skills is the primary focus—toward considering individual's acquisition and use of knowledge as aspects of their participation in social practices.

The situative perspective conceptualizes mathematical knowing as sustained participation in mathematical practices. Situative research on mathematical learning, then, focuses on how students become more successful in participating in mathematical practices and how they develop identities as mathematical knowers and learners. Taking this perspective involves viewing such processes as problem solving, communicating, reasoning, and understanding mathematical connections as aspects of social practices. Students participate in these practices, contributing positively or negatively to those activities of the community—usually the classroom—that involve the processes.

Situative research on classroom activities focuses on patterns of interaction, including ways that people interact with one another as well as ways that individuals or groups interact with resources such as textbooks or computers. Research topics include the study of how teachers and students organize their communication in class discussions or in small group conversations. They also include the study of group or individual activities in

[4] I prefer the term *situative*, as a modifier of *perspective*, *framework*, or *theory*, rather than *situated*, as a modifier of *action*, *cognition*, or *learning*. Phrases such as "situated learning" invite the misconception that there are some kinds of learning that are situated and others that are not situated. Instead, the situative perspective assumes that all learning, cognition, action, and so forth are situated; the differences have to do with where and how these processes are situated, not whether they are.

which students read a textbook, solve problems in an assignment, or use an interactive computer program.

The kinds of events that are most natural to examine as instances of understanding or reasoning are conversations in which conceptual mathematical understanding and inferences are constructed jointly by the participants. Interviews of students by teachers or researchers are a special case of interaction; their purpose is to display what a student is able to contribute to the mutual understanding that the interviewer and student jointly construct. Written tests are an even more unusual type of interaction; the student interacts only with written material that presents statements of problems or questions, constructing representations that are interpreted as answers. In all these cases, situative analyses differ from cognitive or behaviorist analyses. The former focus on interactions in which students participate, whereas the latter focus on an individual student's performance in the context of a test, interview situation, or interaction with other people. Situative analyses can include representations of the informational contents of interactions, and those representations can be used to characterize the mathematical contents of a student's contributions. In the situative perspective, however, it is harder to forget that whatever a student does in an interaction is jointly constructed with other people and material systems, and one is less tempted to make simplistic inferences about how much "knowledge" of mathematics the student "has."

To characterize an individual's knowing, understanding, and learning, we need to identify regularities in the ways in which he or she participates across a variety of situations (Cole, 1999; Shafer & Romberg, 1999). Standard paper-and-pencil tests can be included in those situations, of course, but the common tendency to give dominant emphasis to those performances is misguided, given the peculiarities of the interactions that occur in tests. By considering individuals' trajectories of participation in mathematical activities across time in a community (e.g., a classroom) and across communities, a situative analysis can characterize an individual's *identity* as a mathematical knower and learner.

Some aspects of an individual's mathematical identity involve her or his consistent performances in activities that are attributable to routine skills and informational knowledge. However, the situative perspective raises additional questions about a student's mathematical identity and learning. Other aspects of students' mathematical identities involve their relations with other students, their teachers, and the concepts and methods of mathematical activities as they participate, coming to function in ways that are more or less engaged or alienated and more or less effective or incompetent in their participation. These other aspects of identity include

- students' understandings of what it means to know mathematics;
- expectations that they and others have of their success or failure in performance in mathematical tasks;
- whether they are generally engaged in mathematical activities or alienated from them;
- the ways in which they regularly participate in discussions involving mathematics—for example, whether they regularly contribute explanations to a group's understanding of situations involving mathematical reasoning and inquiry; and
- whether they contribute to mutual understanding by appreciating and explaining assumptions involved in their thinking and that of other participants.

A behaviorist or cognitive characterization of a student's mathematical identity involves a trajectory of that student's skills or knowledge structures. Such a trajectory can be included as part of a student's identity in the situative perspective, which also recognizes that the construction of individual students' reputations for having or lacking skill and knowledge is one of the main activities of mathematical classroom communities. Thus, a situative analysis can include a consideration of how the practices of a school learning community create identities of success and failure, including consideration of the consequences of those identities for different students' opportunities to participate productively in the practices of mathematical learning (e.g., McDermott, 1993; Mehan, Hertwick, & Meihls, 1986; Verenne & McDermott, 1998).

Aspects of Practice in a Situative Perspective

The situative perspective invites consideration of educational practices as systems in which students, teachers, and others participate. This view emphasizes that the ways in which students and teachers go about learning have consequences not only for what they learn to do but also for what kinds of learners they become and how they understand what it means to learn and to know.

Different aspects of learning practices can be understood as depending on different assumptions about what is important in learning (A. L. Brown, 1994; J. S. Brown,

1991). I believe that we can usefully consider three general aspects or orientations of practice in the learning of mathematics: *skill-oriented* aspects, *understanding-oriented* aspects, and *participation-oriented* aspects. These orientations correspond to different aspects of knowing in mathematics—or in any domain. Practices differ in the relative emphasis given to these different aspects of knowing, as well as to the ways in which the different aspects are worked out in relation to one another.

Aspects of learning mathematics that emphasize skill acquisition are well understood. It is advantageous to have clear instructional goals, clear presentations of information and examples of the skills to be acquired, well-organized routines for students to practice the skills, and clear feedback to learners about what they have learned and what they still need to acquire (e.g., Brophy & Good, 1986). Participating successfully in well-organized skill-oriented practices involves attending and practicing carefully to learn the steps of procedures and the appropriate conditions for applying them. Students who succeed in skill-oriented aspects of learning do so by becoming good and careful listeners and conscientious practicers.

Aspects of learning that emphasize understanding focus on the meanings of mathematical concepts and explanations of solutions to problems. Much has been learned in recent years about the difficulties that students have in understanding mathematical concepts and the ways in which learning with understanding can be facilitated (e.g., Siegler, this volume). Recent research has shown that students have significant intuitive understanding that can be the basis of their developing the conceptual understanding they need for progressing in the mathematics curriculum (e.g., Greeno, 1992; Harel & Confrey, 1994; Lesh & Landau, 1983). Participation structures of understanding-oriented learning practices can be very didactic, with the teacher responsible for explaining meanings and relations between concepts clearly, or they can be quite constructivist, with students engaged in activities using materials that embody the concepts they need to come to understand.

Aspects of learning that emphasize participation focus on activities in which students formulate and evaluate questions and problems, as well as propose, evaluate, and explain solutions, answers, examples, and arguments. Metaphorically, learning mathematics is like learning one's way around in an environment, such as a city or a kitchen, in which there are various resources and materials that one learns how to find and use (Greeno, 1991; Schoenfeld, 1998). The activities of learning need to be organized in a sequence that provides a trajectory in which students progress from having informal notions to gaining a more formal and systematic understanding and use mathematical representations and methods more productively. At times, the leading activity may be inquiry into mathematical meaning, with discussions organized so that students' intuitive understandings of number and quantity support their abilities to contribute. At other times, the leading activity may be work on a project in which mathematical concepts and methods play a significant instrumental role. Participation structures in these inquiry-based or project-based activities give students major responsibilities for contributing to the understanding and successful learning of members of their classroom community.

Some Consequences of Learning Practices

In the situative perspective, participating in classroom learning practices is fundamental to students' learning of mathematics. This view is supported in the results of research. In an unusually systematic comparative study, Boaler (1997) spent three years in two schools, studying an age cohort in each school as the students progressed through school mathematics from ages 13 to 16. The students were from very similar, mainly working-class backgrounds, but they had very different mathematics learning experiences. In one school, mathematics was taught in a traditional didactic manner. In the other, the students worked on open-ended projects that typically lasted about three weeks. The projects were designed to engage students in mathematical thinking and reasoning, and the teachers provided mathematical instruction in the form of assistance with the students' projects. In January of the third year of the study, students in the project school were switched to a program of preparation for their GCSE (General Certificate of Secondary Education) examination, with practice on procedures involving textbook questions, worksheets, and past exam questions. At the end of the year, students from both schools took the GCSE examination. Their test scores were about equal overall, but there were some significant differences in the patterns of their performance. In the textbook school, students' scores on procedural items were higher than they were on conceptual items; this discrepancy did not occur for students in the project school. In open-ended assessments, which Boaler designed and administered, students in the project school were more successful than students in the textbook school were.

On the basis of the students' performance and many responses in interviews, Boaler (1997) presented

> a case for two different forms of mathematical knowledge. One of these forms of knowledge ... is inert, inflexible, and tied to the situation or

> context in which it was learned (Whitehead 1962). The other form of knowledge … . is more adaptable, usable, and relational in form (Lave 1993). (p. 96)

More specifically, the mathematical knowledge of students in the textbook school depended on remembering procedures and responding to cues as to which procedure to apply. Students in the project school had learned to think about situations, to understand the available information, and to generate ways to approach problems. About the students in the textbook school, Boaler wrote,

> First, they were not encouraged to think about or understand the methods they used in class, so they were not able to consider the methods they had learned and make informed decisions about the ones they should use. Second, they believed that mathematics was about remembering methods, rather than thinking about questions. In class, they had all been taught to learn methods and to practise them, not to adapt them or think about them. The third problem seemed to be caused by the students' perceptions of the mathematics classroom as a distinct "community of practice" (Lave, 1993, 1996) in which mathematical rules and procedures were learned that were specific to that community of practice. (p. 104)

In contrast, the students in the project school

> did not view mathematics as a formalized and abstract entity that was useful only for school mathematics problems. They had not constructed "boundaries" (Siskin, 1994) around their school mathematical understandings in the way that the [textbook school] students had … . It was the perceiving and interpreting of situations that seemed to characterize the main difference between the students at the two schools. … The [project school] students … were not as well versed in mathematical procedures, but they were able to interpret and develop meaning in the situations they encountered. (pp. 105–106)

Evidence about consequences of didactic teaching also has been provided by Schoenfeld (1988). He reported a study of problem solving by high school geometry students who worked on a construction problem unsuccessfully in spite of the presence on the chalkboard of the proof of the geometric property that was needed to solve the problem. Eventually, Schoenfeld asked the students why they had not used the information on the board as the basis for their approach to the problem. They replied that they believed that proofs are not relevant to construction problems. Schoenfeld (1985) also surveyed students' beliefs about mathematics, asking such questions as how much time it usually takes to solve a problem in mathematics. In their responses, students expressed beliefs that mathematical knowing is mainly a matter of having acquired procedures. These data, along with Boaler's (1997) findings and other results, support the conclusion that in didactic teaching that emphasizes students' learning procedures, most students learn a set of practices in which they understand knowing in mathematics as acquiring a collection of procedures that are applicable to a narrow kind of problem rather than as assembling a set of resources that support thinking and understanding powerfully and generally.

Examples of Successful Participatory Learning Practices

Several projects have developed and studied practices of mathematics education in which students participate more constructively than in traditional mathematics instruction. Other studies have not included the kind of contrasting cases that Boaler (1997) was able to study, but the results demonstrate that the goals of more meaningful learning in mathematics can be adopted in school settings with results that are encouraging for the effort to reform mathematics education significantly.

The QUASAR project (e.g., Stein, Grover, & Henningsen, 1996) included several urban school sites involved in progressive change in middle school mathematics education. Among their findings was that the requirements for student reasoning and understanding in tasks used in the various curricula varied significantly. This finding suggests that standards should emphasize the kinds of reasoning and understanding that students need in order to develop the recommended mathematical capabilities rather than merely suggest the kinds of tasks that can afford these kinds of reasoning and understanding but do not necessarily do so.

Cobb and his colleagues (e.g., Cobb et al., 1991) have developed and studied a curriculum focused on mathematical problem solving and understanding in first and second grade, concluding that for students to participate actively in productive mathematical discussions, classrooms need to develop discourse norms. They distinguish between general *social norms*, which involve politeness in turn-taking and respect in encouraging and considering others' opinions and reasoning; *mathematical norms*, which involve complying with conventional mathematical meanings and reasoning methods; and *sociomathematical norms*, which involve general aspects of discourse that are distinctively mathematical, such as the kinds of considerations that are appropriate for use in mathematical explanations and arguments. This distinction suggests that standards for mathematical communication need to be more specific than merely encouraging students' talking. Instead, standards should specify the characteristics of talk that can be productive and should attend to the need for development of practices, including norms, that support the kinds of talk in which students learn the desired mathematics.

Supporting this suggestion are studies by Ball (e.g., 1993), Lampert (e.g., 1990b), and others who have developed practices in their teaching in which they engage students in significant substantive discussions of mathematics. These discussions emphasize that for teachers to support students' deeper participation in mathematical discourse, they need to develop practices that differ significantly from the customary didactic practices in mathematics education. Standards should include recommendations to support teachers' efforts to accomplish the required changes in their practices.

The Algebra Project, led by Robert Moses (Moses & Cobb, 2001; Moses et al., 1989; Silva et al., 1990), has produced a curriculum and a teacher-development program that emphasize students' participation in discourse that includes developing and evaluating alternative ways of representing quantitative concepts and considering fundamental properties of mathematical symbols. These examples are particularly relevant to the general standard on mathematical representations in the revision of the NCTM (1989, 1991, 1995) *Standards* documents. The curriculum includes arranging an activity in the community—involving a trip on public transportation—that the students experience as a class. This activity provides a shared experience that the class uses in discussions of quantitative properties and in developing symbolic representations. Godfrey and O'Connor (1995) reported an example of discourse in Godfrey's sixth-grade mathematics class in which a group of students constructed a nonstandard pictographic representation, called a vertical handspan, of the amount of difference between heights of students; another group introduced the standard symbols for inequality, which were related to the student-generated symbols in the class discussion.

Another example of nonstandard representation occurred in Lampert's teaching and was analyzed by Hall and Rubin (1998). This representation, which they called a *journey line*, shows the correspondence between distances and times in linear motion at constant velocity. Hall and Rubin traced the use of this representation in a student's journal, in his explanation to another student of why numbers were multiplied in a problem, and then in both of these students' presentation of the explanation to the class. The kinds of discussions that students can have around representations they generate can be valuable in developing their understanding and their agency regarding representational practices, and standards could make a valuable contribution by encouraging such discussions in relation to mathematical communication.

Another curriculum organized by principles of students' participation in mathematical activity is Mathematics in Context (National Center for Research in Mathematical Sciences Education & Freudenthal Institute, 1997). This curriculum embodies a view, called Realistic Mathematics Education, whose "approach to instruction is based on an epistemological view of mathematics as a human activity that reflects the work of mathematicians—finding out why given techniques work, inventing new techniques, justifying assertions, and so forth" (Romberg, 1999, p. 288). This view reflects Freudenthal's (1983) conception of "doing" mathematics, which includes a commitment to engaging students in mathematically authentic activities. In those activities, students can recapitulate the creation of the mathematical discipline and develop practices of understanding and reasoning that include constructing mathematical models, initially *of* problem situations, that then come to be generalized and understood as models *for* mathematical reasoning (Romberg, 1999).

Computational and video technologies have been touted as having strong potential for supporting improvements in education, and several studies have developed and analyzed resources for learning environments to support students' deeper participation in significant mathematical activity. One such project is the Middle-School Mathematics through Applications Project (MMAP), which developed computer programs and curricula for middle school students to work on design projects in architecture, biology modeling, cryptography, or cartography (Goldman et al., 1995). The curriculum materials are used to organize activities with software that functions as a computer-aided design system for students. The curricula also include extension and investigation units in which mathematical concepts and methods become the main topics, but these mathematical materials are related to activities that the students have experienced in their design work. Research studies of processes of teaching and learning in MMAP have provided information about aspects of successful teaching, patterns of student interaction, and students' understanding of mathematical concepts in their activity (Greeno & the MMAP Group, 1997).

In another project, the Cognition and Technology Group at Vanderbilt (e.g., Bransford et al., 1996) developed curricula and studied learning using video dramatizations, called Jasper Adventures, of situations necessitating mathematical reasoning or providing meaningful backgrounds for projects. These learning environments, like the Algebra Project and MMAP, facilitate anchored instruction in mathematics (Cognition and Technology Group at Vanderbilt, 1992)—that is, meaningful and motivating situations that require reasoning and draw on mathematics as a powerful resource—rather than presenting mathematics as a set of concepts and methods to be learned for their own sake. Recently, the Vanderbilt group has used computational technology as a resource for students to develop representational tools, called SMART (Special Multimedia Arenas for Refining Thinking) tools, to help them in their project work. An example of a SMART tool is a graphing function that supports visual estimation of the value of a variable related to another variable by a linear function. These activities of constructing representational tools are another example of learning practices in which students can construct, evaluate, and appreciate properties of mathematical representations rather than simply be taught how to construct representations following the rules of systems delivered to them.

Many people have thought that the promise of technology—especially computers—would be to provide efficient delivery of exercises for learning routine skills. Indeed, some sophisticated systems are superior to standard classroom instruction for helping students learn to solve standard problems in the curriculum (e.g., Anderson et al., 1985). But technology can also be used to support design activities, as with MMAP and the Jasper Project, or activities of exploring a mathematical domain such as geometry and then forming and evaluating conjectures (Schwartz, Yarushalmy, & Wilson, 1993). The uses of these different versions of technological resources serve quite different educational aims and require quite different classroom practices to be effective. The results of research that has examined learning when various kinds of resources are used suggest that standards should advise careful consideration both of the educational aims that technologies are expected to support and of the requirements for classroom organization and teaching if students are to learn successfully in new learning environments.

Aspects of Mathematical Knowing and Learning as Practices

Different ways in which students interact with one another and with a teacher can be considered a kind of design space for learning environments. Teachers, with their students, arrange for the students to participate in interactions in some variety of ways, and these arranged interactions are a selection from the design space of possible ways of interacting. Research in the situative perspective has provided analyses of different kinds of interaction that occur in mathematics classrooms, and these different kinds of interaction are alternatives in the design space.

Because the purpose of mathematics education in schools is for students to acquire mathematical knowledge, the central organizing concepts of school activities have to do with what mathematical knowledge is, how it is represented, how it is used, and how students can show that they have or have not learned successfully. The activity system of a classroom includes the students and the teacher, as well as such material systems as textbooks, a chalkboard or whiteboard, an overhead projector, and computers. Many material systems in mathematics classrooms are used to create representations of mathematical concepts and entities, either directly (as numerals, arithmetic expressions, algebraic formulas, or graphs) or as models or examples of mathematical entities (as with place-value blocks or fraction circles). Of course, the actual physical material of the classroom—the furniture, doorways, windows, and closets—are important as well because they provide locations for individual work, interaction with others, and opportunities to be noticed or hidden in the classroom activity. (Sometimes the physical furnishings of the class are appropriated for the mathematical discussion, for example, by measuring the length of the room to illustrate the meaning of "ten meters.") The classroom activity system also includes the conceptual domain of mathematics, to which students and the teacher relate in different ways as they proceed through the stream of their activities. I show these constituents of the activity system in Figure 21.1.

It is important to remember that the locally situated classroom, with these constituents, exists in a context in which all the constituents are interdependent with constituents of the broader social environment. The school and community provide physical resources and place constraints on educational values and practices of teachers and students. The communities of professional mathematicians, mathematics educators, and influential members of local communities influence the emphasis given to different mathematical topics and aspects of

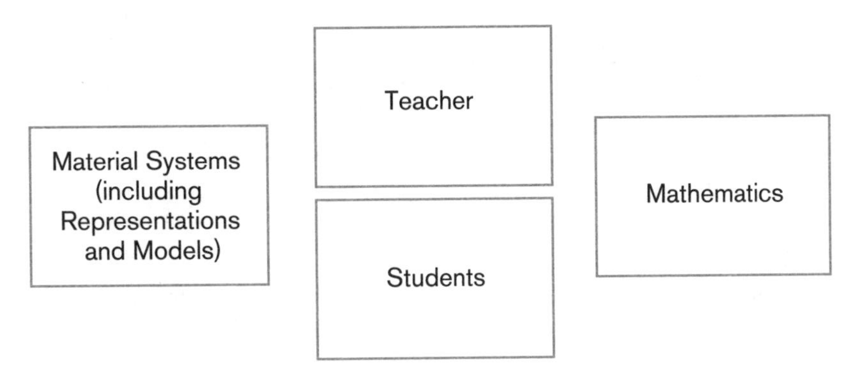

Figure 21.1. Constituents of mathematics learning activity system.

mathematics learning. Textbook and software publishers provide material resources for teachers and students to use. It also is important to remember that the system that includes the constituents of Figure 21.1 is dynamic: The interactions among these constituents and between them and each of the larger systems of the school and society occur over time and evolve.

Any episode of activity in a classroom can be seen as an arrangement of these constituents—the teacher; one or more of the students, usually with some material representation (although sometimes just in talk); and some concepts, principles, and, often, rules that the classroom group takes to be part of mathematics. Research has shown how different patterns of interaction place students in different relations with the teacher, with mathematical representations, and with the concepts and principles of mathematics.

Relations between the students and the conceptual domain of mathematics are especially important. Like material systems, mathematics provides affordances for activities by individuals and groups of people.[5] These affordances depend on people's mathematical practices, as Boaler's (1997) case studies showed. For many people, including the members of many school mathematics classes, mathematics is a collection of propositions and procedures—some of which they know and others they do not. When these people encounter a problem, the question is whether they can tell what procedure to use and remember how to do it. For other people, including members of some school mathematics classes, mathematics is a domain of interrelated concepts and principles that can be explored and used in thinking about problems. When mathematics is treated as a set of things to remember, its main affordance for activity involves showing who has acquired which pieces of knowledge. When mathematics is treated as a domain of interrelated concepts, its affordances are much broader. They include activities of sense making and reasoning within the domain of mathematics and in other domains, with mathematics as a useful resource.

Small-Scale Patterns of Discourse

One of the functions of research is to investigate and explicate the details of processes that need to occur for aims in education to be achieved. When we think of mathematics as an activity in which people participate, rather than as a body of knowledge and skills that people can acquire, a theory needs to address the nature of mathematical activity rather than just the nature of mathematical concepts and procedures.

In this section I discuss an illustrative case: research on patterns of conversational interactions that occur in classroom discourse. Much of what people learn that involves concepts is learned in their conversational interactions—indeed, it can be fruitful to think of conceptual learning as learning patterns of participation in discourse in which people make sense of the events and ideas that are important in their lives (Greeno, Benke, Engle, Lachapelle, & Wiebe, 1998). The importance of discourse in mathematics learning is recognized in the 1989 NCTM Standards in their emphasis on the goal of students' learning to communicate effectively in mathematical discourse. This outcome requires that students have opportunities to talk with one another and their teachers about mathematics. However, just encouraging students to talk in mathematics class probably is not sufficient for them to achieve the goal of becoming effective in mathematical communication. The revised Standards can be helpful in providing more explicit advice about

[5] The term *affordance* was coined by the psychologist J. J. Gibson (1977) to refer to the reciprocal relation between an animal and its environment. An affordance is a resource or support that an object in the environment offers an animal—but only if the animal can perceive and use the object. For example, for people the affordances of a lake are swimming and boating, but for water bugs the lake affords support for walking.

ways that teachers can arrange conversations in which students can participate effectively; some of the properties of these conversations have been identified in research about classroom discourse.

An important example of these research findings, discussed by Forman (this volume), is the distinction between two patterns that occur in brief episodes of interaction. One pattern, found by Mehan (1979), is called IRE for its main sequential constituents: an *initiation* (usually a question by the teacher), a *response* by a student, and an *evaluation* by the teacher. The other pattern, found by O'Connor and Michaels (1996), is called *revoicing*. In revoicing, the initiating presentation is given by a student; the teacher responds to that student's statement, often commenting on its significance in relation to a mathematical concept or principle. The teacher also creates a conversational opening in which either the initiating student or another student can take up the initiating student's idea or question or disagree with it. The initiating student can also agree or disagree with the teacher's interpretation of the meaning of his or her statement.

The difference between these patterns can be represented as variations on a general pattern of discourse interaction, shown in Figure 21.2. This schema is based on ideas about conversation by Clark and Schaefer (1989), whose term *contribution* refers to a joint action of participants in a conversation in which some information becomes part of their conversational common ground.

Figure 21.2 shows two main branches of a contribution. Many contributions have only one, the branch on the right side of Figure 21.2. In the simplest case, a contribution to conversation is just one person's presentation of some statement and another's or others' acceptance of it. This acceptance involves some kind of signal—possibly just allowing the speaker to go ahead—and the speaker and listener take the information as having been mutually understood. Instead of accepting the presentation, a listener may question it, disagree with it, or provide an alternative or elaborated version of it. Then there is a kind of negotiation in which the participants resolve the information they put into the common ground. The resolution may be the intended meaning of the presentation, some modification of the information, or an agreement that they do not mutually accept the information that was presented.

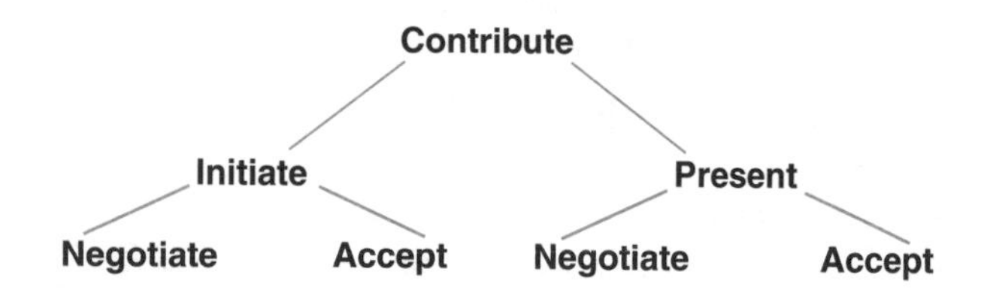

Figure 21.2. Schema of brief discourse interactions.

The IRE and revoicing patterns are more complicated and include both of the branches in Figure 21.2.

The IRE pattern. In the simple, frequently occurring version of IRE shown in Figure 21.3, the teacher initiates the episode by asking a question. Some or all students signal that they understand the question; in many classrooms, they do so by volunteering to answer the question. The teacher selects a student to respond. (In a more detailed analysis, this selection could be represented as beginning another episode embedded in the kind of episode depicted in Figure 21.3.) The selected student then presents an answer to the question, and the teacher accepts the answer by giving a positive evaluation. A negotiation of the question can occur if a student asks what the teacher means or if there is some other indication that students do not understand the question. A negotiation of the response can occur if the teacher gives a negative evaluation and asks another student to respond, or if the teacher is dissatisfied with the answer and rephrases the question. Eventually, if the episode is successful in the conversational sense, some information is understood by the teacher and the students as an acceptable answer to the teacher's initial question, and that information becomes part of the common ground for the class's continuing discourse.

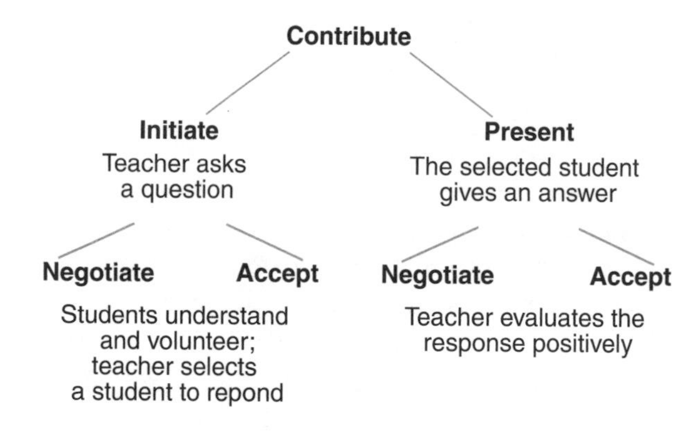

Figure 21.3. Schema of IRE episodes.

The IRE pattern is ubiquitous in many classrooms, as Mehan (1979) found. Here is an excerpt of elementary mathematics teaching dominated by the pattern; it is taken from a paper by McNeal (1993; see also McNeal, 1991). The teacher, Mrs. Rose, followed a suggestion in the teacher's edition of the textbook for teaching the concept of a fact family that involved the inverse relation between addition and subtraction. She began by showing a picture of a domino with five dots on one side and six dots on the other.

01 *Mrs. R:* We have to talk about something brand new today. Nick. I like how Nick is ready with his eyes forward. I like how Abby is ready. OK, boys and girls, we are going to talk about something today. Leif, can I have your attention. Look at this domino, please. Fred, I'm waiting. Notice the domino boys and girls. How many spots do you see on the top of the domino here? (*Pointing*)

02 *Students:* Five.

03 *Mrs. R:* How many do you see on the bottom? (*Pointing*)

04 *Students:* Six.

05 *Mrs. R:* OK. I would like for someone just to give me, ah, an *addition* number sentence for these, for this domino right here. An *addition* number sentence. Who can give me one?

06 *Student:* Five plus six equals eleven.

07 *Mrs. R:* All right. (*Writing* "5 + 6 = 11") Five plus six equals eleven. Who can give me *another* addition number sentence for this? Chuck?

08 *Chuck:* [*Inaudible — perhaps he said nothing*]

09 *Mrs. R:* Up there we have *one* number sentence, 5 + 6 = 11, what else could we do? What else could we use? Use those numbers up there.

10 *Chuck:* Six plus five?

11 *Mrs. R:* Wonderful. Six plus five equals eleven. Who can give me a *subtraction* number sentence using these dominos? Brandy.

12 *Brandy:* Six take away five.

13 *Mrs. R:* How many do we have all together, Brandy?

14 *Brandy:* [*Short pause*] Eleven.

15 *Mrs. R:* Eleven.

16 *Brandy:* Take away five.

17 *Mrs. R:* Eleven take away five equals what, Brandy?

18 *Brandy:* Six.

19 *Mrs. R:* Six. Very good. Who can give me *another* one? Another subtraction number sentence. Karl.

20 *Karl:* Eleven take away six equals five.

21 *Mrs. R:* Now, look here, boys and girls. How many, how many facts do we have there?

22 *Students:* Four/Oh! Four.

23 *Mrs. R:* Four facts. How many numbers, Chuck, are, did we use? How many numbers?

24 *Chuck:* Three. Two.

25 *Mrs. R:* How many numbers did we use?

26 *Chuck:* Three.

27 *Mrs. R:* We used three. We just made what we call a *fact* family.

28 *Student:* A fact? or fat? family. [*Laughter*]

29 *Mrs. R:* A fact family is four facts made out of three numbers. (1994, pp. 12–13)

The excerpt illustrates the IRE pattern and shows how ubiquitous it can be. By my count, the excerpt consisted of nine IRE episodes, preceded by an introductory presentation about the students' attention and the domino and followed by a presentation by Mrs. Rose that introduced the term *fact family*.

Most of the IRE episodes were the simple version not requiring negotiation. In the first two of these, in Turns 01–02 and 03–04, Mrs. Rose used the conversational form of questioning to bring the students' attention to the numbers of dots in her picture of a domino. She accepted the students' choral response in Turn 02 by moving on to her next question, and to their response in Turn 04, she said, "OK," an indication that the students were oriented to the material that they needed for her next question. The exchange of Turns 05, 06, and the first part of 07 is classic IRE. Mrs. Rose's question in 05 used a technical term, *addition number sentence*, in which she emphasized the term *addition*. A student responded in 06 with the answer that Mrs. Rose expected, and she evaluated the answer positively in 07. Note that Mrs. Rose distinguished this answer from the two previous answers by repeating it and writing it on the board, indicating that it had importance for the continuing discussion.

The unit from Mrs. Rose's question in Turn 07 to the beginning of Turn 11 had the IRE form, but Chuck's turn in 08 did not produce the answer, so Mrs. Rose clarified the question—or perhaps induced Chuck's attention—and Chuck answered it satisfactorily in Turn 10. In 11, Mrs. Rose gave her positive evaluation and completed the answer, adding "equals eleven" to Chuck's "six plus five."

The episode from Turn 11 to Turn 19 illustrates negotiation in the IRE schema. Brandy's first answer in Turn 12 was an example of what Mrs. Rose asked for, a subtraction expression using the numbers 6 and 5 from the domino. However, the response was not what Mrs. Rose had in mind. Mrs. Rose redirected Brandy at Turn 13 to produce a subtraction fact involving the total number of dots—11—rather than the difference between the two sets, and Mrs. Rose proceeded to draw the subtraction fact from Brandy in a series of exchanges, ending in Turn 19 with Mrs. Rose's evaluation, "Very good."

Turns 21–23 were an IRE episode that produced the information that the students had a set of four facts. Turns 23–26 were an IRE episode, with uncertainty resolved in 25–26, that produced information that three numbers were involved. In Turns 27–29, Mrs. Rose presented a term that referred to this configuration, specifying its numerical properties.

The revoicing pattern. The revoicing pattern is about the same size as, and follows a scheme similar to, IRE, but the roles of students and teacher are different. Figure 21.4 shows the pattern in a diagram. In this pattern, the teacher establishes an agenda item, as with IRE, but does so by responding to a student's statement or question. This move by the teacher appropriates the student's presentation and makes it the initiation of a conversational episode, setting up a participation structure in which subsequent statements and questions are addressed to the student's initiating statement or question rather than to a statement or question by the teacher. This discourse pattern can be used in a way that encourages participation by many students, so the "negotiate" parts of the interaction can be, and often are, quite extensive.

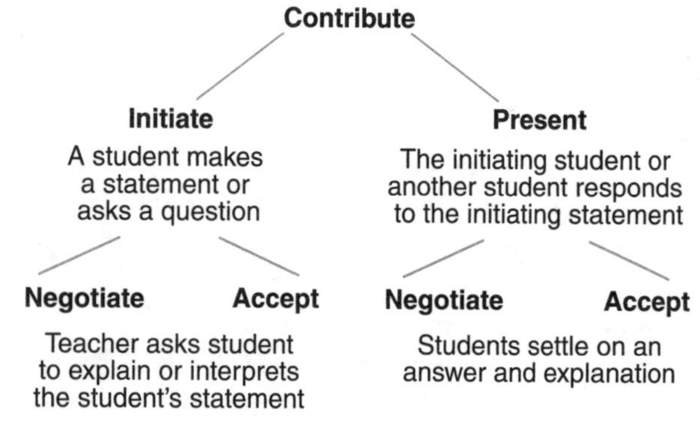

Figure 21.4. Schema of revoicing episodes.

An example of classroom discourse mainly organized by revoicing was reported by Lampert (1990b). Her fifth-grade class was having a discussion of exponents. The students had established that all powers of 5 end in 25 and had considered whether all powers of 6 end in 36, but they had found a counterexample to this conjecture. One of the students had proposed that patterns in the last digits of powers could be investigated without computing all the digits of the products involved; for example, "You don't have to figure out that 5^4 is 625 in order to know that its last digit must be 5." Lampert had asked about 7^4, and the class accepted that its last digit was 1 because "7 times 7 is 49, and 9 times 9 is 81." Lampert then asked the following:

01	*Teacher:*	What about 7 to the fifth power?
02	*Arthur:*	I think it's going to be a 1 again.
03	*Sarah:*	I think it's 9.
04	*Soo Wo:*	I think it's going to be 7.
05	*Sam:*	It is a 7.
06	*Teacher:*	[*Wrote on the board:* "7^5 = *1? 9? 7?*"] You must have a proof in mind, Sam, to be so sure. Arthur, why do you think it's 1?
07	*Arthur:*	Because 7^4 ends in 1, then it's times 1 again.
08	*Gar:*	The answer to 7^4 is 2,401. You multiply that by 7 to get the answer, so it's 7 × 1.
09	*Teacher:*	Why 9, Sarah?
10	*Theresa:*	I think Sarah thought the number should be 49.

11 *Gar:* Maybe they think it goes 9, 1, 9, 1, 9, 1.

12 *Molly:* I know it's 7, 'cause 7 ...

13 *Abdul:* Because 7^4 ends in 1, so if you times it by 7, it'll end in 7.

14 *Martha:* I think it's 7. No, I think it's 8.

15 *Sam:* I don't think it's 8 because, it's odd number times odd number and that's always an odd number.

16 *Carl:* It's 7 because it's like saying $49 \times 49 \times 7$.

17 *Arthur:* I still think it's 1 because you do 7×7 to get 49 and then for 7^4 you do 49×49 and for 7^5, I think you'll do 7^4 times itself and that will end in 1.

18 *Teacher:* What's 49^2?

19 *Soo Wo:* 2,401.

20 *Teacher:* Arthur's theory is that 7^5 should be 2401×2401 and since there's a 1 here and a 1 here ...

21 *Soo Wo:* It's $2{,}401 \times 7$.

22 *Gar:* I have a proof that it won't be a 9. It can't be 9, 1, 9, 1, because 7^3 ends in a 3.

23 *Martha:* I think it goes 1, 7, 9, 1, 7, 9, 1, 7, 9.

24 *Teacher:* What about 7^3 ending in 3? The last number ends in ... 9×7 is 63.

25 *Martha:* Oh.

26 *Carl:* Abdul's thing isn't wrong. 'Cause it works. He said times the last digit by 7 and the last digit is 9, so the last one will be 3. It's 1, 7, 9, 3, 1, 7, 9, 3.

27 *Arthur:* I want to revise my thinking. It would be $7 \times 7 \times 7 \times 7 \times 7$. I was thinking it would be $7 \times 7 \times 7 \times 7 \times 7 \times 7 \times 7 \times 7$. (pp. 50–51)

The interaction structure of this excerpt is considerably more complicated than that of the previous one. A major difference is that students gave three different answers to the question that Lampert used to begin the discussion, and she wrote all of them on the board. Through revoicing, Lampert arranged for these answers to be the initiating turns of several conversational episodes. She commented on Sam's Turn 05 answer, but she did not pursue it—although it was the mathematically correct answer that the class eventually settled on. Instead, she revoiced Arthur's answer, asking him to give an explanation. Arthur's Turn 02, Lampert's Turn 06 question, and Arthur's explanation constituted a revoicing episode that did not terminate with an acceptance. Instead, after hearing an explanation of why 1 was the answer, Lampert revoiced Sarah's presentation, asking for an explanation of the answer 9. Two students, Theresa and Gar, then offered reasons for that possibility, giving a revoicing episode consisting of Turns 03, 09, 10, and 11.

At Turns 12 and 13, Molly and Abdul expressed agreement with the answer 7, which Sam had proposed in Turn 05 and Lampert had revoiced in Turn 06. In Turn 14, Martha also expressed agreement with that answer but then changed her vote to 8. Sam disagreed with that in Turn 15, and the candidate 8 did not become a contender. From Turns 16 through 26, four students and Lampert presented arguments for and against the three candidates. The answer 9 was eliminated in a sequence that Gar initiated in Turn 22; earlier, in Turn 11, he had justified that answer with the possible 9-1-9-1 pattern by noting that the sequence had to include 3. Martha took exception to Gar's proposal, but Lampert restated Gar's point that 3 was required, and Martha accepted that statement. Carl supported the answer 7 in Turn 26 by stating a pattern of digits that was consistent with all the findings that had been established by then. Arthur, who had persisted with the answer 1 in 17, revoiced by Lampert in an exchange with Soo Wo in Turns 18 to 20, agreed with the answer 7 in Turn 27, settling the matter.

Comparison of these patterns. The contrast between these two examples is striking. Even so, in each example are episodes that can be understood as bringing elements of the other pattern into the discourse. For example, Mrs. Rose's writing of the equation "5 + 6 = 11" on the board in her Turn 07, which I interpreted as accepting a student's response to her previous question, also can be considered an instance of revoicing. The student's statement became the reference point of further discussion when Mrs. Rose asked, "Who can give me *another* addition number sentence for this?" Further, Lampert's Turn 24, part of the negotiation process in which the class settled on the answer 7, can be considered an evaluation in which the statement by Martha of the pattern 1-7-9, and so on, was ruled out, albeit on the basis of an earlier statement by Gar. Any classroom interaction—indeed, any conversation—will mix elements of IRE and elements of revoicing. These two examples are instructive because they provide significant contrasts relatively clearly.

My discussions in the two previous sections emphasized differences in the relations between the teacher and students in their participation in the conversations. In IRE episodes, participation is mainly back and forth between the teacher and individual students. The teacher asks a question with an answer in mind, selects some student to answer it, and presents a judgment of whether the student's answer is correct. In revoicing episodes, there is much more interaction between students. The teacher identifies a student's statement or question and makes it a topic for further discussion, so students are responding more to one another's ideas and questions.

Equally important, these two examples illustrate quite different relations between the participants and the subject-matter domain, mathematics. In the IRE discourse, mathematics was treated as a collection of authoritative propositions—a "knowledge base"—and students' statements were evaluated by whether they matched the propositions they were supposed to match to answer Mrs. Rose's questions. Mrs. Rose spoke on behalf of mathematics. In an adaptation of Goffman's (1981) terms, she could be said to *animate* a *principal* who was a personification of the mathematics subject-matter domain. Her main activity was to present questions to determine whether the students knew some of the propositions of mathematics and to do so in a way that would result in students' extending their knowledge to a new concept, the "fact family."

In the revoicing discourse, in contrast, mathematics was treated as a set of interrelated concepts, a kind of territory to be explored. Lampert designed her questions so that multiple answers would be plausible, and in their practice of mathematical discourse, Lampert and her students expected multiple candidates to be proposed and discussed. The students were encouraged to voice their opinions, and they expected to be accountable for giving reasons for their opinions. Indeed, the reasons for opinions were considered at least as carefully as the opinions themselves. In these discussions, the students considered themselves and one another as the principals whose ideas were expressed and debated. Beliefs, opinions, assumptions, and reasons were attributed to members of the class—to the students—rather than to the disembodied agency of mathematics per se. (These matters are discussed in more detail by Forman, this volume.)

In this practice, the relation of students to mathematics is very different from their relation to a set of propositions and procedures that they need to match in their behavior. Here, the students' job is to participate responsibly in discussions of mathematical topics. Students are accountable to one another and to the teacher for arriving at sensible conclusions on mathematical issues. The students, led by the teacher, are accountable to the domain of mathematics, which provides a set of topics, questions, and problems; mathematical practice imposes a set of constraints on acceptable solutions as well as conceptual resources for constructing solutions and arguments.

When students are engaged in constructing their understanding with a teacher as a guide, they do not necessarily construct concepts and representations in their standard form. In guiding her students' discussions, Lampert was mindful of being accountable to the standard meanings of concepts and patterns of reasoning in mathematics. Regarding her activity as a teacher, she has written the following:

> The teacher's social role in these lessons was to make the environment safe for those students who wanted to disagree with their peers and to help everyone clarify his or her position in an argument. The teacher's role as a more experienced knower of the discipline was to supplement students' language, symbols, and representations with those conventionally used by practitioners of mathematics. By enacting these roles, the teacher was contributing to the possibility that classroom communication about mathematics would have some congruence with individual student's thinking and with thinking about the same ideas in the wider mathematical community. (Lampert, 1990a, p. 258)

Lampert's statement presents a clear picture of teaching that is analogous to polycultural farming, rather than monocropping, in Scott's (1998) sense. She did not limit the discussion to mathematical terms and inferences that were consistent with the received forms of academic mathematics. Instead, she encouraged the expression of ideas and understandings that had already grown in the students' minds, and worked to extend these in ways that would move the students' understandings toward standard mathematical practice. Unlike monocropping, which requires maximum uniformity, multiple varieties of thinking were acknowledged and encouraged, and the diversity of opinions was used to nourish the growth of understanding.

Relations to Standards and Practices

In this section, I briefly illustrate a research result that relates to standards and practices in mathematics education. The NCTM (1989, 1991, 1995) *Standards* documents present a vision of learning and teaching mathematics that moves students into a different kind of participation than they typically have in traditional mathematics instruction. This research supports a strong suggestion that students' mathematical dispositions will be shaped by the kind of discourse activity in which they participate. If mathematics is treated as a domain in which students can explore the relationships between concepts and principles, considering alternative ideas and arriving at conclusions based on reasoned arguments, their orientations toward and in mathematics will likely be considerably more positive than if mathematics is treated as a body of knowledge that students must absorb and recite.

But what does the vision of the NCTM (1989, 1991, 1995) *Standards* imply for practice? That is, what kinds of changes in practice will be involved if teachers adopt the aim of changing students' participation in the ways that the *Standards* documents advocate?

The research that distinguishes among different participation structures in discourse provides part of the answer to this question. Changes in the ways that students participate in learning involve significant changes in the patterns of discourse that can be conducted in mathematics classrooms. In principle, these results can be helpful in the efforts to bring about reform in mathematics education because they clarify some of the kinds of changes needed in teaching and learning practices. At the same time, the results also indicate the challenges that need to be met for the reform to be successful. Patterns of conversational interaction such as IRE are deeply ingrained in social practices, and changing them is a deeply challenging undertaking.

Of course, much important detail is omitted from the illustrative examples that I have presented in this section. Perhaps most notably, I have not presented analyses of the conceptual contents of the classroom interactions. Analyses of conceptual contents can be carried out using the methods of cognitive science to show information structures that are constructed by the participants in a conversation (Greeno et al., 1998). Analyses of the information structures involved in mathematical reasoning and problem solving have been developed for individual problem solving and have been used productively in work with teachers in the Cognitively Guided Instruction program (Carpenter, Fennema, Peterson, Chiang, & Loef, 1989). Consequently, the teachers were more attentive to students' thinking processes. The research agenda for the situative perspective will include more detailed studies that combine analyses of informational and interpersonal aspects of students' participation in learning and that identify, in detail, how students' mathematical knowledge and understanding grow through their sustained participation in learning activities, especially discourse in which they play active, generative roles.

Potential Abilities of Students and Teachers for Collaborative Learning Practices

Proposals to change mathematics education to activities involving different kinds of participation by students and teachers presuppose that students and teachers are capable of learning how to engage in that participation. In response to such recommendations as those in the 1989, 1991, and 1995 NCTM *Standards* documents, many people, including students and teachers, believe that only a few students and teachers can develop the ability to participate in such interactions productively. For the *Standards* and other similar proposals to be realistic, we should be able to provide evidence of circumstances in which most, if not all, teachers and students can learn to participate in the kinds of recommended mathematics education practices and to maintain those practices extensively.

The results of research do not support a simple, pessimistic evaluation of this proposition, but neither do they support the idea that adopting these practices is easy for most teachers and students. First, there is substantial evidence that children are generally capable of learning to participate in activities that involve complex quantitative reasoning. In addition to the studies of classroom discourse that I have discussed, this evidence comes from studies of everyday activities in which children are engaged in various kinds of commercial activities, including selling produce, candy, or lottery tickets in the informal economy of a city (Nunes et al., 1993; Saxe, 1990). Curriculum materials in which students engage in a leading activity of design or large-scale problem solving (e.g., Bransford et al., 1996; Goldman et al., 1995; Saxe & Guberman, 1998) are designed to recruit the informal understandings that students bring to school from their general experience so that the learning of mathematics can build from those understandings.

In addition, there is a substantial body of evidence from cognitive studies that students have significant informal understanding of quantitative concepts that are central to the mathematics curriculum. Young children's understandings of early number concepts and of additive operations are substantial and have been thoroughly analyzed (e.g., Carpenter, Moser, & Romberg, 1982; Gelman & Gallistel, 1978; Steffe, Cobb, & von Glasersfeld, 1988). Substantial progress has been made in investigating children's concepts of multiplicative concepts and rational numbers (e.g., Carpenter, Fennema, & Romberg, 1993; Harel & Confrey, 1994), including the discovery that children in the early grades have significant intuitive understanding of processes in which growth or decline occurs by a proportional factor rather than by an additive process (Confrey, 1990). Furthermore, some significant beginning progress has been made in investigating middle school students' informal understanding of functional relations between quantities (e.g., Greeno, 1992; Kaput, 1994). The importance of these results for standards in mathematics education is that they show that students bring significant informal understanding of crucial mathematical ideas to school learning and therefore are equipped to participate meaningfully in discourse in which mathematical concepts are elaborated, refined, and formalized.

Of course, these results do not imply that the development of elaborated, refined, formalized, and abstract mathematical knowledge is easy. Current efforts to reform the curriculum to make better use of students' informal understanding can be expected to result in making that achievement more feasible for many students, but the transitions from more informal to more formal mathematical reasoning and discourse are inherently difficult and challenging for students and in teachers' efforts to foster their students' progress.

The capabilities of teachers to develop practices in which students learn to participate productively in significant mathematical activities are the other side of the question. Of course, this question is not just about teachers; it crucially involves issues of the resources that teachers have available to support the development of more complex practices and to make transitions in their practices.

The most direct evidence concerning teachers' capabilities and resources comes from analytical studies, such as those of the QUASAR project (e.g., Stein et al., 1996) and other evaluations (e.g., Elmore et al., 1996; Stokes, et al., 1997) that I mentioned in the first section. These studies have shown that significant changes in learning and teaching practices can be accomplished and that significant efforts and resources need to be committed to accomplish the changes successfully. The results suggest that in recommending these changes in mathematics education practices, standards should urge schools and teachers to allocate time and material resources to the effort to accomplish changes successfully, including resources for professional development, for collegial planning and reflection among teachers, and for meaningful assessments of student learning that address the variety of proficiencies that the standards identify.

The study of mathematics teaching and teacher development is a growing area of research, in which multiple perspectives are being used productively (Fennema & Scott Nelson, 1997). Research on teacher development in the situative perspective characterizes the process in terms of transformations of teaching practices and in terms of the importance of collegial communities of practice in fostering and supporting teachers' efforts to do so (e.g., Secada & Adajian, 1997; Stein & Brown, 1997; Stein et al., 1998). In its recommendations to the National Educational Research Policy and Priorities Board, the National Academy of Education (1999) emphasized the need for research that considers teacher development as a transformation of the complex, flexible, and generative practices of classroom teaching; this emphasis is an important contrast with much of the discussion of teacher learning that frames the issue as a problem of teachers' simply needing to acquire more adequate knowledge and skills.

Conclusion

The agenda for mathematics education presented in the NCTM (1989, 1991, 1995) *Standards* documents is challenging, as everyone knows. In this chapter, I have discussed results of research relevant to formulating and interpreting a new document that will represent that agenda in light of the experience and knowledge acquired in the 1990s.

The results that I have reviewed include findings that encourage the continued development of resources and practices in support of the agenda of the NCTM *Standards*. Numerous examples of teaching and learning mathematics with constructive student participation in significant aspects of mathematical practice have been documented and analyzed to guide teachers and other educators who adopt the goals of reforming mathematics education. We can be confident that educators can design and implement learning environments in which students are actively and meaningfully involved in mathematical problem solving, communicating, reasoning, and understanding mathematical connections. The consequences of providing these learning environments can be extremely important, leading to forms of mathematical knowing that are much more generative and productive than the knowledge achieved by most students in U.S. mathematics education. The results also show that successful pursuit of these goals requires strong commitments by teachers and other educators as well as the allocation of significant resources to achieve these results.

As the agenda of the NCTM Standards continues to develop, the experience and reflection of practitioners and research studies will continue to provide increased knowledge and understanding of the kinds of conditions and practices that can support successful mathematics learning. The task of educational reform is one of developing a set of practices that will tap the potential of a greater variety of students to be successful learners, thinkers, and users of mathematical concepts, methods, and representations. To return to the analogy of agriculture and Scott's (1998) analysis of projects of legibility, any attempt to accomplish the needed transformations by setting down a list of rules for uniform practices for all students in all schools would be counterproductive, even if those rules specified practices that "work" better than alternatives in a statistical majority of cases. Monocropping in mathematics education seems at least as counterproductive as it is in tropical agriculture. We need, instead, to conduct research that will continue to improve our understanding of the complex, generative practices of teaching and learning in its diverse varieties, and to express the results of that research in terms that will both advance our theoretical understanding and benefit our discourses of practice and policy. The importance of improving mathematics education deserves no less.

ACKNOWLEDGMENTS

This paper was completed while I was a Fellow at the Center for Advanced Study in the Behavioral Sciences, supported in part by a sabbatical leave from Stanford University and in part by fellowship funds provided by the Spencer Foundation through Grant No. 199400132. I am very grateful to participants in the conference organized by Jeremy Kilpatrick who provided comments on a preliminary draft, and to Thomas Romberg for his comments on a subsequent draft.

REFERENCES

Anderson, J. R. (1983). *The architecture of cognition.* Cambridge, MA: Harvard University Press.

Anderson, J. R., Boyle, C. F., & Reiser, B. J. (1985). Intelligent tutoring systems. *Science, 228*, 456–462.

Ball, D. L. (1993). Halves, pieces, and twoths: Constructing and using representational contexts in teaching fractions. In T. P. Carpenter, E. Fennema, & T. A. Romberg (Eds.), *Rational numbers: An integration of research* (pp. 157–196). Hillsdale, NJ: Erlbaum.

Boaler, J. (1997). *Experiencing school mathematics: Teaching styles, sex, and setting*. Buckingham, England: Open University Press.

Bransford, J. D., Zech, L., Schwartz, D., Barron, B., Vye, N., & the Cognition and Technology Group at Vanderbilt. (1996). Fostering mathematical thinking in middle school students: Lessons from research. In R. J. Sternberg & T. Ben-Zeev (Eds.), *The nature of mathematical thinking* (pp. 203–250). Hillsdale, NJ: Erlbaum.

Brophy, J. E., & Good, T. L. (1986). Teacher behavior and student achievement. In M. C. Wittrock (Ed.), *Handbook of research on teaching* (3rd ed., pp. 328–375). New York: Macmillan.

Brown, A. L. (1992). Design experiments: Theoretical and methodological challenges in creating complex interventions in classroom settings. *Journal of the Learning Sciences, 2*, 141–178.

Brown, A. L. (1994). The advancement of learning. *Educational Researcher, 23*(8), 4–12.

Brown, J. S. (1991, January-February). Research that reinvents the corporation. *Harvard Business Review*, pp. 102–111.

California State Board of Education. (1998). *The California mathematics academic content standards*. Sacramento, CA: Author.

Carpenter, T. P., Fennema, E., Peterson, P. L., Chiang, C.-P., & Loef, M. (1989). Using knowledge of children's mathematical thinking in classroom teaching: An experimental study. *American Educational Research Journal, 26*, 499–531.

Carpenter, T. P., Fennema, E., & Romberg, T. A. (Eds.). (1993). *Rational numbers: An integration of research*. Hillsdale, NJ: Erlbaum.

Carpenter, T. P., Moser, J. M., & Romberg, T. A. (Eds.). (1983). *Addition and subtraction: A cognitive perspective*. Hillsdale, NJ: Erlbaum.

Clark, H. H., & Schaefer, E. F. (1989). Contributing to discourse. *Cognitive Science, 13*, 259–294.

Cobb, P., Wood, T., Yackel, E., Nicholls, J., Wheatley, G., Trigatti, B., & Perlwitz, M. (1991). Assessment of a problem-centered second-grade mathematics project. *Journal for Research in Mathematics Education, 22*, 3–29.

Cognition and Technology Group at Vanderbilt. (1992). The Jasper series as an example of anchored instruction: Theory, program description, and assessment data. *Educational Psychologist*, 27, 291–315.

Cole, K. (1999, March). Greasing the wheels: Powerful assessment in a technology-rich math curriculum. *Proceedings of the International Conference on Mathematics/Science Education and Technology* (pp. 84–90). Charlottesville, VA: Association for the Advancement of Computing in Education.

Collins, A. (1992). Toward a design science of education. In. E. Scanlon & T. O'Shea (Eds.), *New directions in educational technology* (pp. 15–22). Berlin: Springer.

Confrey, J. (1990). A review of the research on student conceptions in mathematics, science, and programming. *Review of Research in Education, 16*, 3–56.

Cronon, W. (1991). *Nature's metropolis: Chicago and the Great West*. New York: Norton.

Dewey, J. (1966). *Democracy and education*. New York: Free Press. (Original work published 1916)

Elmore, R. F., Peterson, P. L., & McCarthey, S. J. (1996). *Restructuring in the classroom: Teaching, learning, and school organization*. San Francisco: Jossey-Bass.

Fennema, E., & Scott Nelson, B. (Eds.). (1997). *Mathematics teachers in transition*. Mahwah, NJ: Erlbaum.

Freudenthal, H. (1983). *Didactical phenomenology of mathematical structures*. Dordrecht, The Netherlands: Reidel.

Gagné, R. M. (1968). Learning hierarchies. *Educational Psychologist, 6*, 1–9.

Gelman, R., & Gallistel, C. R. (1978). *The child's understanding of number*. Cambridge, MA: Harvard University Press.

Gibson, J. J. (1977). The theory of affordances. In R. E. Shaw & J. Bransford (Eds.), *Perceiving, acting, and knowing* (pp. 67–82). Hillsdale, NJ: Erlbaum.

Godfrey, L., & O'Connor, M. C. (1995). The vertical hand span: Non-standard units, expressions, and symbols in the classroom. *Journal of Mathematical Behavior, 14*, 327–345.

Goffman, E. (1981). *Forms of talk*. Philadelphia: University of Pennsylvania Press.

Goldman, S., Moschkovich, J., & the Middle-School Mathematics through Applications Project Team. (1995). Environments for collaborating mathematically: The Middle-School Mathematics through Applications Project. In J. L. Schnase & E. L. Cunnius (Eds.), CSCL *'95: The First International Conference on Computer Support for Collaborative Learning* (pp. 1–4). Mahwah, NJ: Erlbaum.

Gravemeijer, K. (1995). Educational development and developmental research in mathematics education. *Journal for Research in Mathematics Education, 25*, 443–471.

Greeno, J. G. (1991). Number sense as situated knowing in a conceptual domain. *Journal for Research in Mathematics Education, 22*, 170–218.

Greeno, J. G. (1992). Mathematical and scientific thinking in classrooms and other situations. In D. F. Halpern (Ed.), *Enhancing thinking skills in the sciences and mathematics* (pp. 39–61). Hillsdale, NJ: Erlbaum.

Greeno, J. G. (1997). Participation as fundamental in learning mathematics. In J. A. Dossey, J. O. Swafford, M. Parmantie, & A. E. Dossey (Eds.), *Proceedings of the 19th Annual Meeting, North American Chapter of the International Group for the Psychology of Mathematics Education* (Vol. 1, pp. 1–14). Columbus OH: ERIC Clearinghouse for Science, Mathematics, and Environmental Education.

Greeno, J. G., Benke, G., Engle, R. A., Lachapelle, C., & Wiebe, M. (1998). Considering conceptual growth as change in discourse practices. In M. A. Gernsbacher & S. J. Derry (Eds.), *Proceedings of the 20th Annual Conference of the Cognitive Science Society* (pp. 442–447). Mahwah, NJ: Erlbaum.

Greeno, J. G., Collins, A. M., & Resnick, L. B. (1996). Cognition and learning. In D. C. Berliner & R. C. Calfee (Eds.), *Handbook of educational psychology* (pp. 15–46). New York: Simon & Schuster Macmillan.

Greeno, J. G., McDermott, R., Cole, K., Engle, R. A., Goldman, S., Knudsen, J., Lauman, B., & Linde, C. (1999). Research, reform, and aims in education: Modes of action in search of each other. In E. C. Lagemann & L. Shulman (Eds.), *Issues in education research* (pp. 299–335). San Francisco: Jossey-Bass.

Greeno, J. G., & the Middle-School Mathematics through Applications Project Group. (1997). Theories and practices of thinking and learning to think. *American Journal of Education, 106*, 85–126.

Greeno, J. G., & the Middle-School Mathematics through Applications Project Group. (1998). The situativity of cognition, learning, and research. *American Psychologist, 53*, 5–26.

Hall, R. P., & Rubin, A. (1998). ... there's five little notches in here: Dilemmas in teaching and learning the conventional structure of rate. In J. G. Greeno & S. V. Goldman (Eds.), *Thinking practices in mathematics and science learning* (pp. 189–236). Mahwah, NJ: Erlbaum.

Harel, G., & Confrey, J. (Eds.). (1994). *The development of multiplicative reasoning in the learning of mathematics*. Albany: State University of New York Press.

Heath, S. B. (1991). "It's about winning!" The language of knowledge in baseball. In L. B. Resnick, J. M. Levine, & S. D. Teasley (Eds.), *Perspectives on socially shared cognition* (pp. 101–124). Washington, DC: American Psychological Association.

Hutchins, E. (1995). *Cognition in the wild*. Cambridge, MA: MIT Press.

Kaput, J. (1994). Democratizing access to calculus: New routes to old roots. In A. H. Schoenfeld (Ed.), *Mathematical thinking and problem solving* (pp. 77–156). Hillsdale, NJ: Erlbaum.

Lampert, M. (1990a). Connecting inventions with conventions. In L. P. Steffe & T. Wood (Eds.), *Transforming children's mathematics education: International perspectives* (pp. 253–265). Hillsdale, NJ: Erlbaum.

Lampert, M. (1990b). When the problem is not the question and the solution is not the answer: Mathematical knowing and teaching. *American Educational Research Journal, 27*, 29-63.

Lave, J. (1988). *Cognition in practice*. Cambridge: Cambridge University Press.

Lave, J. (1993). Situating learning in communities of practice. In L. B. Resnick, J. Levine, & S. Teasley (Eds.), *Perspectives in social shared cognition* (pp. 63–85). Washington, DC: American Psychological Association.

Lave, J. (1996, May). *Situated cognition in mathematics*. Paper presented at a seminar held at Oxford University, Department of Educational Studies.

Lave, J., & Wenger, E. (1991). *Situated learning: Legitimate peripheral participation*. Cambridge: Cambridge University Press.

Lesh, R., & Landau, M. (Eds.). (1983). *Acquisition of mathematics concepts and processes*. New York: Academic Press.

Lieberman, J. C. (1998). Enabling professionalism in high school mathematics departments: The role of generative community (Doctoral dissertation, Stanford University, 1997). *Dissertation Abstracts International, 58*, 375.

Lightman, A. (1993). *Einstein's dreams*. New York: Warner Books.

McDermott, R. P. (1993). The acquisition of a child by a learning disability. In S. Chaiklin & J. Lave (Eds.), *Understanding practice* (pp. 269–305). Cambridge: Cambridge University Press.

McNeal, M. G. (1991). The social context of mathematical development (Doctoral dissertation, Purdue University). *Dissertation Abstracts International, 52*, 2009A.

McNeal, B. (1993, April). *What does it mean to call for more "math talk"? Problem solving in two different mathematics traditions*. Paper presented at the meeting of the American Educational Research Association, Atlanta.

Mead, G. H. (1934). *Mind, self, and society from the standpoint of a social behaviorist*. Chicago: University of Chicago Press.

Mehan, H. (1979). *Learning lessons: Social organization in the classroom*. Cambridge, MA: Harvard University Press.

Mehan, H., Hertwick, A., & Meihls, J. L. (1986). *Handicapping the handicapped*. Stanford, CA: Stanford University Press.

Moses, R. P., & Cobb, C. E., Jr. (2001). *Radical equations: Math literacy and civil rights*. Boston: Beacon Press.

Moses, R. P., Kamii, M., Swap, S. M., & Howard, J. (1989). The Algebra Project: Organizing in the spirit of Ella. *Harvard Educational Review, 59*, 423–443.

National Academy of Education. (1999). *Recommendations regarding research priorities: An advisory report to the National Educational Research Policy and Priorities Board*. New York: Author.

National Center for Research in Mathematical Sciences Education & Freudenthal Institute (Eds.). (1977). *Mathematics in context*. Chicago: Encyclopaedia Britannica.

National Council of Teachers of Mathematics. (1989). *Curriculum and evaluation standards for school mathematics*. Reston, VA: Author.

National Council of Teachers of Mathematics. (1991). *Professional standards for teaching mathematics*. Reston, VA: Author.

National Council of Teachers of Mathematics. (1995). *Assessment standards for school mathematics*. Reston, VA: Author.

National Research Council. (1999). *How people learn: Bridging research and practice*. Washington, DC: National Academy Press.

Nunes, T., Schliemann, A. D., & Carraher, D. W. (1993). *Street mathematics and school mathematics*. Cambridge: Cambridge University Press.

O'Connor, M. C., & Michaels, S. (1996). Shifting participant frameworks: Orchestrating thinking practices in group discussion. In D. Hicks (Ed.), *Discourse, learning and schooling* (pp. 63–103). Cambridge: Cambridge University Press.

Porter, T. M. (1995). *Trust in numbers: The pursuit of objectivity in science and public life*. Princeton, NJ: Princeton University Press.

Reed, E. S. (1996). *Encountering the world: Toward an ecological psychology*. Oxford: Oxford University Press.

Rogoff, B., & Chavajay, P. (1995). What's become of research on the cultural basis of cognitive development? *American Psychologist, 50*, 859–877.

Romberg, T. A. (1993, January). NCTM's *Standards*: A rallying flag for mathematics teachers. *Educational Leadership, 50*(5), 36–41.

Romberg, T. A. (1999). Realistic instruction in mathematics. In J. H. Block, S. T. Everson, & T. R. Guskey (Eds.), *Comprehensive school reform: A program perspective* (pp. 287–314). Dubuque, IA: Kendall-Hunt.

Saxe, G. (1990). *Culture and cognitive development*. Hillsdale, NJ: Erlbaum.

Saxe, G., & Guberman, S. R. (1998). Emergent arithmetical environments in the context of distributed problem solving: Analyses of children playing an educational game. In J. G. Greeno & S. Goldman (Eds.), *Thinking practices in mathematics and science education* (pp. 237–256). Mahwah, NJ: Erlbaum.

Schoenfeld, A. H. (1985). *Mathematical problem solving*. Orlando, FL: Academic Press.

Schoenfeld, A. H. (1988). When good teaching leads to bad results: The disasters of "well-taught" mathematics courses. *Educational Psychologist, 23*, 145–166.

Schoenfeld, A. H. (1998). Making mathematics and making pasta: From cookbook procedures to really cooking. In J. G. Greeno & S. Goldman (Eds.), *Thinking practices in mathematics and science learning* (pp. 299-320). Mahwah, NJ: Erlbaum.

Schwartz, J. L., Yarushalmy, M., & Wilson, B. (Eds.). (1993). *The Geometric Supposer: What is it a case of?* Hillsdale, NJ: Erlbaum.

Scott, J. C. (1998). *Seeing like a state: How certain schemes to improve the human condition have failed*. New Haven, CT: Yale University Press.

Secada, W. G., & Adajian, L. B. (1997). Mathematics teachers' change in the context of their professional communities. In E. Fennema & B. Scott Nelson (Eds.), *Mathematics teachers in transition* (pp. 193–219). Mahwah, NJ: Erlbaum.

Shafer, M., & Romberg, T. A. (1999). Assessment in classrooms that promote understanding. In E. Fennema & T. A. Romberg (Eds.), *Mathematics classrooms that promote understanding* (pp. 159–184). Mahwah, NJ: Erlbaum.

Silva, C. M., Moses, R. P., Rivers, J., & Johnson, P. (1990). The Algebra Project: Making middle school mathematics count. *Journal of Negro Education, 59*, 375–391.

Siskin, L. S. (1994). *Realms of knowledge: Academic departments in secondary schools*. London: Falmer.

Skinner, B. F. (1958). Teaching machines. *Science, 128*, 969–977.

Steffe, L. P., Cobb, P., & von Glasersfeld, E. (1988). *Construction of arithmetical meanings and strategies*. New York: Springer.

Stein, M. K., & Brown, C. (1997). Teacher learning in a social context: Integrating collaborative and institutional processes with the study of teacher change. In E. Fennema & T. A. Romberg (Eds.), *Mathematics classrooms that promote understanding* (pp. 155–192). Mahwah, NJ: Erlbaum.

Stein, M. K., Grover, B. W., & Henningsen, M. (1996). Building student capacity for mathematical thinking and reasoning: An analysis of mathematical tasks used in reform classrooms. *American Educational Research Journal, 33*, 455–488.

Stein, M. K., Silver, E. A., & Smith, M. S. (1998). Mathematics reform and teacher development: A community of practice perspective. In J. G. Greeno & S. Goldman (Eds.), *Thinking practices in mathematics and science learning* (pp. 17–52). Mahwah, NJ: Erlbaum.

Stokes, D. E. (1997). *Pasteur's quadrant: Basic science and technological innovation*. Washington, DC: Brookings Institution Press.

Stokes, L. M., Sato, N. E., McLaughlin, M. W., & Talbert, J. E. (1997). *Theory-based reform and problems of change: Contexts that matter for teachers' learning and community* (Report R97–7). Stanford, CA: Stanford University, School of Education, Center for Research on the Context of Secondary Teaching

Suchman, L. (1987). *Plans and situated action: The problem of human-machine interaction*. Cambridge: Cambridge University Press.

Varenne, H., & McDermott, R. (1998). *Successful failure: The school America builds*. Boulder, CO: Westview Press.

Vygotsky, L. S. (1987). Thinking and speech. In R. W. Rieber & A. S. Cargon (Eds.), *The collected works of L. S. Vygotsky, Volume 1: Problems of general psychology* (pp. 37–285). New York: Plenum. (Original work published 1934)

Wenger, E. (1998). *Communities of practice: Learning, meaning, and identity*. Cambridge: Cambridge University Press.

Whitehead, A. N. (1962). *The aims of education*. London: Ernest Benn.

CHAPTER 22

A Sociocultural Approach to Mathematics Reform: Speaking, Inscribing, and Doing Mathematics Within Communities of Practice

Ellice Ann Forman, University of Pittsburgh

Some people may be surprised to learn that sociocultural theory,[1] a psychological theory developed by Russians in the early 20th century, is being used as one of the guiding frameworks for mathematics reform in the United States. They may wonder why this theory is relevant for mathematics reform in the United States today. The aim of this chapter is to answer that question in at least two general ways. First, I show how sociocultural theory identifies the fundamental link between instructional practices and learning outcomes. For example, if one of a teacher's instructional goals is to enable students to see themselves as a source of mathematical knowledge, then that teacher's instructional practices need to foster the active engagement of students in defining and solving complex problems and debating among themselves the merits of various solutions. This approach to thinking about the relationship between instructional practices and learning outcomes is consistent with some of the themes articulated in *Professional Standards for Teaching Mathematics*, published by the National Council of Teachers of Mathematics (NCTM, 1991).

Second, I demonstrate that mathematics learning occurs through a process of mathematical communication in social contexts. More specifically, mathematical activity involves the use of the speech genres (e.g., argumentation) and the notational systems of the discipline (e.g., numbers, operators, graphs, geometrical figures) (Rotman, 1988, 1993). This second way in which sociocultural theory is relevant is consistent with the emphasis on mathematical discourse in the NCTM (1991) *Standards*. These two themes—how learning is connected with the structures and functions of instructional activity and how learning entails mathematical communication—organize much of my discussion of the impact of sociocultural theory on current and future reform in mathematics education (cf. Lerman, 1996).

Before beginning that discussion, however, I provide an overview of sociocultural theory itself, including its history and its future directions. New developments within sociocultural theory as well as changes in mathematics instruction due to the influence of the NCTM *Standards* documents suggest that numerous fruitful future connections can be made between this theory and educational practice.

Although many people trace the origins of sociocultural theory to the work of three Russian psychologists and their students and colleagues—L.S. Vygotsky, A.N. Leont'ev, and A.R. Luria—its influence had spread widely by the end of the 20th century (Minick, Stone, & Forman, 1993). In particular, sociocultural theory has affected the work of developmental, educational, and cross-cultural psychologists in North and South America, Europe, Asia, and Australia since the 1970s, when more of the work of the Russian psychologists was published and translated. The conceptual and research base for sociocultural theory has always been diverse, by necessity, because Vygotsky, the actual source of many of the basic concepts in the theory, died at an early age and most of his writings remained unpublished until the 1970s. Despite the difficulties of translating and applying early 20th-century Russian psychology to a late 20th-century international psychology, there emerged a core set of principles for sociocultural theory and a distinct

[1] The sociocultural approach has been called by a variety of names (Vygotskian theory, neo-Vygotskian theory, cultural-historical activity theory, cultural psychology, etc.). Some theorists do not trace their intellectual roots directly back to Russians such as Vygotsky. Many of them may align themselves more closely with situated cognition (Greeno, this volume), cultural anthropology (Lave & Wenger, 1991), or other fields. For this chapter, I have simplified somewhat the historical roots and intellectual influences of this approach (see Cole, 1995, 1996; Shweder et al., 1998, for further details).

body of empirical research that are capable of guiding educational reform today.

This chapter sketches the implications of sociocultural theory for conceptualizing educational reform in mathematics and summarizes some of the research literature that can instruct both research and current reform practices. The chapter is organized as follows: First, an overview of sociocultural theory is presented that includes both Vygotsky's original concepts and their more recent formulations; second, the application of these concepts to empirical research of relevance to educational reform in mathematics is critically reviewed; and finally, future directions for sociocultural research and theory are suggested.

An Overview of Sociocultural Theory

Key Concepts From Vygotsky's Writings

Wertsch (1985b) has identified three interrelated themes that unite Vygotsky's original writings from the 1920s and 1930s and tie them to related work by his contemporaries in Russia. The first theme is genetic analysis: To understand a psychological phenomenon, one needs to study it in the process of change. Vygotsky noted that few psychological studies do so; even most of Piaget's work examines the endpoint—what Vygotsky called *fossilized behavior*—not the process of change. The second theme is the social origins of higher mental functions: that conscious, voluntary, and self-regulated mental processes, such as selective attention, voluntary memory, and logical reasoning, are created by social processes. The third theme is that all activity, including thinking, is mediated by signs. These concepts, Wertsch argues, are behind more familiar ideas from Vygotsky, such as the *zone of proximal development* (ZPD)—that is, the difference between assisted and unassisted performance. For example, to understand the ZPD, one would need to study it in a genetic way: to see how a novice's performance becomes less dependent upon *other-regulation* (regulation by others) and becomes more self-regulated over time. In addition, as novices gain in expertise as a result of learning in the ZPD, their performance continues to show evidence of their previous experiences with social regulation. Assisted performance in the ZPD depends upon sign mediation—oral or written language, drawings, notations, gestures—as does expert performance. For example, overt or covert self-directed speech during problem solving may exhibit features of the speech of the person who provided social regulation earlier (e.g., "Make sure you check your answers"). Even experts continue to use mediating devices (e.g., outlines, flowcharts, tables, models, and graphs) to regulate their activities (including their mental activities) that were first employed when they were relative novices.

Three aspects of Vygotsky's framework that are particularly relevant for research on educational reform should be mentioned here. First, Vygotsky (1978) argued for modifications in the methods used to study psychological phenomenon. One modification, mentioned above, is to study the change process and not merely document the outcome of change. Therefore, Vygotsky proposed that when studying children's development, the investigator should attempt to influence that development by providing the material and social conditions thought important in the change process and then should study the outcome of that intervention genetically. Another methodological alteration proposed by Vygotsky was to redefine the unit of analysis in psychological study: from the isolated subject to a unit that "maintains the functional relationship between the individual and the environment" (Minick, 1987, p. 32). This idea was expanded by Vygotsky's colleague Leont'ev in his theory of activity (Wertsch, 1981), in which the focus of research became people engaged in goal-directed, mediated activities. In other words, instead of studying the learning of individuals as if it had occurred without the assistance of others in one or more particular contexts (e.g., a classroom or a psychological laboratory), Leont'ev's approach would be to study changes in problem-solving activities, for example, among a group of people attempting to achieve a particular goal. That goal might include making a suit in a tailor shop (where accurate readings of customer's bodies using tape measures and recording on paper patterns would be required). It might also involve expert tailors supervising the learning of novice tailors. In this tailoring example, the unit of analysis of the study would not be the learning of individual novice tailors but the learning that occurs within the tailor shop as a whole as suits are made and apprentices become tailors.

Second, unlike most psychological theories from the 20th century, Vygotsky's theory stressed the connection between theory and applied research. As van der Veer and Valsiner (1991) argue, Vygotsky saw educational practice as a key way to test psychological theories. They translate the following quotation from a manuscript written by Vygotsky in 1926:

> The most complex contradictions of psychology's methodology are brought to the field of practice and can only be solved there. Here the dispute stops being sterile, it comes to an end. … That is why practice transforms the whole of scientific methodology. (p. 150)

Thus, one distinct advantage of sociocultural theory is its early and continuing commitment to testing psychological theories in educational practice as well as deriving theoretical constructs from practice.

These first two aspects of the Vygotskian framework could have fundamental implications for the study of educational reform that are discussed in some detail here and mentioned again at the end of the chapter (cf. de Corte, Greer, & Verschaffel, 1996). True experiments, in which people (or laboratory rats or materials in a physics experiment) are randomly assigned to treatment conditions, have been seen by many specialists in research methodology (as well as by much of the general public) as the best way to evaluate the results of educational research (Campbell & Stanley, 1963). True experiments may not provide information on the change process that Vygotsky emphasized, however, because they typically rely on comparisons of the different outcomes in the groups studied (such as average achievement scores) but not on comparisons of different processes that occurred during the treatment (such as how instruction was delivered). The assumption of experimental control is also contradictory to Vygotsky's proposal about the unit of analysis because, from his perspective, educational treatments are not composed of separate variables that can be manipulated one at a time. In addition, very few educational researchers are able to conduct anything like a true experiment. It is unlikely that students can be randomly assigned to treatments and that educational experiments can be conducted with the strict control of a laboratory (e.g., where one can manipulate a single variable at a time).

Cole and Means (1981) have noted that, as a result of the difficulties with, and limitations of, experiments, educational research typically relies on designs in which random assignment does not occur and comparison groups are not equivalent in all ways except the experimental treatment. Brown (1992) has argued that educational research usually has a different goal than laboratory research because schools are institutions that attempt to engineer, not just study, particular kinds of change. Finally, Eisenhart (1988) has proposed that the ethnographic research methods of cultural anthropology play an important role in understanding the impact of educational practice on teachers, students, parents, and administrators.[2] It is erroneous, therefore, to assume that one can make an analogy between the classroom, in which reform efforts are typically implemented, and the laboratory, in which experimental control can be accomplished. Vygotsky's suggestions and those of his colleagues have important implications for research methodology in education—implications that have been explored for a number of years by sociocultural researchers and cognitive scientists as well as by research methodologists (e.g., Cole & Means, 1981; Cook & Campbell, 1979; Schoenfeld, 1992; Siegler & Crowley, 1991).

The third aspect of Vygotsky's theory that has relevance for educational reform concerns the importance of social and cultural factors in children's learning and development. The 1991 NCTM Standards were intended to improve mathematics instruction for all students, not just one segment of society. For contemporary U.S. society, that statement means that girls as well as boys; children from low-income as well as middle-income families; children from a wide variety of cultural, racial, and ethnic groups; and children with documented learning difficulties are expected to learn more than the fundamentals of mathematics in school. Teachers are supposed to recognize these students' distinct learning histories, attitudes, motivations, and beliefs about mathematics and to encourage all of them to engage in age-appropriate but also high-level mathematical reasoning and problem solving. Obviously, that goal is a huge challenge for everyone and was a challenge as well in Vygotsky's day, when he and his colleagues addressed the needs of children with handicaps and children from isolated, rural, illiterate, and impoverished families (Luria, 1976; van der Veer & Valsiner, 1994). Nevertheless, Vygotsky's framework, unlike universalistic psychological theories such as that of Piaget, tries to incorporate social and cultural factors into a theory of human learning and development. Contemporary sociocultural researchers have been systematically exploring the implications of Vygotsky's theory for educational reform in communities with diverse or impoverished populations and thus have been able to demonstrate the ways in which his theory can be useful today (e.g., Forman, Minick, & Stone, 1993; Jacob & Jordan, 1993; Moll, 1990; Newman, Griffin, & Cole, 1989; Tharp & Gallimore, 1988).

Contemporary Sociocultural Theory

Vygotsky died in 1934. Because of the political situation in Russia before and during World War II and the cold war immediately afterward, Vygotskian ideas had limited influence in the west until the late 20th century. Concepts such as the ZPD were used in educational research in the 1970s and 1980s by investigators, including Wertsch, Bruner, and Wood, to study how experts (usu-

[2] Boaler's (1977) ethnographic study of educational reform in Great Britain (discussed in more detail later in this chapter) is an excellent example of the value of Eisenhart's recommendation.

ally adults) assisted novices (usually children) by scaffolding (Wood, Bruner, & Ross, 1976) their problem-solving activities. Although this work contributed to our understanding of the teaching-learning process, it was limited in a number of ways. Some of these limitations include an overreliance on the tutorial model of teaching; a tendency to discount other forms of learning, such as peer collaboration and play; a focus on task mastery rather than on task understanding, with little or no attention paid to sign mediation;[3] and a relatively decontextualized model of teaching and learning—as an activity that occurs in a historical vacuum apart from particular cultural practices and institutions, and between people whose emotions, values, and beliefs are irrelevant to the instructional process. (See Minick et al., 1993, for further details.) These limitations are aspects of the early research, however, and not of Vygotsky's theory, because many of the topics ignored by this research (e.g., play, peer collaboration, and cultural and historical contexts) were addressed in Vygotsky's writings.

Beginning in the 1980s and continuing through the 1990s, deeper and richer interpretations of Vygotsky's original theory, as well as those theories of his colleagues and contemporaries like Bakhtin, began to shape sociocultural theory and research on topics like literacy instruction, mathematics problem solving in work settings, and cross-cultural studies of play. Much of this work has been compiled into edited books (e.g., Forman et al., 1993; Goodnow, Miller, & Kessel, 1995; Rogoff & Lave, 1984; van der Veer & Valsiner, 1994; Wertsch, 1981, 1985a), single-author volumes (e.g., Cole, 1996; Kozulin, 1990; Moll, 1990; Rogoff, 1990; van der Veer & Valsiner, 1991; Wertsch, 1985b, 1997), and handbook chapters (e.g., Serpell & Hatano, 1997; Shweder et al., 1998). A complete review of this recent work is beyond the scope of this chapter. Instead, in the following section, I draw from a recent overview by Minick et al. (1993) to provide a brief summary of some theoretical developments. In the body of the chapter, I review the work, both theoretical and empirical, that I believe is most relevant for reform in school mathematics.

There are at least five ways in which contemporary research, at least in North America, goes beyond the work done 15 or more years ago. First, serious attention is being paid to the institutional context of goal-directed activity (in schools, homes, community centers, libraries, etc.). "One cannot develop a viable sociocultural conception of human development without looking carefully at the way these institutions develop, the way they are linked with one another, and the way human social life is organized within them" (Minick et al., 1993, p. 6). Second, goal-directed activity occurs within communities of practice composed of experts and novices (Lave & Wenger, 1991). Thus, peer collaboration and adult-child tutoring are just two forms, among many, of guided participation in communities of practice (Rogoff, 1990). Third, activity is not mediated by *language*—that is, an abstract and generalized semiotic system that can be studied apart from its context of use. Instead, activity is mediated by a variety of speech genres (e.g., narratives, scientific arguments) and symbolic objects (e.g., notations and inscriptions) that are linked to specific institutions and practices (e.g., a biological laboratory, an elementary school classroom). Fourth, teaching and learning activities are typically carried out over time by people who are familiar with one another, who develop particular kinds of relationships, and whose interactions involve complex motives, beliefs, norms, goals, and values. Thus, failure to learn may indicate an unwillingness to identify with a particular teacher; a mismatch among the motives, beliefs, norms, goals, and values of teachers, students, and their families; or cultural or linguistic barriers to effective communication as much as a lack of ability on the part of the learner (Goodnow, 1990). Finally, notions of learning and cognitive development have changed from models that attribute a person's performance on tasks to an internal, individual set of intellectual competencies to models that describe individuals' and groups' goal-directed activities as constituting and constituted by institutionally, culturally, and historically situated communities of practice. Sfard (this volume) has articulated this distinction well in her discussion of the two metaphors for learning: acquisition and participation. I argue that we are moving from a narrow view of learning as something people gain (e.g., knowledge, skills, concepts, strategies for solving problems) to a broader conception of learning that can include what people do (e.g., solve word problems, sell candy, tailor suits, draw maps).

> In brief, we have begun to move beyond the rather decontextualized, universalistic representations of social interaction, language, and cognition characteristic of much educational and psychological research of the recent past ...toward a theory that highlights the rich intercon-

[3] For Vygosky, social and cognitive processes are mediated or regulated by signs (e.g., words, gestures, drawings, models). Studies of learning in social or solitary contexts, therefore, need to take account of the semiotic means by which changes in thinking occur.

nections between cultural institutions, social practices, semiotic mediation, interpersonal relationships, and the developing mind. (Minick et al., 1993, p. 6)

The Application of Sociocultural Theory to Reform in School Mathematics

How can sociocultural theory contribute to our understanding of the teaching and learning of mathematics within reform classrooms? I have organized this section around the two themes identified in the introduction: how the nature of instructional activities influences the outcomes of instruction and how doing mathematics involves communicating mathematically.

Activity Settings for Mathematics Learning

What do we know about the role of the social context in learning and development? First, Vygotsky's proposal to alter our research designs (to emphasize either long-term or short-term change over time) encourages us to look at learning as a dynamic process. Second, his focus on the social origins of higher mental functions makes us look for learning within communities of practice (Lave & Wenger, 1991). One of the implications of examining mathematics learning within communities is that norms for speaking and listening, judging the adequacy of mathematical explanations and solutions, displaying mutual respect, or encouraging competition are as much a part of learning within a community as the mental operations needed to solve problems (Goodnow, 1990; Hatano & Inagaki, 1998; Yackel & Cobb, 1996). Third, his revised unit of analysis (which includes both units of mind and units of social interaction) suggests that we look not merely at individual mental activity but at situated practices. Thus, one would examine not only the achievement outcomes of individuals but also the mathematical activities of students and teachers as they work together in formal educational settings, of parents and children in home and community settings, and of children working and playing with their peers in playgrounds and on street corners (Forman, 1996; Newman et al., 1989; Nunes, Schliemann, & Carraher, 1993; Saxe, 1991; Saxe, Dawson, Fall, & Howard, 1996; Saxe, Guberman, & Gearhart, 1987). In short, sociocultural theory proposes that teachers need to understand the mathematical knowledge that children bring with them to school from the practices outside of school as well as the motives, beliefs, values, norms, and goals developed as a result of those practices. In addition, the assessment of mathematical learning needs to take into account children's out-of-school as well as in-school experiences.

Clearly, that research agenda is ambitious , and studies consistent with that agenda are limited in the field of mathematics education. Rather than provide an exhaustive review of that literature, I highlight a few studies that address some key aspects of this agenda and that employ a sociocultural framework. I begin with the research programs of several groups of investigators who do not restrict their work to formal educational settings. Nevertheless, their research has implications for educational reform in mathematics classrooms. The first research program, by Nunes, Schliemann, and Carraher (1993) on the mathematics of street vendors (and other workers such as cooks), has produced valuable information about the relationship between children's practical and academic knowledge (Nunes & Bryant, 1996). A second set of studies (Saxe, 1991; Saxe et al., 1996; Saxe & Guberman, 1998) examined how children's emergent goals affect their activities and subsequent mathematical learning. A third set of studies focused on reform mathematics classrooms and on the teachers, students, and parents involved. Lehrer and Shumow (1997), for example, investigated the mismatch between teachers' and parents' beliefs about effective instructional activities and the differences between their teaching strategies. Finally, observations of elementary and middle school mathematics classrooms by Stodolsky (1988) and some of my own observations (Forman, 1996) allow an examination of the different activity structures (e.g., whole-class recitation, small-group work) that occur in reform versus traditional settings. Stodolsky and I argue that differences in activity structures are likely to produce differences in students' and teachers' mathematics beliefs, attitudes, and goals. This argument was explored in detail in an ethnographic study of two secondary schools in Great Britain (Boaler, 1997, 1998). Boaler compared the activity structures of the mathematics classrooms in the two schools and found significant relationships between those structures and student attitudes, beliefs, and performance on problem-solving tasks.

Street versus school mathematics. Nunes et al. (1993) conducted a series of studies of mathematical problem solving among Brazilian children who varied in their experience with formal instruction and with selling items on street corners in low-income communities. One of their recurrent findings was that children with limited amounts of schooling could be very successful at "street mathematics" and very unsuccessful at "school mathematics"—even when the same mathematical operations were required in both types of tasks and both tasks were administered outside of school (Nunes & Bryant, 1996).

For example, in one of the first studies by Nunes and her colleagues, children working as street vendors were initially asked for the total price of a specific number of items by an experimenter who posed as a customer. In one instance, the experimenter bought two coconuts, each of which cost 40 cruzieros, with a 500-cruziero bill. Then, after the experimenter identified himself or herself, the street vendors were given a series of problems similar to the first one and were asked about the amount of change necessary from a given bill denomination. After a few days, the experimenter returned and asked each street vendor to solve another series of problems that involved the same numbers and operations as on the previous occasion but that appeared in the format of arithmetic computation or word problems, such as $500 - (40 \times 2) =$ __. The children's success rate on these three types of problems was dramatically different: 98% correct for the street-mathematics format; 74% correct for the word-problem format; 37% correct for the computation-exercise format. Moreover, the children used different strategies to solve the problems. The street mathematics problems were solved orally, whereas the schoollike problems were solved using written algorithms (cited in Nunes & Bryant, 1996).

It is apparent that the street vendors "knew" the mathematics needed to solve problems they typically encountered in their business, but they did not seem to "know" as well the mathematics needed to solve equivalent problems that appeared in a different format. Nevertheless, these performance differences have consequences for the future success of street vendors inside and outside of school—especially if teachers base their judgments of children's mathematical abilities on their success with school-like tasks.

Nunes and Bryant (1996) argue that this body of research calls into question the distinction between practical and academic intelligence. Because school-learned algorithms do not seem to transfer readily to out-of-school tasks (like selling coconuts), academic intelligence may be no more general or abstract than so-called practical intelligence (like street mathematics). Yet educators act as if academic intelligence, as measured by performance on school-like tasks, is an accurate measure of mathematical ability and knowledge. Therefore, at the very least, teachers need to appreciate the potential contribution of out-of-school experiences to students' mathematical understanding and should recognize that school algorithms may not be useful in practical activities. The emphasis put on higher-level reasoning and complex problem solving by the 1989 NCTM Standards may lessen the gap between mathematics practiced within and outside of schools. Nevertheless, as work by Nunes et al. (1993) indicates, all mathematical activities are context bound to some degree, and thus it is unlikely that the gap will or should disappear (see also Lave, 1992).

Emergent goals. Saxe and his colleagues have applied Vygotsky's expanded unit of analysis and Leont'ev's notion of goal-directed, mediated activities to the study of mathematical learning in family, peer, and neighborhood contexts. One of their most valuable findings is that children's goals are emergent in their mathematical activities. That is, goals arise as a result of a complex system of influences from the cultural context (social conventions and artifacts), the social context (interactional practices and activity structures, like rules for buying and selling), and their own prior understandings. As a result, Saxe et al. find that children who are apparently engaged in the "same" mathematical activity may, in fact, be doing different mathematical procedures and learning different mathematical concepts. For example, Brazilian children who sold candy on the street engaged in a variety of price-ratio comparisons, depending upon their age, knowledge, and experience (Saxe, 1991). The youngest children (6- to 11-year-olds) tended to use only one or two ratios, whereas the oldest children (12- to 15-year-olds) were more likely to compare two or three ratios. Saxe also found that younger children tended to get more assistance from adults in their selling—and mathematizing—than older children did. Although the overall practical goal of selling candy was the same for the two groups of children, different mathematical subgoals occurred in their practice. The mathematical subgoal for the youngest children was to avoid ratio comparison (perhaps by getting someone else to do it for them) or to perform a simplified version of it, whereas the subgoal for the older children was to achieve a more complicated ratio comparison (perhaps to offer a more competitive price).

More recently, Saxe and his colleagues (Saxe et al., 1996; Saxe & Guberman, 1998) have studied children's arithmetic calculations using base-ten blocks in a task that resembled an educational game. They found that children 8 to 10 years old who played the game in pairs often revised the rules so as to simplify the mathematical calculations involved. This modification was especially common when younger children played with older children, with the older child acting as a banker so that he or she could do the borrowing and other regrouping calculations for the younger child. Although these simplifications may seem detrimental to children's learning, Saxe and Guberman found that the younger children who had opportunities to play the game with an older partner demonstrated greater learning gains in their understanding of regrouping than those younger children who played with age mates or did not play at all. Thus, by examining the context of children's mathematical activities, Saxe and his colleagues have been able to identify some of the cultural, social, and psychological mechanisms by which learning occurs.

In summary, research on emergent goals in mathematics learning indicates that children, especially while working in a group setting (whether selling candy or doing more school-like tasks), will establish their own priorities for problem solving. Some of those priorities may involve higher-order mathematical reasoning and problem solving; some may not. These findings are consistent with those of other sociocultural investigators who have observed adults and children in school, work, and home settings (e.g., Forman & Cazden, 1985; Forman & Larreamendy-Joerns, 1995; Lave, 1992; Newman et al., 1989). This body of work suggests that teachers need to pay close attention to what students actually do while solving mathematics problems and not assume that a single task will evoke a single set of problem-solving strategies (even a set that simply varies in sophistication). Students who redefine the task goals in ways that simplify the mathematical operations involved may be limiting their learning opportunities.

Obviously, teachers have always been aware that some students will work very hard to get other students to do their work for them, a phenomenon known as the free-rider effect (Salomon & Globerson, 1989). Nevertheless, Saxe and Guberman (1998) found that younger children who relied on the assistance of their older partners eventually learned more than their age mates who worked with a partner of equal status. Thus, these task simplifications could also be seen as a form of peer tutoring or scaffolding and not just as a form of cheating. In my own research, I have found that task modifications may, in fact, raise important questions for students. These questions can lead to discussions about the criteria for defining problems as well as solving them, debates about the effects of experimental error, or attempts to establish social norms for scientific explanations. By paying close attention to students' activities during small-group work, therefore, teachers may discover topics that can be explored in further instructional activities.

Home and school instructional practices. One influential study of the impact of a mismatch between school instructional practices and parental socialization practices in the area of literacy was conducted by Heath (1983). This study showed that not all children arrive at school equally prepared to benefit from the teaching provided. Instead of blaming the students' parents for the academic difficulties they experienced, Heath helped classroom teachers understand something about the community's literacy practices and adapt their instruction accordingly. A similar approach has been successfully used in other literacy interventions (e.g., Moll & Whitmore, 1993; Tharp & Gallimore, 1988).

A recent study by Lehrer and Shumow (1997) directly addressed the potential mismatch between parental practices of, and beliefs and values about, mathematics instruction and those of reform classroom teachers. In the introduction to their report, Lehrer and Shumow discussed the need for teachers and parents to come to a shared understanding of the ZPD for mathematics teaching and learning. Unfortunately, they found a lack of agreement between parents and reform mathematics teachers. For example, they found that parents espoused a variety of beliefs about the instructional practices they felt were the most appropriate ones for their children. Some of those beliefs were consistent with the 1991 NCTM Standards (e.g., students sharing problem-solving strategies), whereas others were not (e.g., opposition to classroom discussion and peer collaboration). In addition, Lehrer and Shumow found that parents differed from teachers in their scaffolding of student problem-solving activity. The teachers were likely to use more indirect forms of assistance (e.g., hints, general encouragement) than the parents did; when teachers were more directive (e.g., demonstrating), they quickly switched to a more indirect form of scaffolding in their next instructional move, whereas parents tended to remain directive throughout the session. As a result of their study, Lehrer and Shumow suggested that reformers of school mathematics cannot afford to ignore potential differences in the instructional goals, beliefs, and practices of teachers and parents—otherwise, seemingly successful interventions can be easily misconstrued by community groups.

Two differences should be noted between the studies of home-literacy practices cited above and the Lehrer and Shumow (1997) study. The former studies focused on low-income families, whereas the latter study focused on middle-income families. The former proposed changing classroom practices to align them, to some degree, with community practices; the latter suggested changing parental beliefs about mathematics instruction to better align them with the beliefs of school mathematics reformers. These studies raise the issue of the appropriate locus of institutional change (the family or the classroom) as well as the issue of possible differences in intervening with students from different socioeconomic communities. These issues are discussed further in later sections of this chapter.

Classroom activity structures. The 1989 NCTM Standards articulate a set of values about mathematics education that, for example, emphasize higher-order thinking and problem solving and de-emphasize the routine application of algorithms. These values, in turn, need to be manifested in different activity structures (e.g., in small-group activities and in student explanations and arguments, with less time spent filling in worksheets) so as to foster changes in students' personal goals, attitudes, and beliefs. The aim is to encourage students to exercise their curiosity and ingenuity and to want to learn from one another as well as from teachers and textbooks.

Have classroom activity structures changed in reform mathematics classrooms? Unfortunately, a complete review of the existing observational studies of mathematics reform is beyond the scope of this chapter. However, some observations in traditional and reform mathematics classrooms may help us explore the hypothesis that changes in activity structures can result in changes in personal goals, attitudes, and beliefs. In my research, I have noticed that the variety of settings available for students in reform classrooms may provide them with more opportunities to actively engage in mathematical problem solving with others than in traditional classrooms (Forman, 1996). In contrast with Stodolsky's (1988) observations in traditional fifth-grade mathematics classrooms in which almost no small-group work was seen and most activities consisted of individual seatwork or whole-class recitations led by the teacher, I have found that middle school reform classrooms often allow students to work in small groups and present their ideas in front of the class. My observations of individual students in those classrooms indicated that their goals, attitudes, and beliefs varied as a function of the activity setting for mathematical practice. For example, some students refused to pay attention to the teacher or other students during whole-class presentations, but those same students interacted readily with other members of their group during problem-solving activities. Other students showed the opposite picture of engagement: being more attentive during whole-class presentations but resistant or anxious during small-group work.

Stodolsky (1988) argued that math anxiety might be a function of the limited and restrictive conditions under which traditional mathematics instruction occurs. Hatano (1988) proposed that small-group work would enhance students' intrinsic motivation to comprehend mathematical problems. I am proposing that apparent anxiety, engagement, or resistance may be a function of individual differences as well as the mathematical activity setting.

Boaler's (1997, 1998) ethnographic study of two secondary schools in Great Britain provides a wealth of information about the relationship among mathematics classroom activity structures, student attitudes, and achievement outcomes. Boaler compared the instructional programs at Amber Hill and Phoenix Park, two schools that served white, working-class neighborhoods. At Amber Hill, the teachers used an approach emphasizing the learning of specified mathematical methods that was common in traditional mathematics instruction in Great Britain when the study was conducted. The students at Amber Hill were tracked into homogeneous ability groups in their mathematics classes. Their lessons tended to involve textbook-based explanations followed by practice exercises that were tailored to their skill level. In contrast, Phoenix Park teachers employed a project-based approach in their mathematics classes. The students were encouraged to work in small, mixed-ability groups on open-ended projects that lasted several weeks.

At the end of her 3-year study of grades 9 through 11 (the students were 13 to 16 years old), Boaler (1997, 1998) found striking differences in student attitudes and achievement in the two schools. Despite the fact that students at Amber Hill were frequently seen working on their exercises and rarely seen being disruptive or disengaged, they did not express many positive attitudes toward mathematics. Her interviews and observations indicated that Amber Hill students found their mathematics classes boring and difficult. She characterized their approach to mathematics as focused on memorizing and applying narrow procedural rules and not on mathematical sense making. Students' attitudes were more positive at Phoenix Park. Students from Phoenix Park were more likely to say that they enjoyed their open-ended work than were students from Amber Hill, and the former were more likely to characterize their work as interesting. At Phoenix Park, some students strongly disliked the instructional approach, whereas others strongly approved of it; at Amber Hill, there was more agreement about the tedium of the approach and the lack of choice in instructional activities. Finally, Boaler found that Phoenix Park students did better than Amber Hill students on open-ended tests of mathematical problem solving and did as well on traditional closed-ended tests of achievement.

Boaler (1997, 1998) concluded that the discrepancies between the activity structures of the two schools resulted in the contrasting beliefs as well as attitudes about mathematics. She claimed that the Phoenix Park students viewed mathematics as a tool that could be applied flexibly in a variety of situations, whereas the Amber Hill students saw mathematics as consisting of specific formulas to be memorized and applied only in limited situations. Thus, the findings of Boaler's study of secondary schools are consistent with observations that Stodolsky (1988) and I (Forman, 1996) have made of elementary and middle schools in terms of the differences between the activity structures of traditional and reform classrooms. Boaler's research goes farther than ours in incorporating measures of students' expressed attitudes and beliefs and of their open-ended and closed-ended problem solving performances.

In these studies of the in-school and out-of-school activity structures for mathematics learning, communication is discussed but not as a central aspect of mathematics learning. Without paying serious attention to communication, one has difficulty understanding the mechanisms by which activity influences and is influenced by learning. In other words, one needs to examine discourse carefully to know how goals emerge in social activity. Otherwise, one could be led to believe that the superficial physical or social characteristics of activity settings (e.g., whether students do more group work and less individual seat work) determine students' learning. This deterministic model of sociocultural influences on learning is consistent neither with basic concepts within the theory (e.g., semiotic mediation) nor with the findings of current research. Jorge Larreamendy-Joerns and I, for example, have found a diversity of learning outcomes as a result of peer collaboration in a group of elementary and middle school students (Forman & Larreamendy-Joerns, 1995). Although each student in our study worked on the same series of mathematical problems with one other classmate, some students ended up learning more about how images of geometric shapes are transformed as a result of being projected as shadows, whereas other students learned more about how to compare static geometric shapes. These differences in learning outcomes could not be attributed to age or initial task-specific expertise but seemed to be a function of the interpersonal dynamics (mediated by discourse) that occurred in each dyad. For example, in some dyads, one student assumed the role of relative expert, and that student exerted a great deal of influence on what got done and what was discussed. In other dyads, each student took turns directing the activities of his or her partner, and, in some cases, both types of divisions of labor occurred. Recent research by Kieran and Sfard (Kieran, 2001; Sfard & Kieran, 2001) provides additional examples of the diversity of outcomes in collaborative problem solving when students solve similiar mathematical tasks. The likelihood that superficial characteristics of activity settings do not determine learning outcomes is crucial to the success of educational reform in mathematics, because one of the aims of reform is to change the interpersonal dynamics within classrooms so as to change what is learned. In the next section, I examine classroom communication practices in mathematics in some detail.

Communication in Mathematics Learning

Research on classroom discourse across a variety of disciplinary settings has documented a recurrent pattern: The teacher tends to initiate discourse (often with a question), one or more students respond (often with an answer), and the teacher evaluates or provides feedback (Mehan, 1979). This pattern is often known as the IRE (or the IRF)—which stands for those three discursive moves: initiate, respond, and evaluate or provide feedback (Cazden, 1986; Wells, 1993). In traditional classrooms, this pattern obviously puts the teacher firmly in control of what is said and done. Reformers of school mathematics have advocated changing classroom discourse practices to allow students to take a more active role in initiating and providing feedback (Cobb, Wood, & Yackel, 1993). Is the IRE (with some modifications) the pattern needed in reform mathematics classrooms? Some educators see a positive role for the IRE if students are encouraged to offer their own explanations and evaluations of their classmates' explanations, as sometimes occurs in Japan but less often in the United States (Inagaki, Morita, & Hatano, 1999). Other educators think that instruction in mathematics should do a better job of introducing students to aspects of the discourse genre of the discipline of mathematics (Lampert, 1990; Lerman, 1996).

Mathematical discourse has been described in a number of ways. One characterization can be found in the work of Halliday and his colleagues (Halliday, 1975, 1988, 1993; Halliday & Martin, 1993). They have investigated the registers of mathematics and science.[4] In brief, Halliday has depicted the mathematics register as exhibiting syntax that is impersonal and outside time and space, as employing specialized vocabulary or new definitions of familiar words (e.g., *similarity*), and as valuing

[4] *Register*, like *genre* and *dialect*, is a linguistic term that refers to language variation. Dialects are typically associated with different groups of speakers (e.g., speakers from a variety of regions and social classes); registers and genres are associated with different situations (e.g., occupational settings and message types). There is much controversy about whether to use *register* or *genre* as the more general or preferred term (Biber, 1994). I use the two terms interchangeably in this chapter.

particular modes of reasoning (e.g., precision, brevity, and logical coherence).

Halliday's perspective on mathematical—and scientific—discourse has been challenged by Bazerman (1988). Bazerman argues that Halliday's approach is limited in three ways. First, it regards the linguistic structures of science and mathematics as static, that is, as eternal and unresponsive to historical changes within scientists' lives and over time in scientific communities. Second, it does not account for the material and symbolic objects to which these linguistic forms refer. And third, it treats the social world as given, not as evolving along with the artifacts and symbolic objects of scientific activity.

According to Bazerman (1988), the central activity of scientists is argumentation within communities of practice: trying to persuade others of the validity of one's ideas as well as those of one's colleagues. The focus of these arguments is on establishing agreement about the truth of symbolic objects, which may or may not refer to objects in the world and may or may not be mathematical objects. For example, the double helix is both a model of a material object, DNA, and a mathematical object, a particular kind of geometric figure. The double helix was the focus of a great deal of argumentation when Watson and Crick first presented it. Nevertheless, other members of the scientific community now accept the model as a scientific "fact" because they have been convinced of its validity by Watson, Crick, and their colleagues.[5]

What Bazerman's analysis provides for understanding discourse in scientific communities, Rotman's (1988, 1993) analysis provides for mathematical discourse. Rotman incorporates three types of discourse in his model of the semiotics of mathematics. In Rotman's account, each mathematical argument contains within it three different voices. The first is that of the Mathematician, also called the Subject, who imagines, inscribes, employs a fragmentary natural language, and addresses the second voice, the Agent, often in imperatives (e.g., assume *Y*, define *X*), using the eternal and impersonal words of Halliday's register, or in Rotman's terms, the Code. The Agent is an automaton who carries out the orders of the Mathematician (e.g., count, integrate) and also uses the Code. The third voice is the Person, who converses in natural languages by using a range of pronouns, including *I*; present, past, and future tenses; and emotional and evaluative expressions, or what Rotman calls the Metacode. The Person also participates fully in a historical and cultural world. The first two voices are present in mathematical texts. The third voice is a participant in mathematical communities and, as such, is concerned with persuading other people of the validity of his or her mathematical arguments. Rotman believes very strongly that the third voice plays an essential role in mathematical activities, even though it is not represented in mathematical texts, because decisions about the worthiness of particular mathematical arguments, their elegance and sophistication, and their empirical validity are negotiated within mathematical communities using this voice.[6]

Thus, Rotman (1988) and Bazerman (1988) view the discourse of science and mathematics as a language of action used informally by communities engaged in all aspects of scientific practice (i.e., generating evidence; producing models of material or symbolic objects; drawing conclusions from the models; and imagining, constructing, and defending logical arguments) and formally by the writers of that practice. Doing mathematics or science may involve different types of linguistic codes than writing about mathematics or science (e.g., Rotman's Code and Metacode). Whereas formal writing about mathematics may exhibit characteristics of Halliday's mathematics register, informal writing and speech about mathematics will resemble the embodied language of everyday life. Both types of language are essential to mathematics, in Rotman's view, because both types are used—for different purposes—in communities of mathematicians.

If teachers want to introduce students to the genre of mathematics discourse, then they need to think about the range of discourse that Rotman has identified: from the very personal, emotional, and evaluative voice of the Person to the very impersonal, logically coherent, but imaginative voice of the Mathematician to the equally impersonal, precise, and tireless voice of the Agent. One could argue that the IRE, as a form of classroom discourse, encourages students to play the role of the Agent (the computational automaton) in response to the commands of the teacher and the textbook (who together would serve as the Mathematician) to compute, prove, draw, and so forth. The voice of the Person seems absent in the IRE and perhaps has been absent in mathematics classrooms in which teachers and students rarely question why certain topics are addressed at length but others are treated superficially, if at all; why some explanations are pre-

[5] More precisely, scientists would say it is the best, most useful model that fits current knowledge about DNA. See Latour (1987) for a more detailed account of the historical context of the double-helix model.

[6] Bazerman (1988) has examined the historical evolution of scientific journals. In his analysis, he traces the disappearance of the more informal types of communication (similar to Rotman's Metacode) from scientific publications and the gradual domination in those publications of the more formal types of communication (similar to Rotman's Code and Halliday's scientific register).

ferred over others; and whether mathematics is frustrating, alienating, or exhilarating. If Rotman's characterization of mathematics is correct, then both students and teachers should be encouraged to assume all three voices and to use both formal and informal mathematical codes.

Some mathematics educators have suggested alternative models of classroom communication—models that do not resemble the IRE and that come closer to that used in mathematical communities. These models, among other things, feature greater student participation in argumentation (Cobb & Bauersfeld, 1995; Lampert, 1990; O'Connor, 2001; Strom, Kemeny, Lehrer, & Forman, 2001). To achieve this goal, Yackel and Cobb (1996) propose that students and teachers need to make explicit the social and sociomathematical norms of discourse in reform classrooms. For example, Yackel and Cobb view a rule like the necessity of adequately explaining one's ideas and evaluating those of others as a social norm (because it could be used in any discipline). In contrast, sociomathematical norms entail the discussion of the rules for evaluating mathematical arguments in terms of elegance, efficiency, and sophistication. These norms, because they are evaluative and personal, would need to be negotiated in the voice of Rotman's Person using the Metacode. Thus, some mathematics educators are proposing that classroom communities attempt to resemble those of mathematicians in which the social and sociomathematical norms for engaging in mathematical argumentation are learned and used.

If an alternative to the IRE is needed for mathematics classrooms, then a number of additional questions need to be raised: Can or should teachers and students change their discourse practices to more closely resemble those of mathematicians? How does classroom discourse differ from that of mathematicians? How can teachers assist students as they begin to appropriate the various linguistic codes and norms of mathematics? In the next section, I discuss the work of researchers who are beginning to explore the ways in which teachers and students negotiate collective arguments in mathematics classrooms. Before examining empirical research on this topic, I first discuss a theoretical framework for the socialization of argumentation in classrooms.

Teacher-orchestrated classroom arguments. One model of classroom argumentation makes an analogy between young children learning their first language and students learning the discourse practices of mathematics and science (Bazerman, 1988). In both these scaffolding situations, the novice and the expert need to interpret or frame a situation in a similar fashion, despite vast differences in linguistic and practical experience (Lerman, 1996). Drawing on Bruner's writings about language learning, Bazerman depicts a situation in which a student's attempts to communicate are interpreted by a teacher, who recasts the students' utterances into forms that would be acceptable to a broader educational or scientific community. These exchanges typically refer to some material or symbolic object (e.g., a geometric figure, a piece of litmus paper) that allows the expert to draw students' attention to relevant perceptual features (e.g., length of side, color change of paper) as well as to the language used to refer to those objects. For example, a student may be explaining how she figured out the area of a geometric figure and concluding that the area is 14.5 cm. Her teacher may recast her conclusion as 14.5 cm^2, thereby emphasizing the correct unit of area measurement. Over time, students' technical vocabularies increase, as does their range of experience with these material and symbolic objects. More important, the students' ability to formulate their own utterances is constrained less by the expert and more by the material and symbolic objects being manipulated. In other words, over time, the teacher's role as orchestrator of discussion changes as he or she becomes a less frequent speaker and as students begin to formulate utterances about the objects that would be seen as more broadly acceptable within the educational or scientific communities.

In Bazerman's (1998) account, classroom discourse would differ from the discourse of scientific communities because of the different distribution of novices and experts in the two communities. Arcavi and his colleagues (Arcavi, Meira, Smith, & Kessel, 1991) argue that the practice of teaching mathematics differs from the doing of mathematics in professional communities in more than the shared knowledge base. In their reflections on Alan Schoenfeld's teaching practices at the postsecondary level, they propose that Schoenfeld had distinct instructional goals in his presentation of mathematical content that resulted in his imposing constraints on student participation in classroom activities. For example, he did not always follow the suggestions of students, choosing instead to make explicit and implicit judgments about productive and unproductive potential solution strategies. In that way, he retained the role of orchestrator of mathematical activities. Another difference between argumentation in classrooms and scientific communities may be the greater need to establish an atmosphere of mutual respect and nurturance in the classroom.

> Extended whole-class discussion is possible only when most students are willing to be good listeners. Requiring students to display their solutions can be productive only when students form

> a caring community for mathematics lessons.... Teachers' fear of damaging a child's self-esteem by public failure is real unless the class is a caring community. (Hatano & Inagaki, 1998, p. 93)

Research by O'Connor and Michaels (1993, 1996) and others (Forman & Ansell, 2001, 2002; Forman, Larreamendy-Joerns, Stein, & Brown, 1998; Strom et al., 2001) has investigated the orchestration of discussion in mathematics classrooms. In this work, the basic research question has been the following: How do the teacher and students create a shared frame or set of expectations for interpreting an argument? O'Connor and Michaels focus on the notion of *revoicing* to illustrate this process. Revoicing involves the repetition, expansion, rephrasing, and reporting by one speaker (often the teacher) of another's (often the student's) utterance. Revoicing can serve a number of purposes: to reformulate an unclear or unacceptable statement, to create alignments and oppositions in an argument, or to redirect the discussion (e.g., "Steven, tell me more about that idea. What part of B would you cut off?" "Does anyone agree with Steven? They think he's got an idea there that, the more they think about it, they think it makes sense?").[7] It is important to note here that revoicing is only one among several strategies that teachers (and students) use to orchestrate discussions. Other strategies include requesting a small-group discussion on a topic or checking additional sources.[8]

Basing their analysis on theoretical work by Goffman (1974, 1981), O'Connor and Michaels (1993, 1996) discuss the mechanism of animation. Expanding the model of conversation beyond the fixed roles of speaker and listener, Goffman argues that sometimes the person making the noise (the animator) is actually voicing the words of someone else (the author) or asking the audience to attribute those words to someone else (the principal). O'Connor and Michaels show that revoicing enables teachers to put claims in students' mouths as well as assign them a particular role in the overall debate. In essence, revoicing allows the teacher to reframe arguments conceptually and socially.

Revoicing can also provide an attitudinal or motivational framework to an argument. Theorists have identified two types of one form of revoicing: direct and indirect quotation. A teacher can revoice through direct quotation (e.g., "Mary says . . .") or through indirect quotation (e.g., "Mary is claiming that . . ."). Direct quotation focuses on the "ideational position of the (other) speaker" (Volosinov, 1929/1986, p. 129). In other words, direct quotation is like a report of someone else's position in an argument. Thus, using direct quotation can demonstrate to students the value of clarifying and authenticating the original speaker's message in some cultural contexts (such as mathematics problem solving). In contrast, indirect quotation (e.g., "Mary is troubled by that") allows one to convey one's own interpretation of the other speaker's attitude; it permits one speaker to ascribe motivations and intentions to another. Through indirect quotation, teachers and students can convey their impression of the emotive quality of other speakers' messages.

My colleagues and I (Forman, Larreamendy-Jones, et al., 1998) have integrated the notion of revoicing with an analysis of argumentation (using notions from Toulmin, 1958, and Krummheuer, 1995) to understand the processes of discussion-orchestration in a reform mathematics classroom located in a low-income, multicultural community.[9] Our analysis showed how a middle school teacher and her students negotiated a shared frame for understanding the norms, values, and conceptual content of a complex mathematical problem through argumentation. One at a time, students presented their explanations of the solution of an area-measurement problem to their classmates. The students reflected upon, evaluated, and integrated the explanations of others in their own arguments as the classroom teacher used revoicing (in all its forms) to recruit students, align students with positions in the collective argument, attribute attitudes and intentions to students, and report or clarify students' explanations.[10]

Rotman's (1998) analysis of the semiotics of mathematics can be useful in helping us connect the previous discussion of mathematical and scientific argumentation with the analysis of classroom discourse. The studies cited in the preceding paragraphs provide multiple instances of teachers' animation of students' utterances through revoicing. That is, a teacher repeats a student's utterance (either exactly or in modified form) to make it more accessible (less ambiguous, better formulated, more canonical) to the other students. In so doing, the teacher is making it clear that this utterance is not his or her own argument. The teacher may also use direct or indirect

[7] These quotes were taken from classroom transcripts that I have used in my research. The classroom videotapes were collected by Richard Lehrer as part of his geometry project. (See Lehrer et al., 1998, and Strom et al., 2001, for further details.)

[8] I am grateful to Giyoo Hatano (personal communication, 24 May 1998) for calling my attention to this issue.

[9] The classroom videotapes were collected by Mary Kay Stein and others as part of the QUASAR project, directed by Ed Silver.

[10] See Strom et al., 2001, for an alternative model of argument analysis and for additional examples of discussion orchestration.

quotes. Direct quotes in revoicing, being more emotionally neutral, more closely resemble the voice of the Mathematician and his or her Agent, whereas indirect quotes, being more personal and emotionally explicit, more closely resemble the voice of the Person—the representative of the mathematics community.

In Rotman's (1988) and Bazerman's (1988) models of the genre of mathematical and scientific argumentation, the aim of an argument is to establish intersubjective agreement about the meaning of symbolic objects (cf. Lerman, 1996). Cobb and his colleagues have referred to these as "taken-to-be-shared-objects of experience" (Cobb et al., 1993, p. 102) in their work on mathematics education.

Changes in the meaning of symbolic objects. Symbolic objects in mathematics can be equations composed of numbers and operators, geometric figures, graphs, or proofs. They may also include oral and written language and other kinds of inscriptions and notations. If learning mathematics involves learning to use symbols in mathematically meaningful ways, how is that accomplished in classrooms? Past research has shown (e.g., Hughes & Grieve, 1980/1993) that children often misunderstand the meaning of the mathematical symbols they learn to manipulate procedurally. I propose that at least two interrelated aspects of symbol use are crucial if students are to learn to use mathematical symbols in a meaningful way. The first aspect of symbol use is the appropriation of mathematical argumentation (as discussed above); the second aspect is the ability to create and modify mathematical symbols. Both these aspects of symbol use are essential to meaningful mathematical and scientific activity in classroom as well as in scientific communities.

Using sociocultural theory as a source, van Oers (1996, 2000) has articulated a perspective on the development of children's knowledge of mathematical symbols. Beginning with Vygotsky's writings about symbolic development and expanding that perspective with notions from Leont'ev's theory of activity and more recent research by Wertsch and others, van Oers argues that children's learning begins with qualitative changes in action patterns during goal-oriented social activity. For example, a young child may count from one to five in the presence of an adult, his mother, who interprets his actions as meaning something about the number 5. As a result, the adult may respond to the child's actions with one word, *five*, a written symbol for that word, 5, and a comment about the cardinality of this quantity. Van Oers suggests that *five* becomes an abbreviated version of the activity of counting—first for the adult and then eventually for the child. So an initial step in the symbolization process involves abbreviation. In this example, the symbol 5 does not refer to some abstract object, divorced from its context of use. Instead, it highlights something about the initial counting situation and predicts something new about that situation (e.g., cardinality). Another crucial step in the process involves reification: the spoken and written symbols *five* and 5 become objects for reflection. Once the initial actions are abbreviated and reified as symbols, their meaning can be remade for new (and initially unforeseen) purposes. For example, this same mother and child may encounter a 5 in a book, which could remind the mother of the earlier counting experience and cause her to comment on the similarity between the five items in the book and the previous shared experience. In this second instance, the number is an object of reflection that connects the two experiences with *five*.[11]

Lehrer, Schauble, Carpenter, and Penner (2000) also discuss symbolization in mathematical activity, but they base their analysis on work from the sociology of science by Latour (1990). Instead of symbolization, they refer to inscription (a form of mathematical symbolization); and instead of abbreviation, reification, and revision, they use Latour's term, *cascade of inscription* (a process by which symbols are abbreviated, reified, and revised repeatedly within the scientific community). In one of their classroom examples, they found that children's initial inscriptions of plant growth were unidimensional and then evolved into two-dimensional and finally three-dimensional representations. Thus, the children's understanding of growth changed as a function of their ability to inscribe it spatially, and these changes in understanding altered their research questions about plants and their arguments about their research findings. In this way, Lehrer and colleagues make an analogy between changes over time in children's symbolization in the classroom and similar changes in adult scientists' symbolization in their professional communities.[12]

In summary, mathematical communication in classrooms has a number of components in common with mathematical communication in scientific communities. At the very least, it involves argumentation and symbolization. Both these aspects of communication are essential for the negotiation of intersubjective meaning in

[11] A similar analysis of symbolization from a sociocultural perspective can be seen in Saxe's writing about form-function shifts in mathematical activity (Saxe, 1991; Saxe et al., 1996).

[12] Forman and Ansell (2002) make a similar argument about the use of inscriptions in communities of mathematicians and in classroom communities.

those two social contexts and are essential for learning, from a sociocultural perspective. The next section addresses criticism of the model we have been developing for understanding mathematical communication in classrooms. Most of this criticism comes from within the community of sociocultural researchers.

A brief critical review of research on mathematical communication. The scaffolding or discussion orchestration model for classroom learning has come under criticism from several fronts. Gee (1996) has wondered whether it is practical to expect one teacher to do an effective job of scaffolding 20 to 40 students. Obviously, this type of teaching requires expertise in both mathematics and pedagogy. One study by my colleagues and me (Forman, McCormick, & Donato, 1998) examined how one teacher tried to encourage multiple solutions to a mathematics problem but experienced difficulty genuinely appreciating students' explanations that differed from hers. We examined, in detail, the scaffolding by this teacher of the explanations of three of her middle school students. Although the teacher claimed to use an approach to instruction that encouraged multiple strategies for solving problems, her espoused instructional goal did not seem to correspond to her instructional practice. We found that two of the three students began their explanations using a solution strategy that differed from the one the teacher had presented the previous day, whereas a third student used the teacher's solution strategy from the beginning. Each time the first two students tried to use an alternative solution strategy, the teacher overlapped her speech with theirs—effectively silencing them. By the end of their presentations, each of those students employed the strategy privileged by the teacher. The third student, who had employed the privileged strategy from the beginning, did not encounter the teacher's overlapping speech.[13] In addition, this student used a speech genre that resembled Rotman's (1988, 1993) Code: The speech was impersonal, was outside time and space, and used mathematical vocabulary. In contrast, the speech of the two other students was more personal, was less precise and clear, employed fewer mathematical expressions, and was more dependent on the immediate spatial context (as in Rotman's Metacode).

We argued that this teacher might have had difficulty understanding the solution strategies that differed from the one she had presented the previous day. Perhaps it is unrealistic to expect a teacher to fully appreciate several legitimate alternative solution paths without time to carefully examine each one or without extensive previous experience with this or similar problems. For example, position-driven discussions allow a teacher to orchestrate a classroom discussion focused on a small number of possible answers (see O'Connor, 2001, for an excellent example of a position-driven discussion). Unfortunately, this teacher's responses to the different solution strategies and the different types of speech may have unintentionally communicated a message to the rest of the class about the competencies of these three students. In essence, she privileged the student who spoke using her notions and who used more abstract terminology and syntax. This message would contradict the recommendation of the 1991 NCTM Standards that all students be encouraged to see themselves as the source of mathematical expertise. Thus, the discourse during this lesson did not resemble the IRE, but the teacher maintained her role as sole initiator and regulator of communication in the classroom.

This lesson raises an important issue about the nature of classroom communication patterns and teachers' goals. How can a teacher simultaneously provide opportunities for active student participation in discussions (e.g., providing explanations, evaluating the explanations of others) while maintaining a clear vision of his or her educational goals for the lesson? Schoenfeld and his colleagues (Arcavi et al., 1991) propose that instructors must keep their goals in mind at all times so that the discussion (and supporting activities) can result in significant learning about mathematics. It is apparent that Schoenfeld's own teaching style depends upon his deep knowledge of the content he is trying to teach as well as his sense of the pedagogical activities that would best communicate to his students the essence of mathematical inquiry. Arcavi et al., in their examination of Schoenfeld's teaching, argue that he tries to maintain a balance between "the dual goals of engaging students' interest and participation and sustaining progress toward important instructional goals" (p. 40). In their view, this balance may involve elements of so-called traditional and nontraditional pedagogy (e.g., lectures, textbooks and other printed materials, small-group work on projects, and teacher-orchestrated whole-class discussions). Thus, the dichotomy introduced previously between reform and traditional classroom practices should be modified to include a wider range of means that might be needed to attain the goals of educational reform.

From another perspective, Walkerdine (1988, 1990, 1997) argues that the rational and impersonal nature

[13] Overlapping speech might be interpreted either positively or negatively—depending upon its context of use. In our analysis, we tried to base our interpretation of the function of overlapping speech in this situation on multiple, convergent sources of information.

of much of mathematical discourse communicates emotional messages as much as intellectual ones. The historical source of these messages can be traced to the Age of the Enlightenment, if not earlier, and involves the notion that nature can be controlled through logic and especially through mathematics. Because of the technological advances that occurred as a result of this harnessing of nature through logic, Western civilization accepted as "natural" the fantasy of control through reason in general and the personification of the agent of control as male, white, and educated. In Walkerdine's analysis, the discourse of mathematics is alienating to those who feel excluded or choose to exclude themselves from the desire for control through logic. If one accepts her view of the fantasy of the power of mathematics, one can easily see why women, people of color, and others have felt ostracized by the powerful group of mathematics users. Rotman's (1988, 1993) analysis suggests a way to counteract that view: highlight the more personal, embodied nature of mathematical discourse. Perhaps a greater awareness of the presence of the community of mathematicians as well as of the classroom community (in which mutual respect needs to be established and maintained) could provide needed balance to the message about the alienating nature of mathematical discourse.

Lampert and her colleagues have expressed another concern about mathematical discourse (Lampert, Rittenhouse, & Crumbaugh, 1996). They noticed that calling attention to disagreements in the classroom could create social tensions because disagreements in traditional classrooms in which only one answer is correct usually mean that someone is wrong and someone else is right. Lampert et al. worry that girls may be more likely than boys to want to please others by seeming to agree so as to preserve a relationship rather than to pursue the argument in the service of greater understanding. Tannen (1993), however, argues that all forms of social influence involve expressions of solidarity and power, albeit in differing degrees. In addition, ethnic and racial differences in argumentation styles have been identified by sociolinguists (Goodwin, 1982; Goodwin & Goodwin, 1987; Schiffrin, 1984). Therefore, teachers need to be aware that students may use different styles to conduct a mathematical argument.

Summary and Conclusions

The overall themes of this chapter concern the nature and impact of instructional activity at the interactional, discursive, and symbolization levels. I have argued, on the basis of sociocultural theory and research, that the forms and functions of instruction (in family, peer, and school contexts) can have a profound influence on children's learning processes and outcomes. This influence is not merely cognitive but also includes motivational, affective, and normative factors and involves beliefs about learning and goals for learning. Several investigators have proposed an analogy between classroom communities of practice and professional communities. If classrooms were to foster some of the essential activities of mathematical communities, then teachers would need to assist students in appropriating mathematical argumentation and symbolization. One means by which mathematical discourse can be learned was proposed: revoicing. I also argued that learning mathematics depends upon a process of symbolization in which students' speech and actions are reinterpreted and abbreviated through negotiation with adults and peers, reified in written inscriptions (e.g., graphs, charts, equations), and revised as a result of argumentation.

This model of learning mathematics was then critically evaluated from several perspectives. A number of additional questions were raised. Can one teacher be expected to orchestrate (in real time) the discussions of 20 to 30 students? How can a teacher keep in mind a set of instructional goals while encouraging active participation from students? Should the beliefs and practices of parents and teachers be aligned in the interest of successful reform in mathematics? If so, who should decide and who should be expected to change?

At this point, perhaps it is appropriate to take a broader perspective on the value of a sociocultural approach to educational reform in mathematics and return to our initial question: What can sociocultural theory contribute to current reform efforts that is not being contributed by other approaches? I propose that a major reason that sociocultural theory is relevant to reform now is that the 1989 and 1991 NCTM *Standards* documents are not just about curriculum revisions (as such documents were in the past) and not just about improving instruction for one segment of the society (students who are middle class, college-bound, male, and white). The NCTM Standards are for *all* students. Sociocultural theory cannot supply a recipe for accomplishing this difficult task. I think, however, that the guiding assumptions of this theory do provide us with a way of conceptualizing and studying educational reform for all students that may not be available from other theoretical perspectives. If Valerie Walkerdine is correct in her assessment of our collective fantasy about the power of reason, then perhaps we should consider seriously a theoretical framework that posits social and cultural activity and not reason as its central organizing premise. In other words, in the interest of educating all students and to overcome our bias toward privileging impersonal rationality over other aspects of human functioning, one would need to begin a theory, such as sociocultural theory or situated cognition (Greeno & the Middle School Mathematics Through Application Project Group, 1998), that sees thought as just one feature of socially organized activity. At the very least, sociocultural theory makes one wonder whether students who fail to learn in reform classrooms are failing not just because of a lack of "ability" or "interest" but also because of resistance to learning the discourse of school mathematics, an alienation from authority figures in classrooms and mathematics comunities, or a misalignment between the beliefs and values of their home community and those of the classroom (cf. Goodnow, 1990). It makes us question assumptions about the locus of blame (individual versus systemic) and the locus of change (students, parents, communities, schools, employers).[14]

A different perspective on the value of sociocultural theory comes from a methodological stance about the close ties between theory and practical application. Vygotsky articulated this position in his writings, as mentioned above, and it is currently influencing educational research in the form of design experiments (Brown, 1992) or teaching experiments (de Corte et al., 1996; Gravemeijer, Cobb, Bowers, & Whitenack, 2000). Like situated cognition researchers (e.g., Greeno & the Middle School Mathematics Through Application Project Group, 1998), who also conduct design experiments, sociocultural researchers (e.g., Newman et al., 1989; Lehrer et al., 2000; Brown, 1992) have argued for an engineering model for educational research. That is, rather than employ a physics model in which controlled experiments are conducted to determine the necessary and sufficient conditions for specific outcomes, the engineer willingly confounds variables to achieve a desired outcome. Although these investigators are still interested in controlling variables in experiments when possible and useful, they also admit that intervening in schools is a complex social activity that involves multiple activity systems (e.g., school boards; teachers' unions; parent groups; federal, state, and local reform initiatives; and state funding decisions). Thus, it may be impossible or unproductive to attempt to conduct anything approaching a controlled experiment in a school setting. If a design experiment can be run successfully, then perhaps further investigations can lead to more information about the necessary and sufficient conditions for change. This methodological innovation is another reason that an approach beginning with the assumptions of sociocultural theory (and of situated cognition) that learning occurs in complex, goal-directed activity systems can be productive in reforming school mathematics.

[14] Some or all of these issues are not specific to mathematics education but appear in the large literature on minority education. A recent review of the history of minority education shows that the deficit model of minority underachievement predominated in the early part of the 20th century and was a major influence in the compensatory education programs of the 1960s such as Head Start. This position was criticized by sociolinguists in the late 1960s and early 1970s who argued that the apparent deficits were really cultural differences due to cross-cultural miscommunication. The cultural difference position, in turn, was further criticized by sociologists and anthropologists for its lack of attention to larger societal forces (class and status hierarchies, differences in economic resources and social mobility, and the different histories of each minority group in terms of colonization, enslavement, or immigration). (See Jacob and Jordan, 1993, for an introduction to the range of current approaches to minority education from anthropological and sociological perspectives.) A full treatment of minority education and its relevance to educational reform in mathematics is beyond the scope of this chapter. The interested reader may want to consult some of the numerous articles, chapters, and books on equity in mathematics education (e.g., Hanna, 1994, 1996; Leder, 1992; Noddings, 1996; Secada, 1992; Secada, Fennema, & Adajian, 1995; Tate, 1995, 1997).

ACKNOWLEDGMENTS

Preparation of this chapter was supported, in part, by grants from the University of Pittsburgh, Central Research Development Fund, and from the U.S. Department of Education, Office of Educational Research and Improvement, to the National Center for Improving Student Learning and Achievement in Mathematics and Science (R305A60007-98). The views expressed do not necessarily reflect the position, policy, or endorsement of the funding agencies. Comments on earlier versions of the chapter from Charles Anderson, Melva Green, Giyoo Hatano, Jorge Larreamendy-Joerns, Eugene Matusov, Martin Packer, Leona Schauble, Gini Stimpson, Pat Tinto, and the participants at the conference organized by Jeremy Kilpatrick are gratefully acknowledged. I would like to thank Deborah Dobransky-Fasiska for her assistance in all phases of the work.

REFERENCES

Arcavi, A., Meira, L., Smith, J. P., & Kessel, C. (1991). Teaching mathematical problem solving: An analysis of an emergent classroom community. In A. H. Schoenfeld, J. Kaput, & E. Dubinsky (Eds.), *Research in collegiate mathematics education* (Vol. 3, pp. 1–70). Providence, RI: American Mathematical Society.

Bazerman, C. (1988). *Shaping written knowledge: The genre and activity of the experimental article in science*. Madison: University of Wisconsin Press.

Biber, D. (1994). An analytical framework for register studies. In D. Biber & E. Finegan (Eds.), *Sociolinguistic perspectives on register* (pp. 31–58). New York: Oxford University Press.

Boaler, J. (1997). *Experiencing school mathematics*. Philadelphia: Open University Press.

Boaler, J. (1998). Open and closed mathematics: Student experiences and understandings. *Journal for Research in Mathematics Education, 29*, 41–62.

Brown, A. L. (1992). Design experiments: Theoretical and methodological challenges in creating complex interventions in classroom settings. *Journal of the Learning Sciences, 2*(2), 141–178.

Campbell, D. T., & Stanley, J. C. (1963). *Experimental and quasi-experimental designs for research*. Chicago: Rand McNally.

Cazden, C. (1986). Classroom discourse. In M. C. Whittrock (Ed.), *Handbook of research on teaching* (pp. 432–674). New York: Macmillan.

Cobb, P., & Bauersfeld, H. (Eds.). (1995). *The emergence of mathematical meaning: Interaction in classroom cultures*. Hillsdale, NJ: Erlbaum.

Cobb, P., Wood, T., & Yackel, E. (1993). Discourse, mathematical thinking, and classroom practice. In E. A. Forman, N. Minick, & C. A. Stone (Eds.), *Contexts for learning: Sociocultural dynamics in children's development* (pp. 91–119). New York: Oxford University Press.

Cole, M. (1995). Socio-cultural-historical psychology: Some general remarks and a proposal for a new kind of cultural-genetic methodology. In J. V. Wertsch, P. del Rio, & A. Alvarez (Eds.), *Socio-cultural studies of mind* (pp. 187–214). New York: Cambridge University Press.

Cole, M. (1996). *Cultural psychology: A once and future discipline*. Cambridge, MA: Harvard University Press.

Cole, M., & Means, B. (1981). *Comparative studies of how people think: An introduction*. Cambridge, MA: Harvard University Press.

Cook, T. D., & Campbell, D. T. (1979). *Quasi-experimentation: Design and analysis issues for field settings*. Chicago: Rand McNally.

De Corte, E., Greer, B., & Verschaffel, L. (1996). Mathematics teaching and learning. In D. C. Berliner & R. C. Calfee (Eds.), *Handbook of educational psychology* (pp. 491–549). New York: Simon & Schuster Macmillan.

Eisenhart, M. A. (1988). The ethnographic research tradition and mathematics education research. *Journal for Research in Mathematics Education, 19*, 99–114.

Forman, E. A. (1996). Learning mathematics as participation in classroom practice: Implications of sociocultural theory for educational reform. In L. Steffe, P. Nesher, P. Cobb, G. A. Goldin, & B. Greer (Eds.), *Theories of mathematical learning* (pp. 115–130). Hillsdale, NJ: Erlbaum.

Forman, E. A., & Ansell, E. (2001). The multiple voices of a mathematics classroom community. *Educational Studies in Mathematics, 46*(1-3), 115–142.

Forman, E. A., & Ansell, E. (2002). Orchestrating the multiple voices and inscriptions of a mathematics classroom. *Journal of the Learning Sciences, 11*(2&3), 251–274.

Forman, E. A., & Cazden, C. B. (1985). Exploring Vygotskian perspectives in education: The cognitive value of peer interaction. In J. Wertsch (Ed.), *Culture, communication, and cognition: Vygotskian perspectives* (pp. 323–347). Cambridge, England: Cambridge University Press.

Forman, E. A., & Larreamendy-Joerns, J. (1995). Learning in the context of peer collaboration: A pluralistic perspective on goals and expertise. *Cognition and Instruction, 13*, 549–565.

Forman, E. A., Larreamendy-Joerns, J., Stein, M. K., & Brown, C. A. (1998). "You're going to want to find out which and prove it": Collective argumentation in a mathematics classroom. *Learning and Instruction, 8*, 527–548.

Forman, E. A., McCormick, D., & Donato, R. (1998). Learning what counts as a mathematical explanation. *Linguistics and Education, 9*, 313–339.

Forman, E. A., & McPhail, J. (1993). Vygotskian perspective on children's collaborative problem-solving activities. In E. A. Forman, N. Minick, & C. A. Stone (Eds.), *Contexts for learning: Sociocultural dynamics in children's development* (pp. 213–229). New York: Oxford University Press.

Forman, E. A., Minick, N., & Stone, C. A. (1993). *Contexts for learning: Sociocultural dynamics in children's development*. New York: Oxford University Press.

Gee, J. P. (1996). Vygotsky and current debates in education: Some dilemmas as afterthoughts to *Discourse, Learning, and Schooling*. In D. Hicks (Ed.), *Discourse, learning and schooling* (pp. 269–282). Cambridge: Cambridge University Press.

Goffman, E. (1974). *Frame analysis*. New York: Harper & Row.

Goffman, E. (1981). *Forms of talk*. Philadelphia: University of Pennsylvania Press.

Goodnow, J. J. (1990). The socialization of cognition: What's involved? In J. W. Stigler, R. A. Shweder, & G. Herdt (Eds.), *Cultural psychology: Essays on comparative human development* (pp. 259–286). Cambridge: Cambridge University Press.

Goodnow, J. J., Miller, P. J., & Kessel, F. (Eds.). (1995). *Cultural practices as contexts for development*. San Francisco: Jossey-Bass.

Goodwin, M. H. (1982). Processes of dispute management among urban black children. *American Ethnologist, 9*, 76–96.

Goodwin, M. H., & Goodwin, C. (1987). Children's arguing. In S. U. Philips, S. Steele, & C. Tanz (Eds.), *Language, gender, and sex in comparative perspective* (pp. 200–249). Cambridge: Cambridge University Press.

Gravemeijer, K., Cobb, P., Bowers, J., & Whitenack, J. (2000). Symbolizing, modeling, and instructional design. In P. Cobb, E. Yackel, & K. McClain (Eds.), *Symbolizing and communicating in mathematics classrooms: Perspectives on discourse, tools, and instructional design* (pp. 225–273). Mahwah, NJ: Erlbaum.

Greeno, J. G., & the Middle School Mathematics Through Application Project Group. (1998). The situativity of knowing, learning, and research. *American Psychologist, 53*, 5–26.

Halliday, M. A. K. (1975). Some aspects of sociolinguistics. In E. Jacobsen (Ed.), *Interactions between linguistics and mathe-*

matical education (UNESCO Report No. ED-74/CONF. 808, pp. 25-52). Paris: UNESCO.

Halliday, M. A. K. (1988). On the language of physical science. In M. Ghadessy (Ed.), *Registers of written English: Situational factors and linguistic features* (pp. 162–178). London: Pinter.

Halliday, M. A. K. (1993). Towards a language-based theory of learning. *Linguistics and Education, 5*, 93–116.

Halliday, M. A. K., & Martin, J. R. (1993). *Writing science*. Pittsburgh: University of Pittsburgh Press.

Hanna, G. (1994). Should girls and boys be taught differently? In R. Biehler, R. W. Scholz, R. Strasser, & B. Winkelmann (Eds.), *Didactics of mathematics as a scientific discipline* (pp. 303–314). Boston: Kluwer.

Hanna, G. (Ed.). (1996). *Towards gender equity in mathematics education*. Boston: Kluwer.

Hatano, G. (1988, Fall). Social and motivational bases for mathematical understanding. *New Directions for Child Development, 41*, 55–70.

Hatano, G., & Inagaki, K. (1998). Cultural contexts of schooling revisited: A review of *The Learning Gap* from a cultural psychology perspective. In S. G. Paris & H. M. Wellman (Eds.), *Global prospects for education: Development, culture, and schooling* (pp. 79–104). Washington, DC: American Psychological Association.

Heath, S. B. (1983). *Ways with words: Language, life, and work in communities and classrooms*. Cambridge: Cambridge University Press.

Hughes, M., & Grieve, R. (1993). On asking children bizarre questions. In M. Gauvain & M. Cole (Eds.), *Readings on the development of children* (pp. 185–191). New York: Scientific American Books. (Original work published 1980)

Inagaki, K., Morita, E., & Hatano, G. (1999). Teaching-learning of evaluative criteria for mathematical arguments through classroom discourse: A cross-national study. *Mathematical Thinking and Learning, 1*(2), 93–111.

Jacob, E., & Jordan, C. (Eds.). (1993). *Minority education: Anthropological perspectives*. Norwood, NJ: Ablex Publishing.

Kieran, C. (2001). The mathematical discourse of 13-year-old partnered problem solving and its relation to the mathematics that emerges. *Educational Studies in Mathematics, 46*, 187–228.

Kirshner, D., & Whitson, J. A. (Eds.). (1997). *Situated cognition: Social, semiotic, and psychological perspectives*. Mahwah, NJ: Erlbaum.

Kozulin, A. (1990). *Vygotsky's psychology: A biography of ideas*. Cambridge, MA: Harvard University Press.

Krummheuer, G. (1995). The ethnography of argumentation. In P. Cobb & H. Bauersfeld (Eds.), *The emergence of mathematical meaning: Interaction in classroom cultures* (pp. 229–269). Hillsdale, NJ: Erlbaum.

Lampert, M. (1990). When the problem is not the question and the solution is not the answer: Mathematical knowing and teaching. *American Educational Research Journal, 27*, 29–63.

Lampert, M., Rittenhouse, P., & Crumbaugh, C. (1996). Agreeing to disagree: Developing sociable mathematical discourse. In D. R. Olson & N. Torrance (Eds.), *The handbook of education and human development: New models of learning, teaching, and schooling* (pp. 731–764). Oxford: Blackwell.

Latour, B. (1987). *Science in action: How to follow scientists and engineers through society*. Cambridge, MA: Harvard University Press.

Latour, B. (1990). Drawing things together. In M. Lynch & S. Woolgar (Eds.), *Representation in scientific practice* (pp. 19–68). Cambridge, MA: MIT Press.

Lave, J. (1992). Word problems: A microcosm of theories of learning. In P. Light & G. Butterworth (Eds.), *Context and cognition: Ways of learning and knowing* (pp. 74–186). Hillsdale, NJ: Erlbaum.

Lave, J. E., & Wenger, E. (1991). *Situated learning: Legitimate peripheral participation*. New York: Cambridge University Press.

Leder, G. C. (1992). Mathematics and gender: Changing perspectives. In D. A. Grouws (Ed.), *Handbook of research on mathematics teaching and learning* (pp. 597–622). New York: Macmillan.

Lehrer, R., Jacobson, C., Thoyre, G., Kemeny, V., Strom, D., Horvath, J., Gance, S., & Koehler, M. (1998). Developing understanding of space and geometry in the primary grades. In R. Lehrer & D. Chazan (Eds.), *Designing learning environments for developing understanding of geometry and space* (pp. 169–200). Mahwah, NJ: Erlbaum.

Lehrer, R., Schauble, L., Carpenter, S., & Penner, D. (2000). The inter-related development of inscriptions and conceptual understanding. In P. Cobb, E. Yackel, & K. McClain (Eds.), *Symbolizing and communicating in mathematics classrooms: Perspectives on discourse, tools, and instructional design* (pp. 325–360). Mahwah, NJ: Erlbaum.

Lehrer, R., & Shumow, L. (1997). Aligning the construction zones of parents and teachers for mathematics reform. *Cognition and Instruction, 15*, 41–83.

Lerman, S. (1996). Intersubjectivity in mathematics learning: A challenge to the radical constructivist paradigm. *Journal for Research in Mathematics Education, 27*, 133–150.

Luria, A. R. (1976). *Cognitive development: Its cultural and social foundations* (M. Lopez-Morillas & L. Solotaroff, Trans.). Cambridge, MA: Harvard University Press.

Mehan, H. (1979). *Learning lessons: Social organization in the classroom*. Cambridge, MA: Harvard University Press.

Mehan, H. (1998). The study of social interaction in educational settings: Accomplishments and unresolved issues. *Human Development, 41*, 245–269.

Minick, N. (1987). The development of Vygotsky's thought: An introduction. In R.W. Rieber & A. S. Carton (Eds.) & N. Minick (Trans.), *The collected works of L. S. Vygotsky: Vol. 1. Problems of general psychology* (pp. 17–36). New York: Plenum.

Minick, N., Stone, C. A., & Forman, E. A. (1993). Introduction: Integration of individual, social, and institutional processes in accounts of children's learning and development. In E. A. Forman, N. Minick, & C. A. Stone (Eds.), *Contexts for learning: Sociocultural dynamics in children's development* (pp. 3–16). New York: Oxford University Press.

Moll, L. C. (1990). Introduction. In L. C. Moll (Ed.), *Vygotsky and education: Instructional implications and applications of sociohistorical psychology* (pp. 1–27). Cambridge: Cambridge University Press.

Moll, L. C., & Whitmore, K. F. (1993). Vygotsky in classroom practice: Moving from individual transmission to social transaction. In E. A. Forman, N. Minick, & C. A. Stone (Eds.), *Contexts for learning: Sociocultural dynamics in children's development* (pp. 19–42). New York: Oxford University Press.

National Council of Teachers of Mathematics. (1989). *Curriculum and evaluation standards for school mathematics*. Reston, VA: Author.

National Council of Teachers of Mathematics. (1991). *Professional standards for teaching mathematics*. Reston, VA: Author.

Newman, D., Griffin, P., & Cole, M. (1989). *The construction zone: Working for cognitive change in school*. New York: Cambridge University Press.

Noddings, N. (1996). Equity and mathematics: Not a simple issue—A review of four new titles. *Journal for Research in Mathematics Education, 27*, 609–615.

Nunes, T., & Bryant, P. (1996). *Children doing mathematics*. Oxford: Blackwell.

Nunes, T., Schliemann, A. D., & Carraher, D. W. (1993). *Street mathematics and school mathematics*. Cambridge: Cambridge University Press.

O'Connor, M. C. (2001). "Can any fraction be turned into a decimal?" A case study of a mathematical group discussion. *Educational Studies in Mathematics, 46*, 143–185.

O'Connor, M. C., & Michaels, S. (1993). Aligning academic task and participation status through revoicing: Analysis of a classroom discourse strategy. *Anthropology and Education Quarterly, 24*, 318–335.

O'Connor, M. C., & Michaels, S. (1996). Shifting participant frameworks: Orchestrating thinking practices in group discussion. In D. Hicks (Ed.), *Discourse, learning, and schooling* (pp. 63–103). New York: Cambridge University Press.

Rogoff, B. (1990). *Apprenticeship in thinking: Cognitive development in social context*. New York: Oxford University Press.

Rogoff, B. (1998). Cognition as a collaborative process. In D. Kuhn & R. S. Siegler (Eds.), *The handbook of child psychology: Cognition, perception, and language* (Vol. 2, pp. 679–744). New York: Wiley.

Rogoff, B., & Lave, J. (Eds.). (1984). *Everyday cognition: Its development in social context*. Cambridge, MA: Harvard University Press.

Rotman, B. (1988). Toward a semiotics of mathematics. *Semiotica*, 72(1–2), 1–35.

Rotman, B. (1993). *Ad infinitum: The ghost in Turing's machine*. Stanford, CA: Stanford University Press.

Salomon, G., & Globerson, T. (1989). When teams do not function the way they ought to. *International Journal of Educational Research, 13*, 89–100.

Saxe, G. B. (1991). *Culture and cognitive development*. Hillsdale, NJ: Erlbaum.

Saxe, G. B., Dawson, V., Fall, R., & Howard, S. (1996). Culture and children's mathematical thinking. In R. J. Sternberg & T. Ben-Zeev (Eds.), *The nature of mathematical thinking* (pp. 119–144). Mahwah, NJ: Erlbaum.

Saxe, G. B., & Guberman, S. R. (1998). Studying mathematics learning in collective activity. *Learning and Instruction, 8*, 489–501.

Saxe, G. B., Guberman, S. R., & Gearhart, M. (1987). Social processes in early number development. *Monographs of the Society for Research in Child Development, 52*.

Schiffrin, D. (1984). Jewish argument as sociability. *Language in Society, 13*, 311–335.

Schoenfeld, A. H. (1992). Research methods in and for the learning sciences. *Journal of the Learning Sciences, 2*(2), 137–139.

Secada, W. G. (1992). Race, ethnicity, social class, language, and achievement in mathematics. In D. A. Grouws (Ed.), *Handbook of research on mathematics teaching and learning* (pp. 623–660). New York: Macmillan.

Secada, W. G., Fennema, E., & Adajian, L. B. (1995). *New directions in equity for mathematics education*. New York: Cambridge University Press.

Serpell, R., & Hatano, G. (1997). Education, schooling, and literacy. In J. W. Berry, P. R. Dasen, & T. S. Saraswathi (Eds.), *Handbook of cross-cultural psychology: Vol. 2. Basic processes and human development* (pp. 339–376). Boston: Allyn & Bacon.

Sfard, A., & Kieran, C. (2001). Cognition as communication: Rethinking learning-by-talking through multi-faceted analysis of students' mathematical interactions. *Mind, Culture, and Activity, 8*(1), 42–76.

Shweder, R. A., Goodnow, J., Hatano, G., LeVine, R. A., Markus, H., & Miller, P. (1998). The cultural psychology of development: One mind, many mentalities. In W. Damon & R. M. Lerner (Eds.), *Handbook of child psychology: Theoretical models of human development* (5th ed., Vol. 1, pp. 865–937). New York: Wiley.

Siegler, R. S., & Crowley, K. (1991). The microgenetic method: A direct means for studying cognitive development. *American Psychologist, 46*, 606–620.

Stodolsky, S. S. (1988). *The subject matters: Classroom activity in math and social studies*. Chicago: University of Chicago Press.

Strom, D., Kemeny, V., Lehrer, R., & Forman, E. A. (2001). Visualizing the emergent structure of children's mathematical argument. *Cognitive Science, 25*, 733–773.

Tannen, D. (1993). The relativity of linguistic strategies: Rethinking power and solidarity in gender and dominance. In D. Tannen (Ed.), *Gender and conversational interaction* (pp. 165–188). New York: Oxford University Press.

Tate, W. F. (1995). Economics, equity, and the national mathematics assessment: Are we creating a national tollroad? In E. Fennema, W. G. Secada, & L. B. Adajian (Eds.), *New directions for equity in mathematics education* (pp. 191–208). New York: Cambridge University Press.

Tate, W. F. (1997). Race-ethnicity, SES, gender, and language proficiency trends in mathematics achievement: An update. *Journal for Research in Mathematics Education, 28*, 652–679.

Tharp, R. G., & Gallimore, R. (1988). *Rousing minds to life: Teaching, learning, and schooling in social context*. Cambridge: Cambridge University Press.

Toulmin, S. E. (1958). *The uses of argument*. Cambridge: Cambridge University Press.

Van der Veer, R., & Valsiner, J. (1991). *Understanding Vygotsky: A quest for synthesis*. Oxford: Blackwell.

Van der Veer, R., & Valsiner, J. (Eds.). (1994). *The Vygotsky Reader*. Oxford: Blackwell.

Van Oers, B. (1996). Learning mathematics as a meaningful activity. In L. Steffe, P. Nesher, P. Cobb, G. A. Goldin, & B. Greer (Eds.) *Theories of mathematical learning* (pp. 91–113). Mahwah, NJ: Erlbaum.

Van Oers, B. (2000). The appropriation of mathematical symbols. A psychosemiotic approach to mathematics learning. In P. Cobb, E. Yackel, & K. McClain (Eds.), *Symbolizing and communicating in mathematics classrooms: Perspectives on discourse, tools, and instructional design* (pp. 133–176). Mahwah, NJ: Erlbaum.

Volosinov, V. N. (1986). *Marxism and the philosophy of language* (L. Matejka & I. R. Titunik, Trans.). Cambridge, MA: Harvard University Press. (Original work published 1929)

Vygotsky, L. S. (1978). *Mind in society: The development of higher psychological processes*. Cambridge, MA: Harvard University Press.

Walkerdine, V. (1988). *The mastery of reason: Cognitive development and the production of rationality*. London: Routledge.

Walkerdine, V. (1990). Difference, cognition, and mathematics education. *For the Learning of Mathematics, 10*(3), 51–56.

Walkerdine, V. (1997). Redefining the subject in situated cognition theory. In D. Kirschner (Ed.), *Situated cognition: Social, semiotic, and psychological perspectives* (pp. 57–70). Mahway, NJ: Erlbaum.

Wells, G. (1993). Reevaluating the IRF sequence: A proposal for the articulation of theories of activity and discourse for the analysis of teaching and learning in the classroom. *Linguistics and Education, 5*, 1–37.

Wenger, E. (1998). *Communities of practice: Learning, meaning, and identity*. New York: Cambridge University Press.

Wertsch, J. V. (Ed.). (1981). *The concept of activity in Soviet psychology*. Armonk, NY: Sharpe.

Wertsch, J. V. (1985). *Vygotsky and the social formation of mind*. Cambridge, MA: Harvard University Press.

Wertsch, J. V. (1990). The voice of rationality in a sociocultural approach to mind. In L. C. Moll (Ed.), *Vygotsky and education: Instructional implications and applications of sociohistorical psychology* (pp. 111–126). New York: Cambridge University Press.

Wertsch, J. V. (1997). *Mind as action*. New York: Oxford University Press.

Wood, D., Bruner, J. S., & Ross, G. (1976). The role of tutoring in problem solving. *Journal of Child Psychology and Psychiatry and Allied Disciplines, 17*(2), 89–100.

Yackel, E., & Cobb, P. (1996). Sociomathematical norms, argumentation, and autonomy in mathematics. *Journal for Research in Mathematics Education, 27*, 458–477.

CHAPTER 23

Balancing the Unbalanceable: The NCTM Standards in Light of Theories of Learning Mathematics

Anna Sfard, University of Haifa

Mathematics is difficult. It is certainly among the most complex of human intellectual endeavors. As a school subject, it is often unmanageable. Much thought has been given over the years to the question of how it can be successfully taught in spite of the difficulty. Educators and mathematicians who try to deal with the problem have found themselves repeatedly thrown from one extreme solution to another (cf. Kilpatrick, 1992). They vacillate between programs such as the famous "new math," which resulted from a concern about the kind of mathematics the child should learn, and projects that, in reaction to the didactic weaknesses of an overlay approach, put the needs and capabilities of the student at the center.

The *Standards* documents of the National Council of Teachers of Mathematics (NCTM, 1989, 1991, 1995) are the result of a serious and comprehensive attempt to teach "mathematics with a human face." This means much care for both mathematics and the student. What is being taught is "mathematics in the making" rather than mathematics as a static body of knowledge. In this process, the needs of the learning child never disappear from the reformers' eyes. It is only relatively recently that these needs became a subject of disciplined study, and it may therefore be the first time in history that we can support an educational project of such impressive dimensions with systematic pedagogical knowledge. There can hardly be any controversy about the humanistic values immanent in the curricular principles intended to guide the current change in mathematics education. Even the most extreme critics agree that "the reform has its merits" (Wu, 1997, p. 946), and even the greatest skeptics applaud the fact that

> it has replaced some of the rote-learning in the traditional curriculum by supplying motivation and heuristic arguments. It has made students aware of the normal process of doing mathematics, such as making conjectures and looking for counter-examples. It has also made mathematics more relevant to the average student by promoting the use of realistic applications in curriculum. (p. 946)

This applause is hardly surprising. The NCTM Standards bring hope of the kind of change we have all been dreaming about for a long time. Ever since it was turned into an obligatory part of school curricula, mathematics has been known as a school bugbear—a subject in which the student is treated as a programmable machine rather than a creative human being. Here, at last, the intellectual and emotional needs of the child are considered. The reform, therefore, stands a reasonable chance of being good for students' mind and soul: Along with intellectual excitement, it returns to the learners of mathematics their long-lost self-esteem. The educational importance of such a change goes well beyond the mathematics classroom itself.

In spite, however, of what appears as a consensus over the general spirit of the NCTM Standards, voices of dissatisfaction can be heard among those who endorse the Standards' goals. Lately, these voices have become only too loud. If not the goals, then what is under fire?

What the Problem Is and How Theories Can Help

The Problem

That the reform pendulum refuses to arrive at the state of equilibrium is anything but surprising. As I try to show throughout this chapter, the needs of mathematics itself and the needs of the child who is supposed to learn it do not necessarily agree. Whatever is done out of a sole concern about mathematics is likely to hurt the child, and whatever is done for the sake of the child invariably compromises some mathematical contents and skills. Any reform movement that tries to make up for the deficiencies of former ways of teaching seems bound to shift the pendulum to the opposite pole. Whenever the balance everybody hoped for appears to be in danger, a backlash to reform is on its way.

According to the critics, the mathematics that the student is supposed to learn has been hurt by the present reform movement—yet little doubt exists that this trend is contrary to what the *Standards* documents' authors had in mind when writing their proposals. The authors' concern about mathematics has been genuine, but their intention was to engage students in what may count as an authentic activity of mathematizing rather than in learning ready-made mathematical facts. The success of educational ideas, however, is never a simple function of the ideas themselves. There is no direct route from general curricular principles to successful instruction.

While listening to the critical voices, one easily notices that, indeed, the true reason for the present controversy lies not so much in the standards as in the ways they are sometimes translated into practice. A certain one-sidedness of this translation may be the principal culprit. When it comes to implementing promising ideas, our enthusiasm is likely to make us guilty of King Midas's mistake: Deeply convinced that our ideas are golden, we are keen to see everything turning into gold. When the alchemy happens, we are left without many vital ingredients. Indeed, educational practices are known for their overwhelming propensity for clear-cut choices and one-for-all solutions. All too often, a new promising idea is embraced to the total exclusion of alternative possibilities. In this way, what was intended as but an ingredient becomes the whole meal; what was supposed to be an optional technique for those who find it helpful and pleasing gradually becomes the only legitimate way of doing things. Such exclusivity is an effective prescription for failure.

My aim in this chapter is to substantiate the preceding claims while answering the question of "what research and theory say about the teaching and learning of mathematics as portrayed in the *Standards* documents and in various criticisms of them" (Kilpatrick, 1997). Although focusing on the learner, I also try to see how the needs of the child and those of mathematics may sometimes be leading in differing directions. Eventually, I attempt to suggest ways of balancing these diverging needs. I do not undertake all this, however, without giving thought to the ways in which theories of learning can support educational practice. In showing strengths and limitations of the theoretical argument, I hope to put the theories to their best use while simultaneously protecting them from being interpreted as saying more than they can possibly say.

What Can Be Done With Theories

The proponents of the NCTM Standards have a strong weapon with which to fight the recent criticism. Their innovative curricular ideas do not result from arbitrary beliefs—many of them can be seen as derivatives of theories of teaching and learning. The theories themselves, the outcome of many decades of impressive development, are grounded in much experience and in empirical research that has shown the ineffectiveness of the old methods and indicated the potential of the new directions. Those who propose the new ideas and those who require the banishment of the traditional approaches seem, therefore, to have good reasons for doing so: They have theories of teaching and learning on their side. However, while using this kind of argument, we will do our job convincingly only if we remain realistic about the power of theories.

To be sure, theories and research are the best tools we have for improving practice and making sensible pedagogical decisions; however, their power is not unrestricted. The claim made above in the context of curricular principles and their classroom implementations will now be repeated with respect to theories and their interpretations: When considering translation from general to concrete and from theoretical to practical, we always have many ways to proceed. Being rooted in metaphors and analogies (Bruner, 1986; Sfard, 1998a; Sheffler, 1990), theories of teaching and learning do not provide unique answers to practical questions. A theory, if well conceived, may lend support to a variety of educational practices without privileging any of them. Thus, theory can only suggest, not dictate; curricular principles and concrete instructional approaches may be implied or supported by theory, but they are certainly not necessitated by theoretical arguments. In a similar vein, we can understand and explain educational success and failure with the help of theoretical argument, but we can hardly predict or control them theoretically.

Other aspects of the discussion that follows should be explained before I approach the task itself. The reader will soon notice that my analysis is somewhat eclectic. I not only promote multiple interpretations of general ideas but also seek inspiration and support in many different theories. Because some of the conceptual frameworks that I will use can be regarded as mutually exclusive, I run the risk of being accused of shifting allegiance—if not incoherence. I therefore declare that I am doing all this deliberately, guided by the same belief that has been presented above in a slightly different context: Exclusivity is an enemy of success. Educational theories, like practical solutions, respond badly to being left alone. They can thrive only in the company of other theories. Although we are quite free in our choices of metaphors, once such a choice is made, we become restricted in our further thinking and decisions. Metaphors and the resulting theories can take us only as far as their entailments allow us to go. As long as the metaphor with which we think is not challenged by an alternative metaphor, its entailments seem natural, self-evident, and unquestionable. Only in the presence of another metaphor and another theory can we become critical toward the seemingly obvious; only when an alternative theory makes us aware of the metaphorical roots of our present beliefs are we able to open ourselves to possibilities that we could not consider before. This plurality of outlooks is extremely important because, in Freudenthal's (1978) words, "Education is a vast field, and even that part which displays a scientific attitude is too vast to be watched with one pair of eyes" (p. 78).

Let me also explain why I am not afraid of becoming incoherent while shifting my gaze from one theory to another or, even more hazardously, from one epistemological framework to another. Multiplicity of theoretical outlooks does not, necessarily, entail a contradiction. Theoretical controversies are very often, if not always, an outcome of differences between underlying metaphors. If so, the theories themselves are only rarely truly incompatible; more often than not, they should rather be viewed as either *complementary*—that is, concerned with different aspects of the same phenomena—or *incommensurable*—that is, speaking different languages rather than really conflicting with each other. In this latter case, no set of common criteria can help to rationally resolve the apparent controversy (Kuhn, 1962; Rorty, 1979). Together with Bruner (1997), I claim that even Piagetian and Vygotskian perspectives, widely acknowledged as incompatible, can in fact be treated as incommensurable rather than mutually exclusive. All this speaks forcefully in favor of the theoretical pluralism. Using multiple theoretical outlooks proves not only possible but also desirable. A combination of theoretical outlooks may have a synergetic effect.

It is not impossible to think about subsuming incommensurable theories under one conceptual super framework. The acceptance of non-Euclidean geometries is good evidence of such a possibility, and today's physics with Bohr's complementarity principle is another. Viewed from the "higher" perspective of the common system, the basic concepts and tenets of the various theories acquire, of course, a slightly different meaning. Thus, for example, after the acceptance of non-Euclidean geometry, the notion of axiom is no longer understood as a "self-evident truth" but rather as a proposition that is chosen to be an atomic element of a theory. When subsumed to a common super framework, the ideas of Piaget and Vygotsky are sure to be understood somewhat differently than in the original context of the respective theories themselves. This process of climbing to ever higher and more comprehensive conceptual systems must, at certain special junctures, bring a reconceptualization of what has been known before, which is fully consistent with what Piaget and Vygotsky said about the way human knowledge develops.

Finally, let me say a few words about the specific theories to which I refer in this chapter. A great many approaches to human cognitive development have been proposed over the last few centuries. Most of the time, the study of thinking has been guided by the metaphor of learning as acquisition: A person who learned something new has been said to *acquire* a new concept or procedure. These new cognitive entities can then be retained in memory thanks to their being incorporated into certain mental schemes. Lately, we have witnessed an emergence of a new outlook. It has been proposed that learning should be regarded as "peripheral participation" (Lave & Wenger, 1991) in a certain type of practice. According to this new approach, the learner of mathematics gradually becomes a member of a mathematizing community. In this case, the theory focuses on the evolving bonds between the individual and the group of practitioners.

We can, therefore, speak roughly about two conceptual super frameworks that divide all these theories into two categories. The "acquisitionist" group consists of those traditional cognitivist approaches that explain learning and knowledge in terms of such mental entities as cognitive schemes, tacit models, concept images, or misconceptions. The "participationist" framework embraces all those relatively new theories that view learning as a reorganization of activity accompanying the integration of an individual learner with a community of practice. In the language of discursive psychology, this latter framework follows the process through which the learner becomes a skillful participant in a given (e.g., mathematical) discourse. Needless to say, the two metaphors cannot be completely separated, and both are usually present in every theory. Nevertheless, in every theory, one metaphor is usually much more prominent than the other. The division of the conceptual frameworks into the two groups can therefore be done according to their dominant ingredient. (For a more comprehensive discussion, see Sfard, 1998a.)

In the following analysis, I support my arguments with both types of theories. In particular, I frequently refer to Piaget and Vygotsky. Piaget can certainly be regarded as a paradigmatic representative of the acquisitionist framework. Vygotsky's theory has been used as an inspiration for the participationist framework, but the language of "internalizing" and concept building, used by Vygotsky himself, still qualifies him as belonging—at least partially—to the acquisitionist framework.

In view of the existing tendency to regard the acquisitionist and participationist approaches as difficult to reconcile, I emphasize again that, faithful to the principle of theoretical pluralism, I do not try to choose one metaphor as generally "better" than the other—and I do not render one of them the status of the exclusive provider of criteria for evaluating the NCTM Standards. As I hope to show, each of the two overall frameworks has something to offer. Moreover, the two outlooks sometimes reinforce each other by providing complementary perspectives on the same basic needs of the learner. Certain important insights about learning would quite obviously be lost if we decided to turn our backs on either of the two epistemologies.

What Will Be Done

In what follows, I survey the available theoretical frameworks, looking for those needs of learners that, according to the theories, are the driving force behind human learning and must be fulfilled if this learning is to be successful. I identify ten such needs. The assumption underlying my analysis is that all of them are quite universal, although they may express themselves differently in different individuals and at different ages. For each of the ten needs, the following four questions will be addressed.

1. *What do we know about this need?* I summarize the theoretical knowledge that explains the nature and the possible ways of satisfying the need in question, citing relevant empirical findings when available.

2. *How do the NCTM Standards address this need?* Here, I discuss the relevant principles in the *Standards* documents.

3. *What can go wrong?* In this section, I consider whether anything in the way the different needs have been addressed may account for some of the recent criticism.

4. *What can be done?* I reflect on possible ways of ensuring that the relevant principles, already existing or still required, are translated into truly effective instructional procedures that answer the need in question without limiting our ability to address other needs.

Along the way, I point to certain dilemmas inherent in the project of teaching mathematics, claiming that although some of the problems do not seem solvable, their impact may perhaps be considerably reduced if we only remain aware of their existence.

The Need for Meaning

What Do We Know About This Need?

"Culture and the quest for meaning within culture are the proper causes of human action," says Bruner (1990, p. 20). Today, few of us would disagree with this conception of human nature. The culturally tinged, but essentially universal, need for meaning and the need to understand ourselves and the world around us have come to be widely recognized as the basic driving force behind all our intellectual activities. Different thinkers try to account for this fundamental human drive in different ways. For Piaget, who built his theory around the metaphor of Darwinian evolution and biological growth, the search for knowledge and meaning is part and parcel of our struggle for survival. Vygotsky views meaning making as the uniquely human activity that stems from the need to communicate our experiences to other human beings. The general recognition of the centrality of the search for understanding makes contemporary philosophers, psychologists, anthropologists, and linguists almost obsessively preoccupied with the question of what meaning is and where it is to be found (see, e.g., Eco, Santambrogio, & Violi, 1988).

If the need for meaning puts in motion all our intellectual activities, it must also be what motivates and guides our learning. Indeed, the search for knowledge is nothing else than an attempt to enhance our understanding of the world. Learners, therefore, can be seen as sense-making creatures who look for order, logic, and causal dependencies behind things, events, and experiences. Mathematics is one of our most important tools in that endeavor. With its help, we often manage to see meaningful relationships between apparently unrelated phenomena. It is also important to note that once created, mathematics turns into an integral part of our world and, as such, becomes a new object of our sense-making efforts. In this chapter, I focus on the need to understand mathematics in its double role of tool and object (Douady, 1985). I see this need as primary, and I present all other needs as its derivatives. Before delving into this issue, however, I make two more remarks about the notion of meaning.

First, it is important to mention that much was written in the twentieth century about the untenability of the long-standing conception of meaning as independent of human mind and residing in symbols. Today's philosophers, psychologists, linguists, and semioticians agree that the meaning of ideas is constructed anew each time anyone learns these ideas (Cobb, Wood, & Yackel, 1991, 1993; Johnson, 1987; Piaget, 1970; Reddy, 1979; Rorty, 1979; von Glasersfeld, 1993). This view entails a rejection of the traditional empiricist stance according to which the learner is but a *tabula rasa* passively absorbing externally generated experiences. It also repudiates the *a priorist* idea of a human mind as a predesigned product of nature that can display only those abilities that were programmed into it in advance. Today, the learner is given the much more exciting and responsible role of an autonomous meaning builder. As explained first in general by Piaget (see, e.g., Piaget, 1953; Piaget & Inhelder, 1969) and then by van Hiele (1985) in terms specific to geometric thinking, children's means for understanding and building meaning change qualitatively with age; thus, the sense-making processes and their products may be quite different at different ages, even for the same child and the same concepts.

My second remark is intended to forestall the doubt with which the sweeping claim about the learner's "hunger" for meaning may be greeted in certain circles. Some teachers may think that the vision of the learner as a determined meaning-maker is somewhat utopian—and that it contradicts their experience. We too often hear complaints about students' thinking as plagued by "misconceptions," and about learners who simply prefer to learn "by rote," viewing the mastery of techniques as the shortest path to success in final examinations. It is easy to show, however, that none of these grievances, however justified, really undermines the assumption about the basic human need for meaning and understanding.

I first tackle the issue of *misconceptions*. This word refers to a phenomenon well known to both teachers and researchers: While learning mathematics, children often tend to "*create their own meanings*—meanings that are not appropriate at all" (Davis, 1988, p. 9). The words *not appropriate* refer not so much to the inner coherence of student's thinking as to possible disparities between students' conceptions and the public versions of the same ideas. Thus, for example, studies have repeatedly shown that an overwhelming majority of high school students tend to believe that any function must have an underlying algorithm; moreover, this conviction persists despite the fact that the definition, which most of the learners can repeat without difficulty, does not require any kind of "regularity" (Malik, 1980; Markovits, Eylon, & Bruckheimer, 1986; Sfard, 1992; Vinner & Dreyfus, 1989). Similarly, young children are known to believe that the operation of multiplication must increase the multiplied number, whereas division must make it smaller (Fischbein, 1987, 1989; Fischbein, Deri, Nello, & Marino, 1985; Harel, Behr, Post, & Lesh, 1989). It is worth emphasizing that children's idiosyncratic notions tend to be consistent, and because of that tendency, these notions are sometimes very difficult to change. All this has been widely documented in misconceptions research, which until recently was probably the most developed type of research in mathematics education (see Confrey, 1990; Smith, diSessa, & Rochelle, 1993; see also studies on such related ideas as *concept images* as in Tall & Vinner, 1991; Vinner, 1983; or *tacit models* as in Fischbein, 1989). Today, our knowledge about the ways in which children think about numbers, functions, proofs, and so forth, is impressively rich. This knowledge may certainly be of great help to those who wish to support children in their meaning-building efforts.

The main point I wish to make, however, is that the appearance of misconceptions does not contradict the vision of learners as sense makers. Indeed, not *in spite* of students' need for meaning but rather *because* of it, students tend to construct their own conceptions. Precisely because of their need to fit new concepts into their former knowledge, their understandings are sometimes at odds with the official definitions. This conflict is evidenced by the seemingly astonishing fact that what we tend to view today as learners' "mistaken" understanding of a concept is often almost identical with what can be found in historical sources (Sfard, 1992, 1994a, 1995; Sfard & Linchevsky, 1994). Like today's students, the mathematicians who were the first to speak about a new concept and could think of this concept only in terms of their former knowledge did not immediately arrive at the version that has survived today. Their imperfect initial conceptualizations, however, often proved a sound basis for innovation. In the same sense, students' misconceptions should be viewed as steppingstones for further development rather than as hurdles to learning. Therefore, the term *misconception*, with its pejorative undertones, may be somehow misleading and should be read as being in quotation marks throughout this paper. For a more extensive critique of theory of misconceptions, see Driver and Easley (1978), Nesher (1987), and Smith et al. (1993).

In sum, the idiosyncrasy of students' conceptions cannot count as evidence of the learners' "thoughtlessness"; on the contrary, it should be regarded as lending support to the thesis of their ever-present need for coherence and meaning. I claim that this thesis remains robust also in the face of some students' apparent preference for rote learning. It is important to understand that a mathematics learner can sometimes abstain from understanding just as a dieting person can abstain from food. In both these cases, the person can temporarily live with the deprivation and in certain situations may even deliberately suppress it. However, sustained preference for rote learning, like the extreme cases of dieting resulting in anorexia, should count as an exception rather than a rule. Instances of voluntary abstention are most likely to happen when the environment seems to encourage the suppression of the need through its reward system. Thus, if one gets the impression that most mathematics students tend to be satisfied with technical mastery and never show any genuine interest in the underlying conceptual structure, this result should be interpreted as evidence of certain basic faults in the learning environment rather than of these students' "natural" thoughtlessness.

How Do the NCTM Standards Address the Need for Learners to Make Meaning?

One glimpse of the NCTM *Standards* documents suffices for one to realize that the new curricular ideas are imbued with the spirit of meaning making. In fact, if one were asked to render the gist of the reform "while standing on one foot," the best thing to say would be that it promotes the vision of a learner as a sense-maker. Of the five major shifts required by the 1991 Standards, the following four are geared toward more meaningful learning:

- Toward logic and mathematical evidence as verification—away from the teacher as the sole authority for right answers
- Toward mathematical reasoning—away from merely memorizing procedures
- Toward conjecturing, inventing, and problem solving—away from an emphasis on mechanistic answer-finding
- Toward connecting mathematics, its ideas, and its applications—away from treating mathematics as a body of isolated concepts and procedures. (NCTM, 1991, p. 3)

This explicit emphasis on the construction of meaning is probably the Standards' most important innovation—and their most important contribution. For too long, the basic human need for meaning has been ignored in mathematics classrooms. Now that it has been explicitly recognized, we seem to have a much better chance for a real improvement in teaching. Because this need makes us learn, instruction that focuses on meaning can be expected to be more effective than instruction that tries to circumvent it. Evidence has begun pouring in that seems to lend support to this claim (see, e.g., Cobb, Boufi, McClain, & Whitenack, 1997; Cobb et al., 1991, 1993; Hiebert & Wearne, 1992, 1996; Yerushalmy, 1997). Moreover, by delegitimizing rote learning, the NCTM Standards convey a refreshing message about the identity of the learner. At long, long last, the mathematics student is pictured as an autonomous human being whose right to independent thought must be respected by everybody, including teachers. The educational message conveyed by this principle is as important inside the mathematics classroom as outside it.

What Can Go Wrong?

I next address the important but elusive issue of what may happen if the NCTM Standards' call for learning with understanding is interpreted in an inappropriate way. Because of the educational importance and beauty of the new way of picturing developing minds that is implicit in the Standards, this call can easily be taken to a dangerous extreme. If not explained, this principle may be misinterpreted as implying the total exclusion of instruction that is not immediately rewarding in terms of understanding. The implementers of the Standards may wrongly conclude that they are expected to shield the student from the vexing experience of insufficient understanding at any cost. Such an extreme interpretation would soon become counterproductive. Effortlessly meaningful mathematics can be only trivial and uninspiring, and any curriculum built around this "principle" is doomed to become "watered down" (Jackson, 1997, p. 695). Above all, however, teachers' and students' fear of lapses in understanding may engender defeatist attitudes that are harmful to both teaching and learning.

Those who believe in the possibility of seamless, uniformly meaningful learning ignore the fact that the quest for meaning, especially in mathematics, is an inherently difficult enterprise. Although in the age of the NCTM Standards, "the goal of many research and implementation efforts in mathematics education has been to promote learning with understanding," more and more people agree that "achieving this goal has been like searching for the Holy Grail" (Hiebert & Carpenter, 1992, p. 65). One explanation for this vexing difficulty is that the way toward understanding is, by its very nature, convoluted and, in a sense, circular. Put more bluntly, it contains a built-in, inevitable, contradiction. I next devote a few sentences to substantiating this claim.

Time and again, many thinkers including Piaget and Vygotsky have emphasized that an understanding of a concept can be achieved only through a subject's activity with that concept. Wittgenstein's (1953) famous declaration that the meaning of a word is tantamount to its use in language implies that one cannot "grasp" or construct the meaning before one has become acquainted with its uses. The situation seems to be relatively simple when the new idea one is trying to learn refers to a physical, perceptually accessible object. In fact, claims Wittgenstein, even when the object in question can be shown, much more complexity arises in the meaning-making process than meets the eyes. Piaget, who argued that what we really learn when dealing with objects are our own actions—actual or mental—on these objects, not our passive impressions of them, would certainly agree with this claim. Doubtlessly, so too would Vygotsky, who regarded the whole activity of meaning making as having its roots in human communication rather than in our direct encounter with nature.

Sufficient or not, the perceptual impression can certainly help. Therefore, when one deals with mathematics, where there are no ready-made mathematical objects to point to, the problem of understanding becomes particularly acute. One may argue that mathematical definitions are a worthy replacement for *ostensive definitions*, that is, definitions by pointing: Although numbers, functions, and sets can be only symbolically *represented* and not really shown, we can understand words and symbols thanks to the precise mathematical description of their meaning.

Can anybody say, however, that students understand a mathematical idea—say, a negative number or a function—through the force of a definition alone, before they are able to work with the concept? Would students declare that they had a good sense of the idea of negative numbers before they can manipulate the symbols –2 and –0.5 in ways characteristic of numbers? Would we agree that they really "got the idea" of function before they could take advantage of the relationships among the different symbols used as "different representations" of functions? No wonder then that empirical research has repeatedly shown that the mathematical conceptions students actually develop may have little to do with the definitions that students have learned (Tall & Vinner, 1981).

This argument implies that in the quest for understanding, one cannot escape a contradiction: If meaning is a function of use, then on the one hand, one must manipulate a concept to understand it; on the other hand, how can one use the concept before one understands it? In discursive terms, all discourses, and mathematical discourse in particular, suffer from an inherent circularity: Signifieds can be built only through discursive activity with signifiers, whereas the existence of the signifieds is a prerequisite for successful use of the signifiers. Or, in simpler language, meaningfulness can arise only from using a concept but at the same time is prerequisite to successful use of that concept. It is important to understand, however, that this circularity, seemingly a serious trap for the learners of mathematics, is actually the driving force behind the incessant growth of knowledge. It fuels the process of learning in which our understanding and our ability to apply mathematical concepts spur each other's development. In this process, the sense of understanding a concept and the ability to apply it are like two legs that make forward movement possible thanks to the fact that they are never in exactly the same place. The salient characteristic of mathematical learning is that at any given moment, one of these two abilities is ahead of the other. (For a more exhaustive treatment of these circularities, see Sfard, 1991, 2000; for a discussion of what is known as the *learning paradox*, which is germane to the present dilemmas, see Bereiter, 1986; Petrie & Oshlag, 1993.)

What Can Be Done?

The previous discussion leaves us with a quandary that has to be resolved if any practical solution is to be found. We should not pretend that such a solution is easy to find, nor should we delude ourselves that the basic dilemma can ever disappear. Rather, we must learn to live with the difficulty while doing our best to find a reasonable path between conflicting needs: the need for understanding and the need for acting even before this understanding has been achieved. I postpone to the following sections any concrete suggestions as to how we can find such a path. Here, I satisfy myself with a remark on the importance of realistic expectations and of constructive attitudes.

The significance of being honest with students about the difficulty of mathematics can hardly be overestimated. From a pedagogical point of view, the utopian idea of frictionless meaningful learning is simply harmful. As research has shown, the student who is not prepared to expect lapses in understanding and thus does not view their occurrences as a natural phenomenon is not likely to make a real effort to overcome a difficulty when it appears (Schoenfeld, 1985, 1989). The awareness of the inevitability of this inherent difficulty is necessary to develop the patience and persistence necessary to cope with it. If learners are unable to defer gratification of their need for meaning, and if they show intolerance toward any lack of clarity, then instances of insufficient understanding are likely to be read as a signal to quit rather than as an exciting challenge and a potential source of future satisfaction. In this situation, learning would simply not take place.

Somebody may oppose this view, saying that being open about difficulties may harm students' belief in the possibility of learning mathematics. To this contention, the response should probably be, once more, that everything is a matter of a measure. Appropriate discussion of the demanding nature of the enterprise will make the learning of mathematics only more challenging and attractive in the eyes of the learners. Of course, the stress in this last sentence is on the word *appropriate*: Simply stating the fact that mathematics is difficult cannot be expected to have the desired effect. Cultivating students' belief in their own ability to cope is also important. This cultivation is certainly in tune with the general spirit of the NCTM Standards, which are "*based on the assumption that all students are capable of learning mathematics*" (NCTM, 1995, p. 1). Learners can be reassured in many different ways that the difficulties they are facing are not insurmountable. The whole history of humankind and of its impressive mathematical achievements can be brought in to show that the inherently paradoxical nature of meaning construction is a source of progress rather than an obstacle. Let us now turn to the long list of other needs, all of which are entailed by the primary need for meaning and understanding.

The Need for Structure

What Do We Know About This Need?

Through the ages, those who tried to capture the elusive nature of meaning have proposed a wide variety of definitions for the term. The motif of meaning as being a matter of relations among concepts rather than just concepts as such—and of understanding as the ability to see structure emerging from these relations—is repeated in one way or another in most of these proposals. It is certainly to be found in the Piagetian idea of meaningful learning as building and reorganizing mental schemes. It is also present in Vygotsky's emphasis on the systemic nature of what he called "scientific" concepts and in his observations on the artificiality of those notions that do not grow out of a well-defined system of other, already known, concepts. Vygotsky's (1962, 1987) rigorous definition of scientific concepts also implied that the system must display a hierarchical organization.

The idea of understanding as almost tantamount to seeing relations was brought into the context of mathematics education more then 60 years ago by Brownell (1935) and became prominent recently thanks to the seminal work by Skemp (1976), who, inspired by Mellin-Olsen, suggested the now well-known distinction between instrumental and relational understanding. *Relational* understanding, according to Skemp, is "knowing both what to do and why," whereas *instrumental* understanding is tantamount to "knowing rules without reasons." To explain his idea, Skemp urges the reader to imagine a person trying to learn his or her way around a foreign city. Basically, there are two ways to accomplish the task: Learn "an increasing number of fixed plans [paths]" or try "to construct in [one's] mind a cognitive map of the town" (p. 25).

Much has been said recently about the advantages of learning that is assisted along the way by an awareness of highly cohesive and well-organized links that turn mathematics into an impressively ordered construction (see, e.g., Hiebert & Carpenter, 1992). Although seeing structure is helpful in any domain of knowledge, in mathematics it may be the very essence of learning. Mathematics, after all, can be viewed as a science of structure, as "the study of ideal constructions (often applicable to real problems), and the discovery thereby of relations between the parts of these constructions, before unknown" (Charles Peirce, quoted in Moritz, 1914/1993, p. 8; see also the description of mathematics as the "science of seeing patterns," National Research Council, 1989; Schoenfeld, 1992). The quest for structure is reflected in mathematics' own well-organized, hierarchic form. Thus, learning mathematics implies seeing structures on many different levels: first the structures that can be extracted directly from concrete things and actions and that constitute the basic mathematical concepts (e.g., natural numbers, simple geometrical forms), then the structures obtained through investigation of relations between these most fundamental mathematical structures, and so on. This process does not have to end, since not even the sky is the limit to the never-stagnant hierarchy of mathematical concepts. The quality of learning, however, is determined by the visibility of the inner logic of this special body of knowledge. If understanding means seeing structure, then the well-organized connections among concepts already learned and those that the students are yet to learn must never disappear from their eyes.

I need to add a remark addressed to those thinkers who, convinced of the essential situatedness of learning, oppose the idea of abstraction, in general, and the mathematicians' idea of abstracting structures from things, in particular (see, e.g., Lave & Wenger, 1991). Such an abstraction, some of them may claim, cannot be learned "as such." In the mind of the learner, a concept will always remain embedded in the context within which it was learned; therefore, we should renounce the hope that the learner might deal with the "pure" mathematical structure or be able to transfer it to a new and different situation. This extreme stance, however, can be shown to result from a certain conceptual confusion (see Sfard, 1998a for a more detailed argument). The reasoning of the opponents of abstraction and transfer is rooted in a participationist framework. Since from the purely analytical point of view, the notions of abstraction and transfer are an integral part of the metaphor of learning-as-acquisition, their metaphorical message simply does not fit into participationist frameworks. When one chooses to conceptualize learning in terms of participation and not of cognitive structures, one excludes oneself from the acquisitionist discourse rather than say that the claims engendered by that approach are false (or true, for that matter). Unfortunately, the proponents of the two frameworks are not always aware of the basic incommensurability of their stances and thus argue with each other. Yet the discussion about abstraction and transfer between acquisitionists and participationists is as senseless as the debate about the sum of angles in a triangle between Euclidean and non-Euclidean geometers. If this somewhat intricate analytic explanation is not enough, let me offer a much more immediate argument: The impressive body of mathematical knowledge constructed throughout ages by generations of mathematicians and successfully learned time and again by many young people seems to be powerful evidence for the possibility of thinking in abstract structures.

How Do the NCTM Standards Address This Need?

The NCTM (1989) *Curriculum and Evaluation Standards* speak extensively about "extracting" mathematical structures from a variety of situations. The idea that mathematical "conceptions are created from objects, events, and relations" (NCTM, 1989, p. 11) is promoted throughout. Also, the need to appreciate the overall structure of mathematics gets a good deal of attention. The fourth standard at each level is titled Mathematical Connections, and the authors of the document explain its rationale as follows:

> This label emphasizes our belief that . . . mathematics must be approached as a whole. Concepts, procedures, and intellectual processes are interrelated. In a significant sense, "the whole is greater than the sum of the parts." Thus, the curriculum should include deliberate attempts, through specific instructional activities, to connect ideas and procedures both among different mathematical topics and with other content areas. (p. 11)

This call engendered much literature on the ways in which different mathematical topics taught in school can be "connected" with one another (and also with the "external" world, a point discussed in subsequent sections of this chapter; see also House, 1995, and focus issues of NCTM journals: "Connections," 1993; "Empowering Students," 1996; "Mathematical Connections," 1993).

What Can Go Wrong?

At times, different curricular ideas aiming at different needs of the learner may clash with one another, which seems to be the case with the emphasis on structure, on the one hand, and on the principle of learning mainly through solving real-life problems, on the other. The latter principle has yet to be discussed and justified. At this point, however, let me show why the two curricular requests may conflict.

First, what the NCTM Standards say may only too easily be translated into the belief that students can do all the structure-building work on their own. This belief has little grounding. It is obviously rooted in the conviction that structure is symbol independent and objective, just waiting to be "discovered." This conviction, however, is almost as unacceptable as Michelangelo's modest claim that in his work as a sculptor, his only role was to expose the forms that were "already there" in the stone. Like sculpting, "extracting structure" is, in fact, a creative work. It is invention rather than discovery. Believing that the student can, on his or her own, extract mathematics from real-life situations is like believing that anyone could have produced Michelangelo's art. The gifted artist's creativity was required to make the famous statues of David and Moses, and mathematicians' ingenuity was required to invent numbers, functions, and derivatives.

A desire for *mathematical closure*, accompanied by a concern that students are unlikely to arrive at it on their own, arises from the current critique of the NCTM *Standards* documents. Some critics (e.g., Wu, 1997) stress the importance of making mathematical abstraction explicit and of helping students see how the result can then be transferred from context to context. An "*over-emphasis* on relevance and 'real-world applications'" (p. 948), says Wu, comes at the expense of structure.

For the same reason, some people express concern about the effectiveness of the call to "connect" mathematics. The call seems to imply that mathematical facts are first to be learned and only then connected, which may be not the best way to proceed. The student may have great difficulty trying to construct an overall organizing structure from the bits and pieces of mathematics scattered among different problems and contexts. The task may be as difficult as putting together a jigsaw puzzle without knowing what the final picture should look like.

Although there are almost as many ways of organizing knowledge as there are potential statues in a piece of stone, only certain types of final outcome seem truly desirable. We certainly want our students to be able to see the whole of mathematics as a hierarchical, cohesive body of knowledge. Left on their own, students may remain convinced that mathematics is nothing but a shapeless, amorphous web of accidental links. Although mathematicians' displeasure with this outcome is dictated mainly by their care for the image of mathematics ("a student coming out of a reform curriculum would not understand why the recent proof of Fermat's Last Theorem is a landmark event in human culture," says Wu, 1997, p. 948), mathematics educators' concern stems from their care for the student: As has been argued, insufficient appreciation of structure means insufficient understanding (cf. Tietze, 1994).

Finally, learning exclusively from "real life" may leave the student at the level of those primary mathematical concepts that are only the beginning of what mathematics is about. Such learning may not let the learner ascend to those layers of mathematical hierarchy where one starts acting in truly mathematical ways. This deficiency may be another reason that some critics complain about a "watering down" of the curriculum.

What Can Be Done?

Because the structure of things lies in the eyes of the beholder rather than in the things themselves, and because the process of constructing the structure cannot be fully controlled "from the outside," it would be unreasonable to demand that it be just "shown" to the learner. Much can be done, however, to help students see it on their own. First and foremost, we can ask learners to be much more explicit about the overall organization of mathematics, as well as about those specific mathematical structures that we would like them to "extract" from the real-life situation. Second, we can keep the real-life context of learning within reasonable limits and enlarge the part that abstract mathematics plays in the curriculum. I elaborate on this possibility below.

Yet another way of helping students is to try to organize at least some parts of learning in ways that reflect the mathematical connections we want students to see. This approach would mean guiding students through the hierarchy of mathematical concepts in as systematic a way as possible without jeopardizing other basic principles of the curriculum. Using Skemp's metaphor, we may imagine mathematics students as tourists in a foreign city trying to understand how the town is laid out. Which alternative seems more plausible: that they would prefer to be shown isolated neighborhoods in an accidental manner or that they would choose to explore the city in a systematic way, guided by a map of the city? Students should have the possibility of referring to something like a "map" while dealing with specific problems, concepts, and procedures. In a subsequent section, I give concrete examples that show how such organizing structures can be built. Here let me add only that within such a general framework, where many of the important links between different elements are readily visible, most ideas do not have to be remembered—they will be easily reconstructed whenever necessary. In the presence of the overall structure, understanding can be built simultaneously in both directions, from particular to general and from general to particular, so that the two attempts meet somewhere in the middle. Of course, every child may have his or her own ways to proceed, which implies only that the greater the assortment of choices, the better the chance that every child will find something for himself or herself. The presence of a "map" does not oblige the child to use it, but it does open up such a possibility.

The Need for Repetitive Action

What Do We Know About This Need?

To recapitulate, according to theorists, the primary need behind the human desire to learn is the need to understand, and understanding, in turn, can be interpreted as being able to see order and structure. When dealing with mathematics, one may rightly ask, Structure of what? What is the primary "setting" of the structures that constitute our mathematical knowledge, and what is the mechanism through which we are able to (re)construct them for ourselves? In this section, I argue that the structures we learn to see while doing mathematics are *structures of our actions*.

This statement is certainly not new, nor is it confined to mathematics. Once again, Piaget and Vygotsky may be acknowledged as proposing the idea, for both stressed the fundamental importance of our actions for learning and understanding. As I already mentioned, Piaget's central claim was that our knowledge has its roots not so much in static objects as in our actions on these objects. A similar message was put forward by Vygotsky when he said, as Kozulin (1990) put it, that "higher mental function is created through activity, is an objectivization of action" (pp. 113–114). Vygotsky's followers, the proponents of what came to be known as Activity Theory, focus their study on "internalization of actions." No wonder then, that Bruner (1990), who defines action as "an intentionally based counterpart of behavior" (p. 19), declares it the focal notion of psychology.

In mathematics, all these claims are certainly in force, but they receive a special meaning. As with learning about the physical world, learning about the "virtual reality" of mathematics implies studying actions. Unlike in "actual reality," however, most of these actions are performed on such intangible mathematical objects as numbers, functions, and sets. The issue of understanding mathematics without trying to answer the question of what these mathematical objects are is difficult to comprehend. Of course, this question is asked here from a psychological, not a philosophical, point of view.

One way to explain the idea of "mathematical object," as indeed has been done by a number of researchers (e.g., Dubinsky, 1991; Greeno, 1983; Harel & Kaput, 1991; Sfard, 1987, 1991; Thompson, 1985), is to say that it is a mental construction created by "extracting" structure from actions. In other words, to declare that one "has learned to see" a mathematical object is to say that this person is aware of a repetitive, constant structure of certain actions and can see it well enough to "reify it." To *reify* means to be able to think and speak about the process in ways in which we think and speak about an object: as if it was a permanent entity whose inner structure does not have to be remembered each time one is dealing with it. The relation between our actions and our ability to "see mathematical objects" and thus structure is, therefore, recursive: The actions are performed on certain objects, but abstracting structure from these actions means constructing new objects, and thus new actions on these new objects, and so on.

It is important to understand that the idea of mathematical object is only a metaphor and that the words *mathematical object* do not really stand alone. They can only come within a phrase, such as "a person has constructed a mathematical object (number, function, set)" or "the student uses the notion of function in an objectified way." The latter expression refers to a very special way of participating in mathematical discourse. This type of discourse may be called "objectified" because it is built in the image of a discourse on tangible objects, such as tables or trees. Indeed, mathematical discourse—like any discourse—is clearly created by recycling linguistic forms taken from the discourse on perceptually accessible reality; this is the mechanism of metaphor. If we transfer the ways we talk about, say, tables and trees into another, more abstract, context, we end up constructing the virtual world of mathematics in the image of the physical world. Moreover, we are often so good at doing this constructing that the putative virtual objects of mathematics eventually become for us experientially real—and we start believing in their independent existence.

This essential interdependence of mathematical actions and objects, and thus of actions and understanding, expresses itself in the process-object duality of mathematical concepts. Such symbols as 5, 2/3, -3, $\sqrt{-1}$, or the function $3x - 2$, although clearly referring to objects, may also be viewed as pointing to certain mathematical processes: counting five objects; dividing an object into three equal parts and taking two of them; subtracting 5 from 2; extracting the square root of -1; and performing the computational procedure of multiplying a number by 3 and subtracting 2, respectively. This duality is the source of both the power and the difficulty of mathematics (see also the idea of *procept* in Gray & Tall, 1994). I build the rest of the argument around the testimony of a mathematician:

> Mathematics is amazingly compressible: you may struggle a long time, step by step, to work through some process or idea from several approaches. But once you really understand it and have the mental perspective to see it as a whole, there is a tremendous mental compression. You can file it away, recall it quickly and completely when you need it, and use it as just one step in some other mental process. The insight that goes with this compression is one of the real joys of mathematics. (Thurston, 1990,p. 847)

If the "compression" is construed as an act of reification—as a transition from an operational (process-oriented) to a structural (objectlike) vision of a concept,[1] this short passage brings into full relief the most important aspects of this intricate process. First, it shows that we have to be well acquainted with a mathematical process to arrive at its structural conception. Second, it implies that reification does much good to our understanding of concepts and to our ability to work with

[1] It does not have to be construed in this way, but this interpretation is consonant with what was said previously about structural conceptions.

them. Third, it says that reification often comes only after a long struggle. Studies in both mathematics education and the history of mathematics (e.g., Breidenbach, Dubinsky, Hawks, & Nichols, 1992; Sfard, 1992; Sfard & Linchevski, 1994) confirm this last claim: They suggest that whether we are talking about functions, numbers, linear spaces, or sets, reification is difficult to attain. The main source of this inherent difficulty is, once again, the circular nature of our understanding (which in the case of mathematics I once called "the vicious circle of reification"): On the one hand, one cannot perform actions (e.g., arithmetic operations) on mathematical objects (e.g., rational numbers) before constructing these objects; on the other hand, one cannot construct the objects before operating on them.

The bottom line is that the structure of actions is the principal object of mathematical inquiry, just as the structure of the human body is the object of research in anatomy. Moreover, because structure means a certain repetitive pattern, we may say that there is no mathematics without repetitive, well-defined actions. Finally, although the only "empirical evidence" I offered was a mathematician's testimony, this claim is equally true of basic mathematics: Some kind of reification—of seeing structure in action—is necessary even at the earliest stages of learning (cf. Nesher, 1986). Indeed, even the transition from counting procedures to the notion of natural number is an act of reification.

How Do the NCTM Standards Address This Need?

The authors of the 1989 NCTM *Standards* document come close to recognizing the importance of action and of reflection on action when they say that "'knowing' mathematics is 'doing' mathematics" and that "a person gathers, discovers, or creates knowledge in the course of some activity having a purpose" (p. 7). However, although this statement can be interpreted in many ways, the interpretation given by the writers of the *Standards* documents themselves is likely to engender practices that are at odds with what has been said previously in this chapter. In the remainder of this section, I focus on a specific aspect of the topic that, in the ongoing debate on the reform, is clearly an object of the most heated controversy—the issue of basic skills, which, being a mastery of well-structured mathematical actions, seem closely related to the needs explained above.

From the point of view of the general public, perhaps the most obvious change brought by the reform is a removal of basic skills from their central place in the curriculum. The new learning process "is different from mastering concepts and procedures" (NCTM, 1989, p. 7), say the *Standards* authors. The rest is done in a series of disclaimers rather than in direct statements: "We do not assert that informational knowledge has no value, only that its value lies in the extent it is useful in the course of some purposeful activity" (p. 7); and "the availability of calculators does not eliminate the need for students to learn algorithms, but such knowledge should grow out of the problem situations that have given rise of such algorithms" (p. 8). If read carefully, none of these statements says that basic skills should be abandoned altogether; rather, they require that the skills be developed in new, more "natural" ways. Because the proposals are stated in negative sentences, however, the 1989 *Standards* document is only too likely to be read as conveying a much more radical message.

What Can Go Wrong?

Judging from what critics are saying, this perception is, indeed, the way the 1989 Standards document is being read. Most people tend to interpret the document as denying the skills any real importance. This tendency is understandable in the light of certain overall trends, quite independently of the NCTM Standards themselves. As Devlin (1997) explained, "We need to reduce drastically the time spent teaching basic skills in middle and high school mathematics classes. . . . *The aim of mathematics education should be to produce an educated citizen, not a poor imitation of a $30 calculator*" (p. 2). But that goal is not the only, and perhaps not even the most important, reason people speak against teaching basic skills. Basic skills—seen as something that can be acquired only through meaningless, dull practice—have been getting bad press for a long time. Our recent stress on understanding has brought intolerance toward anything that does not seem conducive to meaningful learning. Basic skills are deemed a paradigmatic example of what can be learned only by rote repetition. As a consequence, algorithm execution has come to be seen as being in opposition to learning with understanding.

All these claims and beliefs seem rather exaggerated. When one tries to make a case against basic skills, the argument concerning lack of real-life applicability, like the one brought by Devlin, is certainly true and convincing as long as we agree that the only possible reason for learning anything in mathematics is its usability outside mathematics itself. Some types of knowledge, however, may be necessary not because they have practical applications but because of their intramathematical importance—that is, their function as an indispensable element in the process of constructing mathematical knowledge. Basic skills may belong to this category. As paradoxical as it may sound, a reasonable level of mastery of basic algorithms may, in fact, be necessary for understanding mathematics.

The idea I want to promote is simple. In the light of what has been said above, some proficiency in basic mathematical procedures may be necessary if the process of learning is to be possible at all. First, there are empirical studies whose results seem to point in this direction (Fuson & Briars, 1990; Fuson & Kwon, 1992; Hiebert & Wearne, 1996; Siegler, this volume; Stevenson & Stigler, 1992). Second, although this interpretation of the empirical data is not the only possible one, it becomes most plausible on the force of the theoretical argument that goes as follows: If mathematics is a result of reflection on our own actions—first physical and then mental—then the study of actions is the essence of mathematizing. To reflect on the processes we perform, we have first to acquire a certain mastery of these processes, sometimes perhaps even to the degree of automatization. As Vygotsky (1962) said, "In order to subject function to intellectual control, we must first possess it" (p. 90). Children without a reasonable ability to perform basic algorithms may have nothing with which to build their further mathematics. Thus, for example, a child who does not have a reasonable experience with such operations as $3 - 5$ or $8(1 - 4)$ is unlikely to be able to reflect on these operations and reify $3 - 5$ or $1 - 4$ into mathematical objects. This child will therefore be deprived of an understanding of negative numbers, and his or her further progress will be stymied. To recapitulate, we can say that, paradoxically, the emphasis on understanding may have deprived children of something to understand. The claim about the possibility of learning mathematics meaningfully without some mastery of basic procedures may thus be compared to the claim about the possibility of successfully building a brick house without bricks.

What Can Be Done?

The authors of the 1989 NCTM *Standards* document stress *purposeful* activity as a proper setting for learning (p. 7). Indeed, having a purpose is the best motivation for learning. Whether acquiring skills can be regarded as a "purposeful activity" is a matter of what one considers a purpose. From what the *Standards* document says, the authors clearly do not regard the wish to master mathematical procedures as deserving this name. Carefully planned work on mathematical procedures, however, in which the search for meaning goes hand in hand with growing technical mastery (possibly involving the use of calculators!) has a good chance of being received by children as a meaningful, purposeful activity. From my experience as a teacher, many students like practicing because becoming fluent in technique gives them the sense of success and security necessary to proceed. Performing mathematical procedures accompanied by an incessant meaning-making effort[2] will no longer be seen as rote memorization. Rather, it will take the shape of a dialectic process of synchronized increase in proficiency and understanding, in which mastery of an action leads to reflection on the meaning of that action and an increase in understanding leads to new, more complex actions.

I am not trying to say that implementing this suggestion will be easy, or that bringing skills back into schools will be a straightforward move. These measures may be difficult even if what we have to offer is a brand new way of learning skills, one that can be called *reflective practice*. Whether they will be possible depends to great extent on our ability to change the climate—to convince everyone concerned that mastering skills, especially when done through reflective practice, does not mean ignoring the child's basic need to understand but, on the contrary, is a condition for understanding. This effort will require, among other things, that we bring to light some false beliefs about mathematical understanding. First, we must make it clear that treating understanding and repetitive doing (practice) as if they were two separate things, and as if one was possible without the other, is a fundamental mistake. Second, we must explain that persistent doing in situations of only partial understanding is necessary for progress.

A natural ending to this section is the following observation, which Hiebert and Carpenter (1992) made some years ago:

> One of the longstanding debates in mathematics education concerns the relative importance of understanding versus skill. . . . The debate was often carried out in the context of proposing instructional programs emphasizing one kind of knowledge over the other. The prevailing view has seesawed back and forth, weighted by the persuasiveness of the spokesperson for each particular position. Although the arguments may have been convincing at times within the mathematics education community, we have not made

[2] Of course, because not all students learn in the same way, the relative amounts of doing and reflecting may change from learner to learner. The "meaning making" can be done today with the help of computers. Thus, solving equations may be done by applying algebraic and graphical methods simultaneously. As has been shown in several studies (Kieran & Sfard, 1999; Schwartz & Yerushalmy, 1995), this instructional technique may, indeed, have a beneficial effect on students' understanding of the algorithms. Similar results may come from programming activities that require a good sense of procedures (Breidenbach et al., 1992; Sfard, 1992).

> great progress in our understanding of the issue. (p. 77; see also Hiebert & Lafevre, 1986)

In view of the intricate interdependence between doing and understanding, and because of the inherent circularity of the process through which they develop, this zigzagging of instructional approaches is hardly surprising. The time may have arrived, however, to acknowledge that the question of their relative importance is ill posed, as understanding and doing may simply be two sides of the same coin.

The Need for Difficulty

What Do We Know About This Need?

As has been said before in many different ways, true learning implies coping with difficulties. Because people fear difficulty and instinctively try to escape it, I want to stress that when it comes to learning, difficulty is in fact a good thing, provided it is basically manageable. One may say that difficulty is for learning what friction is for movement: It is the condition for its existence. On the one hand, without difficulty, no learning occurs, just as no movement occurs without friction. On the other hand, of course, too great a difficulty makes any learning impossible, just as too much friction stops any movement. To make the argument more "scientific," I next turn to the authority of Piaget and Vygotsky and to their controversy over the relationship between development and learning.

In their discussion, the term *development* referred to the process of natural maturation. Inspired by the metaphor of biological growth, Piaget believed that our mental functions develop naturally, as our body does. However, natural development is not the only possible cause of a change in cognitive processes. These processes can also be molded through learning, that is, through a deliberate attempt to enrich and reorganize our cognitive schemes. Learning, therefore, may advance us toward goals that are not likely to be attained spontaneously. Piaget also had a clear conception of the relationship between learning and development. Guided as always by the metaphor of biological growth, he believed that learning is basically subordinate to development and therefore has well-defined limits. Just as a neonate cannot learn to ride a bicycle, so a child cannot learn things requiring intellectual maturity that he or she does not yet possess.

Vygotsky adopted Piaget's terminology and, like his predecessor, distinguished between development and learning. He took Piaget to task, however, on the issue of the precedence of the former over the latter. More specifically, Vygotsky disagreed with Piaget's extreme claim that development cannot be influenced through learning. His famous idea of the *zone of proximal development* (ZPD) epitomizes the belief that the child's intellectual development can be accelerated by carefully planned learning. Since the purpose of learning should be to take the child across the ZPD—that is, to advance the child from the abilities she or he now possesses to those that, although not yet developed, are already in the process of budding—the best way to teach is to present the learner with tasks that go beyond his or her present developmental level. To be sure, these tasks, although demanding, cannot be too advanced; they have to fit within the ZPD, which also has its limits. One thing, however, is clear from this proposal: Truly substantial learning can occur only when the child experiences a certain difficulty. Thus, Vygotsky can be read as stressing the role of difficulty, whereas Piaget, although it might not have been his intention, has been interpreted as demanding that the child be saved the frustrating challenge of tasks that go beyond his or her present abilities. Quite apart from one's stance on the original issue of the relationship between learning and development, anyone who acknowledges the need for difficulty would find Vygotsky's conception more appealing.

How Do the NCTM Standards Address This Need?

A recognition of the role of difficulty in setting the learning process in motion transpires from several of the curricular principles promoted by the 1989 NCTM *Standards*, but above all from their insistence on problem solving "as a means as well as a goal of instruction" (p. 129). That this suggestion has, indeed, to do with the wish to create the "friction" necessary for the learner to move forward is obvious from the way in which the "genuine problem" is defined. According to the *Standards* document, it "is a situation in which, for the individual or group concerned, one or more appropriate solutions have yet to be developed" (p. 10). The absence of the solution is the difficulty that, combined with student's natural need to understand, is a powerful motivational force. As to the degree of difficulty, the *Standards* authors say that "the situation should be complex enough to offer challenge but not so complex as to be insoluble" (p. 10). This request is certainly reasonable and is in accord with what has been said previously.

What Can Go Wrong?

There are at least two areas in which the good intentions of the 1989 NCTM *Standards* document authors may lead to unhelpful interpretations. First, in spite of the insistence on problem solving, interpreters of the document may undermine the didactic role of difficulty, not simply because of the popular fantasy of "frictionless learning" discussed above but also because of a possible misunderstanding about certain statements in the *Standards*. The document asserts, among other things, that "problem situations must keep pace with the maturity—both mathematical and cultural—and experience of the students" (NCTM, 1989, p. 10). Although this statement sounds very reasonable, without a proper explanation it may be read as consonant with the Piagetian exhortation to protect the child from tasks that are deemed to be slightly above her or his present developmental level but that, for that very reason, would be deemed by Vygotsky as perfectly suitable for "enticing" the learner into her or his ZPD.

Second, much depends on the implementer's understanding of the idea of problematic situation, on the type of problems used to create such situations, and above all, on the extent to which the different situations are related to one another and to the overall direction of learning (if there is learning!). An important discussion of the idea of "problematizing the subject" recently appeared in the *Educational Researcher* (Hiebert et al., 1996; Prawat, 1997; Smith, 1997). Only too easily may the exhortation to "problematize" end up in "watered-down courses filled with cute but mathematically pointless activities" (Jackson, 1997, p. 695), which means that the learning hardly has any general direction. As I argued above, this situation violates students' basic need for structure.

Those accused of introducing such unstructured implementations can provide good arguments to account for their approach. Finding "genuine problems" that would be also good for one's learning turns out to be extremely difficult, and finding ways to interconnect these problems so that they can jointly lead to well-defined, truly important mathematical ideas is even more so. A mathematical problem, to be effective, should confront the student with an authentic but manageable difficulty—meaning that the question itself must both be well understood and seem significant (see the next section for the discussion of the concept of significance) but that the solution must be unknown. If to these requirements, one adds the requests that the problem also be embedded in a real-life context (as, indeed, may be the case with the NCTM Standards; see the next section), that it be tightly connected to some truly important mathematical idea and to a number of other significant problems, and finally, that the problem allow one to satisfy a wide spectrum of students' preferences and abilities, one must concede that the mission of the curriculum developer is extremely difficult if not altogether impossible. No wonder, then, that the question of what should count as a "good problem" has become a topic of lively debate (see, e.g., Nemirovsky, 1996; Yerushalmy, Chazan, & Gordon, 1990). A related topical issue is that of criteria for the assessment of problems constructed by students; see, for example, Silver and Cai (1996).

What Can Be Done?

The first thing that comes to mind in response to the dilemmas outlined above is that rather than try to take difficulty out of curricula, we must simply be more careful about the choice of "problematic situations." As I just noted, doing so is easier said than done. Much thought must be given to the subject if any workable implementation of the certainly appealing, but inherently difficult, idea is to be found. I have every intention of taking a step in this direction. However, since the difficulty of a problem given to the student cannot yet, as such, be regarded as a true challenge—and since to become a powerful driving force for learning, the problem must be perceived as an obstacle to attaining some significant goal—let me turn to an analysis of the notion of significance before I propose a more concrete solution to the dilemmas outlined above.

The Need for Significance and Relevance

What Do We Know About This Need?

When it comes to motivating students' learning by harnessing their hunger for meaning, significance is a necessary complement of difficulty. In this chapter, significance is conceived of as an ability to understand and appreciate the place and importance of what is to be learned within the system of concepts that are already well understood. Or, to put it differently, it is construed as an awareness of the way in which the existing knowledge generates a problem at hand and necessitates its solution. With this interpretation, one can see the sense of significance as a type of understanding concerned with interconceptual relations, as opposed—and complementary—to one that focuses on the inner structure of concepts.

The need for this kind of understanding is already obvious from the foregoing discussion, but it can also be reinforced with the well-known theoretical claim according to which new knowledge can grow only out of existing knowledge. Piaget has translated this almost self-evident truth into a definition of learning as enriching and reorganizing existing mental schemes. The same message is conveyed within Vygotsky's constant emphasis on the "spiral" rather than neatly cumulative nature of knowledge. According to Vygotsky, knowledge develops through our constant dissatisfaction and incessant "reworking" of what we already know. Because significance means linking new to old, this theoretical claim may be interpreted as an assertion of the learner's need for the sense of significance. The already mentioned artificial concepts—a notion coined by Vygotsky—are the ones that, being "parachuted" onto the learner from above rather than spawned by a genuine and well-understood problem, are bound to remain devoid of any significance in the learner's eyes. Only too often, our students leave school with such an image of negative numbers, algebraic operations, and functions.

How Do the NCTM Standards Address This Need?

The stress on significance understood as presented above is conveyed by such statements as "[children] will accept new ideas only if their old ideas do not work or are insufficient" (NCTM, 1989, p. 10). It also emerges from the already mentioned insistence on the shift "toward connecting mathematics, its ideas, and its applications" (NCTM, 1991, p. 3). Applications have become the focal issue. Time and again, the NCTM *Standards* documents urge teachers and learners to relate mathematics to other domains and to pursue the idea that mathematics should be "grown" from familiar, preferably real-life, situations. Finally, the already quoted claim that "a person gathers, discovers, or creates knowledge in the course of some activity having a purpose" (NCTM, 1989, p. 7) may be interpreted as being concerned with significance, provided the words "having a purpose" refer to a situation in which students have a genuine drive to cope and are well aware of the reasons for their actions. Such a drive can arise only from the need to find an answer to a well-understood question deemed truly important.

What Can Go Wrong?

As in the previous cases, the danger lies in an extreme and single-minded interpretation of what is basically a very promising idea. It may well be that because of the NCTM Standards' emphasis on applications, significance is too often confused with real-life relevance. True, the natural need for meaning and significance may be the most important force driving learning, and new knowledge can grow only out of existing knowledge. Moreover, even critics of the Standards are likely to admit the importance of everyday relevance: "The need for applications in school mathematics curriculum is beyond debate," says Wu (1997, p. 948). Who said, however, that the "old knowledge" from which the new ideas are supposed to grow can be only the knowledge of the concrete and the real? Indeed, such a view is a rather limited way of interpreting significance, which, as was already stressed, can also come from within mathematics itself.[3] This last interpretation, nevertheless, is only too often either overlooked or actively rejected. The trend toward equating significance with real-life significance was recently spurred by a body of research that points to the fact that people tend to do significantly better in applying mathematics to real-life problems than in attempting to deal with the same mathematical content in the context of typical school problems (Lave, 1988; Nunes, Schlieman, & Carraher, 1993; Rogoff & Lave, 1984; Saxe, 1991; Schlieman & Carraher, 1996; Walkerdine, 1988). If we add once again our overpowering tendency to seek clear-cut educational solutions, we risk becoming totally preoccupied with finding a realistic context for all mathematical concepts—and openly hostile toward whatever is not being taught that way. By purging the curriculum of the abstract and proposing that mathematics can be learned only through real-life problems, we may be simultaneously committing a number of consequential mistakes.

First, consider the theoretical argument, rooted in Vygotsky's idea of scientific concepts (as opposed to everyday, or spontaneous, concepts) and presented in detail by Davydov, that speaks against an "empirical approach" in teaching mathematics as contradicting the nature of these concepts. I do not repeat Davydov's line of reasoning here but quote only his description (as cited in Kozulin, 1990) of the activities he regarded as most appropriate for learning:

> Such an activity should guide students from *abstract to concrete* [italics added], that is, from the most general relationship characteristic of the given educational subject to its concrete, empirical manifestations. (p. 259)

Second, if we are overeager in our attempts to show the applicability of mathematics in order to motivate stu-

[3]This point is made with particular clarity by the members of the Realistic Mathematics Education project (Gravemeijer, 1994; Streefland, 1991; Treffers, 1987), who require the problems students are supposed to deal with to be *experientially real*, and who then stress that this term "means only that the starting points should be experienced as real by students given their prior participation in both in-school and out-of-school practices, not that they should necessarily involve so-called real-world situations" (Cobb, 1998, p. 189).

dents, we may end up with an opposite effect. As already stated, more advanced mathematics—the kind taught at the secondary level—may be not sufficiently necessary in everyone's life for usability to be the sole argument for learning it. Therefore, our insistence on teaching only what meets some real-life needs will not convince the student that mathematics is really usable; instead, it is likely to make him or her believe that whatever has little practical importance does not have to be learned.

Third, it may be shown that the equation "significance = real-life relevance," often ascribed to the adherents of the participationist perspective, does not really follow from the claims on the essential situatedness of learning. As explained in more detail below, what makes a problem "real" and accessible in the eyes of the solver is not just its content but also, and no less important, the context within which it arises and is being solved. The real-life reason and the real-life purpose of the task motivate the problem solver, give sense to this particular activity, and provide cues for the way the goal can best be accomplished. In other words, the strength of a real-life problem is its being embedded in a real-life context that is organically tied to authentic needs and instincts of the solver. Bringing it to school means stripping it of what gives it this special power.

Finally, we return to what has been said above: The request to teach only what can be taught through real-life applications of mathematics is bound to lead to segmenting and impoverishing the subject matter. It is quite unlikely that a complete, well-organized picture would eventually emerge from the small fragments of the often quite insignificant mathematics that can be abstracted from realistic problems. Adverse effects of this state of affairs on students' understanding were discussed above in the section on the need for structure. Where the issue of significance is concerned, we gain an additional perspective on the same problem (cf. Schlieman, 1995).

What Can Be Done?

Two types of action may be helpful in preventing the good idea of connecting mathematics with life from degenerating into the "hegemony of the applicable." First, we may try to rehabilitate the principle of knowledge and understanding for their own sake. Second, we could try to make it clear that significance may be less likely to come from stand-alone, real-life problems that can produce relatively quick but rather local solutions (and for which the term *puzzle* may therefore be more appropriate) than from problems which, although more abstract, are parts of a system directed at an important overall goal.

Let me put it slightly differently. The 1991 NCTM *Standards* document promotes the idea of a "problematic task" (p. 60) in which students' activity would have a purpose; however, the authors do not explain the exact meaning of the words. They may now wish to add that the adjective *problematic* refers to a situation in which students are helped to see that their present knowledge is as yet insufficient to satisfy certain genuine needs or to answer questions that emerge from what is already known. A real challenge comes from a comprehensible problem whose solution requires a prolonged effort and much work. In that kind of problem, the necessary investment may be considerable, but the reward waiting at the end of the road, and coveted all along, would be proportional. True, questions that are likely to lead to important mathematical ideas require a much greater attention span than the stand-alone puzzles that can be solved during one or two class periods. Such an extension of attention is requested by the Standards—"Students need to work on problems that may take hours, days, and even weeks to solve" (NCTM, 1989, p. 6)—but has evidently been overlooked by many implementers. Also, with comprehensible systems of interrelated problems, students are unlikely to find their way toward solutions on their own; after all, mathematicians often worked on this type of problem for several centuries. Therefore, in this kind of instruction, the demands on the teacher are difficult and many. The teacher has a challenging task to keep students' attention, sustain the sense of a challenge, and preserve the students' awareness of a global direction during the periods when local problems are being solved. In certain classrooms, such an "overall navigating" of learning would seem almost impossible. But because sense of direction is almost tantamount to sense of significance, and because no movement—either forward or backward—is possible without it, sustaining the sense of direction is worth the necessary effort. And as experience shows, with careful planning, this difficult task can be done. Let me mention just two examples.

Let me first turn to the high school calculus course developed in Israel a few years ago.[4] The general idea was to begin the study by helping learners create a conceptual framework for the course. With such a framework, concepts and methods of calculus would appear only when already anticipated, which meant, among other things, that the need for a tool that would

[4]The Israeli education system is centralized, and there is a national curriculum of which mathematics is a central part. The program in question, meant for Grades 10-12, was a product of teamwork in which I took part.

later be called the derivative, the way it might be constructed, and its future uses could be felt and understood in advance. The outline of such a framework is presented in Figure 23.1. At least 20 hours must be devoted to activities through which the questions and observations constituting this framework can be developed. A similar framework has recently been supplemented with ingenious and challenging computer activities (Schwartz & Yerushalmy, 1995).

General goal:

To get acquainted with families of functions.

Problem:

The known methods of investigating functions (build a table, draw a graph) do not seem to work for more complex functions (often not even with computers!).

Subgoal:

To find a new, more reliable, analytical tool for investigating functions.

A promising phenomenon:

Most graphs turn out to be almost linear when watched "closely" (small neighborhood of a point is greatly enlarged)—the existence of a linear approximation.

Questions to be answered:

1. Is the phenomenon above common enough to deserve attention?
2. If the answer to question 1 is yes, how shall we find linear approximations?
3. When we know how to find linear approximations, how shall we use them to investigate functions?

Figure 23.1. A framework for introductory calculus.

My second example comes from a study that Carolyn Kieran and I (Kieran, 1994; Kieran & Sfard, 1999) conducted a few years ago in Montreal. In the experiment, in which introductory algebra was taught to 12-year-olds, the guiding conceptual framework (Figure 23.2) was created at the outset by presenting graphs as objects that require study. A message about the importance of the graphs and their investigation was conveyed in a natural manner by the fact that they were built to model situations that were significant in the eyes of the children, which does not necessarily mean that all the situations were realistic. The aim of the course was therefore recognized from the beginning: The learners' task was to construct ways of investigating graphs. Doing so led them to the "invention" and use of algebraic expressions. Moreover, the learners not only arrived at some of the basic algebraic equivalences on their own but also started solving equations and inequalities before having learned any algebraic method for doing so. The challenge they then faced was to find a method of solving that would be

List of contents

1. *Introducing Cartesian Graphs*
   - 1.1 Discrete graphs, no rule
   - 1.2 Continuous graphs, no rule
   - 1.3 Story-based graphs with a rule
   - 1.4 Abstract graphs with a rule
2. *Algebraic Expressions and the Notion of Function*
   - 2.1 Algebraic expression as a symbolic representation of a rule
   - 2.2 Interplay between story, table, expression, and graph
   - 2.3 The concept of function—an introduction
3. *Exploring Functions*
   - 3.1 Investigating functions with expression, table, and graph
   - 3.2 Matching graphs with expressions
   - 3.3 Linear functions
4. *Operating on Functions: Equivalence of Expressions*
   - 4.1 Operations on graphs: adding and multiplying by a number
   - 4.2 Equivalent expressions
   - 4.3 Adding linear expressions and multiplying an expression by a number
5. *Comparing Functions—Equations and Inequalities*
   - 5.1 Comparing values of two functions using graphs
   - 5.2 Solving linear equations with graphs
   - 5.3 Solving linear inequalities with graphs

Figure 23.2. A framework for introductory algebra (adapted from Kieran & Sfard, 1998).

quicker, more effective, more precise, and fully reliable. This example shows, among other things, that with the help of computers, the issue of sustaining a sense of significance by keeping a clearly defined direction may be not such a difficult task after all.

The Need for Social Interaction

What Do We Know About This Need?

Although all theorists seem to agree on the central importance of human need for knowledge and understanding, it was Vygotsky who alerted us to the essentially social nature of learning and meaning. To quote Bruner (1985):

> One point . . . has usually been overlooked or given second billing in our own achievement-orientated Western culture. It is inherent in [Vygotsky's] conviction that passing on knowledge is like passing on language—his basic belief that social transaction is the fundamental vehicle of education and not, so to speak, solo performance. (p. 25)

Today, the role of social interaction in learning is no longer given second billing. Investigators in mathematics education have written about it extensively (see, e.g., Cobb, 1995, 1999; Cole, 1996; Forman, Minick, & Stone, 1993; Lampert, 1990; O'Connor, 1996; Schoenfeld, 1996). In this section, I therefore limit myself to a brief presentation of a number of central points.

The history of the debate on the relative importance of the individual and the social is probably as old as the history of research on human learning. Mainly because of their relative stance in this debate, Piaget and Vygotsky are regarded today as the leaders of two opposing camps. In the literature, the Piagetian model is depicted as one in which the learner is "a lone organism pitted against nature" (Bruner, 1985, p. 25), whereas Vygotsky is credited with revolutionizing that picture by stating that the way from nature to the child leads through the society. This sharp contrast has sometimes been misinterpreted as hinting at an absence of any reference to the social in Piaget's writings and as Vygotsky's denial of the role of the child's interaction with the natural environment. In fact, neither of these interpretations is true. Piaget declared repeatedly that social interaction is one of the most important factors in learning (see, e.g., Piaget & Inhelder, 1969), and Vygotsky (1962) agreed that nonsocial aspects of the world we live in greatly influence our thinking; that influence is what he certainly had in mind when he spoke, for example, about empirical generalization and about the construction of everyday concepts by creating "family pictures" of similar objects.

Although the two thinkers were indeed divided in their views of the nature of human learning, the gist of the controversy lay not so much in their vision of the mechanism of learning as in their understanding of the nature and origins of human knowledge at large. Whereas Piaget regarded human intellectual development as a biologically determined phenomenon that can be influenced by culture only marginally, Vygotsky gave primacy to sociocultural factors. Moreover, whereas Piaget implied that the knowledge we build is mainly a function of the world around us[5] and is quite literally created anew by every learner, Vygotsky saw both knowledge and meaning as collective creations that are preserved within culture and appropriated time and again by individual children in the process of learning. One may conclude that Piaget and Vygotsky differed in their understanding of the word *social*. Whereas Piaget used this notion almost solely in the context of *ontogenesis*, in relation to the possible techniques of learning, Vygotsky accorded the adjective *social* to concepts as such. He did so not only because of the mechanism of social interaction that is essential for their individual acquisition but also, and perhaps mainly, because of their social *phylogenesis*.[6] The concepts we learn do not simply come directly from nature and are not uniquely determined by nature. Conceptual thinking is a byproduct of human communication and is possible only within language. The language, in turn, is a social creation, and the concepts themselves are therefore essentially social.[7]

Let us now turn to the practical implication of this debate. Vygotsky's insistence on the social nature of knowledge brings to the fore the importance of social interaction in human learning. Unlike Piaget, he implies that without such an interaction, no conceptual learning would be possible. The transactional nature of learning may express itself in many ways. The most obvious form of learning interaction is a student-teacher and student-

---

[5]He did not try to say that the knowledge is an objective picture of this world but only that the world is the crucial factor in creating that knowledge.

[6]Piaget's and Vygotsky's controversy over the exact meaning of *social* is epitomized in their differing interpretations of the phenomenon of "egocentric speech" (little children's well-documented tendency to "talk to themselves") and in the fact that Vygotsky (1962, 1987) objects to Piaget's decision to call nonegocentric speech *social*, saying that *any* speech, egocentric included, is essentially social. Vygotsky prefers therefore to call nonegocentric speech *communicative*.

[7]There is no such thing as "private language," argued Wittgenstein (1953). Language is a vehicle of communication with others, and therefore the idea of private language is, in a sense, an oxymoron.

student exchange. In addition to direct teaching, there are various everyday interactions in which learning occurs, even when unintended. Through an often-imperceptible negotiation of meaning, which is integral to any communication, children learn to understand new words and thus new concepts. Even studying a textbook should count as a form of social interaction. After all, the text being read is a result of an attempt to convey socially constructed meanings, and reading such a text is therefore a form of conversation with others.

Vygotsky's insistence on the essential role of social interaction sounds truly convincing, especially in view of his claim about the possibility of influencing the child's development through an instructional intervention. As to that claim, it immediately implies the central importance of well-planned instruction. The now popular mode of teamwork called *cooperative learning* expresses a general recognition of the need for multifarious forms of learning interactions, including those that do not feature the teacher in the central role (Davidson, 1990; Sutton, 1992). Indeed, social interaction seems necessary if any learning at all is to occur.

How Do the NCTM Standards Address This Need?

The 1989 NCTM Standards, if carefully implemented, would be a substantial step toward Vygotsky's principles. The *Standards* document underscores in a number of ways the principal importance of exposing the learner to different forms of interaction. Among the suggested five modes of learning, three speak explicitly about different types of exchange:

- Group and individual assignments
- Discussion between teacher and students and among students
- Exposition by the teacher (NCTM, 1989, p. 10)

When comparing the different modes, one should stress both the need for diversity in the composition of the groups within which the interaction is to take place and the varying extents of the teacher's and the students' contribution to the exchange. This call for pluralism of form may appear somehow undermined when, in the last of the three *Standards* volumes, the authors issue a call for a "shift in teaching: toward questioning and listening, away from telling" (NCTM, 1995, p. 2).

What Can Go Wrong?

As a reaction to Piaget's underestimation of the role of social interaction, and following Vygotsky's claim about the beneficial effects of the child's interaction with peers, teamwork has come to the fore as the preferred mode of learning. This trend has also been fueled by the growing body of research findings that suggest some advantages of collective effort and its beneficial effect on students' achievement (O'Connor 1998; Siegler, this volume; Webb, 1991, 1994). Despite the Standards' call for plurality of modes, the preference for collaborative learning has pushed aside other, more traditional forms of learning. In some places, individual work and teachers' exposition have disappeared completely. As it turns out, however, the question of interactivity in learning mathematics may be much more complex than any radical follower of Vygotsky would readily admit. For really effective learning, solitary work and the teacher's substantial interventions may be as vital as teamwork. Denying this possibility is a misinterpretation of the theory. Let me substantiate this claim.

Consider the part that the student's solitary effort plays in learning mathematics. For reasons that were explained in previous sections, the dual process of learning and of constructing meaning cannot be seen only as a matter of a collective doing. Rather, it must be understood as an intricate mixture of social interaction and individual reflection. Perhaps because of the central role of this latter ingredient, mathematicians are often reported as preferring solitary work to collaborative problem solving (cf. Sfard, Nesher, Streefland, Cobb, & Mason, 1998). The relative solitude of the working mathematician seems to arise for a good reason, and that reason may also be in force when it comes to students of mathematics. Being extremely demanding in terms of concentration and intellectual effort, mathematical problem solving can sometimes be better practiced in silence, where all the effort may be focused on the problem at hand and on that problem alone. Communication with others, being another strenuous activity, may distract one's attention and make one's problem-solving attempts less effective. That phenomenon may be why a critic of the NCTM Standards observed, "While a bit of group learning . . . is good in the classroom, *too much* of this is happening in the reform classrooms to the detriment of good education" (Wu, 1997, p. 950). Once again, we are facing an excess that may kill a good idea. Besides, as explained previously, those advocates of collaborative learning who refer to Vygotsky for support while demanding exclusivity of this mode are probably taking his intentions to an unintended extreme: Learning is social regardless of the way it occurs.

Indeed, learning does not have to be interactive to be social. Interaction, in turn, does not have to occur among peers; the teacher's intervention is also a form of interaction. Following Vygotsky, I have argued that true learning can occur only in situations in which problems faced by students are somewhat beyond the level of their present competence. If so, one can hardly expect such learning to take place without help of a more knowledgeable person. That claim does not imply that the student needs to be explicitly told what to do or what to think. After Bruner (1985, p. 28), we may compare the activity of a skillful helper to the "constructing of scaffolds" where "the child is permitted to do as much as he can spontaneously do, [and] whatever he cannot do is filled in or 'held up'" by the other person. The need for such help was acknowledged by Vygotsky, who claimed that although a more advanced peer may sometimes play the role of "scaffolder," in many cases the teacher must do the job.

An indirect argument for the need for scaffolding has been made by those who insist that children cannot be expected to reinvent a body of knowledge whose historical construction lasted for thousands of years (see, e.g., Bartolini-Bussi & Pergola, 1996, quoted in Sierpinska & Lerman, 1996). This claim is true whether one is referring to a lone "Piagetian" learner or to a "Vygotskian" team unsupported by a teacher. It may be made more direct if one takes a closer look at the mechanism of knowledge construction.

The greatest need for the teacher's direct intervention must be felt by students at those junctures in learning where their existing knowledge turns out to be less a foundation to build on than an obstacle to progress. Both Piaget and Vygotsky spoke extensively of those special moments when, to proceed, one must reconceptualize what one already knows. Piaget called this reconceptualization "an accommodation of a scheme," whereas Vygotsky elaborated on the intricacies of the transitions from everyday to scientific concepts—transitions that require remaking lower-level generalizations (cf. also Fischbein's, 1989, work on tacit models). All this remaking happens when expectations consistent with the old knowledge prove to be in a conflict with what is to be learned. Such conflict usually occurs even when seemingly straightforward mathematical ideas, such as rational, irrational, or negative numbers, are introduced (see also the notion of epistemological obstacle: Bachelard, 1938; Cornu, 1991; Sierpinska, 1994; also cf. Sfard, 1994).

At such times—times of the greatest confusion but also of the most significant learning—the main difficulty stems from the need to uproot old metamathematical beliefs that obstruct progress. Historically, science and mathematics developed through successive victories over intuition. Tacit beliefs, however, are unlikely to become an object of the learner's reflection in a spontaneous manner; the help of a more knowledgeable person becomes truly indispensable in this regard. Only such a person can make students' hidden beliefs visible to the students themselves. After all, the child who knows only one world cannot be expected to initiate a foray into an "outer space" before the existence of that space is brought to the child's awareness by a more experienced traveler. In light of this observation, it is no wonder that teachers who refrain from "telling" have been found to have a lowered sense of self-efficacy (Smith, 1996).

What Can Be Done?

Once again, exaggeration and a lack of balance seem to be the main sources of the problem. It has a simple solution: Many possible modes of interaction are possible, and all of them should be given a chance. To the question of how much time should be given to each possible form of learning, no one answer suffices. Actual proportions would depend on the specific needs and preferences of each class and each teacher. Clearly, however, the teacher can hardly ever be satisfied with the role of an accommodating, unobtrusive guide; and one can scarcely imagine a successful student who never spends time grappling alone with difficult mathematical ideas. Whenever students have little chance of finding their way toward a new idea or solving a problem on their own, the teacher should not hesitate to enter the scene with concrete proposals; whenever a new complex issue is being dealt with, students must be given a chance to cope with it alone, without the additional stress of simultaneously communicating with others. True, the student should not be given answers to questions that he or she never asked and has not had an opportunity to understand—but if the ground for a new idea is well prepared through collaborative and individual exploratory work, then a bit of teacher telling will be welcome and effective rather than damaging. For that to happen, the "ban on telling" and the imperative to communicate with others must be considerably relaxed.

The Need for Verbal-Symbolic Interaction

What Do We Know About This Need?

Interaction in learning means communication, and communication means using symbols with which one can try to convey to other people his or her own experiences and thoughts. Language is our most developed symbolic system, and speech is therefore our principal mode of communication. Mathematics, with its own special symbols, may be seen as an extension of the natural discourse.

Thus, if mathematics learning is to take place in an interactive setting, students' ability to "talk mathematics" must be fostered. In other words, if a student is acquiring knowledge mainly through discourse, then the student has to talk. For a person who follows Vygotskian ideas of the relation between thought and speech, the reasons for developing the learner's ability to "talk mathematics" go even deeper. Saying that speech is a "tool of thought" and a vehicle for transporting one's thoughts to others implies that speaking is somehow secondary, indeed ancillary, to thinking. However, if one takes seriously the Vygotskian call to try to understand those phenomena that can be labeled "uniquely human," one is likely to conclude that the distinction between speech and thought will become blurred—especially when one realizes that our conceptual systems, and thus both our human selves and the world each of us lives in, are created through and within the activity of speaking. "Speaking" can be an "inner dialogue," of course, and not just a conversation with others. Vygotsky himself questioned the tenability of approaches that try to focus separately either on speech or on thought. He compared such an attempt to trying to understand the properties of water by investigating its components, oxygen and hydrogen.

A rapidly growing community of researchers who call themselves "discursive psychologists" is currently taking these ideas even further (Edwards, 1993; Edwards & Potter, 1992; Harre & Gillet, 1995). In their eyes, faith in the power of conversation is not an isolated opinion but rather a matter of a worldview according to which all our thinking, with mathematical thinking being no exception, is essentially discursive. The discursive psychologists view our conceptual systems, and thus both our human selves and the world in which each of us lives, as created through and within the activity of speaking, either social or private. Being discursive creatures, we cannot just step out of the discourse. Discourse is where all our cognitive activities start, exist, and come to a close. As such, all these activities are essentially social, and even their occasional appearance as leading to universal and mind-independent results is but a discursive byproduct.

The inseparability of speech and thought is demonstrated in the developmental dialectic of word/symbol use and word/symbol meaning—that is, understanding. One just cannot construct the meaning of a concept before introducing a word or a symbol with which one can think about that concept. The sense of understanding then develops through use of the word or symbol. In mathematics, in the absence of the perceptual mediation characteristic of everyday activities, the developmental interdependence between discourse and its objects—as well as the dialectic nature of the relation between signs (i.e., symbols and words) and their understanding—find their clearest expression. As already noted, reflecting on our actions makes new mathematical concepts emerge. Subjecting actions to reflection means turning them into an object of linguistic expressions. The necessary focusing and talking about actions would simply not occur without naming and symbolizing, or without at least describing these actions in so many words and symbols. The act of naming and symbolizing is, in a sense, the act of inception, and using the words and symbols is the activity of constructing meaning. (For more extensive treatments, see Cobb, Gravemeijer, Yackel, McClain, & Whitenack, 1997; Sfard, 2000a.) If so, the more aware we are of the discursive processes that constitute our mathematical activity, the more chance we have to attain good control of these processes; the better our control, the more effective our students' learning. In short, the question is not *whether* to teach through conversation but rather *how*. Because learning mathematics may be equated with the process of entering into a certain well-defined type of discourse, we should give much thought to the ways students' participation in this special type of conversation might be enhanced. (These issues are discussed at length in Forman, Minick, & Stone, 1993; Hicks, 1997a; see also Lampert & Cobb, this volume; Ball, 1991a.)

How Do the NCTM Standards Address This Need?

The authors of the 1989 NCTM *Standards* document speak extensively about the need to give the student opportunities to "speak mathematics":

> The development of a student's power to use mathematics involves learning the signs, symbols, and terms of mathematics. This is best accomplished in problem situations in which students have an opportunity to read, write, and discuss ideas in which the use of the language of mathematics becomes natural. (p. 6)

As students communicate their ideas, explain the *Standards* authors, "they learn to clarify, refine, and consolidate their thinking" (p. 6). This statement clearly implies that the ability to participate in mathematical conversation is promoted not only for its own sake but also for its expected effects on the process of learning and on the quality of the resulting knowledge.[8] It is therefore only natural that many forms of mathematical conversation are promoted. The *Standards* authors ask for peer-instruction, small-group exploration, and whole-class discussion, where verbal exchange is the primary mode of communication. In addition to interacting orally, students are expected to express their ideas in writing.

What Can Go Wrong?

The importance of promoting mathematical conversation seems beyond question. What may, however, be less obvious is that implementing this seemingly straightforward principle is far from simple. Recent experience has shown that there are many ways to turn classroom discussion or group work into learning opportunities (Ball, 1991a; Cobb et al., 1991, 1993; Lampert, 1990; Schoenfeld, 1996)—but even more ways to turn them into a waste of time, or worse, into an obstacle to learning.

As it happens, futile, useless, and even potentially harmful discursive activities can be observed in only too many mathematics classrooms today. One such case (Sfard & Kieran, 2001) occurred in a teaching experiment I conducted with Carolyn Kieran a few years ago in Montreal. In this study, we watched a long series of interactions between 13-year-old children taking their first steps in algebra. While having a close look at a pair of students working together, we realized that the merits of learning-by-talking cannot be taken for granted. Because of the ineffectiveness of the students' communication, the collaboration we had a chance to observe seemed unhelpful and lacking the hoped-for synergetic quality. A number of reasons may exist for the fact that classroom communications of any kind often fail to meet one's expectations.

First, to charge teachers with the responsibility for the effectiveness of the conversation is easy, but to support them in attaining that goal with helpful advice is difficult. Orchestrating a productive mathematical discussion or initiating a genuine exchange between children working in groups turns out to be an extremely demanding and intricate task. The role of discussion coordinator is particularly difficult. Although leading a debate is almost never simple, when that debate has mathematical content, it may be nearly beyond one's ability to manage. Indeed, coping with a mathematical problem may be demanding even when a person's whole attention is given to the problem. Trying to understand other people's reasoning about the problem is often no less strenuous an activity (Arcavi & Schoenfeld, 1992). Combining one's own mathematical thinking with an attempt to listen to many other solvers seems, therefore, the hardest of a teacher's tasks (see, e.g., Ball, 1997). Not many teachers—not even mathematicians—can be expected to do it really well.

Second, the case for leaving space for "noninteractive" learning, made in the previous section, may now be repeated with a new argument. There is certainly much value in attempting to externalize the internal dialogue the student conducts while learning and solving problems. However, insisting too strongly on the "think aloud" mode may interfere with that internal dialogue, damaging the thought itself.

Third, the above-mentioned teaching experiment by Carolyn Kieran and me showed with particular clarity what psychologists have long known: that communication skills can and should be deliberately fostered.

Finally, for many researchers, the idea of learning through conversation is a natural byproduct of the conception of learning as an initiation into mathematical discourse, which is generally regarded as a fairly well defined type of discourse (Lampert, 1990; Lave & Wenger, 1991; Schoenfeld, 1996). A fuller description of this approach and its different implications is given in the next section. Here I remark only that children cannot reinvent the rules and norms that make a discourse mathematical. For that outcome to occur, help from a teacher is indispensable. This point can be overlooked by the implementers of the NCTM Standards, some of whom may believe that telling children to discuss mathematical problems will suffice to engage them in the right kind of activity. However, how can the children play a game whose rules they do not know?

What Can Be Done?

The remarks above imply that perhaps not enough preparatory work has been done to ensure the success of such ideas as small-group exchange and classroom discussion. First, if conversation among children is to be effective for, and conducive to, learning, the art of communicating has to be taught. How to do so and what exactly students should learn remain questions to which the mathematics education community has yet to give much thought. Further, because the teacher mainly determines whether a given mathematical conversation designed for the purpose of learning will be a success or

[8]Journals and books abound in articles whose titles bear a similar message: "On the Learning of Mathematics through Conversation" (Haroutunian-Gordon & Tartakoff, 1996); "Reflection, Communication, and Learning Mathematics" (Wistedt, 1994); "Journal Writing and Learning Mathematics" (Waywood, 1992); "Reading, Writing and Mathematics" (Borasi & Siegel, 1994); and so on.

a failure (cf. Sfard et al., 1998), teachers must be carefully prepared. Without appropriate training, they cannot be expected to implement their role of initiators, moderators, and discussion coordinators in a truly effective way. How such preparation should be carried out, however, is still unclear. Not only the issue of teaching children the art of communication but also the issue of training discussion leaders is seriously underdeveloped. Last but not least, we must think more deeply about activities around which mathematical conversations are to take place. Although some thought has already been given to the nature of mathematical problems that can be trusted to spur useful exchanges, the topic is still far from well understood and, like the others, requires intensive study. (For more ideas about communication in the mathematics classroom, see Steinbring, Bartolini Bussi, & Sierpinska, 1998.)

The Need for a Well-Defined Discourse

What Do We Know About This Need?

The preceding emphasis on conversation may also be presented as a derivative of a new metaphor for knowledge and knowing, the same one that gave rise to the increasingly popular discursive approach to research on thinking and the mind (Edwards & Potter, 1992; Harre & Gillet, 1995). Rather than merely treat conversation as a secure route to knowing, more and more thinkers *equate* knowledge with conversation (Rorty, 1979; for a survey of relevant literature, see Ernest, 1993). Rorty epitomizes the gist of the idea in the following statement:

> If we see knowing not as having an essence, to be described by scientists or philosophers, but rather as a right, by current standards, to believe, then we are well on the way to seeing conversation as the ultimate context within which knowledge is to be understood. (p. 390)

Of course, while talking about knowledge as a "conversation of mankind," Rorty does not necessarily refer to the most immediate, literal meaning of the component terms. For him, as well as for his many colleagues on both sides of the Atlantic, conversation is a broad metaphorical idea that includes all kinds of human communication—from an interactive, real-time oral exchange to the production of written texts (ordinary mail and electronic correspondence, writing articles and books, etc.). *Discourse* is another term widely used these days in a similar sense (Foucault, 1972). In the present context, the word has a very broad meaning and refers to the totality of communicative activities practiced by a given community; to avoid confusion with the everyday narrow sense of the term, some authors such as Gee (1997) propose capitalizing it: *Discourse*. Within the discursive research framework, different communities—the mathematical community being one of them—are understood to be characterized by the distinctive discourses they create. Of course, discourses must also be understood to be dynamic and ever-changing entities; thus, the tasks of determining their exact identities and mapping their boundaries are certainly not as straightforward as a researcher would like. Moreover, discourses of different communities are constantly overlapping, thereby resulting in their incessant crossbreeding. These difficulties notwithstanding, the notion of discourse proves clear enough to spur a steady flow of highly informative research that has the power of eliciting hitherto unnoticed aspects of learning.

In substituting the word *discourse* for *knowledge*, philosophers make salient the central role of speech in human intellectual endeavor. For many researchers, studying mathematical communication has become a task almost tantamount to studying the development of mathematical thinking itself (see, e.g., Bauersfeld, 1995; Forman, this volume; Lampert & Cobb, this volume; Morgan, 1996; Pimm, 1987, 1995). However, the shift of focus that is evident in the renaming goes further than that. First, it may count as an act of "putting body back" into the process of knowledge construction. Knowledge viewed as an aspect of a discursive activity is no longer a disembodied, impersonal set of propositions whose exact nature is a matter of "the true shape" of the real world; rather, it is a human construction. Further, the word *discourse* seems more comprehensive than *knowledge*. Researchers who speak about discourse are concerned not only with those propositions and rules that constitute the proper body of knowledge but also with much less explicit rules of human communicative actions that count as proper ways of constructing and conveying that knowledge. One may therefore speak of *object-level rules* of mathematical discourse and of *metarules*, which are superior to the former, even if only implicitly. Interest in these and related issues can be found in many places in recent literature, even if the authors do not mention metarules explicitly (see, e.g., Bauersfeld, 1995; Bromme & Steinbring, 1994; Cobb et al., 1993; Krummheuer, 1995; Lampert, 1990; Putnam, Lampert, & Peterson, 1990;Voigt, 1985, 1994, 1995, 1996). The term *metarules* seems to include what Cobb and others (1997) call *sociomathematical norms* (for a more elaborate treatment of this subject, see Sfard, 2000b).

For example, object-level rules are involved in the concrete ways of proving from axioms practiced by today's mathematicians, whereas metarules include the belief in proof as a formal derivation from axioms and the mathematicians' right to establish axiomatic systems in any way they wish, provided the systems are free of contradiction. This rule is relatively new; until at least the eighteenth century, only those propositions counted as axioms that were believed to be universally and objectively true. Metarules regulate the ways in which we argue about our mathematical claims (Krummheuer, 1995) and enable us to judge what is and what is not "mathematical" or "mathematically sound." Using these rules, we also decide, usually in an instinctive way, what kind of action on our part would count as proper in the given context and what behavior would look out of place. The way we speak and interact with others conveys these unwritten rules of the game. In other words, students usually learn the "rules of the mathematical game" without conscious effort, simply by participating in the mathematical discourse. As Lampert (1990) put it, they would not learn the rules "simply by being told what to do anymore than one learns how to dance by being told what to do" (p. 58).

Of course, the metarules that characterize mathematical discourse overlap with those that regulate any communication in a classroom (Cazden, 1988) and beyond. Because the metarules are tacit, their grip on our thinking is particularly strong; and yet we are as sensitive to their violation as we are to the violation of explicit laws. This strong influence is probably one reason we have such difficulty with the special "insight" problems that cannot be solved within the confines of the ordinary rules of mathematical discourse; indeed, the rules of discourse may create mindsets. When students believe instinctively that the rules have been changed, they are truly exasperated. Their frustration stems from their inability to account for the uneasiness they feel. Not being fully aware of the exact shape of the violated rules or of the reasons these rules are the way they are, all the students can do is say, "It's not fair." If pressed, they will sometimes add that what they are required to do is, in some inexplicable way, not what they were taught to do. Saying that the game is not "fair" is their way of describing a situation that, from their point of view, is too poor in framing clues to enable further discursive decisions.

Too much rigor is paralyzing, but so is a complete lack thereof. Discursive metarules may be confining, but they are indispensable. In fact, these rules constitute the discourse and must be as much an object of learning as the object-level facts and procedures of mathematics are. Without these rules, the learning of the object-level mathematical content would be seriously impaired, if not altogether impossible. Although they are usually acquired without being directly taught, these rules give the student the sense of understanding the game being played and the ability to take part in it. To learn, the student must believe that the discourse in which she or he is expected to participate is well defined and has an inner logic and clearly delineated borders. In a sense, therefore, the need to know the rules of the discourse to be mastered is a participationist counterpart of the acquisitionist need for structure discussed previously.

The question that should be answered now is where the rules of the classroom mathematical discourse should originate. On the surface, the answer may be simple: Having to do with mathematics, this particular school discourse should be as close as possible to the discourse led by mathematicians. (For insightful essays on different aspects of professional mathematical discourse, see, e.g., Davis & Hersh, 1981.) That the issue is not that simple is evidenced by the experience with the new math movement that, in the late 1950s and the 1960s, tried to transport mathematicians' discourse directly from universities to school classrooms (cf. Brown, 1997). This attempt could not be fully successful simply because its conceivers did not take into consideration the existence and adverse effects of the tension between the professional mathematical discourse and the other discourses in which teachers and children had been participating. Since then, all parties involved recognized the inevitability of *didactic transposition*; that is, of adjusting the school discourse to the needs and possibilities of the learning child.[9]

Nonetheless, the need to preserve certain basic characteristics of the professional discourse in school is unquestionable. After all, the decision to teach mathematics to everyone results from recognizing the importance of this discourse. Today, learning mathematics is often conceptualized as an initiation into the "community of practice" (Lave & Wenger, 1991). According to this vision, mathematics students are beginning practitioners or, as Streefland (see Sfard et al., 1998) calls them, "junior researchers." If we want them to act accordingly, our mission as teachers is to turn the classroom into a "community of inquiry" (Schoenfeld, 1996)

[9]The term *didactic transposition* was coined by Chevallard (1985, 1990; see also Sierpinska & Lerman, 1996) and, in its original version, referred to the fact that professional knowledge must change in accord with the needs of the *institution* in which that knowledge is being practiced; in the language of discourse, we are concerned here with the transformation of discourse in accord with the needs and requirements of different communities.

that would be close in its norms and practices to that of expert practitioners. In this vein, Lampert (1990) claims that the proper goal of teaching should be "to bring the practice of knowing mathematics in school closer to what it means to know mathematics within the discipline" (p. 29). The words "the practice of knowing mathematics" signal an essential difference between the new math approach and the current one. Today the focus is on the discursive process rather than the product; that is, it is on learning to mathematize (Wheeler, 1982) rather than on "acquiring" its result. In other words, the emphasis is on the metadiscursive strata that, in spite of their relative invisibility, give mathematical discourse its unique identity. This approach is germane to what Hicks (1997b) calls "deliberate genre instruction" and to what Cobb et al. (1993) call "talking about talking about mathematics." If this new goal is attained, it will satisfy students' need for a well-defined discourse.

How Do the NCTM Standards Address This Need?

The language of discourse is present in the *Standards* documents and especially in the volume devoted to teaching (NCTM, 1991). The authors explain:

> Discourse refers to the ways of representing, thinking, talking, and agreeing and disagreeing that teachers and students use to engage. . . . The discourse embeds fundamental values about knowledge and authority. Its nature is reflected in what makes an answer right and what counts as legitimate mathematical activity, argument, and thinking. Teachers, through the ways they orchestrate discourse, convey messages about whose knowledge and ways of thinking and knowing are valued, who is considered able to contribute, and who has status in the group. (p. 20)

This definition tries to alert the implementers to the existence of implicit rules and norms, as well as to the indirect discursive ways in which these special contents are being communicated. The NCTM *Standards* documents in their entirety may count as a comprehensive attempt to explicate and define the components of a proper classroom discourse, as understood by the authors. In this context, we should note that quite unlike more traditional documents, the *Standards* make space for certain concrete metalevel rules, such as requiring that the student's own experience and reasoning, rather than the teacher or the textbook, be regarded as the main source of mathematical knowledge and certainty (see, e.g., NCTM, 1989, p. 129).

What Can Go Wrong?

"Mathematics presented with rigor is a systematic deductive science but mathematics in the making is an experimental inductive science," says Polya (1957, p. 11). This statement means, among other things, that the mathematical "discourse of doing" is much more natural and less constraining than the discourse of reporting. Even the discourse of doing, however, is rather disciplined in comparison with other discourses, whether everyday or scientific. The participants hope to attain unquestionable agreement by deliberately and explicitly imposing some unequivocal rules on themselves. By greatly restricting the admissible ways of expression, mathematicians try to ensure that the exact shape and content of their discourse will be treated as externally imposed—that is, independent of the tastes, judgments, and preferences of the interlocutors. Needless to say, one may rightly expect that classroom mathematical discourse will be a greatly relaxed, much less rigorous, and more "popular" version of the mathematicians' discourse. Sometimes, however, the relaxation of rules may be so radical that it starts "redefining what constitutes mathematics" (Wu, 1997, p. 946). A careful analysis of the NCTM Standards' requirements shows that, contrary to the researchers' recommendations quoted above, the new school mathematics may indeed turn out very differently from "what it means to know mathematics within the discipline" (Lampert, 1990, p. 29). While everybody seem to be aware of the inevitability of didactic transposition, what happens now may be going beyond what parents and mathematicians are prepared to tolerate.

I next give a very brief and incomplete account of the ways in which reform-inspired classroom discourse may be different from that of the professional (see also Brown, 1997; Love & Pimm, 1996). As I attempt to show, certain values and norms professed by the NCTM Standards may be at odds with the norms that regulate traditional mathematical discourse. This normative incompatibility can be seen mainly on the discursive metalevel, the level of beliefs about one's rights and obligations as a participant in mathematical discourse.

First, educators and mathematicians are often divided on the issue of what counts as truly mathematical activity. The already mentioned tendency to always look for real-life situations and to eschew dealing with "distilled" mathematical content contradicts what is often believed to be the very essence of mathematization. After all, mathematizing is almost synonymous with abstracting. One may say that in the most fundamental way, mathematics is about "flying high" above the concrete and about classifying things according to features that cut across contexts. Mathematicians would claim (see, e.g., Wu, 1997) that the ability to strip the bones of abstract structures of the flesh of concrete embodiments is the main source of the unique beauty and strength of mathematics. When we restrict ourselves to "contextualized mathematics," we are tying mathematics back to the concrete and particular—and losing the gist of mathematical creation. In addition, the great emphasis on putting mathematics into a real-life context creates a utilitarian atmosphere, foreign to true mathematical discourse. Wu (1997) deplores the disappearance of "the spirit of intellectual inquiry for its own sake" (p. 948).

In a similar way, educators and mathematicians would argue over the admissibility of nonanalytic arguments employing visual means. Although this type of argument is often recognized in schools as sufficient, it is still seen by mathematicians as helpful but far from decisive or final (Davis, 1993; Rotman, 1994; Sfard, 1998b). More generally, the NCTM Standards put a great premium on heuristics, and quite often, the only instruction given to students to engage them in an activity of proving is, "Convince your partner." Because of the practically unlimited freedom in the choice of the ways of "convincing," the unique features of the mathematical discourse of proving may be lost. That loss is probably the reason Wu (1997) speaks of a "cavalier manner in which the reform treats logical argument" (p. 947) while deploring "suppression of precision" (p. 949).

Further, school mathematical discourse engendered by the Standards turns out to be highly personal. Students are invited to speak and write about their mathematical experience in any way they choose using first-person language. The "subjectivization" of the discourse may be taken to an absurd—and, of course, unintended—extreme, however, if the Standards' call for open-ended problems "with no right answers" (NCTM, 1989, p. 6) is misinterpreted as saying that "any solution goes." Such subjectivization stands in a stark contrast with classical mathematical discourse, whose hallmark is an uncompromising impersonality that imbues the discourse with the air of objectivity and "mind independence." A mathematician for whom Platonism is "a working state of mind" if not an outright article of faith (Sfard, 1994b) may find the personal note detrimental to his or her whole project.

In sum, the mathematical discourse that develops in classrooms in which the NCTM Standards are followed is rather different from the professional discourse of the working mathematician. This difference conflicts with the Standards' declared goal of making the student "a legitimate peripheral participant" in true mathematical discourse. Nevertheless, one may deem the state of affairs to be fully justified. After all, the disparities mentioned above are an inevitable result of an attempt to imbue the learning of mathematics with more progressive values—and above all, with respect for the student's ways of thinking. The relaxation of rules is further justified in view of the great diversity in students' needs and capacities. Many would claim that such a change does not have to be acceptable in the eyes of professional mathematicians to be sanctioned as pedagogically sound and necessary. Further, "expert practitioners" must agree that a reasonable compromise is the best possible solution. Indeed, what is the use of trying to teach strict rules of the professional mathematical discourse if almost no one can learn them?

A compromise, on the one hand, can count as reasonable only so long as the discourse we are left with is well defined, intrinsically coherent, and generally convincing. If, on the other hand, we simply reject some of the basic conventions without replacing them by alternative rules, or if the changes we make are accidental and inconsistent, we may end up with a discourse that is rather amorphous and ill defined, and as such simply cannot be learned. Once again, a carelessly exaggerated application of promising ideas would prove detrimental to good pedagogy.

What Can Be Done?

Let me begin by showing that the problem we face is intrinsically complex and that its solution is probably not just a matter of the goodwill of NCTM Standards legislators and implementers. Indeed, mathematicians' and parents' uneasiness about the somewhat paradoxical reincarnation of the most disciplined of discourses as a "fuzzy" school subject (Cheney, 1997) is not a simple outcome of unfortunate proportions. It has a deeper, more elusive reason, one that cannot be either dismissed or easily resolved. On closer examination, it turns out that because of our keen wish to respect students' need to understand, we feel compelled to compromise the very feature of the mathematical discourse that is the basic condition of its comprehensibility: We are sacrificing its inner coherence.

To understand this claim and its didactic implications, we must be more explicit about the reasons that professional mathematical discourse is the way it is. The first thing to realize is that this discourse is geared to the attainment of universal, unquestionable consensus. Therefore, the mathematicians' mathematics may be described as "the science of perfect communication."

Awareness of this goal is necessary to understand and justify the rigorous, confining, and, for too many people, simply contrived metarules mentioned earlier. What mathematicians find motivating, however, does not have to have any appeal for the student. It is highly unlikely that every student will understand and endorse the dream of communicating beyond any doubt. Moreover, even if the learner were able to empathize with mathematicians and their special needs, his or her criteria for judging certainty—or the absence of doubt—are unlikely to be like those of the professionals. If, on the one hand, we add to this argument those against "the ideology of certainty" purportedly inherent in professional mathematical discourse (Borba & Skovsmose, 1997), then motivating students to study school mathematics with the idea of "impeccable communication" becomes a dubious and unlikely project. On the other hand, giving up this kind of motivation means depriving mathematical discourse of its defining feature.

As so often before, we face a seemingly unsolvable didactic dilemma—and the difficulty goes even deeper than that. An ever-present concern for communication leads to a highly structured, inflexible discourse. As a result, mathematics becomes a well-designed, highly organized system that cannot be arbitrarily modified in one place without creating problems in another. Just as the game of chess would become uninspiring and impossible to learn if we arbitrarily replaced or removed some of its rules, so mathematics would become meaningless following too careless a "relaxation." For example, the idea of a negative number cannot be fully understood within a discourse regarded as describing the real physical world; indeed, nothing in the world would require the rule "minus times minus is plus."[10] The only way mathematicians can account for this property in a rational way is to regard it as a dictum of the field axioms for numbers. School discourse, however, is not—and probably cannot be—one that recognizes the supreme authority of axioms and of logical consistency.[11] Similarly, the request for rigorous definitions that may count as "truly mathematical" cannot sound convincing without being related to the idea of mathematical proof; and the mathematical rules of proving, in turn, cannot be understood without an agreement that the ultimate criterion of a proper argumentation is the logical bond between propositions and not the relations between those propositions and physical reality. Evidently sensitive to this interdependence of rules, some educators prefer to view the "relaxation" as a package deal and propose giving up any kind of mathematical rigor. (For a review of educational approaches to the issue of proof and proving, see Hanna & Jahnke, 1996.)

Another metalevel rule of today's mathematics simply cannot be preserved in school mathematics, but without it, the whole discourse becomes problematic. For mathematicians, the overall coherence of mathematics is the ultimate source of its justification and meaningfulness. In this discourse, the meaningfulness of a concept stems from its being an element of a consistent system. For students, who have no means to appreciate this overall coherence, the small isolated pieces successively encountered in the course of learning may always remain somewhat arbitrary.

Because of these insoluble dilemmas, it is extremely difficult to decide on appropriate measures of discipline and rigor in school mathematical discourse. Many solutions have been suggested, and each may be worth some thought. Aware of the basic unsolvability of the problem, some people suggest a radical change in the general approach to school mathematics, or at least to high school mathematics. Building on an analogy with poetry or music, they would propose that, beginning at a certain level of schooling, we teach students about mathematics rather than engage them in doing mathematics. After all, just like poetry and music, mathematical techniques do not have to be fully mastered to be appreciated as part of our culture (see, e.g., Devlin, 1994). It is far from obvious, however, that this proposal is workable: Although one can certainly appreciate and enjoy poetry and music without being able to produce any, the same is probably not true with mathematics. Another, no less radical, solution would be to turn high school mathematics into an elective subject.[12] Perhaps the most realistic suggestion, however, is that we try to be more careful in the choice

[10] On the face of it, this claim may be contested, given that many ideas have been proposed to explain and model negative numbers (e.g. there is the model of movement where time, velocity, and distance can be measured in negative as well as positive numbers; numbers may be represented as vectors, etc.). And yet, on closer examination, all of these explanations and justifications turn out to be derivatives of the same basic decisions about preserving certain former rules of numbers while giving up some others; these fundamental choices are exactly the same as the ones that find their expression in the acceptance of axioms of a numerical field as a basis for any further decision, and they must be (tacitly) accepted prior to any justification.

[11] That is probably the reason behind the once famous "didactic" principle "Minus times minus is plus; the reasons for this we don't discuss." It may also explain the bitter complaint by the French writer Stendhal that his difficulties with negative numbers "simply didn't enter [his teacher's] head" (quoted in Hefendehl-Hebeker, 1991): Without realizing it, Stendhal and his teacher were participating in different discourses.

[12] Some people would claim that the added value of such a move would be an overall improvement in teaching: We all would have to try harder to make the subject attractive in students' eyes. The relationship between popularity and the quality of instruction is, however, far from obvious.

of the rules we modify. While trying to decide, we must be aware of possible pitfalls in the undertaking; we must ensure that the proposed norms are not in conflict with one another and must be careful not to take out ingredients without which the whole construction might collapse. Further, being more explicit with students about the metalevel rules may be of considerable help as well. The fact that this explicitness can be achieved even at the most elementary levels has been shown by Lampert (1990), Ball (1997), and Cobb and his colleagues (Cobb et al., 1991, 1993), all of whom have studied the ways that teachers and children can negotiate the multifarious norms of their classroom discourse.

The Need for Belonging

What Do We Know About This Need?

In the previous section I dealt at length with cognitive challenges likely to motivate learners. Next, I turn to a different type of driving force for learning—motivation, which is brought to the fore by the metaphor of learning as participation.

The term *participation* is almost synonymous with "taking part" and "being a part," and both expressions signal that learning may be viewed as a process of one's integration into a greater social unit. Just as different organs combine to form a living body, so learners contribute to the existence and functioning of the community of those who speak and do mathematics. When a given class of students learns mathematics, it acts as this kind of community; research mathematicians constitute another.[13] The metaphor of learning-as-participation shifts the focus from what "goes into the student's head" to the evolving bonds between an individual and other members of this community, thus giving prominence to the aspect of mutuality that is characteristic of the part-whole relation. Indeed, this metaphor makes salient the dialectic nature of the learning interaction: A community and an individual affect and define each other. On the one hand, the very existence of the community is fully dependent on its members. On the other hand, the identity of an individual, like an identity of a living organ, is a function of her or his being or becoming a part of the greater unit. That belonging is, in fact, what one means when saying that human beings are essentially social creatures. No wonder, then, that the desire to count as a member of a particular social group is a powerful driving force behind most of our actions (cf. Bruner, 1990).

If one thinks in these terms about learning mathematics, then probably the most basic condition for learning is that the student has a desire to belong to the mathematizing community. The question that must be asked now is, What can make the mathematizing community attractive in the eyes of the learner? The most natural response is to say that if children are to be attracted, they must know that being a part of that community can enable them to feel good about themselves. That was certainly not the case for most children in the great majority of traditional mathematics classrooms, but the situation may be changing now.

Other factors to consider in the present context are cultural values. Whether one wishes to be "a member" depends, to great extent, on how much that membership is valued within one's culture. By *culture*, I mean both the local and global levels: the student's immediate social surrounding—her or his family, friends, and school—and the society in which the student lives. Needless to say, there is a strong interdependence between these two cultural environments, and the intricate dialectic between them is a topic for anthropological research.

Because the cultural context changes considerably across time and space, the forces that shape students' desire to learn mathematics vary from place to place and from one historical moment to another. There is little doubt that being mathematically literate was once very highly valued in Western culture and is probably still seen that way in many Far Eastern countries. This esteem may partially explain the consistently high results obtained by students from Singapore, Korea, Japan, and Hong Kong on international comparative tests like the Third International Mathematics and Science Study (Beaton, Mullis, Martin, Gonzalez, Kelly, & Smith, 1996; Mullis, Martin, Beaton, Gonzalez, Kelly, & Smith, 1997). In contrast, the disappointing results of students in many Western countries, including students in the United States, alert us to the possibility that in these parts of the world, the wish to participate in mathematical discourse is in decline.[14] Why that decline might be occurring and how the overall cultural values of Western

[13] For the sake of the present discussion, the "mathematizing community" must be understood broadly as a community of people successful in doing mathematics and not necessarily as the community of professional mathematicians. It is also important to understand that what is called here a *community* is not much more than a group of people who, although they act in similar ways, do not have to function jointly as a team.

[14] That Far Eastern students persistently outperform their Western counterparts has been widely documented; see, for example, Stigler, Lee, and Stevenson (1990), or Stevenson and Stigler (1992). The difference between the cultures of mathematical classrooms in Japan and in America, which may be partly responsible for the disparity in achievements, has been described in Stigler, Fernadez, and Yoshida, 1996.

society may be partially responsible for this phenomenon are discussed later. First, let us consider how the NCTM Standards aim to enhance students' mathematics learning by building on their need to belong.

How Do the NCTM Standards Address This Need?

The NCTM Standards are a major step toward turning the mathematizing classroom into a proper place for children. If the reform is carried out according to the intentions of those who conceived it, a mathematizing community should become an environment in which the child is respected, feels free to speak her or his mind, can succeed on her or his own terms, and has the same chance as anyone else to be creative and make a substantial contribution. Moreover, thanks to the nature of the new type of mathematical activities suggested by the Standards, membership in this community may be a source of much excitement and personal satisfaction.

The issue of cultural importance of mathematics is explicitly addressed in the Standards. "Learning to value mathematics" (NCTM, 1989, p. 5) is one of the principal goals of the mathematics curriculum. This goal is further explained as follows:

> Students should have numerous and varied experiences related to cultural, historical, and scientific evolution of mathematics so that they can appreciate the role of mathematics in the development of our contemporary society and explore relationships among mathematics and the disciplines it serves: the physical and life sciences, and the humanities. . . . Even today, as theoretical mathematics has burgeoned in its diversity and deepened in its complexity and abstraction, it has become more concrete and vital to our technologically oriented society. (pp. 5–6)

Obviously, requiring an appreciation of mathematics is not enough to make it happen. Students' values and opinions are also, and perhaps mainly, shaped by implicit messages about the importance of mathematics; these messages are conveyed by activities that students are supposed to perform and are inherent in the rules of classroom discourse that are expected to develop. In accord with the goals quoted above, the Standards recommend an emphasis on applications of mathematics. The tendency to embed mathematics in an external—everyday and scientific—context means that the usability of mathematics is viewed as the principal reason for students' possible attraction to the "mathematizing community." Other potentially appealing aspects of the community itself, such as the inherently satisfying nature of its activity or certain admirable characteristics of those who manage to become its members, are not explicitly mentioned.

What Can Go Wrong?

Values and desires are only rarely a matter of a deliberate choice, so they cannot be easily influenced by such rational arguments as real-life applicability of the subject we wish to teach. Human beliefs, intentions, and emotions are complex phenomena, dependent on a great many interrelated factors (see McLeod, 1992) and not the result of conscious decisions. Imbuing values and shaping attitudes are, therefore, neither straightforward nor easy, and one can fail in these attempts in far too many ways. Here is a striking example of such a failure. Although it comes from Israel rather than from the United States, its message and implications go well beyond the borders of any particular country.

A few years ago, an Israeli teacher found a letter among the papers that one of her students handed her at the end of a matriculation examination (Figure 23.3).

Dear Examiner,

I cannot possibly pass this test, because I have no idea how to solve all these problems. I have other, more substantial problems in my life, and solving these other problems seems to me much more important. Instead, I am given all these numbers and formulas that are only making my life more complicated and can help with nothing. Five hours of my life every week I have been obliged to devote to memorizing these things. For what? Will I be a better person if I know how to investigate function? Will I become more moral if I prove something by induction? I don't think so.

Five hours of all this every week, but not even one hour on the latest news during the whole year. Not even one conversation about the problems that plague this modern society of ours. I have much to say, but it doesn't find its expression in trigonometric formulas.

It's absolutely clear to me that I will complete my matriculation examinations one day, even though I will do it, in a sense, against my best judgment. For me, this will be a surrender to distorted social conventions, conventions of a society in which one can only succeed by floating with the stream, by getting high grades, and by collecting degrees.

Figure 23.3. A letter to the examiner written by a student on her matriculation exam form.

The message of the letter was unequivocal: The student did not succeed in mathematics—not because she *could not* do it but rather because she had no *genuine interest* in doing it. Mathematics, as such, did not seem to her of any relevance whatsoever, and if she ever decided to learn it, she would do so not for the sake of the knowledge she would acquire but rather to make the best of the social game in which this kind of achievement is a condition for success. This letter might be interpreted as a result of poor teaching, teaching that failed to convey the message about the practical and cultural importance of mathematics. However, that interpretation might be too easy, and it would also be quite unfair toteachers. The student quoted above and the rest of her class did not study in a desert, and the way they were looking at mathematics was a result of both internal and external cultural influences. Today, the values we wish to promote through mathematical discourse, as well as those we *have* to promote to make that discourse possible and successful, may be in conflict with the values of the society within which the learning is taking place.

Membership in a mathematizing community, like any other, may be desired either for its own sake or only as a means to another, more highly rated membership. That the first possibility is a real option is well documented in testimonies of mathematicians (e.g., Hadamard, 1945; Hardy, 1934; Ulam, 1976) who present their career choice as a result of their admiration for the kind of activity in which the mathematizing community is involved. In a culture that values intellectual skills and appreciates knowledge for its own sake, such membership is highly prestigious and gives a person a sense of personal value and satisfaction. When intellectual skills are valued for their own sake, people are prepared to admire mathematicians without having any access to what they are saying or without being shown any immediate evidence that what mathematicians are doing is useful. In such a society, the more esoteric the mathematicians' work, the more respectable they may appear in the eyes of laypeople. Incomprehensibility is taken as evidence of intellectual superiority (cf. Davis & Hersh, 1981).

Whether full-fledged membership in the school mathematizing community will be coveted by children for its own sake depends to great extent on how "child friendly" this community turns out to be. Although not stated explicitly, being in a mathematics classroom may have meant lowered self-esteem and inadequate social positioning for the author of the letter in Figure 23.3, who also apparently did not find doing mathematics either enjoyable or rewarding. In Western society, many consider this latter issue to be crucial and claim that something must be done to make learning mathematics more pleasurable. This basically convincing idea may, however, easily be taken to a dangerous extreme. In a society that lets the need for pleasure speak louder than any other, gratification is expected to come immediately and without effort. As I have argued extensively throughout this chapter, this kind of satisfaction is impossible in mathematics; those who try to turn learning into sweat-free fun are, in fact, obstructing rather than promoting it (see also Thomas, 1996). According to cross-cultural studies, Japanese students, known for their superior achievements in mathematics, display rather negative attitudes toward the subject (McLeod & Ortega, 1993)—a result that implies that one does not have to enjoy studying mathematics to be successful at it.

In a pragmatic, practical minded society, the vision of disinterested pursuit of knowledge may have less allure than in the past. In such a society, people tend to regard learning as a means rather than an end in itself. They value knowledge mainly for its applicability, that is, for what it may give them in other domains of life. Thus, young people may wish to learn mathematics for the sake of the doors it will open for them in the future. Indeed, many of my university students, when asked the reasons for their decision to major in mathematics, give as an argument their belief that "being able to learn mathematics means being generally smart" and "those who succeed in mathematics will succeed in life." These students, therefore, view fluency in mathematical discourse as nothing but a ticket to other, more prestigious "memberships."

This conception is artificially reinforced by the fact that mathematics is widely used today as a selection tool. Mathematical tests serve as a greatly feared sieve at the entrance to universities, constituting a rather arbitrary gateway to many different careers. In a society that simultaneously values critical thinking, an inability to see any other reason for learning mathematics may be fatal to one's mathematical growth; such is certainly the message conveyed by the letter in Figure 23.3. Moreover, the kind of mathematical competence required by the tests is not necessarily the kind of knowing that is cultivated in classrooms that follow the recommendations of the NCTM Standards. A test may be built only on easily measurable skills, whereas the Standards emphasize understanding and creativity. Therefore, students' desire to be members of today's mathematizing community may be seriously reduced by their conviction that it will not be helpful in attaining their future goals. This desire may be diminished even further by old methods of assessment that persist despite the Standards' explicit call for change. Much damage is being done by tests seemingly still aimed at the type of mathematical competence that is no longer viewed as genuinely important.

More generally, in a society that values personal autonomy and critical thinking, the question "Why should I learn this?" must usually be answered for learning to be effective.[15] Convinced of one's right to decide about one's own life, people are no longer ready to accept educational decisions that do not correspond to their norms and beliefs. Norms and beliefs, in turn, are inert, changing at a much slower pace than our educational thinking. This disparity results in a constant tension that may interfere with students' desire to learn the kind of mathematics that we want them to learn. Although today's parents want their children to know mathematics, from the point of view of reformers they may want it for all the wrong reasons. As a result, the mathematics they want their children to learn may be the wrong kind of mathematics.

To summarize, the hidden messages of the many discourses in which the student is participating may be at odds with one another. This kind of conflict may hinder any attempt at improving mathematics learning in Western countries. Some writers go so far as to say that mathematical discourse, as practiced in our society, contradicts the basic values of democracy (Borba & Skovsmose, 1997; Walkerdine, 1994). The difficulty of ensuring interdiscursive consistency may also account for the well-known phenomenon of irreproducibility of educational success. When transplanting educational ideas from one discursive setting into another, one may unwittingly change the context to such an extent that the ideas lose their original meaning. After all, meaning is inherently context dependent; in the most extreme cases, what was perfectly in tune with the general culture of one setting may stand in an opposition to the culture of another setting. The resulting "double bind" effect (Bateson, 1972; Harries-Jones, 1995), or the conflict between implicit and explicit normative messages, turns an originally successful educational approach into something ineffective, if not outright harmful. Thus, for example, a curriculum built on the principle of learning by inquiry may do wonders in a relaxed experimental environment—and then lose all its power in a class whose only concern is to pass external examinations with as high marks as possible. The fact that success lies in the relation between educational ideas and their context rather than in the ideas as such is only too often overlooked by those who wish to bring about change.

What Can Be Done?

In light of the preceding remarks, two obvious conclusions are that students' drive to learn mathematics may be impaired by external influences and that teaching the subject may be difficult in a society that does not value what must be valued if "membership in mathematizing community" is to have any appeal at all. The extent to which the general culture can be influenced from within the mathematics classroom is an open question. Such influence may be particularly difficult to exert in situations in which, according to Comiti and Ball (1996), "mathematics is not a part of the wider culture, or valued as a part of general literacy" (p. 1135). It is also difficult to say how much can be done within a classroom that is not embedded in favorable cultural surroundings.

Even so, one may suggest some promising directions for improvement. Thus, for example, we may try to bring back into the classroom the idea of gaining knowledge as an activity that should be valued for its own sake. After all, this activity is a component of any meaning making, and the need for meaning is the primary cause for our actions. Further, we may try to celebrate mathematics as perhaps the best manifestation of our unique human abilities. We may cultivate it as a part of our culture no less important than poetry or music. Relating the history of the "mathematizing community" and telling life stories of its most prominent members and possible role models may make this community much more appealing in the eyes of the students. Above all, however, we should probably see to it that mathematics is learned not simply because of its use as an often-arbitrary tool of selection. In today's society, this goal may sound utopian but nonetheless is worth trying.

The Need for Balance

What Do We Know About This Need?

To meet learners' multifarious needs, the pedagogy itself must be variegated and rich in possibilities. The learning individual is a complex creature with many needs that must all be satisfied if the learning is to be successful. The principle of a "balanced diet" is therefore as much in force for our minds as it is for our bodies. The need for balance, and the only too frequent lack thereof, has been the leading motif of the previous sections, in which I have tried to show how the hegemony of a single approach is bound to result in damaging rather than helping (cf. Tietze, 1994).

---

[15] That this was not always the case was recently brought to my attention by a university student who wrote in her essay: "As a mathematics learner in school I strived to succeed without asking myself whether and why it is important to know mathematics. I know students are used to asking such questions these days, but we did not ask them at that time." The impact of the need for rationalization on the effectiveness of learning is not as simple a matter as it may appear, and requires thorough investigation. More generally, the question of whether the wish "to belong" has any rational roots is certainly worth pursuing.

How Do the NCTM Standards Address This Need?

The diversity of students' needs and of their ways of thinking is stressed repeatedly in the NCTM *Standards* documents. Because of the numerous and often conflicting needs of each learner, the issue of teaching would be complex enough even if there was only one person to be taught (Arcavi & Schoenfeld, 1992). The Standards, however, are supposed to address the needs of the entire U.S. population, known for its great individual and cultural diversity. Indeed, the issue of equal opportunity lies at the heart of the reform effort, and not just because of its moral importance: "We cannot afford to have the majority of our population mathematically illiterate: Equity has become an economic necessity" (NCTM, 1989, p. 4).

Thus, although the necessity for balance is certainly acknowledged by the *Standards* authors, they must admit that finding this balance has to be left in the hands of the implementers. They repeatedly stress that good teaching cannot be ensured by universal prescriptions: "Because teaching mathematics well is a complex endeavor, it cannot be reduced to a recipe for helping students learn" (NCTM, 1991, p. 22). To enable each person to find his or her own balance, the *Standards* documents try to be inclusive rather than exclusive. The attempt to leave many options open can be felt throughout their pages.

What Can Go Wrong?

Not every reader of the *Standards*, however, may be sensitive enough to notice this openness. That things can go wrong in spite of the authors' good intentions was the theme of all the previous sections. Impatient to put the good ideas to work, interpreters may forget that there is always more than one way to translate general curricular and pedagogical principles into concrete classroom strategies. Zealous to bring about real change, they often start believing that the old and the new are mutually exclusive. This misunderstanding is why the profound constructivist idea of a learner as a builder of his or her own knowledge is sometimes trivialized into a total banishment of "teaching by telling"—and it is how the call to foster mathematical communication may turn into an imperative to make cooperative learning mandatory for all. It is also how the exhortation to teach mathematics through problem solving can bring a complete delegitimatization of instruction that is not problem based—and it is how the request to make mathematics relevant to the student can result in the rejection of mathematics that does not come in a real-life wrapper. Despite the high quality of the ingredients, the meal we cook in this unbalanced way must, sooner or later, prove harmful rather than healthy. Besides, such extremism does not take into account the diversity of students' needs, the very thing the Standards ask us to remember all along the way.

Still, such extremism is a rather natural phenomenon at those junctures in history in which one is trying to make a *real* difference. Also, finding balanced ways of teaching that would likely satisfy the numerous and greatly varying needs of the learner is a dauntingly intricate task. In particular, it is extremely difficult to ensure "the fit between standard-based recommendations and the realities of culturally, linguistically, and economically diverse students" (Tate & D'Ambrosio, 1997, p. 650). Above all, however, as I have tried to show throughout, advocates of a profound change in mathematics face countless dilemmas inherent to the mission itself. The educational process is a very delicate give-and-take between many needs that cannot be satisfied simultaneously. Our project is no exception: It is "riddled with conflicting aims" (Comiti & Ball, 1996, p. 1133) and numerous "vicious circles." On the most general level, our wish to honor students' need for autonomy collides with a certain coercion that is intrinsic to any educational enterprise. More specifically, in the name of democracy the Standards require teaching mathematics to all:

> If all students do not have the opportunity to learn . . . mathematics, we face the danger of creating an intellectual elite and a polarized society. The image of a society in which a few have the mathematical knowledge needed for the control of economic and scientific development is not consistent either with the values of a just democratic system or with its economic needs. (NCTM, 1989, p. 9)

Let me briefly recapitulate some of the dilemmas we face when trying to design balanced instruction. Our problems begin with the circular nature of the process of constructing meaning and continue with "the vicious circle of reification" and the complex dialectic between doing and understanding. The problems are amplified by the fact that mathematics is a tightly knit structure that cannot be easily simplified or taken apart. All these difficulties are inherent in the process of learning. True, only because of the "circles" can we ever move forward. Yet because of them, the learning process is sometimes an intricate bootstrapping that cannot be easily supported "from outside." In addition, we must cope with many obstacles resulting from cultural and societal factors external to mathematics itself. Some of these hurdles can be seen, quite paradoxically, as a product of the democratic spirit of today's culture, which is clearly present in the Standards themselves.

According to the Standards, balancing instruction is the teacher's task. The teacher's role was never as difficult as it is today. The responsibilities and obligations with which the Standards charge their implementers are tremendous: the teacher is expected to plan instruction to fit everybody's needs and abilities, to look for "good problems" likely to stimulate students' thinking and evoke powerful mathematical ideas, to listen to all children and try to understand and assess each one on a daily basis, to orchestrate students' teamwork and lead productive classroom discussions, to cope with technology, to keep pace with educational research and development—and the list goes on and on. As if those expectations were not enough, extreme interpretations of the Standards and the resulting restriction on admissible types of instruction tie teachers' hands and make their task even more difficult. Certainly, numerous beautiful examples of masterful implementations of the principles promoted in the Standards are known from the literature (Ball, 1991b; Cobb et al., 1991, 1993; Lampert, 1990; Schoenfeld, 1996). But all these examples are instances of gifted, exceptional teachers or of teachers massively supported by researchers. Many other teachers, either less able or less willing, may find the task too complex to cope with.

What Can Be Done?

The balance we are looking for is a matter of an extremely delicate tradeoff between potentially clashing needs. It may be spoiled not only by extremism but also by inappropriate proportions. Therefore, if the new proposals are to be appropriate for everybody, the revised Standards should be open rather than constraining, inclusive rather than exclusive. Their recommendations must be diverse and rich enough to ensure that every learner will find in them her or his best route toward knowing, and they must be permissive enough to give every teacher the possibility of doing what she or he knows how to do best. The educator's task is to provide learners opportunities to satisfy all their needs. The Standards' duty is to provide the teacher with as many options as possible for creating such opportunities. There must be a bit of everything in the classroom: Problem solving should not exclude practicing skills, and teamwork should not be recommended at the expense of individual learning and teacher's exposition. Real-life and abstract mathematical problems must be given equal opportunity, and learning by talking should not supplant silent learning.

If the revised Standards are to protect themselves against unintended interpretations, they should probably be more explicit about everything, and in particular about the need for balance and about the difficulty of finding it. They can say in many different ways that there is no one path to success and that all children and teachers can be expected—and should be allowed—to arrive at their personal victories in their personal ways. A plurality of methods should not merely be permitted but recommended.

This approach is extremely important if teachers, who are supposed to translate general principles into deeds, are to do so in a balanced way. Teachers must be attuned to students' needs and capacities as well as to their own. They need a wide assortment of possible methods and approaches on hand. Knowing methods, however, is not enough. Only teachers with a good understanding of the mathematics they are supposed to teach can construct the intramathematical connections that are the basic condition of meaningful learning. Only such teachers can be successful in orchestrating a constructive mathematical discussion. Sufficient knowledge of mathematics gives them the self-assurance necessary to admit, whenever appropriate, that they do not know the answer to a question their problem-solving students are asking. For today's mathematics teacher, charged with more difficult responsibilities then ever before, this kind of knowledge may be indispensable even in the elementary grades.

No Right Answers but Also No Way Back

In this chapter, I have tried to show the unsolvability of the dilemmas we face as mathematics educators. In our attempts to improve the learning of mathematics, we will always remain torn between two concerns: our concern about the learner and our concern about the quality of the mathematics being learned. Because of this ever-present tension, we are repeatedly thrown from one extreme solution to another. It appears impossible to bring the swinging pendulum to a stop. If, on the one hand, we remain adamant in our requests about mathematics, students' achievements will show that we are simply unrealistic. If, on the other hand, we make a considerable compromise with respect to mathematics, we will soon find out that we are impairing the process of learning rather than enhancing it. This point may be illustrated by the story of the poor man who, after smelling a mouth-watering dish in a rich man's house, asked his wife to cook the dish for him, even though some of the ingredients had to be omitted or replaced. The poor fellow could not understand why "the same" dish made with the substitutes his wife could afford proved inedible. Similarly, if we remove too many ingredients from the exquisitely structured system called mathematics, we may be left with a rather tasteless subject that is not conducive to effective learning.

Apparently, therefore, there is no right answer to the question we are all asking. On the other hand, there are certainly many obviously wrong answers. One is that if the new ideas do not work as well as expected, we should think about returning to old ways of doing things. Even if not everything works to our satisfaction, the Standards must be applauded for the values they promote. Withdrawal to former positions seems out of the question if only because today the norms of the mathematics classroom seem more in tune with the norms of a democratic society than they have ever been. The justified criticisms that can sometimes be heard do not mean that the ideas promoted by the present reform movement cannot eventually deliver what has been promised: learning opportunities that, from every possible point of view, have never before been so good. Although the nature of the task is such that we can never expect full success, the situation can certainly be greatly improved. What is basically unbalanceable can still be brought to a near-balance position. If we want that to happen, then rather than try to turn everything into gold, we might instead employ many diverse methods and use them all to find a safe path among numerous conflicting needs.

ACKNOWLEDGMENT

I wish to thank Jeremy Kilpatrick for giving me the reason and the opportunity to write this chapter, and Thomas Dick, Pamela Schram, and the participants of the Conference on Research Foundations for the NCTM *Standards* (Atlanta, March 1998) for their helpful comments.

REFERENCES

Arcavi, A., & Schoenfeld, A. H. (1992). Mathematics tutoring through a constructivist lens: The challenges of sense-making. *Journal of Mathematical Behavior, 11*, 321–335.

Bachelard, G. (1938). *La formation de l'esprit scientifique* [The formation of the scientific spirit]. Paris: J. Vrin.

Ball, D. L. (1991a). What's all this talk about "discourse"? *Arithmetic Teacher, 39*(3), 44–48.

Ball, D. L. (1991b). Research on teaching mathematics: Making subject-matter knowledge part of the equation. In J. Brophy (Ed.), *Advances in research on teaching, Vol. 2: Teachers' subject-matter knowledge* (pp. 1–48). Greenwich, CT: JAI Press.

Ball, D. L. (1997). From the general to particular: Knowing our own students as learners of mathematics. *Mathematics Teacher, 90*, 732–737.

Bartolini-Bussi, M., & Pergola, M. (1996). History in the mathematics classroom: Linkages and kinematic geometry. In H. N. Janke, N. Knoche, & M. Otte (Eds.), *History of mathematics and education: Ideas and experiences* (pp. 39–67). Gottingen: Vandenhoeck & Ruprecht.

Bateson, G. (1972). *Steps to an ecology of mind: Collected essays in anthropology, psychiatry, evolution and epistemology*. San Francisco: Chandler Publishing.

Bauersfeld, H. (1995). "Language games" in the mathematics classroom: Their function and their effects. In P. Cobb & H. Bauersfeld (Eds.), *The emergence of mathematical meaning: Interaction in classroom cultures* (pp. 271–292). Hillsdale, NJ: Erlbaum.

Beaton, A. E., Mullis, I. V. S., Martin, M. O., Gonzalez, E. J., Kelly, D. L., & Smith, T. A. (1996). *Mathematics achievement in the middle school years: IEA's Third International Mathematics and Science Study*. Chestnut Hill, MA: Boston College, TIMSS International Study Center.

Bereiter, C. (1985). Towards the solution of the learning paradox. *Review of Educational Research, 55*, 201–226.

Borasi, R., & Siegel, M. (1994). Reading, writing and mathematics: Rethinking the "basics" and their relationship. In D. F. Robitaille, D. H. Wheeler, & C. Kieran (Eds.), *Selected lectures from the 7th International Congress on Mathematics Education: Québec, 17–23 August 1992* (pp. 35–48). Sainte-Foy, Canada: Les Presses de L'Université Laval.

Borba, M. C., & Skovsmose, O. (1997). The ideology of certainty in mathematics. *For the Learning of Mathematics, 17*(3), 17–23.

Breidenbach, D., Dubinsky, E., Hawks, J., & Nichols, D. (1992). Development of the process conception of function. *Educational Studies in Mathematics, 23*, 247–285.

Bromme, R., & Steinbring, H. (1994). Interactive development of subject matter in the mathematics classroom. *Educational Studies in Mathematics, 27*, 217–248.

Brown, S. I. (1997). Thinking like a mathematician: A problematic perspective. *For the Learning of Mathematics, 17*(2), 35–38.

Brownell, W. A. (1935). Psychological considerations in the learning and teaching of arithmetic. In W. D. Reeve (Ed.), *The teaching of arithmetic* (10th Yearbook of the National Council of Teachers of Mathematics, pp. 1–31). New York: Columbia University, Teachers College, Bureau of Publications.

Bruner, J. (1985). Vygotsky: A historical and conceptual perspective. In J. Wertsch (Ed.), *Culture, communication, and cognition: Vygotskian perspective* (pp. 21–34). Cambridge: Cambridge University Press.

Bruner, J. (1986). *Actual minds, possible worlds*. Cambridge, MA: Harvard University Press.

Bruner, J. (1990). *Acts of meaning*. Cambridge, MA: Harvard University Press.

Bruner, J. (1997). Celebrating divergence: Piaget and Vygotsky. *Human Development, 40*(2), 63–73.

Cazden, C. (1988). *Classroom discourse*. Portsmouth, NH: Heinemann.

Chevallard, Y. (1985). *Transposition didactique: du savoir savant au savoir enseigné* [Didactic transposition: From knowledge learned to knowledge taught]. Grenoble, France: La Pensée Sauvage.

Chevallard, Y. (1990). On mathematics education and culture: Critical afterthoughts. *Educational Studies in Mathematics, 21*, 3–28.

Cheney, L. V. (1997, June 11). President Clinton's mandate for fuzzy math. *The Wall Street Journal*, p. A22.

Cobb, P. (1995). Mathematics learning and small group interactions: Four case studies. In P. Cobb & H. Bauersfeld (Eds.), *Emergence of mathematical meaning: Interaction in classroom cultures* (pp. 25–129). Hillsdale, NJ: Erlbaum.

Cobb, P. (1998). Learning from distributed theories of intelligence. *Mind, Culture, and Activity, 5*, 187–204.

Cobb, P. (1999). Individual and collective mathematical development: The case of statistical data analysis. *Mathematical Thinking and Learning, 1*(1), 5–43.

Cobb, P., Boufi, A., McClain, K., & Whitenack, J. (1997). Reflective discourse and discursive reflection. *Journal for Research in Mathematics Education, 28*, 258–277.

Cobb, P., Gravemeijer, K., Yackel, E., McClain, K., & Whitenack, J. (1997). Mathematizing and symbolizing: The emergence of chains of signification in one first-grade classroom. In D. Kirshner & J. A. Whitson (Eds.), *Situated cognition: Social, semiotic, and psychological perspectives* (pp. 151–234). Mahwah, NJ: Erlbaum.

Cobb, P., Wood, T. & Yackel, E. (1991). A constructivist approach to second grade mathematics. In E. von Glasersfeld (Ed.), *Radical constructivism in mathematics education* (pp. 157–176). Dordrecht, The Netherlands: Kluwer.

Cobb, P., Wood, T. & Yackel, E. (1993). Discourse, mathematical thinking, and classroom practice. In E. Forman, N. Minick, & A. Stone (Eds.), *Contexts for learning: Sociocultural dynamics in children's development* (pp. 91–119). New York: Oxford University Press.

Cole, M. (1996). *Cultural psychology*. Cambridge, MA: Belknap Press.

Comiti, C., & Ball, D. (1996). Preparing teachers to teach mathematics: A comparative perspective. In A. J. Bishop, K. Clements, C. Keitel, J. Kilpatrick, & C. Laborde (Eds.), *International handbook of mathematics education* (pp. 1123–1154). Dordrecht, The Netherlands: Kluwer.

Confrey, J. (1990). A review of research on student misconceptions in mathematics, science, and programming. *Review of Research in Education, 16*, 3–56.

Connections [Special issue]. (1993). *Mathematics Teacher, 86*(9).

Cornu, B. (1991). Limits. In D. Tall (Ed.), *Advanced mathematical thinking* (pp. 153–166). Dordrecht, Netherlands: Kluwer.

Davidson, N. (1990). Small group cooperative learning in mathematics. In T. J. Cooney & C. R. Hirsh (Eds.), *Teaching and learning mathematics in the 1990s* (1990 Yearbook of the National Council of Teachers of Mathematics, pp. 52–61). Reston, VA: National Council of Teachers of Mathematics.

Davis, P. (1993). Visual theorems. *Educational Studies in Mathematics, 24*, 333–344.

Davis, P. J., & Hersh, R. (1981). *The mathematical experience*. London: Penguin Books.

Davis, R. (1988). The interplay of algebra, geometry, and logic. *Journal of Mathematical Behavior, 7*, 9–28.

Devlin, K. (1994). A collegiate mathematical experience for non-science majors. In M. Quigley (Ed.), *Proceedings of the Canadian Mathematics Education Study Group* (pp. 21–35). Regina, Saskatchewan: University of Regina.

Devlin, K. (1997). Editorial: Reduce skills teaching in math class. *Focus, 17*(6), 2–3.

Douady, R. (1985). The interplay between different settings: Tool-object dialectic in the extension of mathematical ability—Examples from elementary school teaching. In L. Streefland (Ed.), *Proceedings of the Ninth International Conference for the Psychology of Mathematics Education* (Vol. 2, pp. 33–52). Utrecht, The Netherlands: State University of Utrecht, Subfaculty of Mathematics, OW&OC.

Driver, R., & Easley, J. (1978). Pupils and paradigms: A review of literature related to concept development in adolescent science students. *Studies in Science Education, 5*, 61–84.

Dubinsky, E. (1991). Reflective abstraction in advanced mathematical thinking. In D. Tall (Ed.), *Advanced mathematical thinking* (pp. 95–123). Dordrecht, The Netherlands: Kluwer.

Eco, U., Santambrogio, M., & Violi, P. (Eds.). (1988). *Meaning and mental representations*. Bloomington: Indiana University Press.

Edwards, D. (1993). But what do children really think? Discourse analysis and conceptual content in children's talk. *Cognition and Instruction, 11*, 207–225.

Edwards, D., & Potter, J. (1992). *Discursive psychology*. Newbury Park, CA: Sage.

Empowering Students Through Connections [Special issue]. (1993). *Arithmetic Teacher, 40*(6).

Ernest, P. (1994). Conversation as a metaphor for mathematics and learning. In *Proceedings of British Society for Research into Learning Mathematics Day Conference, Manchester Metropolitan University, 22 November 1993* (pp. 58–63). Nottingham, England: British Society for Research into Learning Mathematics.

Fischbein, E. (1987). *Intuition in science and mathematics*. Dordrecht, The Netherlands: Reidel.

Fischbein, E. (1989). Tacit models and mathematical reasoning. *For the Learning of Mathematics, 9*(2), 9–14.

Fischbein, E., Deri, M., Nello, M. S., & Marino, M. S. (1985). The role of implicit models in solving verbal problems in multiplication and division. *Journal for Research in Mathematics Education, 16*, 3–17.

Forman, E. A., Minick, N, & Stone, C.A. (Eds.). (1993). *Context for learning: Sociocultural dynamics in children's development*. Oxford: Oxford University Press.

Foucault, M. (1972). *The archaeology of knowledge*. New York: Harper.

Freudenthal, H. (1978). *Weeding and sowing: Preface to a science of mathematics education.* Dordrecht, The Netherlands: Reidel.

Fuson, K. C., & Briars, D. (1990). Using a base-ten blocks learning/teaching approach for first- and second-grade place-value and multidigit addition and subtraction. *Journal for Research in Mathematics Education, 21,* 180–206.

Fuson, K. C., & Kwon, Y. (1992). Korean children's understanding of multidigit addition and subtraction. *Child Development, 63,* 491–506.

Gee, J. P. (1997). Thinking, learning, and reading: The situated sociocultural mind. In D. Kirshner & J. A. Whitson (Eds.), *Situated cognition: Social, semiotic, and psychological perspectives* (pp. 235–260). Mahwah, NJ: Erlbaum.

Gravemeijer, K. E. P. (1994). *Developing realistic mathematics education.* Utrecht, The Netherlands: CD-Beta Press.

Gray, E. M., & Tall, D. O. (1993). Duality, ambiguity, and flexibility: A "proceptual" view of simple arithmetic. *Journal for Research in Mathematics Education, 25,* 116–140.

Greeno, J. (1983). Conceptual entities. In D. Gentner & A. L. Stevens (Eds.), *Mental models* (pp. 227–252). Hillsdale, NJ: Erlbaum.

Hadamard, J. (1945). *The psychology of invention in the mathematical field.* Princeton, NJ: Princeton University Press.

Hanna, G., & Jahnke, H. N. (1996). Proof and proving. In A. J. Bishop, K. Clements, C. Keitel, J. Kilpatrick, & C. Laborde (Eds.), *International handbook of mathematics education* (pp. 877–908). Dordrecht, The Netherlands: Kluwer.

Hardy, G. H. (1967). *A mathematician's apology.* Cambridge: Cambridge University Press. (Original work published 1940)

Harel, G., Behr, M., Post, T., & Lesh, R. (1989). Fischbein's theory: A further consideration. In G. Vergnaud, J. Rogalski, & M. Artigue (Eds.), *Proceedings of the 13th Annual Conference for the Psychology of Mathematics Education* (Vol. 2, pp. 52–59). Paris: University of Paris.

Harel, G., & Kaput, J. (1991). The role of conceptual entities in building advanced mathematical concepts and their symbols. In D. Tall (Ed.), *Advanced mathematical thinking* (pp. 82–94). Dordrecht, The Netherlands: Kluwer.

Haroutunian-Gordon, S., & Tartakoff, D. S. (1996). On the learning of mathematics through conversation. *For the Learning of Mathematics, 16*(2), 2–10.

Harre, R., & Gillet, G. (1995). *The discursive mind.* Thousand Oaks, CA: Sage.

Harries-Jones, P. (1995). *Recursive vision: Ecological understanding and Gregory Bateson.* Toronto: University of Toronto Press.

Hefendehl-Hebeker, L. (1991). Negative numbers: Obstacles in their evolution from intuitive to intellectual constructs. *For the Learning of Mathematics, 11*(1), 26–32.

Hicks, D. (Ed.). (1997a). *Discourse, learning, & schooling.* Cambridge: Cambridge University Press.

Hicks, D. (1997b). Working through discourse genres in school. *Research in the Teaching of English, 31,* 459–485.

Hiebert, J., & Carpenter, T. P. (1992). Learning and teaching with understanding. In D. A. Grouws (Ed.), *Handbook of research on mathematics teaching and learning* (pp. 65–100). New York: Macmillan.

Hiebert, J., Carpenter, T. P., Fennema, E., Fuson, K., Human, P., Murray, H., Olivier, A., & Wearne, D. (1996). Problem solving as a basis for reform in curriculum and instruction: The case of mathematics. *Educational Researcher, 25*(4), 12–21.

Hiebert, J., & Lafevre, P. (1986). Conceptual and procedural knowledge in mathematics: An introductory analysis. In J. Hiebert (Ed.), *Conceptual and procedural knowledge: The case of mathematics* (pp. 1-27). Hillsdale, NJ: Erlbaum.

Hiebert, J., & Wearne, D. (1992). Links between teaching and learning place value with understanding in the first grade. *Journal for Research in Mathematics Education, 23,* 98–122.

Hiebert, J., & Wearne, D. (1996). Instruction, understanding, and skill in multidigit addition and subtraction. *Cognition and Instruction, 14,* 251–284.

House, P. A. (Ed.). (1995). *Connecting mathematics across the curriculum* (1995 Yearbook of the National Council of Teachers of Mathematics). Reston, VA: NCTM.

Jackson, A. (1997). The math wars: California battles it out over mathematics education reform (Parts 1 & 2). *Notices of the American Mathematical Society, 44,* 695–702, 817–823.

Johnson, M. (1987). *The body in the mind: The bodily basis of meaning, imagination, and reason.* Chicago: University of Chicago Press.

Kieran, C. (1994). A functional approach to the introduction of algebra: some pros and cons. In J. P. da Ponte & J. F. Matos (Eds.), *Proceedings of the 18th International Conference for the Psychology of Mathematics Education* (Vol. 1, pp. 157–175). Lisbon: University of Lisbon.

Kieran, C., & Sfard, A. (1999). Seeing through symbols: The case of equivalent equations. *Focus on Learning Problems in Mathematics, 21,* 1–17.

Kilpatrick, J. (1992). A history of research in mathematics education. In D. A. Grouws (Ed.), *Handbook of research on mathematics teaching and learning* (pp. 3–38). New York: Macmillan.

Kilpatrick, J. (1997). *Foundations for school mathematics.* Unpublished manuscript, University of Georgia.

Kozulin, A. (1990). *Vygotsky's psychology: A biography of ideas.* New York: Harvester Wheatsheaf.

Krummheuer, G. (1995). The ethnography of argumentation. In P. Cobb & H. Bauersfeld (Eds.), *The emergence of mathematical meaning: Interactions in classroom culture* (pp. 229–269). Hillsdale, NJ: Erlbaum.

Kuhn, T. (1962). *The structure of scientific revolutions.* Chicago: University of Chicago Press.

Lampert, M. (1990). When the problem is not the question and the solution is not the answer: Mathematical knowing and teaching. *American Educational Research Journal, 27,* 29–63.

Lave, J. (1988). *Cognition in practice.* Cambridge: Cambridge University Press.

Lave, J., & Wenger, E. (1991). *Situated learning: Legitimate peripheral participation.* Cambridge: Cambridge University Press.

Love, E., & Pimm, D. (1996). "This is so": A text on text. In A. J. Bishop, K. Clements, C. Keitel, J. Kilpatrick, & C.

Laborde (Eds.), *International handbook of mathematics education* (pp. 371–409). Dordrecht, The Netherlands: Kluwer.

Malik, M.A. (1980). Historical and pedagogical aspects of definition of function. *International Journal of Mathematics Education in Science and Technology, 1*, 489–492.

Markovits, Z., Eylon, B., & Bruckheimer, M. (1986). Functions today and yesterday. *For the Learning of Mathematics, 6*(2), 18–24.

Mathematical Connections [Special issue]. (1996). *Mathematics Teaching in the Middle School, 1*(9).

McLeod, D. B. (1992). Research on affect in mathematics education: A reconceptualization. In D. A. Grouws (Ed.), *Handbook of research on mathematics teaching and learning* (pp. 575–596). New York: Macmillan.

McLeod, D. B., & Ortega, M. (1993). Affective issues in mathematics education. In P. Wilson (Ed.), *Research ideas for the classroom: High school mathematics* (pp. 21–38). New York: Macmillan.

Morgan, C. (1996). "The language of mathematics": Towards a critical analysis of mathematical text. *For the Learning of Mathematics, 16*(3), 2–10.

Moritz, R. E. (1993). *Memorabilia mathematica: The philomath's quotation book*. Washington, DC: Mathematical Association of America. (Original work published 1914)

Mullis, I. V. S., Martin, M. O., Beaton, A. E., Gonzalez, E. J., Kelly, D. L., & Smith, T. A. (1997). *Mathematics achievement in the primary school years: IEA's Third International Mathematics and Science Study (TIMSS)*. Chestnut Hill, MA: Boston College, TIMSS International Study Center.

National Council of Teachers of Mathematics. (1989). *Curriculum and evaluation standards for school mathematics*. Reston, VA: Author.

National Council of Teachers of Mathematics. (1991). *Professional standards for teaching mathematics*. Reston, VA: Author.

National Council of Teachers of Mathematics. (1995). *Assessment standards for school mathematics*. Reston, VA: Author.

National Research Council. (1989). *Everybody counts: A report to the nation on the future of mathematics education*. Washington, DC: National Academy Press.

Nemirovsky, R. (1996). A functional approach to algebra: Two issues that emerge. In N. Bednarz, C. Kieran, & L. Lee (Eds.), *Approaches to algebra: Perspectives for research and teaching* (pp. 295–313). Dordrecht, The Netherlands: Kluwer.

Nesher, P. (1986). Are mathematical understanding and algorithmic performance related? *For the Learning of Mathematics, 6*(3), 2–9.

Nesher, P. (1987). Toward an instructional theory: The role of student's misconceptions. *For the Learning of Mathematics,* 7(3), 33–40.

Nunes, T., Schlieman, A. D., & Carraher, D. W. (1993). *Street mathematics and school mathematics*. Cambridge: Cambridge University Press.

O'Connor, M. C. (1996). Managing the intermental: Classroom group discussion and the social context of learning. In D. Slobin, J. Gerhardt, A. Kyratzis, & J. Guo (Eds.), *Social interaction, social context, and language* (pp. 495–509). Hillsdale, NJ: Erlbaum.

O'Connor, M. C. (1998). Language socialization in the mathematics classroom: Discourse practices and mathematical thinking. In M. Lampert & M. Blunk (Eds.), *Talking mathematics: Studies of teaching and learning in school* (pp. 17–55). New York: Cambridge University Press.

Petrie, H. G., & Oshlag, R. S. (1993). Metaphor in learning. In A. Ortony (Ed.), *Metaphor and thought* (2nd ed., pp. 579–609). Cambridge: Cambridge University Press.

Piaget, J. (1953). *The origin of intelligence in the child*. London: Routledge & Kegan Paul.

Piaget, J. (1970). *Genetic epistemology*. New York: Columbia University Press.

Piaget, J., & Inhelder, B. (1969). *The psychology of the child*. New York: Basic Books.

Pimm, D. (1987). *Speaking mathematically*. New York: Routledge & Kegan Paul.

Pimm, D. (1995). *Symbols and meanings in school mathematics*. London: Routledge.

Polya, G. (1957). *How to solve it: A new aspect of mathematical method* (2nd ed.). Princeton, NJ: Princeton University Press.

Prawat, R. S. (1997). Problematizing Dewey's views of problem solving: A reply to Hiebert et al. *Educational Researcher, 26*(2), 19–21.

Putnam, R. T., Lampert, M., & Peterson, P. L. (1990). Alternative perspectives on knowing mathematics in elementary school. *Review of Research in Education, 16*, 57–150.

Reddy, M. J. (1983). The conduit metaphor: A case of frame conflict in our language about language. In A. Ortony (Ed.), *Metaphor and thought* (pp. 164–201). Cambridge: Cambridge University Press.

Rogoff, B., & Lave, J. (Eds.). (1984). *Everyday cognition: Its development in social context*. Cambridge, MA: Harvard University Press.

Rorty, R. (1979). *Philosophy and the mirror of nature*. Princeton, NJ: Princeton University Press.

Rotman, B. (1994). Mathematical writing, thinking, and virtual reality. In P. Ernest (Ed.), *Mathematics, education, and philosophy: An international perspective* (pp. 76–86). London: Falmer.

Saxe, G. B. (1991). *Culture and cognitive development: Studies in mathematical understanding*. Hillsdale, NJ: Erlbaum.

Scheffler, I. (1991). *In praise of cognitive emotions*. New York: Routledge.

Schlieman, A. D. (1995). Some concerns about bringing everyday mathematics to mathematics education. In L. Meira & D. Carraher (Eds.), *Proceedings of the 19th International Conference for the Psychology of Mathematics Education* (Vol. 1, pp. 45-60). Recife, Brazil: Universidade Federal de Pernambuco.

Schlieman, A. D., & Carraher, D. W. (1996). Negotiating mathematical meanings in and out of school. In L. P. Steffe, P. Nesher, P. Cobb, G. A. Goldin, & B. Greer (Eds.), *Theories of mathematical learning* (pp. 77–84). Mahwah, NJ: Erlbaum.

Schoenfeld, A. H. (1985). *Mathematical problem solving*. New York: Academic Press.

Schoenfeld, A. H. (1989). Explorations of students' mathematical beliefs and behavior. *Journal for Research in Mathematics Education, 20*, 338–355.

Schoenfeld, A. H. (1992). Learning to think mathematically: Problem solving, metacognition, and sense making in mathematics. In D.A. Grouws (Ed.), *Handbook of research on mathematics teaching and learning* (pp. 334–370). New York: Macmillan.

Schoenfeld, A. H. (1996). In fostering communities of inquiry, must it matter that the teacher knows the "answer"? *For the Learning of Mathematics, 16*(3), 11–16.

Schwartz, J. L., & Yerushalmy, M. (1995). On the need for a bridging language for mathematical modeling. *For the Learning of Mathematics, 15*(2), 29–35.

Sfard, A. (1987). Two conceptions of mathematical notions: Operational and structural. In J. C. Bergeron, N. Herscovics, & C. Kieran (Eds.), *Proceedings of the 11th International Conference on the Psychology of Mathematics Education* (Vol. 3, pp. 162–169). Montreal : Université de Montréal.

Sfard, A. (1991). On the dual nature of mathematical conceptions: Reflections on processes and objects as different sides of the same coin. *Educational Studies in Mathematics, 22*, 1–36.

Sfard, A. (1992). Operational origins of mathematical objects and the quandary of reification: The case of function. In E. Dubinsky & G. Harel (Eds.), *The concept of function: Aspects of epistemology and pedagogy* (MAA Notes 25, pp. 59–84). Washington, DC: Mathematical Association of America.

Sfard, A. (1994a). Mathematical practices, anomalies, and classroom communication problems. In P. Ernest (Ed.), *Constructing mathematical knowledge* (pp. 248–273). London: Falmer.

Sfard, A. (1994b). Reification as a birth of a metaphor. *For the Learning of Mathematics, 14*(1), 44–55.

Sfard, A. (1995). The development of algebra: Confronting historical and psychological perspectives. *Journal of Mathematical Behavior, 14*, 15–39.

Sfard, A. (1998a). Two metaphors for learning mathematics: Acquisition metaphor and participation metaphor. *Educational Researcher, 27*(2), 4–13.

Sfard, A. (1998b). The many faces of mathematics: Do mathematicians and researchers in mathematics education speak about the same thing? In A. Sierpinska & J. Kilpatrick (Eds.), *Mathematics education as a research domain: A search for identity* (pp. 491–512). Dordrecht, The Netherlands: Kluwer.

Sfard, A. (2000a). Symbolizing mathematical reality into being: How mathematical discourse and mathematical objects create each other. In P. Cobb, K. E. Yackel, & K. McClain (Eds), *Symbolizing and communicating: Perspectives on mathematical discourse, tools, and instructional design* (pp. 37–98). Mahwah, NJ: Erlbaum.

Sfard, A. (2000b). On reform movement and the limits of mathematical discourse. *Mathematical Thinking and Learning, 2*(3), 157–189.

Sfard, A., & Kieran, C. (2001). Cognition as communication: Rethinking learning-by-talking through multi-faceted analysis of students' mathematical interactions. *Mind, Culture, and Activity, 8*(1), 42–76.

Sfard, A., & Linchevski, L. (1994). The gains and the pitfalls of reification: The case of algebra. *Educational Studies in Mathematics, 26*, 191–228.

Sfard, A., Nesher, P., Streefland, L., Cobb, P., & Mason, J. (1998). Learning mathematics through conversation: Is it as good as they say? A debate. *For the Learning of Mathematics, 18*(1), 41–51.

Sierpinska, A. (1994). *Understanding in mathematics*. London: Falmer.

Sierpinska, A., & Lerman, S. (1996). Epistemologies of mathematics and of mathematics education. In A. J. Bishop, K. Clements, C. Keitel, J. Kilpatrick, & C. Laborde (Eds.), *International handbook of mathematics education* (pp. 827–876). Dordrecht, The Netherlands: Kluwer.

Silver, E. A., & Cai, J. (1996). An analysis of arithmetic problem posing by middle school students. *Journal for Research in Mathematics Education, 27*, 521–539.

Skemp, R. (1976). Relational understanding and instrumental understanding. *Mathematics Teaching, 77*, 44–49.

Smith, J. P. (1996). Efficacy and teaching mathematics by telling: A challenge for reform. *Journal for Research in Mathematics Education, 27*, 387–402.

Smith, J. P. (1997). Problems with problematizing mathematics: A reply to Hiebert et al. *Educational Researcher, 26*(2), 22–24.

Smith, J. P., diSessa, A. A., & Rochelle, J. (1993). Misconceptions reconceived: A constructivist analysis of knowledge in transition. *Journal of the Learning Sciences, 3*, 115–163.

Steinbring, H., Bartolini Bussi, M. G., & Sierpinska, A. (Eds.). (1998). *Language and communication in the mathematics classroom*. Reston, VA: National Council of Teachers of Mathematics.

Stevenson, H. W., & Stigler, J. W. (1992). *The learning gap: Why our schools are failing and what we can learn from Japanese and Chinese education*. New York: Summit Books.

Stigler, J., Fernandez, C., & Yoshida, M. (1996). Traditions of school mathematics in Japanese and American elementary classrooms. In L. P. Steffe, P. Nesher, P. Cobb, G. A. Goldin, & B. Greer (Eds.), *Theories of mathematical learning* (pp. 149–175). Mahwah, NJ: Erlbaum.

Stigler, J., Lee, S.-Y., & Stevenson, H. W. (1990). *Mathematical knowledge of Japanese, Chinese, and American elementary school children*. Reston, VA: National Council of Teachers of Mathematics.

Streefland, L. (1991). *Fractions in realistic mathematics education: A paradigm of developmental research*. Dordrecht, The Netherlands: Kluwer.

Sutton, G. O. (1992). Cooperative learning works in mathematics. *Mathematics Teacher, 80*(1), 63–66.

Tall, D., & Vinner, S. (1981). Concept image and concept definition in mathematics with particular reference to limits and continuity. *Educational Studies in Mathematics, 12*, 151–169.

Tate, W. F., & D'Ambrosio, B. S. (1997). Guest editorial. *Journal for Research in Mathematics Education, 28*, 650–651.

Thomas, R. (1996). Proto-mathematics and/or real mathematics. *For the Learning of Mathematics, 16*(2), 11–18.

Thompson, P. (1985). Experience, problem solving, and learning mathematics: Considerations in developing mathematics curricula. In E. Silver (Ed.), *Teaching and learning mathematical problem solving: Multiple perspectives* (pp. 189–243). Hillsdale, NJ: Erlbaum.

Thurston, W. P. (1990). Mathematical education. *Notices of the American Mathematical Society, 37*, 844–850.

Tietze, U.-P. (1994). Mathematical curricula and underlying goals. In R. Biehler, R. W. Scholz, R. Strasser, & B. Winkelman (Eds.), *Didactics of mathematics as a scientific discipline* (pp. 41–53). Dordrecht, The Netherlands: Kluwer.

Treffers, A. (1987). *Three dimensions: A model of goal and theory description in mathematics instruction—The Wiskobas Project.* Dordrecht, The Netherlands: Reidel.

Ulam, S. (1976). *Adventures of a mathematician.* New York: Scribners.

Van Hiele, P. M. (1985). The child's thought and geometry. In D. Fuys, D. Geddes, & R. Tischler (Eds.), *English translation of selected writings of Dina van Hiele-Geldof and Pierre M. van Hiele* (pp. 243–252). Brooklyn, NY: Brooklyn College, School of Education.

Vinner, S. (1983). Concept definition, concept image and the notion of function. *International Journal of Mathematical Education in Science and Technology, 14*, 293–305.

Vinner, S., & Dreyfus, T. (1989). Images and definitions for the concept of function. *Journal for Research in Mathematics Education, 20*, 356–366.

Voigt, J. (1985). Patterns and routines in classroom interaction. *Recherches en Didactique des Mathématiques, 6*, 69–118.

Voigt, J. (1994). Negotiation of mathematical meaning and learning mathematics. *Educational Studies in Mathematics, 26*, 275–298.

Voigt, J. (1995). Thematic patterns of interaction and sociomathematical norms. In P. Cobb & H. Bauersfeld (Eds.), *The emergence of mathematical meaning: Interaction in classroom cultures* (pp. 163–201). Hillsdale, NJ: Erlbaum.

Voigt, J. (1996). Negotiation of mathematical meaning in classroom processes: Social interaction and learning mathematics. In L. P. Steffe, P. Nesher, P. Cobb, G. A. Goldin, & B. Greer (Eds.), *Theories of mathematical learning* (pp. 21–50). Mahwah, NJ: Erlbaum.

Von Glasersfeld, E. (1993). *Radical constructivism: A way of knowing and learning.* London: Falmer.

Vygotsky, L. S. (1962). *Thought and langua ge.* Cambridge, MA: MIT Press.

Vygotsky, L. S. (1987). Thinking and speech. In R. W. Rieber & A. C. Carton (Eds.), *The collected works of L. S. Vygotsky* (Vol. 1, pp. 39–285). New York: Plenum.

Walkerdine, V. (1988). *The mastery of reason.* London: Routledge.

Waywood, A. (1992). Journal writing and learning mathematics. *For the Learning of Mathematics, 12*(2), 34–43.

Webb, N. M. (1991). Task-related verbal interaction and mathematics learning in small groups. *Journal for Research in Mathematics Education, 22*, 366–389.

Webb, N. M. & Farivar, S. (1994). Promoting helping behavior in cooperative small groups in middle school mathematics. *American Educational Research Journal, 31*, 369–395.

Wheeler, D. (1982). Mathematization matters. *For the Learning of Mathematics, 3*(1), 45–47.

Wittgenstein, L. (1953). *Philosophical investigations.* Oxford: Blackwell.

Wistedt, I. (1994). Reflection, communication, and learning mathematics. *Learning and Instruction, 4*, 123–138.

Wu, H. (1997). The mathematics education reform: Why you should be concerned and what you can do. *American Mathematical Monthly, 104*, 946–954.

Yerushalmy, M. (1997). Designing representations: Reasoning about functions of two variables. *Journal for Research in Mathematics Education, 28*, 431–466.

Yerushalmy, M., Chazan, D., & Gordon, M. (1990). Mathematical problem posing: Implications for facilitating student inquiry in classrooms. *Instructional Science, 19*, 219–245.

SECTION 3

The Creation of *Principles and Standards for School Mathematics*

CHAPTER 24

Using Research in Policy Development: The Case of the National Council of Teachers of Mathematics' *Principles and Standards for School Mathematics*

Joan Ferrini-Mundy, Michigan State University
W. Gary Martin, Auburn University

For policymakers and practitioners, the question of the role and value of educational research is continually at issue, and various formulations about the relationship of educational research to policy and practice have been advanced (Lagemann, 2000; National Academy of Education, 1999; Shavelson, 1998; Stokes, 1997). Yet few descriptions of actual cases portray the relationship of educational research and its entailments to an instance of policy development and content. Here we present such a case through the description of the recent National Council of Teachers of Mathematics (NCTM) initiative to revise its *Standards* documents (NCTM, 1989a, 1991, 1995, 1998). The organization's latest version of Standards, *Principles and Standards for School Mathematics (Principles and Standards)* (NCTM, 2000a), was released in April 2000.

A formulation provided by Silver (1990) about the role of research in practice—in which he contends that the theoretical constructs and perspectives of research, the methodologies of research, and the findings of research all have implications for practice—will serve as the frame for this case. *Principles and Standards* is a policy document in the sense that it presents national recommendations for changes in practice in prekindergarten through Grade-12 mathematics education. In the first of the three main sections that follow, we provide a brief introduction to NCTM and its *Principles and Standards*. In the next section, we discuss how the development process for *Principles and Standards* was shaped by the theoretical perspectives, methodologies, and findings of research. Finally, we consider the content of *Principles and Standards* from the perspective of how the theoretical perspectives, methodologies, and findings of research influenced and are visible in that content.

Introduction to the National Council of Teachers of Mathematics and *Principles and Standards for School Mathematics*

The NCTM is a professional organization of about 100,000 members that promotes high-quality mathematics teaching and learning in Grades K–12 for all students. In 1989, NCTM led other subject-matter areas by first producing content standards for Grades K–12 mathematics (NCTM, 1989a).[1] This document was followed by standards for teaching (NCTM, 1991) and for assessment (NCTM, 1995). In the mid-1990s, the NCTM Board of Directors began a process for planning the revision of earlier *Standards* documents; in spring 1996, the Writing Group was assembled to plan and ultimately write *Principles and Standards*. The group was composed of mathematicians, mathematics education researchers, classroom teachers, policymakers, curriculum developers, and teacher educators.[2]

Given its composition, the Writing Group was well positioned to ensure that mathematics education research, as well as the considerations of both mathematics

[1] See McLeod, Stake, Schappelle, Mellissinos, and Grierl (1996) for a history of this development process.

[2] Writing Group members included: Jeane Joyner, Angela Andrews, Douglas H. Clements, Alfinio Flores, Carol Midgett, Judith Roitman, Barbara Reys, Francis "Skip" Fennell, Catherine M. Fueglein, Melinda Hamilton, Melissa Manzano-Aleman, Susan Jo Russell, Philip Wagreich, Edward A. Silver, Mary Bouck, Jean Howard, Diana Lambdin, Carol Malloy, James Sandefur, Alan Schoenfeld, Sue Eddins, M. Kathleen Heid, Millie Johnson, Ron Lancaster, Alfred Manaster, and Milton Norman. Joan Ferrini-Mundy served as the chair of this group, and W. Gary Martin served as the project director. Jean Carpenter and Sheila Gorg provided editorial assistance, and Debra Kushner did the layout and design.

and classroom practice, would be prominent in its deliberations. The Commission on the Future of the Standards, similarly composed, was assembled to oversee the development process. The Writing Group was charged to establish standards that built on the foundation of the previous NCTM *Standards* documents; that integrated the classroom-related portions of *Curriculum and Evaluation Standards for School Mathematics* (NCTM, 1989a), *Professional Standards for Teaching Mathematics* (NCTM, 1991), and *Assessment Standards for School Mathematics* (NCTM, 1995); and that were based on four grade bands—pre-K–2, 3–5, 6–8, and 9–12—instead of the three grade bands used in the 1989 *Curriculum and Evaluation Standards.*

Principles and Standards is intended to be a comprehensive and coherent set of goals to improve mathematics teaching and learning in our schools, a resource for all who make decisions about school mathematics, and a tool to stimulate focus and conversation (NCTM, 2000a, p. 5). Building on the 1989 Standards, *Principles and Standards* emphasizes learning mathematics with meaning and understanding by all students. The document includes a chapter that describes six principles intended to guide the formulation of school mathematics decisions, followed by a chapter discussing each of ten standards spanning the mathematical content and process emphases intended for the prekindergarten-through-Grade-12 spectrum. The next four chapters include extended discussions of the standards at each of the four grade bands. The book concludes with a chapter recommending actions by various groups (e.g., teachers, policymakers, teacher educators) positioned to influence the course of mathematics education.

The *Principles and Standards* for School Mathematics Development Process as a Researchlike Endeavor

NCTM supported an extended and elaborate development process for the development of *Principles and Standards*, described in Ferrini-Mundy (2000) and NCTM (2000b). Aspects of that development process are portrayed here to show how theoretical perspectives, methodologies, and research findings all shaped the document's development. The four-year development process involved a large group of people in a variety of roles and was designed to be open and consultative. The process began in 1996; a discussion draft (NCTM, 1998) was released and circulated for comment in fall 1998. The draft was revised in summer 1999, and the final document was released in April 2000.

Retrospectively, we provide commentary on how theoretical perspectives from research shaped the Writing Group's interpretation of its task, the way in which the Writing Group functioned, and the role of the interaction with the field in the development of the document.[3] We then look at how research methodologies influenced one threefold aspect of the development process: the collection, analysis, and use of feedback to the discussion draft. We conclude with a summary of how the outcomes of this researchlike process were used to produce the final version of the document.

Theoretical Perspectives From Research in the Development of Principles and Standards

In reflecting on the processes invented by the Writing Group to address its charge, we can infer the influence of certain theoretical perspectives drawn from research. The 1989 *Curriculum and Evaluation Standards for School Mathematics* (NCTM, 1989a) was widely considered an influential and exemplary set of policy recommendations for mathematics education (Ravitch, 1995). For teachers, the views presented were seen both as somewhat revolutionary and as validation of the practice in which many teachers were trying to engage (e.g., Ferrucci, 1997). In considering the charge to revise, the Writing Group settled on a view that the document would be more evolutionary, designed to take account of the state of activity in mathematics education nationally and to constitute a resource that enabled people to move from their current state of knowledge and practice toward the recommendations of the document. Thus, basic socioconstructivist tenets—the notions of building from the learners' experience, trying to understand the learners' state of knowledge and capacity, and recognizing learning as a socially constructed activity—in a sense could be viewed as shaping the standards as a policy curriculum for the field of learners (teachers, school leaders, policymakers). Saxe and Bermudez (1992) argue, "Children's mathematical environments cannot be understood apart from children's own cognizing activities. . . . Children's construction of mathematical goals and subgoals is interwoven with the socially organized activities in which they are participants" (pp. 2–3). A recasting of these ideas to the context faced in revising the NCTM Standards might be translated as the idea that Grades pre-K–12 mathematics

[3]The authors take full responsibility for this account of the theoretical perspective in the development process and realize that others involved in this process might render it differently.

education practitioners' environments cannot be understood apart from their own practices and activities. Their construction of goals for mathematics education is interwoven with the socially organized contexts and institutions in which they act.

The NCTM leadership, as well as members of the Writing Group, held the view that a solid understanding of the state of the field would be important and that that understanding should shape the document's stance and content. Thus, extensive efforts were made from the beginning of the development process to gather advice and opinions from teachers, mathematicians, policymakers, and mathematics teacher educators; to monitor public debate and discussion about mathematics education; and to consult standards and frameworks from states and from other countries—evidence that this process consciously built on existing knowledge.

The previous NCTM *Standards* documents had been implemented in various ways, perhaps most visibly and directly in the formulation of state standards and frameworks through the early 1990s (Council of Chief State School Officers, 1995). In addition, instructional materials claimed to incorporate the ideas of the NCTM Standards, although the development of processes for analyzing alignment with the NCTM Standards and curricula (e.g., American Association for the Advancement of Science, 2000; U.S. Department of Education, 1999) did not take shape or produce findings until the late 1990s. Before National Science Foundation (NSF)-funded curricula were developed that were based on the Standards in significant and deep ways, teachers, school leaders, and teacher educators worked directly from the *Standards* documents in efforts to implement the Standards. Evidence gathered during the early 1990s suggests that some of these interpretations and implementations, prior to the availability of curriculum materials, were superficial and focused more on pedagogical innovations than on fostering mathematical understanding by all students (Ferrini-Mundy & Johnson, 1997; Peak, 1996). The public controversies that developed over the NCTM Standards began in the mid-1990s (e.g., Allen, 1995; Honest Open Logical Debate [HOLD] on Math Reform, 1996) and may have arisen partially in response to these early and inadequate implementation efforts and as a result of differing values and goals concerning school mathematics.

Thus, the Writing Group, under the guidance of NCTM's Commission on the Future of the Standards, took as part of the development process the need to thoroughly understand all aspects of the national state of school mathematics education and the nature of the debates and disagreements about mathematics education. This task led to a need for methodology that was ethnographic in its perspective but sensitive to Wolcott's (1980) reminder that "the essential ethnographic contribution is interpretive rather than methodological" (p. 59). That is, the emergent theoretical commitments necessitated an information-gathering process that enabled "cultural interpretation" (Geertz, 1973) in the "culture" of mathematics education at a national level.

Information about the audience for *Principles and Standards*—the learners—was crucial in shaping the formulation of the document. The inherent theoretical perspective required building on the existing experience and context in the field in an effort to produce a document that would be accessible enough to all within reach of the field, forward-looking enough to serve teachers who were well along in their Standards-based efforts, and sensible enough to invite constructive discussion from all segments of the diverse community of stakeholders in mathematics education.

In the following section, we address how this theoretical stance was operationalized through a process in which classical qualitative methodologies were employed for gathering feedback to the *Discussion Draft* (NCTM, 1998).

Methodologies From Research in the Development of Principles and Standards

Methodologies drawn from research played a key role in the development of *Principles and Standards*, as seen in the analysis of public input to the process. Attempts were made to gather input from a range of constituencies throughout the development process. In the first phase of reaction, which lasted from 1996 to 1998, formative input was gathered about what form the updated NCTM Standards should take. Member groups of the Conference Board of Mathematical Sciences[4] were asked in 1997 to form Association Review Groups to provide input into the development process. Several formal analyses of this feedback were prepared to guide the Writing Group in their initial formulation and writing of the document.

The second phase began in October 1998 with the release of *Principles and Standards for School Mathematics: Discussion Draft* (NCTM, 1998). This phase of input and analysis is described in the following sections. As was true for the three preceding volumes of NCTM Standards, the release of a draft document (NCTM, 1987,

[4] The Conference Board on the Mathematical Sciences is a consortium of mathematics professional organizations. See http://www.cbmsweb.org for additional information.

1989b, 1993) for public reaction was a central feature of the development process. The difficulties in understanding the issues raised by the feedback became obvious as the first reactions to the discussion draft were distributed to project members. Informal exchanges revealed widely varying interpretations. The project leadership found that when a Writing Group member encountered an issue more than once, he or she tended to identify it as a theme worthy of attention. As different people attended to different parts of the feedback, an overwhelming range of interpretations naturally resulted, implying that a more rigorous process for analyzing the feedback was necessary. Further support for the need for a formal process of analysis was found in the political environment in the late 1990s. The purpose and role of standards were under increasing national discussion, and the process of developing *Principles and Standards* was highly visible. Only a thorough and systematic solicitation and analysis of the feedback could counter the claims that the process was too narrow and not attentive to enough stakeholders.

Because open-ended responses were solicited, methods of qualitative research were naturally found useful to analyze the feedback. Indeed, as the plan for analyzing the feedback unfolded, its designers clearly saw the effort as paralleling a research study's methods and perspectives, with guiding questions framing the effort, methods needed for gathering and analyzing the information, and the results that would affect the substance and form of the final document. The remainder of this section describes the process's elements.

Purposes. The primary purpose of gathering and analyzing reactions to the draft was to better understand the needs and expectations of a diverse set of stakeholders as they might influence the final *Principles and Standards* document. What general approaches should be taken to mathematics education issues? Which issues raised in the draft required attention? What perspectives on these issues should be addressed? To satisfy this purpose, the analysis led to the identification of a set of issues for the Writing Group to consider as general decisions were made for the final document. The process for responding to those issues is considered in the results section below. A secondary purpose of the process was to capture more specific editorial advice.

Feedback collection. Responses to the draft were solicited from a wide range of audiences with a stake in mathematics education (e.g., mathematics teachers, mathematicians, researchers in mathematics education, mathematics supervisors and other administrators, teacher educators, scientists, and science educators). Requests for input were disseminated through articles in various periodicals (e.g., NCTM, 1999); the NCTM Web site; presentations at professional conferences, including mathematics education, mathematical sciences, science, and general education conferences; and mailings to targeted groups, such as leaders of state and national educational associations. Nearly 30,000 copies of the *Discussion Draft* were disseminated, most in response to requests. The *Discussion Draft* was also posted on the NCTM Web site, which received an average of 20,000 hits per month (for all or part of the document) between November 1998 and April 1999. A number of individuals were specially invited to review the document, including 25 commissioned reviewers, past and present members of the NCTM Board of Directors, and members of NCTM standing committees. The Association Review Groups of the mathematical sciences associations were also asked to submit reviews of the *Discussion Draft.*

Prompts calling for open-ended responses to general issues were provided rather than items calling for Likert-scale responses or other check-off responses. The *Discussion Draft* included "Discussion Boxes," a set of questions set off within each chapter. Reaction to specific questions was invited in presentations and articles. The commissioned reviewers were provided with sets of questions that corresponded to the parts of the document they were asked to review. Relatively few of the responses received, however, explicitly addressed the prompts; most dealt with the issues that concerned the writer. Indeed, such responses were actively encouraged in the solicitations for reaction.

Responses were accepted in multiple formats, such as mailed-in responses, e-mailed responses, responses submitted through a form on the NCTM Web site, or responses to a discussion forum co-hosted by the Math Forum. Speakers at conferences submitted discussion summaries. About 700 responses were received, ranging from reactions of a few sentences to reports of more than 60 pages. These responses discussed such diverse matters as specific details of the document, general perspectives about the purposes of mathematics education, and critiques of the viewpoints expressed in the draft. Although many responses were submitted by individuals, others were submitted collaboratively on the basis of substantive discussions about the draft. Table 24.1 lists counts of response by type.

TABLE 24.1. Number of Reviews of the Discussion Draft, and Number Included in the Initial Analysis

Type of review	Number received	Number in the analysis	Percent in the analysis
Noncommissioned individual reviews	561	216	39%
Commissioned reviews	25	18	72%
Reviews by NCTM committees	5	5	100%
Individual reviews by NCTM leaders	18	18	100%
Presentations at NCTM conferences	25	12	48%
Association Review Groups	10	10	100%
Presentations to associations	7	7	100%
Reports from other organizations	50	11	22%
Total	701	297	42%

Analysis of the feedback. The analysis procedure was designed to satisfy the two purposes of the process—to help identify major issues the Writing Group would need to address in the final document and to collect specific editorial comments. The primary focus here is on the first purpose, with a few comments about how the second purpose was simultaneously served. Three phases can be identified in the analysis of the input, beginning with developing a categorization scheme, then coding the responses, and finally reducing the input to a final set of issues to be addressed. The NUD*IST software package (QSR, 1998) was used to manage the feedback received, to facilitate the development of the categorization scheme, and to generate the reports necessary for completing the analysis.

The analysis began with the development of a categorization scheme. Sorting, organizing, and categorizing methods were used, beginning with a rudimentary categorization and moving progressively toward an organizational framework (Glesne & Peshkin, 1992). A five-person team—including the chair of the Commission on the Future of the Standards, the project director, the outreach coordinator, and two graduate students — worked over four months to finalize the scheme. The chair of the Writing Group and other project members offered additional input at critical junctures. To begin the process, Writing Group and Commission members were asked to identify issues they believed were important; these results were used to develop an initial set of categories.

The categorization scheme was then built using a variation of "grounded theory" (Glaser & Strauss, 1967). Several commissioned reviews were selected for use during the initial stages of the process; because they arrived early in the set of responses, they tended to reflect more substantive rather than editorial reactions to the document, and they represented a range of perspectives and concerns. These reviews were coded initially one by one and by the full team of coders. The coding scheme was successively refined and reorganized to better capture the points made by each reviewer. As the scheme became more elaborated, successively larger sets of responses extending beyond commissioned reviews were incorporated to ensure that the coding scheme was effective with the full range of responses that were submitted. The final categorization scheme was based on 12 major iterations.

The NUD*IST software facilitated this process. As each response was considered, it was entered in the database and assigned category codes. The software allowed adding, deleting, or combining categories with ease. Thus, the effects of the changes on previously coded responses could be reviewed. Reports of all responses coded to each category were periodically generated throughout the development process to check for consistency within a category and overlap between categories. Definitions of the categories were subsequently refined to minimize any difficulties found.

When text material is analyzed, the decision on the unit of analysis to use is crucial. In a pilot process, a set of responses was coded using various possible units, such as lines (approximately 70 characters), sentences, and paragraphs. This preliminary work led to the decision to use paragraphs as the unit of analysis. A paragraph generally provided enough context for coding a particular comment within a response but was manageable in size for the production of reports across responses.

The final categorization scheme consisted of three major parts. The need for the first two parts was identified a priori, although their organization was amended somewhat to better fit the data. First, the *demographic* categories identified the respondent's primary job classification and the source of the feedback, such as a commissioned reviewer or a committee report (see Table 24.2, codes beginning with 1 and 2). These codes were generally assigned globally to an entire review, although reports from groups of people required modification when responses of particular individuals within the group could be identified. In those cases, a category of (1 98) was assigned and the codes were adjusted for that section if necessary.

Second, *positional* categories identified the part of the document being addressed, including the chapter (see codes in Table 24.2 beginning with 3) and, for the relevant chapters, either the principle or standard within that chapter (see codes in Table 24.2 beginning with 4 or 5). Special codes identified comments about the document or

TABLE 24.2 Categorization Scheme in the Analysis of Feedback to the Discussion Draft

Code	Category
	Source
(1 1)	Commissioned review
(1 2)	ARG report
(1 3)	Board
(1 4)	Committee
(1 5)	Commission on the Future of the Standards
(1 6)	Writing Group
(1 7)	Presentations at NCTM conferences
(1 8)	Other official presentations
(1 9)	Other group reports
(1 10 *x*)	Reader reaction boxes (*x* = box)
(1 11 *x y*)	Shaping-the-standards articles (*x* = journal; *y* = month)
(1 98)	Individual within a group
(1 99)	Unspecified
	Population
(2 1 *x*)	Math teacher (*x* = grade band)
(2 2)	Mathematician; college math teacher
(2 3)	Math educator
(2 4)	Math supervisor
(2 11)	Science educator
(2 12)	Administrator/policymaker
(2 13)	Educational researcher
(2 19)	Other education
(2 21)	Business
(2 22)	General public
(2 99)	Unspecified
	Chapter
(3 99)	Overall—whole document
(3 9)	Preface
(3 1)	Chapt. 1 Introduction
(3 2)	Chapt. 2 Principles
(3 3)	Chapt. 3 Overviews
(3 4)	Chapt. 4 Grades PreK–2
(3 5)	Chapt. 5 Grades 3–5
(3 6)	Chapt. 6 Grades 6–8
(3 7)	Chapt. 7 Grades 9–12
(3 8)	Chapt. 8
	Principle
(4 99)	General
(4 1)	Equity
(4 2)	Curriculum
(4 3)	Teaching
(4 4)	Learning
(4 5)	Assessment
(4 6)	Technology
	Standard
(5 99)	Intro; general
(5 1)	Number
(5 2)	Algebra
(5 3)	Geometry
(5 4)	Measurement
(5 5)	Data
(5 6)	Problem solving
(5 7)	Reasoning & proof
(5 8)	Communication
(5 9)	Connections
(5 10)	Representation
(5 11)	Content standards
(5 12)	Process standards
	Editorial issues
(6 1)	Writing style
(6 2)	Formatting
(6 3)	Structure
(6 4)	Editorial copyediting
(6 5)	Glossary; terminology
(6 6)	Citations
(6 7)	Electronic format
(6 8)	Inflammatory messages
(6 9)	Clarification of language
(6 10)	Specific examples

(Continued on next page)

TABLE 24.2—*Continued*

Code	Category	Code	Category
	Substantive issues	(7 51)	Core curriculum
(7 1)	General comments about examples	(7 52)	Discrete math
(7 11)	Audience; purpose	(7 53)	Algebra in 8th grade
(7 12)	Global view of math education	(7 54)	Integrated curriculum
(7 14)	Relation to previous standards	(7 55)	Calculus in HS
(7 15)	Visionary	(7 61)	Manipulatives
(7 16)	Linking between standards	(7 62)	Calculators and technology
(7 17)	Articulation of content across grades	(7 71)	Tracking
(7 18)	Global specificity	(7 72)	Gifted
(7 21)	Research base	(7 73)	Special needs
(7 22)	Learning and understanding	(7 74)	Differentiation of needs
(7 31)	Attention to teaching	(7 75)	Diversity and ESL
(7 32)	Attention to assessment	(7 81)	Implementation
(7 33)	Attention to principles	(7 82)	Instructional issues
(7 34)	Relation of content and process standards	(7 83)	Issues about teachers
(7 41)	Basic skills	(7 99)	Other issues
(7 42)	Basic facts		
(7 43)	Computation and algorithms		**Judgments**
(7 44)	Key content issues	(8 1)	Missing message
(7 45)	Proof	(8 2)	Effective message
(7 46)	Skills vs. understanding	(8 3)	Clarify message
(7 47)	Missing math content	(8 4)	Wrong message
(7 48)	Age-appropriateness	(8 5)	Mixed or inconsistent message

a chapter as a whole. Although each coded unit received at least one positional code, in some instances more than one code was assigned when a comment bridged more than one area of the document. The positional categories were particularly important in meeting the second purpose of the research, because they facilitated reporting of comments on particular sections of the document. They were also useful in better understanding the context for a particular comment, because they were included in the reports generated to meet the first purpose.

The final set of categories, the *substantive* categories, was designed to capture the substance of the responses. The categories were reorganized several times, including separating comments of a more editorial nature (see codes beginning with 6 in Table 24.2) from those that addressed the substance of the document (see codes beginning with 7 in Table 24.2). A complete code dictionary was developed, giving the agreed-on definitions for each code and one or more examples of units with that code.

An additional level of coding was introduced (see codes beginning with 8 in Table 24.2) to capture the nature of the reaction. The preceding codes had identified only what issue was being addressed, not whether the reaction was supportive or critical of the stance taken in the document. Following substantial work developing these categories, they did not ultimately prove useful during the data-reduction process.

Initially, the core group that developed the scheme coded the responses. As the scheme neared completion, ten additional people were enlisted to assist with coding the volume of responses. All coders had graduate degrees in mathematics education or were graduate students in the field. Whereas no formal calculation of reliability was undertaken because of time constraints, efforts were made to ensure reliability. First, two members of the initial core group of coders independently coded reviews by the Association Review Groups and all commissioned reviews. The coders resolved any discrepancies with input from the project director as necessary. Second, a core group member coded reports from NCTM committees, NCTM leaders, and conference reports. A second member reviewed the assigned codes, identified problematic codes, and resolved any discrepancies with the initial coder. Finally, the remaining responses were assigned to one coder, and the project director performed spot checks of the coding.

Time constraints prevented full coding of all responses, as the data-reduction phase of the analysis had to begin before the published deadline for submitting responses. Table 24.3 gives the number and percentage of responses coded by category. Most solicited reviews (including commissioned reviews, reviews from the Association Review Groups, and reviews from the NCTM Board and committees) were fully coded, although several commissioned reviews received after the deadline were not. An additional set of responses was purposefully chosen to include constituencies that were less well represented, such as scientists and administrators. A final set of responses was chosen on a random basis as time permitted. The percentage of "Group Reports from Other Organizations" included was relatively low because of time constraints. These responses tended to be very long, meaning that only a small set of them could be chosen to code. An important point to note, however, is that all responses were entered into the NUD*IST database using demographic and positional codes even if they were not completely coded as a part of the issue-development phase of the analysis. This approach allowed for their inclusion in the final section-by-section reports generated for use by the writers during the revision process.

We next examine the process of reducing the feedback to a final set of issues for the Writing Group to consider in answering the first purpose of the research. The goal of this phase of the analysis was to identify a manageable number of issues raised in the feedback and to characterize the range of responses to those issues. The Commission on the Future of the Standards played a significant role in the data reduction. The Commission was NCTM's official oversight body for the Standards project and orchestrated the plan for collecting feedback. It represented a range of audiences with interests in mathematics education, including classroom teachers, two-year-college mathematics teachers, university mathematicians, teacher educators, and educational researchers. Thus, the Commission was particularly well placed to make decisions that were both knowledgeable and organizationally supported. The group met for two days in early May 1999 to reduce the categorized data into a final, organized set of issues.

Prior to that meeting, a subgroup of those involved in the coding met to do initial screening of the categories. The set of 53 *substantive* categories (including all categories in Table 24.2 beginning with 6, 7, or 8) was found to be too large to deal with effectively during a two-day meeting. Furthermore, these categories varied greatly in both volume and consistency of responses included. Whereas some categories included more than 300 units, others had fewer than 20. Some categories had many responses but no recognizable threads of response within them. The subgroup reviewed printouts of all responses in each of the 53 categories and recommended that 34 be given primary consideration by the Commission, on the basis of both the volume of the responses and the existence of identifiable areas of concern within the responses. A selection of responses representing the range of responses was prepared for each of the 34 categories.

At its meeting, the Commission worked in small groups to review these 34 categories, using the summaries produced by the subgroup as well as the full printouts of responses for each category. The group made a final determination of whether a given category represented an issue of sufficient weight to put forward to the Writing Group, again on the basis of the volume of response and the existence of identifiable areas of concern within the reactions. In some instances, they chose to collapse categories that addressed common issues. As a part of the process of deciding to include an issue or to collapse categories, they identified subthemes that represented areas of particular concern within the responses. Several groups requested auxiliary reports from the NUD*IST database, such as a full-text search for a particular concept or the results for a related category, to enhance their analysis. Finally, the subgroups identified characteristic responses for each subtheme, with the intention of showing the range of responses. Attempts to further categorize and quantify the viewpoints expressed proved unsuccessful. The subgroups also reviewed the remaining 19 categories and validated the decision to omit them from the set of issues.

In a subsequent full-group meeting, the Commission discussed the work of the small groups and agreed on a final set of issues to be put forward to the Writing Group. As a part of this process, they further collapsed categories as connections were made among the issues reported by various small groups. Several issues were judged to be of less weight than others on the basis of the number or patterns of response and were not included in the final formulation. The Commission identified one additional area that emerged as an issue within one of the categories but that needed further exploration—the issue of equity. Finally, the Commission clustered the remaining issues into five groups.

Following the Commission meeting, a small group was charged to produce draft reports on each issue, incorporating the work from the subgroups and taking into account the reorganization of issues from the full-group discussion. Each issue was stated in question form along with the identified subthemes. This group also developed the additional issue related to equity, drawing from responses coded as discussing the Equity Principle, code (4 1), and from responses found in a full-text search for terms related to equity. The final list of issues is given in Table 24.3

TABLE 24.3. Issues Identified by the Commission on the Future of Standards

Overarching Issues

Issue 1. Who is the audience, and what is the purpose of the document?
- a. Who is the audience?
- b. What is the purpose?
- c. How are the audience and purpose connected?

Issue 2. In what ways should the document be more specific in its recommendations?
- a. Should there be benchmarks?
- b. Should topics and their sequence be specified for each grade?
- c. Should content be prioritized?
- d. How should the issue of specificity be handled in the process standards?

Issue 3. What global views of mathematics and of mathematics education should be espoused in the document?

Issue 4. What dependencies should there be between the previous *Standards* documents and *Principles and Standards for School Mathematics* (*PSSM*)?
- a. Is *PSSM* intended to be a stand-alone extension of the previous three *Standards* documents?
- b. Does *PSSM* extend the existing *Standards* documents?

Issue 5. Is the vision for school mathematics sufficiently clear and strong to sustain change over the next decade?

Issue 6. Are the needs of teachers addressed?
- a. Are support and resource issues sufficiently addressed?
  - i. Time
  - ii. Resources (texts, technology)
  - iii. Professional development
  - iv. Parents, policymakers, and other stakeholders
- b. Has teachers' knowledge of content been taken into consideration?
  - i. All teachers
  - ii. Specialists

Structure of the Document

Issue 7. How can a more holistic view of classroom instruction be presented that incorporates the process and content standards and the guiding principles?
- a. How can assessment have a greater profile in the document?
- b. Where is the explicit evidence that the principles guided the writing of the standards?
- c. How can the process standards be interwoven into the content standards?
  - i. Which should come first in the discussion, the content or the process standards?
  - ii. Should the process standards be discussed separately in the grade-band chapters?

Issue 8. What should be the role of chapter 3?
- a. Should it be eliminated?
- b. Should it be shortened or reformatted as a part of an executive summary?

Issue 9. How well is research being used to document claims being made?
- a. Are the research citations helpful?
- b. Is the best available evidence cited when needed?
- c. Is research evidence interpreted correctly?

Issue 10. What is the role of examples?
- a. What is the purpose of examples?
  - i. Is there a clear reason for the inclusion of examples?
  - ii. Are the vignettes serving the purpose intended?
  - iii. Is the reason made clear to the reader?
  - iv. Are there sufficient connections across standards?
- b. Is the variety of examples sufficient?
  - i. Is there a balance of contexts, including real-world examples?
  - ii. Is a variety of students used in the examples?
  - iii. Is there a variety of mathematical content?
  - iv. Is a variety of mathematical tools used in the examples?
- c. Is the quality of examples compelling?
  - i. Are they fresh?
  - ii. Are they contrived?
  - iii. Are they distracting?

Issue 11. How can the document be made easier to read, with correct and appropriate terminology?
- a. Will the audience understand all the technical words?

(Continued on next page)

TABLE 24.3—*Continued*

b. Are some words used unclearly or as educational jargon?
c. Are some words used incorrectly?
d. Should there be a glossary?
e. What additional editorial attention is needed?
  i. Wordiness, length?
  ii. Passive/active voice?
  iii. Complexity?
  iv. Single voice?

Content Issues

Issue 12. Are the connections within mathematics addressed adequately?
a. To what extent might the systematic use of unifying mathematical themes structure Grades K–12 mathematics?
b. Should the connections among the content Standards be made more explicit?
c. Should the connections within a content Standard be made more explicit?

Issue 13. How well is articulation across the grades handled?
a. How well was the articulation across the content standards handled?
b. How well was the articulation across the process standards handled?

Issue 14. Is the content appropriate for student in the specified grade bands?
a. Is the content in Grades pre-K–2 appropriate?
b. Is the content in Grades 3–5 appropriate?
c. Is the content in Grades 6–8 appropriate?
  i. How much formal algebra should be included?
d. Is the content in Grades 9–12 appropriate?
  i. Should calculus be a part of the high school curriculum?

Issue 15. Is the role of technology sufficiently addressed?
a. What is the relation to paper-and-pencil computation?
b. Has the power of technology to transform the curriculum been adequately treated?
c. Is the balance between the promise and limitations of technology appropriate?
d. Should more attention be given to technology?
e. What are the main messages about the use of technology?

Issues Related to Learning

Issue 16. Should particular educational theories be promoted explicitly in the document?
a. Should there be more attention to learning theory?
b. Should constructivism be explicitly referenced?
c. Should the van Hiele model be explicitly referenced?

Issue 17. Has the correct balance been struck between skills and understanding?
a. Does the document go too far in emphasizing skills?
b. Is the document clear in its interpretation of "computational fluency"?
c. Does the language used to discuss single-digit arithmetic convey the proper balance between skills and understanding?
d. What should the relationship be between invented and conventional algorithms?
  i. How are the timing of algorithm use and the flow across grades to be orchestrated?
  ii. How do nonconventional (e.g., "low stress") algorithms fit in?
  iii. Has the importance of estimation skills been emphasized sufficiently?
e. What is the relationship between computational proficiency and technology?

Issues Related to Equity

Issue 18. How can the document better address the needs of special student populations?
a. Can the document clarify if and how the curriculum should vary to meet the differing needs of various groups of students?
b. Can the document clarify how instruction should vary to meet the differing needs of various groups of students?

Issue 19. Is the issue of equity addressed clearly and throughout the document?
a. Is sufficient direction given for different audiences?
b. Is the language used to describe equity issues clear?
c. What is the relationship of equity to the other principles and standards?

In addition to finalizing the list of issues, the subgroup developed for each issue a report embedded in a template that included (a) the basic issue; (b) characteristic responses identified by the small groups of the commission, organized by subquestions (where applicable) and presented in a tabular form; and (c) a list of the categories and other data sources on which that issue was developed. An example is given in Figure 24.1. The reports were sent to Commission members for final review and were accepted with minor revisions. The final reports were submitted to the Writing Group for its consideration as the members planned their writing of the final document.

Issue 15. Is the role of technology sufficiently addressed?

Subquestions	Selected Responses
a. What is the relation to paper-and-pencil computation?	Some of us were uncomfortable with the call for the use of technology in the early elementary grades … and were unconvinced by arguments and examples in the [Technology Principle]. (ARG-08 #2)
	[Need to make] a statement [about] the positive educational use of calculators—what calculators do for the curriculum and for learning.... Do not be defensive. (BoD-03 #36)
	[Must] clarify how and when calculators should be used in the primary grades. One long-term objective is for children to select appropriate calculation methods.… Is the recommendation to use calculators [to solve problems in real contexts] in grades K–2, or should that objective be deferred until later grades? There is a much stronger statements in the [grades] 3–5 recommendations. (CR-11 #23)
b. Has the power of technology to transform the curriculum been adequately treated?	How technology influences WHAT we teach has not been convincingly spelled out. [Must specify] which topics in the current curriculum can be de-emphasized because of technology, which new topics need to be addressed, and which current topics need to be emphasized more than they currently are ... .(CR-17 #10)
	… It is important to show that calculators can play a much larger role in the curriculum than just facilitating calculation. It can make many explorations possible which would be much more difficult, if not impossible, without them. (ARG-09 #56)
c. Is the balance between the promise and limitations of technology appropriate?	[Must] communicate clearly and explicitly that many of our treasured "manipulative algebraic techniques" will pass into oblivion.... Students will quickly recognize that computer algebra is a "better mousetrap" when compared to paper and pencil symbolic manipulation procedures. Teachers must be prepared to deal realistically with this fact. (OGR-055 #21)
	I personally hate the "dramatically different world" and "ever increasing technology" arguments for improvements in math teaching.... Schools don't need to change because the world has changed, [but] because we keep finding out more about how to teach better … . (ARG-07 #20)
	It's misleading to suggest that all of modern mathematics is somehow connected with technology. There's plenty of room for wonder without any technology at all. (CR-13 #43)
d. Should more attention be given to technology?	More guidance [about calculator use] needs to be given. (ARG-03 #84)

(Continued on next page)

Figure 24.1. Sample report on an issue raised in the feedback to the discussion draft (from NCTM, 2000b).

Figure 24.1—*Continued*

No guidance is provided how both of these goals [students' access to calculators and becoming familiar with basic arithmetic facts] can be achieved simultaneously. (ARG-04 #123)

While a good deal of "lip service" is paid to technology through the middle grades ... there are almost no specific examples of the use or impact of technology prior to the 9–12 grade span, and, even there, it is almost always a reference instead of a well-developed example of curricular impact or an instructional use of technology. (CR-07 #36)

e. What are the main messages about the use of technology?

This [p. 42, lines 33-41] is perhaps the best discussion of appropriate use of technology. (ARG-10 #146)

[The Technology Principle] needn't be so controversial. The point that you should make ... is that nothing which encourages student growth should be discarded ... .You can then make the argument that computers and calculators do enhance learning ... and that they are important in the Curriculum Principle sense. (Ind-2013 #9)

We think that a clear position has been taken, but a definition of "appropriate use" would be helpful. (Ind-2044 #3)

The ... writers should take a strong, vocal stand on the positive impact of graphing calculators and delineate why they should be continued to be used routinely by grades 9–12 mathematics students. Many more examples should be provided in the final document. (OGR-055 #13)

- Sources:
- Category (7 62)
- Limited use of (7 12)

Figure 24.1. Sample report on an issue raised in the feedback to the discussion draft—*Continued.*

The second purpose for the analysis was to offer specific advice given about particular sections of the document. To address this goal, the full set of responses was sorted by the positional categories to which they pertained and then printed out for the Writing Group to use as they began more detailed editing of the document later in June 1999.

The Use of Findings From the Analysis of the Feedback in the Development of Principles and Standards

In the previous section, we outlined how methodologies adapted from qualitative research were used to analyze the feedback to the *Discussion Draft*. In this section, we consider how the Writing Group made use of those findings when it produced the final document. The Writing Group met in late spring 1999 to consider its responses to the 19 issues developed in the analysis of the feedback. Organized in small groups, the members reviewed the reports produced by the Commission, along with full printouts of the responses of the categories and other data sources underlying each issue. The small groups were charged with developing a general plan for response to the issue, including how the Writing Group would address the issue and a rationale for that response. As time permitted, they were also asked to develop a specific plan to enact that response.

At the end of the first day, the Writing Group reconvened as a whole to review the responses generated by the small groups and to attempt to reach consensus on the proposed plans of response. The full-group discussion continued throughout most of the second day. The leaders of the grade-band writing groups convened following this meeting to finalize a plan of response to each issue; this plan guided the writing of the final document. The template for the plans of response included (a) the basic issue, (b) subissues (where applicable) presented in a tabular form, (c) the intended actions that corresponded to the subissues or to the issue as a whole, and (d) a rationale following each planned action. A sample plan of response to an issue is given in Figure 24.2.

The Commission reviewed the plans for the responses to each of the issues. Further, the issue reports along with the Writing Group's plans of response for the issues

Issue 15. Is the role of technology sufficiently addressed?

Subquestions	Writing Group Response and Rationale
a. What is the relation to paper and-pencil computation? b. Has the power of technology to transform the curriculum been adequately treated? c. Is the balance between the promise and limitations of technology appropriate? d. Should more attention be given to technology? e. What are the main messages about the use of technology?	**General response:** We realize that the treatment of technological matters in this document needs more attention, and we will work toward a more visionary perspective about the potential of technology, both for enhancing mathematical understanding and for reshaping the content and emphases of school mathematics. We also recognize that we have to address some inconsistencies in the document regarding technology. For instance, although we have clear statements about the role of technology in the Technology Principle and elsewhere, the ideas expressed in this statement are not consistently elaborated in the examples in chapters 4 through 7. More consistent attention to appropriate uses of technology needs to be evident in the grade-band sections. We also are continuing to engage the question of how advances in technology can and will transform the curriculum and are rereading in detail the reviews that have addressed this matter for guidance. Finally, with the leadership of the Electronic Format Group, we have arranged a conference titled "The Role of Technology and Examples in *Principles and Standards*" to occur June 6–8, 1999. The agenda has been carefully arranged to ensure that Writing Group members will work with technology experts to consider the best ways of enhancing the document's handling of technology. The electronic version of the document (which was not addressed in the synthesized reviews) provides further illustration of the use of technology and technological tools, and more will be added over the next several months. Particular issues related to the subquestions will be taken up either within chapters 4–7 or in the Technology Principle. **Rationale**: The Electronic Format Group, as well as a number of reviewers, have been concerned about what is perceived to be a conservative stance on technology in mathematics education. On the other hand, some reviewers feel that even this "conservative" stance is too progressive. We have had extensive advice from experts in the role of technology in mathematics education, and we will work to incorporate more of this advice in the interest of making sure the document is forward looking.

Figure 24.2. Sample response by writers on an issue raised in the feedback to the discussion draft.

were presented to a review committee convened by the National Research Council.[5] On the basis of this additional input, the Writing Group prepared a final plan for responding to the issues; this plan guided the subsequent revision of the document.

An important point to note is that the plans for response did not attempt to reflect the consensus of the views of the field, nor was it the intent of the NCTM leadership that they do so. As stated in the final document,

> Although there will never be complete consensus within the mathematics education profession or among the general public about the ideas advanced in any standards document, the Standards provide a guide for focused, sustained efforts to improve students' school mathematics education. *Principles and Standards* supplies guidance and vision while leaving specific curriculum decisions to the local level. (NCTM, 2000a, p. 6)

[5]The NCTM Board of Directors requested a review by the National Research Council of the process of development and review of the *Standards* document. A final report was produced (NRC, 2000) that is available online at http://www.nap.edu/catalog/9870.html.

This point is underscored in the October 1993 Consortium for Policy Research in Education (CPRE) Policy Brief, which stated, "Establishing a broad consensus ... is often in tension with achieving leading-edge standards" (p. 2). The feedback helped the writers identify the issues to which they needed to respond and to better understand the context in which they were writing. The development process for *Principles and Standards* was concluded early in the 2000 calendar year; the document was released in April 2000.

The second component of our commentary about the relationship of research to policy documents focuses directly on the content of *Principles and Standards*. Again, we examine how the theoretical perspectives, methodologies, and findings of research both guided and are visible in the content of this document.

The *Principles and Standards for School Mathematics* Document: Role of Research

In the years following the release of NCTM's 1989 *Curriculum and Evaluation Standards for School Mathematics*, some critics charged that the research base on which the document was built was not explicit (e.g., Allen, n.d.; Wu, 1997). Critics expressed concern that the 1989 Standards, which strongly influenced such important documents as state mathematics curriculum frameworks and federally funded curriculum projects, were proposing an unsubstantiated educational experiment that could not assure a beneficial outcome for student learning. As work on *Principles and Standards* began, national controversy about the substance and direction of Grades K–12 mathematics education gained momentum, often with the NCTM Standards as a central part of the debate (Andrews, 1995; Kilpatrick, 1997; Wu, 1997, 2000).

Partly in response to this climate, the Writing Group made an early decision to more explicitly draw on and include research in the document. Jeremy Kilpatrick, a member of the Commission on the Future of the Standards, organized a major conference (see preface of the current volume) designed to provide research advice to the Writing Group. Scholars from several disciplines came together to address the question "In what ways should research inform the process of developing the Standards?" Aspects of the research base, in terms of both theory and relevant findings, were articulated, and responsible and principled ways in which the findings could be used were discussed. In addition, NCTM's Research Advisory Committee commissioned leading mathematics education researchers to produce white papers synthesizing research on mathematics teaching and learning in their respective areas of expertise. This approach was a major departure from the standard model of report writing and provided the Writing Group a sophisticated and rich set of resources. NCTM subsequently decided to publish the conference proceedings and the white papers in the current volume.

Finally, the NCTM staff assembled a substantial collection of research resources (e.g., journals, handbooks, reports) for the Writing Group to access, and doctoral students in mathematics education[6] reread research articles cited in various drafts and verified their appropriateness. Although members of the Writing Group had varying levels of familiarity with the theoretical perspectives, methodologies, and findings from research that might influence the content of *Principles and Standards*, a collective effort was made to incorporate a research orientation whenever possible. The educational researchers in the Writing Group played a strong part in bringing research perspectives into the work.

The Place of the Theoretical Constructs and Perspectives From Research in the Content of Principles and Standards

In their account of the original NCTM Standards effort, McLeod et al. (1996) describe how ideas from constructivist theories of learning had an impact on the substance of *Curriculum and Evaluation Standards for School Mathematics* (NCTM, 1989a). They cited an interview with Thomas Romberg, the chair of the Writing Group:

> It was very clear that we needed to think about psychological development in ways that were quite different from the behavioral tenets that are behind most education in the United States. Constructivism goes back to Dewey in the early part of this century." (p. 113)

They also cited a researcher in the field of reading as saying, "I view the math *Standards* as the instantiation of a constructivist view of learning within mathematics, parallel to the constructivist instantiation of curriculum in English/language arts and a similar view that's beginning to emerge in science education." (p. 113)

In *Principles and Standards*, this perspective is again prominent, given that its writers were charged to build on the foundations of the original *Standards* documents. Consider, for instance, the statement of the Learning

---

[6] The authors extend many thanks to Dawn Berk, John Beyers, Hope Gerson, Brad Findell, and Todd Grundmeier for their contributions.

Principle: "Students must learn mathematics with understanding, actively building new knowledge from experience and prior knowledge" (NCTM, 2000a, p. 20). Whereas specific invocation of constructivist theory by name was judged by the Writing Group as likely to lead to unnecessary controversy and even confusion, the recommendations are firmly based on a view that "mathematics makes more sense and is easier to remember and to apply when students connect new knowledge to existing knowledge in meaningful ways" (NCTM, 2000a, p. 20). Furthermore, many examples in the document are drawn from classroom-based research projects grounded in constructivist perspectives (see examples given in the next section).

Throughout the document, influences from perspectives in cognitive science are visible as well. For instance, the discussion of the relationship between understanding concepts and executing procedures alerts readers to the robust research finding that "conceptual understanding is an important component of proficiency, along with factual knowledge and procedural facility" (NCTM, 2000a, p. 20), citing the NRC report *How People Learn* (Bransford, Brown, & Cocking, 1999). In addition to the evidence of how contemporary theoretical orientations toward learning are embedded within *Principles and Standards*, the document exhibits a complementary visibility of the influence of some of the current perspectives that guide research and theory about teaching. The Teaching Principle—"Effective mathematics teaching requires understanding what students know and need to learn and then challenging them and supporting them to learn it well" (NCTM, 2000a, p. 16)—is noteworthy in this regard in that it focuses on teaching and its improvement instead of specifying lists of teacher actions and activities. This influence may be attributed in part to the work of such scholars as Ball and Bass (2000) and Stigler and Hiebert (1999), who theorized that improvements in mathematics education are likely to follow from careful analysis of the work of teachers in classrooms and from efforts to better understand its mathematical demands and entailments.

These examples not only illustrate the types of theoretical constructs and perspectives that shaped the content of *Principles and Standards* but also demonstrate how in this document such constructs and perspectives are made explicit, mainly by the extensive use of citations. Less explicit, but equally central to the substance of *Principles and Standards*, are methodologies from research.

The Place of Methodologies From Research in the Content of Principles and Standards

Mathematics education researchers have found in recent years that to better understand students' mathematics learning in the context of classrooms, qualitative research methodologies (such as clinical interviews and ethnographic observation of individuals and groups in classrooms) yield opportunities for deeper looks into student thinking about mathematics and into the interactions among learners, teachers, and curriculum within particular contexts. Mathematics education research has moved from the laboratory into classrooms and schools (Romberg, 1992). Increasingly, research reports include excerpts from transcribed interviews and interactions in class or detailed accounts of classroom practice so that readers have access to empirical grounding for researchers' findings and conclusions. Several groups of researchers, members of the Writing Group among them, have used data from their studies to serve as a basis for cases to be used in teacher education (e.g., Stein, Smith, & Silver, 1999; Schifter, Bastable, & Russell, 1999) or to produce narrative material intended for practitioner audiences (e.g., Ferrini-Mundy, Graham, Johnson, & Mills, 1998).

As a result of this methodological trend in mathematics education research, the Writing Group had available an array of material drawn from real students and classrooms as a resource to illustrate the ideas in *Principles and Standards*. In this sense, the research methodologies are incorporated into the document and made available to the readers. For example, an extended example of mathematical communication in a middle-grades classroom describes student discussion of their work on the following problem: "A certain rectangle has length and width that are whole numbers of inches, and the ratio of its length to its width is 4 to 3. Its area is 300 square inches. What are its length and width?" (NCTM, 2000a, p. 268). The account of classroom discussion that followed was drawn from Silver and Smith's (1997) research, which was conducted as part of the Quantitative Understanding: Amplifying Student Achievement and Reasoning (QUASAR) Project, a major national research endeavor.

In the numbers-and-operations section of the chapter for Grades 3–5, an episode drawn from unpublished classroom-observation notes provides an example of a detailed transcription of student and teacher conversation. The subject of this example is the sharing of student solutions to the division problem of 728 divided by 34 (NCTM, 2000a, p. 153). The page of dialogue among Ricky, Mihaela, Ms. Sparks, Christy, Fanshen, Samir, Henry, Maya, and Rita is included in part to entice readers to undertake the same kind of close examination of the evolving understandings represented in the discourse as a researcher studying children's thinking might make. Research methodologies involve gathering field notes and developing research accounts that are explicitly used in *Principles and Standards.*

The Place of Findings From Research in Principles and Standards

Perhaps the most common expectation of how research will influence policies is that robust research findings will lead clearly to recommendations. Hiebert (1999) offered some useful perspectives about the relationship of research to standards, arguing that "what is 'best' cannot be proven by research" (p. 6), and that "research cannot imagine new ideas" (p. 8). He went on to indicate that research can play a role in shaping standards by serving to document the current situation, and that research can influence the nature of standards. *Principles and Standards* relies on findings from research in both ways.

Data about the state of mathematics education helped frame the line of argument used in *Principles and Standards.* Such sources as the National Assessment of Educational Progress and the Third International Mathematics and Science Study are cited as evidence that students are not learning the mathematics they need or are expected to learn (NCTM, 2000a, p. 5). In more subtle ways, research findings served a purpose by clarifying exactly what *Principles and Standards* is meant to improve. In NCTM's *Curriculum and Evaluation Standards for School Mathematics* (1989a), such notions as "shopkeeper arithmetic," "practice in manipulating expressions and practicing algorithms as a precursor to solving problems," and the need "to reverse existing strategies for teaching" (p. 9) were central characterizations of "traditional mathematics teaching" for which the *Curriculum and Evaluation Standards* were meant to provide a challenging and better alternative. Characterizing the aspects of school mathematics that *Principles and Standards* would try to overcome was a challenge, given that the characterizations offered in 1989 were no longer entirely appropriate for 2000; many teachers had embarked on standards-based change. *Principles and Standards* claims, "Sometimes the changes made in the name of standards have been superficial or incomplete. For example, some of the pedagogical ideas from the NCTM Standards—such as the emphases on discourse, worthwhile mathematical tasks, or learning through problem solving—have been enacted without sufficient attention to students' understanding of the mathematics content" (p. 6). This view, drawn from the research findings as well as the collective experience of the Writing Group, helped guide the document's formulation.

Various sections of the document were designed intentionally to counteract the problem of superficial implementation of earlier standards ideas. Consider the Curriculum Principle, with its strong argument for "curriculum as more than a collection of activities; it must be coherent, focused on important mathematics, and well articulated across the grades" (NCTM, 1989a, p. 14). This principle is meant to help practitioners push beyond the view that using isolated but interesting mathematical tasks constitutes standards-based teaching. The notion of "important mathematics" is meant to point practitioners toward consideration of the mathematical value and implications of their instructional choices, with the idea that standards-based reform is more than content-free instructional trappings and involves significant mathematical depth and understanding. The decision to choose standards that would span the grades is another indication of a commitment to coherent, interconnected mathematics.

In addition to using research that documents the current situation, research findings are used to support claims and recommendations, and to provide detail and elaboration of ideas, throughout the document. In controversial areas in which research was available, several citations were provided in general to illustrate that research supports ideas only as it accumulates. Making an important recommendation based on a single study is less defensible. Consider the following examples from *Principles and Standards* (NCTM, 2000a) of this type of use of research:

- "Students can learn more mathematics more deeply with the appropriate use of technology (Dunham and Dick 1994; Sheets 1993; Boers-van Oosterum 1990; Rojano 1996; Groves 1994)." (p. 25)
- "Research provides evidence that students will rely on their own computational strategies (Cobb et al., 1991). Such inventions contribute to their mathematical development (Gravemeijer, 1994; Steffe, 1994)." (p. 86)
- "Students also develop understanding of place value through the strategies they invent to compute (Fuson et al., 1997)." (p. 82)

In other examples, a single study is cited to support an idea that is quite detailed and perhaps has not been investigated by many researchers: "Making a transition from viewing 'ten' as simply the accumulation of 10 ones to seeing it both as 10 ones and as 1 ten is an important first step for students toward understanding the structure of the base-ten number system. (Cobb and Wheatley 1988)" (NCTM, 2000a, p. 33). Such uses of research appear in all parts of the document.

Hiebert (1999) also discussed how other factors influence choices about standards, including history, current practice, and values. Indeed, the realization that all these influences were central in the work of the Writing Group became clear during the writing of the document. *Principles and Standards* was intended not as a review of research but, rather, as a forward-looking set of goals and ideas. As Silver (1990) noted, "research cannot imagine what ought to be." However, it can play a major role, in several ways, in policy endeavors.

Conclusion

Both the development and substance of NCTM's *Principles and Standards for School Mathematics* have been deeply influenced by the theoretical constructs and perspectives, methodologies, and findings of research in mathematics education and, more broadly, in education. Standards documents such as this one, however, which function in part as statements of goals, are influenced by other important considerations, including the history and the current state of school mathematics, lessons from efforts to change and improve school mathematics, the accumulated wisdom of teachers and other practitioners, the capacity of the field to take up new ideas, the contexts of schools and districts in which mathematics education occurs, value-based judgments about what is important and appropriate for school mathematics, and conceptions of mathematics and school mathematics. Kirst and Bird (1997) summarized these areas as appeals to tradition, science, community, and individual judgment (p. 2). In the document, produced under the aegis of a professional organization, other institutional structures and considerations come into play. Nonetheless, the role of research is prominent in the document, as we have argued and illustrated.

Loveless (1998) takes up "The Use and Misuse of Research in Educational Reform," and in considering policy reforms at the state level, contends that reformers have difficulty interpreting research fairly and accurately (p. 299). We offer *Principles and Standards for School Mathematics* as an example in which the role of research is layered and elaborated beyond strictly interpreting findings. We have argued that the roles of research can be analyzed both as part of the development of, and through its reflection in, the document's content. This image of a dual role for research frames a more complex way of viewing the development of standards and other policy documents from a scholarly standpoint. If we do not overestimate the role of research or overstate claims on its basis, we hope to demonstrate that research can play an essential role in shaping documents that will have significant attention in, and use for, the improvement of educational practice.

ACKNOWLEDGMENTS

We wish to acknowledge the helpful suggestions of Jeane Joyner, Mary Lindquist, Deborah Schifter, and Alan Schoenfeld on earlier drafts of this chapter.

REFERENCES

Allen, F. B. (n.d.). A critical view of NCTM policies with special reference to the Standards reports. Retrieved 2 September 2002 from http://www.mathematicallycorrect.com/allen1.htm

Allen, F. B. (1995). Open letter. Retrieved 2 September 2002 from http://www.mathematicallycorrect.com/allen2.htm

American Association for the Advancement of Science. (2000). *Middle grades mathematics textbooks: A benchmarks-based evaluation.* Washington, DC: Author.

Andrews, G. E. (1995). The irrelevance of calculus reform: Ruminations of a sage-on-the-stage. *UME Trends, 6,* 17, 23.

Ball, D. L., & Bass, H. (2000). Interweaving content and pedagogy in teaching and learning to teach: Knowing and using mathematics. In J. Boaler (Ed.), *Multiple perspectives on the teaching and learning of mathematics* (pp. 83–104). Westport, CT: Ablex.

Bransford, J. D., Brown, A. L., & Cocking, R. R. (Eds). (1999). *How people learn: Brain, mind, experience, and school.* Washington, DC: National Academy Press.

Consortium for Policy Research in Education (CPRE). (1993, October). *Developing content standards: Creating a process for change* (CPRE Policy Brief No. RB-10-10/93). Philadelphia: University of Pennsylvania, Graduate School of Education.

Council of Chief State School Officers. (1995). *State curriculum frameworks in mathematics and science: How are they changing across the states?* Washington, DC: Author.

Ferrini-Mundy, J. (2000). *Principles and standards for school mathematics*: A guide for mathematicians. *Notices of the American Mathematical Society, 47,* 868–876.

Ferrini-Mundy, J., Graham, K., Johnson, L., & Mills, G. (1998). *Making change in mathematics education: Learning from the field.* Reston, VA: National Council of Teachers of Mathematics.

Ferrini-Mundy, J., & Johnson, L. (1997). Highlights and implications. In J. Ferrini-Mundy & T. Schram (Eds.), The recognizing and recording reform in mathematics education project: Insights, issues and implications. *Journal for Research in Mathematics Education Monograph 8*, 87–110.

Ferrucci, B. J. (1997). Institutionalizing mathematics education reform: Vision, leadership, and the standards. In J. Ferrini-Mundy & T. Schram (Eds.), The recognizing and recording reform in mathematics education project: Insights, issues and implications. *Journal for Research in Mathematics Education Monograph 8*, 35–47.

Geertz, C. (1973). The interpretation of cultures. New York: Basic Books.

Glaser, B. G., & Strauss, A. L. (1967). *The discovery of grounded theory: Strategies for qualitative research.* New York: Aldine.

Glesne, C., & Peshkin, A. (1992). *Becoming qualitative researchers:* An introduction. White Plains, NY: Longman.

Hiebert, J. (1999). Relationships between research and the NCTM standards. *Journal for Research in Mathematics Education, 30*, 3–19.

Honest Open Logical Debate (HOLD) on Math Reform. (1996). *HOLD suggestions on NCTM standards.* Retrieved 2 September 2002 from http://www.mathematicallycorrect.com/holdnctm.htm

Kilpatrick, J. (1997). Confronting reform. *American Mathematical Monthly, 104*, 966–962.

Kirst, M. S., & Bird, R. L. (1997). *The politics of developing and maintaining mathematics and science curriculum content standards* (Research Monograph 2). Madison: University of Wisconsin—Madison, National Institute for Science Education.

Lagemann, E. C. (2000). *An elusive science: The troubling history of education research.* Chicago: University of Chicago Press.

Loveless, T. (1998). The use and misuse of research in educational reform. In D. Ravitch (Ed.), *Brookings papers on education policy* (pp. 279–304). Washington, DC: Brookings Institution Press.

McLeod, D. B., Stake, R., Schappelle, B. P., Mellissinos, M., & Gierl, M. J. (1996). Setting the standards: NCTM's role in the reform of mathematics education. In S. Raizen & E. Britton (Eds.), *Bold ventures: Case studies of U. S. innovations in mathematics education* (pp. 13–132). Boston: Kluwer.

National Academy of Education. (1999). *Recommendations regarding research priorities: An advisory report to the National Education Research Policy and Priorities Board.* Downloaded 2 September 2002 from http:// www.nae.nyu.edu/pubs/reportfinal399.doc

National Council of Teachers of Mathematics. (1987). *Curriculum and evaluation standards for school mathematics*, working draft. Reston, VA: Author.

National Council of Teachers of Mathematics. (1989a). *Curriculum and evaluation standards for school mathematics.* Reston, VA: Author.

National Council of Teachers of Mathematics (1989b). *Professional standards for teaching mathematics*, working draft. Reston, VA: Author.

National Council of Teachers of Mathematics (1991). *Professional standards for teaching mathematics.* Reston, VA: Author.

National Council of Teachers of Mathematics (1993). *Assessment standards for school mathematics: Working draft.* Reston, VA: Author.

National Council of Teachers of Mathematics (1995). *Assessment standards for school mathematics.* Reston, VA: Author.

National Council of Teachers of Mathematics (1998). *Principles and standards for school mathematics: Discussion draft.* Reston, VA: Author.

National Council of Teachers of Mathematics (1999). Shaping the standards: Share your thoughts about principles. *Teaching Children Mathematics, 4*, 231.

National Council of Teachers of Mathematics (2000a). *Principles and standards for school mathematics.* Reston, VA: Author.

National Council of Teachers of Mathematics (2000b). *The shaping of "Principles and standards for school mathematics": From discussion draft to final document.* Unpublished technical report, Reston, VA.

National Research Council. (2000). *Final report of the Coordinating Review Committee for the National Council of Teachers of Mathematics: "Principles and standards for school mathematics."* Washington, DC: Author.

Peak, L. (1996). *Pursuing excellence.* Washington, DC: U.S. Department of Education.

QSR. (1998). Non-numerical unstructured data by indexing, searching and theorizing (NUD*IST) [Computer software]. Melbourne, Australia: La Trobe University.

Ravitch, D. (1995). *National standards in American education.* Washington, DC: Brookings Institution Press.

Romberg, T. (1992). Perspectives on scholarships and research methods. In D. A. Grouws (Ed.), *Handbook of research on mathematics teaching and learning* (pp. 49–64). New York: Macmillan.

Saxe, G. B., & Bermudez, T. (1992, August). *Emergent mathematical environments in children's games.* Paper presented at the Seventh International Congress on Mathematical Education, Quebec City, Quebec, Canada.

Schifter, D., Bastable, V., & Russell, S. (1999). *Developing mathematical ideas.* Newton, MA: Dale Seymour Publications.

Shavelson, R. (1998). Contributions of educational research on policy and practice: Constructing, challenging, changing cognition. *Educational Researcher, 17*(7), 4–11, 22.

Silver, E. A. (1990). Contributions of research to practice: Applying findings, methods, and perspectives. In T. Cooney (Ed.), *Teaching and learning mathematics in the 1990s* (pp. 1–11). Reston, VA: National Council of Teachers of Mathematics.

Silver, E. A., & Smith, M. S. (1997). Implementing reform in the mathematics classroom: Creating mathematical discourse communities. In *Reform in math and science education: Issues for teachers*. Columbus, OH: Eisenhower National Clearinghouse. Retrieved 15 November 2002 from Eisenhower National Clearinghouse Web site: http://www.enc.org/professional/learn/research/journal/math

Stein, M. K., Smith, M. S., & Silver, E. A. (1999). The development of professional developers: Learning to assist teachers in new settings in new ways. *Harvard Educational Review*, 69, 237–269.

Stigler, J. W., & Hiebert, J. (1999). *The teaching gap: Best ideas from the world's teachers for improving education in the classroom*. New York: Simon & Schuster.

Stokes, D. (1997). *Pasteur's quadrant: Basic science and technological innovation*. Washington, DC: Brookings Institution Press.

U.S. Department of Education, Mathematics and Science Expert Panel. (1999). *Exemplary and promising mathematics programs*. Washington, DC: U.S. Department of Education.

Wolcott, H. F. (1980). How to look like an anthropologist without really being one. *Practicing Anthropology*, *3*(2), 39.

Wu, H. H. (1997). The mathematics education reform: Why should you be concerned and what you can do. *American Mathematical Monthly*, *104*, 946–954.

Wu, H. H. (2000). The 1997 mathematics standards war in California. In S. Stosky (Ed.), *What's at stake in the K–12 standards war?* (pp. 3–31). New York: Peter Lang.